Functioning of Transmembrane Receptors in Cell Signaling

Functioning of Transmembrane Receptors in Cell Signaling

Editors-in-Chief

Ralph A. Bradshaw
Department of Pharmaceutical Chemistry,
University of California, San Francisco,
San Francisco, California

Edward A. Dennis
Department of Chemistry and Biochemistry,
Department of Pharmacology, School of Medicine,
University of California, San Diego,
La Jolla, California

ELSEVIER

AMSTERDAM • BOSTON • HEIDELBERG • LONDON • NEW YORK • OXFORD
PARIS • SAN DIEGO • SAN FRANCISCO • SINGAPORE • SYDNEY • TOKYO

AP

Academic Press is an imprint of Elsevier

Academic Press is an imprint of Elsevier
525 B Street, Suite 1900, San Diego, CA 92101-4495, USA
30 Corporate Drive, Suite 400, Burlington, MA 01803, USA
32 Jamestown Road, London NW1 7BY, UK
360 Park Avenue South, New York, NY 10010-1710, USA

First edition 2011

Material in the work originally appeared in Handbook of Cell Signaling, Second Edition, edited by Ralph A. Bradshaw and Edward A. Dennis (Elsevier, Inc. 2010)

Library of Congress Cataloging in Publication Data
Application submitted

British Library Cataloging in Publication Data
A catalog record for this book is available from the British Library

ISBN : 978-0-12-382211-6

For information on all Academic Press publications
visit our website at www.elsevierdirect.com

Editorial Advisory Board

Contents

Section A – Overview 1

1. **Transmembrane Receptors and Their Signaling Properties** 3
Ralph A. Bradshaw and Edward A. Dennis

Section B – Biophysical Principles and General Properties 9

2. **Structural and Energetic Basis of Molecular Recognition** 11
Emil Alexov and Barry Honig

3. **Free Energy Landscapes in Protein–Protein Interactions** 15
Jacob Piehler and Gideon Schreiber

4. **Molecular Sociology** 23
Irene M. A. Nooren and Janet M. Thornton

5. **FRET Analysis of Signaling Events in Cells** 29
Peter J. Verveer and Philippe I. H. Bastiaens

6. **Structures of Serine/Threonine and Tyrosine Kinases** 35
Stevan R. Hubbard

7. **Large-Scale Structural Analysis of Protein Tyrosine Phosphatases** 41
Alastair J. Barr and Stefan Knapp

8. **Transmembrane Receptor Oligomerization** 47
Darren R. Tyson and Ralph A. Bradshaw

Section C – Major Receptor Families 53

Part 1
Receptor Tyrosine Kinases 55

9. **Protein Tyrosine Kinase Receptor Signaling Overview** 57
Carl-Henrik Heldin

10. **Receptor Tyrosine Kinase Signaling and Ubiquitination** 65
Daniela Hoeller and Ivan Dikic

11. **Insulin Receptor Complex and Signaling by Insulin** 69
Lindsay G. Sparrow and S. Lance Macaulay

12. **Structure and Mechanism of the Insulin Receptor Tyrosine Kinase** 75
Stevan R. Hubbard

13. **IRS-Protein Scaffolds and Insulin/IGF Action in Central and Peripheral Tissues** 83
Morris F. White

14. **The Epidermal Growth Factor Receptor Family** 95
Wolfgang J. Köstler and Yosef Yarden

15. **Epidermal Growth Factor Kinases and their Activation in Receptor Mediated Signaling** 103
Andrew H. A. Clayton

16. **Role of Lipid Domains in EGF Receptor Signaling** 111
Linda J. Pike

17. **Signaling by the Platelet-Derived Growth Factor Receptor Family** 117
Lars Rönnstrand

18. **The Fibroblast Growth Factor (FGF) Signaling Complex** 125
Wallace L. McKeehan, Fen Wang, and Yongde Luo

19. **The Mechanism of NGF Signaling Suggested by the p75 and TrkA Receptor Complexes** 133
J. Fernando Bazan and Christian Wiesmann

20. **The Mechanism of VEGFR Activation by VEGF** 143
Christian Wiesmann

21. Mechanisms and Functions of Eph Receptor signaling 149
Martin Lackmann

Part 2
Cytokine Receptors 157

22. Overview of Cytokine Receptors 159
Robert M. Stroud

23. Cytokine Receptor Signaling 161
Mojib Javadi Javed, Terri D. Richmond, and Dwayne L. Barber

24. Growth Hormone and Prolactin Family of Hormones and Receptors: The Structural Basis for Receptor Activation and Regulation 177
Anthony A. Kossiakoff and Charles V. Clevenger

25. Erythropoietin Receptor as a Paradigm for Cytokine Signaling 185
Deborah J. Stauber, Minmin Yu, and Ian A. Wilson

26. Structure of IFNγ and its Receptors 193
Mark R. Walter

Part 3
G Protein-Coupled Receptors 197

27. Structures of Heterotrimeric G Proteins and their Complexes 199
Stephen R. Sprang

28. G Protein-Coupled Receptor Structures 209
Veli-Pekka Jaakola and Raymond C. Stevens

29. Heterotrimeric G-Protein Signaling at Atomic Resolution 219
David G. Lambright

30. Structure and Function of G-Protein-Coupled Receptors: Lessons from Recent Crystal Structures 225
Thomas P. Sakmar

31. Chemokines and Chemokine Receptors: Structure and Function 231
Carol J. Raport and Patrick W. Gray

32. The β_2 Adrenergic Receptor as a Model for G-Protein-Coupled Receptor Structure and Activation by Diffusible Hormones 237
Daniel M. Rosenbaum, Søren G. F. Rasmussen, and Brian K. Kobilka

33. Agonist-Induced Desensitization and Endocytosis of G-protein-Coupled Receptors 245
Michael Tanowitz and Mark von Zastrow

34. Functional Role(s) of Dimeric Complexes Formed from G-Protein-Coupled Receptors 253
Raphael Rozenfeld and Lakshmi A. Devi

Part 4
TGFβ Receptors 263

35. Receptor–Ligand Recognition in the TGFβ Superfamily as Suggested by Crystal Structures of their Ectodomain Complexes 265
Matthias K. Dreyer

36. TGFβ Signal Transduction 273
Cristoforo Silvestri, Rohit Bose, Liliana Attisano, and Jeffrey L. Wrana

37. The Smads 285
Malcolm Whitman

Part 5
TNF Receptors 291

38. Structure and Function of Tumor Necrosis Factor (TNF) at the Cell Surface 293
Hao Wu and Sarah G. Hymowitz

39. Tumor Necrosis Factor Receptor-Associated Factors in Immune Receptor Signal Transduction 305
Qian Yin, Su-Chang Lin, Yu-Chih Lo, Steven M. Damo, and Hao Wu

Part 6
Guanylyl Cyclases 313

40. Guanylyl Cyclases 315
Lincoln R. Potter

Section D – Other Transmembrane Signaling Proteins 325

Part 1 Adhesion Molecules 327

41. Mechanistic Features of Cell-Surface Adhesion Receptors 329
Steven C. Almo, Anne R. Bresnick, and Xuewu Zhang

42. Structural Basis of Integrin Signaling 337
Robert C. Liddington

43. Carbohydrate Recognition and Signaling 341
James M. Rini and Hakon Leffler

Part 2 Ion Channels 349

44. An Overview of Ion Channel Structure 351
Daniel L. Minor, Jr.

45. Voltage-Gated Calcium Channels 359
William A. Catterall

46. Store-Operated Calcium Channels 373
James W. Putney, Jr.

47. Intracellular Calcium Signaling 377
Dagmar Harzheim, H. Llewelyn Roderick, and Martin D. Bootman

48. Cyclic Nucleotide-Regulated Cation Channels 383
Martin Biel

Part 3 Immunoglobulin Receptors 389

49. Immunoglobulin–Fc Receptor Interactions 391
Jenny M. Woof

50. T Cell Receptor/pMHC Complexes 399
Markus G. Rudolph, Robyn L. Stanfield, and Ian A. Wilson

51. NK Receptors 407
Roland K. Strong

52. Toll-Like Receptors–Structure and Signaling 415
Istvan Botos and David R. Davies

53. Toll Family Receptors 421
Yann Hyvert and Jean-Luc Imler

Index 427

Preface

Since cell signaling is a major area of biomedical/biological research and continues to advance at a very rapid pace, scientists at all levels, including researchers, teachers, and advanced students, need to stay current with the latest findings, yet maintain a solid foundation and knowledge of the important developments that underpin the field. Carefully selected articles from the 2nd edition of the *Handbook of Cell Signaling* offer the reader numerous, up-to-date views of intracellular signal processing, including membrane receptors, signal transduction mechanisms, the modulation of gene expression/translation, and cellular/organotypic signal responses in both normal and disease states. In addition to material focusing on recent advances, hallmark papers from historical to cutting-edge publications are cited. These references, included in each article, allow the reader a quick navigation route to the major papers in virtually all areas of cell signaling to further enhance his/her expertise.

The Cell Signaling Collection consists of four independent volumes that focus on *Functioning of Transmembrane Receptors in Cell Signaling, Transduction Mechanisms in Cellular Signaling, Regulation of Organelle and Cell Compartment Signaling,* and *Intercellular Signaling in Development and Disease*. They can be used alone, in various combinations or as a set. In each case, an overview article, adapted from our introductory chapter for the Handbook, has been included. These articles, as they appear in each volume, are deliberately overlapping and provide both historical perspectives and brief summaries of the material in the volume in which they are found. These summary sections are not exhaustively referenced since the material to which they refer is.

The individual volumes should appeal to a wide array of researchers interested in the structural biology, biochemistry, molecular biology, pharmacology, and pathophysiology of cellular effectors. This is the ideal go-to books for individuals at every level looking for a quick reference on key aspects of cell signaling or a means for initiating a more in-depth search. Written by authoritative experts in the field, these papers were chosen by the editors as the most important articles for making the Cell Signaling Collection an easy-to-use reference and teaching tool. It should be noted that these volumes focus mainly on higher organisms, a compromise engendered by space limitations.

We wish to thank our Editorial Advisory Committee consisting of the editors of the Handbook of Cell Signaling, 2nd edition, including Marilyn Farquhar, Tony Hunter, Michael Karin, Murray Korc, Suresh Subramani, Brad Thompson, and Jim Wells, for their advice and consultation on the composition of these volumes. Most importantly, we gratefully acknowledge all of the individual authors of the articles taken from the Handbook of Cell Signaling, who are the 'experts' upon which the credibility of this more focused book rests.

Ralph A. Bradshaw, San Francisco, California
Edward A. Dennis, La Jolla, California
January, 2011

Contributors

Emil Alexov (2), Department of Biochemistry and Molecular Biophysics, Howard Hughes Medical Institute, Columbia University, New York

Steven C. Almo (41), Department of Biochemistry, Center for Synchrotron Biosciences, Albert Einstein College of Medicine, Bronx, New York, New York

Liliana Attisano (36), Institute of Medical Sciences, Department of Biochemistry, University of Toronto, Toronto, Ontario, Canada, Donnelly Centre for Cellular and Biomolecular Research, University of Toronto, Toronto, Ontario, Canada

Dwayne L. Barber (23), Division of Stem Cells and Developmental Biology, Ontario Cancer Institute, Departments of Medical Biophysics and Laboratory Medicine & Pathobiology, Faculty of Medicine, University of Toronto, Toronto, Ontario, Canada

Alastair J. Barr (7), Structural Genomics Consortium, Old Road Campus Building, University of Oxford, Oxford, England, UK

Philippe I. H. Bastiaens (5), Cell Biology and Cell Biophysics Program, European Molecular Biology Laboratory, Heidelberg, Germany

J. Fernando Bazan (19), Department of Protein Biology, Genentech, South San Francisco, California

Martin Biel (48), Munich Center for Integrated Protein Science CIPSM and Department of Pharmacy, Center for Drug Research, Ludwig-Maximilians-Universität München, Munich, Germany

Martin D. Bootman (47), The Babraham Institute, Babraham, Cambridge, England, UK and Department of Pharmacology, University of Cambridge, Tennis Court Road, Cambridge, England, UK

Rohit Bose (36), Department of Molecular Genetics, Program in Molecular Biology and Cancer, Samuel Lunenfeld Research Institute, Mt Sinai Hospital, Toronto, Ontario, Canada

Istvan Botos (52), Laboratory of Molecular Biology, National Institute of Diabetes and Digestive and Kidney Diseases, National Institutes of Health, Bethesda, Maryland

Ralph A. Bradshaw (1, 8), Department of Pharmaceutical Chemistry, University of California, San Francisco, San Francisco, CA

Anne R. Bresnick (41), Department of Biochemistry, Albert Einstein College of Medicine, Bronx, New York, New York

William A. Catterall (45), Department of Pharmacology, University of Washington, Seattle, Washington

Andrew H. A. Clayton (15), Ludwig Institute for Cancer Research, Royal Melbourne Hospital, Victoria, Australia

Charles V. Clevenger (24), Department of Pathology, Northwestern University Medical School, Robert H. Lurie Comprehensive Cancer Center, Chicago, Illinois

Steven M. Damo (39), Department of Biochemistry, Weill Medical College of Cornell University, New York, New York

David R. Davies (52), Laboratory of Molecular Biology, National Institute of Diabetes and Digestive and Kidney Diseases, National Institutes of Health, Bethesda, Maryland

Edward A. Dennis (1), Department of Chemistry and Biochemistry and Department of Pharmacology, School of Medicine, University of California, San Diego, La Jolla, CA, USA

Lakshmi A. Devi (34), Department of Pharmacology and Systems Therapeutics, Mount Sinai School of Medicine, New York, New York

Ivan Dikic (10), Institute of Biochemistry II, Goethe University School of Medicine, Frankfurt am Main, Germany

Matthias K. Dreyer (35), Sanofi-Aventis Deutschland GmbH, Structural Biology, Industriepark Höchst, Frankfurt, Germany

Patrick W. Gray (31), Macrogenics, Inc., Seattle, Washington

Dagmar Harzheim (47), The Babraham Institute, Babraham, Cambridge, England, UK and Department of Pharmacology, University of Cambridge, Tennis Court Road, Cambridge, England, UK

Carl-Henrik Heldin (9), Ludwig Institute for Cancer Research, Uppsala University, Sweden

Daniela Hoeller (10), Institute for Medical Biochemistry, Innsbruck Biocentre, Innsbruck, Austria

Barry Honig (2), Department of Biochemistry and Molecular Biophysics, Howard Hughes Medical Institute, Columbia University, New York

Stevan R. Hubbard (6, 12), Kimmel Center for Biology and Medicine of the Skirball Institute of Biomolecular Medicine and Department of Pharmacology, New York University School of Medicine, New York, NY 10016, Skirball Institute of Biomolecular Medicine and Department of Pharmacology, New York University School of Medicine, New York, New York

Sarah G. Hymowitz (38), Department of Structural Biology, Genentech Inc., South San Francisco, California

Yann Hyvert (53), Centre National de la Recherche Scientifique and Faculté des Sciences de la Vie, Université Louis Pasteur, Institut de Biologie Moléculaire et Cellulaire, Strasbourg, France

Jean-Luc Imler (53), Centre National de la Recherche Scientifique and Faculté des Sciences de la Vie, Université Louis Pasteur, Institut de Biologie Moléculaire et Cellulaire, Strasbourg, France

Veli-Pekka Jaakola (28), Department of Molecular Biology, The Scripps Research Institute, La Jolla, California

Mojib Javadi Javed (23), Division of Stem Cells and Developmental Biology, Ontario Cancer Institute, Departments of Medical Biophysics and Laboratory Medicine & Pathobiology, Faculty of Medicine, University of Toronto, Toronto, Ontario, Canada

Stefan Knapp (7), Structural Genomics Consortium, Old Road Campus Building, University of Oxford, Oxford, England, UK

Brian K. Kobilka (32), Department of Molecular and Cellular Physiology, Stanford University School of Medicine, Palo Alto, California

Anthony A. Kossiakoff (24), Department of Biochemistry and Molecular Biology, University of Chicago, Gordon Center for Integrative Sciences, Chicago, Illinois

Wolfgang J. Köstler (14), Department of Biological Regulation, The Weizmann Institute of Science, Rehovot, Israel

Martin Lackmann (21), Department of Biochemistry and Molecular Biology, Monash University, Victoria, Australia

David G. Lambright (29), Program in Molecular Medicine and Department of Biochemistry and Molecular Pharmacology, University of Massachusetts Medical School, Worcester, Massachusetts

Hakon Leffler (43), Section MIG, Department of Laboratory Medicine, University of Lund, Lund, Sweden

Robert C. Liddington (42), Program on Cell Adhesion, The Burnham Institute, La Jolla, California

Su-Chang Lin (39), Department of Biochemistry, Weill Medical College of Cornell University, New York, New York

Yu-Chih Lo (39), Department of Biochemistry, Weill Medical College of Cornell University, New York, New York

Yongde Luo (18), Center for Cancer and Stem Cell Biology, Institute of Biosciences and Technology, Texas A&M Health Science Center, Houston, Texas

S. Lance Macaulay (11), CSIRO Molecular and Health Technologies and Preventative Health Flagship, Parkville, Victoria, Australia

Wallace L. McKeehan (18), Center for Cancer and Stem Cell Biology, Institute of Biosciences and Technology, Texas A&M Health Science Center, Houston, Texas

Daniel L. Minor, Jr. (44), Cardiovascular Research Institute, Departments of Biochemistry & Biophysics and Cellular & Molecular Pharmacology, California Institute for Quantitative Biosciences, University of California, San Francisco

Irene M. A. Nooren (4), European Bioinformatics Institute, Wellcome Trust Genome Campus, Hinxton, Cambridge, UK

Jacob Piehler (3), Division of Biophysics, University of Osnabrück, Osnabrück, Germany

Linda J. Pike (16), Washington University School of Medicine, Department of Biochemistry and Molecular Biophysics, St Louis, Missouri

Lincoln R. Potter (40), Department of Biochemistry, Molecular Biology and Biophysics, University of Minnesota–Twin Cities, Minneapolis, Minnesota

James W. Putney, Jr. (46), Calcium Regulation Section, Laboratory of Signal Transduction, National Institute of Environmental Health Sciences – NIH, Research Triangle Park, North Carolina

Carol J. Raport (31), ICOS Corporation, Bothell, Washington

Søren G. F. Rasmussen (32), Department of Molecular and Cellular Physiology, Stanford University School of Medicine, Palo Alto, California

Terri D. Richmond (23), Division of Stem Cells and Developmental Biology, Ontario Cancer Institute, Departments of Medical Biophysics and Laboratory Medicine & Pathobiology, Faculty of Medicine, University of Toronto, Toronto, Ontario, Canada

James M. Rini (43), Departments of Molecular and Medical Genetics and Biochemistry, University of Toronto, Toronto, Ontario, Canada

H. Llewelyn Roderick (47), The Babraham Institute, Babraham, Cambridge, England, UK and Department of Pharmacology, University of Cambridge, Tennis Court Road, Cambridge, England, UK

Lars Rönnstrand (17), Experimental Clinical Chemistry, Department of Laboratory Medicine, Lund University, Malmö University Hospital, Malmö, Sweden

Daniel M. Rosenbaum (32), Department of Molecular and Cellular Physiology, Stanford University School of Medicine, Palo Alto, California

Raphael Rozenfeld (34), Department of Pharmacology and Systems Therapeutics, Mount Sinai School of Medicine, New York, New York

Markus G. Rudolph (50), Department of Molecular and Structural Biology, Georg-August University, Goettingen, Germany

Thomas P. Sakmar (30), Laboratory of Molecular Biology and Biochemistry, The Rockefeller University, New York, New York

Gideon Schreiber (3), Department of Biological Chemistry, Weizmann Institute of Science, Rehovot, Israel

Cristoforo Silvestri (36), Institute of Medical Sciences, Donnelly Centre for Cellular and Biomolecular Research, University of Toronto, Toronto, Ontario, Canada

Lindsay G. Sparrow (11), CSIRO Molecular and Health Technologies and Preventative Health Flagship, Parkville, Victoria, Australia

Stephen R. Sprang (27), Center for Biomolecular Structure and Dynamics, University of Montana, Missoula, Montana

Robyn L. Stanfield (50), The Scripps Research Institute, Department of Molecular Biology, and The Skaggs Institute for Chemical Biology, La Jolla, California

Deborah J. Stauber (25), Department of Molecular Biology, and The Skaggs Institute for Chemical Biology, The Scripps Research Institute, La Jolla, California

Raymond C. Stevens (28), Department of Molecular Biology, The Scripps Research Institute, La Jolla, California

Roland K. Strong (51), Division of Basic Sciences, Fred Hutchinson Cancer Research Center, Seattle, Washington

Robert M. Stroud (22), Department of Biochemistry and Biophysics, University of California, San Francisco, California

Michael Tanowitz (33), Departments of Psychiatry and Cellular and Molecular Pharmacology, University of California, San Francisco

Janet M. Thornton (4), European Bioinformatics Institute, Wellcome Trust Genome Campus, Hinxton, Cambridge, UK, seconded from University College London (London) and Birkbeck College (London)

Darren R. Tyson (8), Department of Medicine, Vanderbilt University School of Medicine, Nashville, Tennessee

Peter J. Verveer (5), Cell Biology and Cell Biophysics Program, European Molecular Biology Laboratory, Heidelberg, Germany

Mark von Zastrow (33), Departments of Psychiatry and Cellular and Molecular Pharmacology, University of California, San Francisco

Mark R. Walter (26), Department of Microbiology and Center for Macromolecular Crystallography, University of Alabama at Birmingham, Birmingham, Alabama

Fen Wang (18), Center for Cancer and Stem Cell Biology, Institute of Biosciences and Technology, Texas A&M Health Science Center, Houston, Texas

Morris F. White (13), Howard Hughes Medical Institute, Division of Endocrinology, Children's Hospital Boston, Harvard Medical School, Boston, Massachusetts

Malcolm Whitman (37), Department of Developmental Biology, Harvard School of Dental Medicine, Boston, Massachusetts

Christian Wiesmann (19, 20), Department of Structural Engineering, Genentech, South San Francisco, California, Department of Structural Biology Genentech, Inc., 1 DNA Way, South San Francisco, California

Ian A. Wilson (25, 50), Department of Molecular Biology, and The Skaggs Institute for Chemical Biology, The Scripps Research Institute, La Jolla, California, The Scripps Research Institute, Department of Molecular Biology, and The Skaggs Institute for Chemical Biology, La Jolla, California

Jenny M. Woof (49), Division of Medical Sciences, University of Dundee Medical School, Ninewells Hospital, Dundee, Scotland, UK

Jeffrey L. Wrana (36), Department of Molecular Genetics, Program in Molecular Biology and Cancer, Samuel Lunenfeld Research Institute, Mt Sinai Hospital, Toronto, Ontario, Canada

Hao Wu (38, 39), Department of Biochemistry, Weill Cornell Medical College, New York, New York, Department of Biochemistry, Weill Medical College of Cornell University, New York, New York

Yosef Yarden (14), Department of Biological Regulation, The Weizmann Institute of Science, Rehovot, Israel

Qian Yin (39), Department of Biochemistry, Weill Medical College of Cornell University, New York, New York

Minmin Yu (25), Department of Molecular Biology, and The Skaggs Institute for Chemical Biology, The Scripps Research Institute, La Jolla, California

Xuewu Zhang (41), Department of Cell Biology; Albert Einstein College of Medicine, Bronx, New York, New York

Section A

Overview

Chapter 1

Transmembrane Receptors and Their Signaling Properties*

Ralph A. Bradshaw[1] and Edward A. Dennis[2]

[1]*Department of Pharmaceutical Chemistry, University of California, San Francisco, San Francisco, CA*

[2]*Department of Chemistry and Biochemistry and Department of Pharmacology, School of Medicine, University of California, San Diego, La Jolla, CA*

Cell signaling, which is also often referred to as signal transduction or, in more specialized cases, transmembrane signaling, is the process by which cells communicate with their environment and respond temporally to external cues that they sense there. All cells have the capacity to achieve this to some degree, albeit with a wide variation in purpose, mechanism, and response. At the same time, there is a remarkable degree of similarity over quite a range of species, particularly in the eukaryotic kingdom, and comparative physiology has been a useful tool in the development of this field. The central importance of this general phenomenon (sensing of external stimuli by cells) has been appreciated for a long time, but it has truly become a dominant part of cell and molecular biology research in the past three decades, in part because a description of the dynamic responses of cells to external stimuli is, in essence, a description of the life process itself. This approach lies at the core of the developing fields of proteomics and metabolomics, and its importance to human and animal health is already plainly evident.

ORIGINS OF CELL SIGNALING RESEARCH

Although cells from polycellular organisms derive substantial information from interactions with other cells and extracellular structural components, it was humoral components that first were appreciated to be intercellular messengers. This idea was certainly inherent in the 'internal secretions' initially described by Claude Bernard in 1855 and thereafter, as it became understood that ductless glands, such as the spleen, thyroid, and adrenals, secreted material into the bloodstream. However, Bernard did not directly identify hormones as such. This was left to Bayliss and Starling and their description of secretin in 1902 [1].

Recognizing that it was likely representative of a larger group of chemical messengers, the term *hormone* was introduced by Starling in a Croonian Lecture presented in 1905. The word, derived from the Greek word meaning 'to excite or arouse,' was apparently proposed by a colleague, W. B. Hardy, and was adopted, even though it did not particularly connote the messenger role but rather emphasized the positive effects exerted on target organs via cell signaling (see Wright [2] for a general description of these events). The realization that these substances could also produce inhibitory effects, gave rise to a second designation, 'chalones,' introduced by Schaefer in 1913 [3], for the inhibitory elements of these glandular secretions. The word autocoid was similarly coined for the group as a whole (hormones and chalones). Although the designation chalone has occasionally been applied to some growth factors with respect to certain of their activities (e.g., transforming growth factor β), autocoid has essentially disappeared. Thus, if the description of secretin and the introduction of the term hormone are taken to mark the beginnings of molecular endocrinology and the eventual development of cell signaling, then we have passed the hundredth anniversary of this field.

The origins of endocrinology, as the study of the glands that elaborate hormones and the effect of these entities on target cells, naturally gave rise to a definition of hormones as substances produced in one tissue type that traveled

* *Portions of this article were adapted from Bradshaw RA, Dennis EA. Cell signaling: yesterday, today, and tomorrow. In Bradshaw RA, Dennis EA, editors. Handbook of cell signaling. 2nd ed. San Diego, CA: Academic Press; 2008; pp 1–4.*

systemically to another tissue type to exert a characteristic response. Of course, initially these responses were couched in organ and whole animal responses, although they increasingly were defined in terms of metabolic and other chemical changes at the cellular level. The early days of endocrinology were marked by many important discoveries, such as the discovery of insulin [4], to name one, that solidified the definition, and a well-established list of hormones, composed primarily of three chemical classes (polypeptides, steroids, and amino acid derivatives), was eventually developed. Of course, it was appreciated even early on that the responses in the different targets were not the same, particularly with respect to time. For example, adrenalin was known to act very rapidly, while growth hormone required a much longer time frame to exert its full range of effects. However, in the absence of any molecular details of mechanism, the emphasis remained on the distinct nature of the cells of origin versus those responding and on the systemic nature of transport, and this remained the case well into the 1970s. An important shift in endocrinological thinking had its seeds well before that, however, even though it took about 25 years for these 'new' ideas that greatly expanded endocrinology to be enunciated clearly.

Although the discovery of polypeptide growth factors as a new group of biological regulators is generally associated with nerve growth factor (NGF), it can certainly be argued that other members of this broad category were known before NGF. However, NGF was the source of the designation *growth factor* and has been, in many important respects, a Rosetta stone for establishing principles that are now known to underpin much of signal transduction. Thus, considering it to be the progenitor of the field and to be the entity that keyed the expansion of endocrinology, and with it the field of cell signaling, is quite appropriate. The discovery of NGF is well documented [5] as is how this led directly to the identification of epidermal growth factor (EGF) [6], another regulator that has been equally important in providing novel insights into cellular endocrinology, signal transduction, and, more recently, molecular oncology. However, it was not till the sequences of NGF and EGF were determined [7, 8] that the molecular phase of growth factor research began in earnest. Of particular importance was the postulate that NGF and insulin were evolutionarily related entities [9], which suggested a similar molecular action (which, indeed, turned out to be remarkably clairvoyant), and was the first indication that the identified growth factors, which at that time were quite limited in number, were like hormones. This hypothesis led quickly to the identification of receptors for NGF on target neurons, using the tracer binding technology of the time (see Raffioni *et al.* [10] for a summary of these contributions), which further confirmed their hormonal status. Over the next several years, similar observations were recorded for a number of other growth factors, which in turn, led to the redefinition of endocrine mechanisms to include paracrine, autocrine and juxtacrine interactions [11]. These studies were followed by first isolation and molecular characterization using various biophysical methods and then cloning of their cDNAs, initially for the insulin and EGFR receptors [12–14] and then many others. Ultimately, the powerful techniques of molecular biology were applied to all aspects of cell signaling and are largely responsible for the detailed depictions we have today. They have allowed the broad understanding of the myriad of mechanisms and responses employed by cells to assess changes in their environment and to coordinate their functions to be compatible with the other parts of the organism of which they are a part.

TRANSMEMBRANE RECEPTORS

Membranes composed of a lipid bilayer and containing a plethora of proteins that either span or are embedded in it (sometimes by means of lipid anchors), are the structures that allow living cells to organize their intracellular components and organelles and to regulate the passage of molecules and information across it. These membrane-associated proteins constitute upward of one third of the proteins expressed by living organisms and they perform a wide variety of functions. In eukaryotes, they are synthesized on the rough endoplasmic reticulum and are extruded into the lumen of that organelle, where they are variously 'processed.' Some are retained there while the majority are passed through the Golgi apparatus and eventually into vesicles that have various fates, including fusion with the plasma membrane. Those proteins that are found in these bodies but are not associated with the membrane (derived from the ER and Golgi), become part of the secretome (secreted proteins), characteristic of that cell, while those that are embedded in the membrane by one or more transmembrane segments during the extrusion process remain there and become part of the complement of cell surface proteins (as the membrane of the vesicular body becomes part of the plasma membrane as part of the fusion event). The process is similar in prokaryotes, although there is no endoplasmic reticulum or Golgi in these cells. Some organisms also have cell walls in addition to the electrically tight plasma membrane; these also can contain proteins, but they do not serve the same kinds of functions as those found in the lipid-based membranes.

It is in this complex group of proteins inserted into the cellular membrane that are found the receptors and other recognitive molecules that interact with extracellular signals. As with the larger collection of cell surface proteins, this is a diverse group. In it are found various types of receptors for both soluble and membrane-bound ligands, ion channels, and adhesion proteins that participate in junctions of various types with extracellular matrix proteins and other cells. Functionally, these proteins are all involved in cell signaling but in many different ways. The receptors for hormones and growth factors can be grouped into six main types based on their intracellular signaling systems. These are the receptors that contain a tyrosine kinase as an inherent part of the receptor (termed 'receptor tyrosine kinases' or RTKs), cytokine receptors that utilize soluble tyrosine kinases (JAKs) that associate noncovalently with the receptors, serine/threonine kinase-containing receptors (the transforming growth factor β (TGFβ) receptors), the heptameric or G protein-coupled receptors (GPCRs), the tumor necrosis factor (TNF) receptors, and the guanylyl (or guanylate) cyclase receptors. These receptors interact with a broad spectrum of ligands ranging from small molecules (commonly, the heptameric receptors) to proteins and occasionally other macromolecules such as complex polysaccharides, for example, heparin. In addition, important signaling occurs with adhesion receptors (usually involving integrins), receptors that participate in the immune response and by the activation/inactivation of ion channels (some of which are regulated by heptameric receptors). As depicted schematically in Figure 1.1, all nine groups of transmembrane signaling molecules are discussed in this collection of short reviews.

The different classes of receptors have significantly different types of organizational structures. In general, they are divided into three functional domains: an ecto (or extracellular) domain, a transmembrane domain, and an endo (or intracellular) domain.

The ectodomain is the main basis for recognizing the ligands of that receptor, while the endodomain is chiefly responsible for generating the intracellular signal that will, in most cases, be regulated and amplified by a cascade of succeeding reactions/interactions. The transmembrane segments connect the two and are made up of 20–25 largely hydrophobic amino acid segments that occur in an α-helix. Most of the receptors have one such domain per protomer, but the heptameric GPCRs, as the name implies, contain seven such domains per monomer. Multiple membrane-spanning domain proteins are also commonly found among other functional categories such as ion channels.

The single pass transmembrane protomers can be inserted either as type I (N-terminus on the outside) or type II (N-terminus on the inside) proteins, although type I is clearly the most common orientation. In either case, the ectodomains possess the properties of extracellular proteins (reflecting their origins in the lumen of the ER), which

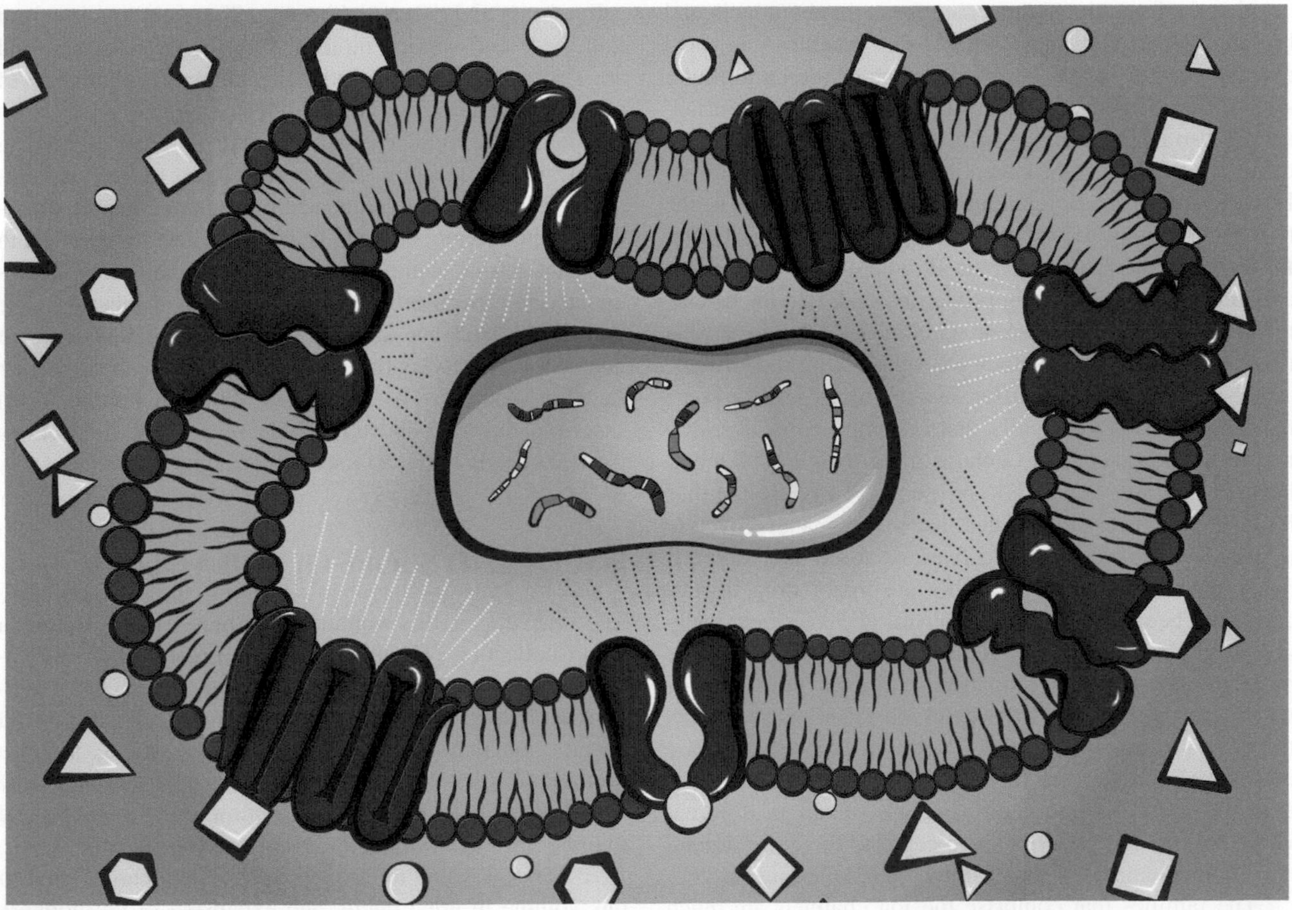

FIGURE 1.1 Schematic representation of nature's wide variety of membrane proteins that interface the cell with its environment, including tyrosine kinase receptors, cytokine receptors, heptameric G-protein coupled receptors, TGFβ receptors, TNF receptors, and guanylyl cyclase receptors as well as adhesion molecules, ion channels, and immunoglobulin receptors.

include complex glycosylation of both the N- and O-linked types and disulfide bonds. In contrast, the endodomains chemically reflect their intracellular origins (this part of the receptor remains in the cytoplasm following extrusion into the ER during translation) and have cysteines (reduced state) and undergo modifications such as phosphorylation and O-GlcNAcylation of serine and threonine residues and, in the former case, tyrosine residues as well, among many others.

Although the cell surface proteins are basically synthesized as single polypeptide chains, various biophysical and modification/mutation experiments have determined that most if not all of the receptor families exist in oligomeric structures in their active states, and a growing body of evidence suggests that these are preformed during biogenesis as opposed to the once strongly prevailing view that the association of the constituent protomers is induced by ligand binding. In the case of the RTKs, these appear to be basically homodimers, and activation (by ligand association) causes a transphosphorylation of several tyrosine residues in the endodomains of each protomer, including those important to stabilize the activation loop of the kinase domain in the open (or active) form. The activation of the kinase domain is apparently achieved by rotational or other conformational movement brought about by ligand binding. Most of the phosphotyrosine residues generated are required for the further perpetration of the intracellular signal. Homodimer (or higher oligomeric structure) formation is also a characteristic of four of the other five classes of receptor families (the exception being the TGFβ family). However, heteromeric complexes are also common in the two subgroups of the cytokine receptors and in the GPCR families. In the latter type, these heteromeric complexes can alter the ligand binding properties such that the heteromeric receptor has a different specificity than either of the two protomers has (when they are in homodimeric forms).

The various families differ in size: the largest (by some margin) is the GPCR family that is composed of over 400 members and potentially contains more than a 1000 members. In contrast, the TGFβ and guanylyl cyclase families have only relatively few. (There are seven guanylyl cyclase transmembrane receptors and five and seven TGFβ type II and type I, respectively). GPCRs are major drug targets, and it has been estimated that upward of 50% of clinically approved pharmaceuticals are so directed with many more targets under active investigation. However, the RTKs, of which there are 58, and the cytokine receptors, of which there are a couple dozen of the two types, collectively, are increasingly the targets for drug development. Both generate tyrosine phosphorylation signals, albeit in different ways. Similar to the cytokine receptors, there are around two dozen members of the TNF receptor superfamily.

The families that comprise the ion channels are much more extensive than most of the transmembrane receptors (with the exception of the GPCRs) and are comparable in number to the full complement of protein kinases (in the human genome) that number in excess of 500. These structures are fundamentally pores that allow the transfer of various ions in or out of cells (thus allowing them to cross the hydrophobic core of the lipid bilayer of the plasma membrane in the process). The resulting changes in ion concentrations control many processes and in the case of calcium ions, is directly connected to other signaling pathways. The kinetics of ion transfer are very rapid and many are, of course, tightly tied to neuronal function, one reason why they have been already extensively investigated even before other forms of signal transduction were appreciated. There are also connections between ion channels and other signaling systems, particularly the GPCRs.

The interaction of cells with other cells, extracellular matrix (ECM), and the substratum are among the most important aspects of cell signaling, particularly as it relates to environmental sensing. Mechanical stress is a major example. The adhesion plaques that form are directly connected to signaling pathways through integrins and other components and involve nonreceptor kinases such as the focal adhesion kinase (or FAK). Similarly, the cadherins, another family of cell surface receptors, participate in the formation of adheren junctions that help in maintaining cell polarity and tissue architecture. The pathways involved link these cell surface molecules to the cytoskeleton and are also active in transcriptional regulation, a hallmark of other cell signaling activities.

Finally, the immune system maintains extensive signaling pathways with receptors that are dedicated to various functions of this essential component of higher vertebrate physiology. The receptors tend to be complex and are involved in both antigen recognition and subsequent antibody generation and in various cell–cell recognition events necessary for immune surveillance and pathogen removal. Key in these interactions, and tying the receptors/ligands involved to other signaling pathways, is the use of the immunoglobin (Ig) superfold that is found throughout recognitive interactions including many of the RTKs, as an example.

FOCUS AND SCOPE OF THIS VOLUME

The chapters of this volume have been selected from a larger collection [15] and have been organized to emphasize the structure and role of cell surface receptors in signaling activities. They have been contributed by recognized experts and they are authoritative to the extent that size limitations allow. It is our intention that this survey will be useful in teaching, particularly in introductory courses, and to more seasoned investigators new to this area.

It is not possible to develop any of the areas covered in this volume in great detail, and expansion of any topic is left to the reader. The references in each chapter provide an excellent starting point, and greater coverage can also be found in the parent work [15]. It is important to realize that this volume does not cover other aspects of cell signaling such as transduction mechanisms and functioning, transcriptional activation and responses in other organelles, and organ-level manifestations, including disease correlates. These can be found in other volumes in this series [16–18].

REFERENCES

1. Bayliss WM, Starling EH. The mechanism of pancreatic secretion. *J Physiol* 1902;**28**:325–53.
2. Wright RD. The origin of the term "hormone". *Trends Biochem Sci* 1978;**3**:275.
3. Schaefer EA. *The endocrine organs.* London: Longman, Green; 1916; p. 6.
4. Banting FG, Best CH. The internal secretion of the pancreas. *J Lab Clin Med* 1922;**7**:251–66.
5. Levi-Montalcini R. The nerve growth factor 35 years later. *Science* 1987;**237**:1154–62.
6. Cohen S. Origins of growth factors: NGF and EGF. *J Biol Chem* 2008;**283**:33793–7.
7. Angeletti RH, Bradshaw RA. Nerve growth factor from mouse submaxillary gland: amino acid sequence. *Proc Natl Acad Sci USA* 1971;**68**:2417–20.
8. Savage CR, Inagami T, Cohen S. The primary structure of epidermal growth factor. *J Biol Chem* 1972;**247**:7612–21.
9. Frazier WA, Angeletti RH, Bradshaw RA. Nerve growth factor and insulin. *Science* 1972;**176**:482–8.
10. Raffioni S, Buxser SE, Bradshaw RA. The receptors for nerve growth factor and other neurotrophins. *Annu Rev Biochem* 1993;**62**: 823–50.
11. Bradshaw RA, Sporn MB. Polypeptide Growth Factors and the Regulation of Cell Growth and Differentiation: Introduction. *Fed Proc* 1983;**42**:2590–1.
12. Ullrich A, Bell JR, Chen EY, Herrera R, Petruzzelli LM, Dull TJ, *et al.* Human insulin receptor and its relationship to the tyrosine kinase family of oncogenes. *Nature* 1985;**313**:756–61.
13. Ullrich A, Coussens L, Hayflick JS, Dull TJ, Gray A, Tam AW, *et al.* Human epidermal growth factor receptor cDNA sequence and aberrant expression of the amplified gene in A431 epidermoid carcinoma cells. *Nature* 1985;**309**:418–25.
14. Ebina Y, Ellis L, Jarnagin K, Edery M, Graf L, Clauser E, *et al.* The human insulin receptor cDNA: the structural basis for hormone transmembrane signalling. *Cell* 1985;**40**:747–58.
15. Bradshaw RA, Dennis EA, editors. *Handbook of cell signaling.* 2nd ed. San Diego, CA: Academic Press; 2008.
16. Dennis EA, Bradshaw RA, editors. *Transduction mechanisms in cellular signaling.* San Diego, CA: Academic Press; 2011.
17. Bradshaw RA, Dennis EA, editors. *Regulation of organelle and cell compartment signaling.* San Diego, CA: Academic Press; 2011.
18. Dennis EA, Bradshaw RA, editors. *Intercellular signaling in development and disease.* San Diego, CA: Academic Press; 2011.

Section B

Biophysical Principles and General Properties

Chapter 2

Structural and Energetic Basis of Molecular Recognition

Emil Alexov and Barry Honig

Department of Biochemistry and Molecular Biophysics, Howard Hughes Medical Institute, Columbia University, New York

INTRODUCTION

Molecular recognition can be thought of as the process by which two or more molecules bind to one another in a specific geometry. Any binding process requires that the associating molecules prefer to interact with each other rather than the alternative, in which the individual binding interfaces interact with the solvent in which they are found. The forces that drive binding are reasonably well understood in a qualitative sense, although the accurate prediction of binding free energies or the structure of a complex given the structures of the interacting subunits remain largely unsolved problems. This chapter will briefly review the physical chemical principles of binding and summarize what has been learned so far from the analysis of the three-dimensional structures of interacting molecules and their complexes. A number of recent reviews should be consulted for more extensive discussion of the topics covered here (see, for example, [1–4]).

PRINCIPLES OF BINDING

What drives proteins to associate with other molecules? The hydrophobic effect clearly plays a central role, and it is possible that close packing at interfaces may allow stronger van der Waals interactions between molecules than either one undergoes with solvent molecules. Both types of forces, in general, will increase as the interfacial surface area increases, and these contributions to binding are often assumed to be proportional to the surface area of both proteins that is buried upon binding. In general, there will always be some factors that oppose binding, including the loss of translational and rotational degrees of freedom as two or more species form a complex [5] and the "strain" induced in each monomer as a result of complex formation [6]. This can involve an increase in the conformational energy of each monomer or entropic losses, such as, for example, side-chains in the interface that lose some configuration freedom upon binding.

Electrostatic interactions [2] also play an important role in binding; however, the magnitude and even sign of the effect are more difficult to predict. The complication in predicting the role of electrostatic interactions is that they generally reflect a balance between two large and opposing forces. For example, the formation of an ion pair as a result of complex formation requires that both charges be removed from the solvent and be completely or partially buried at an interface. For a completely buried ion pair, the loss of solvation is believed to be a larger effect than the gain of Coulomb energy in the complex so that individual ion pairs and hydrogen bonds are believed to oppose complex formation. However, ion pairs close to the surface can remain partially hydrated while still stabilized by Coulomb interactions. In some cases, these may provide a favorable driving force for association [7].

Even when charge–charge interactions oppose binding in a thermodynamic sense, they play a crucial role in specificity, as it would be extremely unfavorable energetically to remove a charge from the solvent and not to form any compensatory interactions. The requirement that buried charges and hydrogen bonding groups be satisfied upon complex formation is fairly strictly observed in known complexes. In some cases, there appear to be interfaces where networks of hydrogen bonds and ion pairs are formed [7]. These can result in a strong enough favorable interaction to compensate for the loss of solvation while at the same time placing fairly stringent specificity requirements on the geometry of the complex.

Handbook of Cell Signaling, Three-Volume Set 2 ed.

Overall, the binding of proteins to other proteins, nucleic acids, and membranes can be thought of as being driving by hydrophobic interactions (including stacking when nucleic acids are involved), constrained by the need to minimize the desolvation of charged and polar groups while optimizing the favorable interaction of these groups at an interface. Within the context of these constraints, as well as that of shape complementarity, the great flexibility in the design of different interfaces allows for the wide range of regulated and highly specific interactions that characterize signaling pathways.

NON-SPECIFIC ASSOCIATION WITH MEMBRANE SURFACES

The interaction of proteins with membrane surfaces provides an example of how different combinations of hydrophobic and electrostatic interactions are combined to achieve various specificities. Many biological membranes contain acidic phospholipids that produce a negative electrostatic potential that can be used to attract positive charges to the membrane surface [8]. A number of membrane-binding motifs are used to anchor proteins to membrane, and these generally consist of some combination of non-polar groups and positively charged amino acids. Some proteins such as Src use unstructured regions for membrane binding and, in the case of Src, this binding involves the N-terminal peptide, which contains basic amino acids and a myristate group [8]. Binding is regulated by phosphorylation, which reduces the electrostatic attraction between the basic amino acids and the acidic phospholipids [9]. Other unstructured regions, such as those of MARCKS and caveolin, use different combinations of aromatic and basic amino acids to effect membrane binding [10].

The same principles operate for structured proteins that bind to membrane surfaces. Many proteins involved in interfacial signaling contain a lipophilic modification (e.g., myristate, farnesyl) that contributes to membrane association by partitioning hydrophobically into the membrane interior. In addition, it appears that peripheral membrane proteins often have positively charged surfaces that provide an additional attraction to the surface of acidic phospholipids. In the case of the $\beta\gamma$ heterodimer of G proteins, the effect appears secondary to that of non-polar penetration [11], while for many C2 domains electrostatics appears to be the dominant interaction [12]. For example, the C2 domain from protein kinase Cβ (PKC-β) and the C2A domain from synaptotagmin I (SytI) associate peripherally with membranes containing anionic phospholipids driven primarily by electrostatic interactions. In contrast, the C2 domain from cytosolic phospholipase A2 (cPLA2) penetrates into the hydrocarbon core of membranes and prefers electrically neutral, zwitterionic phospholipids. Other C2 domains may use a combination of these effects.

PROTEIN–PROTEIN INTERACTIONS

Theoretical calculations of electrostatic interactions can account quantitatively for many of the observed binding properties of peripheral membrane proteins. This is not the case for protein–protein association, in part because the highly specific interactions that characterize protein interfaces place greater demands on the level of theoretical description. In addition, binding is often associated with conformational changes and, possibly, changes in ionization state, and these are extremely difficult to predict. Much of what we know of how protein–protein interfaces are designed has been obtained from the analysis of crystal structures of complexes [13–15]. A somewhat surprising finding has been that protein–protein interfaces are in general very similar in composition to the rest of the protein surface, and they tend to be much less non-polar than the protein core. Some interfaces may be primarily non-polar, while others appear to be characterized by a great deal of electrostatic complementarity [16]. Indeed, the two factors may well be anti-correlated, with some interfaces exploiting hydrophobic interactions and incurring a large electrostatic penalty for binding while others appear to be designed so as to optimize electrostatic interactions and to exploit hydrophobic interactions to a much lesser extent [7].

Given the knowledge of the structures of the isolated monomers it would be extremely useful to be able to predict the structure of the complex they form. This problem is known as the docking problem (see Smith and Sternberg [3] and Camacho and Vajda [4] for recent reviews), and it has been widely studied with the goal of predicting the binding modes of small molecules to proteins. The docking problem is frequently divided into two steps. One involves a geometric matching of the interacting molecules and the other involves "scoring" the model complexes generated in the first step. Scoring functions based on the principles discussed above, maximizing surface area and geometric complementarity while optimizing electrostatic interactions, appears to work quite well. Indeed, assuming that one knows the structure of the monomers as they exist in the complex, it generally appears possible to reproduce the correct complex geometry (see, for example, Norel *et al.* [17]). This suggests that the physical basis of binding is reasonably well understood. Of course, the more meaningful problem is to predict the structure of a complex based on the structures of the free monomers; however, this problem is far from being solved due to unknown conformational changes that accompany complex formation. In general, the larger these changes, the less accurate the result.

PROSPECTS

Although the accurate prediction of binding free energies remains an unsolved problem, the current level of understanding of molecular recognition is such that computational methods can be extremely useful in the design and interpretation of experimental results. Moreover, the use of bioinformatics tools can significantly expand the range of problems that can be addressed. For example, evolutionary information can be used to map regions on a protein surface involved in binding [18,19]. Moreover, once the binding properties of a few members of a protein family have been determined, it should be possible to understand the behavior of many other family members through a combined analysis of sequence, structure, and energetics. Comparing multiple sequence alignments, multiple structure alignments, and the physicochemical properties of protein surfaces can provide a great deal of information as to how binding affinity and specificity are coded onto the three-dimensional structures of proteins.

REFERENCES

1. Elcock A, Sept D, McCammon J. Computer simulation of protein–protein interactions. *J Phys Chem* 2000;**105**:1504–18.
2. Sheinerman F, Norel R, Honig B. Electrostatic aspects of protein–protein interactions. *Curr Opin Struct Biol* 2000;**10**:153–9.
3. Smith GR, Sternberg MJE. Prediction of protein–protein interactions by docking methods. *Curr Opin Struct Biol* 2002;**12**:28–35.
4. Camacho C, Vajda S. Protein–protein association kinetics and protein docking. *Curr Opin Struct Biol* 2002;**12**:36–40.
5. Gilson M, Given J, Bush B, McCammon J. The statistical–thermodynamic basis for computation of binding affinities: a critical review. *Biophys J* 1997;**72**:1047–69.
6. Froloff N, Windemuth A, Honig B. On the calculation of the binding free energies using continuum methods: application to MHC class I protein–protein interactions. *Protein Sci* 1997;**6**:1293–301.
7. Sheinerman F, Honig B. On the role of electrostatic interactions in the design of protein–protein interfaces. *J Mol Biol* 2002;**318**:161–77.
8. McLaughlin S, Aderem A. The myristoyl-electrostatic switch: a modulator of reversible protein-membrane interactions. *Trends Biochem Sci* 1995;**20**:272–6.
9. Murray D, Arbuzova A, Hangyes-Mihalyne G, Gambhir A, Ben Tal. N, Honig B, McLaughlin S. Electrostatic properties of membranes containing acidic lipids and absorbed basic peptides: theory and experiment. *Biophys J* 1999;**77**:3176–88.
10. Arbuzova A, Wang L, Wang J, Hangyas-Mihalyne G, Murray D, Honig B, McLaughlin S. Membrane binding of peptides containing both basic and aromatic residues. Experimental studies with peptides corresponding to the scaffolding region of caveolin and the effector region of MARCKS. *Biochemistry* 2000;**39**:10330–9.
11. Murray D, McLaughlin S, Honig B. The role of electrostatic interactions in the regulation of the membrane associa tion of G protein beta-gamma heterodimers. *J Biol Chem* 2001;**276**:45153–9.
12. Murray D, Honig B. Electrostatic control of the membrane targeting of C2 domains. *Mol Cell* 2002;**9**:145–54.
13. Valdar W, Thornton J. Protein–protein interfaces: analysis of amino acid conservation in homodimers. *Proteins* 2001;**42**:108–24.
14. Tsai C, Lin S, Wolfson H, Nussinov R. Studies of protein–protein interfaces: a statistical analysis of the hydrophobic effect. *Prot Sci* 1997;**6**:53–64.
15. LoConte L, Chothia C, Janin J. The atomic structure of protein–protein recognition sites. *J Mol Biol* 1999;**285**:2177–98.
16. Lawrence M, Colman P. Shape complementarity at protein/protein interfaces. *J Mol Biol* 1993;**234**:946–50.
17. Norel R, Sheinerman F, Petrey D, Honig B. Electrostatic contribution to protein–protein interactions: fast enerrgetic filters for docking and their physical basis. *Prot Sci* 2001;**10**:2147–61.
18. Lichtarge O, Sowa M. Evolutionary predictions of binding surfaces and interactions. *Curr Opin Struct Biol* 2002;**12**:21–37.
19. Armon A, Glaur D, Ben-Tal N. ConSurf: an algorithmic tool for the identification of functional regions in proteins by surface mapping of phylogenetic information. *J Mol Biol* 2001;**307**:447–63.

Chapter 3

Free Energy Landscapes in Protein–Protein Interactions

Jacob Piehler[1] and Gideon Schreiber[2]

[1]*Division of Biophysics, University of Osnabrück, Osnabrück, Germany*

[2]*Department of Biological Chemistry, Weizmann Institute of Science, Rehovot, Israel*

ABSTRACT: Specific protein–protein interactions provide a major part of the basic organization of living cells. Analysis of the structure of the complex provides a high resolution static picture of the complex, while the affinity allows analysis of the equilibrium thermodynamics of the interaction. However, understanding biological processes requires information on the nature of the full energy landscape of the complexation reaction. Analysis of the kinetics of association and dissociation in solution or within membranes allows for characterizing the landscape in more detail. In combination with computational methods, a free energy landscape can be reconstructed based on kinetic data. This includes the transition state and intermediates along the pathway. Of special interest in analyzing the free energy landscape are "hot-spot" residues, the contribution of which towards binding is often cooperative with other residues. The cooperativity can be explained in terms of the modular architecture of binding sites. The thermodynamics and kinetics of protein-protein interactions, and the free energy landscape connecting the free and bound proteins, are the subject of this chapter.

THERMODYNAMICS OF PROTEIN–PROTEIN INTERACTIONS

Specific protein–protein interactions provide a major part of the basic organization of living cells. The structure of a protein complex embeds the information of the relative mutual organization of two proteins in a frozen state. However, it does not intuitively provide information on the affinity between two proteins or the time-dependent process of complex assembly and dissociation. For a mechanistic understanding of biological processes, and for engineering proteins, which fulfill specific therapeutic tasks, we require physicochemical observables, which describe the pathway of protein–protein interaction in detail. The binding affinity between proteins,

$$K_a = \frac{[AB]}{[A][B]}$$

given by the equilibrium concentrations of the proteins [A], [B], and the complex [AB], is directly related to the free energy of interaction $\Delta G^0 = -RT\ln K_a$. Thus, complex formation only takes place if $\Delta G^0 < 0$. The free energy of the complex formation can readily be analyzed by measuring K_a (Figure 3.1). The energetic contributions ($\Delta\Delta G_D$) of individual residues can be determined by measuring changes in the K_a upon mutation. According to the Gibbs–Helmholtz relation $\Delta G^0 = \Delta H^0 - T\Delta S^0$, both the enthalpy ΔH^0 and the entropy ΔS^0 of the complex formation contribute to ΔG^0. ΔH^0 reflects the strength of the interactions between two proteins (e.g., van der Waals, hydrogen bonds, salt bridges) relative to those existing with the solvent molecules, which are excluded upon binding from the interface. ΔS^0, on the other hand, mainly reflects two contributions: changes in solvation entropy, and changes in conformational entropy. Upon binding, the water released from the binding sites leads to a gain in solvent entropy. This gain is particularly important for hydrophobic patches on the protein surface ("hydrophobic effect"). At the same time, the proteins and individual residues within the proteins lose conformational freedom, resulting in a negative change in conformational entropy. The loss of conformational freedom was estimated to be in the order of 15 kcal/mol at 25°C, but values between 0 and 30 kcal/mol have been cited as well [1]. What do we know about the contributions towards entropy and enthalpy on the molecular level? Dehydration of nonpolar residues during association is always entropically

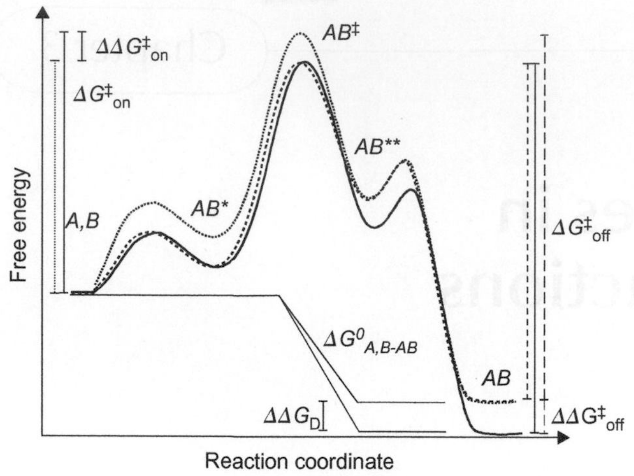

FIGURE 3.1 Free energy profile describing the pathway for the formation of a protein–protein complex (AB) from the free proteins A and B via the encounter complex AB*, the transition state AB‡ and a past transition state intermediate AB**. Comparison of the profiles for the wt proteins (-) with a mutant affecting long-range electrostatic interactions (·····) and a mutant affecting short-range interactions (----), respectively. The free energies ΔG are indicated both for the complex formation ΔG^0 and the transition state, as well as the changes in free energy of the encounter complex.

favorable, while that of polar residues is unfavorable. The enthalpies are nevertheless negative, since they represent the energy of interaction of atoms at the interface relative to their interactions with water. As several partially canceling factors contribute towards the entropy and enthalpy of interaction, it is not surprising that for most mutant complexes the difference in free energy of binding is much smaller than the accompanying changes in ΔH^0 and ΔS^0. This has been emphasized by theoretical studies which show that on forming a cavity in water to accommodate a solvent molecule the change in the enthalpy of water (solvation) is exactly balanced by the entropy of the cavity, thus changes in ΔH^0 and ΔS^0 cancel out each other in ΔG^0 [2]. Thus, enthalpy–entropy compensations seem to be a characteristic of weak non-covalent interactions including protein–protein interactions.

INTERACTION KINETICS

While analysis of the K_a can provide a detailed thermodynamic picture of the complex, it does not allow any conclusion about the pathway that leads to the formation of the complex from the individual proteins. This is entirely determined by the shape of the free energy landscape given by all possible states between the free proteins and the complex, most of which are experimentally not accessible. On this free energy landscape, the reaction itself most likely follows the pathway requiring least free energy. This pathway is called the *reaction coordinate*, and can be experimentally studied through the rates of association and of dissociation (Figure 3.1). Analysis of these kinetic parameters for several structurally and physicochemically well-defined protein–protein interactions allowed establishment of some basic concepts of how protein complexes form. Following is an overview about how kinetics can be used for analyzing the interaction pathway through the free energy landscape, and how this can be understood on the molecular level.

In general, association of a protein complex (AB) from the unbound components (A+B) can be described using a four-state mechanism:

$$A + B \underset{k_{-1}}{\overset{k_1}{\rightleftharpoons}} A:B \underset{k_{-2}}{\overset{k_2}{\rightleftharpoons}} AB^* \underset{k_{-3}}{\overset{k_3}{\rightleftharpoons}} AB$$

Two proteins diffusing in the solution will collide with each other with a rate (k_1) to form A:B solely depending on the translational diffusion constants, as described by the Stokes–Einstein equation [3]. Non-specific collision events may (in a small percent) develop into an encounter complex (AB*) with a rate of k_2, from which, after the transition state, the final complex is formed (k_3). Additional encounter complexes may exist beyond the transition state (AB**), but they do not affect the overall rate of association. The transition state was shown to resemble the final structure, albeit prior to desolvation and structural rearrangements. The transition state is stabilized mainly by long-range electrostatic interactions which, when optimized, lead to a faster association [4,5]. This reaction, which occurs within the milieu of endless competing macromolecules, can be compared to two blind men finding each other in the streets of New York. Yet, it is done rapidly at rates of only one to five orders of magnitude below the Stokes–Einstein diffusion collision limit of $10^{10} m^{-1} s^{-1}$. It has to be emphasized that experimentally one often observes only a single step from A+B to AB, and that the equilibrium association constant (K_A) equals k_{on}/k_{off}). Yet, under certain experimental conditions the pre-complexes are observable. A good example for such case is the interaction between Ras and the Ras binding domain of c-Raf1. Here, a two-step association process was suggested, with an initial rapid equilibrium step followed by an isomerization reaction occurring at several hundreds per second. Increasing the electrostatic steering between these two proteins through mutations did not affect k_3, but did increase k_{on} through increased k_1 and k_2 [6,7].

In recent years a number of theoretical and experimental studies were done to explore the structures of the encounter complex and transition state in order to better understand the mechanism of association. These studies came to define how specific these pre-complexes are, and whether binding is a stochastic or directed event. For the association between the amino-terminal domain of enzyme I and the phosphocarrier protein HPr as measured in very low salt conditions but under very high protein concentrations the encounter complex was found to be diffusive, occupying a large area of the surface and being stabilized by strong

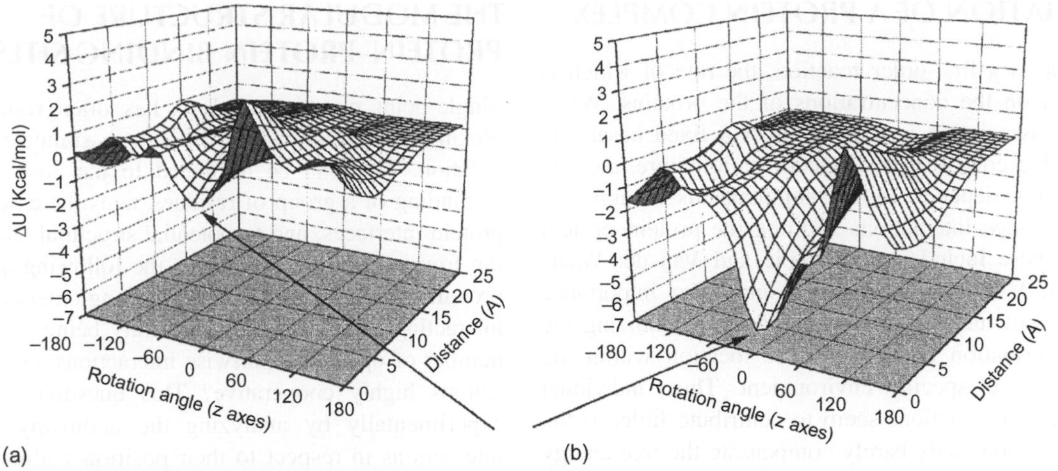

FIGURE 3.2 3D energy landscape of the association between wt TEM1-β–lactamase with wt BLIP (a) and a much faster binding mutant (b). For demonstration, only the z-angle rotation is shown. The magnitude of the Debye Hückel energy of interaction (ΔU) is plotted in 3D versus the distance and the relative rotation angle between the proteins. The arrows point at the 0° rotation angle which is the X-ray crystallographic structure of the TEM1/BLIP complex. For details on the calculations employed, see [11].

electrostatic forces [8]. However, many other experimental and theoretical studies showed the encounter complex and transition state to be much more restricted to the area roughly occupied by the final complex, with the encounter complex occupying a larger volume than the transition state [9–12]. In a recent study we used double-mutant cycle data to measure specific residue–residue interactions existing already at the activated complex for modeling the structure of the transition state. Two types of transition states could be mapped. A specific transition state was observed for the fast associating proteins Barnase and Barstar, while a diffusive one was observed for wild-type TEM1-BLIP and IFNα2 binding IFNAR2. The specific transition state could be structurally modeled while the diffusive transition state lacked preferred structures, which may result in a slower association – as indeed is the case for TEM1-BLIP. The diffusive versus specific transition states are interchangeable, through the introduction of charged mutations by rational design [13]. Figure 3.2 shows the energy landscape during association for a diffusive association (left panel, TEM1-BLIP) and a specific transition state (right panel, engineered TEM1-BLIP for fast association).

Rational computer-based design for faster association of protein complexes was developed based on the direct relation between the electrostatic complementarity between a pair of proteins and the rate of their association. These two quantities are related by eq. 3.1, where U is the electrostatic energy of interaction between the two proteins, $\ln k_{on}^{0}$ s the basal rate of association in the absence of electrostatic forces, κ is the inverse Debye length and a is the minimal distance of approach [7,14]:

$$\ln k_{on} = \ln k_{on}^{0} - \frac{U}{RT}\left(\frac{1}{1+\kappa a}\right) \qquad (3.1)$$

An interesting result of this phenomenon is that one can add charges outside the binding site and still increase the rate of association of the complex without affecting the rate of dissociation [6, 7].

A somewhat different approach to probe association trajectories experimentally is by applying Φ-value analysis ($\Phi=\Delta\Delta G^{\ddagger}_{on}/\Delta\Delta G_D$). A Φ value close to 1 indicates that a specific interaction is formed at the transition state, while a Φ value close to 0 indicates that the interaction is formed after the transition state. In a study of the HyHEL–10 Fab complex, multiple replacements were made in two positions, with most of the replacements having Φ values close to 0. This was interpreted as the transition state being early along the reaction trajectory, before short-range interactions (which have the largest contribution on $\Delta\Delta G_D$) are formed [15]. The notion that short-range interactions affect k_{off} while long-range electrostatic interactions affect k_{on} was directly tested by introducing charged mutations in the vicinity, but outside the binding site, of TEM1-BLIP. These mutations did increase specifically k_{on} 250-fold, but did not affect k_{off} (thus, the increase in k_{on} equals the increase in K_D and $\Phi=1$) [7]. These data suggest that long-range electrostatic interactions increase the rate of association by lowering the free energy of the transition state by the same magnitude as the equilibrium constant (see Figure 3.1). While mutations of non-charged residues do not significantly affect the transition state for association, they can significantly alter k_{off} and K_D. These data implicate that the transition state is stabilized by electrostatic interactions and its structure already resembles that of the final complex, but the proteins do not yet form most of the short-range interactions. The results of the Φ-value analysis and the structural studies of the transition state hence provide a coherent picture of the association process.

DISSOCIATION OF A PROTEIN COMPLEX

Dissociation is a first-order reaction, the rate of which is independent on the concentrations of the proteins. While the rate of association is a function of a fixed basal rate and a variable contribution of electrostatic nature, the rate of dissociation depends on the simultaneous breaking of many short-range interactions forming the protein–protein interface. These include hydrophobic and Van der Waals interactions, H-bonds, and salt bridges. The importance of the different interactions with regard to stabilizing the bound conformation depends on the location within the interface, and its specific environment. Thus, individual charge–charge interactions seem to contribute little, as the gained interactions only barely compensate the free energy required for desolvation of the charged groups. However, if a network of charge–charge interactions is formed, a positive contribution towards binding is regained. H-bonds seem to have a relative small but constant contribution towards binding [16, 17]. The effect of hydrophobic and Van der Waals interactions was estimated from the buried surface area (non-polar and total). However, no good absolute estimations were obtained. The most intriguing question relates to the nature of "hotspots". These are residues which, upon mutation, cause a large shift in complex stability (reducing binding affinity by up to 10,000-fold). While no clear physical definition for "hot-spot" residues could be formulated, they seem to be located at positions that are not water accessible at the complex, and are often located at the center of the interface [18–20]. Thus, the interface can be crudely divided into an outer ring of residues, which forms some kind of a seal, and inner residues, which are fully stripped from solvent molecules. However, only few of these fully buried residues are "hotspots", and the structural and energetic bases of hot-spot residues are not yet completely clear.

THE MODULAR STRUCTURE OF PROTEIN–PROTEIN BINDING SITES

Single-point mutation analysis has often resulted in perplexing results, which are difficult to simulate using computation. This can be attributed to the cooperative effect of binding of a group of residues, as is the case in protein–protein interfaces, and to potential structural changes resulting from mutations. This raises the following question: is a protein–protein binding site just a conglomerate of multiple interacting residues, with the affinity being dictated by the number of optimized pairwise interactions, or are the interactions highly cooperative? This question was explored experimentally by analyzing the additively of pairwise interactions in respect to their position within the binding interface. To define the interface, a contact map of all the interacting residues was built from the physical interactions between the residues, such as H-bonds, electrostatic interactions, van der Waals interactions and aromatic interactions [21, 22]. An example for a contact map is given for the complex between TEM1 and BLIP in Figure 3.3a. This map shows the interface being divided into six individual clusters; each cluster comprised a number of closely interacting residues, with virtually no interactions found between the clusters [22]. The change in binding free energy of pairs of mutations with each mutant being located on a different cluster was found to be always additive, while pairs of mutations located within the same cluster were often cooperative towards one another (Figure 3.3b). Moreover, deleting complete clusters from the interface caused smaller than expected structural and energetic consequences, compared to the additive values of the single mutations constituting the clusters [22]. Constructing interaction maps of many other complexes has shown that the modular architecture is a general design principle of binding sites. Moreover, it

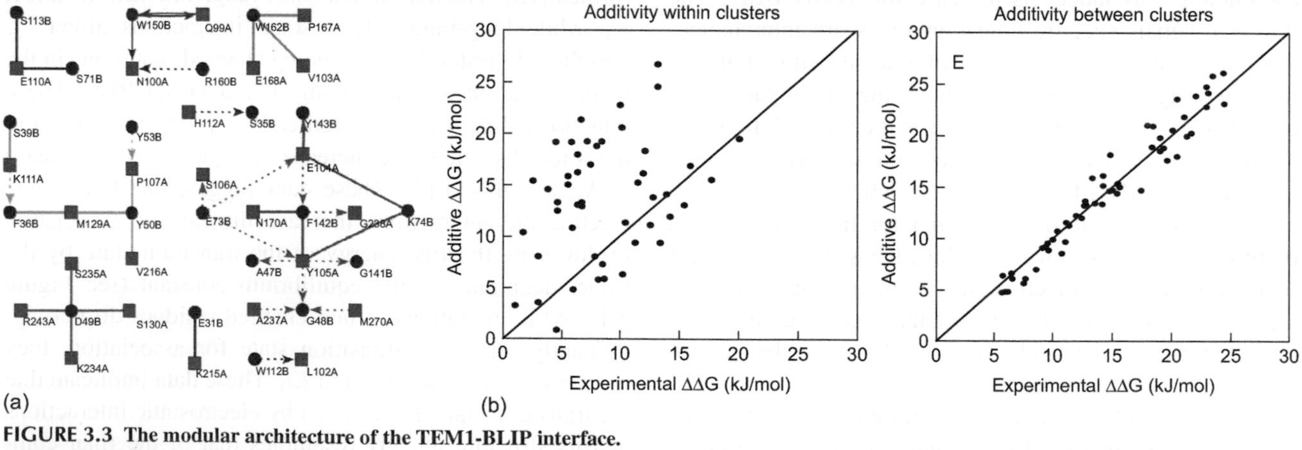

FIGURE 3.3 The modular architecture of the TEM1-BLIP interface.
In (a), a connectivity map was derived for this complex with TEM1 and BLIP residues being nodes in the graph (squares and circles, respectively). Three interaction types are shown in the map: side-chain–side-chain (solid lines), backbone–side-chain (dotted lines) and interactions of both side-chain–side-chain and backbone–side-chain between the same pair of residues (arrows point to the backbone donor atom). (b) shows the degree of additivity of free energy of binding between mutations on TEM1 and BLIP within or between the six clusters shown in (A). Additive $\Delta\Delta G$ is defined as $\Delta\Delta G_{mut1} + \Delta\Delta G_{mut2}$, plotted versus the experimentally determined values of the double-mutant ($\Delta\Delta G_{mut1,mut2}$). Data are taken from [21].

seems that large, highly evolved clusters may result in tight binding interfaces. Contrarily, weak interfaces seem to be characterized by a low complexity of the interaction maps. This view was supported by a bioinformatic analysis of hot-spot residues performed by the groups of Nussinov and Vishveshware, proposing that hot-spots are within densely packed "hot regions." Within these regions, they form networks of interactions, contributing cooperatively to the stability of the complex. However, the contributions of separate, independent hot regions were additive. Since hot-spot residues are also conserved by evolution, proteins binding multiple partners at the same sites are expected to use all or some combination of these regions [23–25].

The modularity of binding sites suggests a roughness of the energy landscape of the interaction surface. This was indeed measured using single-molecule dynamic force spectroscopy for a complex consisting of the small GTPase Ran protein and the nuclear transport receptor Importin-β, indicating a bumpy energy surface, which is consistent with the ability of Importin-β to accommodate multiple conformations and to interact with different, structurally distinct ligands [26].

One of the major challenges in studying proteins is their diversity. While a simple modular architecture is true for some interfaces, for others long-range cooperative effects were detected. One such example is for the T cell receptor variable domain binding a bacterial superantigen, where cooperative binding energies between distinct hot regions that are separated by >20Å were shown [27]. The propagation of these cooperative effects through a dynamic structural network is a complex evolutionary phenomenon which is reminiscent of the allosteric effects found in enzymes. In contrast, the simple additivity found between remote hot-spots in the TEM1-BLIP complex can be viewed as the typical mechanism. Long-range cooperativity may be predicted using statistical coupling analysis (SCA), pioneered by the group of Ranganathan [28]. Using this method, they identified a network of energetically coupled residues that link the functional surfaces of nuclear receptor ligand binding domains, allowing for the long-range cooperativity between the residues.

INTERACTION BETWEEN MEMBRANE-ANCHORED PROTEINS

Key processes responsible for the initiation of transmembrane signaling involve interactions between proteins, which are anchored to the membrane. Thus, signaling through receptor tyrosine kinases, cytokine receptors, and related systems are initiated by ligand-induced interactions between transmembrane proteins. Once the signal is propagated through the membrane, membrane-anchored cytosolic effector proteins such as small or heterotrimeric G-proteins are recruited to activated receptors. Compared to interaction in solution, anchoring of proteins into the membrane severely changes the energetic landscapes of protein–protein interaction. Most importantly, the mobility of the interaction partner is restricted to two dimensions, and the constants of translational and rotational diffusion are reduced by at least two orders of magnitude. Furthermore, the rotational freedom of the interaction partners is restricted by membrane anchoring, thus limiting the range of possible relative orientations during collision. For these reasons, the energy landscapes of membrane-anchored protein complexes substantially differ compared to the landscape of soluble complexes. In general, the reduction in dimensionality as well as potentially favorable pre-orientation has been proposed to promote protein complex formation [29–31]. However, few quantitative experimental studies of two-dimensional protein interactions have been reported. In a seminal paper [32], Whitty and co-workers explored IL-4-induced crosslinking of the IL-4 Rα with the common γ chain on live cells. A surprisingly low efficiency of association of the low-affinity γ-chain subunit with the highly stable complex of IL-4 and IL-4 Rα on the membrane was observed, establishing for the first time affinity-limited complex formation on membranes. Recent systematic studies of ternary cytokine–receptor complexes tethered on artificial membranes have provided some quantitative understanding of two-dimensional interactions. These data confirmed the regulating role of binding affinities for complex formation on membranes. Equilibrium dissociation constants in the order of 1–10 μM for the interaction in solution yielded two-dimensional equilibrium dissociation constants in the order of 10–100 molecules/μm^2 [33]. Kinetic analysis of such complexes revealed that the probability of successful collisions is indeed higher on membranes, which can be ascribed to the pre-orientation and the longer lifetime of the collision complex. This effect is, however, largely compensated by the decreased diffusion rate, which reduces the number of collisions in the same order of magnitude. Probably, the association kinetics of many membrane-anchored proteins is close to diffusion limitation. Orientation of the proteins significantly affects the rate constants of association [34], though protein domains interacting outside the membrane are often probably rather flexible. Overall, the association rates at a given average distance of the molecules in solution and on the surface are surprisingly similar. Also, the dissociation kinetics of protein complexes is substantially affected by protein tethering to the membrane [35]. Here, the reduced diffusion kinetics appears to stabilize the complex, suggesting that the separation of the interacting species by diffusion plays a critical role for the dissociation process. This observation is in line with the notion that the association kinetics is highly diffusion controlled. As a consequence, local membrane fluidity in membrane microdomains may play a critical role for stabilizing membrane protein complexes. Thus, the energy landscapes of cellular membrane protein complexes are highly influenced by the microcompartimentation of the membranes.

SUMMARY

This chapter discusses the energy landscape separating the unbound from the bound state of protein–protein interactions. Along the association pathway, an unstable diffusion encounter complex is formed prior to the transition state. Long-range electrostatic forces play a major role in stabilizing both the encounter complex and the transition state, and thereby effectively steer the association process. In the transition state, the two proteins are correctly orientated towards each other and the interface is just being desolvated, so that subsequently short-range interactions can be formed to stabilize the complex. Accordingly, the free activation energy required for reaching the transition state mostly stems from the energetically costly process of surface desolvation (especially of charged residues). Compared to interaction in solution, anchoring of proteins into the membrane severely changes the energetic landscapes of protein–protein interaction. While the long-range interactions formed during association are a global feature of the protein, the architecture of the interface is modular, with the individual modules being cooperative within themselves but additive between modules.

REFERENCES

1. Karplus M, Janin J. Comment on: "The entropy cost of protein association". *Protein Eng* 1999;**12**:185–6; discussion 187.
2. Yu HA, Karplus M. A thermodynamic analysis of solvation. *J Chem Phys* 1988;**89**:2366–79.
3. Kozer N, Kuttner YY, Haran G, Schreiber G. Protein–protein association in polymer solutions: from dilute to semidilute to concentrated. *Biophys J* 2007;**92**:2139–49.
4. Schreiber G. Kinetic studies of protein–protein interactions. *Curr Opin Struct Biol* 2002;**12**:41–7.
5. Fersht AR. *Structure and mechanism in protein science*. New York: W.H. Freeman; 1999; pp. 132–64.
6. Kiel C, Selzer T, Shaul Y, Schreiber G, Herrmann C. Electrostatically optimized Ras binding Ral guanine dissociation stimulator mutants increase the rate of association by stabilizing the encounter complex. *Proc Natl Acad Sci USA* 2004;**101**:9223–8.
7. Selzer T, Albeck S, Schreiber G. Rational design of faster associating and tighter binding protein complexes. *Nat Struct Biol* 2000;**7**:537–41.
8. Tang C, Iwahara J, Clore GM. Visualization of transient encounter complexes in protein–protein association. *Nature* 2006;**444**:383–6.
9. Miyashita O, Onuchic JN, Okamura MY. Transition state and encounter complex for fast association of cytochrome c2 with bacterial reaction center. *Proc Natl Acad Sci USA* 2004;**101**:16,174–9.
10. Tetreault M, Cusanovich M, Meyer T, Axelrod H, Okamura MY. Double mutant studies identify electrostatic interactions that are important for docking cytochrome c2 onto the bacterial reaction center. *Biochemistry* 2002;**41**:5807–15.
11. Tetreault M, Rongey SH, Feher G, Okamura MY. Interaction between cytochrome c2 and the photosynthetic reaction center from *Rhodobacter sphaeroides*: effects of charge-modifying mutations on binding and electron transfer. *Biochemistry* 2001;**40**:8452–62.
12. Miyashita O, Onuchic JN, Okamura MY. Continuum electrostatic model for the binding of cytochrome c2 to the photosynthetic reaction center from *Rhodobacter sphaeroides*. *Biochemistry* 2003;**42**:11,651–60.
13. Harel H, Cohen M, Schreiber G. On the dynamic nature of the transition state for protein–protein association as determined by double-mutant cycle analysis and simulation. *J Mol Biol* 2007;**371**:180–96.
14. Selzer T, Schreiber G. New insights into the mechanism of protein–protein association. *Proteins* 2001;**45**:190–8.
15. Taylor MG, Rajpal A, Kirsch JF. Kinetic epitope mapping of the chicken lysozyme.HyHEL–10 Fab complex: delineation of docking trajectories. *Protein Sci* 1998;**7**:1857–67.
16. Albeck S, Unger R, Schreiber G. Evaluation of direct and cooperative contributions towards the strength of buried hydrogen bonds and salt bridges. *J Mol Biol* 2000;**298**:503–20.
17. Hendsch ZS, Tidor B. Do salt bridges stabilize proteins? A continuum electrostatic analysis. *Protein Sci* 1994;**3**:211–26.
18. Bogan AA, Thorn KS. Anatomy of hotspots in protein interfaces. *J Mol Biol* 1998;**280**:1–9.
19. Clackson T, Wells JA. A hotspot of binding energy in a hormone–receptor interface. *Science* 1995;**267**:383–6.
20. Halperin I, Wolfson H, Nussinov R. Protein–protein interactions; coupling of structurally conserved residues and of hotspots across interfaces. Implications for docking. *Structure* 2004;**12**:1027–38.
21. Reichmann D, Rahat O, Cohen M, Neuvirth H, Schreiber G. The molecular architecture of protein–protein binding sites. *Curr Opin Struct Biol* 2007;**17**:67–76.
22. Reichmann D, Rahat O, Albeck S, Meged R, Dym O, Schreiber G. The modular architecture of protein–protein binding interfaces. *Proc Natl Acad Sci USA* 2005;**102**:57–62.
23. Keskin O, Ma B, Rogale K, Gunasekaran K, Nussinov R. Protein–protein interactions: organization, cooperativity and mapping in a bottom-up Systems Biology approach. *Phys Biol* 2005;**2**:S24–35.
24. Brinda KV, Vishveshwara S. Oligomeric protein structure networks: insights into protein–protein interactions. *BMC Bioinformatics* 2005;**6**:296.
25. Sohn J, Parks JM, Buhrman G, Brown P, Kristjansdottir K, Safi A, Edelsbrunner H, Yang W, Rudolph J. Experimental validation of the docking orientation of Cdc25 with its Cdk2–CycA protein substrate. *Biochemistry* 2005;**44**:16,563–73.
26. Nevo R, Brumfeld V, Kapon R, Hinterdorfer P, Reich Z. Direct measurement of protein energy landscape roughness. *EMBO Rep* 2005;**6**:482–6.
27. Moza B, Buonpane RA, Zhu P, Herfst CA, Rahman AK, McCormick JK, Kranz DM, Sundberg EJ. Long-range cooperative binding effects in a T cell receptor variable domain. *Proc Natl Acad Sci USA* 2006;**103**:9867–72.
28. Shulman AI, Larson C, Mangelsdorf DJ, Ranganathan R. Structural determinants of allosteric ligand activation in RXR heterodimers. *Cell* 2004;**116**:417–29.
29. Adam G, Delbruck M. Reduction of dimensionality in biological diffusion processes. In: Rich A, Davidson N, editors. *Structural chemistry and molecular biology*. San Francisco: W.H. Freeman and Co. 1968; p. 198–215.
30. Wiegel FW, DeLisi C. Evaluation of reaction rate enhancement by reduction in dimensionality. *Am J Physiol* 1982;**243**:R475–9.

31. Axelrod D, Wang MD. Reduction-of-dimensionality kinetics at reaction-limited cell surface receptors. *Biophys J* 1994;**66**:588–600.
32. Whitty A, Raskin N, Olson DL, Borysenko CW, Ambrose CM, Benjamin CD, Burkly LC. Interaction affinity between cytokine receptor components on the cell surface. *Proc Natl Acad Sci USA* 1998;**95**:13,165–70.
33. Gavutis M, Lata S, Lamken P, Muller P, Piehler J. Lateral ligand–receptor interactions on membranes probed by simultaneous fluorescence-interference detection. *Biophys J* 2005;**88**:4289–302.
34. Lamken P, Gavutis M, Peters I, Van der Heyden J, Uze G, Piehler J. Functional cartography of the ectodomain of the type I interferon receptor subunit ifnar1. *J Mol Biol* 2005;**350**:476–88.
35. Gavutis M, Jaks E, Lamken P, Piehler J. Determination of the two-dimensional interaction rate constants of a cytokine receptor complex. *Biophys J* 2006;**90**:3345–55.

Chapter 4

Molecular Sociology

Irene M. A. Nooren[1] and Janet M. Thornton[1,2]

[1]*European Bioinformatics Institute, Wellcome Trust Genome Campus, Hinxton, Cambridge, UK*

[2]*seconded from University College London (London) and Birkbeck College (London)*

TRANSMEMBRANE SIGNALING PARADIGMS

The initiation of a cell signaling event relies primarily on interactions between molecules in the extracellular and cell-membrane space. Different types of molecules can serve as extracellular signals (Figure 4.1a) [1]: hormones, cytokines, growth factors, and neurotransmitters secreted from distant or neighboring cells; antigens or antibodies free in solution or attached to (migrating) leucocytes or foreign (e.g., virus) cells; small soluble molecules (i.e.,<1000 Da; ions, metabolites); and the extracellular matrix. Except for lipid-soluble signaling molecules that can migrate through the lipid bilayer (e.g., steroid hormones and NO gas), transmembrane proteins are involved in transferring the molecular signal into the target cell. Small soluble molecules can be transported across the plasma membrane by channel and carrier proteins or cell–cell GAP junctions, which provide an electrical and metabolic coupling with the extracellular space and neighboring cells, respectively. Larger soluble or tethered molecules, including filaments from the extracellular matrix, require a specific interaction with a transmembrane (co-) receptor for signal transduction across the membrane.

The transmembrane protein undergoes an intramolecular conformational change or change in the quaternary structure (e.g., dimerization) upon binding of the extracellular molecular signal. It is non-covalently or covalently linked either to an ion channel that allows the change of the ion traffic across the membrane or to intracellular membrane-proximal components that are activated to induce an intracellular signaling cascade. In the latter case, the receptor can contain intrinsic enzyme (e.g., phosphorylation) activity, such as the receptor tyrosine kinases; recruit relevant intracellular enzymes; or associate with G proteins, which in turn activate kinases or ion channels. While the transmembrane signaling process mediated by (ion) channels is immediate and brief, enzyme-linked receptors manifest a slow and more complex molecular mechanism but can achieve a great amplifying signaling effect. Subsequently, gene expression in the nucleus or other cell activities are affected. Recent studies have shown that endocytosis of transmembrane receptor complexes can be used to deliver the complex and affect activities at distant locations in the cell [2].

A careful regulation and coordination of the communication within the molecular signaling society is essential to the initiation of a signaling event, especially in the more complex multicellular organisms. Any molecular interaction, such as that between a receptor or receptor subunits and ligand molecules, is determined by the effective local molecular concentrations and (apparent) dissociation constants. The concentrations of the signaling and receptor molecules can be controlled by various factors at different stages along the path toward an encounter (Figure 4.1b). After synthesis or secretion, enzymatic degradation or temporary storage can influence the concentration of the signaling molecule, whereas lateral capping and endocytosis can alter the density of receptor molecules at the membrane surface. A rapid turnover of signals and receptor molecules is required to respond to fast changes in the environment. Signaling molecules may have to travel far (e.g., endocrine signaling) and depend on fluid streams of the vascular system and diffusion to enable an encounter with their target. The gel-like layer of proteoglycans in the extracellular matrix can serve as a selective molecular sieve to regulate the traffic of migrating cells and signaling molecules.

The local environment at the cell membrane can play an important role in controlling the receptor–ligand interaction (Figure 4.1b). By interacting with the ligands, membrane-associated molecules can localize and/or immobilize extracellular signals for internalization and degradation or

Handbook of Cell Signaling, Three-Volume Set 2 ed.

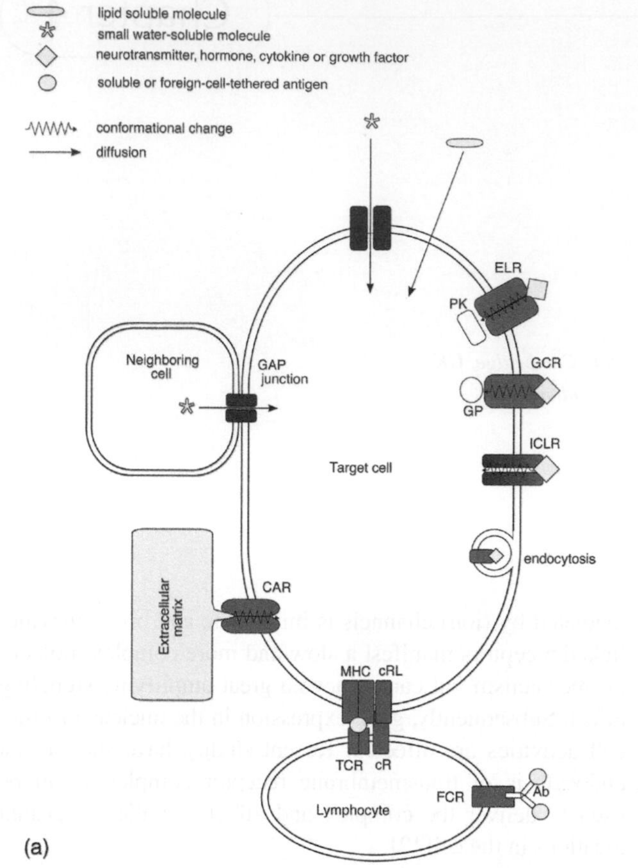

present them for receptor binding (e.g., growth factors and cytokines localized by proteoglycans). The physicochemical and geometrical properties of the molecular interface that determine the (apparent) dissociation constant of the signal binding can be altered by interaction with the local environment. A change in the physiological condition or binding of an effector molecule from the cytosol, membrane, or extracellular space (e.g., ion, metabolite, other protein) can dramatically change the affinity of a receptor–ligand complex by altering the conformation or electrostatic potential at the signal- binding interface. Individual receptor–ligand interactions may be stabilized or activated by effector binding or multiple interactions, (e.g., clustering of receptor–ligand complexes). Some complexes may be very weak on their own and require accessory proteins or co-receptors for stability or activation; for example, CD4 and CD8 co-receptors in the major histocompatibility complex (MHC) multicomponent complexes help to strengthen the adhesion between a T and antigen-presenting cell. Also, low-affinity antibody–antigen interactions may be amplified by multivalent crosslinking. Furthermore, the apparent affinity for the extracellular signaling ligand can be decreased when another molecule (i.e., antagonist) competes for the same target-binding site. These control mechanisms that regulate the signaling interaction network are similar to those that control other biomolecular interactions such as enzyme–substrate interactions. However, altering the affinity of an assembly by covalent modifications (e.g., phosphorylation)

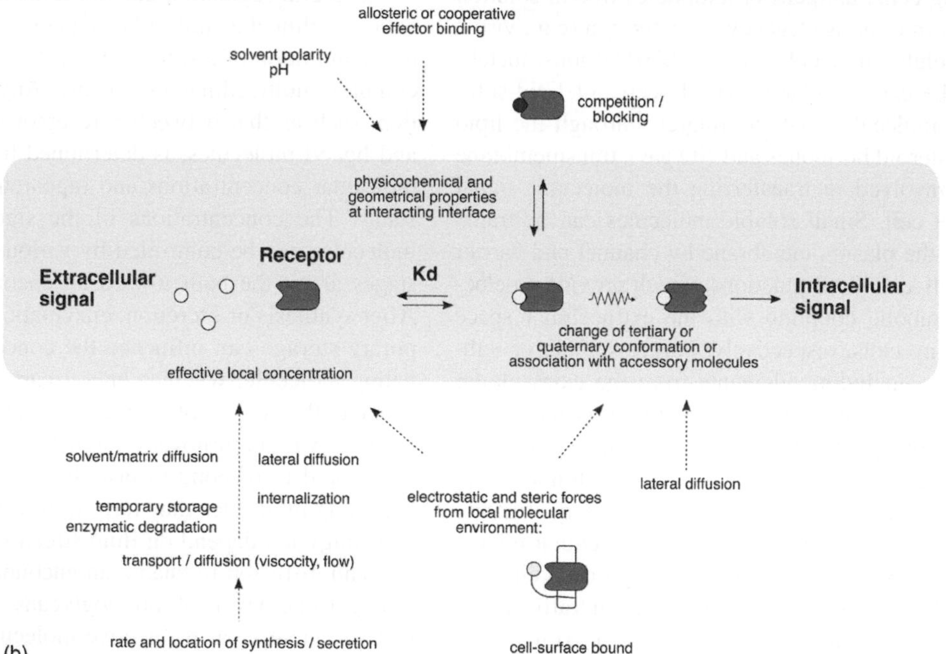

FIGURE 4.1 Transmembrane signaling paradigms.(a) Molecular interactions and mechanisms, and (b) control of an interaction between a (oligomeric) receptor and (oligomeric) ligand. Extracellular signaling molecules are yellow; channels or carrier proteins and receptors are blue and red, respectively (color shown in online version.). Lipid-soluble molecules comprise hydrophobic or small uncharged polar molecules; small water-soluble molecules comprise inorganic ions, sugars, amino acids, nucleotides, and vitamins. ELR, enzyme-linked receptor; GCR, G-coupled receptor; ICLR, ion-channel-linked receptor; PK, protein kinase; GP, G protein; Ab, antibody; MHC, major histocompatibility complex; TCR, T cell receptor; FCR, Fc receptor; cR, co-receptor; cRL, co-receptor ligand; CAR, cell adhesion receptor.

TABLE 4.1 Current receptor–protein signalling complexes in the Protein Data Bank

Protein 1	Protein 2	pdb code(s) (resolution in Å)	Oligomeric state[1]		Receptor Activity	Ref.
Growth hormone receptor	Somatotropin (growth hormone)	3hhr (2.8)[3], lhwg (2.5)	2:1	Receptor dimerization upon ligand binding	Non-protein kinase, associated JAK kinases	[12]
	GH antagonist g130r	1hwh (2.9)[3], 1a22 (2.6)	1:1			
		Remodelled interface: 1axi (2.1)[3]	1:1			
Prolactin receptor	Somatotropin (growth hormone)	1bp3 (2.9)	1:1		Non-protein kinase, associated JAK kinases	[13]
	Placental lactogen	1f6f (2.3)	2:1			
Erythropoietin receptor	Erythropoietin	1cn4 (2.8)[3], 1eer (1.9)	2:1		Non-protein kinase, associated JAK kinases	[14]
Interleukin-1 receptor	Interleukin-1 receptor antagonist	1ira (2.7)	1:1		Non-protein kinase, associated accessory protein	[15]
Interleukin-1 receptor	Interleukin-1 β	1itb (2.5)	1:1		Non-protein kinase, associated accessory protein	[16]
Interleukin-4 receptor α-chain	Interleukin-4	liar (2.3)	1:1		Non-protein kinase, associates with common γ-chain, associated JAK kinase	[17]
Granulocyte colony-stimulating factor receptor	G-csf	1pgr (3.5)[3], 1cd9 (2.8)	2:2-I		Non-protein kinase, associated JAK kinases	
Interleukin-6 receptor GP130 chain	Interleukin-6	1ilr (2.4)	2:2-I		Non-protein kinase, common β-chain (e.g., Gp130), associated JAK kinase	[19]
Trka receptor	Nerve growth factor	1www (2.2)	2:2-II		Tyrosine kinase	[20]
Bone morphogenetic protein receptor 1a	Bone morphogenetic protein-2	1es7 (2.9)	2:2-II		Serine-threonine kinase	[21]
Interferon-γ receptor α	Interferon-γ	1fg9 (2.9)[3], 1fyh (2.0)	2:2-II		Non-protein kinase, associated JAK kinases	[22]
Interleukin-10 receptor	Interleukin-10	1j7v (2.9)	2:2-II or 4:4[2]		Non-protein kinase, associated JAK kinases	[23]
Fibroblast growth factor receptor 1	Fibroblast growth factor-1	1evt (2.8)	2:2-I	Homo- and heterodimerization, interdomain ligand binding; heparin involved	Tyrosine kinase	[24]

(Continued)

TABLE 4.1 (Continued)

Protein 1	Protein 2	pdb code(s) (resolution in Å)	Oligomeric state[1]		Receptor Activity	Ref.
Fibroblast growth factor receptor 1	Fibroblast growth factor-2	1cvs (2.8), 1fq9 (3.0; heparin bound)	2:2-I			
Fibroblast growth factor receptor 2	Fibroblast growth factor-1	1djs (2.4)	2:2-I			
Fibroblast growth factor receptor 2	Fibroblast growth factor-2	1ev2 (2.2)	2:2-1			
	Fibroblast growth factor-2 apert syndrome variant	mutant: 1iil (2.3)[3], 1ii4 (2.7)[3]	2:2-I			
Death receptor 5	TRAIL	1d0g (2.4)[3], 1d4v (2.2)	3:3		Non-rotein kinase, associated TRAF	[25]
Tumor necrosis factor receptor	Tumor necrosis factor β	1tnr (2.9)	3:3	Dimerizes without ligand; trimerizes upon ligand binding	Non-protein kinase, associated TRAF	[26]

[1]There are differences in the literature about the nomenclature used to describe receptor–ligand complexes. Here, we use the stoichiometry of the complex (i.e., the number of protomer chains of the receptor and ligand, respectively, involved in the complex). We identify two types of 2:2 complexes: the 2:2-I type, where each receptor chain contacts both monomeric ligands, and the 2:2-II type, where each receptor chain contacts both protomers of the dimeric ligand.
[2]The 2:2 complex is thought to form an intermediate receptor–ligand complex, whereas the 4:4 is the active receptor–ligand complex. Structural parameters (Figure 4.2) have been computed for the former.
[3]These entries have not been included in the computational analysis shown in Figure 4.2.

is more common in intracellular signaling, where a large repertoire of modifying enzymes is available.

STRUCTURAL BASIS OF PROTEIN–PROTEIN RECOGNITION

A specific interaction between a signaling and membrane receptor molecule is critical to obtain a well-directed signaling event. The molecular recognition process that underlies a specific interaction is provided by the complementarity of the physicochemical and geometrical properties of the two protein surfaces to obtain an energetically favorable complex. This is determined by the hydrophobic effect, close packing with favorable van der Waals interactions, and the formation of hydrogen and ionic bonds. Computational analyses of atomic structures of protein–protein complexes have identified the structural and physiochemical properties of these interfaces [3–5] (see Kleanthous [6] for reviews). Structures of various extracellular molecular signaling complexes have been elucidated so far: the extracellular domains of receptors complexed with hormones and cytokines, the major histocompatibility complex with diverse peptides (pMHC) in association with the T cell receptor (TCR), and antibody (Fab fragments)–antigen complexes [6]. For the majority of these complexes, the individual components form homo- or hetero-oligomers by itself or upon ligand binding. Table 4.1 summarizes the receptor-ligand complexes and their oligomeric disposition [7].

In general, protein–protein interfaces exhibit a mixture of apolar and polar interactions scattered over the binding surface with the polar residues providing fine specificity. The interfaces of non-obligate complexes (i.e., between molecules that also exist on their own), such as extracellular signaling and enzyme–inhibitor complexes are generally more polar than homodimers (Figure 4.2), because of the solubility requirements of the individual molecules. Whereas the percentage of polar atoms in the interface is variable in the receptor–ligand and pMHC–TCR complexes, the antibody–antigen complexes consistently have more than 40 percent polar atoms in the interface and have a relatively small contact area (i.e., interface smaller than 1500 Å^2). The surface area buried in the specific non-obligate protein–protein associations is also highly variable. Large contact areas up to 5000 Å^2 are found for homodimers and various non-obligate complexes, such as multimeric receptor–ligand and large enzyme-inhibitor complexes (Figure 4.2).

Structural rearrangements upon protein–protein association have been identified for many complexes, such as enzyme-inhibitor (e.g., thrombin-hirudin), intracellular signaling

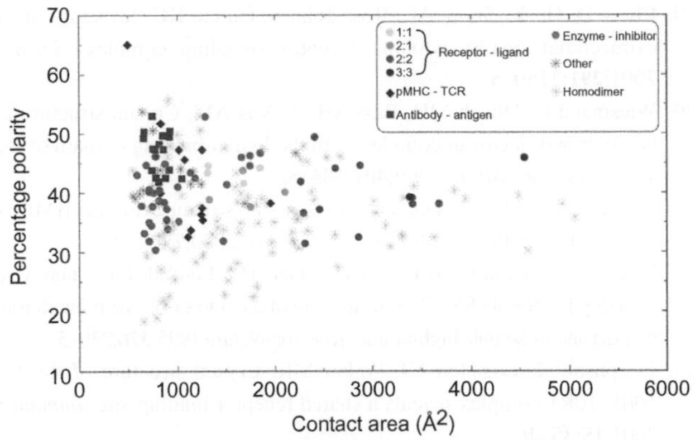

FIGURE 4.2 Correlation between the contact area of the interface and the percentage of the contact area that involves polar atoms for extracellular signaling complexes, diverse other non-obligate complexes, and homodimers.Structures used in this analysis are those studied previously [3,5], and more recent solved structures. Parameters have been calculated as described in Jones *et al.* [27]. When one or both of the proteins involved in the complex are multimers the contact area is summed over all receptor–ligand interfaces to give a total for the complete assembly, and the percentage polarity has been averaged respectively.

(e.g., G_α–$G_{\beta\gamma}$ protein) and receptor–ligand (e.g., receptor–human growth factor) complexes. Remarkably, protein–protein complexes that undergo structural rearrangements usually have large interfaces (i.e.,>1500 Å^2). They involve disorder-to-order transitions, small changes in side-chain conformations (i.e., translational, rotational, and side-chain degrees of freedom), or gross conformational changes such as loop or domain movements. The monomeric human growth factor, for example, shows large helix movements upon binding into a cleft formed by the two subunits of the homodimeric receptor. Conformational changes are expected to play a major role in the transmembrane signaling process. Upon receptor–ligand complexation, side-chain flexibility may facilitate finding the complementary fit [8], whereas larger conformational changes may reveal hydrophobic surfaces or propagate a long-range structural rearrangement of the monomeric or oligomeric transmembrane receptor required for signal transduction. Residue spacing and molecular flexibility have been demonstrated to be important for protein–carbohydrate recognition in signaling, as well [9].

The structural basis of conformational flexibility is difficult to assess experimentally, as the current available experimental methods in structural determination (i.e., X-ray crystallography, nuclear magnetic resonance spectroscopy, electron microscopy) require a stable structure. Capturing the structures under different conditions (e.g., free and bound) or having thermodynamic and mutagenesis data can help to identify these changes or relate them to signal transduction activity. The current structural data in the Protein Data Bank (PDB) and a computational analysis of these protein–protein complexes in the PDB (Figure 4.2) are probably biased toward structures that form stable structures. The atomic structures of many transmembrane domains or proteins have yet to be determined, which leaves the molecular mechanisms responsible for the signal transduction across the membrane still largely unknown.

CONCLUSION

The biology of signaling features a whole range of interactions, from weak to strong. Dissociation constants are found in the nM to mM range [10]. Both the complementarity of the physicochemical and geometrical properties of the interface and the flexibility of the surfaces of the receptor and signaling molecule contribute to the binding free energy. Structural data and currently available computational methods appear inadequate to estimate the binding energy or specificity of an interaction. The interaction between molecules has usually been optimized throughout evolution to tune specificity and affinity to function and physiological environment (Figure 4.1b). For example, high local concentrations of neurotransmitter can be reached in a chemical synapse that allows a low-affinity receptor–neurotransmitter interaction. Also, adjacent helper molecules or multiple interactions allow reduced affinity of isolated complexes by adding up weak interactions, generating a strong multicomponent complex. In comparison to other (cytosolic) protein–protein complexes, the anchoring of molecules in or at the membrane site is advantageous to localizing target subunits and substrate. In contrast, immune response complexes, such as antigen–antibody assemblies lack an extended period of a selective evolutionary optimization toward their physiological environment. Consequently, they exhibit a poorer shape complementarity than other molecular protein–protein associations [11]. In summary, proteins are versatile and the recognition process is different and unique to each molecular complex. Some of these will be discussed in more detail in the following chapters of this section.

REFERENCES

1. Hancock JT. *Cell signalling*. London: Pearson Education Limited; 1997.
2. Di Fiore PP, De Camilli P. Endocytosis and signaling an inseparable partnership. *Cell* 2001;**106**:1–4.
3. Jones S, Thornton JM. Principles of protein–protein interactions. *Proc Natl Acad Sci USA* 1996;**93**:13–20.
4. Larsen TA, Olson AJ, Goodsell DS. Morphology of protein–protein interfaces. *Structure* 1998;**6**:421–7.
5. Conte LL, Chothia C, Janin J. The atomic structure of protein–protein recognition sites. *J Mol Biol* 1999;**285**:2177–98.
6. Kleanthous C. *Frontiers in molecular biology: protein–protein recognition*. Hames BD, Glover DM, editors. London: Oxford University Press; 2000.
7. Meager T. *The molecular biology of cytokines*. London: John Wiley & Sons; 1998.
8. Xu D, Tsai CJ, Nussinov R. Hydrogen bonds and salt bridges across protein–protein interfaces. *Protein Eng* 1997;**10**:999–1012.

9. Tumova S, Woods A, Couchman JR. Heparan sulfate proteoglycans on the cell surface: versatile coordinators of cellular functions. *Int J Biochem Cell Biol* 2000;**32**:269–88.
10. Heldin CH, Purton M. Signal transduction: modular texts in molecular and cell biology. Vol. I, London: Chapman & Hall; 1996.
11. Lawrence MC, Colman PM. Shape complementarity at protein/protein interfaces. *J Mol Biol* 1993;**234**:946–50.
12. de Vos AM, Ultsch M, Kossiakoff AA. Human growth hormone and extracellular domain of its receptor: crystal structure of the complex. *Science* 1992;**255**:306–12.
13. Somers W, Ultsch M, de Vos AM, Kossiakoff AA. The X-ray structure of a growth hormone–prolactin receptor complex. *Nature* 1994;**372**: 478–81.
14. Syed RS, Reid SW, Li C, Cheetham JC, Aoki KH, Liu B, Zhan H, Osslund TD, Chirino AJ, Zhang J, Finer-Moore J, Elliott S, Sitney K, Katz BA, Matthews DJ, Wendoloski JJ, Egrie J, Stroud RM. Efficiency of signalling through cytokine receptors depends critically on receptor orientation. *Nature* 1998;**395**:511–16.
15. Schreuder H, Tardif C, Trump-Kallmeyer S, Soffientini A, Sarubbi E, Akeson A, Bowlin T, Yanofsky S, Barrett RW. A new cytokine-receptor binding mode revealed by the crystal structure of the IL-1 receptor with an antagonist. *Nature* 1997;**386**:194–200.
16. Vigers GP, Anderson LJ, Caffes P, Brandhuber BJ. Crystal structure of the type-I interleukin-1 receptor complexed with interleukin-1beta. *Nature* 1997;**386**:190–4.
17. Hage T, Sebald W, Reinemer P. Crystal structure of the interleukin-4/receptor alpha chain complex reveals a mosaic binding interface. *Cell* 1999;**97**:271–81.
18. Aritomi M, Kunishima N, Okamoto T, Kuroki R, Ota Y, Morikawa K. Atomic structure of the GCSF–receptor complex showing a new cytokine-receptor recognition scheme. *Nature* 1999;**401**:713–17.
19. Chow D, He X, Snow AL, Rose-John S, Garcia KC. Structure of an extracellular gp130 cytokine receptor signaling complex. *Science* 2001;**291**:2150–5.
20. Wiesmann C, Ultsch MH, Bass SH, de Vos AM. Crystal structure of nerve growth factor in complex with the ligand-binding domain of the TrkA receptor. *Nature* 1999;**401**:184–8.
21. Kirsch T, Sebald W, Dreyer MK. Crystal structure of the BMP-2-BRIA ectodomain complex. *Nat Struct Biol* 2000;**7**:492–6.
22. Walter MR, Windsor WT, Nagabhushan TL, Lundell DJ, Lunn CA, Zauodny PJ, Narula SK. Crystal structure of a complex between interferon-gamma and its soluble high-affinity receptor. *Nature* 1995;**376**:230–5.
23. Josephson K, Logsdon NJ, Walter MR. Crystal structure of the IL-10/IL-10R1 complex reveals a shared receptor binding site. *Immunity* 2001;**15**:35–46.
24. Plotnikov AN, Schlessinger J, Hubbard SR, Mohammadi M. Structural basis for FGF receptor dimerization and activation. *Cell* 1999;**98**:641–50.
25. Hymowitz SG, Christinger HW, Fuh G, Ultsch M, O'Connell M, Kelley RF, Ashkenazi A, de Vos AM. Triggering cell death: the crystal structure of Apo2L/TRAIL in a complex with death receptor 5. *Mol Cell* 1999;**4**:563–71.
26. Banner DW, D'Arcy A, Janes W, Gentz R, Schoenfeld HJ, Broger C, Loetscher H, Lesslauer W. Crystal structure of the soluble human 55 kd TNF receptor–human TNF beta complex: implications for TNF receptor activation. *Cell* 1993;**73**:431–45.
27. Jones S, Marin A, Thornton JM. Protein domain interfaces: characterization and comparison with oligomeric protein interfaces. *Protein Eng* 2000;**13**:77–82.

Chapter 5

FRET Analysis of Signaling Events in Cells

Peter J. Verveer and Philippe I. H. Bastiaens
Cell Biology and Cell Biophysics Program, European Molecular Biology Laboratory, Heidelberg, Germany

INTRODUCTION

Detection of fluorescence resonance energy transfer (FRET) by optical techniques provides a sensitive means of detecting and quantifying molecular interactions and protein modifications in cells. Several strategies are available to develop sensors for use in FRET assays based on fluorescent labeling or green fluorescent protein (GFP) fusions. By using these sensors, techniques such as ratio imaging, sensitized emission measurements, photobleaching methods, or fluorescence lifetime imaging can be employed to spatially and temporally resolve the occurrence of FRET in cells. In this contribution, the strengths and weaknesses of the different sensors and measurement methods are discussed and compared and their use illustrated by reviewing the recent literature. We conclude that the spatially and temporally resolved measurement of FRET in cells has opened new opportunities to image biochemistry in intact cells and expect that these techniques will play an increasingly important role in cell biology.

Optical microscopy provides a sensitive, specific, and non-invasive approach to localize fluorescently labeled macromolecules in cells with high spatial and temporal resolution. Moreover, the spectroscopic properties of fluorescence probes can be used to obtain information on their molecular environment. With the advent of genetically encoded variants of green fluorescent proteins, the observation of biochemistry in cells has become feasible *in vivo* [1–3]. Fluorescence resonance energy transfer is one photophysical phenomenon that has been put to good use to detect and quantify molecular interactions and protein modifications [4]. FRET cannot be measured directly, but the resulting changes in the fluorescence properties of the fluorophores can be detected by several optical techniques with spatial and temporal resolution inside cells. FRET is a photophysical effect whereby energy is transferred from an excited donor fluorophore to an acceptor fluorophore. This does not involve the emission and subsequent absorption of a photon but occurs by a direct electromagnetic interaction. The efficiency of transfer depends on the spectral properties of donor and acceptor, and on their relative orientation and distance. An important factor is that the energy transfer efficiency has an inverse sixth order dependence on the distance between the two fluorophores, and typically FRET only occurs when the distance is less than 10 nm, which is generally only achieved when donor and acceptor are attached to the same macromolecule or to interacting molecules. For this reason FRET can be applied to specifically image such events as molecular interactions or conformational changes. FRET can be observed by its consequences, which are reflected by a change in the fluorescence kinetics of both the donor and the acceptor. Due to the transfer of energy the rate at which the donor returns to its ground state increases, and hence its fluorescence lifetime decreases. As a consequence the quantum yield of the donor, and therefore its steady-state fluorescence intensity, also decreases. The steady-state fluorescence of the acceptor, however, is increased by the sensitized emission that is emitted when the acceptor returns to its ground state.

FLUORESCENT PROBES FOR FRET

The design of the fluorescent probes for FRET measurements must match the problem at hand. Single component molecule sensors, consisting of two fluorophores flanking a protein domain or subunit, change FRET efficiency upon a change of conformation or cleavage. These types of sensors are commonly constructed by fusing cyan and yellow variants of GFP to the reporter domains. They can be

Handbook of Cell Signaling, Three-Volume Set 2 ed.

used to detect physiologically relevant ions [5–7] and small organic compounds [8–10], or report on protein activity or conformational or covalent state [11–14]. They can be targeted to specific cellular compartments by incorporating suitable localization signals. Another class of sensors is based upon the interaction between different compounds tagged with a donor or acceptor fluorophore [15–20]. One application is the detection of covalent modification of a protein, in which case the protein is tagged by a donor, and an acceptor-tagged reagent interacting with the conjugated group is present in large excess. The state of donor-tagged proteins can then be probed by FRET. An example is the use of generic phosphoamino-acid-specific antibodies to probe the phosphorylation state of a protein, whereby the protein is tagged with a donor fluorophore (e.g., by fusing a GFP molecule) and the antibody with the acceptor fluorophore (e.g., Cy3). Another application, not necessarily employing an excess of one species, is using cyan and yellow GFP variants to measure interactions between different proteins, or homo- or heterodimerization *in vivo*. The use of GFP variants is attractive for live cell applications, but FRET measurements can also be done via labeled antibodies. This may seem potentially problematic due to the antibody size, but depending on the donor/acceptor configuration and the type of FRET measurement this may in fact be an advantage. Consider an assay to probe the state of a protein based on the observation of the donor fluorophore (e.g. a GFP) only and an antibody labeled with multiple acceptor fluorophores (e.g., Cy3; see [20]). In this case, the increased density of acceptor fluorophores implies a higher probability that upon antibody binding, the donor is in close proximity of an acceptor. However, using an antibody labeled with multiple donor fluorophores would lead to an increased probability of detecting a donor that is too far away from an acceptor for efficient FRET and therefore a lower average signal. In general, the choice of sensor, fluorophore (GFP or fluorescent dye), and where donor and acceptor are located is also determined by the measurement method that is used. The different techniques are described in the next section, where the types of sensors that are appropriate for the respective techniques will also be discussed.

FRET DETECTION TECHNIQUES

Ratio Imaging

Upon the occurrence of FRET the steady-state fluorescence of the donor is quenched, since part of the excited state energy is transferred to the acceptor rather than emitted as photons. Simultaneously, emitting this energy as photons increases the steady-state fluorescence of the acceptor. Therefore, a change in FRET is reflected in an increase in the ratio of sensitized emission over donor emission. This type of measurement is straightforward, as it requires only measurements at two filter settings, and is therefore well suited to live cell imaging.

In an ideal situation, the sensitized emission and donor emission would be directly measured by choosing appropriate combinations of excitation and emission filters, exciting the donor specifically, and detecting the intensities through filters specific for donor and acceptor emissions. Indeed, the donor emission can be imaged specifically; however, in practice the measured acceptor channel contains significant contributions of leak-through of the donor emission and of direct excitation of the acceptor. This implies that the signal in the acceptor channel is strongly dependent on the relative concentrations of donor and acceptor, and that interpretation of the ratio as a diagnostic for FRET is problematic. However, if the relative concentrations of donor and acceptor are fixed, then a change in the donor/acceptor fluorescence ratio can be attributed to a change in FRET. Therefore this approach should mainly be used with sensors that consist of donor and acceptor fluorophores attached to a single molecule. A change in FRET due to a conformational change can then be reliably detected. Another situation is presented by a sensor where donor and acceptor disassociate, for instance by proteolytic activity. In such a case the relative concentrations are known before cleavage and, if no significant differential relocation of the cleaved products is expected, the ratio may still be a good measure of FRET. Also, if one is only interested in the total signal integrated over a cell, the relative total concentrations can be considered to remain the same, although this assumption may be incorrect in a confocal microscope, where molecules may relocate to a position outside the focal plane.

In general, quantification of ratio measurements in terms of the concentrations of the species that exhibit FRET is not straightforward, due to the unknown contributions of leak-through and direct excitation. However, it is possible to externally calibrate the ratio values to physiologically relevant quantities by using reference samples where a known concentration of the species of interest can be related to the measured ratio.

Ratio measurements of FRET have been utilized in a wide spectrum of applications. They have been used as an indicator for adenosine 3′,5′-cyclic monophosphate (cAMP) [8, 10], guanosine 3′,5′-cyclic monophosphate (cGMP) [9], and Ca^{2+} [5, 6]. More recently sensors for protein kinase [11, 12, 14], and GTPase [13] activity have been described.

Sensitized Emission Measurements

Whereas ratio measurements are difficult to quantify, it is possible to make use of reference measurements to further quantify results. Generally three measurements are made: (1) using a filter set where the donor is excited ands the donor emission is measured (donor filter-set); (2) using a filter set where the acceptor is excited and the acceptor emission is measured (acceptor filter-set); and (3) using a filter set where the donor is excited and the acceptor emission is measured (FRET filter-set). The images from the donor and acceptor filter sets are multiplied with correction factors and subtracted from the image taken with the FRET filter-set to obtain the sensitized emission, corrected for contributions of donor leak-through and direct excitation. The correction factors are determined via reference samples that contain only donor or only acceptor molecules. The resulting estimation of the intensity of the sensitized emission is then normalized for donor and/or acceptor concentration by using an expression for apparent energy transfer, for which several approaches have been taken [21–23]. Ideally, quantification should provide the relative fraction of molecules that are exhibiting FRET. Up to a scalar factor, it is indeed possible to determine the fraction of acceptor molecules that are exhibiting FRET, since the acceptor concentration can be directly related to the acceptor fluorescence. Determining the fraction of donor molecules that exhibit FRET is much more difficult, since the donor fluorescence is not proportional to the donor concentration, owing to the quenching of the donor fluorescence by FRET. The donor concentration can, however, be measured using acceptor photobleaching, as described below.

In contrast to ratio measurements, this approach is suited to applications where donor and acceptor are not on the same molecule, due to the correction for leak-through and direct excitation. They are also suitable to live cell imaging since only three measurements need to be made. Mostly these approaches are used when donor- and acceptor-tagged molecules have concentrations that are in the same order of magnitude [17]. This FRET method may become less effective in cases where the acceptor is in large excess, since the corrections for direct excitation become large and the result is more susceptible to noise. Similarly, a large excess of donor leads to large corrections for donor leak-through, with associated problems to estimate the sensitized emission. Thus this approach is likely to be less suitable for sensors where a protein state is determined by a large excess of a probe, although sensors to probe the state of a protein have been reported [18]. Even if donor and acceptor concentrations are similar, the corrections can be large, depending on the spectral properties of the donor/acceptor pair that is employed. The different variants of GFP are examples in which the corrections may be substantial, and in such cases this method works best with a high energy transfer efficiency between donor and acceptor, implying a large relative contribution of sensitized emission.[18]. We note that the correction factors are generally determined from averages of images of reference samples, and are therefore scalar factors. It is not known how much these factors change as a function of the environment or whether scalar correction factors are sufficient.

Sensitized emission measurements have been used to study the interactions between different molecules. Examples are the studies by Sorkin and colleagues, who looked at the interaction of the EGF receptor with Grb2 in living cells [17], and by Mahajan and colleagues, who investigated interaction of Bcl-2 with Bax [24]. The sensitized emission method was also used to localize the activity of the GTPase Rac [18].

Methods Based on Photobleaching

Photobleaching of either the donor or acceptor molecules can be utilized to detect the effects of FRET on the kinetics of the fluorescence of either. Photobleaching is the process whereby a fluorophore is converted to a non-fluorescent species, for instance in the presence of oxygen. This essentially happens only when the donor is in the excited state and therefore the rate at which photobleaching occurs is proportional to the average time it spends in the excited state, which in turn is inversely proportional to the rate at which the molecule returns to its ground state. Hence, an increase in the latter due to FRET can be detected by a decrease in the photobleaching rate. Thus, one approach to measure and quantify FRET is to measure the kinetics of photobleaching of the donor [25, 26]. Generally, the kinetics are described by a sum of exponentials, and it is possible to quantify the fraction of donor molecules exhibiting FRET. Although this is a potentially precise approach, it is not in much use nowadays for several reasons. First, the requirement to photobleach the donor over extended time periods leads to long acquisition times. Therefore the method is mostly useful on fixed samples, although presumably live cells could be used if one is only interested in the integrated response of a complete cell. Second, the mechanisms of photobleaching are not understood very well, although the assumption of a multi-exponential model is reasonable in first approximation.

Rather than examining the photobleaching kinetics of the donor, one can utilize photobleaching of the acceptor [16, 27]. Acceptor molecules can be excited specifically, since their absorption spectrum generally does not overlap with that of the donor. One makes use of the simple property that destroying all acceptor fluorophores abolishes FRET. Thus, after photobleaching the acceptor, the donor intensity should increase, since it is not quenched anymore by FRET. Therefore, comparing the intensity of the donor before and after acceptor photobleaching should indicate whether FRET was occurring. Dividing the difference of the donor intensity before and after photobleaching by the intensity after photobleaching yields an apparent energy transfer measure that is directly proportional to the fraction of donor molecules exhibiting FRET. Obviously this approach is attractive since it is experimentally easy, and also interpretation does not require extensive analysis.

Acceptor photobleaching does suffer from the same drawback as donor photobleaching in that it is a time-consuming approach less suitable for imaging of FRET in live cells. A point of possible concern that has not been addressed much so far is that photobleaching of the acceptor may create a different species of molecule that fluoresces in the donor channel, leading to an overestimation of the unquenched donor signal. Alternatively, a dark species could be created that absorbs light and still acts as an acceptor, leading to an underestimation of the unquenched donor fluorescence.

Acceptor photobleaching has been used as an independent technique to measure FRET. For instance, it was used to study the localization of the A- and B-subunits of cholera toxin [16], and recently to image the three-dimensional distribution of receptor tyrosine kinases interacting with protein tyrosine phosphatase 1B [28]. It is also increasingly being used in combination with other techniques as a standard control for the occurrence of FRET [10, 14, 29, 30].

Fluorescence Lifetime Imaging Microscopy

The kinetics of fluorescence can be measured by fluorescence lifetime imaging microscopy (FLIM). This makes it possible to detect whether the rate at which an excited donor molecule returns to the ground state increases by FRET, since the fluorescence lifetime of a fluorophore is inversely proportional to the sum of rates of all possible pathways by which an excited molecule returns to the ground state [31].

FLIM has been applied in a qualitative fashion where an average lifetime is measured that decreases upon FRET [31]. In such a case, photobleaching the acceptor may serve as an internal control since after destruction of the acceptor fluorophore the lifetime of the donor should attain its normal value [30]. In this case, the acceptor photobleaching can also be applied in live cell imaging by photobleaching after FLIM measurements are made. Since the fluorescence lifetime of the donor is independent of concentration and light path-length, the average donor lifetime can then serve as the control value. FLIM has also been applied quantitatively by resolving the multi-exponential decay kinetics of the donor to determine the fractions of donor molecules exhibiting FRET [20].

So far, FLIM has been applied mostly to donor-only imaging. It is therefore well suited to applications in which the acceptor is present in excess – e.g., to probe the state of a donor-tagged molecule [19, 20]. However, it is also suited to cases in which the donor and acceptor are available in comparable concentrations [28]. FLIM has also been applied by imaging both donor and acceptor simultaneously. In this case, the kinetics of the acceptor come into play, and it becomes possible to use fluorophores that are difficult to separate spectrally, such as GFP and YFP [32]. In principle, it is then possible to quantitatively determine both the fractions of donor and acceptor that are participating in FRET, but this has not been demonstrated yet. FLIM requires the acquisition of multiple images but is rapid enough to enable live cell imaging. One drawback of the method in comparison with other FRET methods is that it is more technically involved and equipment cost is higher.

FLIM has been used in a wide variety of applications, among others to probe the phosphorylation state of proteins such as PKCα [19] and the EGF-receptor [20, 30] and to study the proteolysis of PCKβ1 [15], the oligomerization of EGF-receptors [33], and the interactions between PKCα and ezrin [29]. In a rather different type of application, Murata and colleagues [34] studied the organization of DNA in cell nuclei by visualizing FRET between the AT-specific donor Hoechst 33258 and the GC-specific acceptor 7-aminoactinomycin D.

CONCLUSIONS AND PROSPECTS

Exploitation of the physical phenomenon of FRET in biomolecular systems has opened new ways to image biochemistry in live cells. Several optical techniques to measure FRET have been developed in recent years and the number of applications of these techniques to relevant biological systems has been increasing steadily. All of these techniques have their strengths and weaknesses, and it is to be expected that the technological developments will continue for some time. In addition, the development of novel sensors that are based on FRET measurements promises to open more fields of applications, not in the least due to the large variety of GFPs that is becoming available [35, 36]. FRET is complementary to biochemical approaches for investigating the complex signaling systems that are encountered at the cellular level in the fundamental biological sciences. We therefore expect that FRET measurements will play an increasingly important role in cell biology in the future.

REFERENCES

1. Heim R, Tsien RY. Engineering green fluorescent protein for improved brightness, longer wavelengths and fluorescence resonance energy transfer. *Curr Biol* 1996;**6**:178–82.
2. Matz MV, Fradkov AF, Labas YA, Savitsky AP, Zaraisky AG, Markelov ML, Lukyanov SA. Fluorescent proteins from nonbioluminescent *Anthozoa* species. *Nat Biotechnol* 1999;**17**:969–73.
3. Tsien RY. The green fluorescent protein. *Annu Rev Biochem* 1998;**67**:509–44.
4. Clegg RM. Fluorescence resonance energy transfer. In: Wang and XF, Herman B, editors. *Fluorescence imaging spectroscopy and microscopy*. New York: Wiley; 1996. p. 179–252.
5. Miyawaki A, Llopis J, Heim R, McCaffery JM, Adams JA, Ikura M, Tsien RY. Fluorescent indicators for Ca^{2+} based on green fluorescent proteins and calmodulin. *Nature* 1997;**388**:882–7.

6. Miyawaki A, Griesbeck O, Heim R, Tsien RY. Dynamic and quantitative Ca^{2+} measurements using improved cameleons. *Proc Natl Acad Sci USA* 1999;**96**:2135–40.
7. Truong K, Sawano A, Mizuno H, Hama H, Tong KI, Mal TK, Miyawaki A, Ikura M. FRET-based *in vivo* Ca^{2+} imaging by a new calmodulin-GFP fusion molecule. *Nat Struct Biol* 2001;**8**:1069–73.
8. Adams SR, Harootunian AT, Buechler YJ, Taylor SS, Tsien RY. Fluorescence ratio imaging of cyclic AMP in single cells. *Nature* 1991;**349**:694–7.
9. Honda A, Adams SR, Sawyer CL, Lev Ram VV, Tsien RY, Dostmann WR. Spatiotemporal dynamics of guanosine 3′,5′-cyclic monophosphate revealed by a genetically encoded, fluorescent indicator. *Proc Natl Acad Sci USA* 2001;**98**:2437–42.
10. Zaccolo M, De Giorgi F, Cho CY, Feng L, Knapp T, Negulescu PA, Taylor SS, Tsien RY, Pozzan T. A genetically encoded, fluorescent indicator for cyclic AMP in living cells. *Nat Cell Biol* 1999;**2**:25–9.
11. Nagai Y, Miyazaki M, Aoki R, Zama T, Inouye S, Hirose K, Iino M, Hagiwara M. A fluorescent indicator for visualizing cAMP-induced phosphorylation *in vivo*. *Nat Biotechnol* 2000;**18**:313–16.
12. Ting AY, Kain KH, Klemke RL, Tsien RY. Genetically encode fluorescent reporters of protein tyrosine kinase activities in living cells. *Proc Natl Acad Sci USA* 2001;**18**:15,003–15,008.
13. Mochizuki N, Yamashita S, Kurokawa K, Ohba Y, Nagai T, Miyawaki A, Matsuda M. Spatio-temporal images of growth-factor-induced activation of Ras and Rap1. *Nature* 2001;**411**:1065–8.
14. Zhang J, Ma Y, Taylor SS, Tsien RY. Generically encode reporters of protein kinase A activity reveal impact of substrate thetering. *Proc Natl Acad Sci USA* 2001;**18**:14,997–15,002.
15. Bastiaens PIH, Jovin TM. Microspectroscopic imaging tracks the intracellular processing of a signal transduction protein: fluorescent-labeled protein kinase C βI. *Proc Natl Acad Sci USA* 1996;**93**:8407–12.
16. Bastiaens PIH, Majoul IV, Verveer PJ, Söling H-D, Jovin TM. Imaging the intracelllar trafficking and state of the AB_5 quaternary structure of cholera toxin. *EMBO J* 1996;**15**:4246–53.
17. Sorkin A, McClure M, Huang F, Carter R. Interaction of EGF receptor and Grb2 in living cells visualized by fluorescence resonance energy transfer (FRET) microscopy. *Curr Biol* 2000;**10**:1395–8.
18. Kraynov VS, Chamberlain C, Bokoch GM, Schwartz MA, Slabaugh S, Hahn KM. Localized Rac activation dynamics visualized in living cells. *Science* 2000;**290**:333–7.
19. Ng T, Squire A, Hansra G, Bornancin F, Prevostel C, Hanby A, Harris W, Barnes D, Schmidt S, Mellor H, Bastiaens PIH, Parker PJ. Imaging protein kinase Cα activation in cells. *Science* 1999;**283**:2085–9.
20. Verveer PJ, Wouters FS, Reynolds AR, Bastiaens PIH. Quantitative imaging of lateral ErbB1 receptor signal propagation in the plasma membrane. *Science* 2000;**290**:1567–70.
21. Gordon GW, Berry G, Liang XH, Levine B, Herman B. Quantitative fluorescence energy transfer measurements using fluorescence microscopy. *Biophys J* 1998;**74**:2702–13.
22. Nagy P, Vámosi G, Bodnár A, Locket SJ, Szöllösi J. Intensity-based energy transfer measurements in digital imaging microscopy. *Eur Biophys J* 1998;**27**:377–89.
23. Xia Z, Liu Y. Reliable and global measurement of fluorescence resonance energy transfer using fluorescence microscopes. *Biophys J* 2001;**81**:2395–402.
24. Mahajan N, Linder K, Berry G, Gordon GW, Heim R, Herman B. Bcl-2 and Bax interactions in mitochondria probed with green fluorescenct protein and fluorescence resonance energy transfer. *Nat Biotechnol* 1998;**16**:547–52.
25. Jovin TM, Arndt-Jovin DJ. FRET microscopy: digital imaging of fluorescence resonance energy transfer. In: Kohen E, Hirschberg JG, editors. *Cell structure and function by microspectrofluorometry*. San Diego: Academic Press; 1989. p. 99–115.
26. Jovin TM, Arndt-Jovin DJ. Luminescence digital imaging microscopy. *Annu Rev Biophys Biophys Chem* 1989;**18**:271–308.
27. Bastiaens PIH, Jovin TM. FRET microscopy. In: Celis JE, editor. *Cell biology: a laboratory handbook*. New York: Academic Press; 1998. p. 136–46.
28. Haj FG, Verveer PJ, Squire A, Neel BG, Bastiaens PIH. Imaging sites of receptor dephosphorylation by PTP1B on the surface of the endoplasmic reticulum. *Science* 2002;**295**:1708–11.
29. Ng T, Parsons M, Hughes WE, Monypenny J, Zicha D, Gautreau A, Arpin M, Gschmeissner S, Verveer PJ, Bastiaens PIH, Parker PJ. Ezrin is a downstream effector of trafficking PKC-integrin complexes involved in the control of cell motility. *EMBO J* 2001;**20**:2723–41.
30. Wouters FS, Bastiaens PIH. Fluorescence lifetime imaging of receptor tyrosine kinase activity in cells. *Curr Biol* 1999;**9**:1127–30.
31. Bastiaens PIH, Squire A. Fluorescence lifetime imaging microscopy: spatial resolution of biochemical processes in the cell. *Trends Cell Biol* 1999;**9**:48–52.
32. Harpur AG, Wouters FS, Bastiaens PIH. Imaging FRET between spectrally similar GFP molecules in single cells. *Nat Biotechnol* 2001;**19**:167–9.
33. Gadella Jr TWJ, Jovin TM. Oligomerization of epidermal growth-factor receptors on A431 cells studied by time-resolved fluorescence imaging microscopy – a stereochemical model for tyrosine kinase receptor activation. *J Cell Biol* 1995;**129**:1543–58.
34. Murata S, Herman P, Lin HJ, Lakowicz JR. Fluorescence lifetime imaging of nuclear DNA: effect of fluorescence resonance energy transfer. *Cytometry* 2000;**41**:178–85.
35. Griesbeck O, Baird GS, Campbell RE, Zacharias DA, Tsien RY. Reducing the environmental sensitivity of yellow fluorescent protein. Mechanism and applications. *J Biol Chem* 2001;**276**:29,188–29,194.
36. Nagai T, Ibata K, Park ES, Kubota M, Mikoshiba K, Miyawaki A. A variant of yellow fluorescent protein with fast and efficient maturation for cell-biological applications. *Nat Biotechnol* 2002;**20**:87–90.

Chapter 6

Structures of Serine/Threonine and Tyrosine Kinases

Stevan R. Hubbard
Kimmel Center for Biology and Medicine of the Skirball Institute of Biomolecular Medicine and Department of Pharmacology, New York University School of Medicine, New York, NY 10016

INTRODUCTION

The human genome encodes approximately 500 protein kinases [1], the majority of which adopt a common three-dimensional protein architecture and catalyze phosphoryl transfer to either serine and threonine residues (protein serine/threonine kinases or PSKs) or tyrosine residues (protein tyrosine kinases or PTKs), using ATP as the phosphate donor. Protein phosphorylation is an essential post-translational covalent modification controlling virtually all cellular processes, including proliferation, differentiation, metabolism, migration, and apoptosis. Moreover, through overexpression or gain-of-function mutations, many protein kinases initiate human tumor formation or contribute to tumor progression [2].

The protein kinase domain, whose core size is approximately 275 residues, is present in membrane-spanning receptors and in soluble proteins. Transmembrane receptors include PSKs such as the transforming growth factor-β (TGFβ) receptors and the endoplasmic reticulum-resident IRE1, and also PTKs such as the insulin receptor, fibroblast growth factor receptor, epidermal growth factor (EGF) receptor, and ephrin receptors (Ephs). Examples of soluble PSKs include cyclic AMP-dependent protein kinase (PKA), protein kinase C (PKC), cyclic-dependent protein kinases (CDKs), and mitogen-activated protein kinases (MAPKs). Soluble PTKs include Src-family kinases, Abl, and JAKs, among others. This review highlights the structural features of canonical eukaryotic protein kinases and describes a few of the many molecular mechanisms by which their catalytic activity is regulated.

STRUCTURAL FEATURES OF SERINE/THREONINE AND TYROSINE KINASES

Of the 518 protein kinases that have been identified in the human genome, 478 adopt (or are predicted to adopt) a canonical eukaryotic protein kinase architecture [1], first viewed in the crystal structure of the catalytic subunit of PKA [3]. Canonical protein kinases are two-domain (bi-lobed) enzymes (Figure 6.1a). The N-terminal lobe (N lobe) comprises a five-stranded anti-parallel β-sheet (β1–β5) and an α-helix (αC), and the larger C-terminal lobe (C lobe) comprises seven α-helices (αD–αI) and four short β-strands (β6–β9). For some protein kinases, additional helices in the N and C lobes augment this core architecture (e.g., βB in the N lobe of PKA, αJ in the C lobe of the insulin receptor kinase). The polypeptide segment between β5 and αD connects the N lobe to the C lobe, although a loop N-terminal to this segment, between αC and β4, is anchored in the C lobe, resulting in three pivot points in the rotation of the N lobe relative to the C lobe.

ATP binds in the cleft between the two lobes, coordinated primarily by residues in the N lobe. The loop between β1 and β2 contains the nucleotide binding loop, also known as the glycine-rich loop or P (phosphate) loop, and β3 contains a conserved lysine residue (Lys72 in PKA) which coordinates the α- and β-phosphate groups of ATP. This lysine is oriented for phosphate binding through a salt-bridge interaction with a conserved glutamic acid in αC (Glu91 in PKA).

Residues involved in the phosphoryl transfer mechanism are contained in the eight-residue catalytic loop in the C lobe. This segment, which is remarkably similar in its configuration

Handbook of Cell Signaling, Three-Volume Set 2 ed.

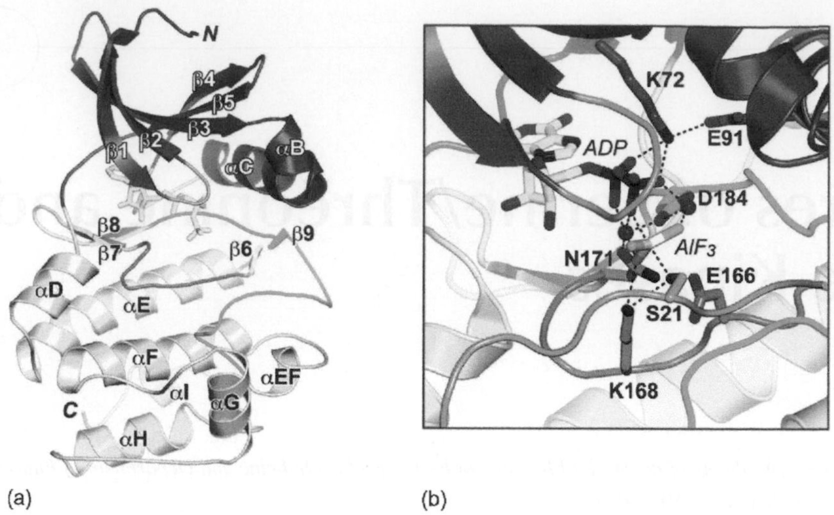

FIGURE 6.1 Crystal structure of the protein kinase domain of PKA.
(a) Ribbon diagram of the core kinase domain of the catalytic subunit of PKA (PDB code 1ATP) [3]. The secondary structural elements are labeled, as are the N- and C-termini. Bound ATP is shown in stick representation. (b) The transition state in PKA (PDB code 1L3R) [5]. The side chains of residues involved in catalysis are labeled. The two Mg^{2+} ions are shown as spheres and hydrogen bonds are shown as dashed lines.

from one protein kinase to the next and between inactive and active states, contains a conserved aspartic acid (Asp166 in PKA) and a conserved asparagine (Asn171), the former of which is directly involved in phosphoryl transfer and the latter coordinates an ATP-associated magnesium ion (Mg^{2+}) (Figure 6.1b). In addition to these two residues, the catalytic loop contains a basic residue (lysine or arginine, Lys168 in PKA), which is positioned either two (PSKs) or four (most PTKs) residues C-terminal to the catalytic aspartic acid. The activation loop/segment in the C lobe, also known as the T loop, provides a critical aspartic acid for Mg^{2+} coordination (Asp184 in PKA, DFG motif) and, for most protein kinases, serves as a steric regulatory "gate" whose conformational state is often controlled by phosphorylation within the loop [4].

PSKs and PTKs catalyze the transfer of the γ phosphate of ATP to the hydroxyl group of their respective substrates. Structural data provide evidence for a direct, in-line mechanism of phosphoryl transfer for PSKs (Figure 6.1b) [5]. Interestingly, for PTKs, biochemical and structural evidence points to a dissociative (meta-phosphate) mechanism [6,7]. This potential difference in phosphoryl transfer mechanism between PSKs and PTKs is likely due to the differing nucleophilicities of the two hydroxyl substrates (serine/threonine and tyrosine).

REGULATION OF SERINE/THREONINE AND TYROSINE KINASE ACTIVITY

Because of the critical roles played by protein kinases in cellular processes such as proliferation, their catalytic activity is tightly regulated, and the mechanisms involved in regulation are highly diverse. In general, protein kinases adopt a multitude of conformations, characterized by differential states of N- and C-lobe closure and positional heterogeneity of αC in the N lobe and the activation loop in the C lobe [8]. The positioning of αC and the activation loop governs, for the most part, the activity state of the kinase. In general, intrasteric (within the polypeptide chain) inhibitory interactions maintain a low-activity state of the kinase (Figure 6.2a), and the proper cellular stimulus results in repositioning and stabilization of αC and the activation loop to achieve an active state, which is structurally very similar among protein kinases (Figure 6.2b).

Receptor PSKs and PTKs are activated by ligand binding to the extracellular region of the receptors, which, in general, results in receptor oligomerization and activation through *trans*-phosphorylation. In the absence of bound ligand, the protein kinase domains in these receptors are maintained in a non-optimized or low-activity state through autoinhibitory interactions that usually involve the activation loop and often the juxtamembrane region (between the transmembrane helix and the kinase domain) [9]. As an example, the receptor PTK for stem cell factor (SCF), c-Kit, is autoinhibited through juxtamembrane interactions with the kinase domain [10], which are released upon SCF-induced dimerization [11] and autophosphorylation (in *trans*) of two tyrosine residues in the juxtamembrane region. Mutations (typically deletions) in the juxtamembrane region of c-Kit that disrupt this autoinhibitory machinery are responsible for ligand-independent hyperactivity and are causative for gastrointestinal stromal tumors [12].

Two other receptors, the EGF receptor (PTK) and IRE1 (PSK), highlight the variety of molecular mechanisms by which receptor protein kinases are regulated. Instead of *trans*-phosphorylation of tyrosine residues in the activation

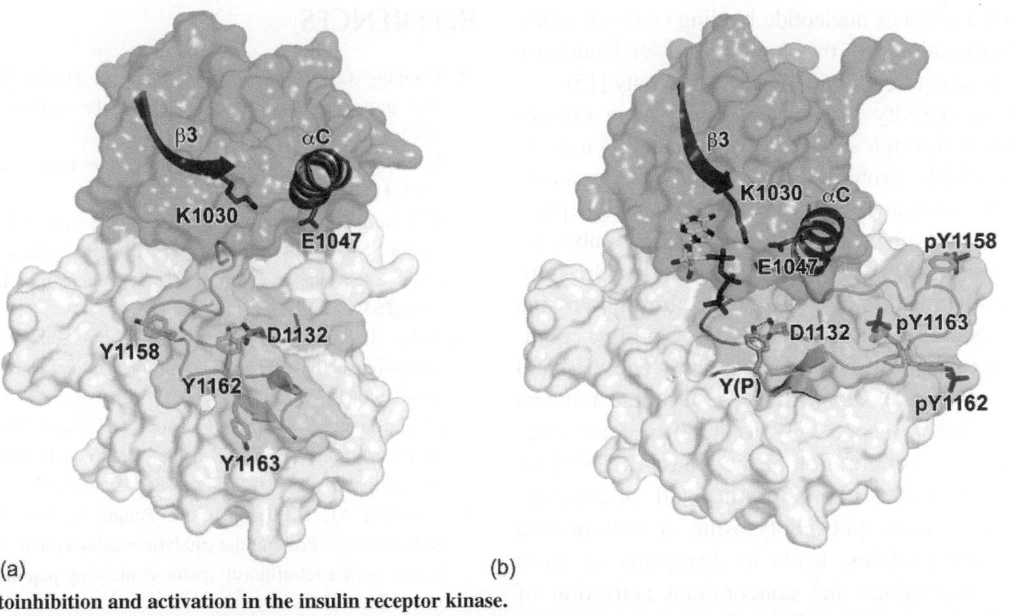

FIGURE 6.2 Autoinhibition and activation in the insulin receptor kinase.
(a) Prior to insulin stimulation, the activation loop adopts an autoinhibitory conformation in which Tyr1162 binds in the kinase active site (hydrogen-bonded to Asp1132), and Glu1047 in αC points away from the ATP binding site (PDB code 1IRK) [23]. A semi-transparent surface of the kinase domain is shown. (b) Insulin binding to the receptor ectodomain promotes autophosphorylation (in *trans*) of the activation loop on Tyr1158, Tyr1162, and Tyr1163, which stabilizes the loop in an active state (PDB code 1IR3) [24]. In addition, αC pivots towards the active site, juxtaposing Glu1047 and Lys1030 for coordination to the ATP phosphate groups (AMP-PNP was co-crystallized). The tyrosine-containing substrate is hydrogen-bonded to Asp1132 and forms a β-strand interaction with the end of the activation loop.

loop or juxtamembrane region, the EGF receptor is activated by an allosteric mechanism in which one kinase domain in the EGF-stabilized dimer [13] activates the other kinase domain through formation of an asymmetric dimer, which favorably positions for catalysis the αC helix (N lobe) and the (unphosphorylated) activation loop of the second kinase (Figure 6.3) [14].

In the endoplasmic reticulum, accumulation of unfolded proteins leads to dimerization of the IRE1 lumenal domain, which results in *trans*-phosphorylation of the kinase activation loop on serine residues. For IRE1, whose cytoplasmic region contains a protein kinase domain followed directly by an endonuclease domain, the only cellular substrate of the kinase is its own activation loop (in *trans*). Activation-loop

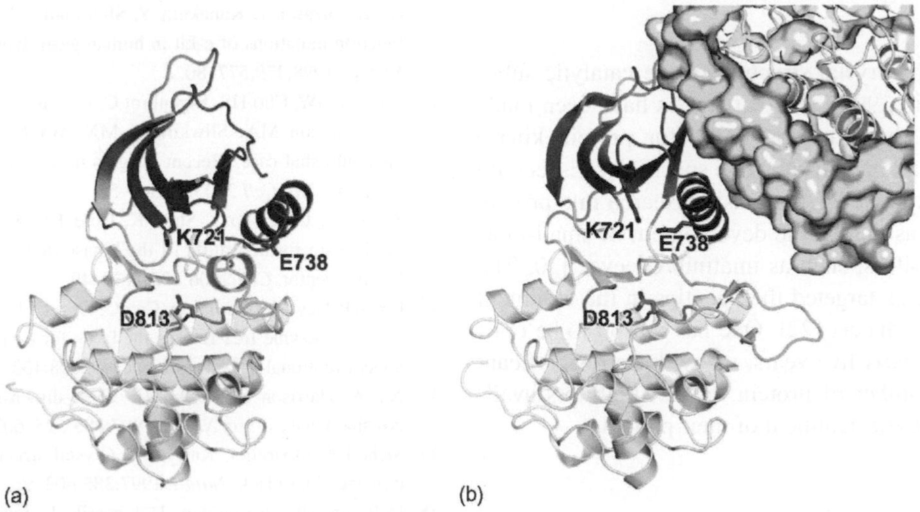

FIGURE 6.3 Autoinhibition and activation in the EGF receptor kinase.
(a) Prior to EGF stimulation, the activation loop adopts an autoinhibitory conformation in which an α-helix in the loop prevents the juxtaposition of Lys721 (β3) and Glu738 (αC) (PDB code 2GS7) [14]. (b) EGF-mediated dimerization (symmetric) of the receptor ectodomain results in formation of an asymmetric kinase dimer, in which the C lobe of one kinase (represented with a cut-away molecular surface) impinges on the N lobe of the other kinase (ribbon representation) (PDB code 1M14) [25]. This interaction repositions αC to bring Glu738 in proximity to Lys721, and also forces the activation loop into an active conformation [14].

phosphorylation facilitates nucleotide binding (ATP or ADP), promoting dimerization of the kinase and endonuclease domains, which is required for endonuclease activity [15].

The catalytic activity of non-receptor protein kinases is also regulated through a variety of molecular mechanisms. Many soluble protein kinases also include multiple non-catalytic domains, either on the same polypeptide chain or on separate chains, which are used for subcellular targeting and/or regulation of kinase activity. Src-family PTKs contain Src homology-2 (SH2, phosphotyrosine binding) and Src-homology-3 (SH3, poly-proline binding) domains upstream of the kinase domain. The SH2 and SH3 domains interact in *cis* with the C-terminal tail and the SH2-kinase linker, respectively, to stabilize a non-catalytically productive configuration of the kinase domain [16, 17]. Dephosphorylation of the C-terminal tail, or competition from higher-affinity phosphotyrosine or poly-proline segments in other proteins, leads to disruption of these autoinhibitory interactions and concomitant activation of the kinase domain, whose activity is further stimulated by activation-loop phosphorylation.

CDKs are soluble PSKs that play essential roles in regulation of the cell cycle. They are maintained in a low basal state of activity through intrasteric interactions that stabilize non-productive (with respect to activity) conformations of αC (known as the PSTAIRE helix in CDKs) and the activation loop (called the T loop in CDKs) [18]. The activation process involves binding of a cognate cyclin to a CDK, which repositions αC and the activation loop, and exposes the activation loop to phosphorylation (Thr160 in CDK2) by a distinct kinase, CDK-activating kinase (CAK), yielding a fully active kinase [19].

PROSPECTS

Beginning with the crystal structure of the catalytic subunit of PKA back in 1991 [3], great strides have been made in understanding the structural bases for protein kinase mechanism, regulation, and substrate specificity. Especially gratifying is the knowledge that basic research into protein kinase structure has guided the development of small-molecule kinase inhibitors, such as imatinib/Gleevec [20, 21], which are serving as targeted therapeutics in the treatment of various human cancers [22]. One has reason to be optimistic that, in the next five years, we will see a significant increase in the number of protein kinase inhibitors available to oncologists for treatment of their patients.

ACKNOWLEDGEMENTS

Support is acknowledged from the National Institutes of Health (DK052916 and NS053414) and from an Irma T. Hirschl-Monique Weill-Caulier Career Scientist Award.

REFERENCES

1. Manning G, Whyte DB, Martinez R, Hunter T, Sudarsanam S. The protein kinase complement of the human genome. *Science* 2002;**298**:1912–34.
2. Blume-Jensen P, Hunter T. Oncogenic kinase signalling. *Nature* 2001;**411**:355–65.
3. Knighton DR, Zheng J, Ten Eyck LF, Ashford VA, Xuong N-h, Taylor SS, Sowadski JM. Crystal structure of the catalytic subunit of cyclic adenosine monophosphate-dependent protein kinase. *Science* 1991;**253**:407–14.
4. Nolen B, Taylor S, Ghosh G. Regulation of protein kinases; controlling activity through activation segment conformation. *Mol Cell* 2004;**15**:661–75.
5. Madhusudan M, Akamine P, Xuong NH, Taylor SS. Crystal structure of a transition state mimic of the catalytic subunit of cAMP-dependent protein kinase. *Nat Struct Biol* 2002;**9**:273–7.
6. Ablooglu AJ, Till JH, Kim K, Parang K, Cole PA, Hubbard SR, Kohanski RA. Probing the catalytic mechanism of the insulin receptor kinase with a tetrafluorotyrosine-containing peptide substrate. *J Biol Chem* 2000;**275**:30,394–8.
7. Parang K, Till JH, Ablooglu AJ, Kohanski RA, Hubbard SR, Cole PA. Mechanism-based design of a protein kinase inhibitor. *Nat Struct Biol* 2001;**8**:37–41.
8. Huse M, Kuriyan J. The conformational plasticity of protein kinases. *Cell* 2002;**109**:275–82.
9. Hubbard SR. Juxtamembrane autoinhibition in receptor tyrosine kinases. *Nat Rev Mol Cell Biol* 2004;**5**:464–71.
10. Mol CD, Dougan DR, Schneider TR, Skene RJ, Kraus ML, Scheibe DN, Snell GP, Zou H, Sang BC, Wilson KP. Structural basis for the autoinhibition and STI-571 inhibition of c-Kit tyrosine kinase. *J Biol Chem* 2004;**279**:31,655–63.
11. Yuzawa S, Opatowsky Y, Zhang Z, Mandiyan V, Lax I, Schlessinger J. Structural basis for activation of the receptor tyrosine kinase KIT by stem cell factor. *Cell* 2007;**130**:323–34.
12. Hirota S, Isozaki K, Moriyama Y, Hashimoto K, Nishida T, Ishiguro S, Kawano K, Hanada M, Kurata A, Takeda M, Muhammad Tunio G, Matsuzawa Y, Kanakura Y, Shinomura Y, Kitamura Y. Gain-of-function mutations of c-kit in human gastrointestinal stromal tumors. *Science* 1998;**279**:577–80.
13. Burgess AW, Cho HS, Eigenbrot C, Ferguson KM, Garrett TP, Leahy DJ, Lemmon MA, Sliwkowski MX, Ward CW, Yokoyama S. An open-and-shut case? Recent insights into the activation of EGF/ErbB receptors. *Mol Cell* 2003;**12**:541–52.
14. Zhang X, Gureasko J, Shen K, Cole PA, Kuriyan J. An allosteric mechanism for activation of the kinase domain of epidermal growth factor receptor. *Cell* 2006;**125**:1137–49.
15. Lee KP, Dey M, Neculai D, Cao C, Dever TE, Sicheri F. Structure of the dual enzyme Ire1 reveals the basis for catalysis and regulation in nonconventional RNA splicing. *Cell* 2008;**132**:89–100.
16. Xu W, Harrison SC, Eck MJ. Three-dimensional structure of the tyrosine kinase c-Src. *Nature* 1997;**385**:595–602.
17. Sicheri F, Moarefi I, Kuriyan J. Crystal structure of the Src family tyrosine kinase Hck. *Nature* 1997;**385**:602–9.
18. DeBondt HL, Rosenblatt J, Jancarik J, Jones HD, Morgan DO, Kim S-H. Crystal structure of cyclin-dependent kinase 2. *Nature* 1993;**363**:595–602.
19. Pavletich NP. Mechanisms of cyclin-dependent kinase regulation: structures of Cdks, their cyclin activators, and Cip and INK4 inhibitors. *J Mol Biol* 1999;**287**:821–8.

20. Schindler T, Bornmann W, Pellicena P, Miller WT, Clarkson B, Kuriyan J. Structural mechanism for STI-571 inhibition of Abelson tyrosine kinase. *Science* 2000;**289**:1938–42.
21. Druker BJ, Talpaz M, Resta DJ, Peng B, Buchdunger E, Ford JM, Lydon NB, Kantarjian H, Capdeville R, Ohno-Jones S, Sawyers CL. Efficacy and safety of a specific inhibitor of the BCR-ABL tyrosine kinase in chronic myeloid leukemia. *N Engl J Med* 2001;**344**: 1031–7.
22. Klein S, McCormick F, Levitzki A. Killing time for cancer cells. *Nat Rev Cancer* 2005;**5**:573–80.
23. Hubbard SR, Wei L, Ellis L, Hendrickson WA. Crystal structure of the tyrosine kinase domain of the human insulin receptor. *Nature* 1994;**372**:746–54.
24. Hubbard SR. Crystal structure of the activated insulin receptor tyrosine kinase in complex with peptide substrate and ATP analog. *EMBO J* 1997;**16**:5572–81.
25. Stamos J, Sliwkowski MX, Eigenbrot C. Structure of the EGF receptor kinase domain alone and in complex with a 4-anilinoquinazoline inhibitor. *J Biol Chem* 2002;**277**:46265–75.

Chapter 7

Large-Scale Structural Analysis of Protein Tyrosine Phosphatases

Alastair J. Barr and Stefan Knapp
Structural Genomics Consortium, Old Road Campus Building, University of Oxford, Oxford, England, UK

OVERVIEW

Protein tyrosine phosphatases (PTPs) are essential regulators of signal transduction pathways, and control, together with protein tyrosine kinases (PTKs), the reversible phosphorylation of tyrosine residues. The human genome contains 107 PTPs grouped into four distinct families [1]. Class I cysteine-based PTPs constitute the largest family. They are divided into 38 "classical" tyrosine-specific PTPs [2] and 61 dual-specificity phosphatases (DUSPs). Class II cysteine-based PTPs are present in all organisms, including prokaryotes. This family is only represented by one member, the low molecular weight PTP LMPTP, in humans. Three CDC25 cell cycle regulators constitute the human class III cysteine-based PTP family. Class I, II, and III cysteine-based PTPs share a similar catalytic mechanism and active site architecture [3, 4, 5, 6]. The high degree of conservation of structural features is surprising, considering that class I PTP and DUSPs share often less than 5 percent sequence identity. The fourth family of PTPs differs from cysteine-based PTPs by catalytic mechanism, and has been reported to hydrolyze either phosphotyrosine- or phosphoserine-containing substrates. This family containing the transcription factor Eyes absent (Eya) proteins has not been extensively characterized, and no representative structure has been reported to date [7, 8, 9].

STRUCTURAL COVERAGE OF THE FAMILY

PTPs are, from a structural biology viewpoint, one of the most comprehensively covered protein families. Currently, 56 structures are publicly available, including structures of the class II enzyme LMPTP [6], two structures of class III PTPs [10], two lipid phosphatases (PTEN [11] and MTMR2 [12]), and 19 DUSP structures. This chapter focuses on the structural comparison of catalytic domains of the major PTP families, and summarizes implications of the available structural data for the understanding of PTP regulation and for the design of specific PTP inhibitors. However, it is important to keep in mind that PTP function is also significantly regulated by modular domains outside the catalytic domain. At least 79 of the 107 PTPs contain one additional domain outside of their phosphatase domain [1] that regulates substrate recruitment, oligomerization, and cellular localization.

All receptor-type phosphatases (RPTPs) except for RPTPα and RPTPε contain large extracellular domains comprising immunoglobulin (IG), fibronectin (FN), meprin/A5/m (MAM), and carbonic anhydrase (CA) domains. The recent crystal structure of RPTPμ revealed an antiparallel dimer of the ectodomain [13]. A large surface area involving residues spanning four domains (MAM, Ig, FN1, and FN2) contributes to this dimer interface. This ectodomain-mediated interaction appears to represent the driving force for correct localization and has been suggested to act as a spacer-clamp which accumulates RPTPμ, in contrast to CD45, at cell contacts [14]. However, more structures of ectodomains are urgently needed to understand the function of these larger extracellular domains of RPTPs.

STRUCTURAL FEATURES

The classical PTP family contains 21 RPTPs, of which 12 contain two catalytic domains (D1 and D2), and 15 are non-receptor type PTPs (excluding PCPTP and STEP, which express both membrane-spanning as well as cytoplasmic isoforms). This family has the best structural coverage to date. Currently, 32 high-resolution structures of the 49 D1

Handbook of Cell Signaling, Three-Volume Set 2 ed.

and D2 phosphatase domains are publicly available, covering all of the major receptor and cytoplasmic subfamilies (Figure 7.1a). The classical PTPs consist of a highly conserved single domain α/β structure of mixed β-sheets flanked by α-helices (Figure 7.1b). The key features of this ~280 residue PTP domain include the PTP signature motif, the WPD loop, and the phosphotyrosine recognition loop. The PTP signature motif, HCSxGxGR(S/T)G, is located at the base of the active site cavity. It constitutes the phosphate-binding loop and also contains the catalytic cysteine

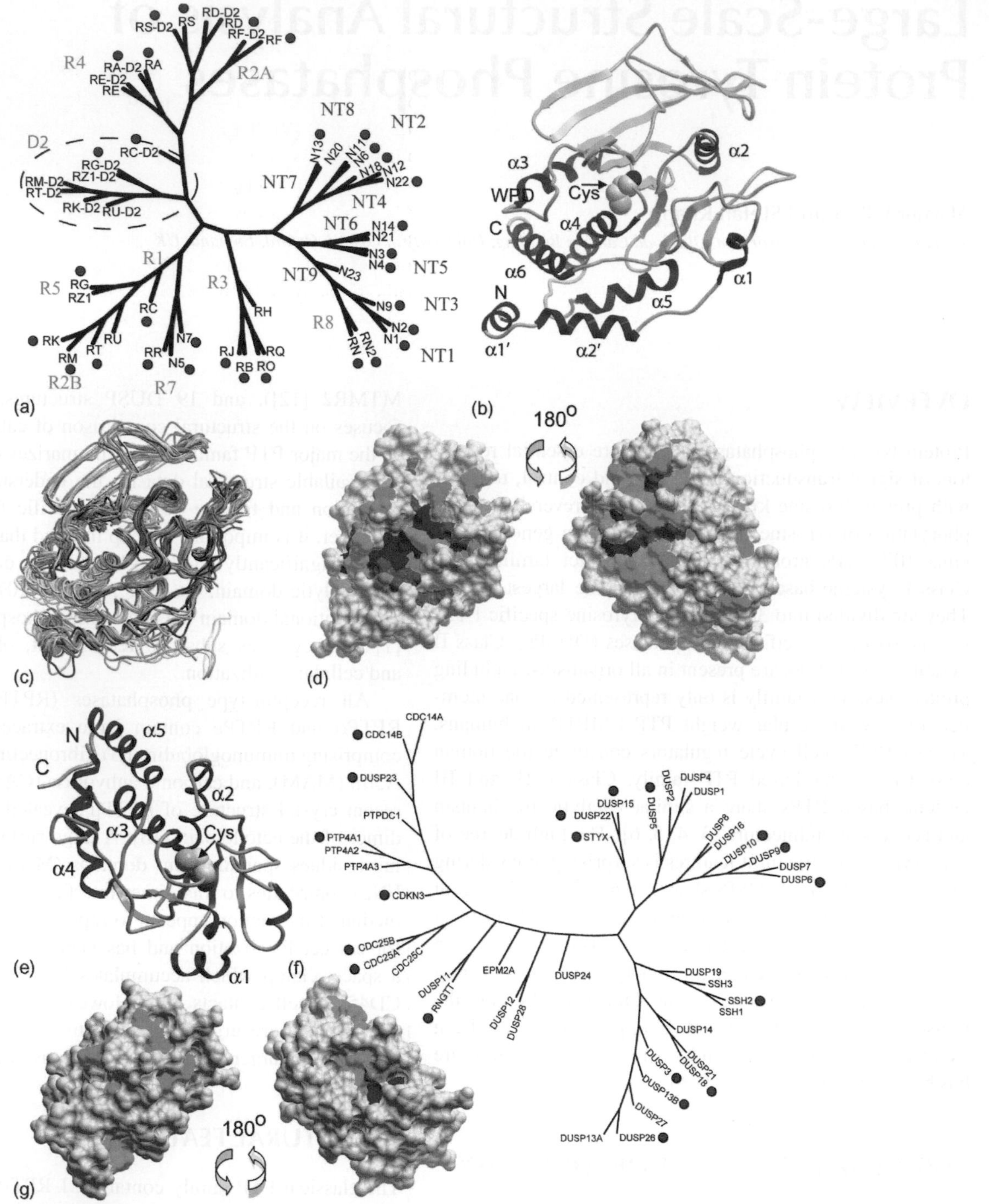

FIGURE 7.1 Overall structure and structural coverage of PTP families.
(a) Phylogenetic tree of the PTP family. Experimental structures are indicated as dots. The HUGO gene name omitting "PTP" has been used. (b) Overall structure of PTP1B with the main structural elements labeled. The active site cysteine is labeled. (c) Superimposition or all PTP structures. (d) Surface conservation of residues mapped on the PTP1B structure. Surface conservation dark gray areas indicate surface patches with high sequence homology. (e) Structural overview of the DUSP family member VHR. The main helices and the active site cysteine are labeled. (f) Phylogenetic tree of the DUSP family (lipid phosphatases and LMPTP have been omitted). (g) Surface conservation of residues mapped on the VHR structure.

(Cys215 PTP1B numbering) and a conserved arginine residue (Arg221) that is involved in substrate binding and stabilization of the phospho-enzyme intermediate. Structural comparison across the family reveals that the overall fold is highly conserved (Figure 7.1c). However, the surface of the PTPs is surprisingly diverse, with only small patches of high residue conservation flanking the active site (Figure 7.1d). Comparing PTP1B with other classical PTPs, one notable region of diversity is the pocket forming the second pTyr binding site, which recognizes pY1163 of the insulin receptor kinase, while the active site binds pY1162 [15]. A shallower pocket is found in most PTPs, with the exception of the NT5, NT6, R7, R2B, and R1 subgroups, in which the loop connecting helix $\alpha2'$ with $\alpha1$ adopts a different conformation, resulting in occlusion of this binding pocket. In the R2A subgroup (RPTPσ, RPTPδ, and LAR), an aspartate residue replaces the arginine found in PTP1B, creating a negatively charged pocket surface. This pocket, where present, together with the so called "gateway" region (258/259 in PTP1B), is likely to contribute to determining substrate specificity. The WPD loop contains the catalytically important aspartic acid residue that functions as a general acid-base. A unique feature of the classical tyrosine-specific PTPs is the phosphotyrosine recognition motif. This loop contributes to the selective recognition of phosphotyrosine over phospho-serine/threonine by stabilizing binding of the tyrosine residue by a phosphotyrosine π–π stacking interaction and by creating a deeper active site pocket, allowing only phosphotyrosine containing substrates to reach the catalytic cysteine residue.

DUSPs share much lower sequence homology between each other, and to classical PTPs, but most secondary structure elements are surprisingly conserved between DUSPs and tyrosine specific PTPs (Figure 7.1e). For example, the DUSP family member VHR (vaccinia H1 related) shares only 10 percent sequence identity with PTP1B [5]. Still, both structures superimpose well (2.3-Å rmsd using 123 of the 173 VHR main chain positions). The main differences between the two structures are the largely diverse N-termini and an extension of the parallel central sheet network that is lacking two β-sheets in VHR as well as the absence of insertion loops between secondary structure elements including the phosphor-tyrosine recognition motif [16]. The structural coverage of the dual-specificity phosphatase family is not as comprehensive as for the classical PTPs, and structural information for some diverse DUSP family members is still lacking (Figure 7.1f). However, 19 structures of the 43 DUSPs compiled in the phylogenetic tree are available to date. Not shown in the phylogenetic tree are the myotubularins, Asp-based PTPs (EYAs, CTDs), the only class II enzyme LMPTP, and the PTEN family. Only the myotubularin MTMR2 [12], PTEN [11], LMPTP [6, 17], and CTDSP1 [18] have been structure-determined in this latter group. As expected from their low sequence homology, DUSP catalytic domains share only small regions of sequence conservation in the vicinity of the active site (Figure 7.1g).

A SHARED CATALYTIC MECHANISM

PTPs and DUSPs share a common catalytic mechanism of phosphate monoester hydrolysis. In contrast to Ser/Thr phosphatases, PTPs require no metal cofactor and catalyze phosphate hydrolysis via a covalent phosphocysteine intermediate [3, 19] (Figure 7.2). Upon binding of the phosphotyrosine substrate, the WPD loop assumes a closed conformation, bringing the WPD aspartate (Asp181 in PTP1B) into position for catalysis (Figure 7.2b). Studies on *Yersinia* PTP showed that WPD aspartate must be protonated, and that substitution with an asparagine reduces catalytic activity by three orders of magnitude, suggesting that this residue acts as a general acid [20]. The phosphotyrosine binds to a conserved arginine residue which is part of the PTP signature motif. This residue reorients upon binding of the substrate to coordinate the phosphate moiety. However, mutagenesis experiments showed that mutations at this site mainly affect k_{cat} and not so much K_m values, indicating that even though this arginine plays a dominant role in substrate binding, it seems to be more important for catalysis [21]. Catalysis is initiated by a nucleophilic attack of the catalytic cysteine, which exists as an unprotonated thiolate ion on the substrate phosphate. The enzymatic reaction proceeds through formation of a phosphor-enzyme intermediate which hydrolyzes with the help of a tightly bound nucleophilic water molecule coordinated by an invariant glutamine. Hydrolysis of the phosphocysteine intermediate is facilitated by the vicinity of a Ser or Thr residue usually present in the PTP signature motif, and by the WPD aspartate now functioning as a general base (Figure 7.3a) [22].

Structures of PTP1B substrate complexes suggested that opening and closing of the WPD loop is triggered by substrate binding. However, many PTP domains have been crystallized in open, closed, and intermediate conformations, regardless if a substrate was present in the active site or not. Unligated structures have been reported for PTP1B in the expected open [4] as well as closed form [23], and the structure of the STEP substrate complex revealed an open conformation of the WPD loop despite the presence of a substrate (pdb code: 2CJZ). The closed apo-structure of PTP1B suggested that water molecules may mimic the binding of a bound substrate. In addition, structures of STEP, LyPTP, and GLEPP1 (pdb codes: 2BIJ, 2P6X, 2GJT) revealed an atypically open WPD loop conformation. Structural information of different conformational states gives valuable insights into the conformation and flexibility of active-site residues. The conformation of the WPD loop changes drastically the size and shape of the active site cavity, and it will be essential to investigate the dynamics of

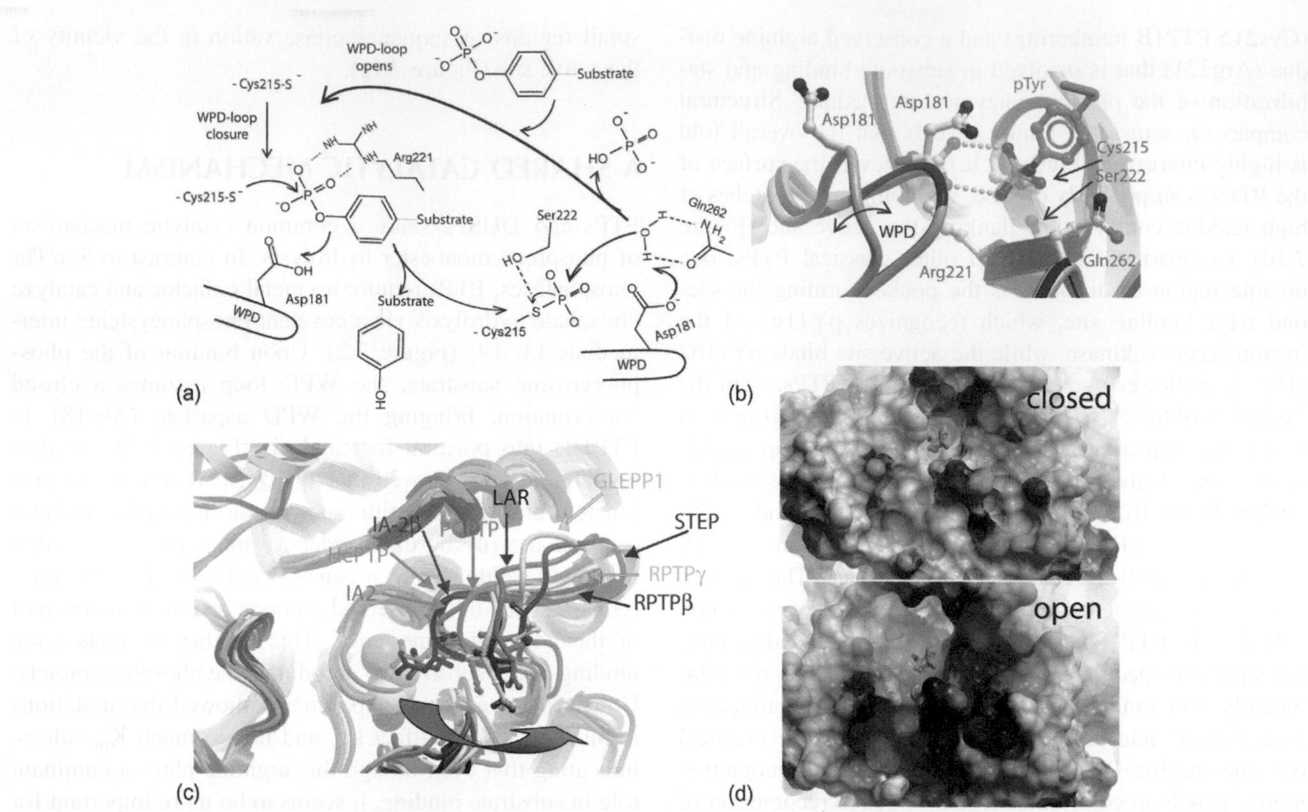

FIGURE 7.2 Superimposition of currently available tandem D1–D2 domains.
(a) The wedge interactions established by the RPTPα crystal structure [35] is highlighted, and a model of a tandem domain dimer has been generated by superimposition of PTPRα on LAR (shown as a gray ribbon). The ribbon model significantly clashes at the two D2 domains, suggesting that the current model is not compatible with the available structural data. (b) RPTPγ "head to toe" dimer observed in the crystal structure.

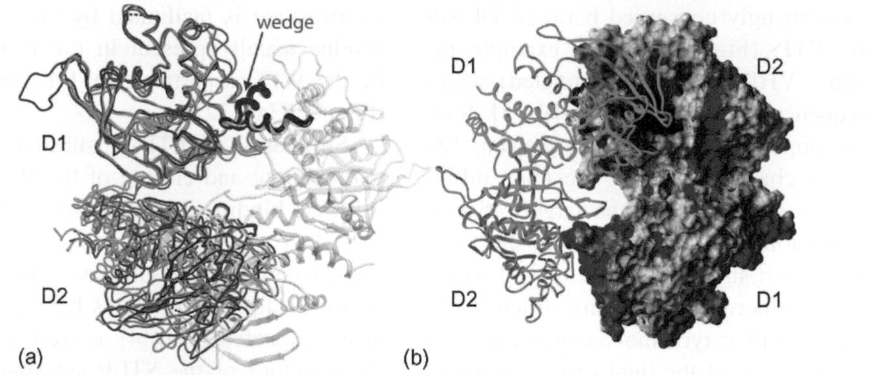

FIGURE 7.3 Catalytic mechanism and active site dynamics of PTPs.
(a) Schematic mechanism of the catalytic cycle. (b) Active site and superimposition on an open and a closed conformation of PTP1B. (c) Superimposition of selected PTP structures showing different conformation of the WPD loop. (d) Changes in active site shape and volume upon closure of the WPD loop in PTP1B. The substrate phosphotyrosine in indicated in both conformations.

this movement in more detail (Figure 7.3c, d). Furthermore, a better understanding of opening and closing movement of the WPD loop is essential for structure-based design of selective inhibitors.

Reversible oxidation of the active site cysteine has been shown also to regulate PTP activity. The active site environment promotes the formation of a cyclic sulfenamide from the oxidized sulfenic acid moiety. The vicinity of His214 to the active site cysteine in PTP1B and other PTPs polarizes the amide bond of Ser216, promoting a nucleophilic attack and ring closure. This reaction causes a significant structural change in the active site, resulting in exposure of the PTP signature motif and Tyr46 of the phospho-binding loop. This conformation protects the active-site cysteine from irreversible oxidation to sulfonic acid, and allows redox regulation of the enzyme by promoting its reversible reduction by reducing agents [24, 25].

Both PTP and DUSP families contain inactive enzymes, and at least 11 PTPs are catalytically inactive. These pseudophosphatases are reminiscent of inactive kinases that constitute about 10 percent of the human kinome [26]. Recently, it has been shown that pseudophosphatases form complexes with active family members – for instance, the inactive myotubularin MTMR5 and MTMR13 bind to active MTMR2, and the active MTMR7 has been shown to interact with inactive MTMR9 [27, 28, 29]. Also, the two inactive RPTPs' major autoantigens in insulin-dependent diabetes mellitus, IA-2 and IA-2β, have been suggested to regulate active PTPs by heterodimerization [30]. Moreover, most D2 domains of RPTPs lack residues that are critical for activity and are either inactive or have negligible catalytic activity. The structure of D2 domains is highly conserved with D1 domains, and catalytic activity can usually be restored to D2 domains by mutating the residues causing inactivation back to residues found at these positions in active phosphatases [31].

An interesting mutation is the exchange of the WPD aspartate for a glutamate residue, which occurs in four human D1 domains (RPTPλ, PTPS31, PTPD1, and HDPTP) as well as in five RPTP D2 domains. Studies on PTPS31 (PTPRQ) showed that PTPS31 has very low activity against phosphotyrosine-containing substrates, but hydrolyzes a broad range of phosphatidylinositol phosphates, including phosphatidylinositol 3,4,5–trisphosphate [32, 33]. Overexpression of PTPS31 in cultured cells inhibits proliferation and induces apoptosis, and the backmutation (E2171D) increases PTPase activity but eliminates PIPase activity and abolishes the inhibitory effects on proliferation and apoptosis. PTPD1 and HDPTP have negligible catalytic activity against a variety of phosphotyrosine substrates, suggesting that these PTPs are either pseudophosphatases or target phospholipids (unpublished data).

RECEPTOR DIMERIZATION

Inactivation of PTPs by receptor dimerization has been discussed as a key regulatory mechanism of PTP activity. This control mechanism was first suggested based on structural studies of the membrane proximal D1 domain of RPTPα [34]. The structure revealed a dimeric assembly in which an inhibitory N-terminal helix–turn–helix wedge motif occluded the active site of the interacting catalytic domain. Subsequently, RPTPα and a number of other PTP domains have been shown to homodimerize at the cell surface, resulting in RPTPα inactivation [35]. Twelve of the 21 RPTPs contain two intracellular catalytic domains, which raises the questions of whether the suggested wedge-packing model is compatible with the structure of the entire cytoplasmic domain, and whether the second commonly inactive D2 domain plays a role in dimerization. The first structure of a cytoplasmic RPTP containing both catalytic domains, LAR, revealed that even though the wedge motif was present in the protein, the cytoplasmic domain did not dimerize, leaving the active sites accessible [31]. Subsequently, the structure of CD45 also revealed a monomeric tandem D1 and D2 domain with largely different crystallographic contacts [36]. To date, two more structures of tandem D1 and D2 domain have been solved (RPTPσ pdb code: 2FH7 and RPTPγ pdb code: 2NLK). All tandem domain structures show a largely conserved domain orientation between the D1 and D2 domains, determined by a short interdomain linker region which allows only limited flexibility (Figure 7.2a). Importantly, reconstruction of a tandem domain dimer of RPTPα based on tandem domain structures reveals that the wedge-packing model is incompatible with the domain orientation found in crystal structures, and would result in a D2 domain that clashes with the D2 domain of the interacting molecule (Figure 7.2a). In addition, analytical ultracentrifugation experiments carried out in our laboratory showed that RPTPα and other single- or dual-domain RPTP domains (except for RPTPγ) have no tendency to dimerize *in vitro* using reducing physiological buffer conditions (unpublished data). The structure of RPTPγ, however, revealed a head-to-toe dimer of the tandem domain with a symmetry related molecule, and biophysical studies showed that RPTPγ dimerizes with a dissociation constant of 3.5 μM (Figure 7.2b). Nevertheless, domains outside the catalytic domains in RPTPα seem to contribute to dimerization, including the extracellular domains and the transmembrane section, and dimerization may occur *in vivo* as a result of a significant structural rearrangement due to oxidation and covalent linkage of the conserved D2 domain cysteine [37, 38, 39]. Also, the dimerization proposed for the inactive major autoantigens IA-2 and IA-2β [30] is not supported by structural data, which revealed monomeric enzymes with largely different crystal contact regions (pdb code: 2I1Y, 2QEP). More detailed studies, and a structure of a covalently linked dimmer, will be needed to clarify these issues. In addition, we also hope that more structural data regarding phosphatases in complex with their substrates and regulators will be available in the near future, which will allow a more detailed insight into the regulation and substrate recognition of this diverse class of enzymes.

ENDNOTE

While this book chapter was in press, we published a related article on large-scale structural analysis of protein tyrosine phosphatases [40].

REFERENCES

1. Alonso A, Sasin J, Bottini N, et al. Protein tyrosine phosphatases in the human genome. *Cell* 2004;**117**:699–711.
2. Andersen JN, Jansen PG, Echwald SM, et al. A genomic perspective on protein tyrosine phosphatases: gene structure, pseudogenes, and genetic disease linkage. *FASEB J* 2004;**18**:8–30.
3. Guan KL, Dixon JE. Evidence for protein-tyrosine-phosphatase catalysis proceeding via a cysteine-phosphate intermediate. *J Biol Chem* 1991;**266**:17,026–17,030.
4. Barford D, Flint AJ, Tonks NK. Crystal structure of human protein tyrosine phosphatase 1B. *Sciece (NY)* 1994;**263**:1397–404.
5. Yuvaniyama J, Denu JM, Dixon JE, Saper MA. Crystal structure of the dual specificity protein phosphatase VHR. *Sciece (NY)* 1996;**272**:1328–31.
6. Su XD, Taddei N, Stefani M, Ramponi G, Nordlund P. The crystal structure of a low-molecular-weight phosphotyrosine protein phosphatase. *Nature* 1994;**370**:575–8.
7. Tootle TL, Silver SJ, Davies EL, et al. The transcription factor Eyes absent is a protein tyrosine phosphatase. *Nature* 2003;**426**:299–302.
8. Jemc J, Rebay I. Identification of transcriptional targets of the dual-function transcription factor/phosphatase eyes absent. *Dev Biol* 2007;**310**:416–29.
9. Jemc J, Rebay I. The eyes absent family of phosphotyrosine phosphatases: properties and roles in developmental regulation of transcription. *Annu Rev Biochem* 2007;**76**:513–38.
10. Reynolds Jr RA, Yem AW, Wolfe CL, Deibel MR, Chidester CG, Watenpaugh KD. Crystal structure of the catalytic subunit of Cdc25B required for G2/M phase transition of the cell cycle. *J Mol Biol* 1999;**293**:559–68.
11. Lee JO, Yang H, Georgescu MM, et al. Crystal structure of the PTEN tumor suppressor: implications for its phosphoinositide phosphatase activity and membrane association. *Cell* 1999;**99**:323–34.
12. Begley MJ, Taylor GS, Kim SA, Veine DM, Dixon JE, Stuckey JA. Crystal structure of a phosphoinositide phosphatase, MTMR2: insights into myotubular myopathy and Charcot-Marie-Tooth syndrome. *Mol cell* 2003;**12**:1391–402.
13. Aricescu AR, Siebold C, Choudhuri K, et al. Structure of a tyrosine phosphatase adhesive interaction reveals a spacer-clamp mechanism. *Science (NY)* 2007;**317**:1217–20.
14. Aricescu AR, Jones EY. Immunoglobulin superfamily cell adhesion molecules: zippers and signals. *Curr Opin Cell Biol* 2007;**19**:543–50.
15. Salmeen A, Andersen JN, Myers MP, Tonks NK, Barford D. Molecular basis for the dephosphorylation of the activation segment of the insulin receptor by protein tyrosine phosphatase 1B. *Mol Cell* 2000;**6**:1401–12.
16. Denu JM, Stuckey JA, Saper MA, Dixon JE. Form and function in protein dephosphorylation. *Cell* 1996;**87**:361–4.
17. Zhang M, Van Etten RL, Stauffacher CV. Crystal structure of bovine heart phosphotyrosyl phosphatase at 2.2-Å resolution. *Biochemistry* 1994;**33**:11,097–105.
18. Zhang Y, Kim Y, Genoud N, et al. Determinants for dephosphorylation of the RNA polymerase II C-terminal domain by Scp1. *Mol Cell* 2006;**24**:759–70.
19. Cho H, Ramer SE, Itoh M, et al. Catalytic domains of the LAR and CD45 protein tyrosine phosphatases from Escherichia coli expression systems: purification and characterization for specificity and mechanism. *Biochemistry* 1992;**31**:133–8.
20. Zhang ZY, Wang Y, Dixon JE. Dissecting the catalytic mechanism of protein-tyrosine phosphatases. *Proc Natl Acad Sci* 1994;**91**:1624–7.
21. Zhang ZY, Wang Y, Wu L, et al. The Cys(X)5Arg catalytic motif in phosphoester hydrolysis. *Biochemistry* 1994;**33**:15,266–70.
22. Zhang ZY. Chemical and mechanistic approaches to the study of protein tyrosine phosphatases. *Acc Chem Res* 2003;**36**:385–92.
23. Pedersen AK, Peters GG, Moller KB, Iversen LF, Kastrup JS. Water-molecule network and active-site flexibility of apo protein tyrosine phosphatase 1B. *Acta Crystallographica* 2004;**60**:1527–34.
24. van Montfort RL, Congreve M, Tisi D, Carr R, Jhoti H. Oxidation state of the active-site cysteine in protein tyrosine phosphatase 1B. *Nature* 2003;**423**:773–7.
25. Salmeen A, Andersen JN, Myers MP, et al. Redox regulation of protein tyrosine phosphatase 1B involves a sulphenyl-amide intermediate. *Nature* 2003;**423**:769–73.
26. Manning G, Whyte DB, Martinez R, Hunter T, Sudarsanam S. The protein kinase complement of the human genome. *Sciece (NY)* 2002;**298**:1912–34.
27. Kim SA, Vacratsis PO, Firestein R, Cleary ML, Dixon JE. Regulation of myotubularin-related (MTMR)2 phosphatidylinositol phosphatase by MTMR5, a catalytically inactive phosphatase. *Proc Natl Acad Sci* 2003;**100**:4492–7.
28. Robinson FL, Dixon JE. The phosphoinositide-3-phosphatase MTMR2 associates with MTMR13, a membrane-associated pseudophosphatase also mutated in type 4B Charcot-Marie-Tooth disease. *J Biol Chem* 2005;**280**:31,699–707.
29. Mochizuki Y, Majerus PW. Characterization of myotubularin-related protein 7 and its binding partner, myotubularin-related protein 9. *Proc Natl Acad Sci* 2003;**100**:9768–73.
30. Gross S, Blanchetot C, Schepens J, et al. Multimerization of the protein-tyrosine phosphatase (PTP)-like insulin-dependent diabetes mellitus autoantigens IA-2 and IA-2β with receptor PTPs (RPTPs). Inhibition of RPTPalpha enzymatic activity. *J Biol Chem* 2002;**277**:48,139–45.
31. Nam HJ, Poy F, Krueger NX, Saito H, Frederick CA. Crystal structure of the tandem phosphatase domains of RPTP LAR. *Cell* 1999;**97**:449–57.
32. Seifert RA, Coats SA, Oganesian A, et al. PTPRQ is a novel phosphatidylinositol phosphatase that can be expressed as a cytoplasmic protein or as a subcellularly localized receptor-like protein. *Exp Cell Res* 2003;**287**:374–86.
33. Oganesian A, Poot M, Daum G, et al. Protein tyrosine phosphatase RQ is a phosphatidylinositol phosphatase that can regulate cell survival and proliferation. *Proc Natl Acad Sci* 2003;**100**:7563–8.
34. Bilwes AM, den Hertog J, Hunter T, Noel JP. Structural basis for inhibition of receptor protein-tyrosine phosphatase-alpha by dimerization. *Nature* 1996;**382**:555–9.
35. Jiang G, den Hertog J, Su J, Noel J, Sap J, Hunter T. Dimerization inhibits the activity of receptor-like protein-tyrosine phosphatase-alpha. *Nature* 1999;**401**:606–10.
36. Nam HJ, Poy F, Saito H, Frederick CA. Structural basis for the function and regulation of the receptor protein tyrosine phosphatase CD45. *J Exp Med* 2005;**201**:441–52.
37. Persson C, Sjoblom T, Groen A, et al. Preferential oxidation of the second phosphatase domain of receptor-like PTP-alpha revealed by an antibody against oxidized protein tyrosine phosphatases. *Proc Natl Acad Sci* 2004;**101**:1886–91.
38. van der Wijk T, Overvoorde J, den Hertog J. H_2O_2-induced intermolecular disulfide bond formation between receptor protein-tyrosine phosphatases. *J Biol Chem* 2004;**279**:44,355–61.
39. Blanchetot C, Tertoolen LG, den Hertog J. Regulation of receptor protein-tyrosine phosphatase alpha by oxidative stress. *EMBO J* 2002;**21**:493–503.
40. Barr AJ, Ugochukwu E, Lee WH, et al. Large-scale structural analysis of the classical human protein tyrosine phosphatome. *Cell* 2009;**136**:352–63.

Chapter 8

Transmembrane Receptor Oligomerization

Darren R. Tyson[1] and Ralph A. Bradshaw[1,2]
[1]*Department of Medicine, Vanderbilt University School of Medicine, Nashville, Tennessee*
[2]*Department of Pharmaceutical Chemistry, University of California, San Francisco, San Francisco, CA*

INTRODUCTION

Signal transduction in eukaryotes is initiated from receptors that are activated by the binding of exogenous ligands. The majority of these receptors is localized in the plasma membrane and can be generally separated into several categories based on structure and signaling properties (see Table 8.1). Although there are many similarities among these families, the most common is the quaternary structure of the activated species. In most (if not all) cases, signaling requires that one or more types of monomers (protomers) assemble themselves in a complex, and the formation of this receptor complex is prerequisite for function. The active complexes can range from simple homodimers to heterotetramers; in some cases, higher order complexes may be required. However, what controls the oligomerization process and the role that ligand plays are, in most cases, poorly defined. While ligand is certainly required to complete the activation, it is less clear whether it is required to induce the formation of the receptor complexes themselves. Evidence supporting both induced and constitutive oligomerized states has been reported for many receptor types, and indeed both mechanisms may be utilized.

The formation of oligomers to generate active (or activatible) species is a common biological mechanism found throughout living systems. It is a primary example of protein–protein interactions, which are a major component of the developing field of proteomics, and it affords several basic physiological advantages [1], including regulatory, kinetic, and specificity opportunities of various types. For example, a simple receptor dimer can easily be viewed as binding its ligand with significantly higher affinity than a monomer due to increased surface binding area. A dimer would also have twice the signaling capacity and provide a simple on/off switch (assuming the monomer is inactive) by

TABLE 8.1 Receptors and classes

Principal receptor classes	Reported oligomeric states	Examples
Tyrosine-kinase-containing receptors (RTK)	Homodimer	EGFR, TrkA, FGFR, CSF1R
	Heterodimer	EGFR/ErbB2, PDGFRα/β
Cytokine receptors, type I	Homodimer	EPOR, GHR
	Heterotrimer	IL-2Rα/β/γχ, IL-4Rα/β/γχ
	Heterotetramer	IL-6R/gp130, IL-11/gp130
Cytokine receptors, type II	Heterotetramer	IFNγR1/2, IL-3R
Guanylyl-cyclase-containing receptors	Homodimer	ANPR
Serine/threonine-kinase-containing receptors	Heterotetramer	TβR1/2, BMPR1/2
TNF receptors	Homotrimer	TNFR1, Fas, CD40
	Homodimer	TNFR1
Heptahelical receptors (G-protein-coupled receptors)	Homodimer	δ-OpioidR, GluR
	Heterodimer	$GABA_BR1/2$, β_2AR/α_2AR

Handbook of Cell Signaling, Three-Volume Set 2 ed.

dissociating. In addition, the formation of heteromeric complexes offers another means to broaden the range of specificity in terms of both ligand binding and the signals produced. Furthermore, a heterodimer of a potentially active protomer with one that cannot be activated could produce a non-functioning complex (even though it can still bind ligands), thereby acting as a negative regulator of receptor signaling. This natural control mechanism has been extensively exploited in the research laboratory in recent years and is referred to as the "dominant-negative" strategy. The sections that follow provide a brief description of how oligomerization is utilized by the major cell transmembrane receptor classes.

TYROSINE KINASE-CONTAINING RECEPTORS

It has long been known that cell-surface *r*eceptors possessing intrinsic *t*yrosine *k*inase activity (RTKs) minimally require a dimeric state for their full activity. It has generally been accepted that RTK dimerization occurs as a result of ligand binding, (i.e., the receptors occur as monomers at the cell surface in the unstimulated state), and the ligand-bound monomer then recruits another monomer into the complex, allowing for the interaction of the kinase domains and their subsequent transphosphorylation. The initial phosphorylation events usually occur on the activation loop and help stabilize the "open" (active) conformation, thereby allowing access of adenosine triphosphate (ATP) and substrate to their respective binding sites within the kinase domain (see Section IIA, Protein Phosphorylation, for more detail). The activated kinases can then carry out further tyrosine phosphorylation events. Both chemical and physical evidence has been obtained to support this model, most of which depends on a lack of evidence for detecting receptor dimers except in cells exposed to stimulation. However, the growing body of evidence that at least some RTKs can exist as preformed dimers in the absence of ligand includes both direct [2–4] and indirect [5] measurements. There are indeed substantial reasons why this would be advantageous to cells. The presence of ligand-independent dimers could provide a precise regulation of the activation of RTKs. Preformed dimers would reduce the time necessary for widely dispersed monomers to associate, thereby reducing the time required to form the active dimeric structure. Additionally, a preformed dimer could present a higher affinity ligand binding site by allowing the interaction of the ligand with both receptors simultaneously. Finally, suppression (inhibition) of the kinase domains in such ligand-independent complexes would provide a means to suppress spurious signaling in the unstimulated state.

A natural example of a ligand-independent dimer is the insulin receptor (IR), which exists as a covalently bound heterotetramer consisting of two α and two β chains joined by disulfide bonds (see Chapters 44 and 45 of Handbook of Cell Signaling, Second Edition). The α and β chains are actually derived from a single precursor, hence the unprocessed insulin receptor resembles the other RTKs except for the covalent links that ensure that the unliganded state will be dimeric. It is also clear that dimerization alone is insufficient for the activation of IR. Somehow, ligand binding to the receptor ectodomain modulates a conformational change of the endodomain even though the plasma membrane is transversed by a single membrane-spanning segment of approximately 20 to 25 residues in each protomer. Although the mechanisms by which this occurs remain largely unknown, it may be surmised that the two protomers (in this case, the protomer is an α/β chain unit) rotate relative to each other. A similar mechanism would be equally effective in the RTKs that are not covalently linked.

The active complexes of dimerized RTKs are stabilized by multiple contacts, such as interactions between monomers and ligand, and between monomers. Interaction between monomers stabilized in a dimeric state has been shown to occur within the ectodomains, transmembrane domains, and endodomains of various receptors. For example, the crystal structure of a ligand-bound domain of TrkA shows direct contacts between the two domains as well as direct contacts to the ligand, nerve growth factor (NGF). In addition, regions within the ectodomains of fibroblast growth factor receptor (FGFR) and platelet-derived growth factor receptor (PDGFR) have been identified that mediate receptor–receptor interactions in the absence of ligand [6,7]. Furthermore, the transmembrane domains of ErbB family members self-associate in cell membranes [8], suggesting that the association may occur in the context of the full-length receptor provided that the ecto- and endodomains do not sterically hinder their association. Residues within the cytoplasmic domains of RTKs also are important for stabilization of dimers [9, 10].

CYTOKINE RECEPTORS

The cytokine receptors are comprised of numerous kinds of transmembrane receptors that are characterized by conserved patterns of amino acid residues in their ectodomains and the lack of enzymatic activity within their endodomains [11, 12]. In lieu of intrinsic kinase activity, they associate with non-receptor tyrosine kinases such as the Janus kinases (JAKs) or Src family kinases. Nevertheless, the active signaling complex induced by ligand is oligomeric and seems to involve the juxtaposition of these non-receptor tyrosine kinases. The quaternary structure of the cytokine receptor superfamily is both complex and varied, ranging from simple homodimers to heterotrimers, and can be separated into two subclasses, type I and type II, with the majority belonging to class I [11, 12]. The class II receptors primarily consist of receptors for interferons (IFNs) and interleukin-10 (IL-10).

Type I Cytokine Receptors

The most-studied class I cytokine receptors are the growth hormone receptor (GHR) and erythropoietin receptor (EPOR) that function as homodimers (reviewed by Frank [13]). Recently, the unliganded and ligand-bound crystal structures of the receptor for erythropoietin have been described and provide substantial insight into the mechanisms of activation of this receptor [14, 15]. EPOR appears to exist at the cell surface as an inactive dimer [14, 15], and this state is mediated by the transmembrane domain [16]. The inactive dimer exists in a conformation in which the transmembrane domains (and presumably the endodomains) are separated by approximately 73 Å, thereby preventing the interaction of the associated JAKs [14, 15]. Upon ligand binding, the separation of the transmembrane domains is reduced to 30 Å, which would likely be sufficient to allow the JAKs to transphosphorylate each other and thus become activated. The ligand-bound structure of GHR, also deduced from X-ray diffraction studies, suggests a similar mechanism of activation in that the endodomains appear to be brought into proximity, allowing activation of the associated JAKs [17]. It is not clear, however, whether the receptors exist in a predimerized state on the cell surface (either loosely or tightly associated) or whether ligand induces the dimerization. Neither is it known whether GHR requires a conformational change upon ligand binding for its activation, although if GHR is predimerized this would almost certainly be required for its activation. Interestingly, there is evidence to suggest that some GHRs, upon ligand stimulation, form covalent dimers through disulfide bridges [13].

Some type I cytokine receptors require three different subunits for their activity. The receptor for interleukin-2 (IL-2R) is one such case, as it has three subunits, α, β, and γ (also known as the common γ chain [γc], as it is used by multiple receptors, including IL-4R). The β and γ receptors together are sufficient for signaling but do not possess sufficient affinity for ligand to be activated by normal levels of IL-2. IL-2Rα is not a member of the cytokine receptor superfamily, as it does not possess the conserved motifs within its ectodomain; however, when present in the complex with the β and γ subunits, it increases the affinity for ligand binding [18] and may induce higher order oligomerization [19]. The ectodomains of the receptor subunits have been shown to interact at the cell surface in the absence of ligand as α/β or β/γ pairs or α/β/γ heterotrimers [20], although the presence of pre-associated, full-length IL-2R subunits remains controversial. In any case, the downstream signaling initiated by IL-2R subunits conforms to the common theme among many cell surface receptors: the juxtaposition of two catalytically active factors allows for their transactivation. With respect to IL-2R, the juxtaposed effectors appear to be JAK1 and JAK3 [21]. It has been shown, however, that the mere juxtaposition of the JAKs is insufficient for signaling and that the correct orientation of these entities is also required [22, 23]. This observation supports the idea that the conformation of receptor endodomains is as important as their proximity within the active receptor complex in the activation process.

Analogous to the use of γc by IL-2R and IL-4R, the IL-6 and IL-11 receptors each require the common gp130 subunit for activity (reviewed in Taga and Kishimoto [24]). IL-6 cannot bind to gp130 alone, and IL-6R alone has a low affinity for IL-6. However, when IL-6R and gp130 are co-expressed, both high- and low-affinity binding sites for IL-6 are generated. IL-6 binds IL-6R on the cell surface or the soluble form of IL-6R (sIL-6R), which then recruits gp130 into the complex. The activity of gp130 requires its association to form homodimers as part of the IL-6/IL–6R/gp130 complex. The stoichiometry of the active receptor complex has been determined to be 2:2:2 [25, 26], suggesting a two-fold symmetry. IL-6R has a short cytoplasmic tail that is dispensable for its function, and gp130 provides all of the required cytoplasmic signaling motifs. Because gp130 has been shown to be constitutively associated with JAK1, JAK2, and Tyk2, the presumed mechanism of activation again involves a juxtaposition of these kinases that allows their transactivation and the phosphorylation of tyrosine residues in the C-terminal region of the endodomain of gp130.

Type II Cytokine Receptors

The type II cytokine receptors consist primarily of receptors for interferons and interleukin-10 (IL-10) [11]. The receptors for interferon-γ (IFNγR) and IL-10 (IL-10R) have a similar structure in that they are each comprised of two type 1 receptors and two type 2 receptors forming a heterotetramer (reviewed by Kotenko and Pestka [27]). As with many of the cytokine receptors, IFNγR subunits are constitutively associated with JAKs: IFNγR1 with JAK1 and IFNγR2 with JAK2. IFNγR1 dimers can bind IFNγ, but this complex is not functional even though the complex contains the associated JAK1. In contrast to the ligand-bound EPOR, which transposes the transmembrane domains within 30 Å in the active conformation, the 27-Å spacing between IFNγR1 molecules acts to prevent downstream signaling [28], exemplifying the importance of the orientation and/or conformation of the cytoplasmic domains. Indeed, it has recently been shown using FRET analysis [S. Pestka, personal communication] that the IFNγR1 and 2 heterotetramer (with the associated JAKs) is, contrary to commonly held opinion, preformed and that binding of IFNγ actually causes the intracellular domains (with their associated JAKs) to move apart. This new model suggests that there are likely inhibitory interactions that keep the JAKs inactive and that displacement is necessary to alleviate this or allow room for substrates (e.g., STATS) to bind. The similarity of this model with that observed for FGFR3 is striking [5].

GUANYLYL CYCLASE-CONTAINING RECEPTORS

The receptors of this type are organized much like the RTKs in that they have an ectodomain, a single transmembrane-spanning domain, and an endodomain composed of a two interacting subdomains: a kinase-homologous regulatory domain that binds ATP and a guanylate cyclase domain that produces cGMP [29]. This group contains seven members, the best studied of which is GC-A, the atrial natriuretic peptide (ANP) receptor [30]. The three-dimensional structure of the ectodomain dimer has been solved by X-ray methods which revealed a bilobal periplasmic binding protein, similar to that found in several other proteins, including several DNA binding proteins [31]. Interestingly, these receptors clearly form unliganded dimers, and activation by ANP occurs through binding to and stimulation of these preformed structures; however, it is currently unclear how this activation is transmitted through the structure. Considering the overall similarity with the RTKs, it is likely that the preformed dimers are in a conformation that results in inhibition of the cyclases, and that binding causes a rotational motion that alleviates this, perhaps by the release of steric hinderances.

SERINE/THREONINE KINASE-CONTAINING RECEPTORS

The serine/threonine kinase receptors are typified by the receptors for transforming growth factor β (TGFβ), a dimeric ligand that exerts its effects through receptors composed of two different subunits designated type I and type II. Each possesses serine/threonine kinase activity. Both classes of receptor protomers are required for mediating the signaling response to ligand binding. The type II TGFβ receptor (TβR-II) exists as a constitutively active dimer and is responsible for the initial interaction with TGFβ, as the type I receptor (TβR-I) cannot bind TGFβ in the absence of TβR-II. Although the exact stoichiometry has not been elucidated, the minimal active signaling complex consists of TGFβ bound to two molecules of TβR-II and two of TβR-I. Two theories exist regarding the association and activation of this complex. The first involves the initial interaction of TGFβ with TβR-II. This complex then actively recruits TβR-I, leading to the phosphorylation and activation of this receptor. The second theory involves an inactive, pre-existing (albeit weakly associated) heterotetramer of TβR-I and TβR-II that undergoes a conformational change upon ligand binding that alters the juxtaposition of TβR-I with TβR-II in such a way as to allow the phosphorylation and activation of TβR-I. In either case, ligand is required to induce or stabilize an active conformation that allows the phosphorylation of TβR-I by TβR-II.

The structure of TβR-II complexed with TGFβ, deduced from X-ray studies, has recently been described [32]. The structure supports the formation of a TGFβ/TβR-II/TβR-I heteropentameric complex in which TβR-I directly contacts both the ligand and TβR-II. These findings do not preclude either theory of receptor association, but they do provide evidence for direct interaction of TβR-I and -II that would be required for association of the receptors in the absence of ligand. It has been shown that TβR-I and -II can associate in the absence of ligand *in vitro* and when co-expressed in mammalian cells [33], two observations giving support for pre-existing heterotetramers in the absence of ligand.

TUMOR NECROSIS FACTOR RECEPTORS

The specific oligomeric nature of active tumor necrosis factor receptor (TNFR) family members has not been definitively proven. The ligands for these receptors are usually trimeric, suggesting that the functional oligomeric complex of the receptors may also be trimeric; however, the receptors may actually function as dimers. At least three different hypotheses have been proposed for the activation of TNFRs. The first, and most widely accepted, involves receptor trimerization. The earliest crystallographic studies of TNFR and TNFR1 bound to TNFβ demonstrate that one receptor molecule binds to each of the three monomer–monomer interfaces of the trimeric TNF ligand [34, 35]. Interestingly, no direct contacts between any of the receptor monomers other than non-specific crystal contacts were detected; however, the receptor endodomains presumably would be sufficiently close to allow interaction. TNF-receptor-associated factors (TRAFs), primary effectors of TNF-Rs, have been crystallized as trimers in association with domains of the TNFR family member CD40, and the structure of trimerized TRAFs supports a model of interaction with trimerized CD40. Furthermore, TRAF activity is greater when induced by trimeric forms of CD40 as compared to monomers or dimers [36]. Biochemical studies of TNFR also suggest that the receptors exist as trimers even in the absence of ligand [37]. There are numerous other studies supporting the notion that the active form of TNFR is trimeric.

Nonetheless, evidence also suggests that active TNFR is assembled as dimers [38]. First, biochemical studies of the ectodomain of TNFR2 determined that two or three TNFR2 molecules bind to each trimer of TNFα or TNFβ ligand, suggesting that a receptor dimer may interact with the trimeric ligand [39]. CD27, a TNFR family member, functions as a disulfide-linked dimer, analogous to the insulin receptor. Chimeric receptors consisting of EPOR or PDGFR ectodomains inframe with the TNFR transmembrane, and cytoplasmic domains function as dimers, either constitutively (the EPOR chimera [40]), or ligand dependently (PDGFR [41]), indicating that two subunits are sufficient for signaling. Furthermore, secreted forms of TNFR have been purified as dimeric proteins. Most substantially, structural studies of crystallized ectodomains of TNFR1 in the absence of ligand indicated that TNFR could exist in at least two different dimeric conformations [42]. These structural studies led to models of dimeric receptor activation.

Two different models exist for the activation of dimeric receptor complexes [43], and both models arise from crystallographic studies of unliganded ectodomains [42]. The first involves a conformational change via an axial rotation that would juxtapose the cytoplasmic domains [44]. The second is the so-called expanding-network hypothesis and involves the formation of higher order oligomeric complexes of dimeric receptors and trimeric ligand [42]. Even though the specific organization of receptor–ligand complexes has not been definitively determined, it is clear that the oligomerization of TNFR family members is required to allow signaling from these receptors.

HEPTAHELICAL RECEPTORS (G-PROTEIN-COUPLED RECEPTORS)

The heptahelical or G-protein-coupled receptors are one of the largest families in the human genome (and likely other mammalian genomes, as well), and with more than 1000 distinct entities they are the most used family of transmembrane signaling receptors. Not surprisingly, as a group, they form one of most popular targets for drug discovery. Their mechanism is well understood in general terms; that is, they interact with trimeric G-protein complexes, of which there are many, which in turn couple them to a wide variety of signaling effector systems, including adenylyl cyclase, the producer of cAMP, which was the first second messenger to be identified.

It has been generally believed that this class of receptors is monomeric and remains so during the binding of ligand and the activation of the trimeric G-protein complex; however, recent evidence suggests that this is not the case [45–47]. Indeed, these receptors appear to exist as homodimers, heterodimers, or even larger oligomers. The identification of these higher order structures has, of course, been aided by improved technology that has allowed detection of the oligomers both in solution and in viable cells [45,46]. The formation of heterodimers is a particularly interesting phenomenon as it allows a greatly increased range of ligand binding and subsequent responses. Apparently, even distantly related receptors have been observed to form complexes and, given the size of this family, the number of possibilities afforded by these interactions is enormous.

CONCLUDING REMARKS

In this brief overview, we have summarized much of the salient observations that support the view that oligomerization is a prevalent and perhaps universal requirement for transmembrane receptor function. There is clearly growing evidence, in part attributable to better technology, that many of these oligomeric structures form independent of ligand and that activation by ligand binding is not due to association of the receptor protomers but rather allows them to seek a different juxtaposition relative to each other, almost certainly at least in part by rotational motions. Such a model eliminates the diffusion control that ligand-induced dimerization (oligomerization) requires and provides additional opportunities for the regulation of signal flux by allowing inhibitory interactions in the unliganded complexes. The reader is directed to the many chapters in this *Handbook* describing individual systems for more specific details.

REFERENCES

1. Klemm JD, Schreiber SL, Crabtree GR. Dimerization as a regulatory mechanism in signal transduction. *Annu Rev Immunol* 1998;**16**:569–92.
2. Gadella Jr. T, Jovin T. Oligomerization of epidermal growth factor receptors on A431 cells studied by time-resolved fluorescence imaging microscopy. A stereochemical model for tyrosine kinase receptor activation. *J Cell Biol* 1995;**129**:1543–58.
3. Mischel PS, Umbach JA, Eskandari S, Smith SG, Gundersen CB, Zampighi GA. Nerve growth factor signals via preexisting TrkA receptor oligomers. *Biophys J* 2002;**83**:968–76.
4. Wiseman PW, Petersen NO. Image correlation spectroscopy. II. Optimization for ultrasensitive detection of preexisting platelet-derived growth factor-beta receptor oligomers on intact cells. *Biophys J* 1999;**76**:963–77.
5. Raffioni S, Zhu YZ, Bradshaw RA, Thompson LM. Effect of transmembrane and kinase domain mutations on fibroblast growth factor receptor 3 chimera signaling in PC12 cells. A model for the control of receptor tyrosine kinase activation. *J Biol Chem* 1998;**273**:35, 250–9.
6. Wang JK, Goldfarb M. Amino acid residues which distinguish the mitogenic potentials of two FGF receptors. *Oncogene* 1997;**14**:1767–78.
7. Omura T, Heldin CH, Ostman A. Immunoglobulin-like domain 4-mediated receptor-receptor interactions contribute to platelet-derived growth factor-induced receptor dimerization. *J Biol Chem* 1997;**272**:12, 676–82.
8. Mendrola JM, Berger MB, King MC, Lemmon MA. The single transmembrane domains of ErbB receptors self-associate in cell membranes. *J Biol Chem* 2002;**277**:4704–12.
9. Tanner KG, Kyte J. Dimerization of the extracellular domain of the receptor for epidermal growth factor containing the membrane-spanning segment in response to treatment with epidermal growth factor. *J Biol Chem* 1999;**274**:35, 985–90.
10. Yu X, Sharma KD, Takahashi T, Iwamoto R, Mekada E. Ligand-independent dimer formation of epidermal growth factor receptor (EGFR) is a step separable from ligand-induced EGFR signaling. *Mol Biol Cell* 2002;**13**:2547–57.

11. Bazan JF. Structural design and molecular evolution of a cytokine receptor superfamily. *Proc Natl Acad Sci USA* 1990;**87**:6934–8.
12. Thoreau E, Petridou B, Kelly PA, Djiane J, Mornon JP. Structural symmetry of the extracellular domain of the cytokine/growth hormone/prolactin receptor family and interferon receptors revealed by hydrophobic cluster analysis. *FEBS Letts* 1991;**282**:26–31.
13. Frank SJ. Receptor dimerization in GH and erythropoietin action-it takes two to tango, but how?. *Endocrinology* 2002;**143**:2–10.
14. Syed RS, Reid SW, Li C, Cheetham JC, Aoki KH, Liu B, Zhan H, Osslund TD, Chirino AJ, Zhang J, Finer-Moore J, Elliott S, Sitney K, Katz BA, Matthews DJ, Wendoloski JJ, Egrie J, Stroud RM. Efficiency of signaling through cytokine receptors depends critically on receptor orientation. *Nature* 1998;**395**:511–16.
15. Livnah O, Stura EA, Middleton SA, Johnson DL, Jolliffe LK, Wilson IA. Crystallographic evidence for preformed dimers of erythropoietin receptor before ligand activation. *Science* 1999;**283**:987–90.
16. Constantinescu SN, Keren T, Socolovsky M, Nam H, Henis YI, Lodish HF. Ligand-independent oligomerization of cell-surface erythropoietin receptor is mediated by the transmembrane domain. *Proc Natl Acad Sci USA* 2001;**98**:4379–84.
17. de Vos AM, Ultsch M, Kossiakoff AA. Human growth hormone and extracellular domain of its receptor: crystal structure of the complex. *Science* 1992;**255**:306–12.
18. Takeshita T, Asao H, Ohtani K, Ishii N, Kumaki S, Tanaka N, Munakata H, Nakamura M, Sugamura K. Cloning of the gamma chain of the human IL-2 receptor. *Science* 1992;**257**:379–82.
19. Eicher DM, Damjanovich S, Waldmann TA. Oligomerization of IL-2Ralpha. *Cytokine* 2002;**17**:82–90.
20. Damjanovich S, Bene L, Matko J, Alileche A, Goldman CK, Sharrow S, Waldmann TA. Preassembly of interleukin 2 (IL-2) receptor subunits on resting Kit 225 K6 T cells and their modulation by IL-2, IL-7, and IL-15: a fluorescence resonance energy transfer study. *Proc Natl Acad Sci USA* 1997;**94**:13, 134–9.
21. Miyazaki T, Kawahara A, Fujii H, Nakagawa Y, Minami Y, Liu ZJ, Oishi I, Silvennoinen O, Witthuhn BA, Ihle JN, Taniguchi T. Functional activation of Jak1 and Jak3 by selective association with IL-2 receptor subunits. *Science* 1994;**266**:1045–7.
22. Hilkens CM, Is'harc H, Lillemeier BF, Strobl B, Bates PA, Behrmann I, Kerr IM. A region encompassing the FERM domain of Jak1 is necessary for binding to the cytokine receptor gp130. *FEBS Letts* 2001;**505**:87–91.
23. Haan C, Heinrich PC, Behrmann I. Structural requirements of the interleukin-6 signal transducer gp130 for its interaction with Janus kinase 1: the receptor is crucial for kinase activation. *Biochem J* 2002;**361**:105–11.
24. Taga T, Kishimoto T. gp130 and the interleukin-6 family of cytokines. *Annu Rev Immunol* 1997;**15**:797–819.
25. Ward LD, Howlett GJ, Discolo G, Yasukawa K, Hammacher A, Moritz RL, Simpson RJ. High affinity interleukin-6 receptor is a hexameric complex consisting of two molecules each of interleukin-6, interleukin-6 receptor, and gp-130. *J Biol Chem* 1994;**269**:23, 286–9.
26. Paonessa G, Graziani R, De-Serio A, Savino R, Ciapponi L, Lahm A, Salvati AL, Toniatti C, Ciliberto G. Two distinct and independent sites on IL-6 trigger gp 130 dimer formation and signaling. *EMBO J* 1995;**14**:1942–51.
27. Kotenko SV, Pestka S. Jak-Stat signal transduction pathway through the eyes of cytokine class II receptor complexes. *Oncogene* 2000;**19**:2557–65.
28. Walter MR, Windsor WT, Nagabhushan TL, Lundell DJ, Lunn CA, Zauodny PJ, Narula SK. Crystal structure of a complex between interferon-gamma and its soluble high-affinity receptor. *Nature* 1995;**376**:230–5.
29. Misono KS. Natriuretic peptide receptor: structure and signaling. *Mol Cell Biochem* 2002;**230**:49–60.
30. van den Akker F. Structural insights into the ligand binding domains of membrane bound guanylyl cyclases and natriuretic peptide receptors. *J Mol Biol* 2001;**311**:923–37.
31. van den Akker F, Zhang X, Miyagi M, Huo X, Misono KS, Yee VC. Structure of the dimerized hormone- binding domain of a guanylyl-cyclase-coupled receptor. *Nature* 2000;**406**:101–4.
32. Hart PJ, Deep S, Taylor AB, Shu Z, Hinck CS, Hinck AP. Crystal structure of the human TβR2 ectodomain-TGF-β3 complex. *Nat Struct Biol* 2002;**9**:203–8.
33. Derynck R, Feng XH. TGF-beta receptor signaling. *Biochim Biophys Acta* 1997;**1333**:F105–50.
34. Banner DW, D'Arcy A, Janes W, Gentz R, Schoenfeld HJ, Broger C, Loetscher H, Lesslauer W. Crystal structure of the soluble human 55 kd TNF receptor-human TNF beta complex: implications for TNF receptor activation. *Cell* 1993;**73**:431–45.
35. D'Arcy A, Banner DW, Janes W, Winkler FK, Loetscher H, Schonfeld HJ, Zulauf M, Gentz R, Lesslauer W. Crystallization and preliminary crystallographic analysis of a TNF-β-55-kDa TNF receptor complex. *J Mol Biol* 1993;**229**:555–7.
36. Werneburg BG, Zoog SJ, Dang TT, Kehry MR, Crute JJ. Molecular characterization of CD40 signaling intermediates. *J Biol Chem* 2001;**276**:43, 334–42.
37. Chan FK, Chun HJ, Zheng L, Siegel RM, Bui KL, Lenardo MJ. A domain in TNF receptors that mediates ligand-independent receptor assembly and signaling. *Science* 2000;**288**:2351–4.
38. Beutler B, Bazzoni F. TNF, apoptosis and autoimmunity: a common thread?. *Blood Cells Mol Dis* 1998;**24**:216–30.
39. Pennica D, Lam VT, Weber RF, Kohr WJ, Basa LJ, Spellman MW, Ashkenazi A, Shire SJ, Goeddel DV. Biochemical characterization of the extracellular domain of the 75-kilodalton tumor necrosis factor receptor. *Biochemistry* 1993;**32**:3131–8.
40. Bazzoni F, Alejos E, Beutler B. Chimeric tumor necrosis factor receptors with constitutive signaling activity. *Proc Natl Acad Sci USA* 1995;**92**:5376–80.
41. Adam D, Kessler U, Kronke M. Cross-linking of the p55 tumor necrosis factor receptor cytoplasmic domain by a dimeric ligand induces nuclear factor-kappa B and mediates cell death. *J Biol Chem* 1995;**270**:17, 482–7.
42. Naismith JH, Devine TQ, Brandhuber BJ, Sprang SR. Crystallographic evidence for dimerization of unliganded tumor necrosis factor receptor. *J Biol Chem* 1995;**270**:13, 303–7.
43. Idriss HT, Naismith JH. TNF alpha and the TNF receptor superfamily: structure-function relationship(s). *Microsc Res Tech* 2000;**50**: 184–95.
44. Bazzoni F, Beutler B. The tumor necrosis factor ligand and receptor families. *N Engl J Med* 1996;**334**:1717–25.
45. Rios CD, Jordan BA, Gomes I, Devi LA. G-protein-coupled receptor dimerization: modulation of receptor function. *Pharmacol Ther* 2001;**92**:71–87.
46. Bouvier M. Oligomerization of G-protein-coupled transmitter receptors. *Nat Rev Neurosci* 2001;**2**:274–86.
47. Devi LA. Heterodimerization of G-protein-coupled receptors: pharmacology, signaling and trafficking. *Trends Pharmacol Sci* 2001;**22**:532–7.

Section C

Major Receptor Families

Part 1

Receptor Tyrosine Kinases

Chapter 9

Protein Tyrosine Kinase Receptor Signaling Overview

Carl-Henrik Heldin
Ludwig Institute for Cancer Research, Uppsala University, Sweden

INTRODUCTION

Protein tyrosine kinase (PTK) receptors constitute an important class of transmembrane receptors that transduce signals regulating cell growth, differentiation, survival, and migration. PTK receptors are also conserved in lower species, and much of our knowledge about their functional properties comes from studies of *Drosophila* and *Caenorhabditis elegans*. Several PTK receptor genes that have been inactivated in mice have revealed the important functional roles of individual PTK receptors in different organs at various stages of the development. Overactivity of PTK receptors has been implicated in a number of diseases, particularly cancer, and several of the PTK receptors were first identified as transforming oncogene products. PTK receptor antagonists are now used clinically for treatment of cancer patients. This chapter reviews the general principles for PTK receptor structure, activation mechanism, and regulation.

PTK SUBFAMILIES

In the human genome, 58 genes encode PTK receptors [1,2]. Each receptor consists of an extracellular ligand binding part, a single transmembrane domain, and an intracellular part with an intrinsic kinase domain. Based on their overall structures, the PTK receptors can be placed into 20 subfamilies (Figure 9.1). Individual subfamilies are characterized by specific structural motifs in their extracellular parts (e.g., Ig-like domains and fibronectin type III domains). Moreover, the sequences of the kinase domains are normally more similar within the subfamilies than between the subfamilies. The major families are briefly introduced below (for reviews, see [1,3]).

The epidermal growth factor (EGF) receptor was the first PTK receptor to be identified. The four members of the family are important for the morphogenesis of epithelial tissues. Members of this family are often amplified or activated through mutations in human malignancies.

The three members of the insulin receptor family are disulfide-bonded dimers that undergo cleavage during processing to generate α and β subunits. In addition to the well-known metabolic effects mediated by the insulin receptor, this family mediates important survival signals.

The platelet-derived growth factor (PDGF) family members are characterized by five Ig-like domains in the extracellular domain, and by the presence of an intervening sequence that splits the kinase into two parts. PDGF receptors are of particular importance for the development of the connective tissue compartments of various organs, as well as for the development of smooth muscle cells of blood vessels. The related receptors for stem cell factor and colony-stimulating factor 1 (CSF-1) are implicated, for example, in the development of hematopoietic cells, germ and neuronal cells, and macrophages.

The vascular endothelial cell growth factor (VEGF) receptor family members have seven Ig-like domains extracellularly and are primarily expressed on endothelial cells; thus, they are implicated in vasculogenesis, angiogenesis, and lymphangiogenesis.

The fibroblast growth factor (FGF) receptor family members are characterized by three Ig-like domains extracellularly, although splice variants with only two Ig-like domains have been described. Like the VEGF receptors, FGF receptors are expressed on endothelial cells and are implicated in angiogenesis; however, these receptors are also expressed in other cell types, and have important roles in the embryonal development of several organs and tissues.

Handbook of Cell Signaling, Three-Volume Set 2 ed.

Members of the neurotrophin receptor family (TrkA, B and C) bind members of the NGF family of neurotrophins, and have important functions during the development and maintenance of the central nervous system.

The two members of the hepatocyte growth factor (HGF) receptor family (Met and Ron) undergo cleavage of their extracellular domains after their syntheses. They have important roles in regulation of cell motility and in organ morphogenesis during embryonal development.

The Eph receptor family, the largest of the PTK receptor subfamilies, has 14 members. Eph receptors are expressed in the nervous system and also in endothelial cells; thus, they are implicated in neuronal guidance and angiogenesis. Interestingly, one class of their ligands, ephrin-Bs, are also transmembrane molecules expressed on the surface of cells; binding of ephrin-Bs to Eph receptors leads not only to activation of the Eph PTK receptor but also to initiation of signaling events at the intracellular part of the ephrin molecules [4].

The remaining PTK subfamilies generally consist of few members and are less well characterized. Interestingly, one of these families, the DDR family, has collagens as ligands [5,6], thus exemplifying the notion that PTK receptors mediate signals not only from soluble or membrane-associated growth factors but also from the surrounding extracellular matrix.

It is notable that four PTK receptor family members have mutations in their kinase domains, rendering them devoid of kinase activity (Figure 9.1); however, they may still have important roles in signaling (see discussion below).

In general, each subfamily binds a family of structurally related ligands. The specificity is not always absolute within the subfamilies; several receptors bind more than one ligand, and several ligands bind more than one receptor. In contrast, high-affinity interactions of individual ligands with more than one subfamily of PTK receptors, or with a PTK receptor and a non-PTK receptor, are rare. One example is Wnt, which, in addition to its non-PTK receptor, can also interact with the PTK receptor Ryk [7].

In addition to tyrosine phosphorylation by the *bona fide* PTK receptors, the TGFβ serine/threonine kinase receptor, which shares some structural characteristics with tyrosine kinases, has been shown to phosphorylate ShcA on tyrosine and serine residues [8]. However, this tyrosine phosphorylation is inefficient compared to that by the conventional PTK receptors, and its physiological significance remains to be determined.

MECHANISM OF ACTIVATION

Ligand-Induced Receptor Dimerization

Protein tyrosine kinase receptors are generally activated by ligand-induced dimerization [3]. This brings the receptor kinase domains close to each other, which results in

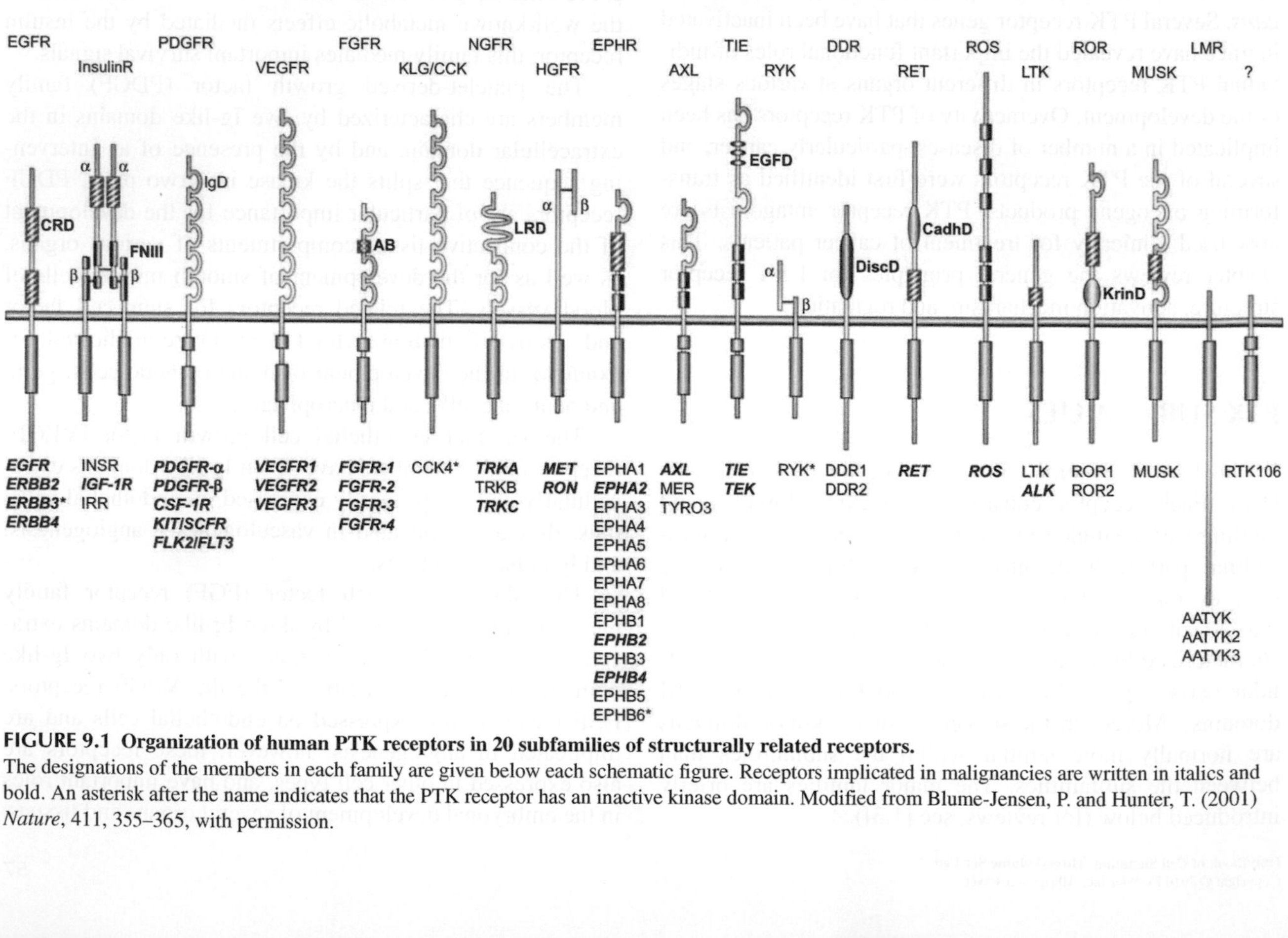

FIGURE 9.1 Organization of human PTK receptors in 20 subfamilies of structurally related receptors.
The designations of the members in each family are given below each schematic figure. Receptors implicated in malignancies are written in italics and bold. An asterisk after the name indicates that the PTK receptor has an inactive kinase domain. Modified from Blume-Jensen, P. and Hunter, T. (2001) *Nature*, 411, 355–365, with permission.

autophosphorylation in *trans* within the intracellular parts of the receptors. The autophosphorylation occurs on tyrosine residues located within or outside the kinase domain of the receptor.

There are, however, many different modes whereby ligand binding induces receptor dimerization [9]. Some ligands are disulfide-bonded dimers, such as PDGFs and VEGFs; the binding of these ligands leads to formation of a symmetric complex consisting of two receptors and one dimeric ligand [10]. In contrast, ephrins are monomeric molecules; after binding of two ephrin molecules to Eph receptors, a dimeric receptor complex is formed in which each ephrin molecule contacts two receptors and each receptor contacts two ligands [11]. Ligand binding to the EGF receptor causes a conformational change allowing direct receptor–receptor interaction, which stabilizes dimerization [12,13]. There are also examples of accessory molecules helping to stabilize a dimeric complex; FGFs are monomeric molecules that interact with receptors at a 1:1 stoichiometry, and receptor dimerization is induced by binding of heparin or heparan sulfate to the complex of FGF and receptor [14]. Finally, members of the insulin receptor family are already disulfide-bonded dimers before ligand binding; binding of ligand presumably induces a conformational change that allows receptor autophosphorylation and activation.

Although receptor dimerization is likely to be necessary for activation of most PTK receptors, it is not always sufficient. Evidence suggests that the orientation of the two intracellular domains in the receptors relative to each other is important [15]. Moreover, there are indications that the initial dimerization of EGF receptors might be followed by further oligomerization, which may be necessary to obtain a fully active receptor [16].

An alternative mechanism of activation of PTK receptors, which does not necessarily involve dimerization, was recently presented for the anaplastic lymphoma kinase (ALK) [17]. The receptor-like protein tyrosine phosphatase β/ζ dephosphorylates ALK and thus keeps it inactive. Binding of the cytokine pleiotrophin to the receptor protein phosphatase β/ζ inactivates its phosphatase activity, thus allowing phosphorylation and activation of ALK.

Homo- and Heterodimerization

In the classical case of ligand-induced dimerization of PTK receptors, two identical receptors form a homodimer; however, two related receptors from the same subfamily may also form a heterodimer. Examples from the PDGF [18] and EGF [19] receptor subfamilies show that heterodimeric receptors may have quantitatively or qualitatively different signaling capacities compared to homodimers of the same receptors.

There are also examples of heteromeric complexes between individual PTK receptors and unrelated receptors. Examples include interactions between PTK receptors from one subfamily with PTK receptors from another subfamily (e.g., PDGF and EGF receptor have been shown to interact [20]). In addition, PTK receptors have been shown to bind to integrins, an interaction that has been shown to enhance integrin signaling [21] and to promote mammary tumor development [22]. Moreover, the hyaluronan receptor CD44 has been shown to interact with certain PTK receptors, including ErbB2 [23] and the PDGF β-receptor [24], and to modulate signaling. In some cases, interactions with non-kinase receptors enhance the affinity for ligand binding (e.g., a long spliced form of VEGF binds to its PTK receptors with higher affinity if the receptor neuropilin is also part of the complex [25]).

Activation of the Receptor Kinase

Ligand-induced PTK receptor dimerization leads to autophosphorylation of the receptors in *trans* within the complex. The autophosphorylation serves two important roles: (1) it causes activation of the kinase domain, and (2) it creates docking sites for downstream SH2- and PTP-domain-containing signaling molecules.

Autophosphorylation may lead to activation of the kinase via several different mechanisms, of which more than one may apply for individual PTK receptors [1]. Tyrosine residues exist within the activation loops of kinases; after phosphorylation, these residues cause the loop to swing out and open up the active site of the kinase [26]. Because most PTK receptors are phosphorylated in this region, this is likely to be a common mechanism for activation of the kinase domain. However, members of the EGF receptor family are not autophosphorylated in the activation loop. In these receptors, it is possible that the activation loops do not efficiently inhibit the kinase of the receptors. Instead, it has been proposed that the long C-terminal tails of these receptors block the active site of the kinase, an inhibition that may be relieved by autophosphorylation and a conformational change of the C-terminal tail; similar mechanisms have been shown for the PTK receptor Tek [27] and the PDGF β-receptor [28]. Moreover, structural studies of the EGF receptor kinase domain have revealed that after ligand-induced dimerization, one kinase domain activates the other through an allosteric mechanism [29]. Finally, the recent elucidation of the three-dimensional structure of the Eph receptor has revealed that in the inactive receptor the juxtamembrane domain forms a helical structure that distorts the small lobe of the kinase domain and prevents access to the active site of the receptor; after autophosphorylation in the juxtamembrane domain, this loop moves away and opens up the active site of the receptor [30]; the juxtamembrane domain may serve similar function also in other PTK receptors [31].

Docking of SH2 and PTB Domain Signaling Proteins

The SH2 domain is a protein module that folds to form a pocket into which a phosphorylated tyrosine residue fits [32]. Genes encoding 110 SH2-domain-containing proteins with a total of 120 SH2 domains are present in the human genome [33]. They interact with phosphorylated tyrosine residues in a specific manner that is directed mainly by the three to six amino acid residues downstream of the phosphorylated tyrosine residue. Also, the PTB domain interacts with phosphorylated tyrosine residues; in this case the specificity is determined by residues N-terminal to the phosphorylated tyrosine.

As an example, Figure 9.2 illustrates the interaction between the autophosphorylated PDGF β-receptor and different SH2-domain-containing molecules. One class of SH2 domain proteins has intrinsic enzymatic activity, e.g., the tyrosine kinase Src, phospholipase Cγ, the tyrosine phosphatase SHP-2, the GTPase-activating protein (GAP) for Ras and the ubiquitin ligase Cbl. The respective enzymatic activities are induced by binding of the SH2 domain to the receptor, or by tyrosine phosphorylation induced by the receptor kinase; alternatively, the enzyme is constitutively active and may, by binding to the receptor, simply be brought to the inner leaflet of the cell membrane, where the next component in the signaling chain is located.

Other SH2 domain proteins are devoid of intrinsic enzymatic activity and serve as adaptors that connect the activated receptors with downstream signaling molecules. Adaptors often have additional domains that mediate interactions with other molecules, such as the SH3 and PH domains [32]. Examples of such adaptors include Nck, Crk, and Shc, as well as Grb2, which forms a complex with Sos, a nucleotide exchange molecule for Ras, and the regulatory subunit p85, which forms a complex with the catalytic subunit p110 of phosphatidylinositol 3′-kinase (PI3-kinase).

The interaction between the activated and autophosphorylated receptor and individual SH2- or PTB-domain-containing molecules initiates signaling pathways that lead to growth stimulation, survival, migration, and actin reorganization. The signaling capacity of a receptor is thus dependent on which such signaling molecules it can dock. Differential autophosphorylation may also be the mechanism by which heterodimeric receptor complexes acquire unique signaling properties [18]. It should be noted that one member of the EGF receptor family, ErbB3, is devoid of kinase activity, yet in heterodimeric configuration with other members of the family it has a potent signaling capacity due to its ability to provide docking sites for SH2 domain proteins [19].

Other Molecules Binding to the Intracellular Parts of PTK Receptors

Signaling via PTK receptors is also modulated by interaction with molecules recognizing other epitopes than

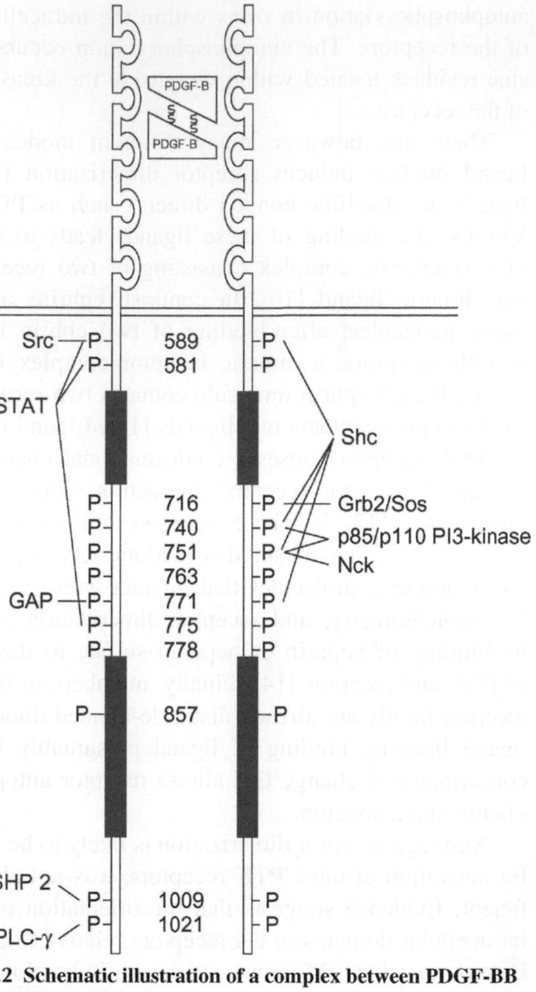

FIGURE 9.2 Schematic illustration of a complex between PDGF-BB and two PDGF β-receptors.
Known autophosphorylated tyrosine residues (P) and their numbers in the receptor sequence are indicated, as well as their interactions with SH2 domain-containing signaling molecules. No PTB domain-containing molecules are known to bind to the PDGF β-receptor. Signaling molecules with intrinsic enzymatic or transcription factor activity are to the left, and adaptors to the right. Note that it is not known how many SH2 domain proteins can bind simultaneously to a dimeric receptor complex.

phosphorylated tyrosines. Examples include the PDZ domain protein NHERF, which binds to the C-terminal tail of the PDGF receptors [34], and the adaptor molecule Alix, which also binds to the PDGF β-receptor in a non-phosphorylation-dependent manner [35].

Inhibition of Phosphatases

The phosphorylation events performed by PTK receptors are counteracted by dephosphorylation by specific tyrosine phosphatases. Recent studies have shown that in order for efficient signaling via PTK receptors, tyrosine phosphatases must be inactivated [36,37]. This may be done by transient, specific oxidation of a cysteine residue in the active site of phosphatases, induced after PTK receptor activation in a PI3-kinase-dependent manner [38].

Nuclear Function of PTK Receptors

Although activation of cytoplasmic signaling pathways by docking of SH2-domain signaling proteins is a major mode of signaling via PTK receptors, an alternative mechanism was recently revealed. The EGF receptor family member ErbB4 was shown to undergo regulated proteolysis in two steps. First, the extracellular domain is cleaved off by a metalloprotease, then another protease, γ-secretase, cleaves within the transmembrane domain and liberates the intracellular domain of ErbB4 for translocation to the nucleus [39], where it regulates transcription by forming a complex with different transcription factors – e.g., the adaptor TAB2 and the co-repressor N-CoR, which represses genes involved in astrocyte differentiation [40,41]. Although regulated intramembrane proteolysis is a well established signaling mechanism for other receptor types (e.g. Notch), its general importance in PTK receptor signaling remains to be elucidated.

There is also evidence that intact EGF receptor [42] or ErbB2 [43] can be translocated to the nucleus, where the receptors are implicated in the induction of cyclin D, iNOS, c-Myb, and COX-2 [44].

CONTROL OF PTK RECEPTOR ACTIVITY

Receptor Internalization and Degradation

After ligand-induced receptor activation, PTK receptors are often accumulated in coated pits and thereafter internalized in endosomes [45], where they are deactivated by several different mechanisms. Upon acidification of the milieu inside the endosomes, the ligand may dissociate from the receptor, which then monomerizes, becoming dephosphorylated by tyrosine phosphatase, and is then recycled back into the membrane. Alternatively, the ligand–receptor complex is degraded after fusion of the endosomes with lysosomes. Moreover, PTK receptors have been shown to become ubiquitinated after activation which enhances internalization [46]. The ubiquitination is often mediated by interaction of the activated receptor with the ubiquitin ligase Cbl, and may also trigger degradation in proteasomes [47,48].

Control of PTK Receptor Signaling

There are several examples of mechanisms that control PTK receptor signaling. When pathways that stimulate certain cellular responses are initiated, signals that inhibit the same responses are often induced. Examples include Ras activation by the PDGF receptor; at the same time as Ras is activated (i.e., converted to its GTP-bound form by the action of the Grb2/Sos complex) it is also inactivated (i.e., converted to the GDP-bound form by RasGAP). The net effect on Ras activation by the PDGF receptor is thus dependent on the stoichiometry in phosphorylation of the tyrosine residues that can bind Grb2/Sos and RasGAP, respectively; evidence suggests that this balance can differ, for example between homo- and heterodimeric receptor complexes [49].

Other examples of such mechanisms are the tyrosine phosphatases SHP-1 and -2, each of which has two SH2 domains through which they can bind to several PTK receptors. The binding to tyrosine-phosphorylated residues activates the enzymatic activities of SHP-1 and -2, which may then counteract signaling by dephosphorylating the receptor or its substrates. It is an interesting possibility that SHP-1 and -2, or other tyrosine phosphatases, may dephosphorylate individual tyrosine residues with different efficiency and thereby modulate signaling not only quantitatively but also qualitatively [50].

To complicate the issue even further, evidence indicates that SHP-2, in addition to its negative modulatory role in signaling, can also promote signaling via the receptors, possibly involving a mechanism that affects phosphorylation of a regulatory C-terminal phosphorylation site in members of the Src family of kinases.

Another mechanism for feedback control of signaling is via activation of protein kinase C (PKC). The classical members of the PKC family are activated by Ca^{2+} and diacylglycerol, which are produced downstream of phospholipase Cγ. For instance, the receptors for EGF, insulin, HGF, and stem cell factor are phosphorylated by PKC in such a way as to inhibit the tyrosine kinase activities of the receptors [51].

CROSS-TALK BETWEEN SIGNALING PATHWAYS

In addition to examples of negative modulation of one signaling pathway by another, as discussed for RasGAP and SHP-1 and -2, components in certain signaling pathways have been found to activate components in other signaling pathways. Examples include Ras and PI3-kinase, which can form a physical complex and activate each other mutually [52,53].

The cross-talk between signaling pathways downstream of PTK receptors may be the reason why the effects of activating one receptor or another in the same cell are rather similar, as illustrated, for example, by the use of microarray analysis of 3T3 cells after activation of PDGF or FGF receptors [54]. Even though there are several indications that the signaling capabilities of PTK receptors overlap extensively, it is likely that qualitative and quantitative differences in signaling capacity occur between PTK receptors, particularly in the situation (common *in vivo*) where the availability of ligand is limiting and only a small fraction of the receptors on the cell surface is activated.

In addition, cross-talk in signaling occurs via PTK receptors and other receptor types. Cytokine receptors, for example, which do not have any intrinsic kinase domain but interact with cytoplasmic tyrosine kinases of the JAK family, exert much of their signaling via activation of members of the STAT family [55]. However, STATs are also activated by certain PTK receptors [3]. Moreover, classical signaling pathways downstream of PTK receptors, such as Ras, PI3-kinase, and phospholipase Cγ, are also activated after ligation of cytokine receptors [56] or integrins [21].

Protein serine/threonine kinase receptors, which mediate growth inhibitory signals, activate SMAD molecules, which after translocation into the nucleus act as transcription factors. The MAP kinase Erk, which is activated by PTK receptors via Ras, has been shown to phosphorylate and inhibit SMADs [57], providing one example of how PTK-receptor induced signals can modulate, in an inhibitory manner, signaling by other receptors.

Another example of cooperation between different receptor types is the finding that the mitogenic activity of certain seven-transmembrane-spanning G-protein-coupled receptors occurs by transactivation of the EGF receptor [58].

PTK RECEPTORS AND DISEASE

Given the importance of PTK receptors in the control of cell proliferation and migration, it is not surprising that overactivity of PTK receptors occurs in cancer and other diseases that involve excess cell proliferation, such as inflammatory and fibrotic conditions, and psoriasis. About half of the PTK receptors are implicated in various human malignancies (Figure 9.1) [1]. Often, the receptors are constitutively activated by amplification or mutational events. Mutations of residues in the extracellular parts of the receptors may affect disulfide binding, thus causing the formation of covalent dimers. Mutations of residues in the transmembrane or juxtamembrane domains may relieve inhibitory mechanisms. In addition, mutation of residues in the kinase domain may change the conformation of the region between the two lobes of the kinase and relieve an inhibitory mechanism [59]. Moreover, the kinase domains of the receptors can form fusion molecules with proteins that normally occur as dimers or oligomers. The end result is a constitutively active kinase that drives cell growth and survival.

Another mechanism of activation of PTK receptors seen in disease is overproduction of the corresponding ligand. If a cell produces a growth factor for which it has the corresponding receptor, autocrine stimulation of growth may result. Alternatively, the growth factor may stimulate cells in the environment in a paracrine manner, which is relevant in tumor progression. Tumor-derived factors (e.g., VEGFs and FGFs) act on angiogenic PTK receptors and cause vascularization of the tumors, which is a prerequisite for tumor growth [60]. Likewise, other growth factors produced by tumor cells (e.g., PDGFs) may promote pericyte coverage of vessels and stimulate the formation of tumor stroma [61].

Since over-activity of PTK receptors is implicated in serious diseases, clinically useful PTK receptor antagonists are warranted. Several types of antagonists are currently used clinically or are in clinical trials for cancer, including monoclonal antibodies and low-molecular-weight selective inhibitors of various tyrosine kinases [62] (Table 9.1). it is

TABLE 9.1 Approved inhibitors for PTK receptors or their ligands

Compound	Target	Indication
Antibodies		
Trastuzumab (Herceptin)	ErbB2	Breast cancer
Cetuximab (Erbitux)	EGFR	Colorectal cancer
Panitumumab (Vectibix)	EGFR	Colorectal cancer
Bevacizumab (Avastin)	VEGF	Colorectal cancer
Low molecular weight compounds		
Imatinib (Gleevec)	Kit, PDGFR	Gastrointestinal stromal tumors
	Bcr-Abl*	Chronic myeloic leukemia
Erlotinib (Tarceva)	EGFR	Lung cancer
Gefitinib (Iressa)	EGFR	Lung cancer
Sorafenib (Nexavar)	VEGFR, PDGFR, c-Raf*	Kidney cancer
Sutinib (Sutent)	VEGFR, PDGFR, c-Raf*	Kidney cancer
	Kit	Gastrointestinal stromal tumors
Lapatinib (Tykerb)	EGFR, ErbB2	Breast cancer

**Brc-Abl is a fusion protein of Bcr and the non-receptor tyrosine kinase Abl, which is characteristic for chronic myeloic leukemia; c-Raf is a serine/threonine kinase which activates the Erk MAP kinase.*

likely that PTK receptor antagonists will be important tools in the treatment of cancer and possibly of other diseases characterized by excessive cell growth.

ACKNOWLEDGEMENTS

Ingegärd Schiller is thanked for her valuable help in the preparation of this manuscript. For space reasons, referencing has been kept to a minimum; I apologize to authors who have not been properly referenced.

REFERENCES

1. Blume-Jensen P, Hunter T. Oncogenic kinase signalling. *Nature* 2001;**411**:355–65.
2. Venter JC, Adams MD, Myers EW, Li PW, Mural RJ, Sutton GG, Smith HO, Yandell M, Evans CA, Holt RA, Gocayne JD, Amanatides P, Ballew RM, Huson DH, Wortman JR, Zhang Q, Kodira CD, Zheng XH, Chen L, Skupski M, Subramanian G, Thomas PD, Zhang J, Gabor Miklos GL, Nelson C, Broder S, Clark AG, Nadeau J, McKusick VA, Zinder N, Levine AJ, Roberts RJ, Simon M, Slayman C, Hunkapiller M, Bolanos R, Delcher A, Dew I, Fasulo D, Flanigan M, Florea L, Halpern A, Hannenhalli S, Kravitz S, Levy S, Mobarry C, Reinert K, Remington K, Abu-Threideh J, Beasley E, Biddick K, Bonazzi V, Brandon R, Cargill M, Chandramouliswaran I, Charlab R, Chaturvedi K, Deng Z, Di Francesco V, Dunn P, Eilbeck K, Evangelista C, Gabrielian AE, Gan W, Ge W, Gong F, Gu Z, Guan P, Heiman TJ, Higgins ME, Ji RR, Ke Z, Ketchum KA, Lai Z, Lei Y, Li Z, Li J, Liang Y, Lin X, Lu F, Merkulov GV, Milshina N, Moore HM, Naik AK, Narayan VA, Neelam B, Nusskern D, Rusch DB, Salzberg S, Shao W, Shue B, Sun J, Wang Z, Wang A, Wang X, Wang J, Wei M, Wides R, Xiao C, Yan C, et al. The sequence of the human genome. *Science* 2001;**291**:1304–51.
3. Schlessinger J. Cell signaling by receptor tyrosine kinases. *Cell* 2000;**103**:211–25.
4. Schmucker D, Zipursky SL. Signaling downstream of Eph receptors and ephrin ligands. *Cell* 2001;**105**:701–4.
5. Shrivastava A, Radziejewski C, Campbell E, Kovac L, McGlynn M, Ryan TE, Davis S, Goldfarb MP, Glass DJ, Lemke G, Yancopoulos GD. An orphan receptor tyrosine kinase family whose members serve as nonintegrin collagen receptors. *Mol Cell* 1997;**1**:25–34.
6. Vogel W, Gish G, Alves F, Pawson T. The discoidin domain receptor tyrosine kinases are activated by collagen. *Mol Cell* 1997;**1**:13–23.
7. Lu W, Yamamoto V, Ortega B, Baltimore D. Mammalian Ryk is a Wnt coreceptor required for stimulation of neurite outgrowth. *Cell* 2004;**119**:97–108.
8. Lee MK, Pardoux C, Hall MC, Lee PS, Warburton D, Qing J, Smith SM, Derynck R. TGFβ activates Erk MAP kinase signalling through direct phosphorylation of ShcA. *EMBO J* 2007;**26**:3957–67.
9. Heldin C-H, Östman A. Ligand-induced dimerization of growth factor receptors: variations on the theme. *Cytokine Growth Factor Rev* 1996;**7**:3–10.
10. Wiesmann C, Fuh G, Christinger HW, Eigenbrot C, Wells JA, de Vos AM. Crystal structure at 1.7 Å resolution of VEGF in complex with domain 2 of the Flt-1 receptor. *Cell* 1997;**91**:695–704.
11. Himanen J-P, Rajashankar KR, Lackmann M, Cowan CA, Henkemeyer M, Nikolov DB. Crystal structure of an Eph receptor–ephrin complex. *Nature* 2001;**414**:933–8.
12. Garrett TP, McKern NM, Lou M, Elleman TC, Adams TE, Lovrecz GO, Zhu HJ, Walker F, Frenkel MJ, Hoyne PA, Jorissen RN, Nice EC, Burgess AW, Ward CW. Crystal structure of a truncated epidermal growth factor receptor extracellular domain bound to transforming growth factor α. *Cell* 2002;**110**:763–73.
13. Ogiso H, Ishitani R, Nureki O, Fukai S, Yamanaka M, Kim JH, Saito K, Sakamoto A, Inoue M, Shirouzu M, Yokoyama S. Crystal structure of the complex of human epidermal growth factor and receptor extracellular domains. *Cell* 2002;**110**:775–87.
14. Schlessinger J, Plotnikov AN, Ibrahimi OA, Eliseenkova AV, Yeh BK, Yayon A, Linhardt RJ, Mohammadi M. Crystal structure of a ternary FGF–FGFR–heparin complex reveals a dual role for heparin in FGFR binding and dimerization. *Mol Cell* 2000;**6**:743–50.
15. Jiang G, Hunter T. Receptor signaling: when dimerization is not enough. *Curr Biol* 1999;**9**:R568–71.
16. Brennan PJ, Kumogai T, Berezov A, Murali R, Greene MI. HER2/Neu: mechanisms of dimerization/oligomerization. *Oncogene* 2000;**19**:6093–101.
17. Perez-Pinera P, Zhang W, Chang Y, Vega JA, Deuel TF. Anaplastic lymphoma kinase is activated through the pleiotrophin/receptor protein-tyrosine phosphatase β/ζ signaling pathway: an alternative mechanism of receptor tyrosine kinase activation. *J Biol Chem* 2007;**282**:28,683–28,690.
18. Heldin C-H, Östman A, Rönnstrand L. Signal transduction via platelet-derived growth factor receptors. *Biochim Biophys Acta* 1998;**1378**:F79–F113.
19. Yarden Y, Sliwkowski MX. Untangling the ErbB signalling network. *Nat Rev Mol Cell Biol* 2001;**2**:127–37.
20. Saito Y, Haendeler J, Hojo Y, Yamamoto K, Berk BC. Receptor heterodimerization: essential mechanism for platelet-derived growth factor-induced epidermal growth factor receptor transactivation. *Mol Cell Biol* 2001;**21**:6387–94.
21. Giancotti FG, Ruoslahti E. Integrin signaling. *Science* 1999;**285**:1028–32.
22. Guo P, Hu B, Gu W, Xu L, Wang D, Huang HJ, Cavenee WK, Cheng SY. Platelet-derived growth factor-B enhances glioma angiogenesis by stimulating vascular endothelial growth factor expression in tumor endothelia and by promoting pericyte recruitment. *Am J Pathol* 2003;**162**:1083–93.
23. Bourguignon LY, Singleton PA, Zhu H, Zhou B. Hyaluronan promotes signaling interaction between CD44 and the transforming growth factor β receptor I in metastatic breast tumor cells. *J Biol Chem* 2002;**277**:39,703–39,712.
24. Li L, Heldin C-H, Heldin P. Inhibition of platelet-derived growth factor-BB-induced receptor activation and fibroblast migration by hyaluronan activation of CD44. *J Biol Chem* 2006;**281**:26,512–26,519.
25. Soker S, Takashima S, Miao HQ, Neufeld G, Klagsbrun M. Neuropilin-1 is expressed by endothelial and tumor cells as an isoform-specific receptor for vascular endothelial growth factor. *Cell* 1998;**92**:735–45.
26. Hubbard SR. Crystal structure of the activated insulin receptor tyrosine kinase in complex with peptide substrate and ATP analog. *EMBO J* 1997;**16**:5572–81.
27. Shewchuk LM, Hassell AM, Ellis B, Holmes WD, Davis R, Horne EL, Kadwell SH, McKee DD, Moore JT. Structure of the Tie2 RTK domain: self-inhibition by the nucleotide binding loop, activation loop, and C-terminal tail. *Struct Fold Des* 2000;**8**:1105–13.
28. Chiara F, Bishayee S, Heldin C-H, Demoulin J-B. Autoinhibition of the platelet-derived growth factor β receptor tyrosine kinase by its C-terminal tail. *J Biol Chem* 2004;**279**:19732–8.

29. Zhang X, Gureasko J, Shen K, Cole PA, Kuriyan J. An allosteric mechanism for activation of the kinase domain of epidermal growth factor receptor. *Cell* 2006;**125**:1137–49.
30. Wybenga-Groot LE, Baskin B, Ong SH, Tong J, Pawson T, Sicheri F. Structural basis for autoinhibition of the EphB2 receptor tyrosine kinase by the unphosphorylated juxtamembrane region. *Cell* 2001;**106**:745–57.
31. Hubbard SR. Juxtamembrane autoinhibition in receptor tyrosine kinases. *Nat Rev Mol Cell Biol* 2004;**5**:464–71.
32. Pawson T, Nash P. Protein-protein interactions define specificity in signal transduction. *Genes Dev* 2000;**14**:1027–47.
33. Liu BA, Jablonowski K, Raina M, Arce M, Pawson T, Nash PD. The human and mouse complement of SH2 domain proteins–establishing the boundaries of phosphotyrosine signaling. *Mol Cell* 2006;**22**:851–68.
34. Maudsley S, Zamah AM, Rahman N, Blitzer JT, Luttrell LM, Lefkowitz RJ, Hall RA. Platelet-derived growth factor receptor association with Na(+)/H(+) exchanger regulatory factor potentiates receptor activity. *Mol Cell Biol* 2000;**20**:8352–63.
35. Lennartsson J, Wardega P, Engström U, Hellman U, Heldin C-H. Alix facilitates the interaction between c-Cbl and platelet-derived growth factor β-receptor and thereby modulates receptor downregulation. *J Biol Chem* 2006;**281**:39,152–39,15.
36. Sundaresan M, Yu ZX, Ferrans VJ, Irani K, Finkel T. Requirement for generation of H_2O_2 for platelet-derived growth factor signal transduction. *Science* 1995;**270**:296–9.
37. Verveer PJ, Wouters FS, Reynolds AR, Bastiaens PIH. Quantitative imaging of lateral ErbB1 receptor signal propagation in the plasma membrane. *Science* 2000;**290**:1567–70.
38. Bae YS, Sung J-Y, Kim O-S, Kim YJ, Hur KC, Kazlauskas A, Rhee SG. Platelet-derived growth factor-induced H_2O_2 production requires the activation of phosphatidylinositol 3-kinase. *J Biol Chem* 2000;**275**:10527–31.
39. Ni C-Y, Murphy MP, Golde TE, Carpenter G. γ-Secretase cleavage and nuclear localization of ErbB-4 receptor tyrosine kinase. *Science* 2001;**294**:2179–81.
40. Linggi B, Carpenter G. ErbB receptors: new insights on mechanisms and biology. *Trends Cell Biol* 2006;**16**:649–56.
41. Sardi SP, Murtie J, Koirala S, Patten BA, Corfas G. Presenilin-dependent ErbB4 nuclear signaling regulates the timing of astrogenesis in the developing brain. *Cell* 2006;**127**:185–97.
42. Lin S-Y, Makino K, Xia W, Matin A, Wen Y, Kwong KY, Bourguignon L, Hung M-C. Nuclear localization of EGF receptor and its potential new role as a transcription factor. *Nat Cell Biol* 2001;**3**:802–8.
43. Giri DK, Ali-Seyed M, Li LY, Lee DF, Ling P, Bartholomeusz G, Wang SC, Hung MC. Endosomal transport of ErbB-2: mechanism for nuclear entry of the cell surface receptor. *Mol Cell Biol* 2005;**25**:11,005–11,018.
44. Lo H-W, Hung MC. Nuclear EGFR signalling network in cancers: linking EGFR pathway to cell cycle progression, nitric oxide pathway and patient survival. *Br J Cancer* 2006;**94**:184–8.
45. Clague MJ, Urbe S. The interface of receptor trafficking and signalling. *J Cell Sci* 2001;**114**:3075–81.
46. Soubeyran P, Kowanetz K, Szymkiewicz I, Langdon WY, Dikic I. Cbl-CIN85 -endophilin complex mediates ligand-induced downregulation of EGF receptors. *Nature* 2002;**416**:183–7.
47. Joazeiro CA, Wing SS, Huang H, Leverson JD, Hunter T, Liu YC. The tyrosine kinase negative regulator c-Cbl as a RING-type, E2-dependent ubiquitin–protein ligase. *Science* 1999;**286**:309–12.
48. Levkowitz G, Waterman H, Zamir E, Kam Z, Oved S, Langdon WY, Beguinot L, Geiger B, Yarden Y. c-Cbl/Sli-1 regulates endocytic sorting and ubiquitination of the epidermal growth factor receptor. *Genes Dev* 1998;**12**:3663–74.
49. Ekman S, Kallin A, Engström U, Heldin C-H, Rönnstrand L. SHP-2 is involved in heterodimer specific loss of phosphorylation of Tyr771 in the PDGF β-receptor. *Oncogene* 2002;**21**:1870–5.
50. Östman A, Böhmer F-D. Regulation of receptor tyrosine kinase signaling by protein tyrosine phosphatases. *Trends Cell Biol* 2001;**11**:258–66.
51. Blume-Jensen P, Siegbahn A, Stabel S, Heldin C-H, Rönnstrand L. Increased Kit/SCF receptor induced mitogenicity but abolished cell motility after inhibition of protein kinase C. *EMBO J* 1993;**12**: 4199–209.
52. Hu Q, Klippel A, Muslin AJ, Fantl WJ, Williams LT. Ras-dependent induction of cellular responses by constitutively active phosphatidylinositol-3 kinase. *Science* 1995;**268**:100–2.
53. Rodriguez-Viciana P, Warne PH, Dhand R, Vanhaesebroeck B, Gout I, Fry MJ, Waterfield MD, Downward J. Phosphatidylinositol-3-OH kinase as a direct target of Ras. *Nature* 1994;**370**:527–32.
54. Fambrough D, McClure K, Kazlauskas A, Lander ES. Diverse signaling pathways activated by growth factor receptors induce broadly overlapping, rather than independent, sets of genes. *Cell* 1999;**97**:727–41.
55. Ihle JN. The Stat family in cytokine signaling. *Curr Opin Cell Biol* 2001;**13**:211–17.
56. Schindler C, Strehlow I. Cytokines and STAT signaling. *Adv Pharmacol* 2000;**47**:113–74.
57. Kretzschmar M, Doody J, Massagué J. Opposing BMP and EGF signalling pathways converge on the TGF-β family mediator Smad1. *Nature* 1997;**389**:618–22.
58. Gschwind A, Zwick E, Prenzel N, Leserer M, Ullrich A. Cell communication networks: epidermal growth factor receptor transactivation as the paradigm for interreceptor signal transmission. *Oncogene* 2001;**20**:1594–600.
59. Chen H, Ma J, Li W, Eliseenkova AV, Xu C, Neubert TA, Miller WT, Mohammadi M. A molecular brake in the kinase hinge region regulates the activity of receptor tyrosine kinases. *Mol Cell* 2007;**27**:717–30.
60. Ferrara N, Alitalo K. Clinical applications of angiogenic growth factors and their inhibitors. *Nat Med* 1999;**5**:1359–64.
61. Östman A, Heldin C-H. Involvement of platelet-derived growth factor in disease: development of specific antagonists. *Adv Cancer Res* 2001;**80**:1–38.
62. Noble ME, Endicott JA, Johnson LN. Protein kinase inhibitors: insights into drug design from structure. *Science* 2004;**303**:1800–5.

Chapter 10

Receptor Tyrosine Kinase Signaling and Ubiquitination

Daniela Hoeller[1] and Ivan Dikic[2]
[1]*Institute for Medical Biochemistry, Innsbruck Biocentre, Innsbruck, Austria*
[2]*Institute of Biochemistry II, Goethe University School of Medicine, Frankfurt am Main, Germany*

INTRODUCTION

Binding of extracellular factors to cell surface receptors initiates intracellular signaling cascades that control biological processes such as proliferation, differentiation, migration, survival, or death. Obviously, both quality and quantity of receptor signaling must be carefully controlled to ensure a rapid and appropriate cellular answer, while avoiding the devastating consequences of aberrant stimulation. Besides protein phosphorylation that serves as the major posttranslational modification to convey signals in the cell, ubiquitination of the receptor as well as several downstream effectors has emerged as an important modification that modulates signaling events. First, the attachment of ubiquitin provides the sorting signal for lysosomal degradation of the receptor leading to permanent termination of signaling. Second, ubiquitination can change the activity, the localization as well as the stability of downstream effectors.

THE UBIQUITIN CONJUGATION SYSTEM

The conjugation of ubiquitin to target proteins is a complex process, which requires successive actions of three enzymes: E1 (Ub-activating enzyme), E2 (Ub-conjugating enzyme), and E3 (Ub-ligase) [1]. In the first step Ub is activated in an ATP-dependent manner by forming a thioester bond with the E1. Ub is then transferred to an E2, which cooperates in the last step with an E3 ligase to couple Ub to the ε-aminogroup of a lysine within the substrate. The interaction of E3 and substrate frequently depends on the phosphorylation of the latter. Ubiquitination is thus an inducible event and molecularly linked to phosphorylation via specific binding domains present in E3 ligases. The attachment of a single Ub molecule is referred to as monoubiquitination. However, Ub contains seven lysines that can serve as acceptor sites in additional rounds of conjugation which gives rise to polyubiquitin chains. Depending on the utilized lysine, distinct linkage types are formed that have different functional consequences for the modified protein [2, 3]. As described in more details below, monoubiquitination is involved in protein sorting and lysosomal degradation. Polyubiquitination via lysine 48 (K48) leads to proteasomal degradation, and lysine 63 (K63)-linked chains have been found to be involved in the assembly of signaling complexes. Other types of chains do exist in cells, but their precise functions remain elusive. Whatever type of ubiquitin modification is attached to a protein it can be removed by the action of de-ubiquitinating enzymes (DUBs), allowing the protein to escape from its fate imposed by the Ub tag. Keeping the right balance between ubiquitination and de-ubiquitination of proteins has proven to be crucial for several cellular processes, as this determines the extent of ubiquitination of a protein. Imbalance in either direction can be associated with human diseases such as cancer [4]. In the following paragraphs, we will describe the molecular basis for the impact of different modes of ubiquitination on multiple steps of receptor signaling.

RTK SIGNALING AND ENDOCYTOSIS ARE MOLECULARLY LINKED

Ligand binding to receptor tyrosine kinases (RTKs), such as the epidermal growth factor receptor (EGFR), leads to receptor dimerization, activation, and phosphorylation at multiple tyrosines within their intracellular part. The phosphorylated tyrosines function as specific docking

Handbook of Cell Signaling, Three-Volume Set 2 ed.

sites for various effector proteins that become themselves phosphorylated by the receptor, and convey the signal to the nucleus and other cellular compartments. Importantly, at the same time as positive signaling is triggered, the receptor initiates its own inactivation as distinct phosphorylation sites are recognized by negative effectors, including phosphatases, ubiquitin ligases, GTPases, and adaptor proteins [5]. For example, the E3-ligases of the Cbl-family are responsible for ubiquitination, endocytosis, and subsequent lysosomal degradation of RTKs (Figure 10.1, reviewed in [6, 7]).

Growing evidence suggests that receptor endocytosis is not only implicated in terminal inactivation of receptor signaling, but rather represents an integral component of signal propagation and diversification [8, 9]. This is particularly clear in the case of neurotrophic factors and their cognate Trk receptors, which need to be internalized and retrogradely transported from nerve terminals to cell bodies in order to promote neuronal survival [10, 11]. Stimulation of nerve terminals with non-internalizable NGF was sufficient to locally induce the growth of distal axons, but failed to activate survival genes in the nucleus. Moreover, in cases where the signaling endpoint is not so distant from the initiation site, the internalization of activated receptor turns out to be critical for the activation of specific signals. For example, activated EGFRs can internalize into a subpopulation of early endosomes carrying the Rab5-effector APPL1. APPL1 undergoes EGF-dependent nucleocytoplasmic shuttling, which is required for cell proliferation [12].

UBIQUITINATION IN RTK ENDOCYTOSIS

Ubiquitin is recognized as a molecular signal playing a dual role in regulation of receptor endocytosis. On one hand, it provides sorting tags on the trafficking cargo that ensure receptor trafficking for degradation [13] (Figure 10.1); on the other, ubiquitination of endocytic sorting proteins may play a role in regulating their function as Ub receptors in the endosome [14].

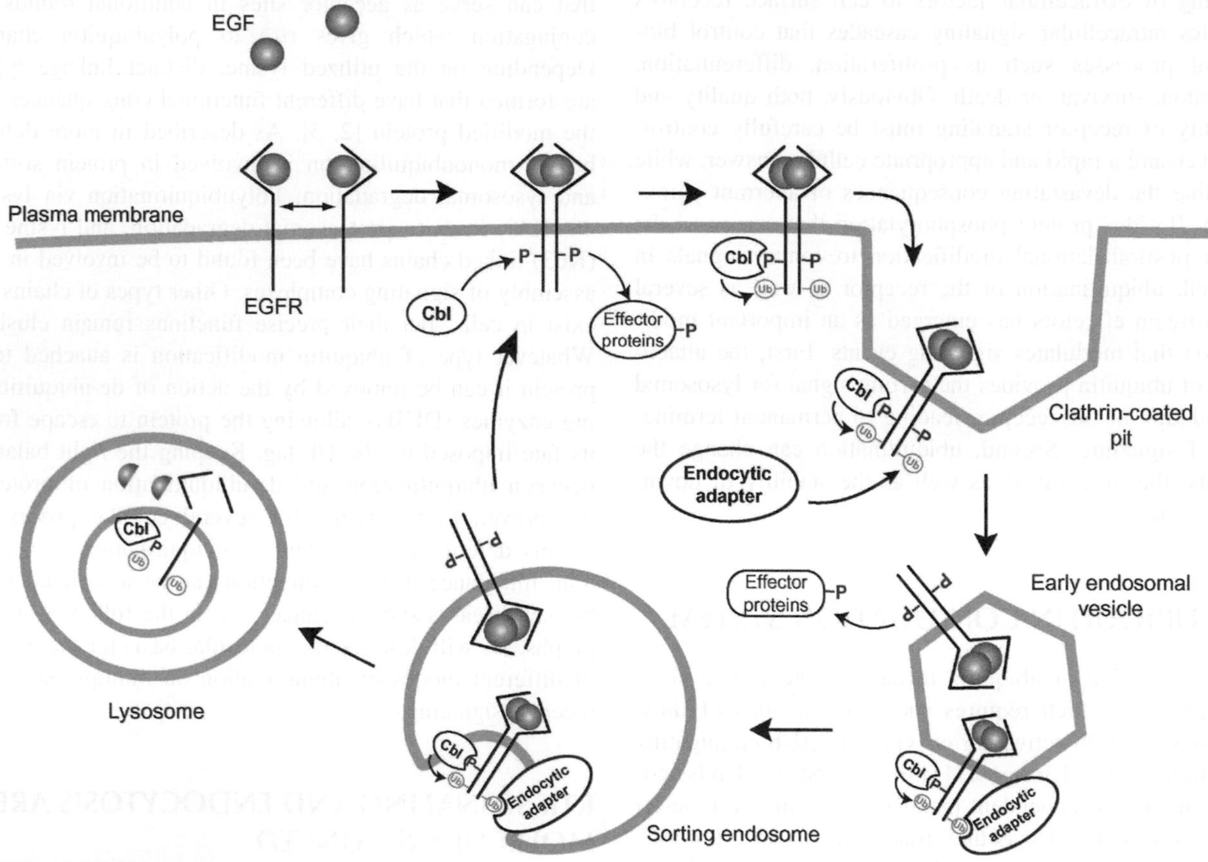

FIGURE 10.1 RTK endocytosis.
Receptor signaling as well as endocytosis is initiated at the plasma membrane by ligand binding, which is followed by receptor dimerization and phosphorylation of the cytosolic tail. Cbl is recruited to the phosphorylated receptor and mediates the attachment of multiple ubiquitin molecules (multiubiquitination) throughout endosomal sorting. While the activated receptor is internalized into endocytic vesicles, it continues to emit signals. Endocytic adaptor proteins such as Eps15, Hrs and Stam2 recognize the ubiquitin tags and link the receptor to membrane domains that deliver the receptor to the lysosome, where it is degraded together with Cbl and the bound ligand. Non-ubiquitinated receptor is recycled to the plasma membrane.

Cbl-mediated attachment of multiple monoubiquitin moieties (multi-ubiquitination) appears to be the key for endosomal sorting of the receptor [15, 16]. Loss of the Ub tags results in recycling of the receptor to the cell membrane, and it has been shown that Cbl needs to stay associated with the receptor along the endosomal route in order to ensure receptor ubiquitination and delivery to the final destination, the lysosome [17, 18]. Moreover, in-frame fusion of monoUb to the EGFR was sufficient to drive its internalization and lysosomal degradation. Though receptor ubiquitination by Cbl starts in the plasma membrane soon after receptor stimulation [17–19], recent reports question a critical role of receptor ubiquitination in early steps of endocytosis in mammalian cells [20, 21]. Though a ubiquitination-deficient (i.e., lysine-less or Cbl-binding deficient) EGFR was not degraded, it was normally internalized into clathrin-coated vesicles. Indeed, there is some experimental evidence that ubiquitinated receptors may utilize a different entry route than non-ubiquitinated receptors [22]. Moreover, besides being multi0ubiquitinated, the EGFR becomes also significantly modified with K63-linked Ub chains upon EGF stimulation [23]. This linkage type has not been implicated in lysosomal sorting in mammalian cells, but rather in the assembly of signaling complexes. Moreover, K63-polyubiquitination of the NGF receptor TrkA has been shown to be required for both its internalization and signaling [24]. Whether, and if so how, this modification contributes to EGFR signaling in a similar manner remains to be determined.

UBIQUITINATION OF EFFECTOR PROTEINS IN RTK SIGNALING

Aside from receptor ubiquintination, ligand stimulation induces the ubiquitination of downstream effectors implicated in signal transduction, as well as endocytic adaptor proteins required for lysosomal sorting of the ubiquitinated receptor. In most cases, K48-linked Ub chains are formed that lead to proteasomal degradation of the substrate.

Proteins that are responsible for the endocytic sorting of the ubiquitinated receptor (such as Eps15, epsins, Hrs, and Stam2) are subjected to a special type of ubiquitination, the so-called *coupled monoubiquitination*. An important feature of these proteins is their ability to interact with ubiquitin via Ub-binding domains (UBDs) [25], which is required not only for their function in receptor sorting but also for their monoubiquitination [26, 27]. The modification triggers intramolecular binding of the UBD to the attached ubiquitin molecule, and leads to the functional inactivation of the protein [14]. Coupled monoubiquitination can be triggered upon growth factor stimulation. It has been shown that receptor activation induces the association of the endocytic adaptor protein Eps15 with two different Ub ligases, the Ring-type E3 ligase Parkin [28] and the HECT-type E3 ligase Nedd4 [29]. The interaction was dependent on the Ub-interacting motif (UIM) of Eps15 and the Ub-like domain (Ubl) of Parkin and the ubiquitin-modification of Nedd4, respectively. On the other hand, several UBD-containing proteins, including Eps15, Stam2, Sts-1/2, and others, can self-ubiquitinate independently of cell stimulation [30]. In this case, the Ub-loaded E2 is directly recruited to the UBD of the substrate and mediates the transfer of Ub without the need for an E3 ligase. This process is considered as an active homeostatic process, which regulates localization, activity, and signaling competence of UBD-containing proteins in a given physiological state. No matter how coupled monoubiquitination is brought about, the created pool of inactive endocytic adaptors could be rapidly reactivated by DUBs.

CONCLUDING REMARKS AND FUTURE PERSPECTIVES

In this chapter, we discussed how ubiquitination regulates receptor tyrosine kinases and how tyrosine kinase signaling pathways control ubiquitination processes. We described an elaborate, although not completely understood, molecular interplay between receptor tyrosine kinase signaling and protein ubiquitination networks that are critical for the regulation of physiological and pathological conditions. In particular, there is a link between constitutive receptor tyrosine kinase (RTK) signaling and the loss of ubiquitin-dependent negative regulation of RTKs, which can cause cell transformation and cancer. Current efforts in the development of cancer therapies aim to induce the downregulation of cancer-associated RTKs. Trastuzumab, a monoclonal antibody directed against the HER2 extracellular domain, was shown to increase Cbl-mediated downregulation of HER-2. However, besides regulating receptor sorting and degradation, ubiquitination emerges as a potent signal regulating localization, activity, and binding-competence of a great variety of proteins, not only in RTK signaling. Ub-binding proteins that recognize ubiquitinated effectors play a key role in the establishment of these signaling networks. Monitoring growth factor-induced Ub binding proteins–Ub interactions in space and time represents a future challenge towards a more comprehensive understanding of Ub-based cellular events.

ACKNOWLEDGEMENTS

We are grateful to Fumiyo Ikeda and Mirko Schmidt for critical comments and suggestions on the book chapter. The work in the authors; labs is supported by grants and fellowships from Deutsche Forschung Gemeinschaft, GIF and EMBO.

REFERENCES

1. Hershko A, Ciechanover A. The ubiquitin system. *Annu Rev Biochem* 1998;**67**:425–79.
2. Schnell JD, Hicke L. Non-traditional functions of ubiquitin and ubiquitin-binding proteins. *J Biol Chem* 2003;**278**:35857–60.
3. Hicke L, Schubert HL, Hill CP. Ubiquitin-binding domains. *Nat Rev* 2005;**6**:610–21.
4. Hoeller D, Hecker CM, Dikic I. Ubiquitin and ubiquitin-like proteins in cancer pathogenesis. *Nat Rev Cancer* 2006;**6**:776–88.
5. Dikic I, Giordano S. Negative receptor signalling. *Curr Opin Cell Biol* 2003;**15**:128–35.
6. Thien CB, Langdon WY. c-Cbl and Cbl-b ubiquitin ligases: substrate diversity and the negative regulation of signalling responses. *Biochem J* 2005;**391**:153–66.
7. Schmidt MH, Dikic I. The Cbl interactome and its functions. *Nat Rev* 2005;**6**:907–19.
8. Miaczynska M, Pelkmans L, Zerial M. Not just a sink: endosomes in control of signal transduction. *Curr Opin Cell Biol* 2004;**16**:400–6.
9. Di Fiore PP, De Camilli P. Endocytosis and signaling. an inseparable partnership. *Cell* 2001;**106**:1–4.
10. Delcroix JD, Valletta JS, Wu C, Hunt SJ, Kowal AS, Mobley WC. NGF signaling in sensory neurons: evidence that early endosomes carry NGF retrograde signals. *Neuron* 2003;**39**:69–84.
11. Ye H, Kuruvilla R, Zweifel LS, Ginty DD. Evidence in support of signaling endosome-based retrograde survival of sympathetic neurons. *Neuron* 2003;**39**:57–68.
12. Miaczynska M, Christoforidis S, Giner A, Shevchenko A, Uttenweiler-Joseph S, Habermann B, Wilm M, Parton RG, Zerial M. APPL proteins link Rab5 to nuclear signal transduction via an endosomal compartment. *Cell* 2004;**116**:445–56.
13. Haglund K, Di Fiore PP, Dikic I. Distinct monoubiquitin signals in receptor endocytosis. *Trends Biochem Sci* 2003;**28**:598–603.
14. Hoeller D, Crosetto N, Blagoev B, Raiborg C, Tikkanen R, Wagner S, Kowanetz K, Breitling R, Mann M, Stenmark H, Dikic I. Regulation of ubiquitin-binding proteins by monoubiquitination. *Nat Cell Biol* 2006;**8**:163–9.
15. Mosesson Y, Shtiegman K, Katz M, Zwang Y, Vereb G, Szollosi J, Yarden Y. Endocytosis of receptor tyrosine kinases is driven by monoubiquitylation, not polyubiquitylation. *J Biol Chem* 2003;**278**:21,323–21,326.
16. Haglund K, Sigismund S, Polo S, Szymkiewicz I, Di Fiore PP, Dikic I. Multiple monoubiquitination of RTKs is sufficient for their endocytosis and degradation. *Nat Cell Biol* 2003;**5**:461–6.
17. de Melker AA, van der Horst G, Calafat J, Jansen H, Borst J. c-Cbl ubiquitinates the EGF receptor at the plasma membrane and remains receptor associated throughout the endocytic route. *J Cell Sci* 2001;**114**:2167–78.
18. Longva KE, Blystad FD, Stang E, Larsen AM, Johannessen LE, Madshus IH. Ubiquitination and proteasomal activity is required for transport of the EGF receptor to inner membranes of multivesicular bodies. *J Cell Biol* 2002;**156**:843–54.
19. de Melker AA, van der Horst G, Borst J. Ubiquitin ligase activity of c-Cbl guides the epidermal growth factor receptor into clathrin-coated pits by two distinct modes of Eps15 recruitment. *J Biol Chem* 2004;**279**:55,465–55,473,.
20. Huang F, Goh LK, Sorkin A. EGF receptor ubiquitination is not necessary for its internalization. *Proc Natl Acad Sci USA* 2007;**104**:16,904–16,909.
21. Duan L, Miura Y, Dimri M, Majumder B, Dodge IL, Reddi AL, Ghosh A, Fernandes N, Zhou P, Mullane-Robinson K, Rao N, Donoghue S, Rogers RA, Bowtell D, Naramura M, Gu H, Band V, Band H. Cbl-mediated ubiquitinylation is required for lysosomal sorting of epidermal growth factor receptor but is dispensable for endocytosis. *J Biol Chem* 2003;**278**:28,950–28,960.
22. Sigismund S, Woelk T, Puri C, Maspero E, Tacchetti C, Transidico P, Di Fiore PP, Polo S. Clathrin-independent endocytosis of ubiquitinated cargos. *Proc Natl Acad Sci USA* 2005;**102**:2760–5.
23. Huang F, Kirkpatrick D, Jiang X, Gygi S, Sorkin A. Differential regulation of EGF receptor internalization and degradation by multiubiquitination within the kinase domain. *Mol Cell* 2006;**21**:737–48.
24. Geetha T, Jiang J, Wooten MW. Lysine 63 polyubiquitination of the nerve growth factor receptor TrkA directs internalization and signaling. *Mol Cell* 2005;**20**:301–12.
25. Hicke L, Dunn R. Regulation of membrane protein transport by ubiquitin and ubiquitin-binding proteins. *Annu Rev Cell Dev Biol* 2003;**19**:141–72.
26. Miller SL, Malotky E, O'Bryan JP. Analysis of the role of ubiquitin-interacting motifs in ubiquitin binding and ubiquitylation. *J Biol Chem* 2004;**279**:33,528–33,537.
27. Polo S, Sigismund S, Faretta M, Guidi M, Capua MR, Bossi G, Chen H, De Camilli P, Di Fiore PP. A single motif responsible for ubiquitin recognition and monoubiquitination in endocytic proteins. *Nature* 2002;**416**:451–5.
28. Fallon L, Belanger CM, Corera AT, Kontogiannea M, Regan-Klapisz E, Moreau F, Voortman J, Haber M, Rouleau G, Thorarinsdottir T, Brice A, van Bergen En Henegouwen PM, Fon EA. A regulated interaction with the UIM protein Eps15 implicates parkin in EGF receptor trafficking and PI(3)K-Akt signalling. *Nature Cell Biol* 2006;**8**:834–42.
29. Woelk T, Oldrini B, Maspero E, Confalonieri S, Cavallaro E, Di Fiore PP, Polo S. Molecular mechanisms of coupled monoubiquitination. *Nature Cell Biol* 2006;**8**:1246–54.
30. Hoeller D, Hecker CM, Wagner S, Rogov V, Dotsch V, Dikic I. E3-independent monoubiquitination of ubiquitin-binding proteins. *Mol Cell* 2007;**26**:891–8.

Chapter 11

Insulin Receptor Complex and Signaling by Insulin

Lindsay G. Sparrow and S. Lance Macaulay
CSIRO Molecular and Health Technologies and Preventative Health Flagship, Parkville, Victoria, Australia

INTRODUCTION

Insulin is an important regulatory hormone that mediates energy uptake by the body. It is the major food storage hormone, and is secreted in response to rising blood sugar levels following a meal. It regulates energy uptake by inhibiting glucose production by the liver and by increasing sugar uptake into muscle and fat, directing this sugar into storage forms as glycogen in liver and muscle, and as fat in adipose cells. Insulin deficiency or resistance to its actions lead to the profound metabolic dysfunctions of type 1 or type 2 diabetes, respectively, which are major diseases of the Western world. There is therefore intense interest in understanding the mechanisms of insulin's actions. Insulin exerts a wide variety of effects on cells that include both metabolic and mitogenic actions, which are triggered by binding to its cell surface receptor. This review briefly explores the nature of this receptor, its interaction with insulin, and its signaling to a major metabolic target, glucose transport.

INSULIN RECEPTOR DOMAIN STRUCTURE

The insulin receptor (IR) is a glycosylated, disulfide-linked homodimer with each monomer being made up of an α chain that is entirely extracellular and a β chain that spans the cell membrane once. The α chain contains the insulin binding determinants of the receptor, while the intracellular portion of the β chain includes a protein tyrosine kinase domain and domains involved in binding signal transduction proteins. The αβ monomer of the IR is encoded by a gene with 22 exons; alternative splicing of the IR pre-mRNA leads to the tissue-specific expression of two isoforms differing by the presence or absence of a 12-residue segment (exon 11) near the C-terminus of the α chain. The receptor is synthesized as a single chain with a 27-residue signal sequence, and is glycosylated, oxidized to the disulfide form, and proteolytically processed to the two-chain form during transport to the cell surface. The mature α chain of the human IR has 731 amino acid residues, while the β chain has 620. Two receptors with close sequence and structural homology to the IR are the receptor for insulin-like growth factor-1 (IGF-1R) and the orphan receptor, the insulin-receptor related receptor (IRR) (see [1–3] for reviews).

Analysis of the sequence of the IR has shown that the molecule can be divided into a number of modules or domains [4]. The ectodomain has two large homologous, leucine-rich domains of approximately 150 residues, L1 and L2, separated by a 150-residue cysteine-rich domain, Cys-rich; in this respect the IR is similar to the epidermal growth factor receptor (EGFR). C-terminal to these, the IR has three fibronectin type III domains (FnIII-1, -2 and -3) of approximately 100–130 residues each. One of these, FnIII-2 has an insertion of 125 residues (the insert domain, ID) which contains the α–β cleavage site, and this results in the N-terminal region of FnIII-2 and part of the insert domain being found at the C-terminus of the α chain, with the remaining portions of these two domains being located at the N-terminus of the β chain. The domain structure of the IR ectodomain is shown schematically in Figure 11.1. The discussion in this section focuses largely on the structure of the ectodomain and its interaction with the ligand, insulin. The structure of the kinase domain is discussed in Chapter 45 of Handbook of Cell Signaling, Second Edition.

The IR ectodomain has been shown to have a single disulfide bond linking the α and β chains, joining residues 647 and 872 (exon 11 plus form), and at least two disulfides symmetrically linking the α chains of the dimer; one at residue 524, and one or more linking the residues 682, 683, and 685 [5]. The human IR has 18 predicted sites for N-linked glycosylation; 14 on the α chain and four

Handbook of Cell Signaling, Three-Volume Set 2 ed.

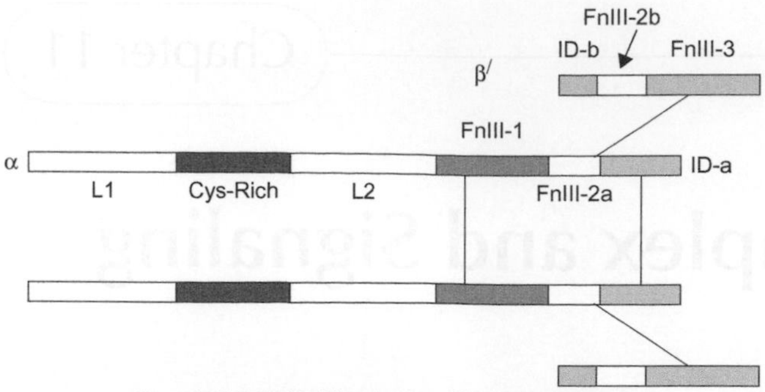

FIGURE 11.1 Domain structure of the IR.
Heavy lines, chain-linking disulfide bonds; β′ refers to that portion of the β-chain that is extracellular. © Elsevier 2003, all rights reserved.

on the β chain. Of the 18 predicted N-linked sites, 16 are occupied by carbohydrate and 1 is unoccupied; O-linked glycans have been shown to occur only on 4 threonine and 2 serine residues in the 22 residues at the N-terminus of the β chain [6, 7].

BINDING DETERMINANTS OF THE INSULIN RECEPTOR

A characteristic property of insulin is its propensity to oligomerize with increasing concentration, first to a dimer and then, in the presence of Zn, to a hexamer which is the storage form of the hormone in the β cells of the pancreas. The surfaces of the insulin molecule used for dimer and hexamer formation also seem to be used for its interaction with the receptor. Insulin is thought to have two distinct receptor binding surfaces, with one site, the "classical" binding site, encompassing at least the residues Gly1, Gln5, Tyr19, and Asn21 of the A chain and residues Val12, Tyr16, Gly23, Phe24, Phe25, and Tyr26 of the B chain, while the second site uses the hexamer surface residues, LeuA13 and LeuB17 [8]; other residues apart from those above are almost certainly involved in this interaction.

Binding of insulin to the native, dimeric IR is characterized by curvilinear Schatchard plots and the phenomenon of negative cooperativity, and these are interpreted as showing that the receptor has both high- and low-affinity binding states. Half receptors, i.e. αβ monomers, formed by mild reduction of the dimeric $(\alpha\beta)_2$ receptor, do not show these phenomena, and bind insulin only with low affinity – as does the expressed, recombinant IR ectodomain. The isolated α chain also binds insulin with low affinity, and appears to have all the insulin binding determinants of the receptor. The IR truncated below the transmembrane domain does show high-affinity binding, as do species where a dimerization moiety such as the IgG Fc or λ domain or a leucine-zipper segment is fused to the C-terminus of the expressed ectodomain [2].

By using alanine-scanning mutagenesis, studies of chimeric receptors and direct crosslinking of insulin to the IR, binding determinants on the receptor have been located in the L1 domain, in the cysteine-rich region, near the N-terminus of the L2 domain and at the C-terminus of the α chain. The fragment L1–Cys-rich–L2 of both the IR and the IGF-1R does not bind the cognate ligand, despite the presence of many of the binding determinants. Low-affinity ligand binding can be restored by adding to the C-terminus of this fragment a 16-residue peptide, designated CT, derived from the C-terminus of the α chain. The peptide can be attached directly to the C-terminus of the L1–Cys-rich–L2 fragment, or by using linkers of varying lengths; indeed, addition of the free peptide to the fragment also restores binding, showing that the mode of attachment is not critical. High-affinity binding is restored when the first fibronectin III domain, FnIII-1, is included in the construct, giving a molecule L1–Cys-rich–L2–FnIII-1–CT, and this suggests that further determinants for binding reside in the FnIII-1 domain [2, 9, 10]. For high-affinity binding the two α chains must be linked by at least one disulfide bond, suggesting that the two monomers are involved in binding a single insulin molecule, thus providing a crosslink between the chains in addition to the disulfide bond.

Structures for the IR ectodomain and for the fragment L1–Cys-rich–L2 from both the IR and the IGF-1R are now available [11, 12, 13]; these were obtained from crystals of expressed, recombinant proteins. The structure of the IR ectodomain was obtained using a construct in which a small section at the N-terminus of the β chain was deleted to reduce heterogeneity and facilitate crystallization; the resulting protein, IRδβ, was crystallized as a complex with Fabs from two IR binding antibodies, 83–7 and 83–14. Though none of these structures include the cognate ligand, they give clear insights into the receptor surfaces that are involved in ligand binding.

The structure of the IR ectodomain construct, IRδβ, is shown in Figure 11.2. Figure 11.2a shows the ectodomain homodimer as an inverted V structure formed by a two-fold rotation of the monomer (seen in Figure 11.2b) about an axis almost parallel to the axis of the inverted V. An important component of the putative insulin binding site, the insert domain (ID) within the FnIII-2 domain, which includes the 16-residue peptide mentioned above, is disordered in this molecule and thus not seen in the structure. In the native IR, the "legs" of the inverted V would extend through the cell membrane. The effect of this juxtaposition of the two monomers is to bring the L1 domain in particular, but also parts of the Cys-rich and L2 domains of one monomer, into the proximity of the FnIII-2 domain of the other monomer and also its insert domain.

Binding of insulin to the IR is thought to be accompanied by a conformational change at each end of the insulin B chain, leading to exposure of hydrophobic residues from the core and formation of one binding site, the classical

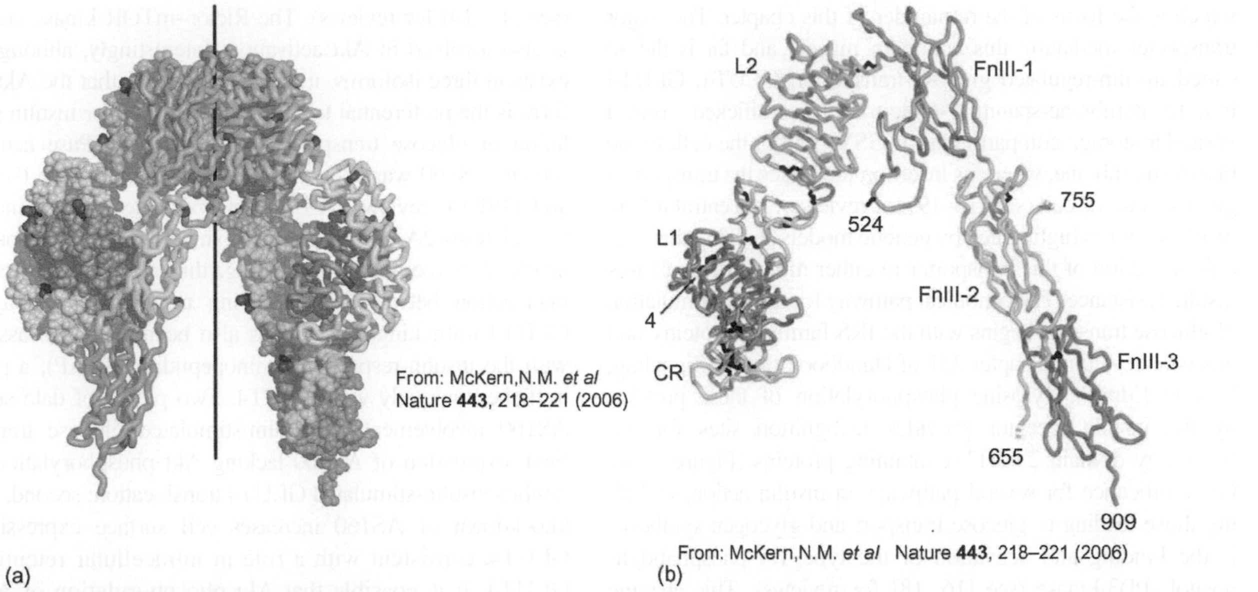

FIGURE 11.2 Structure of the IRδβ ectodomain.
(a) Structure of the homodimer with one monomer shown in tube form and the other in atomic sphere representation; positions of potential N-linked glycosylation sites indicated in black. (b) Structure of a monomer within the homodimer; disulfide bonds are shown in black. From McKern *et al.*, 2006 [11].

binding site. Thus, movement of the C-terminal 8–10 residues would expose key residues, IleA2 and ValA3, while residues IleA10, LeuA13, and LeuB6 are exposed by changes of the N-terminal 8 residues of the B chain [14].

Clearly the receptor also undergoes a major conformation change on insulin binding, resulting in the formation of the high-affinity state. Some physical evidence for the rearrangement is the change in the Stokes radius of the receptor from 9.1 to 7.5 nm [15]. The nature of this change is unclear, but it must in some way bring together the two intracellular kinase domains of the homodimer so that the activation loop of one is accessible to the active site of the other, leading to phosphorylation of the activation loop and activation of the kinase.

The details of the changes in the IR on insulin binding await a structure of the ligand–receptor complex. However, a plausible model for the process has been proposed based on the structures depicted in Figure 11.2 [11]. Figure 11.3 shows a proposed structure for the high-affinity Insulin–IR complex. Here, the classical binding site of insulin is bound to a conserved hydrophobic patch on the surface of the L1 domain of the IR in such a way that the second site, corresponding to the hexamer face of the insulin molecule, is available for further interaction with the receptor (for details, see [11]). As seen in Figure 11.3, this further interaction involves loops from the FnIII-1 domain. Movement of the FnIII-1 domain to make these contacts would change the relative dispositions of the FnIII "legs" of the two monomers, and this in turn should lead to the change in the relative positions of the intracellular kinase domains mentioned above. Activation of the IR kinase is accompanied by autophosphorylation at up to six tyrosines, and this both further activates the kinase and creates binding sites for signaling proteins that in turn become phosphorylated and bind their downstream targets.

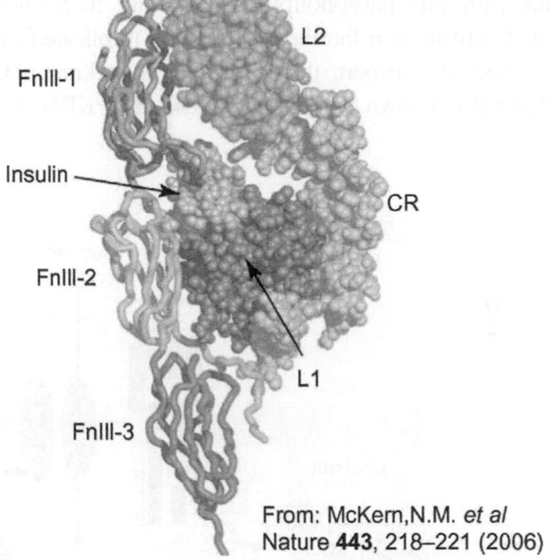

FIGURE 11.3 A possible structure for the Insulin–IR complex.
Domains L1, Cys-rich, and L2 from one monomer are in atomic sphere representation, as is insulin (arrowed), while the fibronectin-3 domains from the other monomer are seen as tubes. From McKern *et al.*, 2006 [11].

INSULIN SIGNALING TO GLUCOSE TRANSPORT

A major (if not the major) endpoint of insulin signaling is the stimulation of glucose uptake (transport) in muscle and fat, which is the focus of the remainder of this chapter. The major transporter mediating this action in muscle and fat is the so called insulin-regulated glucose transporter, GLUT4. GLUT4 is a 12-membrane-spanning protein that is trafficked, from a vesicular storage compartment(s) (GSV) within the cell, to the plasma membrane, where its insertion facilitates the transport of glucose into the cell (see [16–19] for reviews). Its central role in insulin action is highlighted by genetic models showing that specific depletion of this transporter in either muscle or fat causes insulin resistance. The canonical pathway for insulin stimulation of glucose transport begins with the IRS family of proteins, and this is reviewed in Chapter 331 of Handbook of Cell Signaling, Second Edition. Tyrosine phosphorylation of these proteins by the insulin receptor provides recognition sites for Src homology domain 2 (SH2)-containing proteins (Figure 11.4). Of significance for several pathways in insulin action, including those leading to glucose transport and glycogen synthesis, is the binding and activation of the type 1A phosphatidylinositol (PI)3-kinase (see [16, 18] for reviews). This enzyme phosphorylates inositol phospholipids in the plasma membrane, increasing PI3, 4, 5-trisphosphate (PI345P3) levels and enabling the recruitment and activation of a serine/threonine kinase, phosphoinositide-dependent protein kinase (PDK1), via association of its pleckstrin homology (PH) domain with this phospholipid. PDK1 can then phosphorylate and activate two families of proteins implicated in stimulating glucose transport, the atypical protein kinases C ζ/λ, and Akt (also known as protein kinase B (PKB)) isoforms (see [17–19] for reviews). The Rictor–mTOR kinase complex is also involved in Akt activation. Interestingly, although Akt exists in three isoforms, it has become clear that the Akt2 isoform is the preferential target in the pathway for insulin stimulation of glucose transport. Recently, the GTPase activating protein AS160 was shown to be a substrate for Akt (see [20] and [19] for review). AS160 catalyzes the *in vitro* inactivation of Rabs 2A, 8A, 10, and 14, small GTP binding proteins involved in membrane trafficking, thus providing a potential intersection between the signaling trafficking pathway and GLUT4 trafficking. AS160 has also been shown to associate with the insulin-responsive aminopeptidase (IRAP), a protein trafficked uniquely with GLUT4. Two pieces of data support AS160 involvement in insulin-stimulated glucose transport. First, expression of AS160 lacking Akt phosphorylation sites inhibits insulin-stimulated GLUT4 translocation; second, RNAi knockdown of AS160 increases cell surface expression of GLUT4, consistent with a role in intracellular retention of GLUT4. It is possible that Akt phosphorylation of AS160 releases it from IRAP, enabling Rab stimulation of GLUT4 trafficking. However, this has yet to be demonstrated, and it is also likely that other proteins are involved (see, for example, [20]).

Another putative pathway implicated in insulin stimulation of glucose transport is PI3-kinase independent, and involves tyrosine phosphorylation of the protooncogene Cbl by the insulin receptor tyrosine kinase, possibly via association with the Cbl-associated protein (CAP) to the

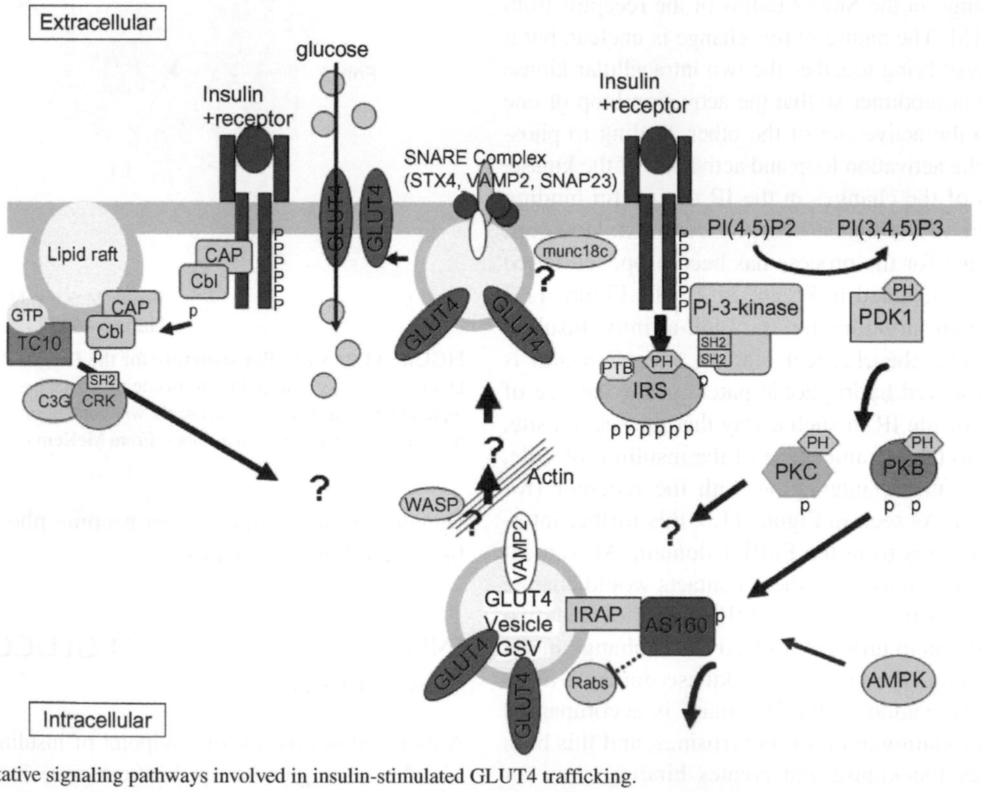

FIGURE 11.4 Putative signaling pathways involved in insulin-stimulated GLUT4 trafficking.

receptor within caveolin-rich lipid rafts (see [16, 17] for reviews). This enables the recruitment of the adaptor protein CrkII and its associated guanine exchange factor, C3G, resulting in the activation of the Rho family GTPase, TC10. However, while CAP association with the insulin receptor is clear, the hypothesis that this pathway participates in insulin-regulated glucose transport has been challenged (see [19] for review) by approaches that include siRNA silencing of intermediates as well as overexpression approaches. Further, ablation of cCbl has been found actually to improve insulin sensitivity.

Muscle contraction and osmotic stress also stimulate GLUT4 translocation, though the signaling pathway is separate, potentially involving the AMP-activated protein kinase (AMPK) (see [21] for review).

How the insulin-responsive PI3-kinase-dependent pathway acts on the vesicular compartments storing GLUT4 to move them to the cell surface is unclear. Recent data suggest involvement of cytoskeleton proteins such as WASP, vimentin, a-tubulin, dynein, and/or cortical actin in moving GLUT4 to the cell surface (see [17, 22] for review). Thus it is clear that the cytoskeleton is involved in trafficking GLUT4, although the detail remains to be determined.

Following translocation of GLUT4 vesicles to regions below the cell surface, the final stages of trafficking and fusion of the GLUT4 vesicles have been shown to involve SNARE proteins (see [17, 22–24] for reviews). These were first described as proteins mediating the fusion of neurotransmitter vesicles at synaptic terminals, and much of the molecular detail of soluble N-ethylmaleimide sensitive factor attachment protein receptor (SNARE) protein interactions was elucidated in this system [25]. In the case of plasma membrane trafficking of GLUT4, the functional vesicle-associated membrane protein (VAMP, also known as synaptobrevin) isoform is VAMP2. This protein binds two plasma-membrane SNARE proteins, syntaxin 4 (STX4) and SNAP23, to form a so-called SNARE complex involving interaction between coiled-coil regions of the three proteins. Although the formation of this complex is possibly sufficient to enable fusion of the GLUT4 vesicle with the plasma membrane, placing the GLUT4 proteins on the cell surface to transport glucose into the cell, effective fusion also requires N-ethylmaleimide-sensitive factor, NSF, and soluble NSF attachment protein (α-SNAP) driven hydrolysis of ATP. The precise role of NSF and α-SNAP in fusion is unclear. Additional accessory proteins have been demonstrated to regulate this fusion process, although the roles of these accessory factors are still unclear (see [17, 22, 23] for reviews). Munc18c is an STX4 binding protein that may have a role as a molecular chaperone for STX4, and in fusion. A structure for munc18c in complex with a STX4 N-terminal peptide was recently reported consistent with a dual role in fusion [26]. Interestingly, insulin-stimulated phosphorylation of these proteins has recently been reported. Other proteins implicated in regulation of fusion events include synip (STX4-interacting protein), synaptotagmin, VAP-33, and Rab4. How these proteins interact with the fusion proteins to regulate this important process remains to be established. Further regulation is provided by the endosomal trafficking of the transporters via clathrin-coated pits for recycling via constititutive or insulin-regulated compartments (see [18–19] for review). Although there is general agreement that the majority of GLUT4 is segregated into a unique insulin-sensitive endosomal compartment (GSV) in the basal state, the nature of this compartment and its regulation remain unclear despite the identification of targeting motifs within the C- and N-terminal regions of GLUT4. These areas offer fruitful areas for investigation over coming years.

ACKNOWLEDGEMENTS

The authors regret that due to space restrictions there was not space to cite the significant contributions of many investigators to the field. We have instead cited representative reviews that cover the respective areas in greater detail.

REFERENCES

1. Marino-Busljе C, Martin-Martinez M, Mizuguchi K, Siddle K, Blundell TL. The insulin receptor: from protein sequence to structure. *Biochem Soc Trans* 1999;**27**:715–26.
2. Adams TE, Epa VC, Garrett TPJ, Ward CW. Structure and function of the type 1 insulin-like growth factor receptor. *Cell Mol Life Sci* 2000;**57**:1050–93.
3. De Meyts P. The structural basis of insulin and insulin-like growth factor-1 receptor binding and negative co-operativity, and its relevance to mitogenic versus metabolic signalling. *Diabetologia* 1994;**37**:S135–48.
4. Bajaj M, Waterfield MD, Schlessinger J, Taylor WR, Blundell T. On the tertiary structure of the extracellular domain of the epidermal growth factor and insulin receptors. *Biochim Biophys Acta* 1987;**916**:220–6.
5. Sparrow LG, McKern NM, Gorman JJ, Strike PM, Robinson CP, Bentley JD, Ward CW. The Disulfide Bonds in the C-terminal Domains of the Human Insulin Receptor Ectodomain. *J Biol Chem* 1997;**272**:29,460–29,467.
6. Sparrow LG, Lawrence MC, Gorman JJ, Strike PM, Robinson CP, McKern NM, Ward CW. N-linked glycans of the human insulin receptor and their distribution over the crystal structure. *Proteins Struct Funct Bioinform*, published online 23 October.
7. Sparrow LG, Gorman JJ, Strike PM, Robinson CP, McKern NM, Epa VC, Ward CW. *Proteins Struct Funct Bioinform* 2007;**66**:261–5.
8. De Meyts P. Insulin and its receptor: structure, function and evolution. *BioEssays* 2004;**26**:1351–62.
9. Kristensen C, Andersen AS, Østergaard S, Hansen PH, Brandt J. Functional reconstitution of insulin receptor binding site from non-binding receptor fragments. *J Biol Chem* 2002;**277**:18,340–45.
10. Surinya KH, Molina L, Soos MA, Brandt J, Kristensen C, Siddle K. Role of insulin receptor dimerization domains in ligand binding, cooperativity and modulation. *J Biol Chem* 2002;**277**:16,718–25.
11. McKern NM, Lawrence MC, Strelsov VA, Lou M-Z, Adamd TE, Lovrecz GO, Elleman TC, Richards KM, Bentley JD, Pilling PA,

Hoyne PA, Cartledge KA, Pham TM, Lewis JL, Sankovich SE, Stoichevska V, Da Silva E, Robinson CP, Frenkel MJ, Sparrow LG, Fernley RT, Epa VC, Ward CW. Structure of the insulin receptor ectodomain reveals a folded-over conformation. *Nature* 2006;**443**:218–21.
12. Lou M-Z, Garrett TPJ, McKern NM, Hoyne PA, Epa VC, Bentley JD, Lovrecz GO, Cosgrove LJ, Frenkel MJ, Ward CW. The first three domains of the insulin receptor differ structurally from the insulin-like growth factor 1 receptor in the regions governing ligand specificity. *Proc Natl Acad Sci USA* 2006;**103**:12,429–12,434.
13. Garrett TPJ, McKern NM, Lou M-Z, Frenkel MJ, Bentley JD, Lovrecz GO, Elleman TC, Cosgrove LJ, Ward CW. Crystal structure of the first three domains of the type-1 insulin-like growth factor receptor. *Nature* 1998;**394**:395–9.
14. Nakagawa SH, Zhao M, Hua Q-X, Hu S-Q, Wan Z-L, Jia W, Weiss MA. Chiral mutagenesis of insulin. Foldability and function are inversely regulated by a stereospecific switch in the B chain. *Biochemistry* 2005;**44**:4984–99.
15. Flörke RR, Klein HW, Reinauer H. Structural requirements for signal transduction of the insulin receptor. *Eur J Biochem* 1990;**191**:473–82.
16. Saltiel AR, Kahn CR. Insulin signalling and the regulation of glucose and lipid metabolism. *Nature* 2001;**414**:799–806.
17. Watson RT, Kanzaki M, Pessin JE. Regulated membrane trafficking of the insulin-responsive glucose transporter 4 in adipocytes. *Endocrine Rev* 2004;**25**:177–204.
18. Simpson F, Whitehead JP, James DE. GLUT4-At the crossroads between membrane trafficking and signal transduction. *Traffic* 2001;**2**:237–45.
19. Huang S, Czech MP. The GLUT4 glucose transporter. *Cell Metab* 2007;**5**:237–52.
20. Bai L, Wang Y, Fan J, Chen Y, Ji W, Qu A, Xu P, James DE. Dissecting multiple steps of GLUT4 trafficking and identifying the sites of insulin action. *Cell Metab* 2007;**15**:47–57.
21. Rose AJ, Richter EA. Skeletal muscle glucose uptake during exercise: how is it regulated? *Physiology* 2005;**20**:260–70.
22. Kanzaki M. Insulin receptor signals regulating GLUT4 translocation and actin dynamics. *Endocrine J* 2006;**53**:267–93.
23. Bryant NJ, Govers R, James DE. Regulated transport of the glucose transporter GLUT4. *Nature Rev* 2002;**3**:267–77.
24. Pessin JE, Thurmond DC, Elmendorf JS, Coker KJ, Okada S. Molecular basis of insulin stimulated vesicle trafficking. *J Biol Chem* 1999;**274**:2593–6.
25. Sollner T, Whiteheart SW, Brunner M, Erdjument-Bromage H, Geromanos S, Tempst P, Rothman JE. SNAP receptors implicated vesicle targeting and fusion. *Nature* 1993;**362**:318–24.
26. Hu SH, Latham CF, Gee CL, James DE, Martin JL. Structure of the Munc18c/Syntaxin4 N-peptide complex defines universal features of the N-peptide binding mode of Sec1/Munc18 proteins. *Proc Natl Acad Sci USA* 2007;**104**:8773–8.

Chapter 12

Structure and Mechanism of the Insulin Receptor Tyrosine Kinase

Stevan R. Hubbard
Skirball Institute of Biomolecular Medicine and Department of Pharmacology, New York University School of Medicine, New York, New York

INTRODUCTION

Insulin activates signaling pathways that control cellular metabolism and growth [1]. The actions of this essential hormone are mediated by the insulin receptor [2, 3], a member of the receptor tyrosine kinase (RTK) family of cell surface receptors. The RTK family also includes, among others, the receptors for insulin-like growth factor-1 (IGF1), epidermal growth factor (EGF), fibroblast growth factors (FGFs), platelet-derived growth factors (PDGFs), nerve growth factor (NGF), and vascular endothelial growth factors (VEGFs). A large family of non-receptor tyrosine kinases also exists, which includes Src-family kinases, Abl kinases, Janus kinases (Jaks), Syk and Zap70, and Fak. Receptor and non-receptor tyrosine kinases are critical components of signaling pathways in metazoans, controlling cellular proliferation, differentiation, migration, and metabolism [4].

In contrast to most RTKs, which are single-chain receptors that are activated by ligand-induced oligomerization, the insulin receptor is an $\alpha_2\beta_2$ heterotetramer (Figure 12.1), with disulfide linkages between the two α-chains and between the α- and β-chains. Insulin binding to the α-chains induces a structural rearrangement within the receptor, resulting in autophosphorylation of specific tyrosines in the β-chain: Tyr972 in the juxtamembrane region, Tyr1158, Tyr1162, and Tyr1163 in the activation loop in the kinase domain, and Tyr1328 and Tyr1334 in the C-terminal region (Figure 12.1). Phosphorylation of Tyr1162 and Tyr1163 (activation loop) stimulates catalytic activity, and phosphorylated Tyr972 (pTyr972; juxtamembrane region), pTyr1158, and pTyr1162 (activation loop) serve as docking sites for downstream signaling proteins. The functional role of tyrosine phosphorylation in the C-terminal region has not been fully established.

Because the insulin receptor is a transmembrane protein, structural studies to understand signaling mechanisms have largely been confined to either the extracellular region or the cytoplasmic region. Recently, the crystal structure of the disulfide-linked dimeric ectodomain of the insulin receptor was determined [5], which reveals an anti-parallel "inverted V" arrangement, and suggests how insulin might cross link the ectodomain by binding to the first L domain of one α chain and to the first fibronectin type III domain on the other α chain. Crystal structures of the insulin-receptor kinase domain (IRK) in the cytoplasmic portion of the β chain have been determined in different states of phosphorylation [6, 7] and for several mutants [8, 9]. IRK shares a similar overall architecture with protein serine/threonine kinases, with an N-terminal lobe (N lobe) comprising a five-stranded anti-parallel β-sheet and one α-helix, and a larger C lobe which is mainly α-helical (Figure 12.1) [10]. ATP binds in the cleft between the two lobes, and the tyrosine-containing segment of a protein substrate binds in the substrate binding groove in the C lobe.

A key mechanism by which the insulin receptor and other tyrosine and serine/threonine kinases regulate catalytic activity is through reversible positioning of their activation loops [11]. The activation loop of the insulin receptor, with three autophosphorylation sites (Tyr1158, Tyr1162, and Tyr1163), begins with the protein kinase-conserved ^{1150}DFG motif and ends with tyrosine kinase-conserved Pro1172. In the crystal structure of unphosphorylated (basal) IRK, the activation loop traverses the cleft between the two kinase lobes, with Tyr1162 bound in the active site [6]. In addition to obstructing the substrate binding site, this conformation of the activation loop also obstructs the nucleotide binding site, ensuring that Tyr1162 cannot be phosphorylated in *cis*. The activation loop is dynamic, however, and upon juxtaposition of the two kinase domains through insulin binding to the

Handbook of Cell Signaling, Three-Volume Set 2 ed.

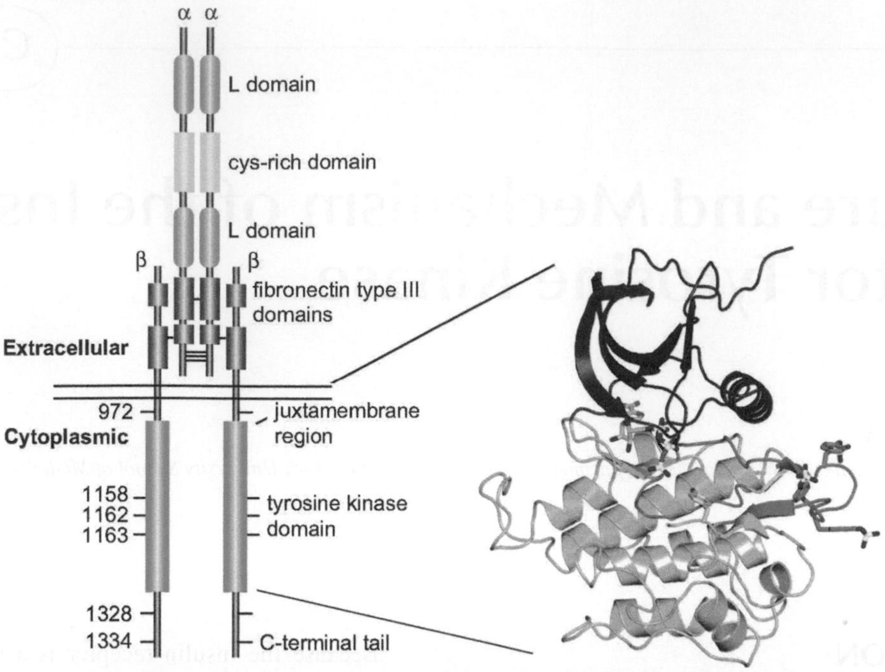

FIGURE 12.1 Overall architecture of the insulin receptor and structure of the tyrosine kinase domain. The α-chains are extracellular and the β-chains pass through the plasma membrane. The major sites of tyrosine autophosphorylation in the cytoplasmic domain are numbered according to Ebina *et al.* [2]. There are three fibronectin type III (FnIII) domains in the extracellular portion of the receptor. FnIII-1 is contained in the α-chain, as is the N-terminal portion of FnIII-2. The C-terminal portion of FnIII-2 is contained in the β-chain, followed by FnIII-3. Solid lines between the α-chains and between the α- and β-chains represent disulfide linkages. On the right, a ribbon diagram of the structure of tris-phosphorylated IRK is shown [7], with the N lobe colored dark gray and the C lobe colored light gray. AMP-PNP is shown in stick representation, as are the three activation-loop phosphotyrosines and the substrate tyrosine.

ectodomain, phosphorylation of the activation loops ensues in *trans*, resulting in phosphotyrosine-mediated stabilization of a loop configuration that is optimal for catalysis [7].

PROTEIN RECRUITMENT TO THE ACTIVATED INSULIN RECEPTOR

Overview

Upon insulin-stimulated activation of its receptor, several signaling proteins are recruited to the receptor, either for signal propagation or for attenuation. Among the proteins recruited for activation of downstream signaling pathways are several adaptor (non-enzymatic) proteins, including the insulin receptor substrate (IRS) proteins, Shc, and SH2B2/APS. Negative regulators recruited to the receptor include the adaptor protein Grb14 and the protein tyrosine phosphatase PTP1B. Recent structural studies have elucidated the modes of binding between these proteins and the cytoplasmic domain of the insulin receptor.

Phosphotyrosine recruitment sites in RTKs usually reside in polypeptide segments that are N-terminal to the kinase domain (juxtamembrane region), or C-terminal to the kinase domain, or in the kinase insert region (between α-helices D and E). These polypeptide segments are typically unstructured (non-α-helix, non-β-strand) and can easily adopt an extended conformation, which is required for phosphorylation, and, upon phosphorylation, for binding to Src homology-2 (SH2) domains and phosphotyrosine-binding (PTB) domains [12]. A subset of SH2 domains, those contained in SH2B2 and Grb14, specifically binds to the phosphorylated activation loop of the insulin receptor, which represents an unusual SH2-domain binding target because it is multiply phosphorylated and held in a relatively stable, turn-containing (non-extended) conformation. As detailed below, these SH2 domains possess three important properties: (1) they are dimeric, (2) they recognize two phosphotyrosines in the IRK activation loop, and (3) they are inhibited from binding canonical phosphotyrosine sequences.

Recruitment of SH2B2 to the Insulin Receptor

SH2B2 is a member of a family of three adaptor proteins, SH2B1–3, previously referred to as SH2B, APS, and Lnk, respectively [13]. These proteins consist of an N-terminal dimerization motif, a pleckstrin homology (PH) domain, an SH2 domain, and polyproline stretches in the N- and C-terminal regions. Recruitment of SH2B2 to the insulin

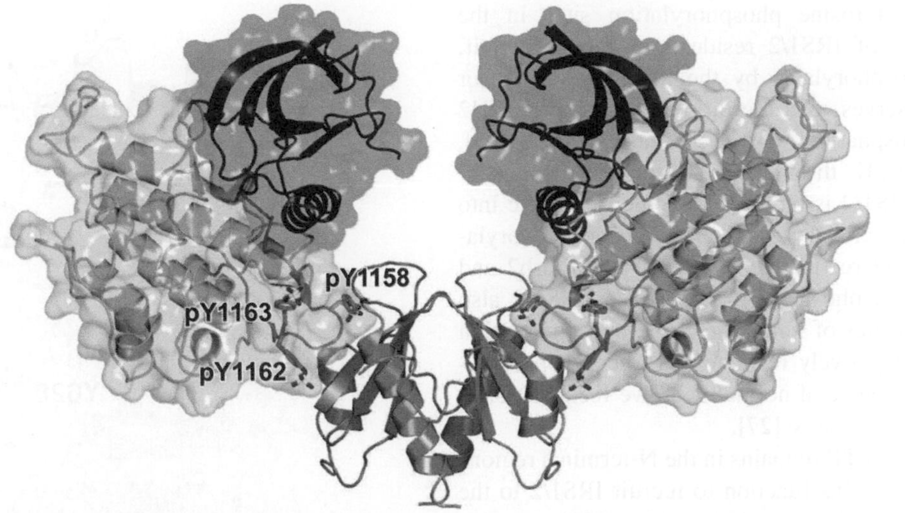

FIGURE 12.2 Structure of the SH2B2 SH2 domain bound to IRK [15].
In this 2:2 complex, IRK is shown with a semi-transparent molecular surface. The phosphotyrosines in the activation loop (pTyr1158/1162/1163) are labeled. The two protomers of the SH2B2 SH2-domain dimer (in the middle of the figure) are shown in ribbon representation.

receptor results in phosphorylation of a C-terminal tyrosine residue in SH2B2, Tyr618, which then serves as a binding site for the tyrosine-kinase binding (TKB) domain of Cbl [14, 15]. This signaling pathway culminates in activation of TC10, a Rho-family GTPase, which is thought to facilitate glucose uptake in adipocytes [1, 16].

The interaction between SH2B2 and the insulin receptor is mediated by the SH2 domain of SH2B2, which, as first shown by yeast two-hybrid studies, interacts with the phosphorylated activation loop of IRK [17, 18]. A crystal structure of the SH2B2 SH2 domain [15] revealed that it possesses a non-canonical architecture, in which the C-terminal half of the domain forms a long α-helix rather than two β-strands (β-strands E and F) and a shorter α-helix (α-helix B) as found in a typical SH2 domain. This structural rearrangement facilitates formation of a novel SH2-domain dimer, which interacts with the activation loops in the two kinase domains of the insulin receptor (Figure 12.2) [15]. Formation of the dimer thwarts binding to canonical phosphotyrosine sequences containing a hydrophobic residue at the +3 position (relative to the phosphotyrosine), and confers specificity for the (turn-containing) phosphorylated IRK activation loop. Rather than coordinating a single phosphotyrosine, each protomer of the SH2B2 SH2 dimer coordinates two phosphotyrosines, the first (pTyr1158) in the canonical (arginine-containing) phosphate binding pocket, and the second (pTyr1162) by two lysines in β-strand D (Figure 12.2). Thus, the SH2B2 SH2 dimer is a quad phosphotyrosine-recognition module. Of note, despite a sequence identity of 79 percent between the SH2 domains of SH2B1 and SH2B2, the SH2B1 SH2 domain is monomeric, which switches its binding preference from the IRK activation loop to pTyr813 of Jak2, a conventional SH2-domain ligand with a +3 hydrophobic residue [19].

Recruitment of SH2B2 to the phosphorylated activation loop of the insulin receptor prolongs receptor activation [20], presumably through protection from dephosphorylation by protein tyrosine phosphatases such as PTP1B, and selects for phosphorylation of Tyr618 near the C-terminus of SH2B2, ~130 residues C-terminal to the SH2 domain. Although there are numerous tyrosine residues in SH2B2 (68-kDa protein), Tyr618 is the predominant site of phosphorylation by the insulin receptor, despite its non-optimal substrate sequence (non-YΦXM, where Φ is hydrophobic and X is any residue). Selectivity for Tyr618 is evidently due to the binding of the SH2B2 SH2 dimer to the kinase activation loops, which facilitates entry of Tyr618, near the C-terminus of SH2B2, into the IRK active site.

Recruitment of IRS1 and IRS2 to the Insulin Receptor

The IRS proteins are a family of four to six adaptor molecules that possess N-terminal PH and PTB domains, followed by more than 900 residues containing multiple sites of tyrosine and serine/threonine phosphorylation [21]. Knockout studies in mice have shown that IRS1 and IRS2 are essential for organismal development and glucose disposal [21]. Despite a conserved domain architecture, common sites of phosphorylation, and overlapping tissue expression, the phenotypes for the *IRS1* and IRS2 knockout mice are distinct. *IRS1*$^{-/-}$ mice are 40 percent smaller than wild-type littermates, and exhibit insulin resistance in peripheral tissues [22, 23]. *IRS2*$^{-/-}$ mice are only slightly smaller (10 percent) than wild-type, but are insulin resistant and develop type 2 diabetes due to insufficient pancreatic β-cell function [24].

Many of the tyrosine phosphorylation sites in the C-terminal region of IRS1/2 reside in a YΦXM motif, which, when phosphorylated by the insulin receptor (or IGF1 receptor), serves as a recruitment site for the SH2 domain(s) of phosphatidylinositol 3-kinase (PI-3K) [25]. Activation of PI-3K through engagement of tyrosine-phosphorylated IRS1/2 is required for glucose uptake into insulin-responsive cells [26]. Other tyrosine phosphorylation sites in IRS1/2 recruit the adaptor protein Grb2 and the protein tyrosine phosphatase SHP2 [21]. IRS1/2 also possess numerous sites of serine/threonine phosphorylation that, in general, negatively regulate tyrosine phosphorylation, either in the course of normal negative feedback or in pathologic insulin resistance [27].

The tandem PH–PTB domains in the N-terminal regions of these adaptor proteins function to recruit IRS1/2 to the insulin receptor for phosphorylation. The PTB domains bind to pTyr972 and adjacent residues (NPXY motif) in the juxtamembrane region of the insulin (and IGF1) receptor [28, 29]. The PH domains are also important for recruitment of IRS1/2 to the receptor [30,31]. Previous yeast two-hybrid studies provided evidence for a second (in addition to the PTB domain) insulin receptor-interacting region in IRS2, which was named the kinase regulatory-loop binding (KRLB) region [32, 33] or receptor binding domain-2 (RBD2) [34]. This region of IRS2 interacts with the kinase domain of the insulin receptor (IRK) in a phosphodependent manner [32, 34]. The KRLB region was roughly mapped to residues 591–733 [32, 34], which starts approximately 300 residues C-terminal to the PTB domain. This region has a high proportion of glycine, serine, and proline residues, i.e., it is likely unstructured, and contains three tyrosine residues in a YΦXM motif. Although the corresponding region in IRS1 contains considerable sequence similarity, including the three YΦXM sites, it does not stably interact with IRK [33, 34]. Mutagenesis studies in the KRLB region identified two non-YΦXM tyrosine residues, Tyr624 and Tyr628 (mouse numbering), which are critical for this region of IRS2 to bind to IRK [33].

The molecular basis for the interaction of the IRS2 KRLB region with the insulin receptor was elucidated through co-crystallization of a 15-residue peptide from the KRLB region (containing Tyr624 and Tyr628) with phosphorylated IRK [35]. The structure revealed that this region of IRS2 binds in the active site of IRK with Tyr628 poised for phosphorylation (Figure 12.3). Biochemical experiments demonstrated that Tyr628 is phosphorylated by IRK, but with a K_m(ATP) that is very high compared to that of a YΦXM substrate (1.6 mM vs 40 μM, respectively). In addition, the phosphorylated Tyr628 peptide retains significant binding affinity, which translates into poor substrate turnover. As a consequence, this segment of IRS2 inhibits phosphorylation by the insulin receptor of other tyrosine sites in IRS2. A possible functional role of the KRLB region is to suppress tyrosine phosphorylation of IRS2, such that only metabolic pathways via PI-3 K are stimulated by insulin and not mitogenic pathways via Grb2/Ras; there are 10 potential PI-3 K recruitment sites in IRS2 versus a single Grb2 site.

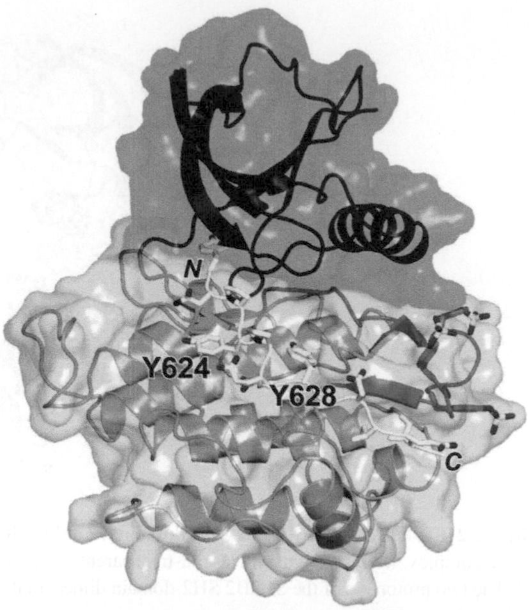

FIGURE 12.3 Structure of the KRLB region of IRS2 bound to IRK [35]. IRK is shown with a semi-transparent molecular surface. The side chains of the KRLB peptide (N- and C-termini are labeled) are shown in stick representation. Tyr624 and Tyr628 of the KRLB region are labeled, the latter of which is bound in the IRK active site.

Recruitment of Grb14 to the Insulin Receptor

Among the negative regulators of insulin signaling, Grb14 is a member of the Grb7/10/14 family of adaptor proteins [36]. *Grb14*$^{-/-}$ mice show tissue-specific (muscle and liver) increases in glucose disposal, and patients with type 2 diabetes have been shown to have increased Grb14 mRNA levels in adipose tissue [37]. Grb7/10/14 proteins possess several signaling modules, including a Ras-associating (RA) domain, a PH domain, and a C-terminal SH2 domain, as well as polyproline sequences. In addition, these proteins contain a ~45-residue region known as BPS (*b*etween *P*H and *S*H2) [38] or PIR (*p*hosphorylated *i*nsulin receptor-interacting *r*egion) [39], which is unique to this adaptor family.

Biochemical studies have demonstrated that the BPS region of Grb7/10/14 is capable of directly inhibiting the catalytic activities of the insulin and IGF1 receptors [40, 41], with a potency rank of Grb14>Grb10>Grb7 [41]. A crystal structure of the Grb14 BPS region in complex with phosphorylated IRK revealed at least one mechanism by which Grb14 inhibits signaling by the insulin receptor [42]. In the structure (Figure 12.4), the N-terminal portion of the BPS region binds in the substrate binding groove in the C lobe of the kinase domain. A leucine (Leu376) is inserted in the active site rather than a tyrosine, and the hydrophobic residues at the +1, +3, and +5 positions relative to Leu376 mimic peptide–substrate binding. Thus, the Grb14 BPS region acts as a pseudosubstrate inhibitor to suppress insulin-receptor signaling.

After the pseudosubstrate segment, the Grb14 BPS region adopts a 16-residue α-helix whose residues make interactions with the phosphorylated activation loop. These interactions fortify the BPS–IRK interaction and also provide specificity; the BPS region of Grb14/10 only inhibits the insulin and IGF1 receptors. The BPS α-helix ends near the kinase N lobe, which positions the C-terminal SH2 domain (~25-residue BPS–SH2 linker) for engagement of pTyr1158 and pTyr1162 in the activation loop [42]. The SH2 domain of Grb14/10 is, like the SH2B2 SH2 domain, dimeric, but in this case the individual protomers possess a canonical architecture. Binding of canonical phosphotyrosine sequences (with +3 hydrophobic residues) is deterred in the Grb14/10 SH2 domain, not by dimer formation (as for SH2B2), but by a valine in the BG loop (BG3) rather than a glycine (as in Src and Abl) [43].

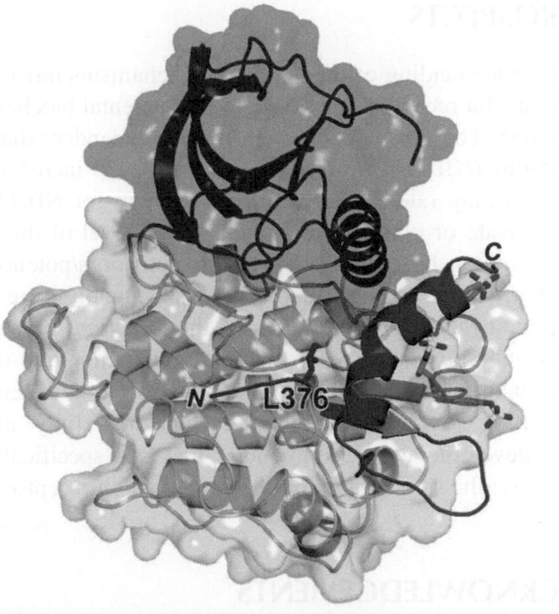

FIGURE 12.4 Structure of the BPS region of Grb14 bound to IRK [35]. IRK is shown with a semi-transparent molecular surface. The BPS region (N- and C-termini are labeled) is shown in ribbon representation. Leu376 (labeled), the pseudosubstrate bound in the IRK active site, is shown in stick representation.

Recruitment of PTP1B to the Insulin Receptor

PTP1B is a key negative regulator of insulin signaling *in vivo*, as shown by gene-deletion studies in mice. Despite the ability of PTP1B to dephosphorylate numerous RTKs *in vitro* and in cells, the *PTP1B*$^{-/-}$ mice are of normal size (i.e., no overgrowth phenotype) and exhibit two apparent phenotypes: a hypersensitivity to insulin, due to prolonged insulin-receptor phosphorylation, and resistance to weight gain on a high-fat diet [44, 45]. The molecular basis for PTP1B's specificity towards the insulin receptor has not been fully elucidated. A crystal structure of PTP1B with a phosphorylated peptide representing the IRK activation loop revealed that a phosphotyrosine (pTyr1163) following the substrate phosphotyrosine (pTyr1162) is a specificity determinant [46]. Because several other RTKs (e.g., IGF1 receptor, TrkA-C, FGF receptors), which do not appear to be targets of PTP1B *in vivo*, contain tandem phosphotyrosines in their activation loops, other specificity determinants must exist.

A crystal structure of a PTP1B–IRK complex was determined that revealed a novel, non-catalytic mode of interaction between the two proteins, in which PTP1B binds not to a phosphotyrosine in the IRK activation loop, but rather to the opposite side of the kinase domain (Figure 12.5). This interaction was evidently facilitated by crystallization in ammonium sulfate, which effectively competes with the activation-loop phosphotyrosines for binding in the PTP1B active site. Although this interaction might be dismissed as a crystallization artifact, the PTP1B–IRK interface comprises residues that are highly characteristic for these two proteins, including Tyr152 and Tyr153 of PTP1B, which had been previously implicated in the interaction between PTB1B and the insulin receptor [47–49]. Moreover, the highly related IGF1 receptor would be incapable of engaging PTP1B in this manner. Thus, it remains an attractive hypothesis that the interaction observed in the crystal structure represents a recruitment mode of binding to localize PTP1B to the insulin receptor, although biochemical evidence for this hypothesis is still outstanding.

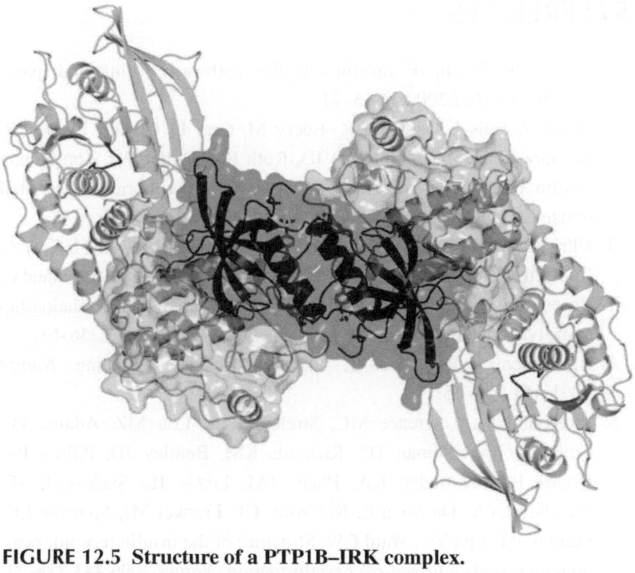

FIGURE 12.5 Structure of a PTP1B–IRK complex. In this 2 : 2 complex, the IRK dimer is in the center of the figure, and the two PTP1B molecules are on the outside, shown in ribbon representation. The molecular (non-crystallographic) two-fold axis is perpendicular to the page.

PROSPECTS

An understanding of the signaling mechanisms intrinsic to the insulin receptor is not only of fundamental biochemical interest. The prevalence of non-insulin-dependent diabetes mellitus (NIDDM) in developed countries is increasing at an alarming rate. One therapeutic strategy for NIDDM is to activate or potentiate signaling at the level of the insulin receptor. Indeed, small-molecule activators/potentiators of the insulin receptor that act cytoplasmically have been reported [50, 51]. Knowledge of the structural transitions that underlie insulin-receptor kinase activation, and of the mechanisms by which downstream positive and negative signaling proteins are recruited, will hopefully aid in the development of compounds that act specifically to increase the signaling output from the insulin receptor.

ACKNOWLEDGEMENTS

Support is acknowledged from the National Institutes of Health (DK052916, NS053414) and from an Irma T. Hirschl-Monique Weill-Caulier Career Scientist Award.

REFERENCES

1. Saltiel AR, Pessin JE. Insulin signaling pathways in time and space. *Trends Cell Biol* 2002;**12**:65–71.
2. Ebina Y, Ellis L, Jarnagin K, Edery M, Graf L, Clauser E, Ou JH, Masiarz F, Kan YW, Goldfine ID, Roth RA, Rutter WJ. The human insulin receptor cDNA: the structural basis for hormone-activated transmembrane signaling. *Cell* 1985;**40**:747–58.
3. Ullrich A, Bell JR, Chen EY, Herrera R, Petruzzelli LM, Dull TJ, Gray A, Coussens L, Liao YC, Tsubokawa M, Mason A, Seeburg PH, Grunfeld C, Rosen OM, Ramachandran J. Human insulin receptor and its relationship to the tyrosine kinase family of oncogenes. *Nature* 1985;**313**:756–61.
4. Blume-Jensen P, Hunter T. Oncogenic kinase signalling. *Nature* 2001;**411**:355–65.
5. McKern NM, Lawrence MC, Streltsov VA, Lou MZ, Adams TE, Lovrecz GO, Elleman TC, Richards KM, Bentley JD, Pilling PA, Hoyne PA, Cartledge KA, Pham TM, Lewis JL, Sankovich SE, Stoichevska V, Da Silva E, Robinson CP, Frenkel MJ, Sparrow LG, Fernley RT, Epa VC, Ward CW. Structure of the insulin receptor ectodomain reveals a folded-over conformation. *Nature* 2006;**443**:218–21.
6. Hubbard SR, Wei L, Ellis L, Hendrickson WA. Crystal structure of the tyrosine kinase domain of the human insulin receptor. *Nature* 1994;**372**:746–54.
7. Hubbard SR. Crystal structure of the activated insulin receptor tyrosine kinase in complex with peptide substrate and ATP analog. *EMBO J* 1997;**16**:5572–81.
8. Till JH, Ablooglu AJ, Frankel M, Bishop SM, Kohanski RA, Hubbard SR. Crystallographic and solution studies of an activation loop mutant of the insulin receptor tyrosine kinase: insights into kinase mechanism. *J Biol Chem* 2001;**276**:10,049–55.
9. Li S, Covino ND, Stein EG, Till JH, Hubbard SR. Structural and biochemical evidence for an autoinhibitory role for tyrosine 984 in the juxtamembrane region of the insulin receptor. *J Biol Chem* 2003;**278**:26,007–14.
10. Hubbard SR, Till JH. Protein tyrosine kinase structure and function. *Annu Rev Biochem* 2000;**69**:373–98.
11. Johnson LN, Noble MEM, Owen DJ. Active and inactive protein kinases: structural basis for regulation. *Cell* 1996;**85**:149–58.
12. Kuriyan J, Cowburn D. Modular peptide recognition domains in eukaryotic signaling. *Annu Rev Biophys Biomol Struct* 1997;**26**:259–88.
13. Ahmed Z, Pillay TS. Functional effects of APS and SH2-B on insulin receptor signalling. *Biochem Soc Trans* 2001;**29**:529–34.
14. Liu J, Kimura A, Baumann CA, Saltiel AR. APS facilitates c-Cbl tyrosine phosphorylation and GLUT4 translocation in response to insulin in 3T3-L1 adipocytes. *Mol Cell Biol* 2002;**22**:3599–609.
15. Hu J, Liu J, Ghirlando R, Saltiel AR, Hubbard SR. Structural basis for recruitment of the adaptor protein APS to the activated insulin receptor. *Mol Cell* 2003;**12**:1379–89.
16. Chiang SH, Baumann CA, Kanzaki M, Thurmond DC, Watson RT, Neudauer CL, Macara IG, Pessin JE, Saltiel AR. Insulin-stimulated GLUT4 translocation requires the CAP-dependent activation of TC10. *Nature* 2001;**410**:944–8.
17. Moodie SA, Alleman-Sposeto J, Gustafson TA. Identification of the APS protein as a novel insulin receptor substrate. *J Biol Chem* 1999;**274**:11,186–11,193,.
18. Ahmed Z, Smith BJ, Kotani K, Wilden P, Pillay TS. APS, an adapter protein with a PH and SH2 domain, is a substrate for the insulin receptor kinase. *Biochem J* 1999;**341**:665–8.
19. Hu J, Hubbard SR. Structural basis for phosphotyrosine recognition by the Src homology-2 domains of the adapter proteins SH2-B and APS. *J Mol Biol* 2006;**361**:69–79.
20. Ahmed Z, Pillay TS. Adapter protein with a pleckstrin homology (PH) and an Src homology 2 (SH2) domain (APS) and SH2-B enhance insulin-receptor autophosphorylation, extracellular-signal-regulated kinase and phosphoinositide 3-kinase-dependent signalling. *Biochem J* 2003;**371**:405–12.
21. White MF. IRS proteins and the common path to diabetes. *Am J Physiol Endocrinol Metab* 2002;**283**:E413–22.
22. Araki E, Lipes MA, Patti ME, Bruning JC, Haag B, 3rd, Johnson RS, Kahn CR. Alternative pathway of insulin signalling in mice with targeted disruption of the IRS-1 gene. *Nature* 1994;**372**:186–90.
23. Tamemoto H, Kadowaki T, Tobe K, Yagi T, Sakura H, Hayakawa T, Terauchi Y, Ueki K, Kaburagi Y, Satoh S, et al. Insulin resistance and growth retardation in mice lacking insulin receptor substrate-1. *Nature* 1994;**372**:182–6.
24. Withers DJ, Gutierrez JS, Towery H, Burks DJ, Ren JM, Previs S, Zhang Y, Bernal D, Pons S, Shulman GI, Bonner-Weir S, White MF. Disruption of IRS-2 causes type 2 diabetes in mice. *Nature* 1998;**391**:900–4.
25. Myers Jr. MG, Backer JM, Sun XJ, Shoelson S, Hu P, Schlessinger J, Yoakim M, Schaffhausen B, White MF. IRS-1 activates phosphatidylinositol 3′-kinase by associating with src homology 2 domains of p85. *Proc Natl Acad Sci USA* 1992;**89**:10,350–4.
26. Okada T, Kawano Y, Sakakibara T, Hazeki O, Ui M. Essential role of phosphatidylinositol 3-kinase in insulin-induced glucose transport and antilipolysis in rat adipocytes. Studies with a selective inhibitor wortmannin. *J Biol Chem* 1994;**269**:3568–73.
27. Pirola L, Johnston AM, Van Obberghen E. Modulation of insulin action. *Diabetologia* 2004;**47**:170–84.
28. Craparo A, O'Neill TJ, Gustafson TA. Non-SH2 domains within insulin receptor substrate-1 and SHC mediate their phosphotyrosine-dependent interaction with the NPEY motif of the insulin-like growth factor I receptor. *J Biol Chem* 1995;**270**:15,639–43.
29. Eck MJ, Dhe-Paganon S, Trub T, Nolte RT, Shoelson SE. Structure of the IRS-1 PTB domain bound to the juxtamembrane region of the insulin receptor. *Cell* 1996;**85**:695–705.

30. Yenush L, Makati KJ, Smith-Hall J, Ishibashi O, Myers MG, Jr. White MF. The pleckstrin homology domain is the principal link between the insulin receptor and IRS-1. *J Biol Chem* 1996;**271**:24,300–6.
31. Burks DJ, Pons S, Towery H, Smith-Hall J, Myers MG, Jr. Yenush L, White MF. Heterologous pleckstrin homology domains do not couple IRS-1 to the insulin receptor. *J Biol Chem* 1997;**272**:27,716–21.
32. Sawka-Verhelle D, Tartare-Deckert S, White MF, Van Obberghen E. Insulin receptor substrate-2 binds to the insulin receptor through its phosphotyrosine-binding domain and through a newly identified domain comprising amino acids 591–786. *J Biol Chem* 1996;**271**:5980–3.
33. Sawka-Verhelle D, Baron V, Mothe I, Filloux C, White MF, Van Obberghen E. Tyr624 and Tyr628 in insulin receptor substrate-2 mediate its association with the insulin receptor. *J Biol Chem* 1997;**272**:16,414–20.
34. He W, Craparo A, Zhu Y, O'Neill TJ, Wang L-M, Pierce J, Gustafson TA. Interaction of insulin receptor substrate-2 (IRS-2) with the insulin and insulin-like growth factor I receptors. *J Biol Chem* 1996;**271**:11,641–5.
35. Wu J, Tseng YD, Xu C-F, Neubert TA, White FM, Hubbard SR. Structural and biochemical characterization of the KRLB region in insulin receptor substrate-2. *Nat Struct Mol Biol* 2008, in press.
36. Holt LJ, Siddle K. Grb10 and Grb14: enigmatic regulators of insulin action – and more? *Biochem J* 2005;**388**:393–406.
37. Cariou B, Capitaine N, Le Marcis V, Vega N, Bereziat V, Kergoat M, Laville M, Girard J, Vidal H, Burnol AF. Increased adipose tissue expression of Grb14 in several models of insulin resistance. *FASEB J* 2004;**18**:965–7.
38. He W, Rose DW, Olefsky JM, Gustafson TA. Grb10 interacts differentially with the insulin receptor, insulin-like growth factor I receptor, and epidermal growth factor receptor via the Grb10 Src homology 2 (SH2) domain and a second novel domain located between the pleckstrin homology and SH2 domains. *J Biol Chem* 1998;**273**:6860–7.
39. Kasus-Jacobi A, Perdereau D, Auzan C, Clauser E, Van Obberghen E, Mauvais-Jarvis F, Girard J, Burnol AF. Identification of the rat adapter Grb14 as an inhibitor of insulin actions. *J Biol Chem* 1998;**273**:26,026–35.
40. Stein EG, Gustafson TA, Hubbard SR. The BPS domain of Grb10 inhibits the catalytic activity of the insulin and IGF1 receptors. *FEBS Letts* 2001;**493**:106–11.
41. Bereziat V, Kasus-Jacobi A, Perdereau D, Cariou B, Girard J, Burnol AF. Inhibition of insulin receptor catalytic activity by the molecular adapter Grb14. *J Biol Chem* 2002;**277**:4845–52.
42. Depetris RS, Hu J, Gimpelevich I, Holt LJ, Hubbard SR. Structural basis for inhibition of the insulin receptor by the adaptor protein Grb14. *Mol Cell* 2005;**20**:325–33.
43. Stein EG, Ghirlando R, Hubbard SR. Structural basis for dimerization of the Grb10 Src homology 2 domain, Implications for ligand specificity. *J Biol Chem* 2003;**278**:13,257–64.
44. Elchebly M, Payette P, Michaliszyn E, Cromlish W, Collins S, Loy AL, Normandin D, Cheng A, Himms-Hagen J, Chan CC, Ramachandran C, Gresser MJ, Tremblay ML, Kennedy BP. Increased insulin sensitivity and obesity resistance in mice lacking the protein tyrosine phosphatase-1B gene. *Science* 1999;**283**:1544–8.
45. Klaman LD, Boss O, Peroni OD, Kim JK, Martino JL, Zabolotny JM, Moghal N, Lubkin M, Kim YB, Sharpe AH, Stricker-Krongrad A, Shulman GI, Neel BG, Kahn BB. Increased energy expenditure, decreased adiposity, and tissue-specific insulin sensitivity in protein-tyrosine phosphatase 1B-deficient mice. *Mol Cell Biol* 2000;**20**:5479–89.
46. Salmeen A, Andersen JN, Myers MP, Tonks NK, Barford D. Molecular basis for the dephosphorylation of the activation segment of the insulin receptor by protein tyrosine phosphatase 1B. *Mol Cell* 2000;**6**:1401–12.
47. Bandyopadhyay D, Kusari A, Kenner KA, Liu F, Chernoff J, Gustafson TA, Kusari J. Protein-tyrosine phosphatase 1B complexes with the insulin receptor in vivo and is tyrosine-phosphorylated in the presence of insulin. *J Biol Chem* 1997;**272**:1639–45.
48. Dadke S, Kusari J, Chernoff J. Down-regulation of insulin signaling by protein-tyrosine phosphatase 1B is mediated by an N-terminal binding region. *J Biol Chem* 2000;**275**:23,642–7.
49. Dadke S, Chernoff J. Interaction of protein tyrosine phosphatase (PTP) 1B with its substrates is influenced by two distinct binding domains. *Biochem J* 2002;**364**:377–83.
50. Zhang B, Salituro G, Szalkowski D, Li Z, Zhang Y, Royo I, Vilella D, Diez MT, Pelaez F, Ruby C, Kendall RL, Mao X, Griffin P, Calaycay J, Zierath JR, Heck JV, Smith RG, Moller DE. Discovery of a small molecule insulin mimetic with antidiabetic activity in mice. *Science* 1999;**284**:974–7.
51. Manchem VP, Goldfine ID, Kohanski RA, Cristobal CP, Lum RT, Schow SR, Shi S, Spevak WR, Laborde E, Toavs DK, Villar HO, Wick MM, Kozlowski MR. A novel small molecule that directly sensitizes the insulin receptor in vitro and in vivo. *Diabetes* 2001; **50**:824–30.

Chapter 13

IRS-Protein Scaffolds and Insulin/IGF Action in Central and Peripheral Tissues

Morris F. White
Howard Hughes Medical Institute, Division of Endocrinology, Children's Hospital Boston, Harvard Medical School, Boston, Massachusetts

INTRODUCTION

The insulin-like signaling system is integrated throughout the body to coordinate systemic growth and development with peripheral and central nutrient homeostasis, fertility, and lifespan [1]. Dysregulated insulin signaling is associated with a cohort of common systemic disorders, including obesity and dyslipidemia, cardiovascular disease and hypertension, infertility, and neurodegeneration. Each of these disorders is also associated with Type 2 diabetes, which usually occurs at middle age, afflicts more than 30 million people over the age of 65, but is appearing with greater frequency in children and adolescents as the obesity epidemic spreads to younger patients [2]. The conventional view of Type 2 diabetes places a significant emphasis upon insulin action in liver, muscle, and adipose tissues. However, the failure of pancreatic β-cells to secrete sufficient insulin to compensate for peripheral insulin resistance plays a major role [3]. The insulin receptor substrates proteins – mainly Irs1 and Irs2 – coordinate insulin and insulin-like growth factor signals in all tissues. The insulin-like signaling cascade through Irs2 is critical for β-cell growth, function, and survival, so insulin secretion and action are coordinated by similar signaling cascades [4]. Moreover, IRS-proteins are modulated by heterologous mechanisms – including serine/threonine phosphorylation, ubiquitinylation and acetylation – which integrates the insulin-like responses to the prevailing physiological state.

INSULIN, IGFS, AND THEIR RECEPTORS

Mammals produce three insulin-like peptides, insulin, insulin-like growth factor-1 (Igf1), and insulin like growth factor-2 (Igf2), which activate five homologous insulin-like receptor tyrosine kinases encoded by two genes, the insulin receptor (IR) and the insulin-like growth factor-1 receptor (IGF1R). Insulin is produced in pancreatic β-cells in response to circulating glucose concentrations, whereas endocrine IGF1 is largely secreted from hepatocytes stimulated by nutrients and growth hormone; IGF1 and IGF2 are also produced locally in many tissues and cells, including the central nervous system [5]. The IGF1 can work coordinately with insulin to regulate nutrient homeostasis, insulin sensitivity, and pancreatic β-cell function [5, 6]. The insulin receptor and IGF receptor genes encode homologous precursors that form covalently linked dimers that are cleaved by proteolysis to generate a tetramer with two extracellular α-subunits and two transmembrane β-subunits [7, 8]. The extracellular α-subunits create the ligand binding domain that regulates the activity of the tyrosine kinase on the intracellular portion of the transmembrane β-subunits [9–11].

The IGF1R binds IGF1 and IGF2 with relatively high affinity, whereas insulin binds with about 100-fold lower affinity. By comparison, the IR gene encodes two isoforms: IRb binds insulin strongly and predominates in adult liver, muscle, and adipose tissues (Figure 13.1), and IRa binds both IGF2 and insulin with moderate affinity and predominates in fetal tissues, the adult CNS, and hematopoietic cells [11, 12]. Hybrid receptors form randomly in cells depending upon the concentration of the IR and IGF1R [13]: only liver and adipose lack IGF1R expression, making them purely insulin responsive tissues [5]. The hybrid IGF1R::Irb, composed of αβ-subunits of the IGF1R and Irb, selectively binds IGF1; the hybrid IGF1R::Ira, composed of αβ-subunits of the IGF1R and Ira, binds insulin, IGF1, and IGF2 with similar moderate affinities (Figure 13.1). Whether different binding affinities produce unique receptor activity states requires more details of ligand binding at the atomic level [11]. More work is needed to determine whether each receptor isoform has a specialized role in neurons, pancreatic β-cells, muscle, and other tissues where the IR and IGF1R are expression together.

Handbook of Cell Signaling, Three-Volume Set 2 ed.

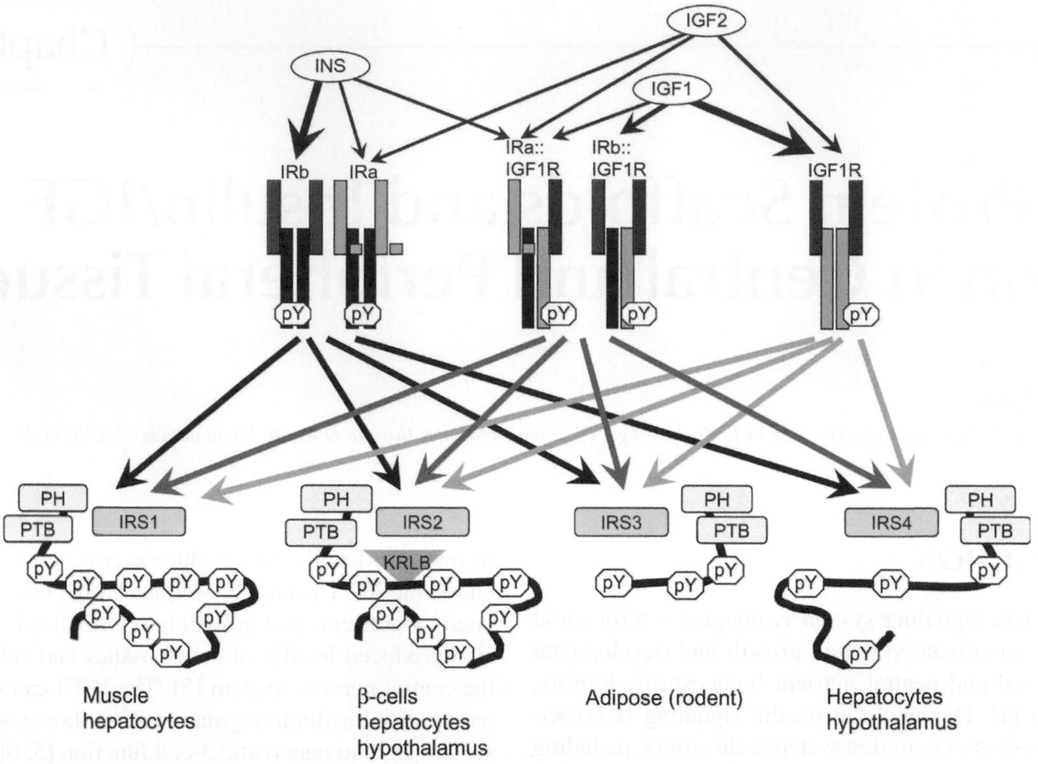

FIGURE 13.1 The insulin-like network.
Insulin like signaling consists of three homologous peptides: insulin, insulin-like growth factor-1 (IGF1), and insulin-like growth factor-2 (IGF2). These peptide ligands bind to five distinct receptor isoforms that generate cytoplasmic signals: two insulin receptor isoforms (IRa and IRb); the insulin-like growth factor receptor, IGF1R; and two hybrid receptors, IRa::IGF1R and IRb::IGF1R. The insulin receptor is the primary target for insulin throughout development and life. The IGF-1 receptor is the primary target for IGF-1. IGF-2 binds to IRa during embryonic development, and binds the IGF-1 receptor throughout life. The receptors are tyrosine kinases that are activated by ligand binding and phosphorylate the insulin receptor substrates, which are required for the insulin signal in various tissues.

High affinity ligand binding induces structural transitions in the catalytic domain of the β-subunit that promote phosphorylation of three tyrosine residues in the kinase regulatory loop – for example, Tyr_{1158}, Tyr_{1162}, and Tyr_{1163} in IRa [14]. Autophosphorylation releases the regulatory loop from its inhibitory position, which opens the catalytic site to phosphorylate other proteins [15]. Moreover, the phosphorylated regulatory loop interacts with other signaling proteins that modulate kinase activity, including Grb10, APS, and SH2B [16]. A fourth tyrosine residue within an NPEY motif located outside the kinase domain and near the plasma membrane (Tyr_{960} in Irb, Tyr_{972} in Ira, Tyr_{950} in IGF1R) is also phosphorylated, which recruits insulin receptor substrates (IRS-proteins) for tyrosine phosphorylation by the activated receptor kinase [17, 18].

INSULIN RECEPTOR SUBSTRATES

Cell based and mouse based experiments show that most if not all insulin signals are produced or modulated through tyrosine phosphorylation of IRS1, IRS2, or its homologs; or other scaffold proteins including SHC, CBL, APS and SH2B, GAB1, GAB2, DOCK1, and DOCK2 [19–25]. Although the role of each of these substrates merits attention, work with transgenic mice suggests that many insulin responses – especially those that are associated with somatic growth and nutrient homeostasis – are mediated through IRS1 or IRS2 [26].

The IRS-proteins are adapter molecules that link the insulin-like receptors to common downstream signaling cascades (Figure 13.1). Four Irs-protein genes have been identified in rodents, but only three of these genes (IRS1, IRS2, and IRS4) are expressed in humans [27]. IRS1 and IRS2 are broadly expressed in mammalian tissues, whereas IRS4 is largely restricted to the hypothalamus [28]. Each of these proteins is targeted to the activated insulin-like receptors through an NH_2-terminal pleckstrin homology (PH) domain and a phosphotyrosine binding (PTB) domain (Figure 13.2a). The PTB domain binds specifically to the phosphorylated NPEY motif in the activated receptor kinases [29]. This interaction is important, as deletion of the PTB domain or deletion of $Tyr_{960/972}$ in the IR impairs signal transduction [17, 30]. The PH domain also promotes the interaction between IRS-proteins and the IR, but the mechanism is poorly understood. Although PH domains are generally

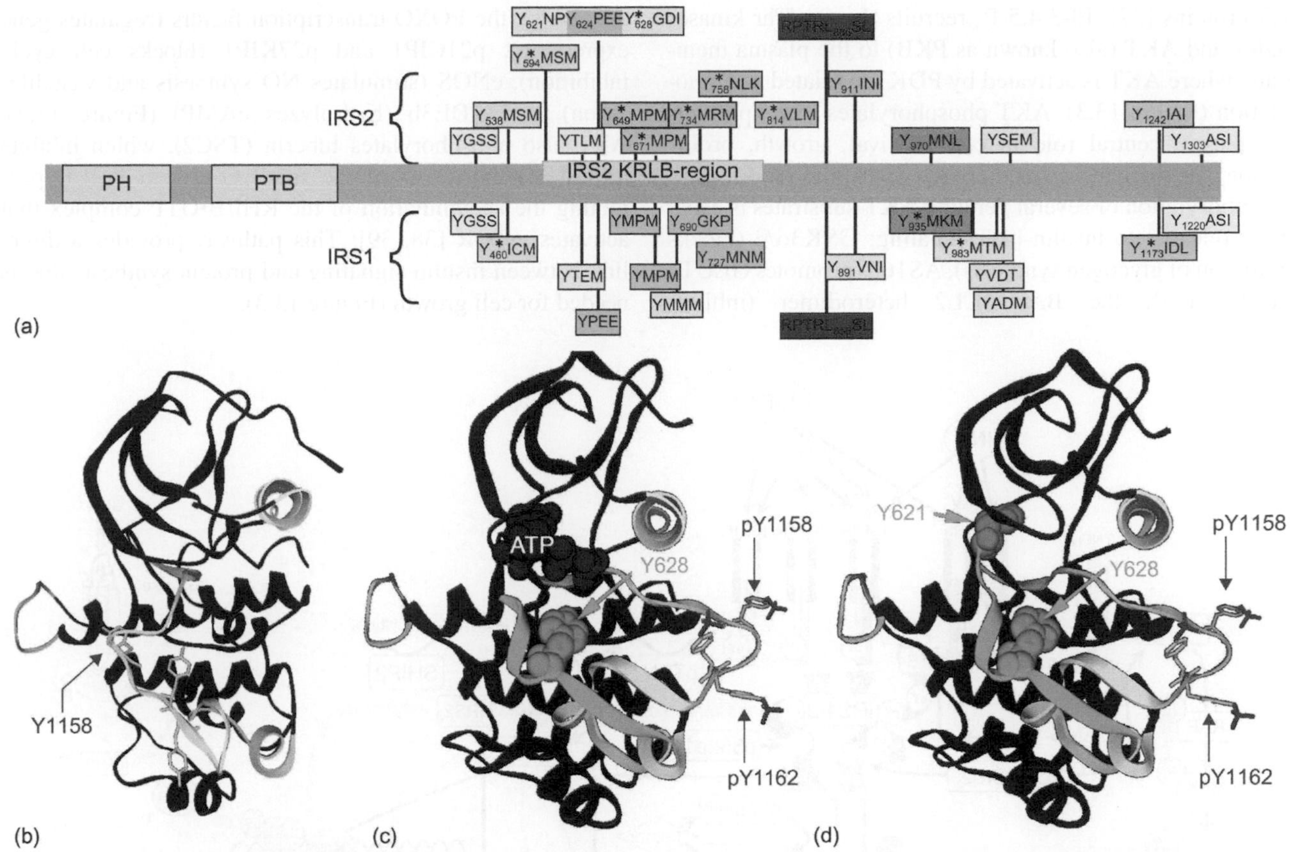

FIGURE 13.2 Structure of IRS1 and IRS2.
(a) Schematic comparison of the composition of IRS1 and IRS2, including the location of the PH and PTB domains, and putative and validated (*) tyrosine phosphorylation sites. The position of the kinase regulatory loop binding domain (KRLB) in IRS2 is indicated. (b) The backbone of the insulin receptor catalytic domain (blue) including the location of the regulatory loop in the inactive kinase (yellow) with Tyr_{1158} indicated for orientation. (c) The backbone of the insulin receptor catalytic domain (blue) including the location of Tyr_{1158} and Tyr_{1162} in the regulatory loop reoriented away from the catalytic pocket of the active kinase (yellow). The interaction of the IRS2 peptide surrounding Tyr_{628} – but lacking Tyr_{621} – is shown (green). (d) The same diagram as (c), but Tyr_{621} is included to show how it interacts with the ATP binding pocket of the insulin receptor.

thought to bind phospholipids, the PH domains in IRS1 and IRS2 are poor examples of this specificity [31, 32]. Regardless, the PH domain in the IRS-protein plays an important and specific role as it can be interchanged among the IRS-proteins without noticeable loss of bioactivity, whereas heterologous PH domains inhibit IRS1 function when substituted for the normal PH domain [33].

In addition to the PH and PTB domains, IRS2 utilizes an additional mechanism to interact with the activated insulin receptor [34]. This IRS2 specific interaction was originally localized between amino acid residues 591 and 786 in IRS2, and subsequently shown to require Tyr_{624} and Tyr_{628} in IRS2 [35]. This binding region in IRS2 was originally called the kinase regulatory loop binding (KRLB) domain because the interaction was shown to require the three tyrosine autophosphorylation sites in the regulatory loop of the functional receptor kinase [34]. However, Wu *et al.* showed recently that receptor autophosphorylation was required to move the regulatory loop out of the catalytic site so the functional part of the KRLB domain residues 620–634 in murine IRS2 can bind in the catalytic site with Tyr_{621} inserted into the ATP binding pocket and Tyr_{628} positioned for phosphorylation [36] (Figure 13.2b). Tyr_{628} might be a critical phosphorylation site that needs to be oriented properly by the KRLB domain for immediate phosphorylation upon insulin binding. Alternatively, the KRLB domain might attenuate signaling through IRS2 by blocking ATP access to the catalytic site. Further work is needed to establish the physiological significance of this unique motif in IRS2.

IRS→PI3K→AKT Cascade

One of the best studied insulin-like signaling cascades involves the production of PI-3,4,5-P_3 by the phosphatidylinositol 3-kinase (PI 3-kinase). The type 1 PI 3-kinase is composed of a regulatory subunit that contains two src-homology-2 (SH2) domains and a catalytic subunit that is inhibited by the regulatory subunit until its SH2 domains are occupied by phosphorylated tyrosine residues in the

IRS-proteins [37]. PI-3,4,5-P_3 recruits the Ser/Thr kinases PDK1 and AKT (also known as PKB) to the plasma membrane where AKT is activated by PDK1 mediated phosphorylation (Figure 13.3). AKT phosphorylates many proteins that play a central role in cell survival, growth, proliferation, angiogenesis, metabolism, and migration [38]. Phosphorylation of several genuine AKT substrates is especially relevant to insulin-like signaling: GSK3α/β (blocks inhibition of glycogen synthesis), AS160 (promotes GLUT4 translocation), the BAD•BCL2 heterodimer (inhibits apoptosis), the FOXO transcription factors (regulates gene expression), p21CIP1 and p27KIP1 (blocks cell cycle inhibition), eNOS (stimulates NO synthesis and vasodilatation), and PDE3b (hydrolyzes cAMP) (Figure 13.3). AKT also phosphorylates tuberin (TSC2), which inhibits its GAP activity toward the small G-protein RHEB promoting the accumulation of the RHEB•GTP complex that activates mTOR [38, 39]: This pathway provides a direct link between insulin signaling and protein synthesis that is needed for cell growth (Figure 13.3).

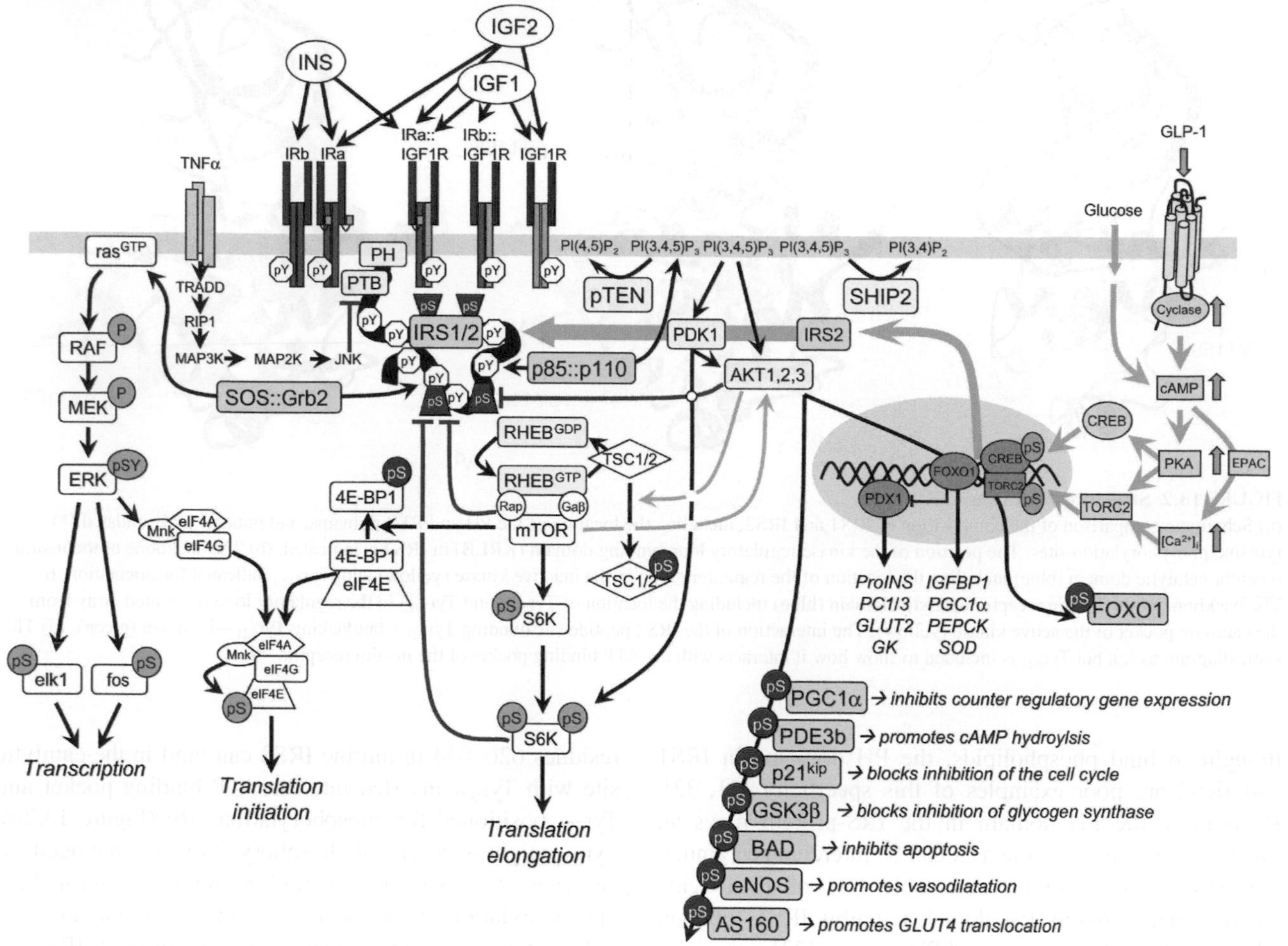

FIGURE 13.3 Insulin-like signaling cascade.

There are two main limbs that propagate the signal generated through the IRS-proteins: the PI 3-kinase and the Grb2/Sos→ras cascade. Activation of the receptors for insulin and IGF-1 results in tyrosine phosphorylation of the IRS-proteins, which bind PI 3-kinase and Grb2/SOS. The GRB2/SOS complex promotes GDP/GTP exchange on $p21^{ras}$, which activates the ras→raf→MEK→ERK1/2 cascade. The activated ERK stimulates transcriptional activity by direct phosphorylation of elk1 and by phosphorylation of fos through $p90^{rsk}$. The activation of PI 3-kinase by IRS-protein recruitment produces PI 3,4P_2 and PI 3,4,5P_3 (antagonized by the action of PTEN or SHIP2), which recruit PDK1 and AKT to the plasma membrane, where AKT is activated by PDK- and mTOR mediated phosphorylation. The mTOR kinase is activated by $Rheb^{GTP}$, which accumulates upon inhibition of the GAP activity of the TSC1::TSC2 complex by PKB mediated phosphorylation. The $p70^{s6k}$ is primed through mTOR mediated phosphorylation for activation by PDK1. AKT phosphorylates many cellular proteins: inactivating PGC1α, $p21^{kip}$, GSK3β, BAD, and AS160; and activating PDE3b, eNOS. The AKT mediated phosphorylation of the forkhead proteins results in their sequestration in the cytoplasm, which inhibits their influence upon transcriptional activity. Insulin stimulates protein synthesis by altering the intrinsic activity or binding properties of key translation initiation and elongation factors (eIFs and eEFs, respectively) as well as critical ribosomal proteins. This occurs via phosphorylation and/or sequestration of repressive factors into inactive complexes. Components of the translational machinery that are targets of insulin regulation include eIF2B, eIF4E, eEF1, eEF2, and the S6 ribosomal protein [98]. TNFα activates JNK, which can phosphorylate IRS1 inhibiting its interaction with the insulin receptor and subsequent tyrosine phosphorylation. IRS2 expression is promoted by nuclear FOXO, which increases IRS2 expression during fasting conditions. CREB : TORC2 complex also promotes IRS2 expression especially in β-cell cells, placing IRS2 under the control of glucose and GLP1.

The role of IRS-proteins in the PI3K→AKT signaling cascade is validated by a wide array of cell based and mouse based experiments. Although IRS1 was originally purified and cloned from rat hepatocytes, the principle role of Irs1 and Irs2 during insulin signaling in hepatocytes *in vivo* was verified only recently [40–43]. The simplest experiments employ an intraperitoneal injection of insulin into ordinary mice, or mice lacking hepatic IRS1, IRS2, or both substrates. In ordinary mice, insulin rapidly stimulates Akt phosphorylation, and the phosphorylation of its downstream substrates Foxo1 and Gsk3α/β. However, both IRS1 and IRS2 must be deleted before insulin receptors are uncoupled from the PI3K→AKT cascade [42]. These results confirm the shared but absolute requirement for IRS1 or IRS2 for the hepatic insulin signaling.

The IRS→PI3K→AKT cascade is very robust owing to the expression of multiple isoforms of each component that confers both redundancy and specificity upon the system [8]. For example, the type 1_A PI 3-kinase is a dimer assembled from one catalytic subunit, p110α or p110β, and one of eight regulatory subunit isoforms encoded by three different genes, Pik3r1, Pik3r2, and Pik3r3 [38]. In most cells including the liver, p85α and its two shorter isoforms p55α and p50α encoded by Pik3r1 constitute most of the regulatory subunits; however, p85β encoded by Pik3r2 also contributes to the total pool of regulatory subunits [44, 45]. Thus, both genes must be inactivated to reduce the cellular concentration of PIP3 by more than 90 percent and confirm the role of PI3K in the hepatic insulin signaling cascade [46].

AKT is also encoded by thee genes. However, mouse based experiments clearly show that these isoforms are not redundant components of the insulin-like signaling cascade. AKT1 deficient mice display developmental differences and are generally smaller than littermate controls [47]. By comparison, AKT2 deficient mice display metabolic defects; and AKT3 deficient mice display neural defects [38]. AKT2 rather than AKT1 is primarily responsible for insulin stimulated GLUT4 translocation in adipose tissues [48]. More work is needed to understand the beneficial and detrimental effects of the redundant signaling, how AKT isoforms generate selective metabolic signals, and whether this diversity can be exploited to treat metabolic disease and cancer.

DYSREGULATION OF IRS-PROTEIN SIGNALING

Insulin resistance is a common pathological state that is associated with many health disorders – obesity, hypertension, chronic infection, dysregulated female reproduction, and kidney and cardiovascular diseases. Over the past 15 years, mouse based experiments have revealed how mutations in genes that mediate the insulin signal, modulate the insulin signal, or respond to the insulin signal, contribute to insulin resistance and diabetes. Whereas genetic mutations are obvious sources of life-long insulin resistance, they are usually associated with rare metabolic disorders. Environmental, physiological, and immunological stress causes insulin resistance through heterologous signaling cascades coordinated by complex genetic backgrounds [26].

Obesity is always associated with peripheral insulin resistance. Recent studies reveal a variety of factors secreted from adipose tissue that inhibit insulin signaling—FFAs, tumor necrosis factor-alpha (TNFα), and resistin—or factors that promote insulin signaling—adipocyte complement related protein of 30 kDa (adiponectin) and leptin. Each of these factors has specific effects upon gene expression patterns that can alter the response of cell to insulin. However, the effect of these factors upon the expression or function of the IRS-proteins could contribute to the mechanism for insulin resistance [49, 50]. Signaling cascades activated during acute trauma or chronic metabolic or inflammatory stress dysregulate IRS-proteins through various mechanisms, including phosphatase mediated dephosphorylation, proteasome mediated degradation, and Ser/Thr phosphorylation (Figure 13.3). Dysregulation of IRS-protein function also provides a plausible framework to understand the loss of compensatory β-cell function while peripheral insulin resistance emerges.

Experiments with TNFα revealed one of the first mechanisms linking inflammatory cytokines to insulin resistance [51]. TNFα activates the NH_2-terminal JUN kinase (JNK), which phosphorylates IRS1 on serine residues that inhibit the activation of the PI 3-kinase/Akt pathway in response to insulin [52]. Phosphorylation of Ser^{307} in the rodent Irs1 (Ser^{312} in human IRS1) is one of the residues thought to mediate the effect of JNK agonists, but others are clearly involved [49, 53]. JNK mediated phosphorylation of IRS1 may also mediate the effects of cellular stress, including endoplasmic reticulum stress [50]. Insulin itself promotes Ser^{307} phosphorylation through activation of the PI 3-kinase, revealing feedback regulation that might be mediated by many kinases – AKT, PKCζ, IKKβ, JNK, mTOR, and S6K1 (Figure 13.3) [49].

The Dominant Role of the IRS→FOXO Cascade

Hyperglycemia and dyslipidemia owing to hepatic insulin resistance are key pathologic features of Type 2 diabetes [2, 54]. In mice, near total hepatic insulin resistance can be introduced via the systemic or liver specific knockout of key insulin signaling genes [41, 55–58]. Among these approaches, the compound suppression or deletion of the insulin receptor substrates, Irs1 and Irs2, is the least complicated by defective insulin clearance or liver failure [40, 41]. The program of gene expression directed by Foxo1 and its cofactors ordinarily protects cells, as well as whole organisms, from the life threatening consequences of nutrient deprivation, and oxidative and genotoxic stresses [59]. During prolonged starvation, hepatic Foxo1 ensures the production of sufficient glucose to prevent life threatening hypoglycemia [60]. In healthy animals, the decreased insulin concentration during fasting promotes the nuclear localization of Foxo1, where it interacts with Ppargc1a and Creb/Crtc2 to increase the expression of the key gluconeogenic enzymes G6pc and Pck1 [61–66] (Figure 13.3). Foxo1 also coordinates decreased nutrient availability with reduced somatic growth by increasing the hepatic expression of Igfbp1 – a secreted factor that limits the bioavailability of Igf1 [65]. Finally, in conjunction with Creb/Crtc2, Foxo1 increases the expression of Irs2 and reduces the expression of the Akt inhibitor Trib3, which together can enhance fasting insulin sensitivity and augment the insulin response upon eventual feeding [67, 68].

Hepatic gene expression is dramatically altered by fasting and feeding owing to the effects of insulin during the postprandial state and the action of counter regulatory hormones including glucagon and glucocorticoids during fasting. Foxo1 is a dominant regulator of hepatic gene expression that is ordinarily inactivated through the Irs1 or Irs2 branch of the insulin signaling system [42]. Some reports suggest that hepatic Foxo1 is preferentially regulated through the Irs2 branch of the insulin signaling cascade [69]. However, in murine hepatocytes either Irs1 or Irs2 is sufficient for insulin regulated phosphorylation of Foxo1 [42]. Many hepatic genes are dysregulated in mice lacking both hepatic Irs1 and Irs2, confirming that insulin signaling plays a critical role. Upon deletion of *Foxo1* gene expression is largely normalized in both the fasted and postprandial states. Remarkably, acute or chronic inactivation of Foxo1 also restores nearly normal glucose homeostasis even in the absence of the IRS1 and IRS2 – branches of the insulin signaling cascade [42]. Although Foxo1 can be regulated by many mechanisms, its inactivation in postprandial liver appears to be under the exclusive control of the Irs1/2→Akt cascade (Figure 13.3).

Hepatic deficiency of Irs1 and Irs2 leads to a significant growth retardation in both prediabetic and diabetic mice. Consistent with this phenotype, the expression of several hepatic genes that promote organismal growth (*Igf1*, *Igfals*, and *Ghr*) are significantly reduced in mice lacking Irs1/2 [42]. Moreover, the expression of *Igfbp1* increases significantly in Irs1/2 deficient liver, which leads to increased circulating concentrations of Igfpb1 that inactivates Igf1 [5]. Again, the deletion of hepatic *Foxo1* restores normal expression of these genes in the Irs1/2 deficient liver, which restores growth of the mice to a normal size. Stat5b is well known to regulate *Igf1* and *Igfals* expression, but whether Foxo1 directly modulates their expression requires additional work (Figure 13.4) [70]. Regardless, the postprandial inactivation of Foxo1 by the IRS→PI3K→AKT cascade is essential for the usual systemic effect of growth hormone and hormonal Igf1 upon body growth (Figure 13.4).

The Central Role of IRS2 Signaling in Pancreatic β-cells

Mice lacking the gene for Irs1 or Irs2 are insulin resistant, with impaired peripheral glucose utilization [4, 71, 72]. Both

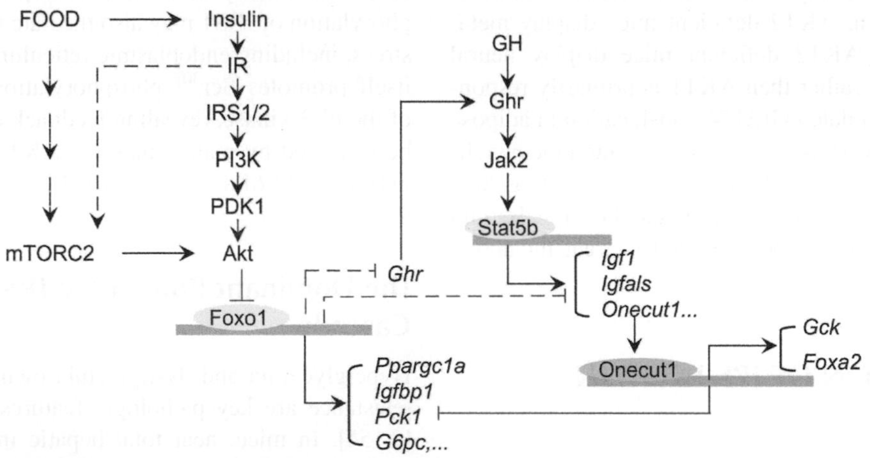

FIGURE 13.4 The role of Foxo1 in the insulin and growth hormone signaling pathways.
Insulin can inhibit Foxo1 transcriptional activity through IRS1/2 mediated PI3K→Akt pathway to phosphorylate Foxo1. As a transcription factor, Foxo1 can activate one set of genes including *Ppargc1a*, *Pck1*, *G6pc*, and *Igfbp1*, and directly or indirectly suppress another set of genes including *Ghr*, *Igf1*, *Onecut1*, and *Gck*. Through regulation of *Ghr* gene expression, Foxo1 also influences expression of growth hormone regulated genes including *Igf1*, *Igfals*, and *Onecut1*. Moreover, in response to insulin and growth hormone signals, Onecut1 can impact on nutrient metabolism through activation of *Foxa2* and *Gck* gene expression and suppression of *Pck1* gene expression. Fonts in bold indicate protein molecules and fonts in italic indicate mRNA molecules. Arrows indicate activation and blunted lines represent inhibition. Solid lines or arrows indicate reported links in the literature and dotted lines or arrows indicate implicated links from this current study.

types of knockout mice display metabolic dysregulation, but only the Irs2$^{-/-}$ mice develop diabetes between 10 and 15 weeks of age owing to a near complete loss of pancreatic β-cells [4]. This result positions the insulin-like signaling cascade through IRS2 at the center of β-cell function. The Igf1r→Irs2 pathway has a major role in adult β-cell growth, function and survival [73]. Restricting our attention to 4-week old mice, β-cell mass is nearly normal in Irs2$^{-/-}$ mice, whereas it is reduced by 50 percent in Igf1r$^{+/-}$::Irs2$^{+/-}$ mice, and nearly undetectable in Igf1r$^{+/-}$::Irs2$^{-/-}$ mice [73]. However, the targeted deletion of the Igf1r in β-cells has insignificant effects upon β-cell growth and survival, suggesting that Igf1→Irs2 signaling might not be β-cell autonomous [74, 75]. By comparison, the β-cell specific deletion of both IR and IGF1R in β-cells causes loss of β-cell mass by 2 weeks of age, suggesting that both receptor kinases – and possibly receptor hybrids – promote the IRS2 signaling needed for β-cell growth and survival [76].

Many factors are required for proper β-cell function, including the homeodomain transcription factor Pdx1. Pdx1 regulates downstream genes needed for β-cell growth and function, and mutations in PDX1 cause autosomal forms of early onset diabetes in people (MODY) [77, 78] (Figure 13.5). Pdx1 is reduced in Irs2$^{-/-}$ islets and Pdx1 haploinsufficiency further diminishes the function of β-cells lacking Irs2. By comparison, transgenic Pdx1 expressed in Irs2$^{-/-}$ mice restores β-cell function and normalizes glucose tolerance, which links the insulin-like signaling cascade to the network of β-cell transcription factors [79, 80]. Transgenic upregulation of Irs2 or suppression of Foxo1 increases Pdx1 concentrations in Irs2$^{-/-}$ mice, supporting the hypothesis that Pdx1 is modulated by Irs2→Foxo1 cascade in β-cells [80, 81] (Figure 13.5). Haploinsufficiency for PTEN also prevents β-cell failure in Irs2$^{-/-}$ mice, owing at least in part to the simulation of the PI 3-kinase→PKB/AKT cascade that inhibits Foxo1 (Figure 13.5).

Glucose and glucagon-like peptide-1 have strong effects upon β-cell growth, which depend upon the Irs2 signaling cascade. In β-cells, Irs2 is strongly upregulated by cAMP and Ca^{2+} agonists – including glucose and glucagon-like peptide-1 (GLP1) – which activate cAMP responsive element binding protein (CREB) and the CREB regulated transcription coactivator 2 (TORC2) [3]. While many cAMP mediated pathways oppose the action of insulin, the upregulation of IRS2 by glucose and GLP1 reveals an unexpected intersection of these important signals (Figure 13.5). Thus, hyperglycemia resulting from the daily consumption of high caloric food promotes β-cell growth by increasing IRS2 expression [82]. Mice expressing low concentrations of glucokinase in β-cells display impaired intracellular calcium responses to glucose, which suppresses the phosphorylation of CREB that induces Irs2 expression. Similarly, GLP1 couples into this mechanism by directly increasing the concentration of cAMP in β-cells [3]. Thus, glucose and GLP1 have no effect upon β-cell growth in Irs2$^{-/-}$ mice [83]. Together, these results suggest that the Irs2 branch of the insulin-like signaling cascade – rather than Irs1 or the upstream receptor kinases – is the "ordinary gatekeeper" for β-cell plasticity and function (Figure 13.5).

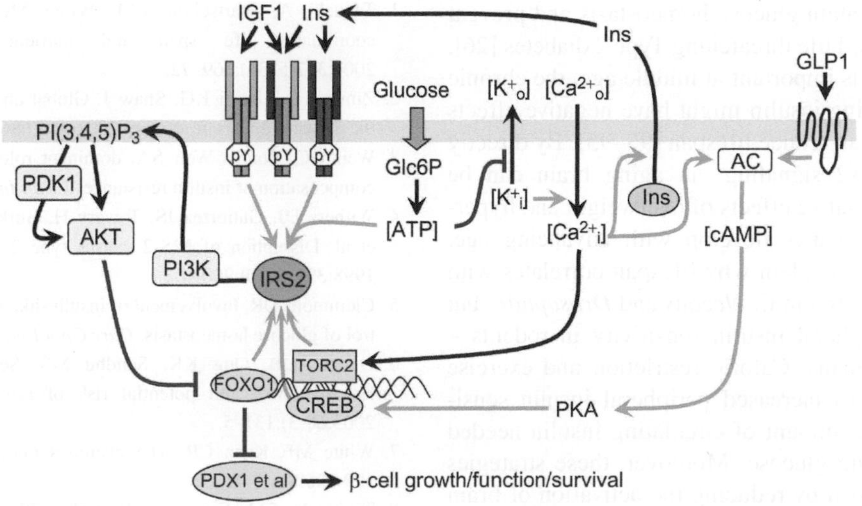

FIGURE 13.5 The integrative role of Irs2 in pancreatic β-cells.
The diagram shows the relation between the MODY genes, especially Pdx1, and the IRS2-branch of the insulin signaling pathway (363). Drugs that promote IRS2 signaling are expected to promote PDX1 function in β-cells, including the phosphorylation of BAD and FOXO1, which will promote β-cell growth, function, and survival. Induction of PDX1 promotes the expression of genes products that enhance glucose sensing and insulin secretion. Activation of CREB by GLP1→cAMP, and TORC2 by glucose→Ca^{2+} induces IRS2 expression in β-cells, revealing a mechanism that promotes β-cell growth, function, and survival.

IRS2 Signaling and Central Control of Nutrient Homeostasis and Lifespan

Work with lower metazoans – mainly *C. elegans* and *Drosophila* – has drawn attention to the idea that less insulin-like signaling can extend the healthy lifespan [84]. Female fruit flies live up to 80 percent longer when the activity of the *Drosophila* insulin-like receptor is reduced by 80 percent [85]. Moreover, ablation of neurosecretory cells in the adult *Drosophila* that produce three of the seven insulin-like peptides also extends lifespan. Although the long-lived flies accumulate excess lipids and carbohydrate, they also show resistance to oxidative stress [86]. Thus, reduced insulin-like signaling can extend lifespan in lower metazoans, while causing physiological alterations that are associated with life threatening diseases in rodents and humans.

Recent results point to the brain as the site where reduced insulin-like signaling might have a consistent effect to extend mammalian lifespan – as it does in *C. elegans* and *D. melanogaster* [84, 87]. Irs1 and Irs2 are expressed in neurons throughout the brain – and Irs4 is expressed in the hypothalamus. However, Irs2 appears to play a major role for brain growth, fertility [88, 89], and nutrient homeostasis [90–92]. Proper regulation of Irs2 signaling in the brain might also play a role in mouse lifespan [1]. At 22 months of age, brain Irs2 knockout mice are slightly hyperphagic, overweight, hyperinsulinemic, and glucose intolerant; however, compared to control mice, brain Irs2 knockout mice display stable superoxide dismutase-2 (SOD) concentration in the hypothalamus during meals. However, reduced Irs2 signaling in the brain increases by nearly 6 months the lifespan of mice maintained on a high energy diet [1].

As mammals age, compensatory hyperinsulinemia usually develops to maintain glucose homeostasis and prevent the progression toward life threatening Type 2 diabetes [26]. While this response is important at middle age, the chronic elevation of circulating insulin might have negative effects upon the brain that can reduce lifespan [93–95]. By directly attenuating brain Irs2 signaling, an aging brain can be shielded from the negative effects of overweight and hyperinsulinemia that ordinarily develop with advancing age. This hypothesis helps explain why lifespan correlates with reduced insulin signaling in *C. elegans* and *Drosophila*, but correlates with peripheral insulin sensitivity in rodents – and probably in humans. Calorie restriction and exercise are the usual ways to increased peripheral insulin sensitivity and reduce the amount of circulating insulin needed to regulate circulating glucose. Moreover, these strategies might increase lifespan by reducing the activation of brain IRS2. Other genetic strategies that improve peripheral insulin sensitivity – including reduced growth hormone or adipocyte insulin signaling – can be explained by a similar mechanism [96, 97]. Consistent with this hypothesis, human centenarians display increased peripheral insulin sensitivity and reduced circulating insulin levels [93]. Hence, less insulin-like signaling in the brain increases mammalian lifespan, just as reduced insulin-like signaling extends the lifespan of *C. elegans* and *Drosophila*.

SUMMARY AND PERSPECTIVES

The investigation of the insulin signaling cascade has revealed a broad physiologic role that extends far beyond the classical insulin target tissues – liver, muscle, and fat. It is now clear that insulin-like signaling plays a major role in pancreatic β-cells and in the central nervous system. Peripheral insulin resistance is ordinarily opposed by increased insulin secretion from the β-cells; however, compensatory hyperinsulinemia might have negative consequences in the brain that shorten lifespan. The tools now available to probe the insulin-like signaling cascades in healthy and diabetic tissues provides a rational platform to develop new strategies to treat insulin resistance and prevent its progression to Type 2 diabetes and neurodegeneration. Understanding how the IRS-proteins negotiate the conflicting signals generated during insulin and cytokines action might be a valuable starting point. Whether better management of inflammatory responses can attenuate insulin resistance and diminish its consequences is an important area of investigation. Future work must better resolve the network of insulin responses that are generated in various tissues, because too much insulin action might also shorten our lives.

REFERENCES

1. Taguchi A, Wartschow LM, White MF. Brain IRS2 signaling coordinates life span and nutrient homeostasis. *Science* 2007;**317**(5836):369–72.
2. Zimmet P, Alberti KG, Shaw J. Global and societal implications of the diabetes epidemic. *Nature* 2001;**414**(6865):782–7.
3. Weir GC, Bonner-Weir SA. dominant role for glucose in beta cell compensation of insulin resistance. *J Clin Invest* 2007;**117**(1):81–3.
4. Withers DJ, Gutierrez JS, Towery H, Burks DJ, Ren JM, Previs S, et al. Disruption of IRS-2 causes type 2 diabetes in mice. *Nature* 1998;**391**(6670):900–4.
5. Clemmons DR. Involvement of insulin-like growth factor-I in the control of glucose homeostasis. *Curr Opin Pharmacol* 2006;**6**(6):620–5.
6. Dunger DB, Ong KK, Sandhu MS. Serum insulin-like growth factor-I levels and potential risk of type 2 diabetes. *Horm Res* 2003;**60**(3):131–5.
7. White MF, Kahn CR. The insulin signaling system. *J Biol Chem* 1994;**269**(1):1–4.
8. Taniguchi CM, Emanuelli B, Kahn CR. Critical nodes in signalling pathways: insights into insulin action. *Nat Rev Mol Cell Biol* 2006;**7**(2):85–96.
9. Ward CW, Lawrence MC, Streltsov VA, Adams TE, McKern NM. The insulin and EGF receptor structures: new insights into ligand-induced receptor activation. *Trends Biochem Sci* 2007;**32**(3):129–37.

10. McKern NM, Lawrence MC, Streltsov VA, Lou MZ, Adams TE, Lovrecz GO, et al. Structure of the insulin receptor ectodomain reveals a folded-over conformation. *Nature* 2006;**443**(7108):218–21.
11. Lawrence MC, McKern NM, Ward CW. Insulin receptor structure and its implications for the IGF-1 receptor. *Curr Opin Struct Biol* 2007;**17**(6):699–705.
12. Sciacca L, Prisco M, Wu A, Belfiore A, Vigneri R, Baserga R. Signaling differences from the A and B isoforms of the insulin receptor (IR) in 32D cells in the presence or absence of IR substrate-1. *Endocrinology* 2003;**144**(6):2650–8.
13. Blanquart C, Achi J, Issad T. Characterization of IRA/IRB hybrid receptors using bioluminescence resonance energy transfer. *Biochem Pharmacol* 2008;**76**(7):873–83.
14. White MF, Shoelson SE, Keutmann H, Kahn CRA. cascade of tyrosine autophosphorylation in the beta-subunit activates the phosphotransferase of the insulin receptor. *J Biol Chem* 1988;**263**(6):2969–80.
15. Till JH, Ablooglu AJ, Frankel M, Bishop SM, Kohanski RA, Hubbard SR. Crystallographic and solution studies of an activation loop mutant of the insulin receptor tyrosine kinase: insights into kinase mechanism. *J Biol Chem* 2001;**276**(13):10,049–55.
16. Hubbard SR. Crystal structure of the activated insulin receptor tyrosine kinase in complex with peptide substrate and ATP analog. *EMBO J* 1997;**16**(18):5572–81.
17. White MF, Livingston JN, Backer JM, Lauris V, Dull TJ, Ullrich A, et al. Mutation of the insulin receptor at tyrosine 960 inhibits signal transmission but does not affect its tyrosine kinase activity. *Cell* 1988;**54**:641–9.
18. Hsu D, Knudson PE, Zapf A, Rolband GC. Olefsky JM. NPXY motif in the insulin-like growth factor-I receptor is required for efficient ligand-mediated receptor internalization and biological signaling. *Endocrinology* 1994;**134**(2):744–50.
19. Yenush L, White MF. The IRS-signaling system during insulin and cytokine action. *Bio Essays* 1997;**19**(5):491–500.
20. Pawson T, Scott JD. Signaling through scaffold, anchoring, and adaptor proteins. *Science* 1997;**278**(5346):2075–80.
21. Kotani K, Wilden P, Pillay TS. SH2-Balpha is an insulin-receptor adapter protein and substrate that interacts with the activation loop of the insulin-receptor kinase. *Biochem J* 1998;**335**(Pt 1):103–9.
22. Lock P, Casagranda F, Dunn AR. Independent SH2-binding sites mediate interaction of Dok-related protein with RasGTPase-activating protein and Nck. *J Biol Chem* 1999;**274**(32):22,775–84.
23. Noguchi T, Matozaki T, Inagaki K, Tsuda M, Fukunaga K, Kitamura Y, et al. Tyrosine phosphorylation of p62(Dok) induced by cell adhesion and insulin: possible role in cell migration. *EMBO J* 1999;**18**(7):1748–60.
24. Chiang SH, Baumann CA, Kanzaki M, Thurmond DC, Watson RT, Neudauer CL, et al. Insulin-stimulated GLUT4 translocation requires the CAP-dependent activation of TC10. *Nature* 2001;**410**(6831):944–8.
25. Baumann CA, Ribon V, Kanzaki M, Thurmond DC, Mora S, Shigematsu S, et al. CAP defines a second signalling pathway required for insulin-stimulated glucose transport. *Nature* 2000;**407**(6801):202–7.
26. White MF. Insulin signaling in health and disease. *Science* 2003;**302**(5651):1710–1.
27. Bjornholm M, He AR, Attersand A, Lake S, Liu SC, Lienhard GE, et al. Absence of functional insulin receptor substrate-3 (IRS-3) gene in humans. *Diabetologia* 2002;**45**(12):1697–702.
28. Numan S, Russell DS. Discrete expression of insulin receptor substrate-4 mRNA in adult rat brain. *Brain Res Mol Brain Res* 1999;**72**(1):97–102.
29. Wolf G, Trub T, Ottinger E, Groninga L, Lynch A, White MF, et al. The PTB domains of IRS-1 and Shc have distinct but overlapping specificities. *J Biol Chem* 1995;**270**:27,407–10.
30. Yenush L, Zanella C, Uchida T, Bernal D, White MF. The pleckstrin homology and phosphotyrosine binding domains of insulin receptor substrate 1 mediate inhibition of apoptosis by insulin. *Mol Cell Biol* 1998;**18**(11):6784–94.
31. Lemmon MA, Ferguson KM, Schlessinger J. PH domains: diverse sequences with a common fold recruit signaling molecules to the cell surface. *Cell* 1996;**85**(5):621–4.
32. Lemmon MA, Ferguson KM, Abrams CS. Pleckstrin homology domains and the cytoskeleton. *Growth Regul* 2002;**513**(1):71–6.
33. Burks DJ, Wang J, Towery H, Ishibashi O, Lowe D, Riedel H, et al. IRS pleckstrin homology domains bind to acidic motifs in proteins. *J Biol Chem* 1998;**273**(47):31,061–7.
34. Sawka-Verhelle D, Tartare-Deckert S, White MF, Van Obberghen E. Insulin receptor substrate-2 binds to the insulin receptor through its phopshotyrosine-binding domain and through a newly identified domain comprising amino acids 591-786. *J Biol Chem* 1996;**271**(11):5980–3.
35. Sawka-Verhelle D, Baron V, Mothe I, Filloux C, White MF, Van Obberghen E. Tyr624 and Tyr628 in insulin receptor substrate-2 mediate its association with the insulin receptor. *J Biol Chem* 1997;**272**(26):16,414–20.
36. Wu J, Tseng YD, Xu CF, Neubert TA, White MF, Hubbard SR. Structural and biochemical characterization of the KRLB region in insulin receptor substrate-2. *Nat Struct Mol Biol* 2008;**15**(3):251–8.
37. Backer JM, Myers Jr MG, Shoelson SE, Chin DJ, Sun XJ, Miralpeix M, et al. Phosphatidylinositol 3'-kinase is activated by association with IRS-1 during insulin stimulation. *EMBO J* 1992;**11**: 3469–79.
38. Manning BD, Cantley LC. AKT/PKB signaling: navigating downstream. *Cell* 2007;**129**(7):1261–74.
39. Astrinidis A, Henske EP. Tuberous sclerosis complex: linking growth and energy signaling pathways with human disease. *Oncogene* 2005;**24**(50):7475–81.
40. Taniguchi CM, Ueki K, Kahn CR. Complementary roles of IRS-1 and IRS-2 in the hepatic regulation of metabolism. *J Clin Invest* 2005;**115**(3):718–27.
41. Dong X, Park S, Lin X, Copps K, Yi X, White MF. Irs1 and Irs2 signaling is essential for hepatic glucose homeostasis and systemic growth. *J Clin Invest* 2006;**116**(1):101–4.
42. Dong XC, Copps KD, Guo S, Li Y, Kollipara R, DePinho RA, et al. Inactivation of hepatic foxo1 by insulin signaling is required for adaptive nutrient homeostasis and endocrine growth regulation. *Cell Metab* 2008;**8**(1):65–76.
43. Kubota N, Kubota T, Itoh S, Kumagai H, Kozono H, Takamoto I, et al. Dynamic functional relay between insulin receptor substrate 1 and 2 in hepatic insulin signaling during fasting and feeding. *Cell Metab* 2008;**8**(1):49–64.
44. Ueki K, Fruman DA, Yballe CM, Fasshauer M, Klein J, Asano T, et al. Positive and negative roles of p85 alpha and p85 beta regulatory subunits of phosphoinositide 3-kinase in insulin signaling. *J Biol Chem* 2003;**278**(48):48,453–66.
45. Ueki K, Algenstaedt P, Mauvais-Jarvis F, Kahn CR. Positive and negative regulation of phosphoinositide 3-kinase-dependent signaling pathways by three different gene products of the p85alpha regulatory subunit. *Mol Cell Biol* 2000;**20**(21):8035–46.
46. Taniguchi CM, Kondo T, Sajan M, Luo J, Bronson R, Asano T, et al. Divergent regulation of hepatic glucose and lipid metabolism by

phosphoinositide 3-kinase via Akt and PKClambda/zeta. *Cell Metab* 2006;**3**(5):343–53.

47. Cho H, Thorvaldsen JL, Chu Q, Feng F, Birnbaum MJ. Akt1/PKBalpha is required for normal growth but dispensable for maintenance of glucose homeostasis in mice. *J Biol Chem* 2001;**276**(42):38,349–52.
48. Bae SS, Cho H, Mu J, Birnbaum MJ. Isoform-specific regulation of insulin-dependent glucose uptake by Akt/protein kinase B. *J Biol Chem* 2003;**278**(49):49,530–6.
49. Zick Y. Ser/Thr phosphorylation of IRS proteins: a molecular basis for insulin resistance. *Sci STKE* 2005;**268**:e4.
50. Ozcan U, Cao Q, Yilmaz E, Lee AH, Iwakoshi NN, Ozdelen E, et al. Endoplasmic reticulum stress links obesity, insulin action, and type 2 diabetes. *Science* 2004;**306**(5695):457–61.
51. Wellen KE, Hotamisligil GS. Inflammation, stress, and diabetes. *J Clin Invest* 2005;**115**(5):1111–9.
52. Aguirre V, Uchida T, Yenush L, Davis R, White MF. The c-Jun NH(2)-terminal kinase promotes insulin resistance during association with insulin receptor substrate-1 and phosphorylation of Ser(307). *J Biol Chem* 2000;**275**(12):9047–54.
53. Giraud J, Haas M, Feener EP, Copps KD, Dong X, Dunn SL, et al. Phosphorylation of Irs1 at SER-522 inhibits insulin signaling. *Mol Endocrinol* 2007;**21**(9):2294–302.
54. Brown MS, Goldstein JL. Selective versus total insulin resistance: a pathogenic paradox. *Cell Metab* 2008;**7**(2):95–6.
55. Michael MD, Kulkarni RN, Postic C, Previs SF, Shulman GI, Magnuson MA, et al. Loss of insulin signaling in hepatocytes leads to severe insulin resistance and progressive hepatic dysfunction. *Mol Cell* 2000;**6**(1):87–97.
56. Cho H, Mu J, Kim JK, Thorvaldsen JL, Chu Q, Crenshaw III EB, et al. Insulin resistance and a diabetes mellitus-like syndrome in mice lacking the protein kinase Akt2 (PKB beta). *Science* 2001;**292**(5522):1728–31.
57. Okamoto Y, Ogawa W, Nishizawa A, Inoue H, Teshigawara K, Kinoshita S, et al. Restoration of glucokinase expression in the liver normalizes postprandial glucose disposal in mice with hepatic deficiency of PDK1. *Diabetes* 2007;**56**(4):1000–9.
58. Mora A, Lipina C, Tronche F, Sutherland C, Alessi DR. Deficiency of PDK1 in liver results in glucose intolerance, impairment of insulin-regulated gene expression and liver failure. *Biochem J* 2005;**385** (Pt 3):639–48.
59. van der Horst A, Burgering BM. Stressing the role of FoxO proteins in lifespan and disease. *Nat Rev Mol Cell Biol* 2007;**8**(6):440–50.
60. Matsumoto M, Pocai A, Rossetti L, DePinho RA, Accili D. Impaired regulation of hepatic glucose production in mice lacking the forkhead transcription factor foxo1 in liver. *Cell Metab* 2007;**6**(3):208–16.
61. Dentin R, Liu Y, Koo SH, Hedrick S, Vargas T, Heredia J, et al. Insulin modulates gluconeogenesis by inhibition of the coactivator TORC2. *Nature* 2007;**449**(7160):366–9.
62. Koo SH, Flechner L, Qi L, Zhang X, Screaton RA, Jeffries S, et al. The CREB coactivator TORC2 is a key regulator of fasting glucose metabolism. *Nature* 2005;**437**(7062):1109–11.
63. Puigserver P, Rhee J, Donovan J, Walkey CJ, Yoon JC, Oriente F, et al. Insulin-regulated hepatic gluconeogenesis through FOXO1-PGC-1alpha interaction. *Nature* 2003;**423**(6939):550–5.
64. Schilling MM, Oeser JK, Boustead JN, Flemming BP, O'Brien RM. Gluconeogenesis: re-evaluating the FOXO1-PGC-1alpha connection. *Nature* 2006;**443**(7111):E10–1.
65. Barthel A, Schmoll D, Unterman TG. FoxO proteins in insulin action and metabolism. *Trends Endocrinol Metab* 2005;**16**(4): 183–9.
66. Mounier C, Posner BI. Transcriptional regulation by insulin: from the receptor to the gene. *Can J Physiol Pharmacol* 2006;**84**(7):713–24.
67. Canettieri G, Koo SH, Berdeaux R, Heredia J, Hedrick S, Zhang X, et al. Dual role of the coactivator TORC2 in modulating hepatic glucose output and insulin signaling. *Cell Metab* 2005;**2**(5):331–8.
68. Matsumoto M, Han S, Kitamura T, Accili D. Dual role of transcription factor FoxO1 in controlling hepatic insulin sensitivity and lipid metabolism. *J Clin Invest* 2006;**116**(9):2464–72.
69. Wolfrum C, Asilmaz E, Luca E, Friedman JM, Stoffel M. Foxa2 regulates lipid metabolism and ketogenesis in the liver during fasting and in diabetes. *Nature* 2004;**432**(7020):1027–32.
70. Le RD, Bondy C, Yakar S, Liu JL, Butler A. The somatomedin hypothesis. *Endocr Rev* 2001;**22**(1):53–74.
71. Kubota N, Tobe K, Terauchi Y, Eto K, Yamauchi T, Suzuki R, et al. Disruption of insulin receptor substrate 2 causes type 2 diabetes because of liver insulin resistance and lack of compensatory beta-cell hyperplasia. *Diabetes* 2000;**49**(11):1880–9.
72. Previs SF, Withers DJ, Ren JM, White MF, Shulman GI. Contrasting effects of IRS-1 vs IRS-2 gene disruption on carbohydrate and lipid metabolism in vivo. *J Biol Chem* 2000;**275**(50):38,990–4.
73. Withers DJ, Burks DJ, Towery HH, Altamuro SL, Flint CL, White MF. Irs-2 coordinates Igf-1 receptor-mediated beta-cell development and peripheral insulin signalling. *Nat Genet* 1999;**23**(1):32–40.
74. Xuan S, Kitamura T, Nakae J, Politi K, Kido Y, Fisher PE, et al. Defective insulin secretion in pancreatic beta cells lacking type 1 IGF receptor. *J Clin Invest* 2002;**110**(7):1011–9.
75. Kulkarni RN, Holzenberger M, Shih DQ, Ozcan U, Stoffel M, Magnuson MA, et al. beta-cell-specific deletion of the Igf1 receptor leads to hyperinsulinemia and glucose intolerance but does not alter beta-cell mass. *Nat Genet* 2002;**31**(1):111–5.
76. Ueki K, Okada T, Hu J, Liew CW, Assmann A, Dahlgren GM, et al. Total insulin and IGF-I resistance in pancreatic beta cells causes overt diabetes. *Nat Genet* 2006;**38**(5):583–8.
77. Jonsson J, Carlsson L, Edlund T, Edlund H. Insulin-promoter-factor 1 is required for pancreas developement in mice. *Nature* 1994; **371**(6498):606–9.
78. Stoffers DA, Zinkin NT, Stanojevic V, Clarke WL, Habener JF. Pancreatic agenesis attributable to a single nucleotide deletion in the human IPF1 gene coding sequence. *Nat Genet* 1997;**15**(1):106–10.
79. Kushner JA, Ye J, Schubert M, Burks DJ, Dow MA, Flint CL, et al. Pdx1 restores beta cell function in Irs2 knockout mice. *J Clin Invest* 2002;**109**(9):1193–201.
80. Kitamura T, Nakae J, Kitamura Y, Kido Y, Biggs III WH, Wright CV, et al. The forkhead transcription factor Foxo1 links insulin signaling to Pdx1 regulation of pancreatic beta cell growth. *J Clin Invest* 2002;**110**(12):1839–47.
81. Hennige AM, Burks DJ, Ozcan U, Kulkarni RN, Ye J, Park S, et al. Upregulation of insulin receptor substrate-2 in pancreatic beta cells prevents diabetes. *J Clin Invest* 2003;**112**(10):1521–32.
82. Terauchi Y, Takamoto I, Kubota N, Matsui J, Suzuki R, Komeda K, et al. Glucokinase and IRS-2 are required for compensatory beta cell hyperplasia in response to high-fat diet-induced insulin resistance. *J Clin Invest* 2007;**117**(1):246–57.
83. Park S, Dong X, Fisher TL, Dunn S, Omer AK, Weir G, et al. Exendin-4 uses irs2 signaling to mediate pancreatic Beta cell growth and function. *J Biol Chem* 2006;**281**(2):1159–68.
84. Kenyon C. The plasticity of aging: insights from long-lived mutants. *Cell* 2005;**120**(4):449–60.
85. Tatar M, Kopelman A, Epstein D, Tu MP, Yin CM, Garofalo RS. A mutant *Drosophila* insulin receptor homolog that extends life-span and impairs neuroendocrine function. *Science* 2001;**292**(5514):107–10.

86. Broughton SJ, Piper MD, Ikeya T, Bass TM, Jacobson J, Driege Y, et al. Longer lifespan, altered metabolism, and stress resistance in *Drosophila* from ablation of cells making insulin-like ligands. *Proc Natl Acad Sci U S A* 2005;**102**(8):3105–10.
87. Hughes KA, Reynolds RM. Evolutionary and mechanistic theories of aging. *Annu Rev Entomol* 2005;**50**:421–45.
88. Schubert M, Brazil DP, Burks DJ, Kushner JA, Ye J, Flint CL, et al. Insulin receptor substrate-2 deficiency impairs brain growth and promotes tau phosphorylation. *J Neurosci* 2003;**23**(18):7084–92.
89. Burks DJ, de Mora JF, Schubert M, Withers DJ, Myers MG, Towery HH, et al. IRS-2 pathways integrate female reproduction and energy homeostasis. *Nature* 2000;**407**(6802):377–82.
90. Lin X, Taguchi A, Park S, Kushner JA, Li F, Li Y, et al. Dysregulation of insulin receptor substrate 2 in beta cells and brain causes obesity and diabetes. *J Clin Invest* 2004;**114**(7):908–16.
91. Choudhury AI, Heffron H, Smith MA, Al-Qassab H, Xu AW, Selman C, et al. The role of insulin receptor substrate 2 in hypothalamic and beta cell function. *J Clin Invest* 2005;**115**(4):940–50.
92. Masaki T, Chiba S, Noguchi H, Yasuda T, Tobe K, Suzuki R, et al. Obesity in insulin receptor substrate-2-deficient mice: disrupted control of arcuate nucleus neuropeptides. *Obes Res* 2004;**12**(5):878–85.
93. Barbieri M, Rizzo MR, Manzella D, Grella R, Ragno E, Carbonella M, et al. Glucose regulation and oxidative stress in healthy centenarians. *Exp Gerontol* 2003;**38**(1–2):137–43.
94. Schriner SE, Linford NJ, Martin GM, Treuting P, Ogburn CE, Emond M, et al. Extension of murine life span by overexpression of catalase targeted to mitochondria. *Science* 2005;**308**(5730):1909–11.
95. Hekimi S, Guarente L. Genetics and the specificity of the aging process. *Science* 2003;**299**(5611):1351–4.
96. Bonkowski MS, Rocha JS, Masternak MM, Al Regaiey KA, Bartke A. Targeted disruption of growth hormone receptor interferes with the beneficial actions of calorie restriction. *Proc Natl Acad Sci U S A* 2006;**103**(20):7901–5.
97. Bluher M, Kahn BB, Kahn CR. Extended longevity in mice lacking the insulin receptor in adipose tissue. *Science* 2003;**299**(5606):572–4.
98. Rhoads RE. Signal transduction pathways that regulate eukaryotic protein synthesis. *J Biol Chem* 1999;**274**(43):30,337–40.

Chapter 14

The Epidermal Growth Factor Receptor Family

Wolfgang J. Köstler and Yosef Yarden
Department of Biological Regulation, The Weizmann Institute of Science, Rehovot, Israel

INTRODUCTION

Receptor tyrosine kinases (RTKs), transmembrane molecules whose intracellular portions exhibit a catalytic kinase activity specific for tyrosine residues, couple binding of growth factor ligands to the intracellular signaling pathways that orchestrate a variety of cellular processes, such as migration, proliferation, differentiation, and survival. The receptor for the epidermal growth factor (EGFR, ERBB-1) represents a prototypical RTK, and was the first cell-surface signaling protein characterized by molecular and genetic methods [1].

The complexity of ERBB signaling has increased considerably during evolution. In invertebrates such as *Caenorhabditis elegans*, a single worm ERBB ortholog, LET-23, binding to a single ligand, LIN-3, induces the development of the vulva, the male tail, and posterior ectoderm [2]. In *Drosophila melanogaster*, the interactions of a single fly homolog of ERBB, DER, with its five stimulatory ligands, regulate several stages of development, including oogenesis, embryogenesis, and wing and eye development [3]. The present review focuses on the ERBB signaling network in humans, which comprises 11 stimulatory ligands and 4 ERBB receptors – EGFR, ERBB-2 (or HER2/neu), ERBB-3 (HER3), and ERBB-4 (HER4). These ERBB receptors are both widely expressed and intricately involved in the development and function of epithelial, mesenchymal, and neuronal tissues. Analyses of the function of ERBBs have revealed their critical developmental role in inductive cell-fate determination in mammals (see Table 14.1). Due to their highly redundant function, lack of individual ligands results in milder phenotypes.

STRUCTURE AND ACTIVATION OF ERBB RECEPTORS AND THEIR LIGANDS

ERBB family ligands are polypeptides that include an EGF-like consensus sequence consisting of three disulfide-bonded intramolecular loops. Ligands are generated upon the regulated cleavage of glycosylated transmembrane precursor molecules by GPCR (G-protein-coupled receptor) activated zinc-dependent proteases of the ADAM (*a disintegrin and metalloprotease*) family. This inter-receptor cross-talk (GPCR to RTK) couples the potent signaling capacities of the ERBB network to the largest group of cell surface receptors, which can sense a broad variety of physical and chemical stimuli [4,5]. ERBB ligands differ with respect to their receptor specificity and binding affinity. Epidermal growth factor (EGF), transforming growth factor-α (TGF-α), β-cellulin (BTC), and amphiregulin (AR) are specific ligands for ERBB-1; the neuregulins (NRG) 2, 3, and 4 bind only ERBB-4; whereas the ligand specificities of heparin-binding EGF (HB-EGF) and Epiregulin (EPR; both binding ERBB-1 and ERBB-4), neuregulin-1 (NRG1; binding ERBB-3 and ERBB-4), and epigen (EPG; binding ERBB-1 and ERBB-3 or -4 in the presence of ERBB-2) are more relaxed [6].

ERBB receptors are composed of a large extracellular ligand binding domain, which has four subdomains (I–IV) and a single hydrophobic transmembrane domain. The intracellular portion consists of a small intracellular juxtamembrane domain, a bilobular tyrosine kinase domain and a carboxyl-terminal tail containing tyrosine autophosphorylation sites (Figure 14.1). With the exception of the ligand-less ERBB-2 and the kinase-dead ERBB-3, the activation

Handbook of Cell Signaling, Three-Volume Set 2 ed.

TABLE 14.1 Phenotypes of mice deficient in ERBB proteins

EGFR$^{-/-}$	Strain-dependent phenotypes (embryonic or perinatal lethality): death *in utero* due to defects in placental spongioblasts. After birth: progressive neurodegeneration and abnormalities in multiple organs, including brain, skin, lung, eyes, GI tract, and hair follicles.
ERBB-2$^{-/-}$ (lethal E10.5)	Insufficient heart development with trabeculae malformation in ventricles. Introduction of transgenic ERBB-2 into myocardial cells results in perinatal lethality and nervous system defects, including lack of Schwann cells, defects in motor and sensory neurons and neuromuscular junctions.
ERBB-3$^{-/-}$ (lethal E13.5)	Defects in valve formation in heart development and nervous system defects, including neural crest defects, lack of Schwann cells, and degeneration of the peripheral nervous system.
ERBB-4$^{-/-}$ (lethal E10.5)	Insufficient heart development and trabeculae malformation in ventricles and central nervous system defects, including hindbrain innervation.

of the other two ERBB family receptors can occur through formation of ligand-bound receptor homo- and heterodimers. Recent structural data have provided novel insights into how all four members of the EGFR family differ with respect to the mode of activation:

In the absence of ligand binding, the majority ERBB-1 exists in a tethered, autoinhibited form in which the dimerization loop of subdomain II is buried by an intermolecular interaction with domain IV. Ligands bind bivalently to the leucine-rich repeats in subdomains I and III. Thereby, ERBB-1 ligands trap the receptor in an untethered conformation, which dimerizes with other receptors through the exposed dimerization arm in a 2:2 receptor/ligand stoichiometry, thus driving the equilibrium

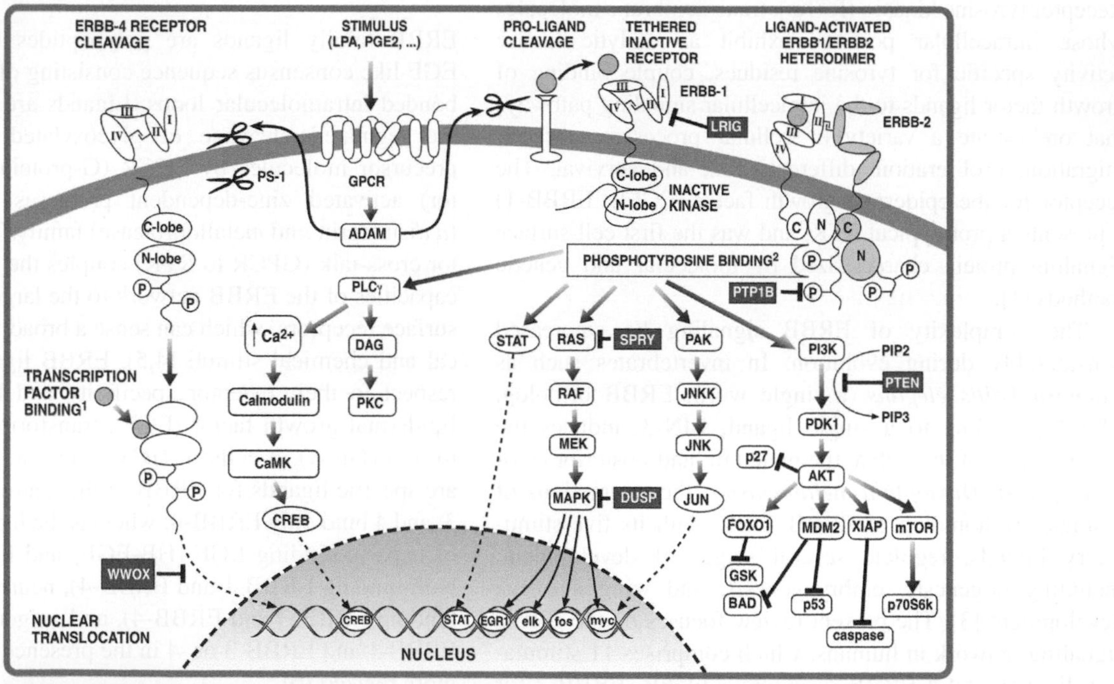

FIGURE 14.1 Mechanisms of ERBB family receptor activation and main routes of signal transduction.
ERBB receptor activation occurs predominantly through ligand binding (top right), but proteolytic cleavage of the extracellular domain (top left) resulting in nuclear translocation of receptors (bottom left) represents an alternative mechanism. ERBB activation induces phosphorylation of tyrosine residues in the cytoplasmatic domain (top right), which serve as docking sites for phosphotyrosine - binding proteins (see below) that link ERBB receptors to pathways essential for cellular motility (e.g., PLC-γ, PKC), proliferation (e.g., RAS–RAF–MAPK, JNKK–JNK–JUN) and cell survival (e.g., PI3K–PDK1–AKT). The main inhibitors of ERBB-induced signaling are depicted in dark gray boxes. They include inhibitors transcriptionally induced by ERBB signaling (e.g., LRIG, SPRY and DUSPs).
[1]Nuclear translocation of receptors induces nuclear import of transcription factors YAP1 (ERBB-4) and STAT (EGFR, ERBB-4).
[2]Phosphotyrosine binding proteins (with SH2 or PTB domains) linking specific phosphorylated tyrosine residues on ERBB receptors to signal transduction pathways include PLC-γ, CBL, GRB2, GAP1, SHC, SHP1, CHK1, p85 subunit of PI3K, and GRB7.

of receptors to a state of dimerization [7]. Whereas ligand-induced activation of ERBB-3 and ERBB-4 resembles that of ERBB-1, the ligand-less ERBB-2 constitutively boasts a structure resembling the active conformation of ERBB-1, with an exposed dimerization loop poised to interact with heterodimerization partners [8].

The conformational transition of extracellular domains is relayed to the intracellular receptor portion, leading to removal of the inhibitory C-terminal tail from the ATP binding cleft, juxtaposition of the C-lobe of the kinase domain of one receptor molecule of the receptor dimer with the N-lobe of the other dimerization partner, and consecutive auto- or transphosphorylation of terminal tyrosine residues [9]. ERBB-3 is devoid of intrinsic kinase activity, and thus its activation is dependent upon heterodimerization with other members of the EGFR family.

Phosphorylated tyrosine residues on the C-terminal tail of ERBB receptors serve as docking points for signaling effectors or adaptor proteins containing Src homology 2 (SH2) or phosphotyrosine binding (PTB) domains, thereby linking the receptors to intracellular signaling pathways and to the endocytotic machinery (see below).

In addition to ligand-mediated activation of the ERBB network, ligand-independent ERBB signaling can occur by two different mechanisms: First, metalloprotease-mediated cleavage of the extracellular receptor domain can lead to either truncated or constitutively activated receptors [10], or nuclear translocation and signaling of cytoplasmic receptor domains [11–13]. Second, unphysiological stimuli (e.g., oxidative and mechanical stress, UV light or gamma-radiation) and activation of chemokine receptors can induce EGFR transmodulation mostly by stimulating non-receptor kinases (e.g. FAK, p38, PKC, PYK2, or SRC) or by inactivating inhibitory phosphatases, thereby shifting the basal activation/deactivation equilibrium of ERBB receptors [5,14].

ERBB-INDUCED SIGNALING PATHWAYS

Signaling pathways induced through ERBBs are dictated by the phosphorylation pattern of cytoplasmic receptor tyrosine residues, as the specificity of SH2 and PTB domain binding is conferred by amino acids surrounding the tyrosine phosphorylation site. Because the members of the ERBB family differ with respect to their tyrosine phosphorylation sites, the identity of recruited pathways critically depends upon the identity of ligand and receptor complex composition. This layered information processing module allows for the combinatorial interactions of ligands and receptors to phosphorylate specific tyrosine sites, thereby determining which signaling proteins are engaged [15].

All ERBB ligands and receptors couple to activation of the RAS-MAPK pathway through recruitment of adaptor proteins such as growth factor-receptor bound-2 (GRB-2), SRC-homology-2 containing (SHC), and the GRB-2-bound exchange factor son of sevenless (SOS). Activation of phosphatidylinositol 3-kinase (PI3K) is differentially induced: ERBB-3 contains six putative binding sites for the SH2 domain of the PI3K p85 regulatory subunit, while ERBB-4 contains one binding site. EGFR and ERBB-2 couple to PI3K indirectly through adaptor proteins such as GAB1 and c-CBL. Thus, ERBBs induce the proliferative and survival signals resulting from the activation of PI3K and its downstream effectors, such as AKT and p70 S6 kinase, with differing potencies and kinetics. Other signaling effectors are recruited only to some ERBB family members, such as the recruitment of phospholipase C (PLC-gamma) to EGFR and ERBB-2. PLC-gamma activation results in the hydrolysis of phosphatidylinositol-4,5-bisphosphate and the generation of the second messengers diacylglycerol and inositol trisphosphate, leading to activation of protein kinase C (PKC) and increases in intracellular calcium concentrations. Signaling through EGFR involves multiple additional pathways: c-SRC is activated upon stimulation with EGF and phosphorylates two tyrosine residues of EGFR implicated in mitogenesis. Furthermore, c-SRC phosphorylates and activates signal transducer and activator of transcription (STAT) family transcription factors, cytoskeletal proteins, and proteins involved in endocytosis (see below). The FAK and PYK2 kinases are implicated in EGF-induced cell migration, and link EGFR to integrin signaling pathways. The JNK pathway is also activated by ERBBs, likely through the adaptor protein CRK (v-CRK sarcoma virus CT10 oncogene homolog), DBL family exchange factors, and the small GTPases RAC1 and CDC42 (cell division cycle-42). Further, after nuclear translocation, ERBB receptors may function as transcription factors, or contranscriptional regualators, respectively (Figure 14.1) [11,12].

These pathways funnel into the rapid induction of transcriptional activation of a set of immediate early genes including prototype oncogenes, such as the AP-1 components FOS and JUN and the zinc finger transcription factor EGR1 that peak 20–40 minutes after stimulation with EGF [16].

Recent phosphoproteomic approaches have identified potential additional tyrosine phosphorylation sites on intracellular residues of ERBB receptors, and demonstrated that the binding specificity of tyrosine residues with synthetic SH2 and PTB domains may be more relaxed than previously anticipated [17]. Awaiting further validation, these results may imply that the promiscuity of tyrosine-containing domains may increase in parallel with receptor abundance, thereby potentially linking disease states involving enhanced ERBB receptor activation (see below) to more potent and broad recruitment of secondary signaling pathways.

SPECIFICITY OF SIGNALING THROUGH THE ERBB NETWORK

The specificity of signaling through ERBB receptors is regulated at multiple levels. The distinct organ- and developmental stage-specific expression profiles of ERBB receptors and ligands regulate biological responses throughout development and adulthood by influencing ERBB homo- or heterodimer formation, as well as the identity of the phosphorylation sites within individual ERBBs [18]. Ligand affinity influences signal strength and duration: low affinity ligands (EPG, EPR, AR, and virally encoded low-affinity ligands) gain high potency by evading physiological mechanisms of signaling desensitization due to receptor downregulation [6] (see below). Further, the pH stability of the ligand–receptor interaction influences receptor trafficking; the pH-resistant interaction of EGFR with EGF targets the receptor to the lysosome, whereas TGF-α and NRG-1 dissociate from their receptors in early endosomes, thus favoring receptor recycling and signal potentiation.

Dimerization of ERBBs adds an additional level of signal diversification. Homodimeric receptor combinations are less potent and mitogenic than heterodimers. The critical role of heterodimers is most clearly demonstrated in mice with targeted deletions in individual ERBBs (see Table 14.1). The defective heart formation phenotype of both ERBB-2 and ERBB-4 knockout mice demonstrates that ERBB homodimers cannot functionally substitute for receptor heterodimers. Although ERBB-2 does not bind directly to known ligands, it is the preferred heterodimer partner for other ERBBs. ERBB2-containing heterodimers are the most potent complexes, due to an increased affinity of ligand binding, relaxed ligand specificity, decreased ligand dissociation, decreased rate of endocytosis, and increased receptor recycling, as well as highly promiscuous PTB and SH2 binding sites [15,17].

ATTENUATION OF THE ERBB SIGNALING NETWORK

Concomitant with receptor activation, signal attenuation is initiated by both transcription-independent and transcription-based negative feedback initiated by both preexisting and transscriptionally-induced negative regulators, thereby defining the specificity, duration, and amplitude of RTK signaling. The major route of transcription-independent signal attenuation, receptor endocytosis, is reviewed below and depicted in Figure 14.2. In addition, transcription-independent signal attenuation is also achieved by dephosphorylation, including recruitment of tyrosine phosphatases such as density-enhanced phosphatase-1 (DEP1) and protein tyrosine phosphatase-1B (PTP1B).

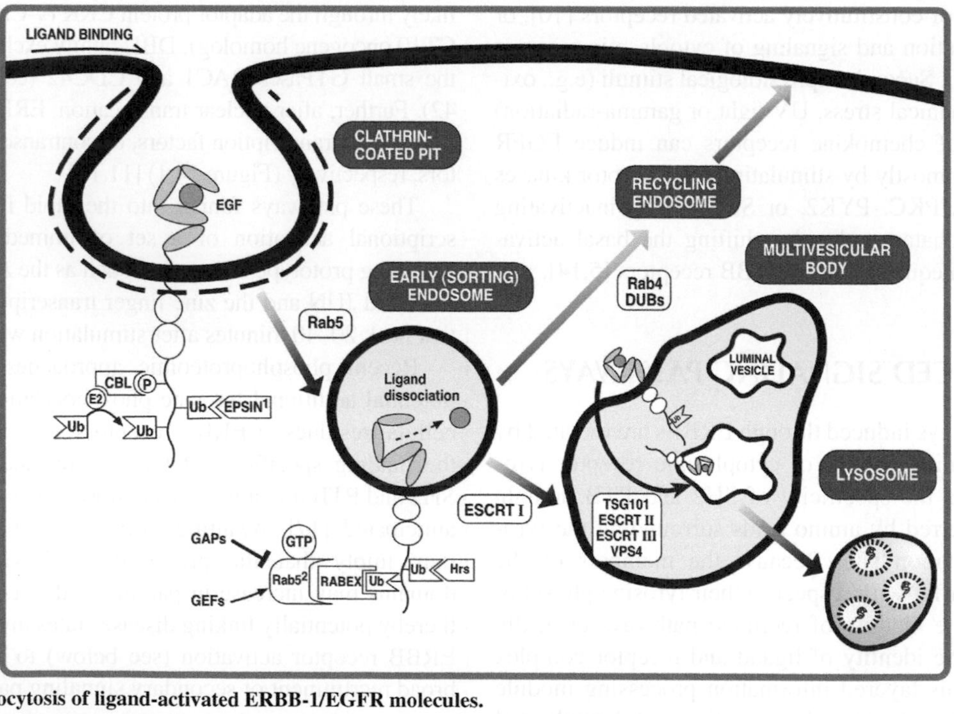

FIGURE 14.2 Endocytosis of ligand-activated ERBB-1/EGFR molecules.
Ligand binding induces receptor segregation in clathrin-coated membrane domains, which invaginate to form clathrin-coated pits and clathrin-coated vesicles (top left). In parallel, CBL binds to phosphorylated tyrosine residue 1045 and induces receptor ubiquitylation, which is required for receptor sorting by the endocytotic machinery in both early endosomes (middle) and the multivesicular body (MVB; bottom right). Ligand-dissociation in the mildly acidic pH of early endosomes and receptor de-ubiquitylation by de-ubiquitylating enzymes (DUBs) can lead to receptor recycling (top right). Alternately, receptors are sorted from early endosomes to the luminal vesicles in the MVB leading to lysosomal ERBB receptor degradation.
[1]Epsin is chosen as an example for several activated and recruited proteins including the AP-2 complex, Endophilin, Dynamin, and Eps15 that link ubiquitylated receptors to early steps of endocytosis.
[2]Rab5 complex: Rab5, EEA1, Rabex5, Rabaptin-5.

Receptor endocytosis

RTK endocytosis represents a highly regulated process in which activated receptors are internalized, directed from the early endosome to the multivesicular body (MVB) and, ultimately, degraded in lysosomes. Alternately, recycling of RTKs to the plasma membrane by a default recycling pathway can occur after de-ubiquitylation or dissociation of ligands in the acidic endosomal environment. For a detailed overview on ERBB endocytosis see [19,20], and ligand-induced ERBB-1 endocytosis will be briefly exemplified herein (see Figure 14.2). Notably, the other members of the ERBB family are less efficiently coupled to the degradative endocytotic machinery, which contributes to the increased signaling potency of homo- and heterodimers containing ERBB-2, ERBB-3, and ERBB-4. In addition, ERBB-1 activated by ligand-independent mechanisms exhibits a different endocytotic fate [14,19].

In resting cells, a large proportion of EGFR is localized in plasma membrane microdomains termed caveolae, which are lipid rafts enriched in caveolin proteins, glycosphingolipids, and cholesterol. Caveolin-1 interacts with EGFR and inhibits its catalytic activity. Activation of RTKs and receptor-mediated activation of SRC family kinases allows EGFR to migrate out of caveolae and aggregate over clathrin-coated regions of the plasma membrane (clathrin-coated pits); however, some ERBB-1-induced signaling pathways and non-clathrin-mediated internalization also occur in caveolae. At the plasma membrane, ligand-activated ERBB-1 directly (via phosphorylation of tyrosine residue 1045) or indirectly (via GRB2) recruits the ubiquitin E3 ligase CBL [21], which attaches ubiquitin monomers as well as oligomers to lysine residues of the receptor, thereby tagging the receptor for internalization and linking it to dynamin-binding endophilins. In parallel, ERBB-1 activates components of the endocytotic machinery by various mechanisms, including coupling to NEDD4/AIP4 ubiquitin ligases (which ubiquitylate adaptors such as Epsin and EPS15 necessary for the assembly of a clathrin coat) and SRC kinase-induced phosphorylation of coat proteins such as clathrin heavy chain and dynamin. Ubiquitylated ERBB-1 is sorted by a set of ubiquitin-interacting motif (UIM) containing adaptors (including HRS/HGS and STAM/EAST) at the early endosome, and later into the luminal vesicles of the MVB (a prelysosomal compartment), where sorting is executed by Tsg101, Tal, Vps4 and endosomal complexes required for transport (ESCRTs) [19,22].

Transcription-Induced Negative Regulators

Stimulation of ERBB signaling induces upregulation of positive and negative regulatory genes, including a transcriptional module of delayed early genes responsible for signal attenuation, as well as later acting transcriptional modules that decouple the ERBB network from repeated stimulation. The transcriptional module of delayed early genes (peaking 40–240 minutes after stimulation), includes nucleotide binding proteins (e.g., NAB2, FOSL1, and JUNB) that are recruited into existing transcriptional complexes of immediate early genes to attenuate their transcriptional activity, and RNA binding proteins (e.g., ZFP36) specifically terminating translation of mRNAs induced by immediate early genes. In addition, delayed early genes comprise newly transcribed dual-specificity phosphatases (DUSPs), which inhibit core signaling pathways downstream of ERBB receptors, including p38 MAPK, JNK, and ERK [16].

The induction of transcriptional modules that desensitize the ERBB network from repeated stimulation peaks 2–3 hours after stimulation, and includes signal attenuators predominantly acting at the level of receptors or their interface with recruited signaling pathways: Sprouty-2 (SPRY), mitogen-inducible gene-6/receptor-associated late transducer (MIG6/RALT), leucine-rich repeats and immunoglobulin-like domains-1 (LRIG1) and suppressor of cytokine signaling-5 (SOCS-5) proteins downregulate ERBB signaling by linking receptors to the degradative endocytotic pathway, as well as by uncoupling receptors from secondary signaling pathways [23,24].

ERBB PROTEINS AND PATHOLOGICAL CONDITIONS

Due to their widespread expression and signaling potency, ERBB molecules are involved in a variety of physiological processes (e.g., myelination, implantation, wound healing, mammary development, and angiogenesis). Accordingly, dysregulated ERBB signaling has been implicated in such different pathological states as respiratory disease (e.g., airway inflammation and asthma), gastric ulcer disease, cardiomyopathy, inflammatory hyperproliferative diseases such as psoriasis, and demyelinating diseases such as multiple sclerosis and leprosy [25]. However, the best studied is the oncogenic aspect of the ERBB network in human malignancies. ERBBs were first implicated in cancer upon the characterization of an aberrant form of EGFR encoded by the avian erythroblastosis tumor virus. Particularly, EGFR and ERBB-2 have since been causally implicated in most forms of human cancers, and therefore represent *bona fide* oncogenes. Abnormal activation of these receptors occurs through various mechanisms: Overexpression by transcriptional upregulation or gene amplification (e.g., ERBB-2 gene amplification occurs in approximately one-fourth of breast cancers [26]), constitutive activation of mutant receptors (e.g., activating mutations in the EGFR tyrosine kinase domain in non-small cell lung cancers or deletion of most of the extracellular domain in glioblastomas), or autocrine and paracrine growth factor loops. These perturbations have not only been implicated in resistance to conventional therapeutic modalities (e.g., resistance of EGFR- overexpressing head and neck cancers to radiation therapy, and resistance of ERBB-2 overexpressing breast cancers to endocrine therapy), but have also been shown to be promising targets for specific therapeutic interventions ("targeted therapies"): Addition of monoclonal antibodies targeting EGFR has been shown to improve cure rates of radiotherapy in patients with head and neck cancer, and to improve the therapeutic efficacy of cytotoxic chemotherapy in patients with colorectal cancer, respectively. Likewise, combinations of a monoclonal antibody targeting HER2/neu with chemotherapy have been shown to improve cure rates and to prolong survival of patients with early-stage and metastatic breast cancer, respectively [27,28]. In the same vein, small molecule inhibitors of the tyrosine kinase domain of EGFR have enriched the therapeutic armamentarium available for patients with lung or pancreatic cancer. Likewise, a dual specificity kinase inhibitor (targeting both EGFR and HER2/*neu*) has shown promising efficacy in patients with ERBB-2 overexpressing breast cancer.

A more complete understanding of the ERBB network may allow for the development of additional therapeutic strategies that might be of clinical benefit in a variety of pathological conditions.

ACKNOWLEDGEMENTS

Our laboratory is supported by research grants from the National Cancer Institute, the Israel Science Foundation, the Israel Cancer Research Fund, the Prostate Cancer Foundation and the German-Israel Foundation. Yosef Yarden is the incumbent of the Harold and Zelda Goldenberg Professorial Chair.

REFERENCES

1. Ullrich A, Coussens L, Hayflick JS, Dull TJ, Gray A, Tam AW, et al. Human epidermal growth factor receptor cDNA sequence and aberrant expression of the amplified gene in A431 epidermoid carcinoma cells. *Nature* 1984;**309**(5967):418–25.
2. Moghal N, Sternberg PW. The epidermal growth factor system in caenorhabditis elegans. *Exp Cell Res* 2003;**284**(1):150–9.
3. Shilo BZ. Signaling by the drosophila epidermal growth factor receptor pathway during development. *Exp Cell Res* 2003;**284**(1):140–9.
4. Prenzel N, Zwick E, Daub H, Leserer M, Abraham R, Wallasch C, et al. EGF receptor transactivation by G-protein-coupled receptors requires metalloproteinase cleavage of proHB-EGF. *Nature* 1999;**402**(6764):884–8.
5. Fischer OM, Hart S, Gschwind A, Ullrich A. EGFR signal transactivation in cancer cells. *Biochem Soc Trans* 2003;**31**(Pt 6):1203–8.
6. Kochupurakkal BS, Harari D, Di-Segni A, Maik-Rachline G, Lyass L, Gur G, et al. Epigen, the last ligand of ErbB receptors, reveals intricate relationships between affinity and mitogenicity. *J Biol Chem* 2005;**280**(9):8503–12.
7. Burgess AW, Cho HS, Eigenbrot C, Ferguson KM, Garrett TP, Leahy DJ, et al. An open-and-shut case? Recent insights into the activation of EGF/ErbB receptors. *Mol Cell* 2003;**12**(3):541–52.
8. Garrett TP, McKern NM, Lou M, Elleman TC, Adams TE, Lovrecz GO, et al. The crystal structure of a truncated ErbB2 ectodomain reveals an active conformation, poised to interact with other ErbB receptors. *Mol Cell* 2003;**11**(2):495–505.
9. Zhang X, Gureasko J, Shen K, Cole PA, Kuriyan J. An allosteric mechanism for activation of the kinase domain of epidermal growth factor receptor. *Cell* 2006;**125**(6):1137–49.
10. Codony-Servat J, Albanell J, Lopez-Talavera JC, Arribas J, Baselga J. Cleavage of the HER2 ectodomain is a pervanadate-activable process that is inhibited by the tissue inhibitor of metalloproteases-1 in breast cancer cells. *Cancer Res* 1999;**59**(6):1196–201.
11. Lin SY, Makino K, Xia W, Matin A, Wen Y, Kwong KY, et al. Nuclear localization of EGF receptor and its potential new role as a transcription factor. *Nat Cell Biol* 2001;**3**(9):802–8.
12. Ni CY, Murphy MP, Golde TE, Carpenter G. gamma -Secretase cleavage and nuclear localization of ErbB-4 receptor tyrosine kinase. *Science* 2001;**294**(5549):2179–81.
13. Giri DK, Ali-Seyed M, Li LY, Lee DF, Ling P, Bartholomeusz G, et al. Endosomal transport of ErbB-2: mechanism for nuclear entry of the cell surface receptor. *Mol Cell Biol* 2005;**25**(24):11,005–11,018.
14. Zwang Y, Yarden Y. p38 MAP kinase mediates stress-induced internalization of EGFR: implications for cancer chemotherapy. *EMBO J* 2006;**25**(18):4195–206.
15. Yarden Y, Sliwkowski MX. Untangling the ErbB signaling network. *Nat Rev Mol Cell Biol* 2001;**2**(2):127–37.
16. Amit I, Citri A, Shay T, Lu Y, Katz M, Zhang F, et al. A module of negative feedback regulators defines growth factor signaling. *Nat Genet* 2007;**39**(4):503–12.
17. Jones RB, Gordus A, Krall JA, MacBeath G, et al. A quantitative protein interaction network for the ErbB receptors using protein microarrays. *Nature* 2006;**439**(7073):168–74.
18. Olayioye MA, Graus-Porta D, Beerli RR, Rohrer J, Gay B, Hynes NE, et al. ErbB-1 and ErbB-2 acquire distinct signaling properties dependent upon their dimerization partner. *Mol Cell Biol* 1998;**18**(9):5042–51.
19. Gur G, Zwang Y, Yarden Y. Endocytosis of receptor tyrosine kinases: implications forsignal transduction by growth factors. In: Dikic I, editor. *Endosomes*: Eurekah Bioscience Database; 2006.
20. Waterman H, Yarden Y. Molecular mechanisms underlying endocytosis and sorting of ErbB receptor tyrosine kinases. *FEBS Letts* 2001;**490**(3):142–52.
21. Levkowitz G, Waterman H, Ettenberg SA, Katz M, Tsygankov AY, Alroy I, et al. Ubiquitin ligase activity and tyrosine phosphorylation underlie suppression of growth factor signaling by c-Cbl/Sli-1. *Mol Cell* 1999;**4**(6):1029–40.
22. Amit I, Yakir L, Katz M, Zwang Y, Marmor MD, Citri A, et al. Tal, a Tsg101-specific E3 ubiquitin ligase, regulates receptor endocytosis and retrovirus budding. *Genes Dev* 2004;**18**(14):1737–52.

23. Kario E, Marmor MD, Adamsky K, Citri A, Amit I, Amariglio N, et al. Suppressors of cytokine signaling 4 and 5 regulate epidermal growth factor receptor signaling. *J Biol Chem* 2005;**280**(8):7038–48.
24. Gur G, Rubin C, Katz M, Amit I, Citri A, Nilsson J, et al. LRIG1 restricts growth factor signaling by enhancing receptor ubiquitylation and degradation. *EMBO J* 2004;**23**(16):3270–81.
25. Tapinos N, Ohnishi M, Rambukkana AE. ErbB2 receptor tyrosine kinase signaling mediates early demyelination induced by leprosy bacilli. *Nat Med* 2006;**12**(8):961–6.
26. Slamon DJ, Godolphin W, Jones LA, Holt JA, Wong SG, Keith DE, et al. Studies of the HER-2/neu proto-oncogene in human breast and ovarian cancer. *Science* 1989;**244**(4905):707–12.
27. Piccart-Gebhart MJ, Procter M, Leyland-Jones B, Goldhirsch A, Untch M, Smith I, et al. Trastuzumab after adjuvant chemotherapy in HER2-positive breast cancer. *N Engl J Med* 2005;**353**(16):1659–72.
28. Slamon DJ, Leyland-Jones B, Shak S, Fuchs H, Paton V, Bajamonde A, et al. Use of chemotherapy plus a monoclonal antibody against HER2 for metastatic breast cancer that overexpresses HER2. *N Engl J Med* 2001;**344**(11):783–92.

Chapter 15

Epidermal Growth Factor Kinases and their Activation in Receptor Mediated Signaling

Andrew H. A. Clayton
Ludwig Institute for Cancer Research, Royal Melbourne Hospital, Victoria, Australia

INTRODUCTION

The receptor for epidermal growth factor (EGFR or HER1 or erbB1) [1, 2] is a member of the receptor tyrosine kinase (RTK) family and has been the subject of intensive investigation over the last three decades [3, 4], in part as a prototype for RTK activation and signaling [5], and in part because of its involvement in cancer initiation and progression [6]. It was recognized over two decades ago that an oncogenic chicken virus encodes a constitutively activated form (v-erbB) of EGFR [7]. Since that time, three other ErbB receptor family members have been discovered (Table 15.1) – ErbB2 [8–11], ErbB3 [12, 13], and ErbB4 [14, 15] – together with 11 ligands (peptide ligands; Table 15.2) that together generate a complex signaling network which plays a central role in development, homeostasis, and diseases such as cancer [6], inflammation [16], psoriasis [17], heart disease [18], and Alzheimer's disease [19]. The purpose of this chapter is to give an overview of EGFR signaling and activation, with particular emphasis on structural biology and cell biophysical studies that have emerged in the past 6 years.

The domain architecture of the EGFR is analogous to the other three members (Figure 15.1) [4]. Each receptor consists of a glycosylated and disulfide-bonded ectodomain that recognizes ligand, a single transmembrane domain, and a large cytoplasmic region that contains a juxtamembrane region, a tyrosine kinase domain, and a C-terminal cytoplasmic tail with multiple phosphorylation sites. Binding of ligand to the EGFR initiates oligomerization, which is required for activation and signaling. ErbB2 differs from EGFR in that it has no known ligand but can function in an oligomeric complex with the other ErbB members [20]. ErbB3 binds a different subset of ligands and is enzymatically inactive because of point mutations, and requires heteroligomerization with erbB2, erbB1 or erbB4 to activate signaling [21]. ErbB4 is a ligand-activated kinase, but is cleaved by membrane proteases, and the activated intracellular kinase domain moves into the cell nuclear membrane where it is reported to signal [22].

TABLE 15.1 EGFR (or ErbB) family of receptors

Receptor	Comment	Original reference	Recent Review(s)
EGFR (erbB1,HER1)	Ligand activated tyrosine kinase	Carpenter *et al.* (1975) [1]	Burgess (2008) [3]
		Ullrich *et al.* (1984) [2]	Ferguson (2008) [4]
erbB2 (HER2, neu)	No known ligand	Stern (1986) [8]	Landgraf (2007) [10]
		Bargmann and Weinberg (1988) [9]	Warren and Landgraf (2006) [11]
erbB3 (HER3)	Inactive kinase	Plowman *et al.* (1990) [12]	Sithanandam and Anderson (2008) [13]
erbB4 (HER4)	Kinase cleaved after activation	Plowman *et al.* (1993) [14]	Carpenter (2003) [15]

Handbook of Cell Signaling, Three-Volume Set 2 ed.

TABLE 15.2 Binding of ligands to ErbB receptors

	Receptor			
Ligand	**EGFR**	**ErbB2**	**ErbB3**	**ErbB4**
EGF	Yes	No	No	No
TGF-α	Yes	No	No	No
HB-EGF	Yes	No	No	Yes
Amphiregulin	Yes	No	No	No
Betacellulin	Yes	No	No	Yes
Epigen	Yes	No	No	No
Epiregulin	Yes	No	No	Yes
Neuregulin-1	No	No	Yes	Yes
Neuregulin-2	No	No	Yes	Yes
Neuregulin-3	No	No	No	Yes
Neuregulin-4	No	No	No	Yes

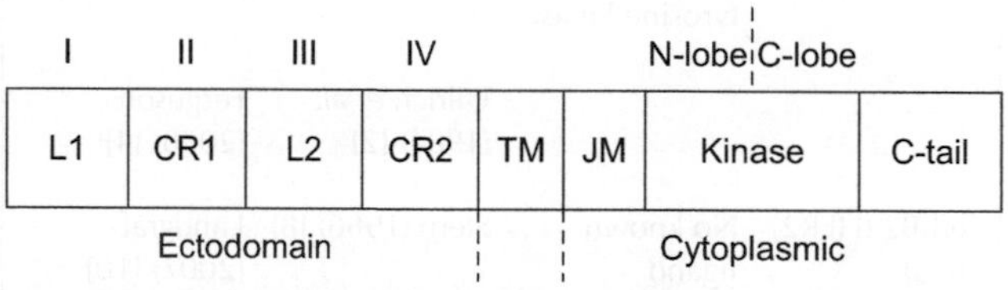

FIGURE 15.1 Domain structure of EGFR family.
The receptors have an ectodomain characterized by two cysteine-rich regions (CR1 and CR2 or domains I and II). The ligand binding site is formed by domains I and III (or L1 and L2). There is a single transmembrane helix domain (TM), followed by a juxtamembrane region (JM), a tyrosine kinase domain (with N- and C-lobes), and a C-terminal (CT) region containing autophosphorylation sites.

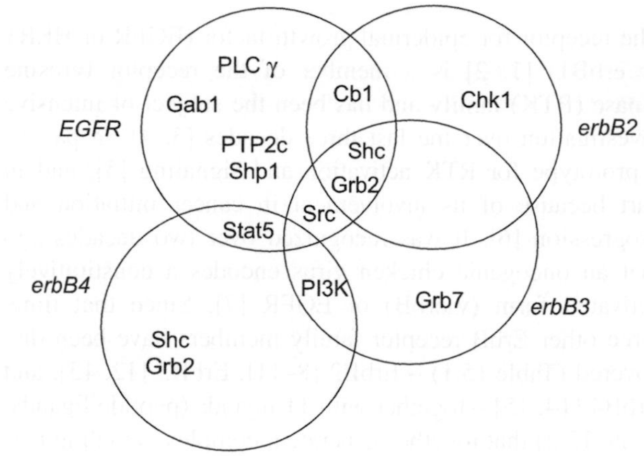

FIGURE 15.2 Major proteins recruited to phosphorylation sites of erbB intracellular domains.
All erbBs have shc/grb2 binding sites to activate MAPK pathway. ErbB3 has the most binding sites for PI3 kinase and is an important co-receptor (with EGFR or erbB2) to sustain Akt signaling resulting in cell survival.

EGFR SIGNALING NETWORK PATHWAYS

Ligand binding to the ectodomain ultimately leads to activation of the cytoplasmic kinase domain and "auto"-phosphorylation of tyrosines at the C-terminal tail. These phosphorylated tyrosines act as docking sites for cytoplasmic adaptors, and second messengers that direct transcriptional program that alter cellular responses. The major pathways [3, 23, 24] induced by activated ErbBs in normal cells are the MAP/Ras kinase, phospholipase Cγ, *s*ignal *t*ransducer and *a*ctivator of *t*ranscription (STAT) and phosphatidylinositol (PtdIns)-3-kinase. Figure 15.2 contains a summary of the major proteins that are recruited to ErbB receptors. All ErbBs couple to the Ras-MAPK pathway through binding to adaptor protein Grb-2 and/or Shc [23].

ErbB3 has numerous PI3-kinase binding sites to initiate Akt/mTOR signaling, while EGFR/erbB2 can couple to the PI3-kinase pathway indirectly through binding to adaptors Gab1 and c-Cbl [25]. Some signaling pathways are controlled by effector binding to specific receptors, such as STAT5 to EGFR/erbB4, src to erbB2/EGFR, and PLCγ to EGFR. These different binding preferences for intracellular adaptors give each receptor a different propensity to activate particular pathways in the cell.

Several other signaling systems are modulated by EGFR, or can modulate EGFR signaling [3] – for example, the wnt [26], steroid receptors [27], and PKC-associated pathways [28, 3]. Signaling regulation is also controlled by interactions and cross-talk with other EGFR family members, signaling from other receptor tyrosine kinase systems (e.g. PDGFR [29]) or receptor-associated kinases). A host of multiple negative regulators [30] (phosphatases, ubiquitin ligases, adaptors, RALT (Mig6, gene 33)) can also act to modulate EGFR signaling, and inhibition of their activity may well underlie some aspects of EGFR activation itself.

After activation, erbB receptors are internalized and degraded in the cell [31]. EGFR is often internalized and degraded relatively quickly. However, erbB2 and erbB3 appear to internalize more slowly, and may recycle for further participation in signaling [32, 33].

STRUCTURAL BIOLOGY OF RECEPTOR FRAGMENTS

At present there are no high-resolution structural data on the intact full-length EGFR in a lipid bilayer environment. However, in the past decade there has been a wealth of structural data [4] on the extracellular domain fragment of all four monomeric erbB receptors in the unliganded state, the dimeric extracellular domain of EGFR in complex with EGF or TGFα, and monomeric extracellular domains of erbB2 or EGFR in complex with antibody fragments of therapeutic interest (see [4] and references therein). The structure of the intracellular kinase domain in different activation states has also been determined [4].

CONFORMATIONS OF THE ECD FRAGMENTS OF ErbB RECEPTORS

The ECD regions of EGFR family members share common features – two ligand binding domains (domains I and III or L1 and L2) and two cysteine-rich domains (domains II and IV or CR1 and CR2). EGFR, erbB3, and erbB4 appear to have two conformations in solution – a tethered conformation and an untethered one (Figure 15.3a). In the unliganded state of EGFR [34], erbB3 [35], and erbB4 [36], the monomeric receptors exist in the tethered intramolecular conformation in which a dimerization motif is prevented from mediating receptor–receptor interaction by a "tether" between domains II and IV [37]. Unliganded erbB2 [38] and liganded EGFR [34] exist in the untethered conformation in which the domain II/IV tether is released and a dimerization arm is exposed ready to oligomerize with erbB2 or the other three liganded ErbB members. In the liganded EGFR back-to-back dimer [39, 40], the interaction is mediated

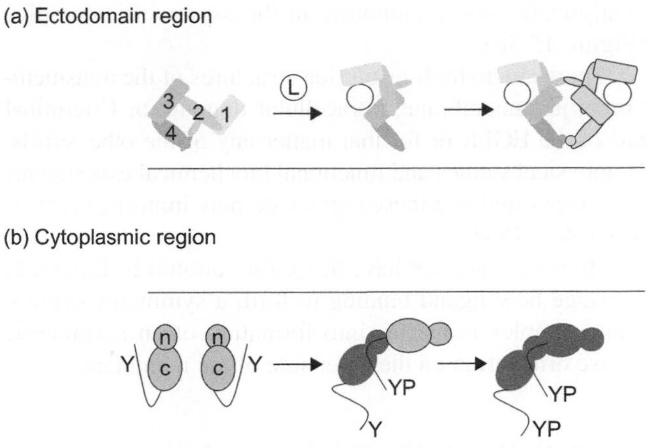

FIGURE 15.3 Structures of ectodomain and kinase domains of EGFR. Schematic representation of the major structural forms of the ectodomain (tethered monomer, untethered monomer, back-to-back dimer) (a) and kinase domain fragments (inactive symmetric dimer, active asymmetric dimer) (b) of erbB receptors, as deduced from high-resolution X-ray crystallography studies.

exclusively through receptor–receptor contacts (between dimerization arms on each receptor monomer) and held in a defined orientation by the simultaneous binding of domains I and III in each receptor to ligand. This liganded dimer is symmetric in the sense that the two liganded monomer halves of the receptor are identical. The receptor-mediated dimer structure can be contrasted with other RTK and cytokine receptors which show that the ligand effectively crosslinks the two monomers into the dimeric complex [41].

Neither the conformational status of the EGFR ectodomain (tethered or untethered) in full length receptors on the cell surface nor the dynamics of interconversion between the conformational states in living cells have been measured directly as yet [3].

KINASE DOMAIN FRAGMENT STRUCTURES

Ligand-induced dimerization or oligomerization was originally proposed to facilitate a high local concentration of cytoplasmic tails, thereby promoting "auto"/transphosphorylation [42]. Structural studies with and without inhibitor, as well as modeling studies performed a decade earlier, suggest that oligomerization facilitates kinase activation via formation of an asymmetric kinase dimer [43, 44]. In this mechanism, two EGFR tyrosine kinase domains form an asymmetric dimer in which the C-lobe of a catalytically inactive monomer contacts the N-lobe of the other monomer and stabilizes it in a catalytically active conformation through displacement of the kinase domain activation loop. Transphosphorylation of tyrosine residues on the catalytically inactive monomer is facilitated by interaction with the kinase domain of the catalytically active monomer in the asymmetric complex (Figure 15.3b).

There are no high-resolution structures of the transmembrane, juxtamembrane, intracellular domain, or C-terminal tail of the EGFR or for that matter any of the other erbBs. Biophysical studies and functional biochemical experiments have suggested that these regions do play important regulatory roles [45–48].

However, until we have this information it is difficult to envisage how ligand binding to form a symmetric ectodomain complex translates into formation of an asymmetric kinase orientation on the other side of the membrane.

BIOPHYSICAL STUDIES OF ErbB ACTIVATION AT THE CELL SURFACE

Developments in microscopy methods and probes make it possible to measure conformation [49], oligomerization [50], activation [51], dynamics [52], and internalization [53] of EGFR family members on the cell surface. Studies directed at understanding the association state of the EGFR and the other erbBs in unliganded and liganded form on the cell surface lead to some interesting and perhaps surprising results that are providing challenges to our understanding of erbB activation in the cellular context.

ErbB RECEPTORS EXIST AS PREDOMINANTLY PRE-FORMED DIMERS IN CELLS

It is now recognized that the unliganded, inactive EGFR is present in a predominantly pre-dimerized (or pre-oligomerized) state on the surface of cells. Image correlation spectroscopy on a biologically active GFP derivative of EGFR expressed at normal levels determined an average oligomeric state of two in the absence of ligands or other erbB family members [50]. Another study [54], using image cross-correlation spectroscopy, determined that the majority of receptors were in preformed dimers (80 percent dimers) by measuring the coincidence of green (GFP) tagged and red (RFP) tagged receptors as they diffuse through a focused laser beam. Interestingly, the degree of dimerization was found to be independent of receptor expression level over the physiological to pathological range (20,000–260,000 receptors/cell). This observation can be understood in the context of a recent fluorescence complementation study which demonstrated that dimers (or oligomers) of EGFR, erbB2, erbB3, and erbB4 form in the endoplasmic reticulum, and are trafficked and presented to the cell surface as preformed complexes [55]. Dimerization is required, but is clearly not sufficient, for activation of the erbBs.

BEYOND DIMERS: A LIGAND-INDUCED EGFR TETRAMER IS FORMED DURING ACTIVATION

Biophysical studies on cells stably expressing normal levels of EGFR and in the absence of other erbB family members have found another level of complexity in EGFR organization on the cell surface [50, 56]. Image correlation spectroscopy measurements show that EGF induces a dimer-to-tetramer transition under conditions in which the receptor is activated and kept at the cell surface (prior to internalization and signal attenuation) [50]. Measurements of the distance between fluorescently-tagged ligands using Förster Resonance Energy Transfer (FRET)[50] and Förster Resonance homotransfer [56] has revealed an unexpectedly close separation between ligands of approximately 5 nm – a distance that is too short to be consistent with the 11-nm separation of ligands in the back-to-back EGFR dimer structure. Taken together, these studies are consistent with a tetramer formed from two liganded back-to-back structures that are in a side-by-side (or slightly staggered [50, 57]) arrangement (see Figure 15.4). Head-to-head and head-to-tails models have also been invoked [56, 57]. Mutational analysis shows that the head-to-tail contact is not required for signaling [40], and a head-to-head tetramer would be inconsistent with the high fraction of molecules undergoing FRET (the maximum is 50 percent for the head-to-head model compared with >80 percent observed in BaF/3 cells [50]).

Tetramers are formed after ligand binding, and under conditions where internalization is inhibited. However it is important to determine whether tetramers are required for

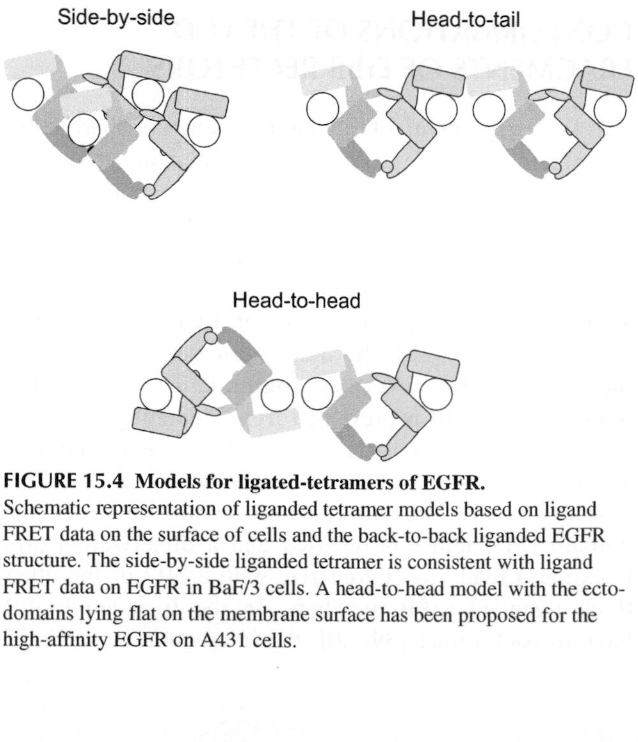

FIGURE 15.4 Models for ligated-tetramers of EGFR. Schematic representation of liganded tetramer models based on ligand FRET data on the surface of cells and the back-to-back liganded EGFR structure. The side-by-side liganded tetramer is consistent with ligand FRET data on EGFR in BaF/3 cells. A head-to-head model with the ectodomains lying flat on the membrane surface has been proposed for the high-affinity EGFR on A431 cells.

initial activation, are required for sustained positive signaling, or are involved in the early steps of signal attenuation. Two recent studies have attempted to answer these questions. In one study, Sako [58] used a single molecule technique to measure the ratio of EGF to activated receptor within clusters on the cell surface. Values greater than two were found, indicating that liganded EGFR complexes could transactivate unliganded EGFR via formation of dynamic higher-order clusters. Another used a hybrid microscopy that combines FRET to detect active (phosphorylated) EGFR-GFP based on transfer to an acceptor-conjugated antibody and image correlation microscopy to measure the relative oligomeric state of EGFR-eGFP, with the result that the activated oligomers are approximately four times larger than the inactive ones [59]. The presence of exposed phosphotyrosines of these oligomers attests to their role in early signaling, prior to full engagement by adaptors. From these studies, it is now clear that the predominant phosphorylated form of the EGFR on the surface of intact cells is the tetramer (and/or higher-order oligomers). Understanding the role of the various oligomeric forms in kinase activation and the quality and magnitude of signaling is an important but presently unanswered question.

ACTIVATION-DEPENDENT HIGHER-ORDER ErbB OLIGOMERS IN CANCER CELLS

Studies of receptor activation and oligomerization in cancer cell lines can be complicated by multiple factors; the presence of mutated receptors, overexpressed receptors, secreted ligands, or combinations, all of which result in constitutive or basal erbB activation. Nevertheless, changes in higher-order oligomerization as a function of serum conditions (level and starvation), added ligand, and added kinase inhibitors have been reported (Table 15.3). In A431 cells which express 1 million–3 million EGFRs on the cell surface, tetramers [57, 60, 61] and higher-order oligomers containing 10–30 receptors per cluster have been observed [60, 52]. The tetramers appear to be preformed, since they are present when a ligand-blocking antibody [60] or serum starvation is used [57, 61]. Addition of ligand increases the cluster size [61], while addition of a kinase inhibitor decreases the cluster size [60], partially disperses the EGFR clusters [60], and alters the conformation of the extracellular domain [60, 62]. Studies of erbB2 in the human breast cancer cell line SKBR-3 (containing 800,000 ErbB2 receptors) reveal larger clusters containing 110 erbB2 chains in low serum medium, and that cluster sizes decrease with

TABLE 15.3 Sizes of erbB oligomers observed on the cell surface

Cell	Type	Receptor	Condition	Oligomer Size	Ref(s)
BaF/3	Hemopoetic	5×10^4 EGFR	No ligand	2	[50]
BaF/3	Hemopoetic	5×10^4 EGFR	EGF	4	[50]
CHO	Fibroblast	EGFR	No ligand	2	[54]
		erbB2	No ligand	2	[54]
		EGFR-erbB2	No ligand	heterodimer	[54]
		2×10^4–2.6×10^5 receptors/cell			
A431	Carcinoma	1–3×10^6 EGFR	Ligand blocked or stripped	4	[60, 61] [52, 60]
				10–30	
A431	As above	As above	EGF added	11	[61]
SKBR3	Carcinoma	8×10^5 erbB2	No ligand	111	[61]
			erbB1,2,3		
SKBR3	As above	As above	EGF	71	[61]
SKBR3	As above	As above	heregulin	32	[61]
MCF7	Carcinoma	2.5×10^4 erbB3	inactive	1, 2 ,3, >3	[63]

either EGF stimulation or neuregulin stimulation [61]. This is consistent with erbB2 forming ligand-dependent heteromers with EGFR (upon EGF stimulation) and erbB3 (with neuregulin stimulation). In stark contrast to EGFR, erbB2 displays an inverse relationship between homo-oligomer cluster size and activation level (as assessed by erbB2 phosphorylation) [61]. In MCF7 breast-cancer cells (expressing 25,000 receptors per cell), higher-order inactive erbB3 oligomers were trapped using photo-crosslinkable aptamers [63]. These inactive erbB3 oligomers were found to be spatially segregated from inactive erbB2, activated erbB2, and activated erbB3 [63]. No cellular data are presently available reporting on the cluster sizes of erbB4 at the cell surface or inside the cell.

NEW PARADIGM IN ErbB ACTIVATION AND SIGNALING

We are not at a stage yet where we can describe all the changes occurring from ligand binding to the ectodomain that lead to activation of the kinase domain on the inside of the cell. However, the picture that emerges from integrating current structural and biophysical data is that erbB activation at the cell surface is a complex process involving multiple conformational (tethered/untethered ectodomain; symmetric/asymmetric kinase) and oligomeric states (varying ligand/receptor stoichiometry and oligomer size). In particular, the existence of preformed dimers and the presence of activated high-order EGFR homo-oligomers and inactive higher-order erbB2 or erbB3 homo-oligomers forces us to rethink erbB activation and signaling beyond the textbook inactive monomer–active dimer model. ErbB oligomers should perhaps be considered in a broader context as dynamic recognition modules whose size and stoichiometry appear to control the quality, timing, and magnitude of the signal being sent to the cell. In that sense the erbB family serve as archetypal horizontal receptors, where the lateral organization on the cell surface provides the scaffold for enzyme activation and effector coupling. Further structural studies and high-resolution fluorescence microscopy approaches will no doubt add further details to this emerging view, but because the complexity of the problem is greatly increased, systems biology approaches will also be needed if we are to unravel the exciting details surrounding the regulation of EGFR family signaling.

REFERENCES

1. Carpenter G, Lembach KJ, Morrison MM, Cohen S. Characterization of the binding of 125-I-labeled epidermal growth factor to human fibroblasts. *J Biol Chem* 1975;**250**(11):4297–304.
2. Ullrich A, Coussens L, Hayflick JS, Dull TJ, Gray A, Tam AW, Lee J, Yarden Y, Libermann TA, Schlessinger J, et al. Human epidermal growth factor receptor cDNA sequence and aberrant expression of the amplified gene in A431 epidermoid carcinoma cells. *Nature* 1984;**309**(5967):418–25.
3. Burgess AW. EGFR family: structure physiology signalling and therapeutic targets. *Growth Factors* 2008;**26**(5):263–74.
4. Ferguson KM. Structure-based view of epidermal growth factor regulation. *Annu Rev Biophys* 2008;**37**:353–73.
5. Schlessinger J. Cell signaling by receptor tyrosine kinases. *Cell* 2000;**103**:211–25.
6. Blume-Jensen P, Hunter T. Oncogenic kinase signaling. *Nature* 2001;**411**:355.
7. Downward J, Yarden Y, Mayes E, Scrace G, Totty N, Stockwell P, Ullrich A, Schlessinger J, Waterfield MD. Close similarity of epidermal growth factor receptor and v-erb-B oncogene protein sequences. *Nature* 1984;**307**:521–7.
8. Stern DF, Heffernan PA, Weinberg RA. p185, a product of the neu proto-oncogene, is a receptor-like protein associated with tyrosine kinase activity. *Mol Cell Biol* 1986;**6**(5):1729–40.
9. Bargmann CI, Weinberg RA. Increased tyrosine kinase activity associated with the protein encoded by the activated neu oncogene. *Proc Natl Acad Sci USA* 1988;**85**(15):5394–8.
10. Landgraf R. HER2 therapy. HER2 (ERBB2): functional diversity from structurally conserved building blocks. *Breast Cancer Res* 2007;**9**(1):202.
11. Warren CM, Landgraf R. Signaling through ERBB receptors: multiple layers of diversity and control. *Cell Signal* 2007;**18**(7):923–33.
12. Plowman GD, Whitney GS, Neubauer MG, Green JM, McDonald VL, Todaro GJ, Shoyab M. Molecular cloning and expression of an additional epidermal growth factor receptor-related gene. *Proc Natl Acad Sci USA* 1990;**87**(13):4905–9.
13. Sithanandam G, Anderson LM. The ERBB3 receptor in cancer and cancer gene therapy. *Cancer Gene Ther* 2008;**15**(7):413–48.
14. Plowman GD, Culouscou JM, Whitney GS, Green JM, Carlton GW, Foy L, Neubauer MG, Shoyab M. Ligand-specific activation of HER4/p180erbB4, a fourth member of the epidermal growth factor receptor family. *Proc Natl Acad Sci USA* 1993;**90**(5):1746–50.
15. Carpenter G. ErbB-4: mechanism of action and biology. *Exp Cell Res* 2003;**284**(1):66–77.
16. Bueter W, Dammann O, Zscheppang K, Korenbaum E, Dammann CE. ErbB receptors in fetal endothelium–a potential linkage point for inflammation-associated neonatal disorders. *Cytokinem* 2006;**36**(5–6):267–75.
17. Wierzbicka E, Tourani JM, Guillet G. Improvement of psoriasis and cutaneous side-effects during tyrosine kinase inhibitor therapy for renal metastatic adenocarcinoma. A role for epidermal growth factor receptor (EGFR) inhibitors in psoriasis? *Br J Dermatol* 2006;**155**(1):213–14.
18. Negro A, Brar BK, Lee KF. Essential roles of Her2/erbB2 in cardiac development and function. *Recent Prog Horm Res* 2004;**59**:1–12.
19. Sardi SP, Murtie J, Koirala S, Patten BA, Corfas G. Presenilin-dependent ErbB4 nuclear signaling regulates the timing of astrogenesis in the developing brain. *Cell* 2006;**127**(1):185–97.
20. Graus-Porta D, Beerli RR, Daly JM, Hynes NE. ErbB-2, the preferred heterodimerization partner of all ErbB receptors, is a mediator of lateral signaling. *EMBO J* 1997;**16**(7):1647–55.
21. Guy PM, Platko JV, Cantley LC, Cerione RA, Carraway KL. III. Insect cell-expressed p180erbB3 possesses an impaired tyrosine kinase activity. *Proc Natl Acad Sci USA* 1994;**191**(17):8132–6.
22. Ni CY, Murphy MP, Golde TE, Carpenter G. Gamma-secretase cleavage and nuclear localization of ErbB-4 receptor tyrosine kinase. *Science* 2001;**294**(5549):2179–81.

23. Schulze WX, Deng L, Mann M. Phosphotyrosine interactome of the ErbB-receptor kinase family. *Mol Syst Biol* 2005;**1**. 2005.0008.
24. Yarden Y, Sliwkowski MX. Untangling the ErbB signalling network. *Nature Rev Mol Cell Biol* 2001;**2**(2):127–37.
25. Soltoff SP, Cantley LC. p120cbl is a cytosolic adapter protein that associates with phosphoinositide 3-kinase in response to epidermal growth factor in PC12 and other cells. *J Biol Chem* 1996;**271**(1):563–7.
26. Kim SE, Choi KY. EGF receptor is involved in WNT3a-mediated proliferation and motility of NIH3T3 cells via ERK pathway activation. *Cell Signal* 2007;**19**(7):1554–64.
27. Silva CM, Shupnik MA. Integration of steroid and growth factor pathways in breast cancer: focus on signal transducers and activators of transcription and their potential role in resistance. *Mol Endocrinol* 2007;**21**(7):1499–512.
28. Santiskulvong C, Rozengurt E. Protein kinase C-alpha mediates feedback inhibition of EGF receptor transactivation induced by Gq-coupled receptor agonists. *Cell Signal* 2007;**19**(6):1348–57.
29. Walker F, Burgess AW. Transmodulation of the epidermal-growth-factor receptor in permeabilized 3T3 cells. *Biochem J* 1988;**256**(1):109–15.
30. Fry WH, Kotelawala L, Sweeney C, Carraway KL. III. Mechanisms of ErbB receptor negative regulation and relevance in cancer. *Exp Cell Res* 2008. Jul 31.
31. Cohen S, Fava RA. Internalization of functional epidermal growth factor:receptor/kinase complexes in A–431 cells. *J Biol Chem* 1985;**260**(22):12,351–12,358.
32. Baulida J, Kraus MH, Alimandi M, Di Fiore PP, Carpenter G. All ErbB receptors other than the epidermal growth factor receptor are endocytosis impaired. *J Biol Chem* 1996;**271**(9):5251–7.
33. Sorkin A, Goh LK. Endocytosis and intracellular trafficking of ErbBs. *Exp Cell Res* 2008;**314**(17):3093–106.
34. Ferguson KM, Berger MB, Mendrola JM, Cho HS, Leahy DJ, Lemmon MA. EGF activates its receptor by removing interactions that autoinhibit ectodomain dimerization. *Mol Cell* 2003;**11**(2):507–17.
35. Cho HS, Leahy DJ. Structure of the extracellular region of HER3 reveals an interdomain tether. *Science* 2002;**297**(5585):1330–3.
36. Bouyain S, Longo PA, Li S, Ferguson KM, Leahy DJ. The extracellular region of ErbB4 adopts a tethered conformation in the absence of ligand. *Proc Natl Acad Sci USA* 2005;**102**(42):15,024–15,029.
37. Burgess AW, Cho HS, Eigenbrot C, Ferguson KM, Garrett TP, Leahy DJ, Lemmon MA, Sliwkowski MX, Ward CW, Yokoyama S. An open-and-shut case? Recent insights into the activation of EGF/ErbB receptors. *Mol Cell* 2003;**12**(3):541–52.
38. Garrett TP, McKern NM, Lou M, Elleman TC, Adams TE, Lovrecz GO, Kofler M, Jorissen RN, Nice EC, Burgess AW, Ward CW. The crystal structure of a truncated ErbB2 ectodomain reveals an active conformation, poised to interact with other ErbB receptors. *Mol Cell* 2003;**11**(2):495–505.
39. Ogiso H, Ishitani R, Nureki O, Fukai S, Yamanaka M, Kim JH, Saito K, Sakamoto A, Inoue M, Shirouzu M, Yokoyama S. Crystal structure of the complex of human epidermal growth factor and receptor extracellular domains. *Cell* 2002;**110**(6):775–87.
40. Garrett TP, McKern NM, Lou M, Elleman TC, Adams TE, Lovrecz GO, Zhu HJ, Walker F, Frenkel MJ, Hoyne PA, Jorissen RN, Nice EC, Burgess AW, Ward CW. Crystal structure of a truncated epidermal growth factor receptor extracellular domain bound to transforming growth factor alpha. *Cell* 2002;**110**(6):763–73.
41. Boulanger MJ, Garcia KC. Shared cytokine signaling receptors: structural insights from the gp130 system. *Adv Protein Chem* 2004;**68**:107–46.
42. Honegger AM, Schmidt A, Ullrich A, Schlessinger J. Evidence for epidermal growth factor (EGF)-induced intermolecular autophosphorylation of the EGF receptors in living cells. *Mol Cell Biol* 1990;**10**(8):4035–44.
43. Groenen LC, Walker F, Burgess AW, Treutlein HR. A model for the activation of the epidermal growth factor receptor kinase involvement of an asymmetric dimer? *Biochemistry* 1997;**36**(13):3826–36.
44. Zhang X, Gureasko J, Shen K, Cole PA, Kuriyan J. An allosteric mechanism for activation of the kinase domain of epidermal growth factor receptor. *Cell* 2006;**125**(6):1137–49.
45. Moriki T, Maruyama H, Maruyama IN. Activation of preformed EGF receptor dimers by ligand-induced rotation of the transmembrane domain. *J Mol Biol* 2001;**311**(5):1011–26.
46. Sato T, Pallavi P, Golebiewska U, McLaughlin S, Smith SO. Structure of the membrane reconstituted transmembrane-juxtamembrane peptide EGFR(622–660) and its interaction with Ca^{2+}/calmodulin. *Biochemistry* 2006;**45**(42):12,704–12,714.
47. Lee NY, Hazlett TL, Koland JG. Structure and dynamics of the epidermal growth factor receptor C-terminal phosphorylation domain. *Protein Sci* 2006;**15**(5):1142–52.
48. Lee NY, Koland JG. Conformational changes accompany phosphorylation of the epidermal growth factor receptor C-terminal domain. *Protein Sci* 2005;**14**(11):2793–803.
49. Jares-Erijman EA, Jovin TM. Imaging molecular interactions in living cells by FRET microscopy. *Curr Opin Chem Biol* 2006;**10**(5):409–16.
50. Clayton AH, Walker F, Orchard SG, Henderson C, Fuchs D, Rothacker J, Nice EC, Burgess AW. Ligand-induced dimer-tetramer transition during the activation of the cell surface epidermal growth factor receptor-A multidimensional microscopy analysis. *J Biol Chem* 2005;**280**(34):30,392–30,399.
51. Wouters FS, Verveer PJ, Bastiaens PI. Imaging biochemistry inside cells. *Trends Cell Biol* 2001;**11**(5):203–11.
52. Keating E, Nohe A, Petersen NO. Studies of distribution, location and dynamic properties of EGFR on the cell surface measured by image correlation spectroscopy. *Eur Biophys J* 2008;**37**(4):469–81.
53. Carter RE, Sorkin A. Endocytosis of functional epidermal growth factor receptor-green fluorescent protein chimera. *J Biol Chem* 1998;**273**(52):35,000–35,007.
54. Liu P, Sudhaharan T, Koh RM, Hwang LC, Ahmed S, Maruyama IN, Wohland T. Investigation of the dimerization of proteins from the epidermal growth factor receptor family by single wavelength fluorescence cross-correlation spectroscopy. *Biophys J* 2007;**93**(2):684–98.
55. Tao RH, Maruyama IN. All EGF(ErbB) receptors have preformed homo- and heterodimeric structures in living cells. *J Cell Sci* 2008;**121**(Pt 19):3207–17.
56. Whitson KB, Beechem JM, Beth AH, Staros JV. Preparation and characterization of Alexa Fluor 594-labeled epidermal growth factor for fluorescence resonance energy transfer studies: application to the epidermal growth factor receptor. *Anal Biochem* 2004;**324**(2):227–36.
57. Webb SE, Roberts SK, Needham SR, Tynan CJ, Rolfe DJ, Winn MD, Clarke DT, Barraclough R, Martin-Fernandez ML. Single-molecule imaging and fluorescence lifetime imaging microscopy show different structures for high- and low-affinity epidermal growth factor receptors in A431 cells. *Biophys J* 2008;**94**(3):803–19.
58. Ichinose J, Murata M, Yanagida T, Sako Y. EGF signalling amplification induced by dynamic clustering of EGFR. *Biochem Biophys Res Commun* 2004;**324**(3):1143–9.
59. Clayton AH, Orchard SG, Nice EC, Posner RG, Burgess AW. Predominance of activated EGFR higher-order oligomers on the cell surface. *Growth Factors* 2008. October 20:1 (epub ahead of press).

60. Clayton AH, Tavarnesi ML, Johns TG. Unligated epidermal growth factor receptor forms higher order oligomers within microclusters on A431 cells that are sensitive to tyrosine kinase inhibitor binding. *Biochemistry* 2007;**46**(15):4589–97.
61. Szabó A, Horváth G, Szöllosi J, Nagy P. Quantitative characterization of the large-scale association of ErbB1 and ErbB2 by flow cytometric homo-FRET measurements. *Biophys J* 2008;**95**(4):2086–96.
62. Gan HK, Walker F, Burgess AW, Rigopoulos A, Scott AM, Johns TG. The epidermal growth factor receptor (EGFR) tyrosine kinase inhibitor AG1478 increases the formation of inactive untethered EGFR dimers. Implications for combination therapy with monoclonal antibody 806. *J Biol Chem* 2007;**282**(5):2840–50.
63. Park E, Baron R, Landgraf R. Higher-order association states of cellular ERBB3 probed with photo-cross-linkable aptamers. *Biochemistry* 2008. October 23 (epub ahead of press).

Chapter 16

Role of Lipid Domains in EGF Receptor Signaling

Linda J. Pike

Washington University School of Medicine, Department of Biochemistry and Molecular Biophysics, St Louis, Missouri

INTRODUCTION

Epidermal growth factor (EGF) is a 53 amino acid polypeptide that promotes growth and differentiation of a wide variety of cells. Its existence was first recognized almost 50 years ago, when it was identified as a factor in mouse submaxillary glands that induced precocious eyelid opening when injected into newborn mice [1].

EGF elicits its cellular effects through binding to the cell surface EGF receptor. The EGF receptor is a transmembrane tyrosine kinase composed of an ~620 amino acid extracellular ligand binding domain, a single-pass transmembrane domain, a tyrosine kinase domain and an ~200 amino acid C-terminal tail [2]. Upon binding ligand, the EGF receptor dimerizes and undergoes inter-chain autophosphorylation, primarily in the C-terminal tail of the receptor [3, 4]. This promotes complex formation with SH2 and PTB domain-containing proteins and initiates intracellular downstream signaling cascades.

X-ray crystallography studies of the EGF receptor in its unliganded [5] and ligand-bound state [6, 7] have provided insight into the mechanism of EGF receptor homodimer formation. The extracellular domain of the EGF receptor contains four subdomains, numbered I through IV. Domains I, II, and III form a crescent-shaped structure that comprises the EGF binding site. In the absence of EGF, the receptor is held in a closed, tethered conformation through the interaction of loops or arms that extend out from the back of domains II and IV. In the presence of ligand, the interaction between domains II and IV is broken, allowing the adoption of an open configuration that is stabilized by strong interactions between EGF and subdomains I and III. In this open configuration, the "dimerization" arm in subdomain II that was previously involved in holding the receptor in the closed configuration is now available for interaction with the dimerization arm from the second EGF receptor. A back-to-back receptor dimer is formed that allows for activation of the intracellular tyrosine kinase activity.

Activation of the tyrosine kinase appears to occur through the formation of a dimer of the cytoplasmic kinase domains. Zhang *et al.* [8] reported the X-ray crystallographic structure of an asymmetric EGF receptor kinase dimer in which the N-terminal lobe of monomer A interacts with the C-terminal lobe of monomer B. This leads to the activation of monomer A and the phosphorylation of the C-terminal tail of monomer B. Presumably, a conformational change in the kinase domains allows for a shift in the position of each monomer, allowing activation of both kinase domains in the receptor dimer.

The EGF receptor is a member of the ErbB receptor family that has four members: the EGF receptor (ErbB1), HER2/neu (ErbB2), ErbB3, and ErbB4. The receptors are structurally similar; however, in ErbB3 the kinase domain is inactive. The ErbB receptors bind a family of ligands, with each receptor having a different ligand selectivity. ErbB2 is the exception, in that there is no known ligand for this receptor (for review, see [9, 10]). The EGF receptor can form heterodimers with other ErbB family members. Essentially, all combinations of ErbB receptor heterodimers have been shown to exist [11, 12]. However, in all cases, ErbB2 appears to be the preferred dimerization partner [11, 12].

Both the EGF receptor and ErbB2 are frequently overexpressed in tumors, including breast, colon, and lung carcinomas [9, 13]. Several novel cancer therapeutics are directed against the EGF receptor or ErbB2, including the monoclonal antibodies cetuximab, trastuzumab, and pertuzumab, and the small molecule tyrosine kinase inhibitors gefitinib, erlotinib, and lapatinib. Thus, the function and regulation of the EGF receptor and its homologous family members is important both biologically and medically.

Handbook of Cell Signaling, Three-Volume Set 2 ed.

A variety of studies have demonstrated that the EGF receptor partitions into low-density, cholesterol-enriched regions of the membrane known as lipid rafts, and that this localization is important in receptor function [14–18]. This review will briefly describe lipid rafts and how they are studied, and then discuss what is known regarding the mechanism through which the EGF receptor is targeted to lipid rafts, and the effect of lipid rafts on EGF receptor-mediated signaling.

STUDYING LIPID RAFTS

According to the consensus definition developed at the 2006 Keystone Symposium on Lipid Rafts and Cell Function [19], "Lipid rafts are small (10–200 nm), heterogeneous, highly dynamic, sterol- and sphingolipid-enriched domains that compartmentalize cellular processes. Small rafts can sometimes be stabilized to form larger platforms through protein–protein and protein–lipid interactions." The small size and heterogeneity of lipid rafts has made the analysis of their properties and the characterization of their function difficult.

The most common method employed for the biochemical analysis of lipid rafts is the use of cell fractionation procedures to isolate low-density, cholesterol-enriched membrane fractions for characterization. Solubilization in Triton X-100 – or, more specifically, resistance to solubilization in Triton X-100 – was the original method of choice for isolating lipid rafts and their close cousins, caveolae [20]. However, it is now recognized that the use of detergents can introduce artifacts through the fusion of domains from the same membrane or even from two different cellular membranes into a single hybrid "raft" [21]. Thus, the use of detergents is no longer deemed an acceptable method for demonstrating the localization of proteins to lipid rafts.

Several procedures for the preparation of lipid raft membranes in the absence of detergents have been devised [17, 22, 23]. One involves the use of high-pH carbonate buffer to extract extrinsic membrane proteins and includes extensive sonication of the membranes [23]. The resulting raft fractions are contaminated with internal membrane fragments [22], and hence the preparation is not optimal for producing purified raft membranes. The method of Smart *et al.* [17] yields highly purified rafts, but is time consuming; the preparation of Macdonald and Pike produces equally pure fractions but is more rapid [22]. Both have the advantage of generating raft fractions that are likely derived primarily from the plasma membrane and represent domains that existed in the cells prior to their disruption with isotonic buffers.

A persistent problem in the study of lipid rafts is demonstrating their involvement in a specific physiological process. For this purpose, methyl-β-cyclodextrin, an agent that extracts cholesterol from membranes, is often used to disrupt cholesterol-based lipid rafts. The rationale is that if rafts are involved in a particular process, then disrupting the rafts by removing cholesterol should inhibit the phenomenon under study. The absence of a substantially better alternative makes this an acceptable approach. However, the possibility that a particular raft may be less susceptible to disruption by cholesterol extraction, or that cholesterol depletion may have global effects on the membrane [24], should always be borne in mind. Optimally, evidence for the involvement of lipid rafts in a specific process should be supported by data obtained using other biochemical or biophysical approaches.

A key feature of the definition of lipid rafts is that they are, in general, quite small. Thus, rafts cannot be effectively visualized using light microscopy. Immunofluorescence approaches lack sufficient resolution to provide information on the co-localization of proteins in lipid rafts. Therefore, the results of such studies should be viewed with caution. Recently, single-particle methods such as fluorescence resonance energy transfer (FRET), fluorescence lifetime imaging (FLIM), single particle tracking, or fluorescence correlation spectroscopy have been applied to the study of lipid rafts. These approaches show great promise for unraveling the physical nature of the micro- or nanodomains referred to as lipid rafts [25].

LOCALIZATION OF THE EGF RECEPTOR IN LIPID RAFTS

Data from early studies suggested that the EGF receptor resided in a low-density membrane fraction that also contained the marker protein, caveolin-1 [14, 16–18]. As a result, the EGF receptor was initially thought to reside in the caveolin-1-coated membrane invaginations known as caveolae. However, later studies demonstrated that the EGF receptor is present in flattened, low-density lipid rafts that lack caveolin-1, rather than in caveolae proper [26, 27]. The initial confusion was due to the fact that caveolae are isolated along with EGF-containing lipid rafts in most subcellular fractionation protocols. Although no specific data are available, it appears from several studies that in most cells, the majority of EGF receptors are present in the low-density lipid raft fraction [28].

The EGF receptor localizes to lipid rafts under basal conditions. Using a non-detergent method for preparing lipid raft membranes, Mineo *et al.* [29] reported that upon stimulation with EGF, the receptor rapidly exits lipid rafts. However, Ringerike *et al.* [26] failed to see movement of EGF receptors out of lipid rafts using electron microscopy. Similarly, based on studies using image cross-correlation spectroscopy, Keating *et al.* [28] concluded that more than half of the EGF receptors in cells were present in lipid rafts in both the presence and absence of EGF. Hofman *et al.*

[30] also reported that the partitioning of the EGF receptor into lipid raft membrane fractions was not altered by the addition of agonist, and that EGF actually induced the coalescence of two different types of lipid rafts to form a larger signaling platform. Other workers have reported the recruitment of EGF receptors into lipid rafts following receptor activation [31, 32]. Thus, while there is agreement with regard to the association of the EGF receptor with lipid rafts, there is no consensus on the effect of EGF on the localization of its receptor. The differences may be due to differences in methodology, or to cell type-specific differences. The weight of evidence seems to support a neutral or positive effect of EGF on the partitioning of its receptor into lipid rafts.

Several structural features of proteins have been identified that serve to target the proteins to lipid rafts. These include palmitoylation, myristoylation, and prenylation of the protein, as well as the addition of a GPI anchor [20, 33, 34]. However, none of these mechanisms appear to function in the case of the EGF receptor, which has never been shown to acquire any such post-translational lipid modifications. Mutational analyses of the EGF receptor have demonstrated that a receptor in which the entire cytoplasmic domain has been deleted still localizes to lipid rafts, indicating that the intracellular domain is not required for localization of the receptor to these low-density domains [29, 35]. Similarly, a receptor lacking the first 270 amino acids of the extracellular domain continued to localize to rafts, indicating that this portion of the receptor was not required for localization [29]. Recent work has demonstrated that a 60 amino acid segment in the most membrane-proximal region of the extracellular domain of the EGF receptor is responsible for targeting the receptor to lipid rafts [36]. This suggests that the EGF receptor may localize to lipid rafts as a result of the interaction of its extracellular domain with a resident raft protein or lipid. The observation that the extracellular domain of the EGF receptor binds to gangliosides [37, 38], and that receptor kinase activity is regulated by these important components of lipid rafts [39–41], suggests the possibility that this interaction may play a role in localizing the receptor to these membrane domains. Consistent with this possibility, FLIM analysis of the EGF receptor has recently shown that it co-localizes with GM1 in a cholesterol-independent fashion in membrane nanodomains [30].

Lipidomics studies of the lipid rafts into which the EGF receptors partition indicate that they exhibit the expected enrichment in cholesterol and sphingomyelin [15, 42]. Non-detergent rafts that contained EGF receptors exhibited a balance of inner (phosphatidylethanolamine) and outer (phosphatidylcholine and sphingomyelin) leaflet lipids, suggesting that the rafts were derived from lipid bilayers. By contrast, rafts isolated on the basis of resistance to Triton X-100 and lacking the EGF receptor exhibited a large excess of outer leaflet lipids [15]. This indicates that detergent extraction procedures preferentially solubilize inner leaflet lipids, and therefore do not truly reflect the original membrane bilayer.

Unexpectedly, non-detergent lipid rafts were also enriched in phosphatidylserine and plasmenylethanolamine [42]. A recent study suggests that the former may serve a role in the function of these domains. Al-Nedawi *et al.* [43] reported that an oncogenic form of the EGF receptor is shed from cells in vesicles derived from lipid rafts. These microvesicles are capable of fusing with other cells, endowing the recipient cells with oncogenic characteristics. The fusion process was found to be phosphatidylserine-dependent, suggesting that the enrichment of this lipid in rafts from which these vesicles are derived contributes to the ability of the shed vesicles to fuse with cells.

RAFTS AND EGF RECEPTOR-MEDIATED SIGNALING

Because lipid rafts harbor a variety of proteins that are involved in signal transduction, it has been hypothesized that these low-density domains serve to organize cell signaling, enhancing the specificity or efficiency of the processes [44]. For the EGF receptor, the data are more consistent with the hypothesis that lipid rafts suppress receptor signaling by preventing unregulated interaction between EGF receptors.

Depletion of cholesterol, and hence disruption of lipid rafts, induces ligand-independent tyrosine phosphorylation of the EGF receptor as well as activation of MAP kinase [45, 46]. Studies using fluorescence correlation spectroscopy and an analysis of brightness have shown that depletion of cholesterol leads to an increase in the clustering of the EGF receptor, while enrichment of cholesterol decreases clustering of the receptor [47]. Consistent with this observation, Orr *et al.* [48] reported that cholesterol depletion reduced the mobility of the EGF receptor while cholesterol enrichment increased the mobility of the EGF receptor. These findings suggest that in the absence of ligand, the aggregation state of the EGF receptor is modulated by the cholesterol content of cells, with high cholesterol levels blocking receptor–receptor interactions. Given that the EGF receptor must dimerize in order to signal, inhibition of receptor aggregation would reduce the likelihood of spurious activation of EGF receptor signaling.

Indeed, lipid rafts appear to exert a suppressive effect on both receptor binding and kinase activity. Several groups have reported that disruption of rafts by treatment of cells with methyl-β-cyclodextrin leads to an increase in ^{125}I-EGF binding [27, 49, 50], while cholesterol loading decreases EGF binding [27]. Similarly, depletion of cholesterol is associated with an increase in receptor autophosphorylation [26, 49, 50], while cholesterol enrichment decreases autophosphorylation [26, 47]. Surprisingly, not all sites of autophosphorylation are equally affected by changes in cholesterol content [50]. It is possible that cholesterol-dependent changes in receptor conformation may specifically favor the phosphorylation of certain sites.

Experiments with enantiomers of cholesterol showed that depletion of cholesterol followed by repletion with either natural cholesterol or its enantiomer resulted in similar effects on receptor kinase activity [50]. Thus, the effect of cholesterol on EGF receptor function is not stereoselective. Since enzymes with specific binding sites for cholesterol distinguish between the two enantiomers, these data suggest that cholesterol modulates EGF receptor autophosphorylation via a membrane-level effect.

The effect of cholesterol on EGF receptor kinase activity is traceable to its effect on the ability of EGF to induce receptor dimer formation. Cholesterol depletion enhanced receptor dimerization, whereas cholesterol loading impaired dimerization [26, 47]. This is consistent with the effects of cholesterol on the aggregation of the EGF receptor in the absence of ligand [47, 48]. Interestingly, elevation of circulating cholesterol in SCID mice increased the cholesterol content of lipid rafts isolated from xenograft tumors, and this was associated with an increase in basal protein tyrosine phosphorylation [51]. Thus, alterations in plasma cholesterol levels may be a physiologically relevant mechanism for regulating EGF receptor signaling.

Other aspects of EGF receptor signaling also involve lipid rafts. Both EGF- and bradykinin-stimulated phosphatidylinositol turnover appear to occur exclusively in lipid rafts [52]. Disruption of lipid rafts by depletion of cholesterol completely ablated the ability of both hormones to stimulate phosphatidylinositol turnover, implying that intact rafts are absolutely required for this response [16]. EGF stimulation of the initial steps in MAP kinase activation, namely the recruitment of ras and raf to plasma membranes, also appears to take place within the lipid raft compartment [14]. However, lipid rafts are not essential for the EGF-induced activation of MAP kinase. Depletion of cholesterol from cells leads to hyperactivation of MAP kinase by EGF [53]. As cholesterol depletion results in the disruption of lipid rafts, these data suggest that activation of MAP kinase can occur in the absence of intact rafts. This is consistent with studies that indicate that GTP binding causes H-ras to move out of lipid rafts, and that activation of Raf by H-ras is less efficient when it occurs in rafts as compared to bulk plasma membrane [54]. Thus, the events leading to the activation of the kinases upstream of MAP kinase can apparently occur outside of lipid rafts.

THE EGF RECEPTOR AND CAVEOLIN

Caveolin is a 21-kDa membrane protein that was first identified as a substrate for pp60^{src} [55], and appears to be responsible for stabilizing the invaginated structure of caveolae [56, 57]. Cells lacking caveolin-1 do not exhibit plasma membrane caveolae, yet these cells still have low-density, cholesterol-enriched lipid rafts that contain high levels of signaling proteins. The relationship between caveolae proper and lipid rafts is not clear, but they appear to be distinct entities based on the ability to separate caveolin-containing low-density membranes from other similar membranes that contain raft markers such as GPI-linked proteins [58]. The EGF receptor is thought to be present in lipid rafts rather than caveolae [26, 27].

Despite the fact that the two proteins reside in different compartments, numerous studies have reported evidence for a physical or functional relationship between caveolin-1 and the EGF receptor. Upregulation of caveolin-1 expression was shown to inhibit EGF-stimulated MAP kinase activation in human diploid fibroblasts [59]. This has been attributed to the ability of caveolin to interact directly with the EGF receptor and inhibit its activity. Using GST-caveolin-1 and synthetic peptides corresponding to the "scaffolding domain" of caveolin-1 (residues 81–101), Couet *et al.* [60] reported a physical association of the EGF receptor with caveolin-1 as well as an inhibition of EGF receptor kinase activity by caveolin-1 *in vitro*. Similar findings were reported regarding the effect of caveolin-1 expression on the activity of the EGF receptor homolog, ErbB2 [61]. However, EGF-mediated MAP kinase activation appeared to be normal in caveolin-1 knockout mice [62], suggesting that in the *in vivo* situation, caveolin-1 does not suppress EGF receptor signaling. The discrepancy may be due to the use of recombinant proteins and synthetic peptides in the *in vitro* studies that do not accurately recapitulate the relationship of caveolin-1 and the EGF receptor *in vivo*.

Studies from other investigators have provided data consistent with the results from the caveolin-1 knockout mice. Immunodepletion of low-density membrane fractions with an antibody to caveolin resulted in the physical separation of EGF receptors from caveolin-containing membranes [18]. In addition, these investigators reported that co-immunoprecipitation of the EGF receptor and caveolin was "irreproducible," consistent with the absence of a stable association between the two proteins. Thus, direct interaction between the EGF receptor and caveolin-1 *in vivo* seems unlikely.

EGF does affect caveolin-1 phosphorylation and trafficking. Overexpression of the wild-type EGF receptor or a C-terminally truncated form of the EGF receptor results in an EGF-dependent tyrosine phosphorylation of caveolin-1 that seems to be mediated by pp60^{src} [63]. Furthermore, EGF induces the redistribution of caveolin-1 from the plasma membrane to an early endocytic compartment [64]. The observation that EGF stimulates the phosphorylation and endocytosis of phospholipid scramblase 1, a resident lipid raft protein [65], is consistent with the notion that EGF may regulate the trafficking of a variety of proteins present in low-density membrane domains.

Whether the EGF receptor itself is internalized via lipid rafts or caveolae remains an open question. Sigismund *et al.* [66] reported that while a low dose of EGF was internalized exclusively through the clathrin pathway, a substantial fraction of a high dose of EGF was internalized via a lipid raft-dependent pathway. The mechanism may involve the recruitment of endocytic machinery to lipid rafts, enabling the endocytosis of the EGF receptor directly from a raft membrane without requiring movement of the receptor into a coated pit [32]. Kazazic *et al.* [67] reported no difference in the fraction of bound EGF localizing to caveolae at low and high doses of EGF, and hence did not confirm the findings of Sigismund *et al.* [66]. They found that high concentrations of EGF did not increase the mobility of caveolae, and concluded that these domains were not involved in EGF receptor endocytosis. By contrast, Orlichenko *et al.* [68] reported that treatment of cells with EGF stimulated the assembly and internalization of caveolae via a mechanism that involves phosphorylation of caveolin on tyrosine-14. The reason for these differences is not clear, but it suggests that there may be cell-type specific differences in the utilization of caveolae and rafts for EGF receptor internalization.

REFERENCES

1. Cohen S. Purification and metabolic effects of a nerve growth-promoting protein from snake venom. *J Biol Chem* 1959;**234**:1129–37.
2. Ullrich A, Coussens L, Hayflick JS, et al. Human epidermal growth factor receptor cDNA sequence and aberrant expression of the amplified gene in A431 epidermoid carcinoma cells. *Nature* 1984;**309**:418–25.
3. Yarden Y, Schlessinger J. Epidermal growth factor induces rapid, reversible aggregation of the purified epidermal growth factor receptor. *Biochemisty* 1987;**26**:1443–51.
4. Yarden Y, Schlessinger J. Self-phosphorylation of epidermal growth factor receptor: evidence for a model of intermolecular allosteric activation. *Biochemistry* 1987;**26**:1434–42.
5. Ferguson KM, Berger MB, Mendrola JM, Cho H-S, Leahy DJ, Lemmon MA. EGF activates its receptor by removing interactions that autoinhibit ectodomain dimerization. *Mol Cell* 2003;**11**:507–17.
6. Garrett TPJ, McKern NM, Lou M, et al. Crystal structure of a truncated epidermal growth factor receptor extracellular domain bound to transforming growth factor a. *Cell* 2002;**110**:763–73.
7. Ogiso H, Ishitani R, Nureki O, et al. Crystal structure of the complex of human epidermal growth factor and receptor extracellular domains. *Cell* 2002;**110**:775–87.
8. Zhang X, Gureasko J, Shen K, Cole PA, Kuriyan J. An allosteric mechanism for activation of the kinase domain of epidermal growth factor receptor. *Cell* 2006;**125**:1137–49.
9. Olayioye MA, Neve RM, Lane HA, Hynes NE. The ErbB signaling network: receptor heterodimerization in development and cancer. *EMBO J* 2000;**19**:3159–67.
10. Yarden Y, Sliwkowski MS. Untangling the ErbB signalling network. *Nature Rev Mol Cell Biol* 2001;**2**:127–37.
11. Graus PD, Beerli RR, Daly JM, Hynes NE. ErbB2, the preferred heterodimerization partner of all ErbB receptors, is a mediator of lateral signaling. *EMBO J* 1997;**16**:1647–55.
12. Tzahar E, Waterman H, Chen X, et al. A hierarchical network of interreceptor interactions determines signal transduction by Neu differentiation factor/neuregulin and epidermal growth factor. *Mol Cell Biol* 1996;**16**:5276–87.
13. Normanno N, De Luca A, Bianco C, et al. Epidermal growth factor receptor (EGFR) signaling in cancer. *Gene* 2006;**366**:2–16.
14. Mineo C, James GL, Smart EJ, Anderson RGW. Localization of epidermal growth factor-stimulated Ras/Raf-1 interaction to caveolae membrane. *J Biol Chem* 1996;**271**:11930–5.
15. Pike LJ, Han X, Gross RW. Epidermal growth factor receptors are localized to lipid rafts that contain a balance of inner and outer leaflet lipids: a shotgun lipidomics study. *J Biol Chem* 2005;**280**:26,796–804.
16. Pike LJ, Miller JM. Cholesterol depletion de-localizes PIP2 and inhibits hormone-stimulated phosphatidylinositol turnover. *J Biol Chem* 1998;**273**:22,298–304.
17. Smart EJ, Ying Y-S, Mineo C, Anderson RGW. A detergent-free method for purifying caveolae membrane from tissue culture cells. *Proc Natl Acad Sci USA* 1995;**92**:10,104–8.
18. Waugh MG, Lawson D, Hsuan JJ. Epidermal growth factor receptor activation is localized within low-buoyant density, non-caveolar membrane domains. *Biochem J* 1999;**337**:591–7.
19. Pike LJ. Rafts defined: a report on the keystone symposium on lipid rafts and cell function. *J Lipid Res* 2006;**47**:1597–8.
20. Brown DA, Rose JK. Sorting of GPI-anchored proteins to glycolipid-enriched membrane subdomains during transport to the apical cell surface. *Cell* 1992;**68**:533–44.
21. Shogomori H, Brown DA. Use of detergents to study membrane rafts: the good, the bad, and the ugly. *Biol Chem* 2003;**384**:1259–63.
22. Macdonald JL, Pike LJ. A simplified method for the preparation of detergent-free lipid rafts. *J Lipid Res* 2005;**46**:1061–7.
23. Song KS, Li S, Okamoto T, Quilliam LA, Sargiacomo M, Lisanti MP. Co-purification and direct interaction of ras with caveolin, an integral membrane protein of caveolae microdomains. *J Biol Chem* 1996;**271**:9690–7.
24. Kwik J, Boyle S, Fooksman D, Margolis L, Sheetz MP, Edidin M. Membrane cholesterol, lateral mobility, and the phosphatidylinositol 4,5-bisphosphate-dependent organization of cell actin. *Proc Natl Acad Sci USA* 2003;**100**:13,964–9.
25. Jacobon K, Mouritsen OG, Anderson RGW. Lipid rafts: at a crossroad between cell biology and physics. *Nature Cell Biol* 2007;**9**:7–14.
26. Ringerike T, Glystad FD, Levy FO, Madshus IH, Stang E. Cholesterol is important in control of EGF receptor kinase activity but EGF receptors are not concentrated in caveolae. *J Cell Sci* 2002;**115**:1331–40.
27. Roepstorff K, Thomsen P, Sandvig K, van Deurs B. Sequestration of EGF receptors in non-caveolar lipid rafts inhibits ligand binding. *J Biol Chem* 2002;**277**:18,954–60.
28. Keating E, Nohe A, Petersen NO. Studies of distribution, location and dynamic properties of EGFR on the cell surface measured by image correlation spectroscopy. *Eur Biophys J* 2008;**37**:469–81.
29. Mineo C, Gill GN, Anderson RGW. Regulated migration of epidermal growth factor receptor from caveolae. *J Biol Chem* 1999;**274**:30,636–43.
30. Hofman EG, Ruonala MO, Bader AN, et al. EGF induces coalescence of different lipid rafts. *J Cell Sci* 2008;**121**:2519–28.
31. Kasai A, Shima T, Okada M. Role of Src family tyrosine kinases in the down-regulation of epidermal Growth factor signaling in PC12 cells. *Genes to Cells* 2005;**10**:1175–87.
32. Puri C, Tosoni D, Comai R, et al. Relationships between EGFR-signaling-competent and endocytosis-competent membrane microdomains. *Mol Biol Cell* 2005;**16**:2704–18.

33. Melkonian Ka, Ostermeyer AG, Chen JZ, Roth MG, Brown DA. Role of lipid modifications in targeting proteins to detergent-resistant membrane rafts. *J Biol Chem* 1999;**274**:3910–7.
34. Moffett S, Brown DA, Linder ME. Lipid-dependent targeting of G proteins into rafts. *J Biol Chem* 2000;**275**:2191–8.
35. Macdonald JL, Li Z, Wu W, Pike LJ. The membrane proximal disulfides of the EGF receptor extracellular domain are required for high affinity binding and signal transduction but do not play a role in the localization of the receptor to lipid rafts. *Biochim Biophys Acta Mol Cell Res* 2006:870–8.
36. Yamabhai M, Anderson RGW. Second cysteine-rich region of EGFR contains targeting information for caveolae/rafts. *J Biol Chem* 2002;**277**:24,843–6.
37. Miljan EA, Meuillet EJ, Mania-Farnell GD, Yamamoto H, Simon H-G, Bremer EG. Interaction of the extracellular domain of the epidermal growth factor receptor with gangliosides. *J Biol Chem* 2002;**277**:10,108–3.
38. Yoon S-J, Nakayama K, Hikita T, Handa K, Hakomori S-i. Epidermal growth factor receptor tyrosine kinase is modulated by GM3 interaction with N-linked GlcNac termini of the receptor. *Proc Natl Acad Sci USA* 2007;**103**:18,987–91.
39. Hanai N, Nores GA, MacLeod C, Torres-Mendex CR, Hakomori S. Ganglioside-mediated modulation of cell growth, specific effects of GM3 and lyso-GM3 in tyrosine phosphorylation of the epidermal growth factor receptor. *J Biol Chem* 1988;**263**:10,915–21.
40. Meuillet EJ, Mania-farnell B, George D, Inokuchi J-I, Bremer EG. Modulation of EGF receptor activity by changes in the GM3 content in a human epidermoid carcinoma cell line, A431. *Exp Cell Res* 2000;**256**:74–82.
41. Zhou Q, Hakomori S, Kitamura K, Igarashi Y. GM3 directly inhibits tyrosine phosphorylation and De-N-acetyl-G_{M3} directly enhances serine phosphorylation of epidermal growth factor receptor, independently of receptor–receptor interaction. *J Biol Chem* 1994;**269**:1959–65.
42. Pike LJ, Han X, Chung K-N, Gross R. Lipid rafts are enriched in plasmalogens and arachidonate-containing phospholipids and the expression of caveolin does not alter the lipid composition of these domains. *Biochem* 2002;**41**:2075–88.
43. Al-Nedawi K, Meehan B, Micallef J, et al. Intercellular transfer of the oncogenic receptor EGFRvIII by microvesicles derived from tumor cells. *Nature Cell Biol* 2008;**10**:619–24.
44. Simons K, Toomre D. Lipid rafts and signal transduction. *Nature Rev Mol Cell Biol* 2000;**1**:31–41.
45. Chen X, Resh MD. Cholesterol depletion from the plasma membrane triggers ligand-independent activation of the epidermal growth factor receptor. *J Biol Chem* 2002;**277**:49,631–7.
46. Lambert S, Vind-Kezunovic D, Karvinen S, Gniadecki R. Ligand-independent activation of the EGFR by lipid raft disruption. *J Invest Dermatol* 2006;**126**:954–62.
47. Saffarian S, Li Y, Elson EL, Pike LJ. Oligomerization of the EGF receptor investigated by live cell fluorescence intensity distribution analysis. *Biophys J* 2007;**93**:1021–31.
48. Orr G, Hu D, Ozcelik S, Opresko LK, Wiley HS, Colson SD. Cholesterol dictates the freedom of EGF receptors and HER2 in the plane of the membrane. *Biophys J* 2005;**89**:1362–73.
49. Pike LJ, Casey L. Cholesterol levels modulate EGF receptor-mediated signaling by altering receptor function and trafficking. *Biochem* 2002;**41**:10,315–22.
50. Westover EJ, Covey DF, Brockman HL, Brown RE, Pike LJ. Cholesterol depletion results in site-specific increases in EGF receptor phosphorylation due to membrane level effects: studies with cholesterol enantiomers. *J Biol Chem* 2003;**278**:51,125–33.
51. Zhuang L, Kim J, Adam RM, Solomon KR, Freeman MR. Cholesterol targeting alters lipid raft composition and cell survival in prostate cancer cells and xenographs. *J Clin Invest* 2005;**115**:959–68.
52. Pike LJ, Casey L. Localization and turnover of phosphatidylinositol 4,5-bisphosphate in caveolin-enriched membrane domains. *J Biol Chem* 1996;**271**:26,453–6.
53. Furuchi T, Anderson RGW. Cholesterol depletion of caveolae causes hyperactivation of extracellular signal-related kinase (ERK). *J Biol Chem* 1998;**273**:21,099–104.
54. Prior IA, Harding A, Yan J, Sluimer J, Parton RG, Hancock JF. GTP-dependent segregation of H-ras from lipid rafts is required for biological activity. *Nature Cell Biol* 2001;**3**:368–75.
55. Glenney JR, Zokas L. Novel tyrosine kinase substrates from rous sarcoma virus transformed cells are present in the membrane skeleton. *J Cell Biol* 1989;**108**:2401–8.
56. Fra AM, Williamson E, Simons K, Parton RG. De novo formation of caveolae in lymphocytes by expression of VIP21-caveolin. *Proc Natl Acad Sci USA* 1995;**92**:8655–9.
57. Le PU, Guay G, Altschuler Y, Nabi IR. Caveolin-1 is a negative regulator of caveolae-mediated endocytosis to the endoplasmic reticulum. *J Biol Chem* 2002;**277**:3371–9.
58. Schnitzer JE, McIntosh DP, Dvorak AM, Liu J, Oh P. Separation of caveolae from associated microdomains of GPI-anchored proteins. *Science* 1995;**269**:1435–9.
59. Park W-Y, Park J-S, Cho K-A, et al. Up-regulation of caveolin attenuates epidermal growth factor signaling in senescent cells. *J Biol Chem* 2000;**275**:20,847–52.
60. Couet J, Sargiacomo M, Lisanti MP. Interaction of a receptor tyrosine kinase, EGF-R, with caveolins. Caveolin binding negatively regulates tyrosine and serine/threonine kinase activities. *J Biol Chem* 1997;**272**:30,429–30,438.
61. Engelman JA, Lee RJ, Karnezis A, et al. Reciprocal regulation of neu tyrosine kinase activity and caveolin-1 protein expression *in vitro* and *in vivo*. *J Biol Chem* 1998;**273**:20,448–55.
62. Razani B, Engelman JA, Wang XB, et al. Caveolin-1 null mice are viable but show evidence of hyperproliferative and vascular abnormalities. *J Biol Chem* 2001;**275**:38,121–38.
63. Kim Y-N, Wiepz GJ, Guadarrama AG, Bertics PJ. Epidermal growth factor-stimulated tyrosine phosphorylation of caveolin-1. *J Biol Chem* 2000;**275**:7481–91.
64. Pol A, Lu A, Pons M, Peiro S, Enrich C. Epidermal growth factor-mediated caveolin recruitment to early endosomes and MAPK activation. *J Biol Chem* 2000;**275**:30,566–72.
65. Sun J, Nanjundan M, Pike LJ, Wiedmer T, Sims PJ. Plasma membrane phospholipid scramblase 1 is enriched in lipid rafts and interacts with the epidermal growth factor receptor. *Biochem* 2002;**41**:6338–45.
66. Sigismund S, Woelk T, Puri C, et al. Clathrin-independent endocytosis of ubiquitinated cargos. *Proc Natl Acad Sci USA* 2005;**102**:2760–5.
67. Kazazic M, Roepstorff K, Johannessen LE, et al. EGF-induced activation of the EGF receptor does not trigger mobilization of caveolae. *Traffic* 2006;**7**:1518–27.
68. Orlichenko L, Huang B, Krueger E, McNiven MA. epithelial growth factor-induced phosphorylation of caveolin 1 at tyrosine 14 stimulates caveolae formation in epithelial cells. *J Biol Chem* 2006;**281**:4570–9.

Chapter 17

Signaling by the Platelet-Derived Growth Factor Receptor Family

Lars Rönnstrand

Experimental Clinical Chemistry, Department of Laboratory Medicine, Lund University, Malmö University Hospital, Malmö, Sweden

PLATELET-DERIVED GROWTH FACTOR ISOFORMS

The history of PDGF dates back to the original observation by Balk [1] that serum provides a better growth support for fibroblasts than plasma does. Platelets were identified as the source of the growth factor, and the name platelet-derived growth factor was coined. PDGF is a family of disulfide-bonded dimeric proteins, made up of A, B, C, or D chains. Five dimeric combinations exist: PDGF-AA, PDGF-AB, PDGF-BB, PDGF-CC, and PDGF-DD (reviewed in [2], Figure 17.1). Although originally found in platelets, PDGF is widely expressed in many different cell types. The traditional isoforms of PDGF exist in homodimeric and heterodimeric configurations, i.e., PDGF-AA, PDGF-AB and PDGF-BB. A few years ago, two additional members of the family were identified through searching the databases for homologs of the closely related VEGF proteins, namely PDGF-C and PDGF-D. They both differ from the classical PDGFs in that they have an N-terminal CUB domain and a C-terminal PDGF/VEGF domain, separated by a hinge region. In contrast to the classical PDGFs, they do not form heterodimers. In order for PDGF-C and PDGF-D to be active they need to be cleaved off the CUB domain, which releases them and renders them fully functional.

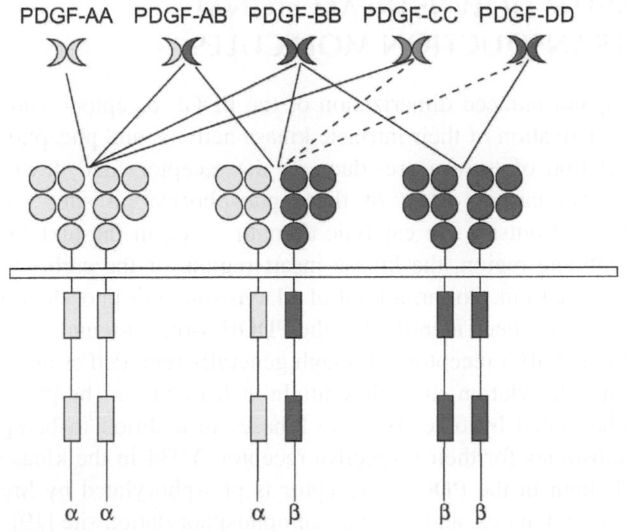

FIGURE 17.1 Receptor binding specificity of various isoforms of PDGF.
The ability of the five different PDGF isoforms to induce the formation of homo- and heterodimeric receptor is depicted. Ig domains in the receptors are indicated by circles.

PHYSIOLOGICAL FUNCTION OF PDGF

Although originally regarded as a growth stimulator of fibroblasts and smooth muscle cells, many other cell types express PDGF receptors, including neurons, astrocytes, mesenchymal stem cells, pericytes, hepatic stellate cells, etc. The function of PDGF in development has been thoroughly studied (reviewed in [3]). The two PDGF receptors have distinct functions. While the PDGF β-receptor is involved in pericyte recruitment to capillaries, development of smooth muscle cells in vessels, and development of mesangial cells in the kidney, the PDGF α-receptors are required for formation of alveolar smooth muscle cells, hair follicle development, proper villus formation in the gut, and oligodendrocyte development. In the adult organism, PDGF signaling contributes to wound healing through stimulation of, for example, fibroblasts, smooth muscle cells, and different inflammatory cells. PDGF β-receptors are also involved in regulation of the interstitial fluid pressure, and can thus control transport of fluids from the vessels to surrounding tissues [4].

Handbook of Cell Signaling, Three-Volume Set 2 ed.

ACTIVATION OF PLATELET-DERIVED GROWTH FACTOR RECEPTORS AND REGULATION OF KINASE ACTIVITY

PDGFs mediate their biological function through binding to two types of PDGF receptors, denoted α and β, which belong to the type III family of receptor tyrosine kinases (RTKs also including the M-CSF receptor, c-Kit, and Flt3). They both share a similar layout, in that they possess an extracellular domain of five immunoglobulin-like (Ig-like) domains, a transmembrane domain, a juxtamembrane domain, a split kinase domain interrupted by a stretch of amino acids denoted the kinase insert, and a carboxyterminal tail. Ig-like domains 1–3 are involved in ligand binding [5, 6], while the fourth Ig-like domain is involved in dimerization of receptors [7]. There is no crystal structure of the PDGF receptors to date, but in the closely related stem cell factor receptor/c-Kit Ig-like domain 5 is also involved in the dimerization of receptors [8]. Whether this is the case for the PDGF receptors remains to be seen.

The four different PDGF chains show differences in binding specificity to the two PDGF receptors. PDGF-A and PDGF-C bind exclusively to PDGF α-receptors [9, 10], while PDGF-B binds to both PDGF α-receptors and PDGF β-receptors. PDGF-D is a PDGF β-receptor agonist, but has also been reported to be able to induce PDGF α–β heterodimers (for review, see [11]). For a summary of the interaction between the PDGF isoforms and their ligands, see Figure 17.1.

As mentioned, PDGF-A and PDGF-C bind to the PDGF α-receptor, and they seem to induce identical signal transduction responses (Ulf Eriksson, personal communication). This is in contrast to the ErbB family of RTKs, where different ligands induce distinctive signal transduction patterns [12]. Each PDGF molecule binds two receptors, which leads to dimerizations of receptors. Due to the lack of structural information on the PDGF receptors, it is not known exactly how the ligand binding leads to receptor activation. Structural data from the ectodomain of the RTK c-Kit shows that the ligand, SCF, binds to Ig-like domains 1–3. The induced dimer is further stabilized by interactions between the Ig-like domains 4 on two receptors, and also between Ig-like domains 5. Since the dimerization motif found in Ig-like domain 4 of c-Kit is conserved in the PDGF receptors [8], it is likely that the PDGF receptors are activated by a similar mechanism.

Dimerization of receptors brings them in close proximity to each other and promotes their mutual phosphorylation on tyrosine residues. One of these residues in the second part of the catalytic domain (Y849 in the PDGF α-receptor and Y857 in the PDGF β-receptor) is conserved in most tyrosine kinases. The exact function of these tyrosine residues in the PDGF receptor is not known, but in the thoroughly studied insulin receptor the activation loop is folded over the kinase domain, and suppresses its activity when these tyrosine residues are not phosphorylated. Upon phosphorylation the activation loop is repositioned, leading to a concomitant increase in catalytic activity of the kinase domain (reviewed in [13]). In the unphosphorylated state, one of the tyrosines, Y1162, acts as a pseudosubstrate inhibitor. Due to the lack of information regarding the structure of the PDGF receptors, it is not known whether a similar mechanism is acting in the PDGF receptors. Studies have shown that mutation of Y857 to phenylalanine in the PDGF β-receptor leads to reduction in the kinase activity of the receptor, suggesting that it has a function in regulating kinase activity [14, 15].

In order to tightly control the RTKs from accidental activation by, for example, random dimerization of receptors in the plasma membrane, several mechanisms keep the non-ligand-bound RTKs in an inactive state. The juxtamembrane region has an important function in regulating the kinase activity of many RTKs (reviewed in [13]). The crystal structure of the type III RTK Flt3 has revealed a mechanism whereby the juxtamembrane region makes extensive contacts with the amino- and carboxyterminal lobes of the kinase, which function to stabilize an inactive configuration of the activation loop [16]. The activation loop adopt an autoinhibitory conformation in which Y842 (corresponding to Y857 of the PDGF β-receptor) is positioned in the active site as a pseudosubstrate. It is highly likely that a similar mechanism is acting on the PDGF receptors, since several activating mutations of the PDGF receptors reside in the juxtamembrane region [17, 18].

INTERACTION OF THE PDGF RECEPTORS WITH DOWNSTREAM SIGNAL TRANSDUCTION MOLECULES

Ligand-induced dimerization of the PDGF receptors leads to activation of their intrinsic kinase activity, and phosphorylation of tyrosine residues on the receptors and downstream targets. Most of these phosphorylation sites are located outside the catalytic domain – i.e., in the juxtamembrane region, the kinase insert region, or the carboxyterminal tail. So far, a total of 11 tyrosine phosphorylation sites have been identified in the PDGF α-receptor and 13 in the PDGF β-receptor. Although generally regarded as autophosphorylation sites, they might at least in part be phosphorylated by other tyrosine kinases in addition to being substrates for their respective receptor. Y934 in the kinase domain in the PDGF β-receptor is phosphorylated by Src family kinases, and is not an autophosphorylation site [19].

Many of the signal transduction molecules that are activated by the receptors bind to receptors through Src homology 2 (SH2) domains and phosphotyrosine binding (PTB) domains, which recognize phosphorylated tyrosine residues in a specific sequence environment. Other signal transduction molecules associate indirectly through adaptor proteins that provide a bridge to the receptor. Finally, PDZ domain-containing proteins can interact with receptors through carboxyterminal valine residues in a specific sequence context (for review, see [20]).

A summary of protein associating to the activated PDGF α- and β-receptors is found in Figure 17.2. For a more detailed overview of the signal transduction molecules activated by the PDGF receptors, see [21]. Several of the tyrosine residues in the PDGF receptors share binding partners, and thus the signaling outcome will at least in part be determined by the expression level of individual signal transduction molecules. Many of these proteins are involved in more than one biological response, and the signal transduction pathways activated by each of them frequently overlap. It is also striking that both positive and negative effectors of a certain pathway can be activated by the PDGF receptor. One example is the Ras/Erk pathway, which is activated by binding of Grb2–Sos complexes either directly to the PDGF receptors, or indirectly through adaptor and scaffolding proteins (such as Shc). RasGAP, binding to Y771 in the PDGF β-receptor, is a negative regulator of Ras activity. The protein tyrosine phosphatase SHP2 gives mostly a positive signal in Ras/Erk signaling, despite the fact that it is a phosphatase. However, SHP2 can also, under certain circumstances, dephosphorylate Y771 in the PDGF β-receptor, and thereby interfere with binding of RasGAP [22]. It should also be remembered that the pathways activated by the PDGF receptors are not linear; there is a considerable cross-talk between the pathways, and also regulation by feedback mechanisms. In the case of the non-receptor tyrosine kinase Abl, there is a requirement for PLC-γ1 activity, leading to reduced levels of phosphatidylinositol 4,5 bisphosphate in the cell, for full activation of Abl. Conversely, Abl phosphorylates PLC-γ1 at key residues leading to attenuation of its lipase activity [23]. There are several examples of downstream signal transduction targets that provide a negative feedback loop to attenuate signaling. Erk kinases are known to phosphorylate Sos, which leads to reduced binding of Grb2–Sos complex to phosphorylated tyrosine residues. This in turn leads to attenuation of the Ras/Erk pathway [24].

The biological outcomes of PDGF-induced signaling include proliferation, survival, cytoskeletal rearrangement, chemotaxis, and, in some cases, differentiation.

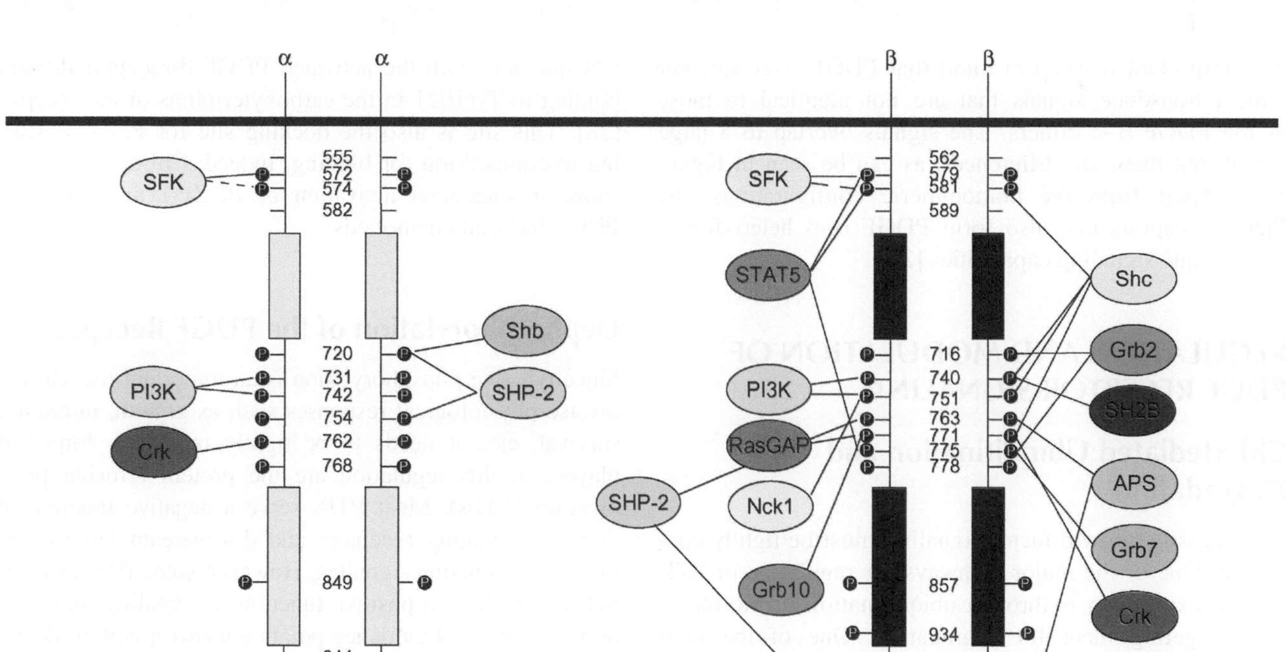

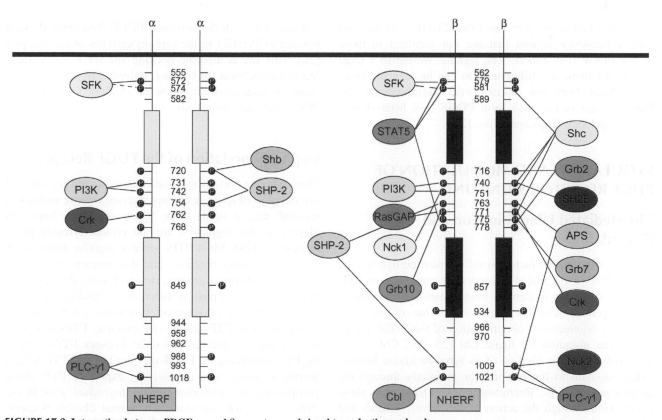

FIGURE 17.2 Interaction between PDGF α- and β-receptors and signal transduction molecules.
The intracellular parts of the receptors are depicted. All tyrosine residues outside the catalytic domains and their numbers are indicated; known phosphorylation sites are indicated by an encircled P. The known interactions between individual phosphorylated tyrosine residues and different SH2 domain containing signal transduction molecules are shown. NHERF interacts with the carboxyterminal sequence of the PDGF receptors in a phosphorylation independent manner.

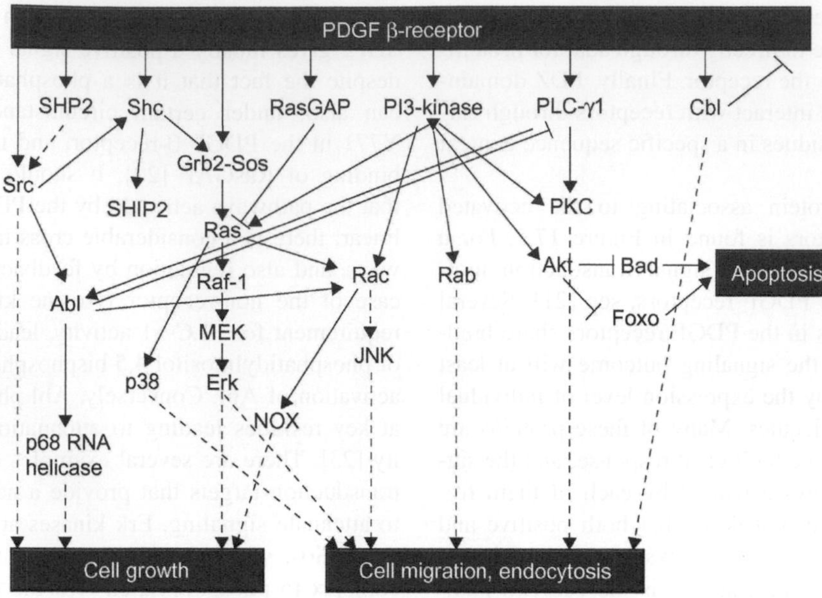

FIGURE 17.3 Schematic illustration of selected signaling pathways that are initiated by the PDGF β-receptor and which lead to cell growth, cytoskeletal reorganization and protection from apoptosis.
Arrows indicate experimentally verified stimulatory interactions. Indirect interactions are indicated with stippled lines. Inhibitory actions are indicated with-.

It is important to keep in mind that PDGF α–α receptor dimers transduce signals that are not identical to those of the PDGF β–β dimers. The signals overlap to a large extent, but there are differences, as can be seen in Figure 17.3. Apart from the homodimeric configurations, the PDGF receptors can also form PDGF α–β heterodimers with unique signaling capabilities [22].

REGULATION AND MODULATION OF PDGF RECEPTOR SIGNALING

Cbl Mediated Ubiquitination and Degradation

By necessity, growth factor signaling must be tightly controlled. One of the major pathways for regulation of RTK protein expression is through ubiquitination of the receptors, targeting them for degradation. One of the most important regulators of ubiquitination of the PDGF receptors is the ubiquitin E3 ligase Cbl [25, 26]. Cbl associates with RTKs via binding of its tyrosine kinase binding (TKB) domain to the receptor, or indirectly through the adaptor protein Grb2. Recruitment of Cbl leads to its phosphorylation through the action of Src family kinases and ubiquitination of lysine residues on target proteins, which tags the proteins to lysosomal degradation via the endosomal sorting complex (reviewed in [27]). It seems that direct binding of Cbl to the PDGF receptors through the TKB domain is the main route of recruiting of Cbl [25]. Cbl interacts with the activated PDGF β-receptor through binding to Tyr1021 in the carboxyterminus of the receptor [28]. This site is also the docking site for PLC-γ1, leading to competition for binding. Indeed, Cbl-deficient cells show an enhanced activation of PLC-γ1 and enhanced PDGF-induced chemotaxis.

Dephosphorylation of the PDGF Receptor

Since tyrosine phosphorylation is an important mediator of diverse physiological responses such as growth, migration, survival, etc., it needs to be tightly regulated. Important players in this regulation are the protein tyrosine phosphatases (PTPs). Most PTPs serve a negative function by dephosphorylating receptors and downstream targets, and thereby terminating signaling. However, some PTPs, such as SHP2, also have a positive function in signaling. An accumulating mass of evidence points towards a certain degree of specificity in PTP action – for example, PTPs recognize specific target sequences that differ between PTPs. One of the PTPs regulating the PDGF receptor is T cell PTP, which, despite its name, is a ubiquitously expressed PTP. Using phosphospecific antibodies against individual sites in the PDGF β-receptor and mouse embryonal fibroblasts carrying a targeted deletion of T cell PTP, Y1021 in the PDGF β-receptor was identified as a target for T cell PTP [29]. Increased phosphorylation of PLC-γ1, the docking partner of Y1021, was seen in cells lacking T cell PTP, as well as a hyperchemotactic response to PDGF-BB.

Stabilization of PDGF Receptor Dimer Formation by NHERF

The Na^+/H^+ exchanger regulatory factor (NHERF) interacts with the carboxyterminal end of the PDGF receptors through one of its PDZ domains [30]. This interaction is ligand dependent but phosphorylation independent, and leads to stabilization of PDGF receptor dimers, thus enhancing PDGF receptor signaling [31]. It links the N-cadherin/catenin complex to the PDGF receptors, and thereby regulates the actin cytoskeleton and cell motility [32].

Modulation of Receptor Signaling by Serine/Threonine Phosphorylation

Apart from dephosphorylation of tyrosine residues on RTKs and their degradation, several RTKs have been shown to be regulated through serine/threonine phosphorylation, including the insulin receptor, the EGF receptor and the stem cell factor receptor/c-Kit [33–35]. Bioukar and colleagues studied PDGF-BB-induced phosphorylation of the PDGF β-receptor and found a rapid increase in serine phoshorylation [36]. An inhibitor of the serine/threonine kinase casein kinase 1 (CK1) blocked this effect. Phosphorylation on serine residues by CK1 was found to attenuate receptor tyrosine kinase activity. Another serine/threonine kinase that regulates PDGF receptor kinase activity is the serine/threonine kinase G-protein-coupled receptor kinase-2 (GRK2), which is mainly known for its role in phosphorylation and desensitization of a large number of G-protein-coupled receptors (reviewed in [37]). GRK2 phosphorylates the PDGF β-receptor at Ser1104 in the carboxyterminal end of the protein, thereby interfering with its interaction with NHERF [38]. Due to the ability of NHERF to facilitate PDGF-induced receptor dimerization, this leads to lower degree of PDGF β-receptor activation and less activation of downstream targets such as Akt.

A-Raf is a less well studied member of the same family of serine/threonine kinase as Raf-1. A-Raf exists in a preformed complex with the PDGF β-receptor and regulates receptor activity through a not fully understood mechanism [39]. Expression of a partially active mutant of A-Raf led to decreased phosphorylation of the PDGF β-receptor selectively at Y857 and Y1021, the activation loop site and the site of interaction with PLC-γ1, respectively.

Modulation of PDGF α-Receptor Signaling by Heparin

Heparan sulfate and heparin are essential components for full ligand binding in a number of receptor systems, including vascular endothelial cell growth factor (VEGF) and fibroblast growth factor (FGF) [40, 41]. Heparin also potentiates the PDGF-BB induced activity of the PDGF α-receptor but not the PDGF β-receptor [42] by a mechanism that is not fully understood.

Transactivation of G-Protein-Coupled Receptors

Contrary to what was originally thought, RTKs and G-protein-coupled receptors (GPCRs) do not signal through distinct cellular systems in a linear manner. Instead, the different receptor systems cross-talk with each other in various ways (reviewed in [43]). This means that RTKs can activate GPCRs and *vice versa*. One of the most well-characterized examples is transactivation of the PDGF β-receptor by angiotensin II. Interestingly, the PDGF α-receptor does not seem to be transactivated by angiotensin II. The mechanism of transactivation is not fully understood, but is dependent on reactive oxygen species (ROS) and independent of Ca^{2+} levels, and occurs independently of the extracellular domain of the PDGF receptor [44, 45]. In some cases, the PDGF β-receptor exists as a preformed complex with GPCRs. EDG1 (synonymous to the sphingosine 1-phosphate receptor) exists in a physically tethered complex with the PDGF β-receptor. G-protein-coupled receptor kinase 2 (GRK2) and β-arrestin were also found in this complex [46].

Signaling does not, however, necessarily go from the GPCR to the PDGF receptor; there are also examples of the reciprocal. In smooth muscle cells, PDGF-induced production of ROS by NADPH oxidase was mediated through the G-protein receptor subunit Gα1,2 and the PDGF α-receptor [47].

Activation of the PDGF β-Receptor by Urokinase

It has been known for quite some time that urokinase-induced signaling in human vascular smooth muscle cells stimulates cell migration and proliferation. Recently, it was found that this is mediated through urokinase-dependent but PDGF-independent activation of the PDGF β-receptor [48]. Urokinase binds to its glycolipid-anchored receptor, which leads to association with the PDGF β-receptor. This interaction leads to dimerization of the PDGF receptor and PDGF-independent activation of its intrinsic kinase activity. Thus, urokinase-induced migration and proliferation can be inhibited by inhibitors of PDGF receptor kinase activity.

Cross-Talk with the TGFβ Receptor System

Transforming growth factor β is a potent cytokine that is involved in numerous physiological processes (for review, see [49]). It exerts most, if not all, of its actions by phosphorylating and activating a class of transcription factors, the Smads. TGFβ mediated phosphorylation of Smad isoforms on serine residues leads to their dimerization and translocation to the nucleus, where they influence gene transcription. In contrast to these classical modes of activating Smads, it was recently demonstrated that receptor tyrosine kinases can also activate Smads in an indirect manner. Yoshida and colleagues [50] clearly showed that PDGF stimulation can phosphorylate and activate Smad2 and Smad3 via the action of the serine/threonine kinase JNK. Furthermore, it was shown that the TGFβ and PDGF signals were synergistic.

Taken together, this shows us that there are no watertight barriers between the signal transduction systems used by various growth factors and cytokines. Cytokine receptors were first found to activate JAK kinases, leading to STAT phosphorylation and transcriptional activation, but also RTKs can phosphorylate STATs. Likewise, the Smad proteins can be phosphorylated by TGFβ but also by RTKs such as the PDGF receptors. It remains to be shown whether cytokine receptors also activate Smads, and whether TGFβ receptors can modulate STAT signaling.

The Role of Reactive Oxygen Species (ROS) in PDGF Signaling

PDGF stimulation leads to the generation of ROS through activation of NADPH oxidase (NOX) in a PI3-kinase- and Rac-dependent manner (reviewed in [51]). The cellular targets for PDGF-induced ROS seem to be the PTPs. In the active center of all PTPs resides an invariate cysteine residue that is essential for the catalytic activity. It participates in formation of a reaction intermediate with target phosphate groups through formation of a thioester. ROS oxidizes this cysteine residue. Several PTPs are known to be deactivated by PDGF-induced ROS, including PTP-PEST, T cell PTP, PTP1B, SHP2, and LMW-PTP. Inactivation of PTPs leads to potentiation of PDGF receptor signaling. Whether additional signal transduction targets for PDGF-induced ROS exist is not known to date.

CONCLUSIONS

Since the PDGF receptor was identified as a ligand-activated tyrosine kinase in 1982, a vast amount of information has accumulated regarding the mechanisms of PDGF receptor signaling. Most of the tyrosine phosphorylation sites and their corresponding docking partners have been identified. However, many issues remain to be solved. This will involve more detailed studies of the role of additional posttranslational modifications on receptor signaling. What is the role of other tyrosine kinases in mediating phosphorylation of the receptors and downstream signalng? What impact does serine/threonine phosphorylation have on receptor function? In recent years, cross-talk between the PDGF receptors and several other receptor systems has emerged, but largely the details remain unknown. The issues of cell type specificity in signaling also need to be addressed. How is specificity in receptor signaling achieved despite the fact that there is considerable overlap in the molecules activated by different receptor tyrosine kinases? Most studies involving PDGF receptors have dealt with the homodimers of PDGF α- and β-receptors, respectively, but data suggest that the heterodimers mediate specific signals, which deserve to be investigated in more detail. Are the different receptor ligands equivalent in their ability to mediate downstream signaling? With the advent of novel technologies, we will in the future be able to study the signal transduction processes at the subcellular level to a higher degree than we do today, and this will likely enlighten and broaden our knowledge and understanding of growth factor signaling.

ACKNOWLEDGEMENTS

My apologies to authors of many relevant references that I was unable to cite due to space restriction.

REFERENCES

1. Balk SD. Stimulation of the proliferation of chicken fibroblasts by folic acid or a serum factor(s) in a plasma-containing medium. *Proc Natl Acad Sci USA* 1971;**68**:1689–92.
2. Heldin CH, Eriksson U, Östman A. New members of the platelet-derived growth factor family of mitogens. *Arch Biochem Biophys* 2002;**398**:284–90.
3. Betsholtz C. Insight into the physiological functions of PDGF through genetic studies in mice. *Cytokine Growth Factor Rev* 2004;**15**:215–28.
4. Heuchel R, Berg A, Tallquist M, Åhlén K, Reed RK, Rubin K, Claesson-Welsh L, Heldin CH, Soriano P. Platelet-derived growth factor beta receptor regulates interstitial fluid homeostasis through phosphatidylinositol-3′ kinase signaling. *Proc Natl Acad Sci USA* 1999;**96**:11,410–11,415.
5. Lokker NA, O'Hare JP, Barsoumian A, Tomlinson JE, Ramakrishnan V, Fretto LJ, Giese NA. Functional importance of platelet-derived growth factor (PDGF) receptor extracellular immunoglobulin-like domains. Identification of PDGF binding site and neutralizing monoclonal antibodies. *J Biol Chem* 1997;**272**:33,037–33,044.
6. Heidaran MA, Pierce JH, Jensen RA, Matsui T, Aaronson SA. Chimeric α- and β-platelet-derived growth factor (PDGF) receptors define three immunoglobulin-like domains of the α-PDGF receptor that determine PDGF-AA binding specificity. *J Biol Chem* 1990;**265**:18,741–18,744,.
7. Omura T, Heldin CH, Östman A. Immunoglobulin-like domain 4-mediated receptor-receptor interactions contribute to platelet-derived growth factor-induced receptor dimerization. *J Biol Chem* 1997;**272**:12,676–12,682.
8. Yuzawa S, Opatowsky Y, Zhang Z, Mandiyan V, Lax I, Schlessinger J. Structural basis for activation of the receptor tyrosine kinase KIT by stem cell factor. *Cell* 2007;**130**:323–34.
9. Claesson-Welsh L, Eriksson A, Westermark B, Heldin CH. cDNA cloning and expression of the human A-type platelet-derived growth

factor (PDGF) receptor establishes structural similarity to the B-type PDGF receptor. *Proc Natl Acad Sci USA* 1989;**86**:4917–21.

10. Li X, Pontén A, Aase K, Karlsson L, Abramsson A, Uutela M, Bäckström G, Hellström M, Boström H, Li H, Soriano P, Betsholtz C, Heldin CH, Alitalo K, Östman A, Eriksson U. PDGF-C is a new protease-activated ligand for the PDGF alpha-receptor. *Nat Cell Biol* 2000;**2**:302–9.
11. Fredriksson L, Li H, Eriksson U. The PDGF family: four gene products form five dimeric isoforms. *Cytokine Growth Factor Rev* 2004;**15**:197–204.
12. Sweeney II. C, Lai C, Riese DJ, Diamonti III. AJ, Cantley LC, Carraway KL Ligand discrimination in signaling through an ErbB4 receptor homodimer. *J Biol Chem* 2000;**275**:19,803–19,807.
13. Hubbard SR. Juxtamembrane autoinhibition in receptor tyrosine kinases. *Nat Rev Mol Cell Biol* 2004;**5**:464–71.
14. Fantl WJ, Escobedo JA, Williams LT. Mutations of the platelet-derived growth factor receptor that cause a loss of ligand-induced conformational change, subtle changes in kinase activity, and impaired ability to stimulate DNA synthesis. *Mol Cell Biol* 1989;**9**:4473–8.
15. Kazlauskas A, Cooper JA. Autophosphorylation of the PDGF receptor in the kinase insert region regulates interactions with cell proteins. *Cell* 1989;**58**:1121–33.
16. Griffith J, Black J, Faerman C, Swenson L, Wynn M, Lu F, Lippke J, Saxena K. The structural basis for autoinhibition of FLT3 by the juxtamembrane domain. *Mol Cell* 2004;**13**:169–78.
17. Heinrich MC, Corless CL, Duensing A, McGreevey L, Chen CJ, Joseph N, Singer S, Griffith DJ, Haley A, Town A, Demetri GD, Fletcher CD, Fletcher JA. PDGFRA activating mutations in gastrointestinal stromal tumors. *Science* 2003;**299**:708–10.
18. Stover EH, Chen J, Folens C, Lee BH, Mentens N, Marynen P, Williams IR, Gilliland DG, Cools J. Activation of FIP1L1-PDGFRalpha requires disruption of the juxtamembrane domain of PDGFRalpha and is FIP1L1-independent. *Proc Natl Acad Sci USA* 2006;**103**:8078–83.
19. Hansen K, Johnell M, Siegbahn A, Rorsman C, Engström U, Wernstedt C, Heldin CH, Rönnstrand L. Mutation of a Src phosphorylation site in the PDGF β-receptor leads to increased PDGF-stimulated chemotaxis but decreased mitogenesis. *EMBO J* 1996;**15**:5299–313.
20. Pawson T, Nash P. Assembly of cell regulatory systems through protein interaction domains. *Science* 2003;**300**:445–52.
21. Heldin CH, Östman A, Rönnstrand L. Signal transduction via platelet-derived growth factor receptors. *Biochim Biophys Acta* 1998; **1378**:F79–F113.
22. Ekman S, Kallin A, Engström U, Heldin CH, Rönnstrand L. SHP-2 is involved in heterodimer specific loss of phosphorylation of Tyr771 in the PDGF β-receptor. *Oncogene* 2002;**21**:1870–5.
23. Plattner R, Irvin BJ, Guo S, Blackburn K, Kazlauskas A, Abraham RT, York JD, Pendergast AM. A new link between the c-Abl tyrosine kinase and phosphoinositide signalling through PLC-gamma1. *Nat Cell Biol* 2003;**5**:309–19.
24. Buday L, Warne PH, Downward J. Downregulation of the Ras activation pathway by MAP kinase phosphorylation of Sos. *Oncogene* 1995;**11**:1327–31.
25. Bonita Jr. DP, Miyake S, Lupher ML, Langdon WY, Band H Phosphotyrosine binding domain-dependent upregulation of the platelet-derived growth factor receptor alpha signaling cascade by transforming mutants of Cbl: implications for Cbl's function and oncogenicity. *Mol Cell Biol* 1997;**17**:4597–610.
26. Miyake S, Mullane-Robinson KP, Lill NL, Douillard P, Band H. Cbl-mediated negative regulation of platelet-derived growth factor receptor-dependent cell proliferation. A critical role for Cbl tyrosine kinase-binding domain. *J Biol Chem* 1999;**274**: 16,619–16,628.
27. Haglund K, Di Fiore PP, Dikic I. Distinct monoubiquitin signals in receptor endocytosis. *Trends Biochem Sci* 2003;**28**:598–603.
28. Reddi AL, Ying G, Duan L, Chen G, Dimri M, Douillard P, Druker BJ, Naramura M, Band V, Band H. Binding of Cbl to a phospholipase Cgamma1-docking site on platelet-derived growth factor receptor beta provides a dual mechanism of negative regulation. *J Biol Chem* 2007;**282**:29,336–29,347.
29. Persson C, Sävenhed C, Bourdeau A, Tremblay ML, Markova B, Böhmer FD, Haj FG, Neel BG, Elson A, Heldin CH, Rönnstrand L, Östman A, Hellberg C. Site-selective regulation of platelet-derived growth factor beta receptor tyrosine phosphorylation by T-cell protein tyrosine phosphatase. *Mol Cell Biol* 2004;**24**:2190–201.
30. Maudsley S, Zamah AM, Rahman N, Blitzer JT, Luttrell LM, Lefkowitz RJ, Hall RA. Platelet-derived growth factor receptor association with Na(+)/H(+) exchanger regulatory factor potentiates receptor activity. *Mol Cell Biol* 2000;**20**:8352–63.
31. Demoulin JB, Seo JK, Ekman S, Grapengiesser E, Hellman U, Rönnstrand L, Heldin CH. Ligand-induced recruitment of Na+/H+-exchanger regulatory factor to the PDGF (platelet-derived growth factor) receptor regulates actin cytoskeleton reorganization by PDGF. *Biochem J* 2003;**376**:505–10.
32. Theisen III. CS, Wahl JK, Johnson KR, Wheelock MJ NHERF links the N-cadherin/catenin complex to the platelet-derived growth factor receptor to modulate the actin cytoskeleton and regulate cell motility. *Mol Biol Cell* 2007;**18**:1220–32.
33. Takayama S, White MF, Kahn CR. Phorbol ester-induced serine phosphorylation of the insulin receptor decreases its tyrosine kinase activity. *J Biol Chem* 1988;**263**:3440–7.
34. Countaway JL, Nairn AC, Davis RJ. Mechanism of desensitization of the epidermal growth factor receptor protein-tyrosine kinase. *J Biol Chem* 1992;**267**:1129–40.
35. Blume-Jensen P, Rönnstrand L, Gout I, Waterfield MD, Heldin CH. Modulation of Kit/stem cell factor receptor-induced signaling by protein kinase C. *J Biol Chem* 1994;**269**:21,793–21,802.
36. Bioukar EB, Marricco NC, Zuo D, Larose L. Serine phosphorylation of the ligand-activated β-platelet-derived growth factor receptor by casein kinase I-γ2 inhibits the receptor's autophosphorylating activity. *J Biol Chem* 1999;**274**:21,457–21,463.
37. Ribas Jr. C, Penela P, Murga C, Salcedo A, Garcia-Hoz C, Jurado-Pueyo M, Aymerich I, Mayor F. The G-protein-coupled receptor kinase (GRK) interactome: role of GRKs in GPCR regulation and signaling. *Biochim Biophys Acta* 2007;**1768**:913–22.
38. Hildreth KL, Wu JH, Barak LS, Exum ST, Kim LK, Peppel K, Freedman NJ. Phosphorylation of the platelet-derived growth factor receptor-beta by G-protein-coupled receptor kinase-2 reduces receptor signaling and interaction with the Na(+)/H(+) exchanger regulatory factor. *J Biol Chem* 2004;**279**:41,775–41,782.
39. Mahon ES, Hawrysh AD, Chagpar RB, Johnson LM, Anderson DH. A-Raf associates with and regulates platelet-derived growth factor receptor signalling. *Cell Signal* 2005;**17**:857–68.
40. Jakobsson L, Kreuger J, Holmborn K, Lundin L, Eriksson I, Kjellén L, Claesson-Welsh L. Heparan sulfate in trans potentiates VEGFR-mediated angiogenesis. *Dev Cell* 2006;**10**:625–34.
41. Schlessinger J, Plotnikov AN, Ibrahimi OA, Eliseenkova AV, Yeh BK, Yayon A, Linhardt RJ, Mohammadi M. Crystal structure of a ternary FGF–FGFR–heparin complex reveals a dual role for heparin in FGFR binding and dimerization. *Mol Cell* 2000;**6**:743–50.

42. Rolny C, Spillmann D, Lindahl U, Claesson-Welsh L. Heparin amplifies platelet-derived growth factor (PDGF)-BB-induced PDGF alpha -receptor but not PDGF β-receptor tyrosine phosphorylation in heparan sulfate-deficient cells. Effects on signal transduction and biological responses. *J Biol Chem* 2002;**277**:19,315–19,321.
43. Natarajan K, Berk BC. Crosstalk coregulation mechanisms of G-protein-coupled receptors and receptor tyrosine kinases. *Methods Mol Biol* 2006;**332**:51–77.
44. Heeneman S, Haendeler J, Saito Y, Ishida M, Berk BC. Angiotensin II induces transactivation of two different populations of the platelet–derived growth factor beta receptor. Key role for the p66 adaptor protein Shc. *J Biol Chem* 2000;**275**:15,926–15,932.
45. Gao BB, Hansen H, Chen HC, Feener EP. Angiotensin II stimulates phosphorylation of an ectodomain-truncated platelet-derived growth factor receptor-β and its binding to class IA PI3K in vascular smooth muscle cells. *Biochem J* 2006;**397**:337–44.
46. Alderton F, Rakhit S, Kong KC, Palmer T, Sambi B, Pyne S, Pyne NJ. Tethering of the platelet-derived growth factor beta receptor to G-protein-coupled receptors. A novel platform for integrative signaling by these receptor classes in mammalian cells. *J Biol Chem* 2001;**276**:28,578–28,585.
47. Kreuzer J, Viedt C, Brandes RP, Seeger F, Rosenkranz AS, Sauer H, Babich A, Nurnberg B, Kather H, Krieger-Brauer HI. Platelet-derived growth factor activates production of reactive oxygen species by NAD(P)H oxidase in smooth muscle cells through Gi1,2. *FASEB J* 2003;**17**:38–40.
48. Kiyan J, Kiyan R, Haller H, Dumler I. Urokinase-induced signaling in human vascular smooth muscle cells is mediated by PDGFR-β. *EMBO J* 2005;**24**:1787–97.
49. Moustakas A, Pardali K, Gaal A, Heldin CH. Mechanisms of TGF-β signaling in regulation of cell growth and differentiation. *Immunol Letts* 2002;**82**:85–91.
50. Yoshida K, Matsuzaki K, Mori S, Tahashi Y, Yamagata H, Furukawa F, Seki T, Nishizawa M, Fujisawa J, Okazaki K. Transforming growth factor-beta and platelet-derived growth factor signal via c-Jun N-terminal kinase-dependent Smad2/3 phosphorylation in rat hepatic stellate cells after acute liver injury. *Am J Pathol* 2005;**166**:1029–39.
51. Kang SW. Two axes in platelet-derived growth factor signaling: tyrosine phosphorylation and reactive oxygen species. *Cell Mol Life Sci* 2007;**64**:533–41.

Chapter 18

The Fibroblast Growth Factor (FGF) Signaling Complex

Wallace L. McKeehan, Fen Wang, and Yongde Luo

Center for Cancer and Stem Cell Biology, Institute of Biosciences and Technology, Texas A&M Health Science Center, Houston, Texas

INTRODUCTION

The fibroblast growth factor (FGF) signaling system is a ubiquitous fast-acting cellular sensor of local environmental changes and mediator of cell-to-cell communication with broad roles in cellular homeostasis in development and adult organs. Through the interaction of heparan sulfate (HS) with both activating FGF polypeptides and transmembrane FGF receptor (FGFR) tyrosine kinases, the system is rigorously modulated by the extracellular matrix and tissue architecture. Diversity and cell and tissue specificity of signaling result from the combinatorial oligomerization of a family of 18 receptor binding FGF homologs, diverse oligosaccharide motifs within HS chains of proteoglycans, and ectodomain variants resulting from splice variations from four genes coding for four transmembrane tyrosine kinases. Most recently FGFR signaling has been implicated in endocrine control of metabolic homeostasis that includes cholesterol/bile acid metabolism [1], lipid and glucose metabolism [2] and calcium and phosphate metabolism [3]. The endocrine activities of FGF signaling are regulated by membrane-anchored co-factors klotho (αklotho) and βklotho [4].

FGF POLYPEPTIDES

Activating FGF ligands are single heparin binding polypeptides sharing a homologous core split and flanked by areas of less homology [5, 6]. The heparin-binding property reflects the intimate association of the FGF polypeptides with motifs within HS chains of proteoglycans in the extracellular matrix and in the FGFR signaling complex [7, 8]. Most FGFs are translated with a conventional secretory signal peptide at the N-terminus; however, FGF1, FGF2, and several other homologs are not, although they appear in the external environment. This has fueled questions concerning novel mechanisms of FGF exit from cells as well as potential intracellular signaling [9]. Crystal structures indicate that the FGFs share a similar three-dimensional structure characterized by a conserved hydrophobic patch that interacts with FGFR and a unique heparin binding surface that differs among FGFs [10]. The HS and FGFR binding domains in FGFs are comprised of secondary and tertiary structure formed from distal sequence domains. Functional analysis by site-directed mutagenesis indicates that heparin binding, FGFR binding, and biological activity are intimately associated [5, 10, 11]. The specificity of an individual FGF is likely determined by a composite of the unique heparin binding domain and side-chain interactions with the ectodomain of a specific oligomeric HS-FGFR tyrosine kinase complex.

A subset of the human FGF homologs, FGF19, FGF21 and FGF23, appears to play endocrine roles in metabolic homeostasis [4, 12], in contrast to the majority of FGFs that act short-range within tissues mediating cell-to-cell communication in control of cellular homeostasis. Murine FGF15 is thought to be the interspecies ortholog of human FGF19 [13] although it exhibits only a 53 percent sequence homology [14]. The three "endocrine" FGFs exhibit a reduced affinity for heparin that is thought to facilitate movement through the blood from a distal origin to the target cells displaying FGFR [15]. Consistent with their roles in either cellular or metabolic homeostasis, expression of diverse members of the FGF polypeptide family is usually partitioned within a different cell type or tissue, respectively, than the target FGFR signaling complex [16–18]. This promotes a controlled cell-specific cell-to-cell paracrine communication or a tissue-to-tissue endocrine communication via the partitioned signal and receptor. A major level of control of cell-specific expression of diverse FGFs both in development and adult homeostasis is by cell-specific

Handbook of Cell Signaling, Three-Volume Set 2 ed.

transcription factors, among them nuclear transcription factor receptors [19–21].

FGFR TYROSINE KINASES

FGFs elicit activity by activation of an oligomeric complex of transmembrane receptor tyrosine kinases and HS proteoglycans. The FGFR kinase monomer is composed of a single polypeptide chain that has a glycosylated extracellular ligand binding domain, a transmembrane domain flanked by juxtamembrane sequences, and a core tyrosine kinase with distinct structural subdomains followed by a C-terminal domain (Figure 18.1). The intracellular juxtamembrane region is where the most important intracellular signal transmission relay adaptor (FRS2) and tyrosine kinase substrate associates with FGFR [22]. The region is also critical in relay of conformational changes in the ectodomain to the intracellular domain of the FGFR kinase [23]. The C-terminal to the intracellular juxtamembrane domain is the relatively conserved core tyrosine kinase sequence domain comprised of adenosine triphosphate (ATP) binding and catalytic subdomains. Within the kinase domain is a mobile kinase repression/de-repression sequence containing two conserved tyrosine autophosphorylation sites that when phosphorylated allow access of substrates to the FGFR kinase [24]. Extending from the C-terminus of the kinase domain is a less-conserved region among the four FGFRs containing a third conserved tyrosine autophosphorylation site that recruits phospholipase C gamma (PLCγ) [25].

The extracellular domain of the transmembrane FGFR kinase consists of two or three immunoglobulin (Ig)-like modules. Ig modules II and III of FGFR independently bind heparin or HS, and, in complex with affinity-selected heparin or HS, exhibit independent binding sites for an FGF [26]. Heparin-module II complexes bind a wider spectrum of FGFs than Ig module III-heparin complexes [26, 27]. However, only one FGF binds to a monomeric ectodomain composed of Ig modules II and III, indicating that the independent binding sites on modules II and III likely cooperate in affinity and specificity for a single FGF [26, 28, 29]. Ig modules II and III cooperate both within monomers and across dimers with cellular HS to confer specificity for FGF [28, 29]. The FGFR forms ligand-independent dimers or higher-order oligomers on the cell surface that are dependent on the sequence between Ig modules II and III [30]. A highly conserved sequence domain rich in basic amino acids within the N-terminus of Ig module II is required for the interaction with heparin and HS [31]. Mutations in this domain significantly affect both HS and FGF binding and activities of the FGFR complex. The N-terminal to Ig module II is an alternatively spliced sequence rich in serines and acidic residues called the *acidic box*. Alternatively spliced Ig module I modifies the affinity for both heparin and FGF [32]. The exon coding for the acidic box sequence is always included whenever Ig module I is present. The acidic box sequence between Ig modules I and II is a structural element that contributes to reduction of affinity for both HS and FGF by Ig module I [32].

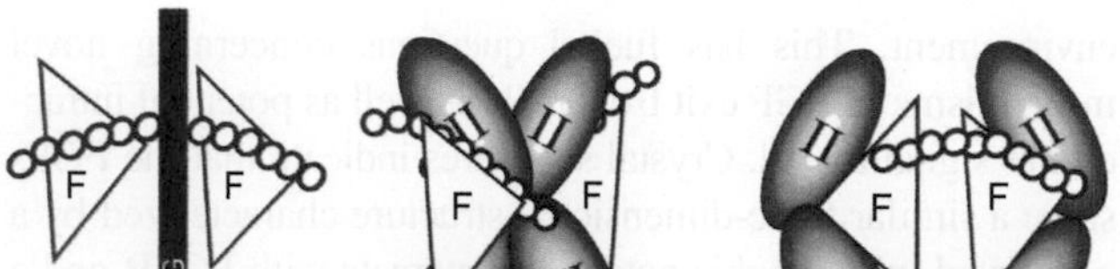

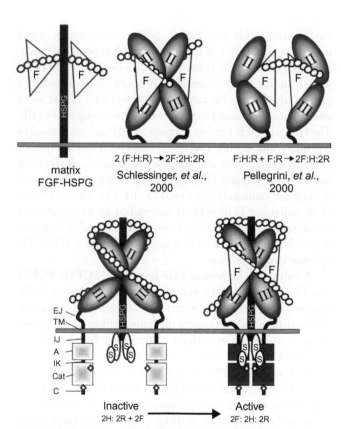

FIGURE 18.1 Models of the FGFR signaling complex.Independent binding of FGF to matrix heparin-sulfate chains of a membrane-anchored proteoglycan (HSPG) in both orientations confirmed by crystal structures of heparin–FGF complexes is indicated. Also shown are the two divergent models of active dimeric complexes of heparin oligosaccharide (H) (open circles), FGFR (R) comprised of Ig modules II and III, and FGF (F), with the order of assembly proposed by the authors. At the bottom, an alternative conformational model in membrane context is depicted. EJ, extracellular juxtamembrane; TM, transmembrane; A, kinase ATP binding domain; IK, interkinase domain sequence; Cat, kinase catalytic domain; C, C-terminal sequence; S, membrane-anchored FGFR kinase substrates. Functional tyrosine phosphorylation sites are indicated by diamonds.

HEPARAN SULFATE AND KLOTHOS

Pericellular matrix HS participates in rigorous control of the FGFR signaling system at multiple levels that exceeds the classical role of HS motifs in anticoagulation in complexity and impact on biological processes. The independent interaction of matrix HS with FGFs affects location and trafficking within tissues, access to the FGFR signaling complex, and lifetime and stability of FGFs by protection against proteases [5, 7, 8, 10]. Structure–function analyses and co-crystal structures of FGF1 and FGF2 with heparin oligosaccharides indicate that the heparin binding domain of FGF is a composite domain contributed by distal sequence residues and formed by secondary and three-dimensional structures [10, 33]. Of the 18 FGFR-mediated FGF homologs, co-crystal structures of only FGF1 and FGF2 with fully sulfated heparin oligosaccharides have been determined. Although a variety of carbohydrate-like electrolytes interact, affinity purification and structural analysis indicate that a hexameric heparin oligosaccharide exhibiting at least 2-O sulfation and both 2-O and 6-O sulfation is required for highest-affinity interaction of heparin oligosaccharides to FGF2 and FGF1, respectively [10, 34]. Crystal structures of other FGFs have been determined, but attempts at co-crystallization with simple 2-O and 6-O sulfated 6- to 12-mer oligosaccharides have failed [10]. Differing heparin binding domains among FGFs suggest specific requirements in respect to composition and length of interactive oligosaccharide [7, 10]. As predicted by the unique heparin binding domain of FGF7 relative to FGF1 and FGF2, a longer heparin oligosaccharide that exhibits anticoagulant activity and the presence of a 3-O sulfate is required to interact with and protect FGF7 against protease [7, 10]. FGF19, FGF21, and FGF23 exhibit the least affinity for heparin coincident with the sparse distribution of basic amino acid residues and in some cases the presence of acidic residues on the HS binding face homologous to FGF1 and FGF2 [15].

In the absence of FGF, self-associated dimers of the FGFR ectodomain form a complex with crude heparin with a K_d that exceeds 10 nM that is competent to bind FGF in the absence of soluble heparin or cellular HS [5, 26, 35–37]. The strict dependence of FGF binding on heparin/HS and the high-affinity formation of a functional, specific complex of FGFR with heparin/HS requires the presence of physiological levels of extracellular divalent cations [35] and assembly of dimers of FGFR anchored in a membrane by its transmembrane domain [7]. Similar to FGF7, only the fraction of heparin or HS that binds to antithrombin and has anti-Factor Xa activity (anticoagulant heparin/HS) exhibited functional high-affinity binding to dimers of FGFR in the absence of FGF [10, 37]. The independent interaction of unliganded dimeric complexes of the FGFR ectodomain with heparin or HS protects them against proteolysis, stabilizes them, and likely represents the composite oligomeric complex that is activated upon the binding of FGF [7, 26, 35, 36]. A partnership between the FGFR isotype and the rare and specific motif within an HS chain that interacts with the FGFR ectodomain dimer determines the specificity of the overall FGFR signaling complex for a particular FGF [7, 8, 17, 26, 36].

αklotho is a type I membrane-anchored protein with anti-aging effects that is preferentially expressed in kidney, bone, and parathyroid, and involved in vitamin D, calcium, and phosphate homeostasis parallel to similar biological effects of FGF23 [4, 38]. βklotho is a homolog of αklotho that is expressed predominantly in the liver and white adipose tissue – key organs in cholesterol, lipid, and glucose metabolism. A deficiency of βklotho in mice caused similar effects on cholesterol and bile acid metabolism as a deficiency in FGF15, and hepatocyte FGFR4. αklotho and βklotho are required for high-affinity binding and activation of selected FGFR isoforms by FGF21, FGF19, and FGF23, and, similar to HS for other FGFs, have been reported to interact with both the three FGFs and FGFR independently.

THE OLIGOMERIC FGF–FGFR–HS SIGNALING COMPLEX

The FGFR tyrosine kinase is limited by a flexible structural domain for which tyrosine phosphorylation by a neighboring FGFR kinase within a dimer or higher-order oligomer releases the repression [24]. The order of assembly; the stoichiometry; the conformation of inactive and active complexes of FGFR; and how rearrangement of the interactions by FGF binding signals transactivation of the kinases and interaction and activation of intracellular mediators of downstream signaling pathways is largely unclear. One monomeric unit of the FGFR complex is generally agreed to be a composite ternary complex of one FGF, one FGFR, and one HS chain in which the sugar chain in a specific orientation concurrently interacts with heparin binding domains from both FGFR and FGF [5, 28, 29]. However, prediction of how monomeric units come together to transactivate the intracellular kinases is complicated by irreconcilable models from dimeric crystal structures derived from the same FGF, FGFR ectodomain, and artificial heparin-derived oligosaccharide in which the second unit of the dimer is fundamentally different [28, 29] (Figure 18.1). One structure suggests a symmetrical complex that arises from back-to-back interaction of two identical ternary complexes of FGF, FGFR, and heparin/HS [28]. This dimer, consisting of FGF, FGFR, and heparin/HS with stoichiometry of 2 : 2 : 2, is stabilized by secondary interactions between FGF in one ternary unit and FGFR in the other unit, FGFR–FGFR contacts, and each heparin chain that spans both FGFRs. Another structure [29] suggests that the ternary complex of FGF, FGFR, and a single heparin/HS chain recruits a second complex of FGF and FGFR. The single heparin/HS chain binds asymmetrically to an FGF bound independent of heparin/HS to a second FGFR. The net result is a heteropentameric complex with overall stoichiometry of two FGFs to two FGFRs to one HS chain, with minimal direct contacts between the two face-to-face FGFRs. The dimeric complex is stabilized simply by the sugar chain bridge between one ternary unit and FGF bound to the second unit. In both models, sustained de-repression of the active sites of the monomeric FGFR kinases by transphosphorylation was viewed to be limited by concentration, proximity, and stochastic interactions of ectodomains of FGFR, short, soluble, relatively non-specific, HS chains, and FGF which are stabilized by both HS and FGF. The irreconcilable differences in crystal structures have been attributed to different conditions of crystallization and therefore crystallographic artifact [39]. Although the structures provide clues to elements of the FGFR complex, they are both likely subject to artifact due to use of non-specific fully sulfated heparin oligosaccharides, the sensitivity of specific HS interactions with FGF and FGFR to electrolyte conditions, and difficulty in dissection of specific and non-specific interactions among heparin, FGFR, and FGF at high concentrations of the fully sulfated heparin polyelectrolyte.

An alternative model of the oligomeric FGFR signaling complex that unifies biochemical, structural, and functional data has been proposed [5, 23, 26, 35–37]. This model proposes a pre-existing, inactive, unliganded symmetric complex of two FGFR ectodomains interacting back-to-back upstream of Ig module III, most likely at the hinge-like connecter sequence between Ig modules II and III, while anchored to a proteoglycan core through two HS chains. HS chains interact with a primary heparin binding domain on Ig module II and extend across the dimer to interact with Ig module III on the adjacent partner (Figure 18.1). Specific functional HS binding to FGFR in contrast to non-specific electrolyte interaction is divalent cation dependent, and requires rare oligosaccharide motifs properly spaced within the HS chain [7, 8]. Divalent cations and HS in a 2 FGFR : 2 HS unliganded complex cooperate in the ectodomain to conformationally restrict the kinase-to-kinase substrate relationship that is in relative proximity in the intracellular domain. This maintains dependence on but allows rapid and graded transactivation of the kinases by the binding of FGF or other perturbations of the external matrix HS–FGFR relationship. Docking of activating FGF into composite HS–protein sites formed by Ig modules II and III of the preexisting inactive complex or other perturbations of the FGFR–HS relationship is transmitted to the intracellular kinase domains to overcome the conformational restrictions that limit sustained transactivation of the kinases and relationship with membrane-bound substrates requiring phosphorylation or conformational reshuffling for activity. Transmission of conformational change across dimers and to the intracellular juxtamembrane and ATP binding site of the kinase is enhanced by the bivalent contact of FGF and HS with both FGFRs within the dimer [23]. In this model, HS chains of matrix proteoglycans play a central role in negative conformational control of pre-existing unliganded FGFR kinase complexes, the requirement and specificity for FGF, the cell context specificity of FGF signaling, and integration of signaling with tissue matrix remodeling. Consistent with the idea that the cellular FGFs and FGFR ectodomains are unlikely to exist as free components in the pericellular environment, total HS both in the extracellular matrix and at the cell surface is in large excess over concentrations of FGF and FGFR [7]. FGF ligands exhibit a wide range of affinities and specificity for HS motifs, ranging from widely overlapping interactions largely dominated by non-specific charge density to highly specific requirements independent of charge density requiring specific disposition of side-chains on the HS motif [7, 40]. The structural requirements in the HS motif for high affinity and specific interaction with FGFR appear to be more restrictive, and requirements for the ternary interaction between FGFR and FGF are likely even more restrictive [36, 37]. In the case of FGF7 and FGFR2IIIb, the rare HS motif with highest affinity and exhibiting structural specificity for FGF7 and FGFR2IIIb independently converges with the one required for activation of the FGFR signaling complex [7, 8]. These observations suggest a model where non-receptor bound HS sites compete with HS–FGFR complexes for FGF in overall control of access of FGFs to the unliganded, inactive HS–FGFR complex.

Structural-level mechanisms underlying the interaction and modification of HS–FGFR oligomeric complexes by αklotho and βklotho to confer high-affinity binding and activation of FGFR signaling by FGF19, FGF21, and FGF23 remain to be determined.

INTRACELLULAR SIGNAL TRANSDUCTION BY THE FGFR COMPLEX

Overall, FGF tyrosine kinase receptors appear to work with the same canonical intracellular signal transduction networks as other tyrosine receptor kinases (RTKs) [41, 42]. Yet FGF signaling elicits unique co-factor and context-dependent endpoints that differ not only from other RTK families but also among the four different tyrosine kinases within the intracellular domains of FGFR isotypes. How this specificity relatively to other RTKs and diversity of endpoints within the FGFR family arises remains a major question.

There are seven phosphotyrosines in the intracellular domain of FGFR1, five of which appear temporally in specific order upon activation of the FGFR complex with FGF [43]. Only four phosphotyrosine sites (counterparts of tyr653, tyr654, tyr730 and tyr766 in FGFR1) are conserved across all four isotypes of the FGFR family. Only three of the conserved sites clearly play a direct role in FGFR signaling. Phosphorylation of the additional tyrosines appear to play indirect conformational roles in control of the magnitude of FGFR kinase activity [43]. Tyr653 plays an essential role in the conformational blockade of the kinase site that prevents sustained activating transphosphorylation between FGFR monomers and then access to the site of signal-mediating substrates [24]. Tyr766 appears to be the single site that plays a role in direct recruitment of a phosphotyrosine-binding signaling molecule (PLCγ).

The single phosphotyrosine recruitment site PLCγ distinguishes FGFR signaling from most other RTKs which recruit downstream signaling mediators via diverse phosphotyrosine sites distributed across the intracellular domain sequence. FGFR signaling is in the largest part mediated by two isotypes of a single membrane-bound organizer of signal relay molecules, FGF receptor substrate 2 (FRS2) [22]. Both isotypes, FRS2α and FRS2β, are anchored to the cell membrane via the N-terminus by a lipid-binding myristyl group in constitutive association through an N-terminal PTB domain with an unphosphorylated sequence in the intracellular juxtamembrane domain in FGFRs. The interaction is stabilized indirectly by phosphorylation of FGFR phosphotyrosine sites, but unaffected by phosphorylation of tyrosine sites on FRS2α [44]. FRS2α and FRS2β exhibit six and five tyrosine phosphorylation sites, respectively. Four and three sites, respectively, are involved in the interaction with Grb2 that carries its partners SOS, Cbl, and Gab1 [22]. Guanine nucleotide exchange factor (GEF) provides the conduit to the canonical ras–raf–ERK signaling pathway. Cbl is an E3 ubiquitin ligase, and Gab1 interfaces with the PI–3 kinase pathway. Two other phosphotyrosine sites interface with tyrosine phosphatase Shp2. Other intracellular signal modifiers interact with FRS2 independent of its phosphorylation [22]. These include Rnd, which interfaces with G-protein signaling; cell-cycle modifiers Cks1/Cks2; and Sprouty, a negative regulator of the Ras/ERK pathway.

In summary, similar to the extracellular domain, the membrane-proximal intracellular domain of unliganded FGFR is part of a multi-subunit complex of largely pre-assembled membrane-anchored subunits poised for activation by the FGFR kinase signaled through the cell membrane upon binding of FGF in the ectodomain. The quantity and quality of phosphorylation of FRS2 is both FGFR isotype and cell-type specific [45, 46]. Similar to affinity and specificity for FGF [17, 18] and HS oligosaccharide motif [7, 36, 37], dependence of the FGFR kinases on HS and FGF for transactivation [35], and specific structural relationships in the unliganded HS–FGFR dimer [23, 26, 30], the differential phosphorylation of FRS2 by FGFR appears only to occur in intact cells and is dependent on cell membrane context [46]. The composition of the FRS2 complex and differential phosphorylation of both FRS2 and assembled substrates determined by proximity to the FGFR kinase site may be involved in determination of the signaling specificity of FGFR isotypes. Diverse conformational relationships determined by cell membrane and cytoskeletal context among the intracellular FGFR kinases and membrane-anchored substrates and molecules complexed with them that are transmitted by conformational arrangements among FGFR ectodomains, pericellular HS, and FGF in the extracellular environment likely account for the diversity of FGFR isotype- and cell-type-specific FGF signaling. In some systems, the four FGFR isotypes elicit similar and redundant effects on cell responses [47, 48]. In others, individual isotypes exhibit dramatically different effects on cell phenotype, some of which are in opposition. Whether FGFR signaling is transient or sustained from a single isotype can also affect quality of the response. For example, although FGFR1 and FGFR3 intracellular domains appear redundant in eliciting neurite outgrowth in PC12 neural cells [48], only the FGFR1 kinase elicits neurite outgrowth when the FGFR ectodomain is utilized [49]. The FGFR3 kinase failed to sustain outgrowth and Ras-dependent gene expression, but instead induced neural-specific gene expression pathways that were Ras independent [50]. In bladder [51], prostate [52, 53], and salivary tumor epithelial cells [54], the resident FGFR2 kinase promotes homeostasis between cellular compartments and suppresses progression toward malignancy. Resident hepatocyte FGFR4 that is a key regulator of cholesterol/bile acid and lipid metabolism in hepatocytes is a hepatoma tumor suppressor [55]. This is in contrast to normally mesenchymal FGFR1, which drives malignancy when it appears ectopically in parenchymal cells [52, 56–58].

As mentioned earlier, membrane-bound co-factors αklotho and βklotho convert the HS–FGFR1 signaling complex that normally responds to tissue FGFs to mediate cellular homeostasis in development and response to injury into an FGF21- and FGF23-controlled mediator of tissue-specific metabolic homeostasis [59, 60]. Remarkably, FGF21 and FGF23 do not appear to promote mitogenic and other phenotypes associated with cellular homeostasis that are characteristic of FGFR1. The cell- and tissue-specific partnership of HS motif, FGFR isotype, and klothos in the ectodomain, FGF-specific conformational perturbations that are transmitted to the intracellular domain to sustain de-repression of the conformational blockade of the FGFR kinase, and access of components of the multi-subunit FRS2 complex to the kinase are candidates underlying this diversity of cellular responses.

For more details on the role of the FGF family in specific biological systems, the diverse signaling pathways impacted by the FGF signaling system and the irreconcilable differences in structural models of the FGF signaling complex, readers are referred to specialized reviews [4–6, 12, 14, 22, 39, 42, 61–65].

REFERENCES

1. Yu C, Wang F, Kan M, et al. Elevated cholesterol metabolism and bile acid synthesis in mice lacking membrane tyrosine kinase receptor FGFR4. *J Biol Chem* 2000;**275**(20):15,482–9.
2. Huang X, Yang C, Luo Y, Jin C, Wang F, McKeehan WL. FGFR4 prevents hyperlipidemia and insulin resistance but underlies high-fat diet induced fatty liver. *Diabetes* 2007;**56**(10):2501–10.
3. Razzaque MS, Lanske B. The emerging role of the fibroblast growth factor-23-klotho axis in renal regulation of phosphate homeostasis. *J Endocrinol* 2007;**194**(1):1–10.
4. Kuro-o M. Endocrine FGFs and klothos: emerging concepts. *Trends Endocrinol Metab* 2008;**19**(7):239–45.
5. McKeehan WL, Wang F, Kan M. The heparan sulfate–fibroblast growth factor family: diversity of structure and function. *Prog Nucleic Acid Res Mol Biol* 1998;**59**:135–76.
6. Itoh N, Ornitz DM. Evolution of the Ff and Fgfr gene families. *Trends Genet* 2004;**20**(11):563–9.
7. Luo Y, Ye S, Kan M, McKeehan WL. Structural specificity in a FGF7-affinity purified heparin octasaccharide required for formation

of a complex with FGF7 and FGFR2IIIb. *J Cell Biochem* 2006;**97**(6):1241–58.

8. Luo Y, Ye S, Kan M, McKeehan WL. Control of fibroblast growth factor (FGF) 7- and FGF1-induced mitogenesis and downstream signaling by distinct heparin octasaccharide motifs. *J Biol Chem* 2006;**281**(30):21,052–61.
9. Friesel R, Maciag T. Fibroblast growth factor prototype release and fibroblast growth factor receptor signaling. *Thromb Haemost* 1999;**82**(2):748–54.
10. Ye S, Luo Y, Lu W, et al. Structural basis for interaction of FGF-1, FGF-2, and FGF-7 with different heparan sulfate motifs. *Biochemistry* 2001;**40**(48):14,429–39.
11. Luo Y, Lu W, Mohamedali KA, et al. The glycine box: a determinant of specificity for fibroblast growth factor. *Biochemistry* 1998;**37**(47):16,506–15.
12. Fukumoto S. Actions and mode of actions of FGF19 subfamily members. *Endocr J* 2008;**55**(1):23–31.
13. Wright TJ, Ladher R, McWhirter J, Murre C, Schoenwolf GC, Mansour SL. Mouse FGF15 is the ortholog of human and chick FGF19, but is not uniquely required for otic induction. *Dev Biol* 2004;**269**(1):264–75.
14. Jones S. Mini-review: endocrine actions of fibroblast growth factor 19. *Mol Pharm* 2008;**5**(1):42–8.
15. Goetz R, Beenken A, Ibrahimi OA, et al. Molecular insights into the klotho-dependent, endocrine mode of action of fibroblast growth factor 19 subfamily members. *Mol Cell Biol* 2007;**27**(9):3417–28.
16. Yan G, Fukabori Y, McBride G, Nikolaropolous S, McKeehan WL. Exon switching and activation of stromal and embryonic fibroblast growth factor (FGF)–FGF receptor genes in prostate epithelial cells accompany stromal independence and malignancy. *Mol Cell Biol* 1993;**13**(8):4513–22.
17. Kan M, Uematsu F, Wu X, Wang F. Directional specificity of prostate stromal to epithelial cell communication via FGF7/FGFR2 is set by cell- and FGFR2 isoform-specific heparan sulfate. *In Vitro Cell Dev Biol Anim* 2001;**37**(9):575–7.
18. Jin C, Wang F, Wu X, Yu C, Luo Y, McKeehan WL. Directionally specific paracrine communication mediated by epithelial FGF9 to stromal FGFR3 in two-compartment premalignant prostate tumors. *Cancer Res* 2004;**64**(13):4555–62.
19. Yan G, Fukabori Y, Nikolaropoulos S, Wang F, McKeehan WL. Heparin-binding keratinocyte growth factor is a candidate stromal-to-epithelial-cell andromedin. *Mol Endocrinol* 1992;**6**(12):2123–8.
20. Lu W, Luo Y, Kan M, McKeehan WL. Fibroblast growth factor-10. A second candidate stromal to epithelial cell andromedin in prostate [published erratum appears in *J Biol Chem.* 1999;**274**(39):28,058; 1999;**274**(18):12,827–34.
21. Moore DD. Physiology. Sister act. *Science* 2007;**316**(5830):1436–8.
22. Gotoh N. Regulation of growth factor signaling by FRS2 family docking/scaffold adaptor proteins. *Cancer Sci* 2008;**99**(7):1319–25.
23. Uematsu F, Jang JH, Kan M, Wang F, Luo Y, McKeehan WL. Evidence that the intracellular domain of FGF receptor 2IIIb affects contact of the ectodomain with two FGF7 ligands. *Biochem Biophys Res Commun* 2001;**283**(4):791–7.
24. Mohammadi M, Schlessinger J, Hubbard SR. Structure of the FGF receptor tyrosine kinase domain reveals a novel autoinhibitory mechanism. *Cell* 1996;**86**(4):577–87.
25. Wang F, McKeehan K, Yu C, McKeehan WL. Fibroblast growth factor receptor 1 phosphotyrosine 766: molecular target for prevention of progression of prostate tumors to malignancy. *Cancer Res* 2002;**62**(6):1898–903.
26. Uematsu F, Kan M, Wang F, Jang JH, Luo Y, McKeehan WL. Ligand binding properties of binary complexes of heparin and immunoglobulin-like modules of FGF receptor 2. *Biochem Biophys Res Commun* 2000;**272**(3):830–6.
27. Wang F, Lu W, McKeehan K, et al. Common and specific determinants for fibroblast growth factors in the ectodomain of the receptor kinase complex. *Biochemistry* 1999;**38**(1):160–71.
28. Schlessinger J, Plotnikov AN, Ibrahimi OA, et al. Crystal structure of a ternary FGF–FGFR–heparin complex reveals a dual role for heparin in FGFR binding and dimerization [in process citation]. *Mol Cell* 2000;**6**(3):743–50.
29. Pellegrini L, Burke DF, von Delft F, Mulloy B, Blundell TL. Crystal structure of fibroblast growth factor receptor ectodomain bound to ligand and heparin. *Nature* 2000;**407**(6807):1029–34.
30. Wang F, Kan M, McKeehan K, Jang JH, Feng S, McKeehan WL. A homeo-interaction sequence in the ectodomain of the fibroblast growth factor receptor. *J Biol Chem* 1997;**272**(38):23,887–95.
31. Kan M, Wang F, Xu J, Crabb JW, Hou J, McKeehan WL. An essential heparin-binding domain in the fibroblast growth factor receptor kinase. *Science* 1993;**259**(5103):1918–21.
32. Wang F, Kan M, Yan G, Xu J, McKeehan WL. Alternately spliced NH2-terminal immunoglobulin-like Loop I in the ectodomain of the fibroblast growth factor (FGF) receptor 1 lowers affinity for both heparin and FGF-1. *J Biol Chem* 1995;**270**(17):10,231–5.
33. Faham S, Hileman RE, Fromm JR, Linhardt RJ, Rees DC. Heparin structure and interactions with basic fibroblast growth factor. *Science* 1996;**271**(5252):1116–20.
34. Pye DA, Vives RR, Hyde P, Gallagher JT. Regulation of FGF-1 mitogenic activity by heparan sulfate oligosaccharides is dependent on specific structural features: differential requirements for the modulation of FGF-1 and FGF-2. *Glycobiology* 2000;**10**(11):1183–92.
35. Kan M, Wang F, To B, Gabriel JL, McKeehan WL. Divalent cations and heparin/heparan sulfate cooperate to control assembly and activity of the fibroblast growth factor receptor complex. *J Biol Chem* 1996;**271**(42):26,143–8.
36. Kan M, Wu X, Wang F, McKeehan WL. Specificity for fibroblast growth factors determined by heparan sulfate in a binary complex with the receptor kinase. *J Biol Chem* 1999;**274**(22):15,947–52.
37. McKeehan WL, Wu X, Kan M. Requirement for anticoagulant heparan sulfate in the fibroblast growth factor receptor complex. *J Biol Chem* 1999;**274**(31):21,511–14.
38. Kuro-o M. Klotho as a regulator of oxidative stress and senescence. *Biol Chem* 2008;**389**(3):233–41.
39. Mohammadi M, Olsen SK, Ibrahimi OA. Structural basis for fibroblast growth factor receptor activation. *Cytokine Growth Factor Rev* 2005;**16**(2):107–37.
40. Kreuger J, Jemth P, Sanders-Lindberg E et al. Fibroblast growth factors share binding sites in heparan sulphate. *Biochem J* 2005;**389**(Pt 1):145–50.
41. Schlessinger J. Cell signaling by receptor tyrosine kinases. *Cell* 2000;**103**(2):211–25.
42. Schlessinger J. Common and distinct elements in cellular signaling via EGF and FGF receptors. *Science* 2004;**306**(5701):1506–7.
43. Furdui CM, Lew ED, Schlessinger J, Anderson KS. Autophosphorylation of FGFR1 kinase is mediated by a sequential and precisely ordered reaction. *Mol Cell* 2006;**21**(5):711–17.
44. Zhang Y, McKeehan K, Lin Y, Zhang J, Wang F. Fibroblast growth factor receptor 1 (FGFR1) tyrosine phosphorylation regulates binding of FGFR substrate 2alpha (FRS2alpha) but not FRS2 to the receptor. *Mol Endocrinol* 2008;**22**(1):167–75.

45. Jin CL, McKeehan K, Lambert D, Wang F. The intracellular microenvironment in host cell-specific phosphorylation of SNT1 by the FGFR1 tyrosine kinase. *Shi Yan Sheng Wu Xue Bao* 2002;**35**(3):184–90.
46. Wang F. Cell- and receptor isotype-specific phosphorylation of SNT1 by fibroblast growth factor receptor tyrosine kinases. *In Vitro Cell Dev Biol Anim* 2002;**38**(3):178–83.
47. Wang Q, Green RP, Zhao G, Ornitz DM. Differential regulation of endochondral bone growth and joint development by FGFR1 and FGFR3 tyrosine kinase domains. *Development* 2001;**128**(19):3867–76.
48. Raffioni S, Thomas D, Foehr ED, Thompson LM, Bradshaw RA. Comparison of the intracellular signaling responses by three chimeric fibroblast growth factor receptors in PC12 cells. *Proc Natl Acad Sci USA* 1999;**96**(13):7178–83.
49. Lin HY, Xu J, Ischenko I, Ornitz DM, Halegoua S, Hayman MJ. Identification of the cytoplasmic regions of fibroblast growth factor (FGF) receptor 1 which play important roles in induction of neurite outgrowth in PC12 cells by FGF-1. *Mol Cell Biol* 1998;**18**(7):3762–70.
50. Choi DY, Toledo-Aral JJ, Lin HY, et al. Fibroblast growth factor receptor 3 induces gene expression primarily through Ras-independent signal transduction pathways. *J Biol Chem* 2001;**276**(7):5116–22.
51. Ricol D, Cappellen D, El Marjou A, et al. Tumour suppressive properties of fibroblast growth factor receptor 2-IIIb in human bladder cancer. *Oncogene* 1999;**18**(51):7234–43.
52. Feng S, Wang F, Matsubara A, Kan M, McKeehan WL. Fibroblast growth factor receptor 2 limits and receptor 1 accelerates tumorigenicity of prostate epithelial cells. *Cancer Res* 1997;**57**(23):5369–78.
53. Matsubara A, Kan M, Feng S, McKeehan WL. Inhibition of growth of malignant rat prostate tumor cells by restoration of fibroblast growth factor receptor 2. *Cancer Res* 1998;**58**(7):1509–14.
54. Zhang Y, Wang H, Toratani S, et al. Growth inhibition by keratinocyte growth factor receptor of human salivary adenocarcinoma cells through induction of differentiation and apoptosis. *Proc Natl Acad Sci USA* 2001;**98**(20):11,336–40.
55. Huang X, Yang C, Jin C, Luo Y, Wang F, McKeehan WL. Resident hepatocyte fibroblast growth factor receptor 4 limits hepatocarcinogenesis. *Mol Carcinogenesis* 2008. in press.
56. Jang JH. Reciprocal relationship in gene expression between FGFR1 and FGFR3: implication for tumorigenesis. *Oncogene* 2005;**24**(5):945–8.
57. Huang X, Yu C, Jin C, et al. Ectopic activity of fibroblast growth factor receptor 1 in hepatocytes accelerates hepatocarcinogenesis by driving proliferation and vascular endothelial growth factor-induced angiogenesis. *Cancer Res* 2006;**66**(3):1481–90.
58. Acevedo VD, Gangula RD, Freeman KW, et al. Inducible FGFR-1 activation leads to irreversible prostate adenocarcinoma and an epithelial-to-mesenchymal transition. *Cancer Cell* 2007;**12**(6):559–71.
59. Ogawa Y, Kurosu H, Yamamoto M, et al. BetaKlotho is required for metabolic activity of fibroblast growth factor 21. *Proc Natl Acad Sci USA* 2007;**104**(18):7432–7.
60. Suzuki M, Uehara Y, Motomura-Matsuzaka K, et al. betaKlotho is required for fibroblast growth factor (FGF) 21 signaling through FGF receptor (FGFR) 1c and FGFR3c. *Mol Endocrinol* 2008;**22**(4):1006–14.
61. Klint P, Claesson-Welsh L. Signal transduction by fibroblast growth factor receptors. *Front Biosci* 1999;**4**:D165–77.
62. Powers CJ, McLeskey SW, Wellstein A. Fibroblast growth factors, their receptors and signaling. *Endocr Relat Cancer* 2000;**7**(3):165–97.
63. Ornitz DM, Itoh N. Fibroblast growth factors. *Genome Biol* 2001;**2**(3). REVIEWS3005.
64. Pellegrini L. Role of heparan sulfate in fibroblast growth factor signalling: a structural view. *Curr Opin Struct Biol* 2001;**11**(5):629–34.
65. Harmer NJ, Ilag LL, Mulloy B, Pellegrini L, Robinson CV, Blundell TL. Towards a resolution of the stoichiometry of the fibroblast growth factor (FGF)–FGF receptor–heparin complex. *J Mol Biol* 2004;**339**(4):821–34.

Chapter 19

The Mechanism of NGF Signaling Suggested by the p75 and TrkA Receptor Complexes

J. Fernando Bazan[1] and Christian Wiesmann[2]

[1] *Department of Protein Biology, Genentech, South San Francisco, California*

[2] *Department of Structural Engineering, Genentech, South San Francisco, California*

INTRODUCTION

Nerve growth factor (NGF) is the founding member of the neurotrophins, a family of secreted growth factors responsible for the growth, survival, and developmental plasticity of neuronal populations in the vertebrate peripheral and central nervous system [1, 2]. In addition to NGF, the neurotrophin family comprises brain-derived neurotrophic factor (BDNF), neurotrophin-3 (NT-3), neurotrophin-4/5 (NT-4/5), and neurotrophin-6 (NT-6). These molecules exert their biological activities through binding to two unrelated classes of receptors in a remarkable signaling relationship that may predate the emergence of primitive chordates [3, 4]. One of these, the p75 neurotrophin receptor ($p75^{NTR}$), also referred to as the low-affinity receptor, binds to all neurotrophins with nanomolar affinity and is responsible for their apoptotic activities [5, 6]. The Trks, or so-called high-affinity receptors, are more specific in their engagements and mediate the trophic effects of neurotrophins through dimerization-induced tyrosine kinase activity [7–9]. There is considerable interest in therapeutic targeting of this neurotrophin ligand/receptor system in neurodegenerative disorders such as Alzheimer's, Huntington's and Parkinson's diseases [10–12], for the treatment of neuroinflammatory pain [13, 14], and in other immune-related pathologies outside of the nervous system [14, 15]. In the present chapter, we focus on new structures of neurotrophin receptor complexes that have emerged since 2004 [16–18], and which have significantly reshaped our molecular understanding of engagements between these important ligands and their binding receptors.

NEUROTROPHINS

The neurotrophins share a pairwise sequence identity of about 50 percent and belong to a group of the large and diverse superfamily of cystine-knot-containing growth factors [19, 20]. They adopt very similar three-dimensional structures, encompassing a core fold with a pair of irregular, anti-parallel, two-stranded β-sheets that are tied together by a distinctive disulfide bond array characterized as a cystine-knot motif [21] (Figure 19.1). Two neurotrophin monomers assemble side-by-side in a parallel fashion to yield a non-covalent but tightly packed homodimer with a dumbbell-like shape, a dimer configuration that recurs only in interleukin (IL)-17 and *Drosophila* Spatzle structures [22, 23]. The conserved residues are clustered in distinct chain segments that cover about half of the molecule, and share a greater sequence identity among all neurotrophins of about 70 percent. Three of these segments map onto three of the β-strands that form the "handle" of the dumbbell as well as contribute to a large portion of the dimer interface. To the present date (and summarized in Table 19.1), crystal structures have been reported for the uncomplexed NGF, NT-3, and NT-4/5 homodimers, as well as the BDNF/NT-3 and BDNF/NT-4/5 heterodimers [21, 24–27]; the Trk receptor ligand-binding immunoglobulin (Ig) domains have been captured in uncomplexed fashion [28, 29], and otherwise engaged to NGF (TrkA; see [30]) and NT4/5 (TrkB, see [31]). More recent reports present the structures of NGF in complex with the entire ectodomain of TrkA in a 2:2 stoichiometry [17], of an NGF dimer in an asymmetric complex binding to a single $p75^{NTR}$

Handbook of Cell Signaling, Three-Volume Set 2 ed.

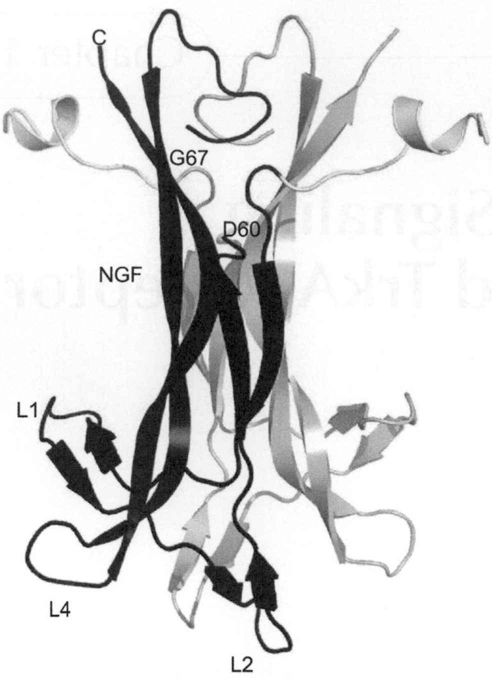

FIGURE 19.1 Structure of the NGF dimmer, taken from the NGF–TrkA-D5 complex [31].
One NGF monomer is depicted in light gray, the other in dark gray. Note the N-terminus of NGF adopting a helical conformation upon TrkA binding. One loop, bounded by residues 60 and 67, is poorly ordered or disordered in most neurotrophin structures.

receptor [16], as well as an NT-3 dimer in a symmetric complex pinned between two p75^{NTR} receptors [18].

The experimentally derived patterns of receptor binding in the neurotrophin system reveal that TrkA binds to NGF (and weakly to NT-3), TrkB interacts with BDNF and NT-4/5 (but weakly to NT-3), and TrkC binds to NT-3. The breakthrough structure of an NGF dimer in complex with a minimal pair of TrkA-D5 Ig domains revealed for the first time in 1999 the precise details of how neurotrophins specifically recognize their primary tyrosine kinase receptors [30] – findings that were bolstered by the later complex structure of NT4/TrkB-D5 [31]. These structures, along with a wealth of mutagenic, computational, and protein engineering experiments, have led to the successful design of multi- and pan-specific neurotrophin variants (reviewed in [26,32]). Nonetheless, the highest (picomolar) affinity binding events on cell surfaces occur in the presence of both p75 and Trk receptors [1], and it is these full-length ternary complexes – neutrotrophin, p75, and Trk chains – about which many questions remain regarding assembly, architecture, and signaling mechanism [17, 33, 34].

Trks

The Trks are transmembrane receptors with cytoplasmic tyrosine kinase domains and a molecular weight of about 85 kDa. Their extracellular portions share a sequence identity of about 35 percent, and have five distinct globular domains [35]. Domain 2 (D2) consists of three leucine-rich repeats (LRRs), and is sandwiched between two cysteine-rich modules (D1 and D3). D4 and D5 in turn have an Ig-like fold, and are followed by a 40-residue-long linker segment connecting the ectodomains to the transmembrane helix. Domain deletion studies, experiments with truncated and chimeric versions of the Trks, and Ala-scanning mutagenesis experiments have singled out D5 as being most critical for high-affinity neurotrophin binding (see references in [34]). While some studies indicate that D4 enhances complex formation between the neurotrophins and the Trks, others indicate that D4 is required (at least *in vitro*) for proper folding of the receptor; D4 has also been suggested to be necessary to prevent ligand-independent formation of receptor complexes. If only the membrane-proximal D4–D5 domain pair is critical for neurotrophin binding and receptor stability, what are the roles of the D1, D2, and D3 modules? In contrast to earlier studies indicating that the LRRs are somehow involved in neurotrophin binding, the 2007 crystal structure of NGF in complex with the complete extracellular domain of TrkA [17] reaffirms the supremacy of the TrkA-D5 Ig module in the recognition of NGF as glimpsed in the 2001 X-ray study [30]. In the new complex, D4 as well as the N-terminal trio of domains D1–D3 curl away from the NGF dimer at the center of the signaling complex and reveal that the Trk receptor has an elongated shape with a length of about 105 Å. The LRRs (D2) together with the cysteine-rich domains (D1 and D3) form a relatively compact structural unit that does not participate in receptor : receptor contacts.

The crystal structures of the minimal D5 Ig folds of TrkA (TrkA-D5), TrkB (TrkB-D5), and TrkC (TrkC-D5) are all domain-swapped dimers, where β-strand A of each molecule is replaced by the analogous β-strand of a symmetry-related neighbor [28, 30]. This domain swapping is believed to be a folding artifact caused by the absence of D4 (that anchors the N-terminal end of D5 β-strand A), but the correctly folded structures can be modeled with little difficulty. The crystallization of domains extracted from larger, multimodular protein chains has been observed to give rise to inopportune swapped dimers; for instance, the central GTPase domain from the LRRK2 chain (a kinase implicated in Parkinson's disease) has been recently captured in both solo (swapped) and accompanied (non-swapped) forms [36, 37]. The D5 structures show a β-sandwich fold of the I-set of the Ig superfamily; however, they possess a number of distinct and unusual features that deviate from the canonical I-set fold [28]. For example, the usual buried disulfide bridge, connecting both β-sheets, is absent in D5; instead, D5 possesses a solvent-exposed disulfide bond formed between cysteine residues in two adjacent β-strands. Another noteworthy difference concerns the loop connecting strands A and B which is three residues shorter than in other I-set members, and it is likely

TABLE 19.1 Structures of neurotrophins

PDB file	Protein(s)	Species	Resolution (Å)	Release Date	Reference
1BET	NGF	Mouse	2.30	5/31/94	21
1BTG	NGF	Mouse	2.50	3/8/96	24
1BND	BDNF/NT3	Human	2.75	4/4/96	25
1B8K	NT3	Human	2.15	2/9/99	25
1B8M	BDNF/NT4	Human	2.75	2/9/99	26
1B98	NT4	Human	2.75	2/26/99	26
1NT3	NT3	Human	2.40	6/6/99	27
1WWA	TrkA-D5	Human	2.50	7/7/99	28
1WWB	TrkB-D5	Human	2.10	7/7/99	28
1WWC	TrkC-D5	Human	1.90	7/7/99	28
1WWW	NGF/TrkA-D5	Human	2.20	15/9/99	30
1HE7	TrkA-D5	Human	2.00	4/2/01	29
1HCF	NT4/TrkB-D5	Human	2.70	12/6/01	31
1SG1	NGF/p75	Human/rat	2.40	6/1/04	16
2IFG	NGF/TrkA	Human	3.40	2/13/07	17
3BUK	NT3/p75	Human/rat	2.60	7/15/08	18
1ZAN	Anti-NGF MAb	Human	1.70	4/4/06	93

that the economy of this loop increases the propensity of these domains to misfold when singly expressed.

NGF–TrkA COMPLEXES

The crystal structure of the complex between NGF and TrkA-D5 [30] revealed for the first time the structural basis for specificity and cross-reactivity among the neurotrophins and their Trks, and intimated the orientation of the complex with respect to the membrane. The overall shape of this minimal complex resembles a bat, with the wings formed by the two copies of TrkA-D5 book-ended around the NGF dimer in the center (Figure 19.2). By contrast, the silhouette of the entire TrkA ectodomain complexed to NGF was described as reminiscent of a crab, with the NGF dimer body flanked on either side by pincers cast by the curving Trk-A D1–D5 modules [17]. The heart of the complex formed by the NGF dimer and two TrkA-D5 Ig folds shows virtually no structural changes (rmsd of 0.2 Å for 409 common C_α positions) between minimal and extended architectures [17, 30].

Because the C-terminus of D5 is expected to steer towards the membrane, NGF is likely oriented with its two-fold axis perpendicular to the membrane, with its termini and the cystine-knot motif at the top and the three hairpin loops L1, L2, and L4 closer to the membrane plane (Figures 19.1, 19.2). In a key contact, the A/B loop of TrkA-D5 is tightly packed against the waist of the NGF dumbbell [30]. While ligand binding does not induce any significant differences in D5, NGF undergoes an important conformational change when compared to its unbound form: the N-terminal residues of NGF, which are known to be critical for receptor binding and are flexible in all crystal structures of neurotrophins either in their unbound state or captive by $p75^{NTR}$ [16, 18], become well-ordered in the NGF–TrkA complexes [17, 30]. The NGF N-terminus now forms a helical turn (Figures 19.1, 19.2), which packs tightly into a hydrophobic pocket on the surface of D5; interestingly, the unusual disulfide bridge found in D5 but absent from other I-set Ig

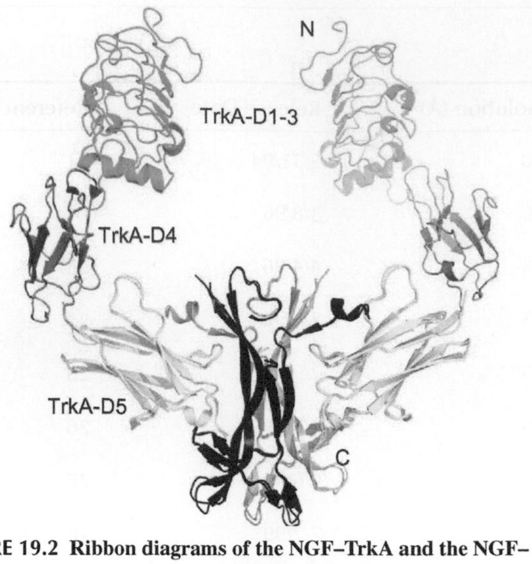

FIGURE 19.2 Ribbon diagrams of the NGF–TrkA and the NGF–TrkA-D5 complexes superimposed onto each other.
Domains 1–3 of TrkA are at the top of the figure, followed by D4 and D5 binding on either side of the NGF dimer. The NGFs of both complexes exclusively form interactions with domain 5 of TrkA.

members forms the floor of this pocket. A comparison of the D5 structures of TrkA, TrkB, and TrkC reveals that the residues lining the walls of the pocket are not conserved; therefore, the interactions in this patch are likely to confer specificity between neurotrophin ligands and their Trk family receptors [28]. A second binding patch involves residues from the central portion of the NGF dimer. Most of the residues involved in this part of the interface are preserved among both ligands and receptors, and similar interactions are predicted for all neurotrophin–Trk complexes [30], as confirmed by the homologous NT4/5/TrkB-D5 complex structure [31]. Finally, hairpin loops L2 and L4 at the bottom of the NGF dimer (Figure 19.1) show poor sequence conservation and are known from swapping experiments to be important for specificity [12]. While neither of these loops is in contact with TrkA D5, the C-terminus of D5 is aligned with a groove between the packed loops [17, 30]. Therefore, in the native complex with intact TrkA, the upper portion of the linker segment connecting D5 to the transmembrane helix could interact with these loops and form a third contact point, consistent with the observation that deletions or certain point mutations in this segment result in decreases in neurotrophin binding affinity.

$p75^{NTR}$

$p75^{NTR}$ enhances survival signals induced by NGF when in the presence of TrkA, while it activates apoptotic signals in a TrkA-negative background (reviewed in [33, 38, 39]). $p75^{NTR}$ lacks intrinsic catalytic activity and is a member of the tumor necrosis factor receptor (TNFR) superfamily [40], where it can be functionally grouped with other CNS-expressed, death-domain containing TNFRs [41]. Members of this superfamily contain between two and six cysteine-rich domains (CRDs) in their extracellular portion, identified by six signature cysteine residues that form a ladder of three disulfide bridges in each CRD fold [42]. These modules are fairly rigid building blocks, and stacking of multiple (or half) repeats gives rise to the elongated, rod-like profile of TNFR family receptors.

$p75^{NTR}$ has four CRDs followed by a long 60 amino-acid linker to the transmembrane helix. Crystal structures of complexes in this superfamily have shown that CRD2 and CRD3 are most intimately involved in canonical TNF trimer interactions [43–45], while non-canonical pairings – most notably HVEM in complex with the B-cell Ig fold molecule BTLA – utilize other CRDs [46, 47]. In the case of p75, while all four CRDs are required for NGF binding, CRD2 appears to be most critical; in contrast to the Trks, the linker segment C-terminal of the CRDs is dispensable for binding [34]. The binding site for $p75^{NTR}$ on the neurotrophins has been mapped through mutational analyses onto two spatially separated regions; one patch at the top of NGF involves a number of positively charged residues in loop L3 and the C-terminal chain, while the second patch, at the bottom of NGF, features positively charged residues from loops L1 and L4. The crystal structure of the NGF–TrkA-D5 complex shows that neither of these patches is buried by contact with TrkA-D5, in principle sanctioning the formation of ternary 2:1:1 NGF dimer–TrkA-$p75^{NTR}$ complexes at the cell surface. Based on these observations, a possible model for such a ternary complex has been proposed [34]. How have these predictions fared with the X-ray crystallographic solutions of NGF–p75 complex structures [16, 18]?

NGF–$p75^{NTR}$ COMPLEXES

The first crystal structure of a neurotrophin bound to p75 was the 2:1 complex of a human NGF dimer surprisingly clasped by only one copy of a rat p75 chain [16, 32]. The rigid TNFR-like body of p75 makes two contacts along the seam between packed NGF monomers; a "top" patch involving CRD2, and a "bottom" patch at the interface between CRDs 3 and 4 (Figure 19.3a). This mode of ligand interaction for p75 is distinct from TNFR orthodoxy, where the opposite face of the slender CRD repeat structure is involved in trimeric TNF family ligand recognition [16]. For this asymmetric complex, the authors critically argue that the "pinching" of the NGF dimer (across the concave "waist") by the single bound p75 chain allosterically distorts the equivalent site on the other face of the ligand, creating an inactive pair of pseudosites that are suffer a respective 6–10° movement, sufficient to exclude a second p75 chain binding [16]. Residues at the productive p75 binding interface are largely preserved among neurotrophin

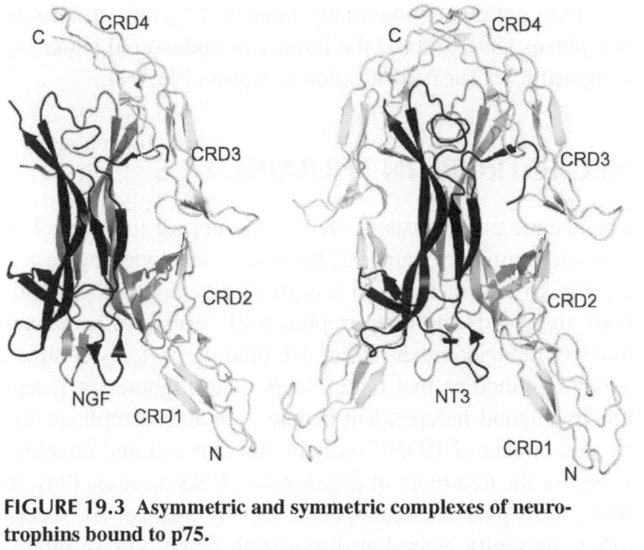

FIGURE 19.3 Asymmetric and symmetric complexes of neurotrophins bound to p75.
Left: crystal structure of NGF bound to p75 in an asymmetric fashion; right: the symmetric complex of p75 bound to NT-3. Note that the C-termini of the p75 are located at the top of the page.

sequences, offering an explanation for the pan-specific nature of p75. There is a significant electrostatic contribution to binding, as the NGF dimer presents a largely basic face to a very acidic p75 interface. Comparison of the NGF contacts between the minimal TrkA-D5 complex that buries Ig-fold loops into the NGF dimer waist [30] and the asymmetric p75 complex (where the NGF dimer is crimped at the edges of the waist) strongly suggests that both receptor chains compete for essentially the same binding face – with the caveat that the ligand orientations between p75- and Trk-engaged NGF dimers are reversed with respect to the membrane plane [32]. This latter finding would not necessarily hinder the formation of hypothetical cell surface 1 : 2 : 1 p75–NGF dimer–Trk heterocomplexes, but just require flexibility of the juxtamembrane linkers from one or both receptor chains in order to accommodate the antiparallel binding orientations [16, 32, 33].

The matter of whether NGF could nucleate a symmetric signaling complex was fully resolved with the full length TrkA ectodomain co-crystal structure where the NGF dimer was sandwiched between two TrkA chains [18]. A scant year later, a new neurotrophin–p75 complex suggested that p75 could also be dimerized by ligand binding (Figure 19.3b). The new structure of the human NT3 dimer bound symmetrically to a pair of rat p75 chains posed the question of whether glycosylation of the p75 receptors could have helped capture the 2 : 2 complex (by perhaps favoring another crystal lattice), counter to the earlier, unglycosylated structure [18]. There are few other differences to be seen between the NGF and NT3 complexes where the binding mode of p75 is essentially the same, save for an additional site-3 contact between the C-terminal end of the receptors and the base of the NT3 dimer [18]. It is very likely that the 2 : 1 asymmetric complex is not an artifact but rather reflects a midpoint in the assembly of the 2 : 2 NGF–p75 complex as ligand concentrations build on the cell surface, in competition for available p75 and Trk receptors.

NEUROTROPHIN SIGNALING EXCURSIONS

The available neurotrophin structures, either alone or in concert with p75 or Trk receptors, uniformly represent the C-terminal mature halves (i.e., cystine-knot growth factor folds) of larger, secreted precursors that include cleavable N-terminal pro-domains [48–52]. For instance, human proNGF comprises a 102-residue pro-domain (ending in a dibasic KR motif that marks the cleavage site by plasmin and matrix metalloproteases) followed by the 119 amino-acid mature NGF sequence [48, 49]; significantly, there are clear biophysical signals that the NGF pro-domain assists in folding of the mature cystine-knot module and is itself partly structured [53]. Unprocessed proneurotrophins have started to garner attention as active molecular entities with distinct biological functions from their mature counterparts; notably, proneurotrophins trigger cell death by preferentially binding p75 with high affinity, an opposite action to the mature factors that are disposed to signal survival through Trks [48–51].

Proneurotrophins appear to form larger complexes with p75 and another cell surface molecule called sortilin, a Vps10-domain-class trafficking or sorting receptor, further potentiating the cell death signal of p75 [54–59]. While the exact mechanism and geometry of ternary complex formation is unknown, both proBDNF and proNGF have been shown to directly interact with both p75 and sortilin [55, 60], and a computational model has been proposed for a proneurotrophin–sortilin complex that invokes a beta-propeller fold for the Vps10 binding domain [61]. Further complicating the interaction landscape, p75 has separately been found to consort with two other LRR array cell surface molecules, Nogo receptor (NgR) and LINGO-1 [62–66]. Intriguingly, TROY, an orphan TNF receptor superfamily member, can supplant p75 in ternary complexes with NgR and LINGO-1 in the regulation of axonal growth and regeneration [67–70].

An additional layer of complexity and regulation of p75 function – and (pro)neurotrophin binding – is revealed by juxta- and intramembrane processing of this tall receptor by membrane metalloprotease sheddases and γ-secretase, respectively, that give rise to truncated forms of p75 that either lack most of the ectodomain (p75-CTF), or free a signaling-capable death domain into the cytosol (p75-ICD) [71–75]. In mice, there is a naturally occurring variant gene encoding a short, membrane-bound form of p75 (akin to p75-CTF), variously called PLAIDD, NRH2, or NRADD, that forms a signaling complex with p75 [76–80]; the syntenic gene in humans has degraded to a non-functional pseudogene.

PROSPECTS FOR TERNARY RECEPTOR COMPLEXES

Even with seminal structures in hand for full-length p75 and Trk ectodomains individually tethered to neurotrophins [16–18], the path to a p75•Trk•neurotrophin ternary complex (which could arguably take a 1 : 1 : 2 or 2 : 2 : 2 stoichiometry [16–18, 32, 33]) is surprisingly elusive. Two key structural issues, the conflicting NGF dimer orientations of p75 and Trk complexes (viewed with respect to their membrane postures) and the steric overlap of p75 and Trk binding sites on NGF, stand in the way of assembling a productive receptor signaling complex with simultaneous p75 and Trk participants. The experimental evidence for such a ternary complex is not clear cut and, as a result, groups in the field have not settled on a uniform interpretation of the structural results (see, for example, the appended comments to ref. [17] at the journal website http://www.neuron.org/content/article/comments?uid=PIIS0896627306007720). At a minimum, the symmetric p75 and Trk complexes show that neurotrophins are capable of creating two functionally orthogonal 2 : 2 transmembrane signaling complexes, one with paired death domains, and the other with tyrosine kinase modules [17, 18, 33].

A number of earlier cytokine receptor complex structures have conditioned us to expect the unobstructed juxtaposition of receptor chains around a bound cytokine in an embrace that draws intracellular domains into signaling proximity below the membrane plane [81]. Because of their distinctive dimer architectures, cystine-knot growth factors tend to assemble extended, two-fold symmetric receptor signaling complexes. Three telling examples include (1) cytokines of the TGF-β family that nucleate 2 : 2 : 2 receptor complexes, with each ligand subunit in contact with a heterodimer of type I and II tyrosine kinase receptors [82, 83]; (2) Glial-derived neurotrophic factors (GDNFs) have been crystallographically captured in complex with non-signaling GFRa co-receptors [84, 85], forming a symmetric scaffold for a 2 : 2 : 2 RET tyrosine kinase complex; and (3) VEGFs also nucleate a 2 : 2 : 2 cell surface complex of neuropilin and VEGFR tyrosine kinase chains, with a recent neuropilin ectodomain structure indicating the likely binding epitope for VEGF [86]. It will be further interesting to see what types of receptor contacts are formed by IL17 and Spatzle family cytokines, since they show such close monomer likeness and dimer symmetry with neurotrophin folds [22, 23, 87, 88]. Still, the notion that neurotrophins will likewise converge on a similar strategy with dueling p75 and Trk binding chains is likely to be resolved only by additional biochemical and biophysical work on complexes that adds significantly new information about (1) neurotrophin prodomain functions, (2) alternative p75 binding partners like sortilin or PLAIDD/NRH2/NRADD-like variants, (3) the interdependence of functionally antagonistic Trk (kinase-activated) and p75 (death domain-triggered) signaling pathways potentially launched by mixed, ternary complexes [89], and (4) the impact of endosomal signaling compartments for neurotrophin receptors [90, 91].

NEUROTROPHIN THERAPEUTICS

Therapeutic antagonists of NGF signaling for the control of neuroinflammatory pain [92] have been developed by targeting the growth factor with a high-affinity monoclonal antibody that binds the neurotrophin fold "waist" in a manner that would block either p75 or Trk binding [93]. By contrast, agonist antibodies that target TrkB (and oligomerize receptors in a ligand-independent manner) are able to replicate the trophic effects of BDNF on neuronal survival and development, for the treatment of degenerative CNS diseases [94]. In truth, most pharmacotherapeutic activity against Trk receptors is presently geared at discovering potent kinase inhibitors of Trk signaling for indications in cancer [95], but these discussions are beyond the scope of the current review. The antibody-based therapeutics – and the Trk kinase drugging efforts – are still very much in their initial phases of development, yet already benefit from the architectural plans of the two neurotrophin signaling complex structures. At an even earlier, exploratory stage, other groups are using the available complex structures (and their interaction "hotspots") as templates for the design and selection of small molecule modulators of neurotrophin or receptor function [96–98]; nonetheless, the success rate of targeting protein–protein interfaces with drug-like compounds remains a difficult task [99]. The growing involvement of the neurotrophin ligand–receptor system in the pathophysiology of human disease, coupled with a better functional understanding of the molecular intricacies and excursions of the neurotrophin ligand–receptor system, assure a continuing wave of complex structures and therapeutics in this field [100–102].

REFERENCES

1. Bibel M, Barde YA. Neurotrophins: key regulators of cell fate and cell shape in the vertebrate nervous system. *Genes Dev* 2000;**14**:2919–37.
2. von Bartheld CS, Fritzsch B. Comparative analysis of neurotrophin receptors and ligands in vertebrate neurons: tools for evolutionary stability or changes in neural circuits? *Brain Behav Evol* 2006;**68**:157–72.
3. Bothwell M. Evolution of the neurotrophin signaling system in invertebrates. *Brain Behav Evol* 2006;**68**:124–32.
4. Halböök F, Wilson K, Thorndyke M, Olinski RP. Formation and evolution of the chordate neurotrophin and Trk receptor genes. *Brain Behav Evol* 2006;**68**:133–44.
5. Dechant G, Barde YA. The neurotrophin receptor p75(NTR): novel functions and implications for diseases of the nervous system. *Nat Neurosci* 2002;**5**:1131–6.
6. Kalb R. The protean actions of neurotrophins and their receptors on the life and death of neurons. *Trends Neurosci* 2005;**28**:5–11.

7. Chao MV. Neurotrophins and their receptors: a convergence point for many signalling pathways. *Nat Rev Neurosci* 2003;**4**:299–309.
8. Segal RA. Selectivity in neurotrophin signaling: theme and variations. *Annu Rev Neurosci* 2003;**26**:299–330.
9. Reichard LF. Neurotrophin-regulated signaling pathways. *Philos Trans R Soc Lond B Biol Sci* 2006;**361**:1545–64.
10. Siegel GJ, Chauhan NB. Neurotrophic factors in Alzheimer's and Parkinson's disease brain. *Brain Res Rev* 2000;**33**:199–227.
11. Blesch A. Neurotrophic factors in neurodegeneration. *Brain Pathol* 2006;**16**:295–303.
12. Evans JR, Barker RA. Neurotrophic factors as a therapeutic target for Parkinson's disease. *Expert Opin Ther Targets* 2008;**12**:437–47.
13. Pezet S, McMahon SB. Neurotrophins: mediators and modulators of pain. *Annu Rev Neurosci* 2006;**29**:507–38.
14. Allen SJ, Dawbarn D. Clinical relevance of the neurotrophins and their receptors. *Clin Sci* 2006;**110**:175–91.
15. Ayyadhury S, Heese K. Neurotrophins – more than neurotrophic. *Curr Immunol Rev* 2007;**3**:189–215.
16. He XL, Garcia KC. Structure of nerve growth factor complexed with the shared neurotrophin receptor p75. *Science* 2004;**304**:870–5.
17. Wehrman T, He X, Raab B, Dukipatti A, Blau H, Garcia KC. Structural and mechanistic insights into nerve growth factor interactions with the TrkA and p75 receptors. *Neuron* 2007;**53**:25–38.
18. Gong Y, Cao P, Yu HJ, Jiang T. Crystal structure of the neurotrophin-3 and p75NTR symmetrical complex. *Nature* 2008;**454**:789–93.
19. Sun PD, Davies DR. The cystine-knot containing growth-factor superfamily. *Annu Rev Biophys Biomol Struct* 1995;**24**:269–91.
20. Isaacs NW. Cystine knots. *Curr Opin Struct Biol* 1995;**5**:391–5.
21. McDonald NQ, Lapatto R, Murray-Rust J, Gunning J, Wlodawer A, Blundell TL. New protein fold revealed by a 2.3-Å resolution crystal structure of nerve growth factor. *Nature* 1991;**354**:411–14.
22. Hymowitz SG, Filvaroff EH, Yin JP, Lee J, Cai L, Risser P, Maruoka M, Mao W, Foster J, Kelley RF, Pan G, Gurney AL, de Vos AM, Starovasnik MA. IL-17s adopt a cystine knot fold: structure and activity of a novel cytokine IL-17F, and implications for receptor binding. *EMBO J* 2001;**20**:5332–41.
23. Hoffmann A, Funkner A, Neumann P, Juhnke S, Walther M, Schierhorn A, Weininger U, Balbach J, Reuter G, Stubbs MT. Biophysical characterization of refolded Drosophila Spatzle, a cystine knot protein, reveals distinct properties of three isoforms. *J Biol Chem* 2008. Epub ahead of print September 12, 2008, doi: 10.1074/jbc.M801815200.
24. Holland DR, Cousens LS, Meng W, Matthews BW. Nerve growth factor in different crystal forms displays structural flexibility and reveals zinc binding sites. *J Mol Biol* 1994;**239**:385–400.
25. Robinson RC, Radziejewski C, Stuart DI, Jones EY. Structure of the brain-derived neurotrophic factor/neurotrophin 3 heterodimer. *Biochemistry* 1995;**34**:4139–46.
26. Robinson RC, Radziejewski C, Spraggon G, Greenwald J, Kostura MR, Burtnick LD, Stuart DI, Choe S, Jones EY. The structures of the neurotrophin 4 homodimer and the brain-derived neurotrophic factor/neurotrophin 4 heterodimer reveal a common Trk-binding site. *Protein Sci* 1999;**8**:2589–97.
27. Butte MJ, Hwang PK, Mobley WC, Fletterick RJ. Crystal structure of neurotrophin-3 homodimer shows distinct regions are used to bind its receptors. *Biochemistry* 1998;**37**:16,846–16,852.
28. Ultsch MH, Wiesmann C, Simmons LC, Henrich J, Yang M, Reilly D, Bass SH, de Vos AM. Crystal structures of the neurotrophin-binding domain of TrkA, TrkB and TrkC. *J Mol Biol* 1999;**290**:149–59.
29. Robertson AG, Banfield MJ, Allen SJ, Dando JA, Mason GG, Tyler SJ, Bennett GS, Brain SD, Clarke AR, Naylor RL, Wilcock GK, Brady RL, Dawbarn D. Identification and structure of the nerve growth factor binding site on TrkA. *Biochem Biophys Res Commun* 2001;**282**:131–41.
30. Wiesmann C, Ultsch MH, Bass SH, de Vos AM. Crystal structure of nerve growth factor in complex with the ligand-binding domain of the TrkA receptor. *Nature* 1999;**401**:184–8.
31. Banfield MJ, Naylor RL, Robertson AG, Allen SJ, Dawbarn D, Brady RL. Specificity in Trk receptor:neurotrophin interactions: the crystal structure of TrkB-d5 in complex with neurotrophin-4/5.. *Structure* 2001;**9**:1191–9.
32. Wiesmann C, de Vos AM. Nerve growth factor: structure and function. *Cell Mol Life Sci* 2001;**58**:748–59.
33. Barker PA. High affinity not in the vicinity. *Neuron* 2007;**53**:1–4.
34. Zampieri N, Chao MV. The p75 NGF receptor exposed. *Science* 2004;**304**:833–4.
35. Schneider R, Schweiger M. A novel modular mosaic of cell adhesion motifs in the extracellular domains of the neurogenic Trk and TrkB tyrosine kinase receptors. *Oncogene* 1991;**6**:1807–11.
36. Deng J, Lewis PA, Greggio E, Sluch E, Beilina A, Cookson MR. Structure of the ROC domain from the Parkinson's disease-associated leucine-rich repeat kinase 2 reveals a dimeric GTPase. *Proc Natl Acad Sci U S A* 2008;**105**:1499–504.
37. Gotthardt K, Weyand M, Kortholt A, Van Haaster PJ, Wittinghofer A. Structure of the Roc–COR domain tandem of C. tepidum, a prokaryotic homologue of the human LRRK2 Parkinson kinase. *EMBO J* 2008;**27**:2239–49.
38. Barker PA. p75^{NTR}: a study in contrasts. *Cell Death Differ* 1998;**5**:346–56.
39. Barker PA. p75^{NTR} is positively promiscous: novel partners and new insights. *Neuron* 2004;**42**:529–33.
40. Bodmer JL, Schneider P, Tschopp J. The molecular architecture of the TNF superfamily. *Trends Biochem Sci* 2002;**27**:19–26.
41. Haase G, Pettmann B, Raoul C, Henderson CE. Signaling by death receptors in the nervous system. *Curr Opin Neurobiol* 2008. Epub ahead of print September 18, 2008, doi: 10.1016/jconb.2008.07.013.
42. Naismith JH, Sprang SR. Modularity in the TNF-receptor family. *Trends Biochem Sci* 1998;**23**:74–9.
43. Zhang G. Tumor necrosis factor family ligand-receptor binding. *Curr Opin Struct Biol* 2004;**14**:154–60.
44. Hymowitz SG, Patel DR, Wallweber HJ, Runyon S, Yan M, Yin J, Shriver SK, Gordon NC, Skelton NJ, Kelley RF, Starovasnik MA. Structures of APRIL–receptor complexes: like BCMA, TACI employs only a single cysteine-rich domain for high affinity ligand binding. *J Biol Chem* 2005;**280**:7218–27.
45. Compaan DM, Hymowitz SG. The crystal structure of the costimulatory OX40–OX40L complex. *Structure* 2006;**14**:1321–30.
46. Compaan DM, Gonzalez LC, Tom I, Loyet KM, Eaton D, Hymowitz SG. Attenuating lymphocyte activity: the crystal structure of the BTLA–HVEM complex. *J Biol Chem* 2005;**280**:39,553–39,561,.
47. Nelson CA, Fremont MD, Sedy JR, Norris PS, Ware CF, Murphy KM, Fremont DH. Structural determinants of herpesvirus entry mediator recognition by murine B and T lymphocyte attenuator. *J Immunol* 2008;**180**:940–7.
48. Lee R, Kermani P, Teng KK, Hempstead BL. Regulation of cell survival by secreted proneurotrophins. *Science* 2001;**294**:1945–8.
49. Ibañez CF. Jekyll-Hyde neurotrophins: the story of proNGF. *Trends Neurosci* 2002;**25**:284–6.

50. Lu B, Pang PT, Woo NH. The yin and yang of neurotrophin action. *Nat Rev Neurosci* 2005;**6**:603–14.
51. Hempstead BL. Dissecting the diverse actions of pro- and mature neurotrophins. *Curr Alzheimer Res* 2006;**3**:19–24.
52. Schweigreiter R. The dual nature of neurotrophins. *Bioessays* 2006;**28**:583–94.
53. Kliemannel M, Golbik R, Rudolph R, Schwarz E, Lilie H. The propeptide of proNGF: Structure formation and intramolecular association with NGF. *Protein Sci* 2007;**16**:411–19.
54. Nykjaer A, Lee R, Teng KK, Jansen P, Madsen P, Nielsen MS, Jacobsen C, Kliemmanel M, Schwarz E, Willnow TE, Hempstead BL, Petersen CM. Sortilin is essential for proNGF-induced neuronal cell death. *Nature* 2004;**427**:843–8.
55. Teng HK, Teng KK, Lee R, Wright S, Tevar S, Almeida RD, Kermani P, Torkin R, Chen ZY, Lee FS, Kraemer RT, Nykjaer A, Hempstead BL. ProBDNF induces neuronal apoptosis via activation of a receptor complex of p75NTR and sortilin. *J Neurosci* 2005;**25**:5455–63.
56. Kaplan DR, Miller FD. Neurobiology: a move to sort life from death. *Nature* 2004;**427**:798–9.
57. Bronfman FC, Fainzilber M. Multi-tasking by the p75 neurotrophin receptor: sortilin things out?. *EMBO Rep* 2004;**5**:867–71.
58. Nykjaer A, Willnow TE, Petersen CM. p75NTR–live or let die. *Curr Opin Neurobiol* 2005;**15**:49–57.
59. Arevalo JC, Wu SH. Neurotrophin signaling: many exciting surprises!. *Cell Mol Life Sci* 2006;**63**:1523–37.
60. Clewes O, Fahey MS, Tyler SJ, Watson JJ, Seok H, Catania C, Cho K, Dawbarn D, Allen SJ. Human proNGF: biological effects and binding profiles at TrkA, p75(NTR) and sortilin. *J. Neurochem.* 2008; **107**:1124–35.
61. Paiardini A, Caputo V. Insights into the interaction of sortilin with proneurotrophins: a computational approach. *Neuropeptides* 2008;**42**:205–14.
62. Wang KC, Kim JA, Sivasankaran R, Segal R, He Z. p75 interacts with the Nogo receptor as a co-receptor for Nogo, MAG and OMgp. *Nature* 2002;**420**:74–8.
63. Wong ST, Henley JR, Kanning KC, Huang KH, Bothwell M, Poo MM. A p75(NTR) and Nogo receptor complex mediates repulsive signaling by myelin-associated glycoprotein. *Nat Neurosci* 2002;**5**:1302–8.
64. Dechant G, Barde YA. The neurotrophin receptor p75(NTR): novel functions and implications for diseases of the nervous system. *Nat Neurosci* 2002;**5**:1131–6.
65. Hempstead BL. The many faces of p75^{NTR}. *Curr Opin Neurobiol* 2002;**12**:260–7.
66. Mi S, Lee X, Shao Z, Thill G, Ji B, Relton J, Levesque M, Allaire N, Perrin S, Sands B, Crowell T, Cate RL, McCoy JM, Pepinsky RB. LINGO-1 is a component of the Nogo-66 receptor/p75 signaling complex. *Nat Neurosci* 2004;**7**:221–8.
67. Shao Z, Browning JL, Lee X, Scott ML, Shulga-Morskaya S, Allaire N, Thill G, Levesque M, Sah D, McCoy JM, Murray B, Jung V, Pepinsky RB, Mi S. TAJ/TROY, an orphan TNF receptor family member, binds Nogo-66 receptor 1 and regulates axonal regeneration. *Neuron* 2005;**45**:353–9.
68. Park JB, Yiu G, Kaneko S, Wang J, Chang J, He XL, Garcia KC, He Z. A TNF receptor family member, TROY, is a co-receptor with Nogo receptor in mediating the inhibitory activity of myelin inhibitors. *Neuron* 2005;**45**:345–51.
69. Mandemakers WJ, Barres BA. Axon regeneration: it's getting crowded at the gates of TROY. *Curr Biol* 2005;**15**:R302–5.
70. Mi S. Troy/Taj and its role in CNS axon regeneration. *Cytokine Growth Factor Rev* 2008;**19**:245–51.
71. Kanning KC, Hudson M, Amieux PS, Wiley JC, Bothwell M, Schecterson LC. Proteolytic processing of the p75 neurotrophin receptor and two homologs generates C-terminal fragments with signaling capability. *J Neurosci* 2003;**23**:5425–36.
72. Paul CE, Vereker E, Dickson KM, Barker PA. A pro-apoptotic fragment of the p75 neurotrophin receptor is expressed in p75NTR^{ExonIV} null mice. *J Neurosci* 2004;**24**:1917–23.
73. Kenchappa RS, Zampieri N, Chao MV, Barker PA, Teng HK, Hempstead BL, Carter BD. Ligand-dependent cleavage of the p75 neurotrophin receptor is necessary for NRIF nuclear translocation and apoptosis in sympathetic neurons. *Neuron* 2006;**50**:219–32.
74. Uma S, Escudero CA, Ramos P, Lisbona F, Allende E, Covarrubias P, Parraguez JL, Zampieri N, Chao MV, Annaert W, Bronfman FC. TrkA receptor activation by nerve growth factor induces shedding of the p75 neurotrophin receptor followed by endosomal gamma-secretase-mediated release of the p75 intracellular domain. *J Biol Chem* 2007;**282**:7606–15.
75. Bronfman FC. Metalloproteases and gamma-secretase: new membrane partners regulating p75 neurotrophin receptor signaling?. *J Neurochem* 2007;**103**(Suppl. 1):91–100.
76. Frankowski H, Castro-Obregon S, del Rio G, Rao RV, Bredesen DE. PLAIDD, a type II death domain protein that interacts with p75 neurotrophin receptor. *Neuromolecular Med* 2002;**1**:153–70.
77. Wang X, Shao Z, Zetoune FS, Zeidler MG, Gowrishankar K, Vincenz C. NRADD, a novel membrane protein with a death domain involved in mediating apoptosis in response to ER stress. *Cell Death Differ* 2003;**10**:580–91.
78. Gowrishankar K, Zeidler MG, Vincenz C. Release of a membrane-bound death domain by γ-secretase processing of the p75^{NTR} homolog NRADD. *J Cell Sci* 2004;**117**:4099–111.
79. Murray SS, Perez P, Lee R, Hempstead BL, Chao MV. A novel p75 neurotrophin receptor-related protein, NRH2, regulates nerve growth factor binding to the TrkA receptor. *J Neurosci* 2004;**24**:2742–9.
80. Rabizadeh S, Bredesen DE. Ten years on: mediation of cell death by the common neurotrophin receptor p75^{NTR}. *Cytokine Growth Factor Rev* 2003;**14**:225–39.
81. Stroud RM, Wells JA. Mechanistic diversity of cytokine receptor signaling across cell membranes. *Sci STKE* 2004;**2004**(231):re7.
82. Groppe J, Hinck CS, Samavarchi-Tehrani P, Zubieta C, Schuermann JP, Taylor AB, Schwarz PM, Wrana JL, Hinck AP. Cooperative assembly of TGF-β superfamily signaling complexes is mediated by two disparate mechanisms and distinct modes of receptor binding. *Mol Cell* 2008;**29**:157–68.
83. Massague J. A very private TGF-β receptor embrace. *Mol Cell* 2008;**29**:149–50.
84. Wang X, Baloh RH, Milbrandt J, Garcia KC. Structure of artemin complexed with its receptor GFRα3: convergent recognition of glial cell line-derived neurotrophic factors. *Structure* 2006;**14**:1083–92.
85. Prakash V, Lepänen VM, Virtanen H, Jurvansuu JM, Bespalov MM, Sidorova YA, Runeberg-Roos P, Saarma M, Goldman A. The structure of the GDNF:co-receptor complex: insights into RET signalling and heparin binding. *J Biol Chem* 2008. Epub ahead of print, October 8, 2008; doi.
86. Appleton BA, Wu P, Maloney J, Yin J, Liang WC, Stawicki S, Mortara K, Bowman KK, Elliott JM, Desmarais W, Bazan JF, Bagri A, Tessier-Lavigne M, Koch AW, Wu Y, Watts RJ, Wiesmann C. Structural studies of neuropilin/antibody complexes provide insights into semaphorin and VEGF binding. *EMBO J* 2007;**26**:4902–12.

87. Shen F, Gaffen SL. Structure–function relationships in the IL-17 receptor: implications for signal transduction and therapy. *Cytokine* 2008;**41**:92–104.
88. Gangloff M, Murali A, Xiong J, Arnot CJ, Weber AN, Sandercock AM, Robinson CV, Sarisky R, Holzenburg A, Kao C, Gay NJ. Structural insight into the mechanism of activation of the Toll receptor by the dimeric ligand Spätzle. *J Biol Chem* 2008;**283**:14,629–14,635,.
89. Twiss JL, Chang JH, Schanen NC. Pathophysiological mechanisms for actions of the neurotrophins. *Brain Pathol* 2006;**16**:320–32.
90. Moises T, Dreier A, Flohr S, Esser M, Brauers E, Reiss K, Merken D, Weis J, Krüttgen A. Tracking TrkA's trafficking: NGF receptor trafficking controls NGF receptor signaling. *Mol Neurobiol* 2007;**35**:151–9.
91. Bronfman FC, Escudero CA, Weis J, Krüttgen A. Endosomal transport of neurotrophins: roles in signaling and neurodegenerative diseases. *Dev Neurobiol* 2007;**67**:1183–203.
92. Hefti FF, Rosenthal A, Walicke PA, Wyatt S, Vergara G, Shelton DL, Davies AM. Novel class of pain drugs based on antagonism of NGF. *Trends Pharmacol Sci* 2006;**27**:85–91.
93. Covaceuszach S, Cassetta A, Konarev PV, Gonfloni S, Rudolf R, Svergun DI, Lamba D, Cattaneo A. Dissecting NGF interactions with TrkA and p75 receptors by structural and functional studies of an anti-NGF neutralizing antibody. *J Mol Biol* 2008;**381**:881–96.
94. Qian MD, Zhang J, Tan XY, Wood A, Gill D, Cho S. Novel agonist monoclonal antibodies activate TrkB receptors and demonstrate potent neurotrophic activities. *J Neurosci* 2006;**26**:9394–403.
95. Geiger TR, Peeper DS. Critical role for TrkB kinase function in anoikis suppression, tumorigenesis, and metastasis. *Cancer Res* 2007;**67**:6221–9.
96. Price RD, Milne SA, Sharkey J, Matsuoka N. Advances in small molecules promoting neurotrophin function. *Pharmacol Ther* 2007;**115**:292–306.
97. Longo FM, Massa SM. Small molecule modulation of p75 neurotrophin receptor functions. *CNS Neurol Disord Drug Targets* 2008;**7**:63–70.
98. Skaper SD. The biology of neurotrophins, signalling pathways, and functional peptide mimetics of neurotrophins and their receptors. *CNS Neurol Disord Drug Targets* 2008;**7**:46–62.
99. Wells JA, McClendon CL. Reaching for high-hanging fruit in drug discovery at protein–protein interfaces. *Nature* 2007;**450**:1001–9.
100. Thoenen H, Sendtner M. Neurotrophins: from enthusiastic expectations through sobering experiences to rational therapeutic approaches. *Nat Neurosci* 2002;**5**:1046–50.
101. Chao MV, Rajagopal R, Lee FS. Neurotrophin signalling in health and disease. *Clin Sci* 2006;**110**:167–73.
102. Schulte-Herbrüggen O, Braun A, Rochlitzer S, Jockers-Scherübl MC, Hellweg R. Neurotrophic factors – a tool for therapeutic strategies in neurological, neuropsychiatric and neuroimmunological diseases?. *Curr Med Chem* 2007;**14**:2318–29.

Chapter 20

The Mechanism of VEGFR Activation by VEGF

Christian Wiesmann

Department of Structural Biology Genentech, Inc., 1 DNA Way, South San Francisco, California

Vascular endothelial growth factor (VEGF-A) is a secreted cytokine that plays a central role in angiogenesis, the process of new blood vessel formation, and is essential for numerous physiological processes such as embryonic development and wound healing (reviewed in [1]). In addition, VEGF-A is involved in a number of human diseases that depend on dysregulated angiogenesis, including cancer, rheumatoid arthritis, and age-related macular degeneration. For this reason, VEGF-A and its receptors have proven excellent therapeutic targets for treatment of many of these disorders [2].

Solved over 10 years ago, the crystal structure of VEGF-A bound to the second immunoglobulin domain of its receptor VEGFR1 represented the first structure of a receptor tyrosine kinase ectodomain in complex with its ligand [3]. Supported by numerous biochemical experiments, the structure strongly suggests a general model of VEGFR activation by VEGF-type cytokines, whereby the dimeric VEGF molecule brings together two VEGFRs on the cell surface, presumably allowing the receptor kinase domains to trigger an intracellular signaling cascade [4]. While precise structural information of a complete VEGFR ectodomain in complex with VEGF continues to elude crystallographic solution, a number of structures of related receptors and growth factors, together with electron microscopy studies, have added to our understanding of the core VEGF-mediated receptor activation and signaling module. However, a number of questions remain unanswered about the function of auxiliary molecules present in the signaling complex. For example, many isoforms of VEGF, in addition to binding to a subset of the VEGFRs, are also capable of forming complexes with heparin, a negatively-charged glycan, or neuropilins (Nrp). Neuropilins are co-receptors for VEGF and semaphorins, and essential for not only the development of the vasculature, but also the formation of the nervous system (reviewed in [5]). Our understanding of the composition and architecture of these larger signaling complexes containing multiple receptors remains limited.

STRUCTURAL CHARACTERIZATION OF VEGF FAMILY MEMBERS

VEGF-A, with its distinctive cystine-knot growth factor fold [6], is an obligatory dimeric glycoprotein and the prototype for the VEGF family of growth factors. This family includes placental growth factor (PlGF), VEGF-B, VEGF-C, VEGF-D, (reviewed in [7]), highly homologous proteins encoded by Orf-viruses collectively called VEGF-E [8], and the more recently identified VEGF-F from snake venom [9]. Members of the VEGF family can be expressed as multiple gene splice variants, and are additionally subject to post-translational modification through glycosylation. The core cystine-knot fold, common to all VEGF family members, is composed of approximately 100 amino acids and has a sequence identity of between 30 and 50 percent among family members. This architectural motif is also present in the closely related platelet-derived growth factors (PDGFs), as well as more distant cytokines such as nerve growth factor (NGF) and members of the transforming growth factor (TGF) family [6]. This core domain determines the binding specificity of VEGFs towards one or more of three receptor tyrosine kinases, VEGFR-1 (flt-1, fms-like tyrosine kinase-1), VEGFR-2 (also called kinase insert domain containing receptor or KDR), and VEGFR-3 (flt-4, fms-like tyrosine kinase-4). VEGF-B and PlGF are specific for VEGFR-1; VEGF-E and VEGF-F selectively bind to VEGFR-2; while VEGF-A is capable of binding VEGFR-1 as well as VEGFR-2. VEGF-C and VEGF-D preferably bind to VEGFR-3, but also recognize VEGFR-2 (reviewed in [10]).

Handbook of Cell Signaling, Three-Volume Set 2 ed.

The currently available crystal structures of VEGF-family receptor-binding modules include VEGF-A [11], PlGF [12], VEGF-B [13], VEGF-E [14], and VEGF-F [9]. These structures confirm that VEGF adopts a cystine-knot fold. The structures of VEGF family members reported so far superimpose very well with root mean square deviations (rmsd) of about 1.0–1.6 Å for 175 to 190 common Cα positions within each dimer (Figure 20.1). Each VEGF monomer contains an N-terminal α-helix, followed by two pairs of anti-parallel two-stranded β-sheets with a short helical insertion L1 connecting β1 and β2, and loops L2 and L3 connecting β2 to β3, and β3 to β4, respectively. The cystine-knot motif and the N-terminal helix are located on the same end of the β-sheet. The largest structural differences occur in the loop regions that have a relatively high inherent degree of flexibility (11) and carry many of the VEGFR binding determinants. In the active dimer, two VEGF monomers are assembled in an anti-parallel fashion with the N-terminal helix of one VEGF monomer packing atop its partner to form a flat sheet-like structure that is 65 Å long and 40 Å wide, but in its center only about 10 Å thick. The two monomers forming the dimer are covalently connected via two disulfide bonds formed by cysteine residues that stem from the central β1-strands. Interestingly, the mature forms of VEGF-C and VEGF-D receptor-binding domains have been reported to form non-covalent dimers [15, 16], even though the cysteine residues responsible for covalent dimer formation within other VEGF family members are conserved. Unfortunately, the structures of VEGF-C and VEGF-D are not yet available. In addition to the receptor-binding domain, many isoforms of VEGF family members contain a heparin binding domain [17], and numerous VEGF isoforms bind to neuropilins, presumably via their C-terminus [18, 19].

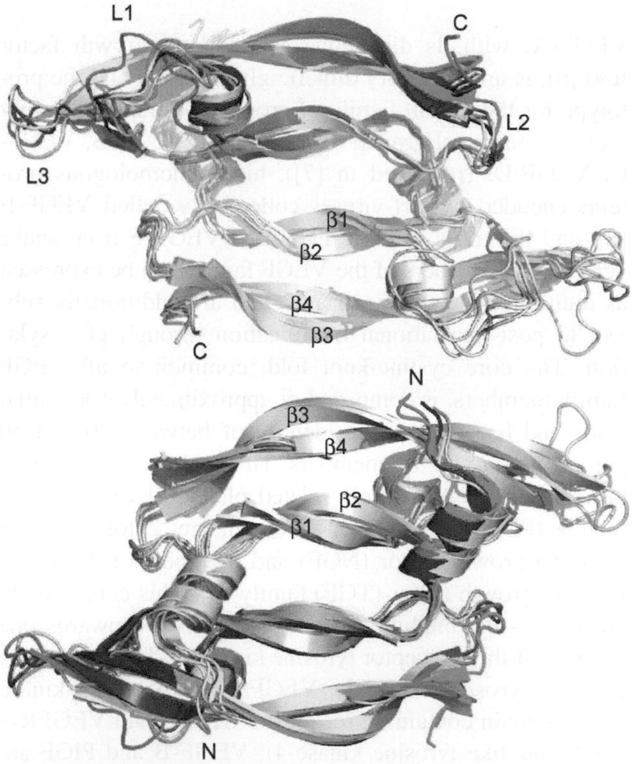

FIGURE 20.1 Two different views of VEGF family members in ribbon representation superimposed on top of each other.
One monomer of each VEGF dimer is shown in white, the other is shown in gray, from light to dark in the order VEGF-A, PlGF, VEGF-B, VEGF-E, and VEGF-F. The disulfide bonds connecting the monomers are shown as sticks; the termini and secondary structure elements are labeled.

STRUCTURAL CHARACTERIZATION OF VEGFRs

The three receptors that exert most functions of the VEGF family members belong to the class of receptor tyrosine kinases and can be regarded as a subfamily of the platelet-derived growth factor receptors (PDGFRs) [20]. The global architecture of the three VEGF-receptors is identical; they are all composed of seven immunoglobulin-like (Ig-like) repeats in their extracellular portion, labeled domain-1 through domain-7, beginning at the N-terminus. Interestingly, VEGFR-3 is the only one of the three receptors that possesses a proteolytic cleavage site within the fifth Ig-like domain, although, after cleavage, the two receptor parts remain linked via a disulfide bond [21]. The ectodomain of the VEGFRs connects via a transmembrane helix to the intracellular portion, which carries a tyrosine kinase domain. Overall, the sequence homology between human VEGFR-1 and the other two human VEGFRs is about 50 percent within the ectodomain and approximately 80 percent within the intracellular domain.

Currently, high-resolution structures of the VEGFRs are limited to domain-2 of VEGFR1 in its unliganded form [22] or in complex with VEGF-A [3] or PlGF [23] (Figure 20.2). These structures established the central importance of domain-2 for the ligand–receptor interactions within the

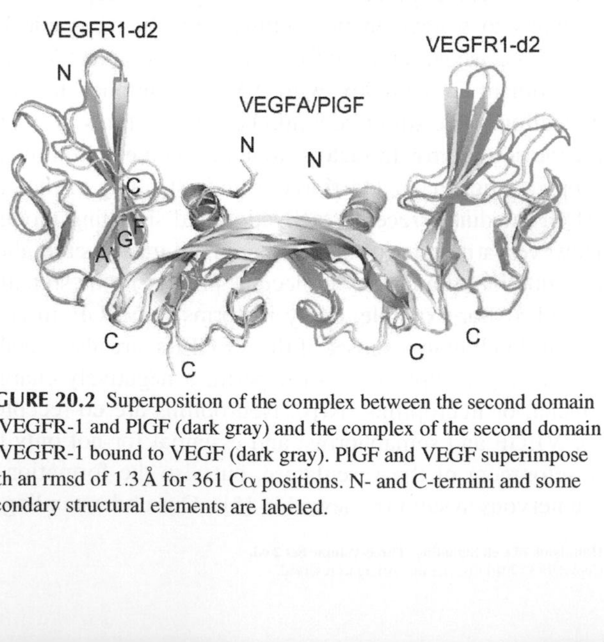

FIGURE 20.2 Superposition of the complex between the second domain of VEGFR-1 and PlGF (dark gray) and the complex of the second domain of VEGFR-1 bound to VEGF (dark gray). PlGF and VEGF superimpose with an rmsd of 1.3 Å for 361 Cα positions. N- and C-termini and some secondary structural elements are labeled.

VEGFR family. Domain-2 of VEGFR-1 is a distorted member of the I-set of Ig-like domains [24], and contacts the N-terminal alpha helix and with loops L1, L2, and L3 of PlGF or VEGF-A via the C-terminal end of its A′GFC β-sheet. The residues that link domains-2 and -3 of the receptor point towards a groove at the bottom of VEGF-A, which, in agreement with mutagenesis studies, would allow the VEGFR domain-3 to contact the bottom face of the growth factor dimers [25].

Among the residues specifically forming the contact area on the ligand for the receptor, VEGF-A and PlGF share about 52 percent sequence conservation, similar to the overall sequence homology for their receptor-binding domains. It has been speculated that such a low degree of sequence conservation within the binding site might produce differences in the general architecture of the signaling complexes [26]. However, the comparison of VEGFR-1 domain-2 bound to either VEGF-A or PlGF demonstrates that the architecture of both complexes is the same, and many of the structural details of the interaction are conserved [23] (Figure 20.2). For example, in both structures, the interfaces between ligand and receptor are rather flat and dominated by hydrophobic contacts. Also, the same set of VEGFR-1 domain-2 receptor residues contacts equivalent residues in both PlGF and VEGF-A. Residues that are located in the core of the interface and that are not conserved between VEGF-A and PlGF have compensatory exchanges nearby that allow for the receptor domain to bind both ligands in a very similar way. Interestingly, based on structural alignments, PlGF complexed to VEGFR-1 adopts a conformation more similar to the structure of VEGF-A bound to the same receptor domain, than to PlGF in its unbound form [23]. Although structures of VEGFR-2 and VEGFR-3 are currently unavailable, additional functional studies using VEGF mutants and chimeras identify receptor-selective VEGF variants [14, 27] and pinpoint residues that mediate receptor selectivity within the loop regions L1 and L3 of VEGF.

While high-resolution structures of the entire VEGFR ectodomain are still unavailable, recent electron microscopy studies revealed the general architecture of VEGFR-2 in its free form as well as bound to VEGF-A [28]. These electron micrographs identify the receptor as a highly flexible, monomeric molecule adopting random conformations in the absence of ligand. Upon addition of VEGF-A, the receptors dimerize and, while maintaining some flexibility, align in a parallel fashion. The electron micrographs further verify domains-2 and -3 of the receptor as responsible for ligand binding, corroborating truncation studies that identified these two domains as the ligand-binding module [29, 30]. The study shows domain-7 as the module that is most often engaged in receptor : receptor contacts, and, in agreement with earlier findings [31], suggests that domain 4 of VEGFR-2 also forms interactions between two receptor molecules.

PDGFR AND ANALOGIES TO VEGFRs

The VEGFR family is closely related to the family of PDGF receptors, which includes PDGFR-α, PDGFR-β, colony stimulating factor-1 (CSF-1) receptor, the stem cell factor (SCF) receptor (also called KIT), and Flt-3 [32]. In contrast to the VEGFRs, the ectodomain of the PDGFRs is composed of only five Ig-like domains. Interestingly, the PDGFR receptor family binds ligands that belong to two distinct structural classes. The first class includes the PDGFs, cystine-knot containing growth factors that are homologous and structurally very similar to the members of the VEGF family; the second class comprises MCSF, SCF, and Flt3-ligand (Flt3-L), proteins that belong to the family of four-helix cytokines, a fold that is unrelated to VEGF [33]. Sequence comparisons, together with insights provided by the genomic organization of the receptors [34] and their evolutionary relationship [35], suggest that the architecture of the N-terminal five domains and the areas of the receptors involved in ligand binding for the four-helix bundle-type growth factors and the cystine-knot containing growth factors are conserved between the VEGF- and PDGF-receptor families.

Recent reports revealed the crystal structures of the human and murine SCFR ectodomain in complex with their SCF ligands [36, 37]. SCF is structurally unrelated to VEGF and possesses a different dimerization mode, as SCF monomers form head-to-head dimers, in contrast to the side-by-side arrangement within PDGF or VEGF dimers (Figure 20.3). As a consequence, these cytokine dimers engage their receptors in different manners. While VEGF-A and PlGF contact the VEGFR-1 with both molecules of the growth factor dimer, the SCF : SCFR receptor structure shows each molecule within the dimer in contact with only one receptor. The nature of the binding interface is also very dissimilar, as VEGF and PlGF bind to VEGFR-1 mainly via hydrophobic residues, while SCF engages SCFR predominantly via hydrophilic and charged interactions. Aside from these local differences, the overall architecture of the molecules within the ligand : receptor complexes is surprisingly similar. Analogous to the reports on the VEGF : VEGFR interactions, the crystal structures of the SCFR : SCF complex identify the three N-terminal receptor domains as important for ligand binding, with the interactions of Ig modules 2 and 3 dominating the binding interface. Domain 2 of the SCFR shares a sequence identity of only 22 percent with the respective domain of the VEGFR-1, but is also a distorted member of the I-set of Ig-like domains. Like domain 2 of VEGFR-1, it engages the ligand with residues that, albeit not conserved, are in equivalent positions on the C-terminal end of the A′GFC β-sheet. When superimposing the second receptor domains of both receptor complexes, only small conformational changes are required to position domain-3 of SCFR at the bottom face of VEGF, further supporting a role for domain-3 of VEGFR-1 in VEGF binding as expected from functional

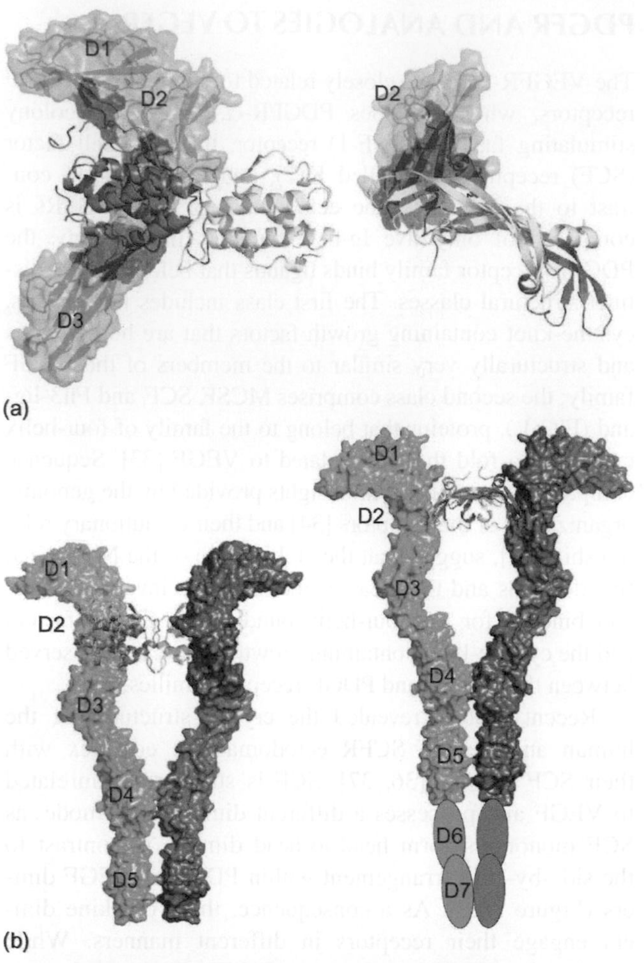

FIGURE 20.3 (a) Left side: SCF binding to KIT. The N-terminal three receptor domains are depicted in white, with all residues in contact with the ligand colored dark gray. The two molecules of the SCF dimer are shown in light gray. Right side: complex between VEGF-A and VEGFR-1. Domain-2 of the receptor is shown in white and in the same orientation as domain-2 of the KIT receptor on the left. Residues in contact with VEGF-A are shown in dark gray. The two molecules of the VEGFA dimer are depicted in light gray. Note that for both ligand : receptor complexes the same general area of the second receptor domain is forming numerous contacts to the ligand. (b) Structure of SCF (shown in ribbon diagram) in complex with KIT (surface representation) (37) and model of a complex between VEGF-A (ribbon diagram) and VEGFR-1 (surface representation).

studies [25]. The same superposition places domain-1 of the SCF receptor relatively far from the ligand, corroborating the observation that this domain in VEGFR-1 has only a minor role for VEGF binding.

The structure of the SCF : SCFR complex confirms functional studies which show that, after ligand binding, domains 4 and 5 of the two receptors come into close vicinity with one another. Especially intriguing is the presence of a small receptor : receptor interface within the fourth domain that contains two salt-bridges forming around the two-fold axis of the dimer [37]. Although the fourth domains of the PDGFR and VEGFR families share a number of features, such as the lack of the hallmark disulfide bridge that is present in all other domains within the receptor families, the residues engaged in formation of receptor : receptor contacts in domain 4 are only conserved within the PDGFR family. Curiously, a homologous sequence motif to this binding interface is found in the membrane proximal domain-7 of VEGFRs, corroborating the importance of domain-7 for receptor : receptor interactions [28] within the VEGFR family.

VEGF CO-RECEPTORS: NEUROPILINS

Neuropilins are a family of class I transmembrane glycoproteins that were discovered as neuron-specific cell-surface antigens and identified as important for axon guidance [38, 39]. Later, it was revealed that, in addition to the development of the embryonic nervous system, neuropilins are also crucial for the development of cardiovascular development [40]. The two members of the neuropilin family, neuropilin-1 and -2, share a sequence identity of approximately 45 percent, and both receptors are also expressed as soluble, secreted proteins as a result of differential splicing events. Neuropilins are unable to induce cellular signaling in the absence of other signaling molecules. Instead, they function as co-receptors to enhance the angiogenic and mitogenic potency of VEGF [41] or the neurogenic effects of the semaphorins [38]. They mediate their biological function not only by binding to the heparin-binding isoforms of VEGF and semaphorins, both of which are soluble extracellular ligands, but also by interacting with a number of cell-surface receptors, such as VEGFR-1 [42], members of the plexin family, [43] and heparin.

Neuropilins possess a large extracellular domain of about 830 residues, followed by a transmembrane helix and a short, approximately 40-residue cytoplasmic tail. The extracellular portion of the receptors folds into five domains referred to as a1, a2, b1, b2, and c [5]. The crystal structures of the Nrp-2 a1, a2, b1, b2 in complex with neuropilin-binding antibody fragments [44], and of Nrp-1 b1, b2 in complex with Tuftsin, a short peptide homologous to the C-terminus of the VEGF, have been reported [18]. These crystal structures describe the general architecture of the neuropilin extracellular region and suggest how VEGF and heparin might interact with neuropilin, but do not reveal what role neuropilins play in the organization of larger signaling modules containing VEGF, VEGFRs, or semaphorins and plexins, and how its participation in these complexes might modify intracellular signaling events.

CONCLUSION

With numerous studies elucidating structural aspects of ligand : receptor interactions of the VEGFR and the PDGFR families and their ligands, we now have a solid understanding of the general architecture and mechanism of receptor activation within these families. In light of these studies, it is probable that all VEGFR and PDGFR family members utilize their N-terminal three domains, or a subset thereof, to bind their respective dimeric ligand. These resulting complexes containing two receptor molecules and one dimeric ligand are further stabilized via receptor:receptor interactions mediated by Ig-like domains 4 and 5 for the PDGFR family and 4–7 within the VEGFR family. The interaction of the receptor domains proximal to the membrane ensures that the intracellular kinase domains of the receptor molecules are in close proximity and able to transphosphorylate each other to initiate the signaling cascade.

However, the picture of VEGR activation is still incomplete, as only limited information is available on the composition and architecture of larger signaling complexes that contain co-receptors like heparin or neuropilins. Future studies will hopefully shed light on structural and functional aspects of the interactions between VEGF, VEGFRs, heparin, neuropilin, and potential other co-receptors.

REFERENCES

1. Breen EC. VEGF in biological control. *J Cell Biochem* 2007;**102**(6):1358–67.
2. Ferrara N, Mass RD, Campa C, Kim R. Targeting VEGF-A to treat cancer and age-related macular degeneration. *Annu Rev Med* 2007;**58**:491–504.
3. Wiesmann C, Fuh G, Christinger HW, Eigenbrot C, Wells JA, de Vos AM. Crystal structure at 1.7 A resolution of VEGF in complex with domain 2 of the Flt-1 receptor. *Cell* 1997;**91**(5):695–704.
4. Wiesmann C, de Vos AM. The mechanism of VEGFR activation suggested by the complex of VEGF-flt1-D2. In: *Handbook of Cell Signaling*. Amsterdam: Elsevier Science; 2003, p. 285–9.
5. Pellet-Many C, Frankel P, Jia H, Zachary I. Neuropilins: structure, function and role in disease. *Biochem J* 2008;**411**(2):211–26.
6. Sun PD, Davies DR. The cystine-knot growth-factor superfamily. *Annu Rev Biophys Biomol Struct* 1995;**24**:269–91.
7. Tammela T, Enholm B, Alitalo K, Paavonen K. The biology of vascular endothelial growth factors. *Cardiovasc Res* 2005;**65**(3): 550–63.
8. Ogawa S, Oku A, Sawano A, Yamaguchi S, Yazaki Y, Shibuya M. A novel type of vascular endothelial growth factor, VEGF-E (NZ-7 VEGF), preferentially utilizes KDR/Flk-1 receptor and carries a potent mitotic activity without heparin-binding domain. *J Biol Chem* 1998;**273**(47):31,273–2.
9. Yamazaki Y, Takani K, Atoda H, Morita T. Snake venom vascular endothelial growth factors (VEGFs) exhibit potent activity through their specific recognition of KDR (VEGF receptor 2). *J Biol Chem* 2003;**278**(52):51985–8.
10. Otrock ZK, Makarem JA, Shamseddine AI. Vascular endothelial growth factor family of ligands and receptors: review. *Blood Cells Mol Dis* 2007;**38**(3):258–68.
11. Muller YA, Christinger HW, Keyt BA, de Vos AM. The crystal structure of vascular endothelial growth factor (VEGF) refined to 1.93 A resolution: multiple copy flexibility and receptor binding. *Structure* 1997;**5**(10):1325–38.
12. Iyer S, Leonidas DD, Swaminathan GJ, Maglione D, Battisti M, Tucci M, Persico MG, Acharya KR. The crystal structure of human placenta growth factor-1 (PlGF-1), an angiogenic protein, at 2.0-Å resolution. *J Biol Chem* 2001;**276**(15):12,153–61.
13. Iyer S, Scotney PD, Nash AD, Ravi Acharya K. Crystal structure of human vascular endothelial growth factor-B: identification of amino acids important for receptor binding. *J Mol Biol* 2006;**359**(1): 76–85.
14. Pieren M, Prota AE, Ruch C, Kostrewa D, Wagner A, Biedermann K, Winkler FK, Ballmer-Hofer K. Crystal structure of the Orf virus NZ2 variant of vascular endothelial growth factor-E. Implications for receptor specificity. *J Biol Chem* 2006;**281**(28):19,578–19,587,.
15. Joukov V, Sorsa T, Kumar V, Jeltsch M, Claesson-Welsh L, Cao Y, Saksela O, Kalkkinen N, Alitalo K. Proteolytic processing regulates receptor specificity and activity of VEGF-C. *EMBO J* 1997; **16**(13):3898–911.
16. Stacker SA, Stenvers K, Caesar C, Vitali A, Domagala T, Nice E, Roufail S, Simpson RJ, Moritz R, Karpanen T, Alitalo K, Achen MG. Biosynthesis of vascular endothelial growth factor-D involves proteolytic processing which generates non-covalent homodimers. *J Biol Chem* 1999;**274**(45):32,127–32,136.
17. Fairbrother WJ, Champe MA, Christinger HW, Keyt BA, Starovasnik MA. Solution structure of the heparin-binding domain of vascular endothelial growth factor. *Structure* 1998;**6**(5):637–48.
18. Vander Kooi CW, Jusino MA, Perman B, Neau DB, Bellamy HD, Leahy DJ. Structural basis for ligand and heparin binding to neuropilin B domains. *Proc Natl Acad Sci U S A* 2007;**104**(15):6152–7.
19. Pan Q, Chathery Y, Wu Y, Rathore N, Tong RK, Peale F, Bagri A, Tessier-Lavigne M, Koch AW, Watts RJ. Neuropilin-1 binds to VEGF121 and regulates endothelial cell migration and sprouting. *J Biol Chem* 2007;**282**(33):24,049–56.
20. van der Geer P, Hunter T, Lindberg RA. Receptor protein-tyrosine kinases and their signal transduction pathways. *Annu Rev Cell Biol* 1994;**10**:251–337.
21. Pajusola K, Aprelikova O, Pelicci G, Weich H, Claesson-Welsh L, Alitalo K. Signalling properties of FLT4, a proteolytically processed receptor tyrosine kinase related to two VEGF receptors. *Oncogene* 1994;**9**(12):3545–55.
22. Starovasnik MA, Christinger HW, Wiesmann C, Champe MA, de Vos AM, Skelton NJ. Solution structure of the VEGF-binding domain of Flt-1: comparison of its free and bound states. *J Mol Biol* 1999;**293**(3):531–44.
23. Christinger HW, Fuh G, de Vos AM, Wiesmann C. The crystal structure of placental growth factor in complex with domain 2 of vascular endothelial growth factor receptor-1. *J Biol Chem* 2004;**279**(11):10,382–8.
24. Harpaz Y, Chothia C. Many of the immunoglobulin superfamily domains in cell adhesion molecules and surface receptors belong to a new structural set which is close to that containing variable domains. *J Mol Biol* 1994;**238**(4):528–39.
25. Muller YA, Li B, Christinger HW, Wells JA, Cunningham BC, de Vos AM. Vascular endothelial growth factor: crystal structure and functional mapping of the kinase domain receptor binding site. *Proc Natl Acad Sci U S A* 1997;**94**(14):7192–7.
26. Autiero M, Waltenberger J, Communi D, Kranz A, Moons L, Lambrechts D, Kroll J, Plaisance S, De Mol M, Bono F, Kliche S, Fellbrich G, Ballmer-Hofer K, Maglione D, Mayr-Beyrle U, Dewerchin M, Dombrowski S, Stanimirovic D, Van Hummelen P, Dehio C, Hicklin DJ, Persico G, Herbert JM, Communi D, Shibuya M, Collen D, Conway EM, Carmeliet P. Role of PlGF in the intra- and

intermolecular cross talk between the VEGF receptors Flt1 and Flk1. *Nat Med* 2003;**9**(7):936–43.
27. Li B, Fuh G, Meng G, Xin X, Gerritsen ME, Cunningham B, de Vos AM. Receptor-selective variants of human vascular endothelial growth factor. Generation and characterization. *J Biol Chem* 2000;**275**(38):29,823–8.
28. Ruch C, Skiniotis G, Steinmetz MO, Walz T, Ballmer-Hofer K. Structure of a VEGF-VEGF receptor complex determined by electron microscopy. *Nat Struct Mol Biol* 2007;**14**(3):249–50.
29. Shinkai A, Ito M, Anazawa H, Yamaguchi S, Shitara K, Shibuya M. Mapping of the sites involved in ligand association and dissociation at the extracellular domain of the kinase insert domain-containing receptor for vascular endothelial growth factor. *J Biol Chem* 1998;**273**(47):31,283–8.
30. Davis-Smyth T, Chen H, Park J, Presta LG, Ferrara N. The second immunoglobulin-like domain of the VEGF tyrosine kinase receptor Flt-1 determines ligand binding and may initiate a signal transduction cascade. *EMBO J* 1996;**15**(18):4919–27.
31. Barleon B, Totzke F, Herzog C, Blanke S, Kremmer E, Siemeister G, Marme D, Martiny-Baron G. Mapping of the sites for ligand binding and receptor dimerization at the extracellular domain of the vascular endothelial growth factor receptor FLT-1. *J Biol Chem* 1997;**272**(16):10,382–8.
32. Blume-Jensen P, Hunter T. Oncogenic kinase signalling. *Nature* 2001;**411**(6835):355–65.
33. Wiesmann C, Muller YA, de Vos AM. Ligand-binding sites in Ig-like domains of receptor tyrosine kinases. *J Mol Med (Ber)* 2000;**78**(5):247–60.
34. Kondo K, Hiratsuka S, Subbalakshmi E, Matsushime H, Shibuya M. Genomic organization of the flt-1 gene encoding for vascular endothelial growth factor (VEGF) receptor-1 suggests an intimate evolutionary relationship between the 7-Ig and the 5-Ig tyrosine kinase receptors. *Gene* 1998;**208**(2):297–305.
35. Shibuya M. Vascular endothelial growth factor receptor family genes: when did the three genes phylogenetically segregate? *Biol Chem* 2002;**383**(10):1573–9.
36. Liu H, Chen X, Focia PJ, He X. Structural basis for stem cell factor-KIT signaling and activation of class III receptor tyrosine kinases. *EMBO J* 2007;**26**(3):891–901.
37. Yuzawa S, Opatowsky Y, Zhang Z, Mandiyan V, Lax I, Schlessinger J. Structural basis for activation of the receptor tyrosine kinase KIT by stem cell factor. *Cell* 2007;**130**(2):323–34.
38. He Z, Tessier-Lavigne M. Neuropilin is a receptor for the axonal chemorepellent Semaphorin III. *Cell* 1997;**90**(4):739–51.
39. Kolodkin AL, Levengood DV, Rowe EG, Tai YT, Giger RJ, Ginty DD. Neuropilin is a semaphorin III receptor. *Cell* 1997;**90**(4):753–62.
40. Gu C, Rodriguez ER, Reimert DV, Shu T, Fritzsch B, Richards LJ, Kolodkin AL, Ginty DD. Neuropilin-1 conveys semaphorin and VEGF signaling during neural and cardiovascular development. *Dev Cell* 2003;**5**(1):45–57.
41. Soker S, Takashima S, Miao HQ, Neufeld G, Klagsbrun M. Neuropilin-1 is expressed by endothelial and tumor cells as an isoform-specific receptor for vascular endothelial growth factor. *Cell* 1998;**92**(6):735–45.
42. Fuh G, Garcia KC, de Vos AM. The interaction of neuropilin-1 with vascular endothelial growth factor and its receptor flt-1. *J Biol Chem* 2000;**275**(35):26690–5.
43. Takahashi T, Fournier A, Nakamura F, Wang LH, Murakami Y, Kalb RG, Fujisawa H, Strittmatter SM. Plexin-neuropilin-1 complexes form functional semaphorin-3A receptors. *Cell* 1999;**99**(1):59–69.
44. Appleton BA, Wu P, Maloney J, Yin J, Liang WC, Stawicki S, Mortara K, Bowman KK, Elliott JM, Desmarais W, Bazan JF, Bagri A, Tessier-Lavigne M, Koch AW, Wu Y, Watts RJ, Wiesmann C. Structural studies of neuropilin/antibody complexes provide insights into semaphorin and VEGF binding. *EMBO J* 2007;**26**(23): 4902–12.

Chapter 21

Mechanisms and Functions of Eph Receptor signaling

Martin Lackmann
Department of Biochemistry and Molecular Biology, Monash University, Victoria, Australia

INTRODUCTION

Eph receptor tyrosine kinases (RTKs) [1] together with their cell-surface-bound ephrin ligands are amongst the principle cell-cell communication systems that control positioning and morphology of cells and to some degree affect cell proliferation and survival during normal and oncogenic development [2]. A single ancestral Eph gene in sponges and in *C. elegans* has expanded to become the largest vertebrate RTK family, seemingly reflecting increasing need for accurate cell positioning during vertebrate development. Vertebrate Ephs and ephrins are dynamically expressed in complementary or overlapping domains, and control developmental programs ranging from germ layer formation [3] and somitogenesis [4] to patterning of skeletal [5] neural [6] and vascular [7] networks. In adults, Eph and ephrin are implicated in regulating neural plasticity [8], bone homeostasis [9], and insulin release [10], and their de-regulated function contributes to the progression and metastasis of epithelial and mesenchymal tumors [11].

Signaling between mesenchymal- and epithelial/endothelial-expressed Eph and ephrin interaction partners [12, 13] is an emerging paradigm underlying many of these developmental tasks. In general, Ephs as well as ephrins are signal transducers, which function on a molecular level by responding to the graded expression pattern of their interaction partner with fine-tuned cellular responses ranging from cell rounding, cell-cell repulsion, and increased motility, to cell spreading and tight cell-substrate and cell-cell adhesion. These functions are controlled by a range of parameters, including Eph and ephrin cell surface densities and signaling capacities, and the composition, size, and persistence of their signaling clusters, as well as cross-talk with other principle cell signaling systems, such as the RAS/Erk/mitogen-activated kinase (MAPK), phosphoinositol-3-kinase (PI3K), and Wnt signaling pathways [2].

EPH/EPHRIN PROTEIN STRUCTURES AND SIGNALING CONCEPTS

Eph RTKs (Ephs) and ephrins are highly-conserved proteins: the 16 vertebrate Ephs are comprised of modular extracellular and intracellular subdomains, which, apart from the unique N-terminal ephrin binding domain and a cytoplasmic sterile alpha motif (SAM) protein–protein interaction domain, are common to other receptor tyrosine kinase families (Figure 21.1). The domain structure and overall function of Ephs is conserved from the single *Drosophila* and *C. elegans* genes to all of the vertebrate family members. Likewise, all of the nine vertebrate ephrins are closely related in their core domains (30–70 percent similarity), and highly conserved between orthologs (80–95 percent identity). Structural features and binding preferences classify 6 glycophosphatidyl inositol (GPI) membrane-anchored A-type ephrins that can interact with 10 type-A Ephs, and 3 transmembrane type-B ephrins that bind to 6 type-B Ephs ([14], Figure 21.1). Some Eph family members can be activated by A- and B-type ephrins, including EphA4, which also interacts with ephrin-B2 and ephrin-B3; and ephrin-A5, which can also activate EphB2.

In contrast to a conventional concept of kinase activation through ligand-induced receptor dimerization, multimeric Eph/ephrin complexes that are induced by cell membrane-bound ephrins are necessary to induce robust Eph phosphorylation and biological responses, whereas non-clustered soluble ephrins act as functional antagonists *in vitro* and *in vivo* [2]. Contact between Eph and ephrin-expressing cells invokes formation of a 2:2 Eph/ephrin hetero-tetramer, which is tethered by two contact surfaces within the Eph and ephrin binding domains and a third interface within the Eph cysteine-rich linker region [14]. In agreement with Eph/ephrin complexes assembling on opposing cell membranes *in trans,* this configuration

Handbook of Cell Signaling, Three-Volume Set 2 ed.

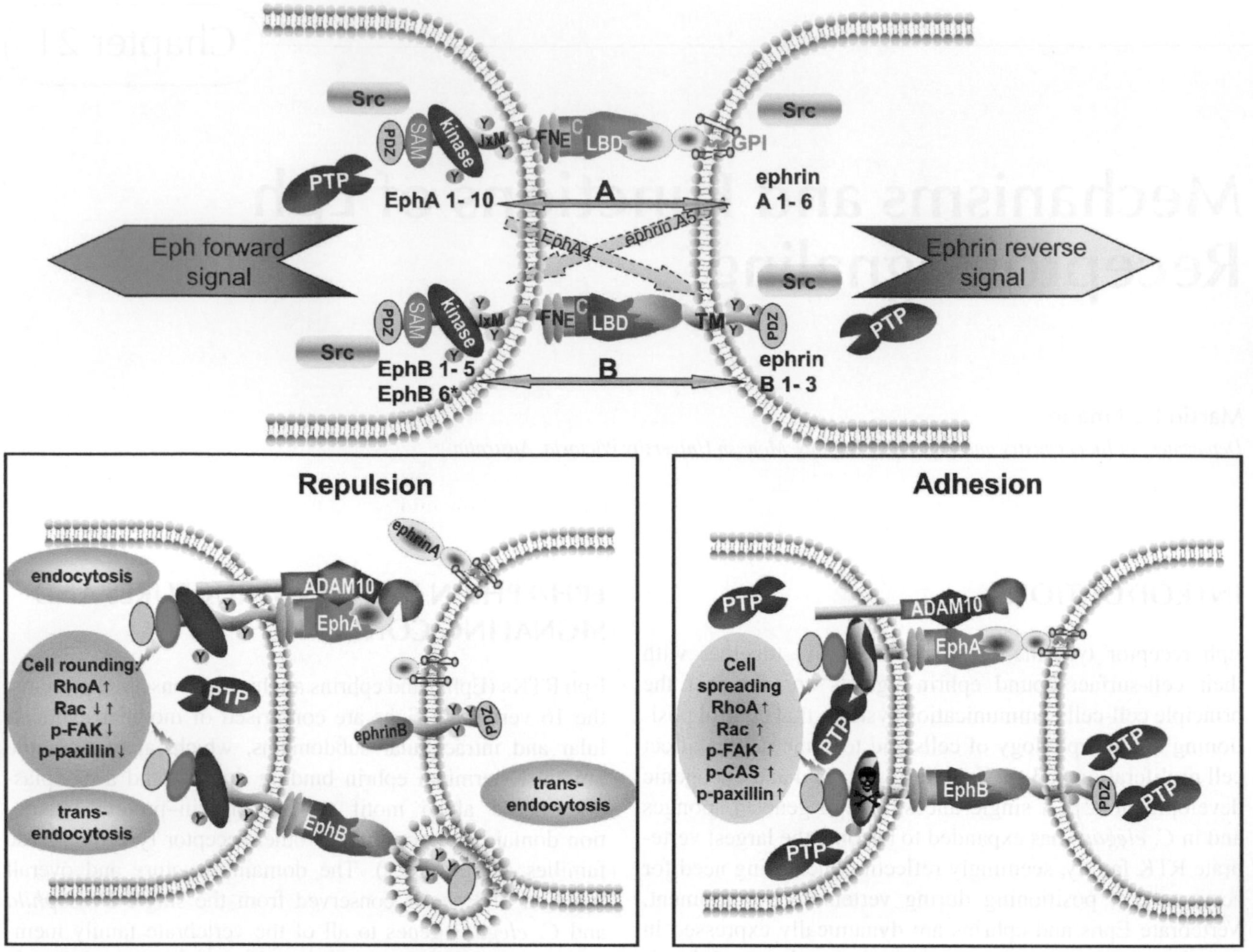

FIGURE 21.1 Mechanisms of Eph and ephrin signaling.
Structural modules of EphA receptors (EphA1-EphA10) include the N-terminal ligand binding domain (LBD), cysteine-rich domain (C), EGF-like motif (E), fibronectin-type III motifs (FN), regulatory juxtamembrane domains (JxM) containing two tyrosine (Y) phosphorylation/Src homology 2 (SH2) domain binding sites, kinase domain, steril-alpha-motif interaction domain (SAM), and PDZ binding motif. EphB6 is catalytically inactive (EphB6*). Ephrins (ephrin A1-ephrin A6) are membrane anchored via a glycosyl-phosphatidylinositol (GPI), or are transmembrane (TM) proteins with SH2 docking sites and PDZ binding motifs (ephrin B1- phrin B3). Solid and broken arrows indicate Eph/ephrin interactions within and across Eph/ephrin subfamilies, respectively. Disruption of Eph-ephrin complexes via ADAM10-mediated ephrin shedding or transendocytosis of intact EphB/ephrin B complexes is illustrated. Eph kinase signaling is abrogated by inactivating mutations (X) or elevated PTPs, and, via kinase-independent downstream signaling, leads to cell–cell adhesion and spreading. Eph downstream signaling components affecting cell morphology include Src kinase (Src), protein tyrosine phosphatase (PTP), RhoA, Rac, focal adhesion kinase (pFAK), paxillin, Crk-associated substrate (CAS); (p-), phosphorylated, (↑), increase, (↕), transient increase, (↓), decrease. From [2], Lackmann M, Boyd AW. Eph, a protein family coming of age: more confusion, insight, or complexity? *Sci Signal* 2008; 1: re2.

positions the ephrin and Eph C-termini to opposite sides of a ring-like structure ([15], Figure 21.1). The initial complex triggers progressive recruitment of further Ephs via homotypic interactions into large signaling clusters, which are rapidly internalized into cytoplasmic cell compartments. Because the interacting Ephs and ephrins are tethered to the cell surface, this critical step of the signaling cascade relies on disruption of the multivalent molecular tethers between cells - likely involving regulated metalloprotease activity - which provides the switch between cell-cell repulsion and adhesion [16–19]. Importantly, Eph clustering is essential for phosphotyrosine-dependent Eph/ephrin signaling, but also triggers tyrosine-independent functions, in particular cell adhesion and migration. In both settings, the composition and dynamic regulation of Eph/ephrin signaling clusters determine the nature and strength of cell-biological responses [2, 20, 21].

Unlike most RTK ligands, membrane-bound ephrins also have signaling functions, such that upon interaction, ligand- and receptor-expressing cells both respond with intracellular signaling by Ephs (forward signaling) and ephrins (reverse signaling) [22, 23]. The conserved ephrin-B cytoplasmic domains contain SH2- and PDZ-binding motifs acting as nucleation sites for signaling complexes, and while circumstantial evidence also indicates ephrin-A signaling, lack of a cytoplasmic domain makes the involved mechanisms less well understood.

REGULATION OF EPH/EPHRIN SIGNALING ACTIVITY

Cellular responses to Eph/ephrin interactions necessarily depend on the enzymatic capacity of the Eph kinase, and the ratio of kinase active and inactive receptors determines whether Eph/ephrin contact leads to cell-cell repulsion or adhesion. The default response to Eph kinase activation is cell-cell repulsion, but attenuation of cell contraction signals through co-expression of signaling-compromised receptor splice variants [24] or because of low ephrin surface concentrations results in integrin-dependent or -independent adhesion [2, 21, 25].

The signaling function of the intact Eph kinase is tightly regulated by phosphorylation of a conserved tyrosine in the kinase activation loop as well as two additional juxtamembrane (JM) domain tyrosines, which together modulate the conformation, accessibility, and activity of the kinase domain, and at the same time provide docking sites for downstream signaling proteins [26]. In the inactive state of the kinase, the activation loop tyrosine blocks ATP binding by protruding into the active site. Similar to the auto-inhibition mechanism of the TGFβR1 receptor kinase, where the non-phosphorylated JM segment keeps the kinase in an inactive conformation, the Eph JM domain folds into an ordered α-helix that clamps onto the small kinase domain lobe and prevents the activation loop from attaining an active conformation [27]. Phosphorylation of the two JM tyrosines releases the ordered α-helical JM domain into a dynamic, disordered protein fold, releasing the constraint on the activation loop and allowing activation to occur [26]. Accordingly, protein tyrosine phosphatases (PTPs) can directly regulate Eph tyrosine phosphorylation levels and kinase activity, and thereby modulate biological functions such as endothelial capillary-like tube assembly, retinal axon repulsion, cell attachment, and polarization. PTPs thought to directly modulate Eph functions include LMW-PTP [28], "PTP receptor type O" (Ptpro) [29], and PTP-PEST [30], while PTP-BL regulates ephrin-B phosphorylation and activity [31].

In addition to enzymatic regulation, the density of Ephs and ephrins on the cell surface directly influences the signaling outcome, likely by impacting on the size and stability of signaling clusters [2, 28, 32]. Thus, assembly of A- and B-type ephrins in highly ordered membrane microdomains [33] is thought to effect partitioning of signaling molecules such as Src family kinases into these signaling complexes [34]. Localization to membrane microdomains probably also facilitates mutual clustering of Ephs and ephrins via homotypic interactions, and stabilizes their multivalent clusters at cell–cell contacts in a manner similar to that reported for other receptor systems, such as the T cell receptor. Importantly, Ephs and ephrins that are co-expressed on the same cell membrane laterally segregate into distinct lipid microdomains, partition their respective signaling complexes, and prevent Eph/ephrin interactions in *cis* from occurring [35].

DISRUPTION OF CELL–CELL CONTACTS AND INTERNALIZATION OF SIGNALING COMPLEXES

The default response to Eph activation and phosphotyrosine-dependent signaling is cell-cell repulsion. However, cell contact-dependent assembly of clustered surface-attached Ephs and ephrins initially tightly joins the interacting cells, and regulated disruption of these Eph/ephrin clusters and their endocytosis is critical for ensuing signaling pathways and cell-biological outcomes. Currently, two different molecular mechanisms that may control the termination of Eph/ephrin-mediated cell-cell contacts are being explored:

In EphB and ephrin-B overexpressing fibroblasts, Rac-mediated ruffling of the opposing cell membranes appears to trigger "trans-endocytosis" of the whole, intact EphB/ephrin-B complex (including the adjacent plasma membrane components) into one of the interacting cells [17, 36]. The direction of trans-endocytosis seems to be regulated by the presence of the Eph and ephrin intracellular domains, as overexpression of cytoplasmic truncated forms leads to preferential endocytosis into ephrin- or Eph-expressing cells, respectively [17, 36]. It remains unclear how bidirectional signaling can proceed when one of the participating cells loses its Eph or ephrin during trans-endocytosis.

In the case of EphA/ephrin-A-mediated cell-cell contacts, shedding of Eph-bound ephrins via the transmembrane metalloprotease ADAM10 releases the molecular tether between the opposing cells [16, 18]. This ephrin-shedding is tightly regulated, and only intact EphA/ephrin-A complexes provide the high-affinity recognition site that is required for ADAM10 to efficiently cleave Eph-bound ephrins [18]. Similar to ADAM-shedding of other RTK ligands [37], ephrin cleavage also underlies feedback control by Eph kinase activity and restricts disruption of the Eph/ephrin tethers to conditions of ongoing "repulsive" Eph/ephrin signaling. Shedding of B-type ephrins by a metalloprotease has been demonstrated, and is part of a γ-secretase-mediated cleavage mechanism thought to trigger Src activation of ephrin-B signaling [19].

EPH (FORWARD) SIGNALING

Ephrin-induced Eph activation leads to phosphorylation of several conserved tyrosine residues, including the kinase activation loop tyrosine and two juxtamembrane region tyrosines which provide interaction sites for various SH2 domain-containing adaptor and scaffold proteins, including Src and the Abelson (Abl) kinases and various regulators of small guanine nucleotide triphosphatases (GTPases) of the Rho family [38, 39]. One exception is EphB6, which, due to several critical amino acid substitutions in its kinase domain, is catalytically inactive, and instead is phosphorylated by the Src kinase Fyn [40].

Key signaling proteins downstream of activated Ephs have well-established functions in controlling cytoskeletal plasticity, actin fiber and focal contact assembly, cell adhesion and cell motility (Figure 21.1), and include, amongst others: PI3K, Crk, Nck, Src and Fyn, Rho-GTPases, R-ras, Ras GTPase-activating protein (RasGAP), Grb2, 7 and 10, p130CAS, FAK, Abl and Arg. Consistent with the concept that Eph receptors communicate with pathways regulating directional migration, they provide critical building blocks for a guidance system that steers moving cells and controls their adhesive properties. Considering the attention Eph-ephrin biology has attracted, and the number of proteins known to partake in downstream signaling, the understanding of the pathways executing cell-biological responses to Eph-ephrin signaling is surprisingly limited. To some extent, this may reflect the challenge of dissecting pathways that rely on tyrosine kinase activation and on the assembly of multimeric receptor clusters for productive, SH2 domain-mediated signaling pathways [28, 41].

For example, while Rho GTPase activation is critical for actin-myosin based growth cone collapse and axon retraction downstream of activated Ephs [39], activation of RhoA- and inhibition of Rac- and Cdc42 is also required for ephrin-B-triggered inhibition of cell migration [42], for EphB4/ephrin-B2 triggered membrane ruffling, and for internalization of Eph/ephrin complexes [17]. However, little is known about the effectors linking Ephs and Rho GTPases, and while the Rho-guanine nucleotide-exchange factor (Rho-GEF) ephexin-1 is implicated in Eph-A growth cone signaling, its overall biological relevance is unclear, as ephexin1-knockout (KO) mice have no obvious phenotype [43]. On the other hand, inactivation of the Rac GTPase-activating protein (GAP) α-chimerin phenocopies the pathfinding defects of EphA4 and ephrinB3 KO mice, suggesting its essential role in EphA4 signaling [44].

EPHRIN (REVERSE) SIGNALING

Interaction of ephrins with Eph receptors triggers "reverse" signaling pathways that generally induce cell biological responses similar to those seen during Eph forward signaling [22, 23, 45]. The highly conserved ephrin-B cytoplasmic domains contain several protein interaction motifs, including five tyrosine residues as potential SH2-domain docking sites, a polyproline stretch, and a **P**ostsynaptic **d**ensity protein/**D**isc large/**Z**ona occludens (PDZ) binding motif at the extreme C-terminus. Ephrin clustering promotes the recruitment and activation of Src family kinases, which, in the case of B-type ephrins, phosphorylate specific cytoplasmic tyrosine residues [31], likely generating docking sites for various SH2-containing adaptor proteins. While the critical role of ephrin reverse signaling in axon guidance, segmentation, cell migration, and angiogenesis is well-established, the current knowledge about downstream signaling mechanisms is limited to very few examples [23]. Binding of the GTPase activating protein PDZ-RGS3, which catalyzes GTP hydrolysis in heterotrimeric G proteins, to the PDZ binding motif of ephrin-B leads to inhibition of CXCR4 chemokine receptor-mediated cerebellar granule cell chemotaxis. Phosphotyrosine-dependent recruitment of the SH2-SH3 adaptor protein Grb4 results in increased FAK activity and ultimately leads to changes in the actin cytoskeleton and focal adhesions resulting in cell rounding [46]. Ephrin-B reverse signaling is down-modulated by recruitment of the PDZ domain-containing phosphatase PTP-BL, which inactivates Src-family kinases and dephosphorylates ephrin-B phosphotyrosines [31].

Despite the lack of an intracellular domain, GPI-anchored A-type ephrins also signal upon interactions with Ephs, but the involved mechanisms remain largely unexplored [23]. It is thought that A-type ephrins assemble into membrane microdomains or rafts [33] into which signaling molecules such as Src family kinases partition [34]. However, downstream signaling adaptors that relay ephrin-A signals have remained elusive until now. On the other hand, genetic studies in mice and *C. elegans* have unambiguously confirmed ephrins as ligands and signal transducers, and suggest that Eph-ephrin mediated cell positioning can be viewed conclusively only by considering forward and reverse signaling as being equally important components [23].

CROSS-TALK WITH OTHER SIGNAL PATHWAYS

In line with their prominent role in controlling cell morphology and positioning, there is extensive cross-talk of Ephs with integrin-, focal adhesion and PI3K-signaling pathways with established functions in the regulation of cell adhesion and cytoskeletal plasticity. However, in addition to direct involvement in cell positioning, emerging evidence confirms that Eph signaling also modulates pathways that control cell proliferation, cell survival, and cell fate, including particularly the RAS/Erk/MAPK and Wnt signaling cascades. In contrast to other RTKs, which respond to ligand activation and autophosphorylation by activation of MAPK signaling [47], Ephs and ephrins generally down-modulate MAPK pathways [2]. For example, the absence of ephrin A2 reverse signaling in *ephrin* $A2^{-/-}$ mice leads to an increased number of proliferating neural progenitor cells without affecting cell migration [48], and demonstrates a direct anti-mitogenic role for ephrin A2 signaling. The cross-talk between Eph and MAPK signaling is highly conserved, and in *C. elegans* plays a critical role in the regulation of oocyte maturation [49]. The cross-talk between Eph and Wnt signaling, on the other hand, has considerable impact during gut development in vertebrates, where graded expression of B-type Ephs and ephrins is programmed by the Wnt/β-catenin/Tcf4 transcription system and controls colonic epithelial cell positioning during normal and oncogenic development of colonic crypts [50].

EPH/EPHRIN FACILITATED CELL-CELL COMMUNICATION DURING VERTEBRATE DEVELOPMENT

Eph and ephrin family members are expressed and function throughout vertebrate development. Ephs and ephrins operate as a "chemotactic" guidance system [51] that directs motile cells into genetically-predetermined positions. An increasing range of normal and oncogenic developmental programs, including gastrulation, somitogenesis, patterning of the skeletal, nervous, and vascular systems, tumor invasion, neoangiogenesis, and metastasis [11, 21, 52], rely on chemoattractive/chemorepulsive cell guidance by Ephs and ephrins. Accurate cell positioning relies on genetically predetermined multidimensional gradients and counter-gradients of Eph and ephrin interaction partners that are transformed into a dynamic range of fine-tuned cellular responses ranging from cell-cell repulsion to adhesion, and from increased motility to tight adhesion [6].

Eph/ephrin cell guidance is possibly explored most extensively in the development of the vertebrate optical system. A three-dimensional array of Eph and ephrin expression gradients and counter-gradients, involving most of the vertebrate family members, on retinal axons and collicular target domains controls the accurate positioning of axon arbors. On a cellular level, this is accomplished by promotion and inhibition of axon (back-) branching toward the correct and aberrant termination zones, respectively [20]. On a molecular level, axonal Ephs respond by eliciting repulsive or attractive responses according to the abundance, type, and combination of ephrins that they encounter on the interacting cell surface [20]. Importantly, as the incoming axons compete with all other axons for their collicular targets, the relative rather than the absolute abundance of Ephs and ephrins, together with their signaling activities, determines the accurate position of retinal axons [6].

Forward genetic approaches in zebrafish and *C. elegans* developmental models reveal that, in addition to controlling cell intermingling during gastrulation and somitogenesis [53], Ephs and ephrins facilitate critical communication between epithelial and mesenchymal cell populations, and orchestrate the mesenchymal to epithelial transition that underlies epithelialization at somite boundaries [54-56]. Furthermore, signaling by B-type Ephs and ephrins is essential for the development of the vertebrate cardiovascular system, and, in addition to critical cell guidance functions, expression of Ephs and ephrins on endothelial and mesenchymal cells facilitates the essential cell-cell contacts and communication between pericytes, smooth muscle cells, and the underlying endothelial cell layer [13, 57].

EPHS IN ONCOGENESIS: DE-REGULATED CELL POSITIONING DURING INVASION AND METASTASIS

Ephs and ephrins, distinctively less prevalent in adult than in embryonal tissues, are overexpressed in a broad range of cancers, including breast, colon, lung, and kidney carcinomas, melanomas, sarcomas, neuroblastomas, and ovarian and prostate cancers [11]. The complexity of Eph/ephrin function during vertebrate development, with seemingly conflicting roles to facilitate accurate cell positioning by eliciting cell-cell adhesive or repulsive mechanisms, is mirrored during cancer progression. Both of these contrary biological effects, as well as direct effects on proliferation, are implicated in the disease progression of a variety of cancer types, and it is likely that Ephs and ephrins regulate different steps of tumor formation, growth, neo-angiogenesis, invasion, and metastasis [2, 58]. Thus, while in some cases of breast and colon cancer elevated EphB/ephrin-B levels are correlated with tumor progression, in other cases EphB signaling is tumor suppressive *in vitro* [59], in mouse colon cancer models [60] and in human colorectal cancer [61].

In non-epithelial tumors there is a more consistent link between increased expression and tumor progression, perhaps best exemplified in malignant melanoma, where one or more of at least three Eph genes are increased during melanoma progression and metastasis. In this case, Eph-mediated cell-cell repulsive responses are likely responsible for tumor cell dislodgement and spreading from primary tumor sites, while the shift to Eph/ephrin-mediated adhesion may provide a docking mechanism for metastasizing cancer cells [41]. Moreover, Eph-induced neo-angiogenesis, discussed above, is also likely to promote metastasis via increased exposure of tumors to the vascular and lymphatic systems. EphA2 expression in melanoma is correlated with both poor clinical outcome and the ability for vascular mimicry [62], a unique ability of highly malignant melanoma cells to form vasculogenic networks. Thus, Eph-facilitated vascular mimicry may represent an additional mechanism which promotes tumor progression by reactivating EphA2 signaling pathways that promote, in addition to increased invasion and migration, vasculogenic remodeling in tumor cells [62].

REFERENCES

1. Hirai H, Maru Y, Hagiwara K, Nishida J, Takaku F. A novel putative tyrosine kinase receptor encoded by the eph gene. *Science* 1987;**238**:1717–20.
2. Lackmann M, Boyd AW. Eph, a protein family coming of age: more confusion, insight, or complexity? *Sci Signal* 2008;**1**:re2.
3. Oates AC, Lackmann M, Power MA, Brennan C, Down LM, Do C, Evans B, Holder N, Boyd AW. An early developmental role for eph-ephrin interaction during vertebrate gastrulation. *Mech Dev* 1999;**83**:77-94.
4. Barrios A, Poole RJ, Durbin L, Brennan C, Holder N, Wilson SW. Eph/Ephrin signaling regulates the mesenchymal-to-epithelial transition of the paraxial mesoderm during somite morphogenesis. *Curr Biol* 2003;**13**:1571-82.
5. Compagni A, Logan M, Klein R, Adams RH. Control of skeletal patterning by ephrinB1-EphB interactions. *Dev Cell* 2003;**5**:217-30.
6. Lemke G, Reber M. Retinotectal mapping: new insights from molecular genetics. *Annu Rev Cell Dev Biol* 2005;**21**:551-80.
7. Adams RH. Vascular patterning by Eph receptor tyrosine kinases and ephrins. *Semin Cell Dev Biol* 2002;**13**:55-60.
8. Gerlai R. Eph receptors and neural plasticity. *Nat Rev Neurosci* 2001;**2**:205-9.
9. Mundy GR, Elefteriou F. Boning up on ephrin signaling. *Cell* 2006;**126**:441-3.
10. Konstantinova I, Nikolova G, Ohara-Imaizumi M, Meda P, Kucera T, Zarbalis K, Wurst W, Nagamatsu S, Lammert E. EphA-Ephrin-A-mediated beta cell communication regulates insulin secretion from pancreatic islets. *Cell* 2007;**129**:359-70.
11. Wimmer-Kleikamp SH, Lackmann M. Eph-modulated cell morphology, adhesion and motility in carcinogenesis. *IUBMB Life* 2005;**57**:421-31.
12. Oike Y, Ito Y, Hamada K, Zhang XQ, Miyata K, Arai F, Inada T, Araki K, Nakagata N, Takeya M, Kisanuki YY, Yanagisawa M, Gale NW, Suda T. Regulation of vasculogenesis and angiogenesis by EphB/ephrin-B2 signaling between endothelial cells and surrounding mesenchymal cells. *Blood* 2002;**100**:1326-33.
13. Foo SS, Turner CJ, Adams S, Compagni A, Aubyn D, Kogata N, Lindblom P, Shani M, Zicha D, Adams RH. Ephrin-B2 controls cell motility and adhesion during blood-vessel-wall assembly. *Cell* 2006;**124**:161-73.
14. Himanen JP, Nikolov DB. Eph signaling: a structural view. *Trends Neurosci* 2003;**26**:46-51.
15. Himanen JP, Rajashankar KR, Lackmann M, Cowan CA, Henkemeyer M, Nikolov DB. Crystal structure of an Eph receptor–ephrin complex. *Nature* 2001;**414**:933-8.
16. Hattori M, Osterfield M, Flanagan JG. Regulated cleavage of a contact-mediated axon repellent. *Science* 2000;**289**:1360-5.
17. Marston DJ, Dickinson S, Nobes CD. Rac-dependent trans-endocytosis of ephrinBs regulates Eph-ephrin contact repulsion. *Nat Cell Biol* 2003;**5**:879-88.
18. Janes PW, Saha N, Barton WA, Kolev MV, Wimmer-Kleikamp SH, Nievergall E, Blobel CP, Himanen JP, Lackmann M, Nikolov DB. Adam meets Eph: an ADAM substrate recognition module acts as a molecular switch for ephrin cleavage in trans. *Cell* 2005;**123**:291-304.
19. Georgakopoulos A, Litterst C, Ghersi E, Baki L, Xu C, Serban G, Robakis NK. Metalloproteinase/Presenilin1 processing of ephrinB regulates EphB-induced Src phosphorylation and signaling. *EMBO J* 2006;**25**:1242-52.
20. McLaughlin T, O'Leary DD. Molecular gradients and development of retinotopic maps. *Annu Rev Neurosci* 2005;**28**:327-55.
21. Pasquale EB. Eph receptor signalling casts a wide net on cell behaviour. *Nat Rev Mol Cell Biol* 2005;**6**:462-75.
22. Cowan CA, Henkemeyer M. Ephrins in reverse, park and drive. *Trends Cell Biol* 2002;**12**:339-46.
23. Davy A, Soriano P. Ephrin signaling *in vivo*: look both ways. *Dev Dyn* 2005;**232**:1-10.
24. Holmberg J, Clarke DL, Frisen J. Regulation of repulsion versus adhesion by different splice forms of an Eph receptor. *Nature* 2000;**408**:203-6.
25. Holmberg J, Frisen J. Ephrins are not only unattractive. *Trends Neurosci* 2002;**25**:239-43.
26. Wiesner S, Wybenga-Groot LE, Warner N, Lin H, Pawson T, Forman-Kay JD, Sicheri F. A change in conformational dynamics underlies the activation of Eph receptor tyrosine kinases. *EMBO J* 2006;**25**:4686-96.
27. Wybenga-Groot LE, Baskin B, Ong SH, Tong J, Pawson T, Sicheri F. Structural basis for autoinhibition of the Ephb2 receptor tyrosine kinase by the unphosphorylated juxtamembrane region. *Cell* 2001;**106**:745-57.
28. Stein E, Lane AA, Cerretti DP, Schoecklmann HO, Schroff AD, Van Etten RL, Daniel TO. Eph receptors discriminate specific ligand oligomers to determine alternative signaling complexes, attachment, and assembly responses. *Genes Dev* 1998;**12**:667-78.
29. Shintani T, Ihara M, Sakuta H, Takahashi H, Watakabe I, Noda M. Eph receptors are negatively controlled by protein tyrosine phosphatase receptor type O. *Nat Neurosci* 2006;**9**:761-9.
30. Wimmer-Kleikamp SH, Nievergall E, Gegenbauer K, Adikari S, Mansour M, Yeadon T, Boyd AW, Patani NR, Lackmann M. Elevated protein tyrosine phosphatase activity provokes Eph/ephrin-facilitated adhesion of pre-B leukemia cells. *Blood* 2008. in press.
31. Palmer A, Zimmer M, Erdmann KS, Eulenburg V, Porthin A, Heumann R, Deutsch U, Klein R. EphrinB phosphorylation and reverse signaling: regulation by Src kinases and PTP-BL phosphatase. *Mol Cell* 2002;**9**:725-37.
32. Hansen MJ, Dallal GE, Flanagan JG. Retinal axon response to ephrin-as shows a graded, concentration-dependent transition from growth promotion to inhibition. *Neuron* 2004;**42**:717-30.
33. Davy A, Gale NW, Murray EW, Klinghoffer RA, Soriano P, Feuerstein C, Robbins SM. Compartmentalized signaling by GPI-anchored ephrin-A5 requires the Fyn tyrosine kinase to regulate cellular adhesion. *Genes Dev* 1999;**13**:3125-35.
34. Gauthier LR, Robbins SM. Ephrin signaling: One raft to rule them all? One raft to sort them? One raft to spread their call and in signaling bind them?. *Life Sci* 2003;**74**:207-16.
35. Marquardt T, Shirasaki R, Ghosh S, Andrews SE, Carter N, Hunter T, Pfaff SL. Coexpressed EphA receptors and ephrin-A ligands mediate opposing actions on growth cone navigation from distinct membrane domains. *Cell* 2005;**121**:127-39.
36. Zimmer M, Palmer A, Kohler J, Klein R. EphB-ephrinB bi-directional endocytosis terminates adhesion allowing contact mediated repulsion. *Nat Cell Biol* 2003;**5**:869-78.
37. Blobel CP. ADAMs: key components in EGFR signalling and development. *Nat Rev Mol Cell Biol* 2005;**6**:32-43.
38. Kullander K, Klein R. Mechanisms and functions of Eph and ephrin signalling. *Nat Rev Mol Cell Biol* 2002;**3**:475-86.
39. Noren NK, Pasquale EB. Eph receptor–ephrin bidirectional signals that target Ras and Rho proteins. *Cell Signal* 2004;**16**:655-66.
40. Matsuoka H, Obama H, Kelly ML, Matsui T, Nakamoto M. Biphasic functions of the kinase-defective Ephb6 receptor in cell adhesion and migration. *J Biol Chem* 2005;**280**(29):355-63.

41. Wimmer-Kleikamp SH, Janes PW, Squire A, Bastiaens PI, Lackmann M. Recruitment of Eph receptors into signaling clusters does not require ephrin contact. *J Cell Biol* 2004;**164**:661-6.
42. Miao H, Strebhardt K, Pasquale EB, Shen TL, Guan JL, Wang B. Inhibition of integrin-mediated cell adhesion but not directional cell migration requires catalytic activity of EphB3 receptor tyrosine kinase: Role of Rho family small GTPases. *J Biol Chem* 2005;**280**:923-32.
43. Sahin M, Greer PL, Lin MZ, Poucher H, Eberhart J, Schmidt S, Wright TM, Shamah SM, O'Connell S, Cowan CW, Hu L, Goldberg JL, Debant A, Corfas G, Krull CE, Greenberg ME. Eph-dependent tyrosine phosphorylation of ephexin1 modulates growth cone collapse. *Neuron* 2005;**46**:191-204.
44. Iwasato T, Katoh H, Nishimaru H, Ishikawa Y, Inoue H, Saito YM, Ando R, Iwama M, Takahashi R, Negishi M, Itohara S. Rac-GAP alpha-chimerin regulates motor-circuit formation as a key mediator of EphrinB3/EphA4 forward signaling. *Cell* 2007;**130**:742-53.
45. Murai KK, Pasquale EB. 'Eph'ective signaling: forward, reverse and crosstalk. *J Cell Sci* 2003;**116**:2823-32.
46. Cowan CA, Henkemeyer M. The SH2/SH3 adaptor Grb4 transduces B-ephrin reverse signals. *Nature* 2001;**413**:174-9.
47. Chang L, Karin M. Mammalian MAP kinase signalling cascades. *Nature* 2001;**410**:37-40.
48. Holmberg J, Armulik A, Senti KA, Edoff K, Spalding K, Momma S, Cassidy R, Flanagan JG, Frisen J. Ephrin-A2 reverse signaling negatively regulates neural progenitor proliferation and neurogenesis. *Genes Dev* 2005;**19**:462-71.
49. Miller MA, Ruest PJ, Kosinski M, Hanks SK, Greenstein D. An Eph receptor sperm-sensing control mechanism for oocyte meiotic maturation in Caenorhabditis elegans. *Genes Dev* 2003;**17**:187-200.
50. Batlle E, Henderson JT, Beghtel H, van den Born MM, Sancho E, Huls G, Meeldijk J, Robertson J, van de Wetering M, Pawson T, Clevers H. Beta-catenin and TCF mediate cell positioning in the intestinal epithelium by controlling the expression of EphB/ephrinB. *Cell* 2002;**111**:251-63.
51. Van Haastert PJ, Devreotes PN. Chemotaxis: signalling the way forward. *Nat Rev Mol Cell Biol* 2004;**5**:626-34.
52. Poliakov A, Cotrina M, Wilkinson DG. Diverse roles of Eph receptors and ephrins in the regulation of cell migration and tissue assembly. *Dev Cell* 2004;**7**:465-80.
53. Xu Q, Mellitzer G, Wilkinson DG. Roles of Eph receptors and ephrins in segmental patterning. *Philos Trans R Soc Lond B Biol Sci* 2000;**355**:993-1002.
54. George SE, Simokat K, Hardin J, Chisholm AD. The VAB-1 Eph receptor tyrosine kinase functions in neural and epithelial morphogenesis in *C. elegans*. *Cell* 1998;**92**:633-43.
55. Chong SW, Jiang YJ. Off limits – integrins holding boundaries in somitogenesis. *Trends Cell Biol* 2005;**15**:453-7.
56. Marston DJ, Goldstein B. Actin-based forces driving embryonic morphogenesis in *Caenorhabditis elegans*. *Curr Opin Genet Dev* 2006;**16**:392-8.
57. Adams RH. Molecular control of arterial-venous blood vessel identity. *J Anatomy* 2003;**202**:105-12.
58. Noren NK, Pasquale EB. Paradoxes of the EphB4 receptor in cancer. *Cancer Res* 2007;**67**:3994-7.
59. Noren NK, Foos G, Hauser CA, Pasquale EB. The EphB4 receptor suppresses breast cancer cell tumorigenicity through an Abl-Crk pathway. *Nat Cell Biol* 2006;**8**:815-25.
60. Batlle E, Bacani J, Begthel H, Jonkheer S, Gregorieff A, van de Born M, Malats N, Sancho E, Boon E, Pawson T, Gallinger S, Pals S, Clevers H. EphB receptor activity suppresses colorectal cancer progression. *Nature* 2005;**435**:1126-30.
61. Davalos V, Dopeso H, Castano J, Wilson AJ, Vilardell F, Romero-Gimenez J, Espin E, Armengol M, Capella G, Mariadason JM, Aaltonen LA, Schwartz S Jr, Arango D. EPHB4 and survival of colorectal cancer patients. *Cancer Res* 2006;**66**:8943-8.
62. Hess AR, Margaryan NV, Seftor EA, Hendrix MJ. Deciphering the signaling events that promote melanoma tumor cell vasculogenic mimicry and their link to embryonic vasculogenesis: Role of the Eph receptors. *Dev Dyn* 2007;**236**:3283-96.

Part 2

Cytokine Receptors

Chapter 22

Overview of Cytokine Receptors

Robert M. Stroud
Department of Biochemistry and Biophysics, University of California, San Francisco, California

Cytokine and cell adhesion receptors are generally single-crossing receptors that are activated by binding a ligand on one side of the membrane that initiates a response on the other side. The ligands associate with binding sites that may lie within one single receptor molecule or may involve bridging interactions between multiple receptors. The relative disposition of transmembrane segments and their associated domains induces a signal transduction cascade on the other side of the membrane. In all cases, this is mediated by changing lateral associations of receptors by horizontal signaling.

Horizontal receptor signaling is primarily found in multicellular organisms. Downstream signaling pathways inside the cell generally control changes in cell metabolism, such as those that lead to changes in transcription, translation, or replication, or changes that result in apoptosis of the cell. The horizontal receptors are activated by binding a protein ligand that can be monomeric or multimeric and induces a reordering of quaternary interactions between receptors in the cell membrane.

Most cytokines can be grouped into four groups (I to IV) based on the preponderance of α-helical β-sheet, mixed α/β, or mosaic substructures. Reordering of receptor associations as a result of binding is a common theme in horizontal signaling, with enormous diversity in the protein folds, binding sites, and stoichiometries of these signaling complexes. Much less accessible, but of increasing significance, is the evidence that cytokine receptors are maintained in an inactive state by their ordered associations prior to binding the cytokines. The first such inactive but pre-associated states have now been established; however, because inactive species are more difficult to detect than activated ones, they have only recently been sought. In some cases, these are poised to bind cytokine, as for members of the tumor necrosis factor (TNF) receptor class. In other cases, such as the erythropoietin (EPO) receptor, the receptors are associated in an "off" state that requires them to dissociate before a productive complex can be made.

In many cases, the intracellular domains of receptors are pre-associated with protein kinases, either non-covalently, as is the case with EPO receptors, or by covalent construction on the same gene, as is the case with the epidermal growth factor (EGF) class of receptors. In other cases, the intracellular domains associate with kinases after they bind cytokine, as is the case for human growth hormone (hGH) receptors. Once the receptors are appropriately oriented, the kinases act intermolecularly to phosphorylate each other, regions of the intracellular domains, or other proteins. These in turn act as docking sites for binding and activating other signaling factors. Thus, the membrane surface serves as the nexus for a plethora of pathways within the cell. In their role as the "mailbox" of the cell, they will be the key to intervention by therapeutic drugs as the ability to target protein-protein interfaces reaches maturity.

Cytokine–receptor complexes include those where two identical receptor molecules are dimerized by binding to two different sites on a single cytokine to produce a 2:1 complex, as seen for growth hormone and erythropoietin. Other cytokines form complexes where two cytokine molecules bind two identical receptors, such as in the gp130-interleukin-6 (IL-6), granulocyte colony-stimulating factor (GCSF)-GCSF receptor (GCSFR), and fibroblast growth factor (FGF)-FGF receptor (FGFR) complexes to induce allosteric changes that lead to back-to-back associations of the receptors without any contact between cytokines themselves. Still other cytokines act as monomers to bind two different receptors simultaneously such those found in the IL-4 system and gamma-interferon (γ-IFN). Still higher order trimeric complexes are seen for the TNF receptors class.

Overall, our study of the mechanisms of cytokine signaling via single crossing receptors has been driven by

Handbook of Cell Signaling, Three-Volume Set 2 ed.

defining the activated complex structures and their function in recruiting molecular complexes, but this research has also led to determining what constitutes the "off" state of receptors. In the case of erythropoietin receptors, only 50 receptor dimers, oriented by binding erythropoietin on the cell surface, are required to evoke ≈50% signaling. Therefore, our understanding of the reference "off" states is key to understanding the horizontal signaling receptors. The chapters of this section detail some of the most pertinent examples of horizontal signaling mechanisms, which serve to define the pathway to the future.

Chapter 23

Cytokine Receptor Signaling

Mojib Javadi Javed, Terri D. Richmond, and Dwayne L. Barber
Division of Stem Cells and Developmental Biology, Ontario Cancer Institute, Departments of Medical Biophysics and Laboratory Medicine & Pathobiology, Faculty of Medicine, University of Toronto, Toronto, Ontario, Canada

INTRODUCTION

Cytokines regulate many steps of proliferation, cell survival, and differentiation within the bone marrow and extrahematopoietic tissues. These growth factors generally fall into a superfamily of helical proteins that contain four α-helices that possess a unique up–up–down–down architecture. Cytokines exert their function by binding to cognate receptors that are expressed in the appropriate target cell. The advent of expression cloning results in the isolation of cytokines or their receptors, due to the availability of a powerful assay based on proliferation.

The focus of this chapter is to summarize signaling pathways activated by cytokine receptors. In addition, developmental regulation of the cytokine signaling pathway and the involvement of this pathway in human disease will be addressed. For reasons of simplicity, the Erythropoietin Receptor (EPO-R) will be utilized as a model. Additional details can be gleaned from Richmond *et al.*, 2005 [1]. Comparisons to other receptors will be utilized where appropriate.

GENERATION OF HIGH-AFFINITY CYTOKINE–RECEPTOR COMPLEXES

Cytokines bind to their cognate receptor with high affinity, usually in the picomolar range. Receptors are activated when a high-affinity ligand–receptor complex is generated. Receptor oligomerization generally falls into three classes. Some cytokines, including EPO, prolactin, and growth hormone, bind to a receptor homodimer. In these cases, both the ligand binding and signaling requirements can be found on the same receptor. In other cases, such as the interleukin-3 family of cytokines (consisting of IL-3, IL-5, and granulocyte-macrophage colony stimulating factor (GM-CSF)), a heterodimeric receptor complex is formed. This consists of a unique α chain for each receptor and a shared "common" β chain that couples to downstream signal transduction. The situation is more complex for IL-2, which has a trimeric complex consisting of a unique IL-2 R α chain, an IL-2 R β chain (shared with IL-15), and an IL-2 R γ chain. The IL-2 R γ chain is the "common" chain (IL-2 Rγc) in the IL-2 family of cytokines, shared with receptors for IL-2, IL-7, IL-9, IL-15, and IL-21. The composition of high-affinity receptors and specific JAK and STAT activation downstream of cytokine stimulation are presented in Table 23.1.

Somewhat surprisingly, there are few conserved features amongst the cytokine receptor superfamily. Four cysteine residues are conserved within the extracellular domain, which form two intrachain disulfide bonds. A "WSXWS" motif is located close to the transmembrane domain, and is thought to be important in receptor trafficking and/or surface localization. All cytokine receptors that deliver intracellular signals contain Box-1 and Box-2 motifs. These proline-rich sequences are responsible for association of the Janus (or JAK) tyrosine kinases. Selected JAK kinases bind to certain cytokine receptors due to the precise sequences of the Box-1 and Box-2 domains. Cytokine receptors express several tyrosine residues, which can become tyrosine phosphorylated by JAK kinases and/or other associated tyrosine kinase enzymes downstream of receptor activation.

ARCHITECTURE OF EXTRACELLULAR DOMAIN

Ultrastructural analysis has provided tremendous insight regarding how the EPO-R is activated. Because of the significant commercial interest in EPO as a biotechnology product, three-dimensional structures have been described for the unliganded EPO-R [2], an EPO-R antagonist [3], a mimetic peptide [4], and the EPO–EPO-R complex [5].

Handbook of Cell Signaling, Three-Volume Set 2 ed.

TABLE 23.1 Cytokine receptor architecture and downstream JAK/STAT pathways

Cytokine receptor	Architecture (number of chains)	Associated JAKs	Activated STATs
Type I receptors			
Erythropoietin (EPO)	Single-chain (1)	JAK2	STAT1, STAT3, STAT5
Thrombopoietin (TPO)	Single-chain (1)	JAK2	STAT3, STAT5
Growth hormone (GH)	Single-chain (1)	JAK2	STAT1, STAT3, STAT5
Prolactin (PRL)	Single-chain (1)	JAK2	STAT5
Interleukin-3	Common β chain (2)	JAK1, JAK2	STAT5
Granulocyte-macrophage CSF	Common β chain (2)	JAK1, JAK2	STAT5
Interleukin-5	Common β chain (2)	JAK1, JAK2	STAT5
Interleukin-2	Common γ or γ-like (3)	JAK1, JAK3	STAT1, STAT3, STAT5
Interleukin-4	Common γ or γ-like (2)	JAK1, JAK3	STAT3, STAT5, STAT6
Interleukin-7	Common γ or γ-like (2)	JAK1, JAK3	STAT3, STAT5
Interleukin-9	Common γ or γ-like (2)	JAK1, JAK3	STAT1, STAT3, STAT5
Interleukin-13	Common γ or γ-like (2)	JAK1, Tyk2	STAT6
Interleukin-15	Common γ or γ-like (2)	JAK1, JAK2, JAK3	STAT1, STAT3, STAT5
Interleukin-21	Common γ or γ-like (2)	JAK1, JAK3	STAT1, STAT3, STAT5
Thymic stromal lymphopoietin	Common γ or γ-like (2)	JAK1, JAK2	STAT5
Granulocyte-CSF	IL-6 family (1)	JAK1, JAK2, Tyk2	STAT1, STAT3, STAT5
Interleukin-6	IL-6 family (2)	JAK1	STAT1, STAT3
Leukemia inhibitor factor (LIF)	IL-6 family (1)	JAK1	STAT1, STAT3
Ciliary neutrotrophic factor	IL-6 family (3)	JAK1	STAT1, STAT3
Cardiotrophin I	IL-6 family (2)	JAK1	STAT1, STAT3
Oncostatin M	IL-6 family (2)	JAK1	STAT1, STAT3
Interleukin-11	IL-6 family (2)	JAK1	STAT1, STAT3
Interleukin-12	IL-6 family (1)	JAK2, Tyk2	STAT4
Type II receptor			
Interferon-α,β	(2)	JAK1, Tyk2	STAT1, STAT2, STAT3
Interferon-γ	(2)	JAK1, JAK2	STAT1
Interleukin-10	(2)	JAK1, JAK2	STAT1, STAT3, STAT5

The extracellular domain of the EPO-R consists of two fibronectin type III repeats, referred to as domains 1 (D1) and 2 (D2). Analysis of the unliganded extracellular domain of the EPO-R revealed that this region of the receptor exists as a preformed dimer [2]. The D1 domains are oriented 160° apart and the D2 domains are 135° apart. The distance between the ends of the D2 domain is 73Å. Assuming that the transmembrane domain forms an α-helix, JAK2 would be associated with the Box-1 and Box-2 regions of the EPO-R.

Comparison with the structure of the EPO-R extracellular domain with bound EPO revealed that the D1 domains are now 120° apart, whereas the D2 domains are parallel to each other. This reduces the spacing between the D2 domains to 30Å. It is believed that this conformational change is adequate to bring the associate JAK2 proteins in close proximity to facilitate activation.

In support of these hypotheses, a crystal structure of an EPO mimetic, EPO mimetic peptide-1 (EMP1), bound to the extracellular domain of the EPO-R has also been solved.

The D1 domains are 180° apart, whereas the D2 domains are now 45° apart. This increases the spacing between the two domains to 39 Å. Biochemical data supports this model, as EMP1 is partially active in the mM range. Additional details regarding ultrastructural approaches in understanding EPO-R function can be gleaned from Wilson and Jolliffe, 1999 [6] and Jian and Hunter, 1999 [7].

RECEPTOR SIGNALING-UTILIZING EPO-R AS A MODEL

Upon generation of a high-affinity receptor complex, JAK kinase activation leads to tyrosine phosphorylation of the cytoplasmic tail of the receptor. The tyrosine phosphorylated receptor presents discrete binding motifs for SH2 domain binding proteins. Cytokine receptors activate three major intracellular signaling pathways: JAK-STAT, phosphatidylinositol 3′ kinase, and MAP kinase enzymes. All three signaling pathways will be considered within this chapter.

Mouse modeling experiments have been completed for the proximal steps in the EPO signaling pathway. EPO [8] and EPO-R [8–10] knockout mice have been generated. Both mouse models suffer from embryonic lethality due to a fatal anemia that occurs at E12.5–E13.5, a time corresponding to massive expansion of the erythroid compartment in the fetal liver.

Early experiments completed in transfected hematopoietic cell lines suggested that a negative regulatory region was contained within the distal EPO-R cytoplasmic tail [11]. Ihle and colleagues generated two truncated EPO-R knock-in alleles to test this in greater detail [12]. EPO-R H mice express only the proximal tyrosine residue, Y343, whereas EPO-R HM mice have a tyrosine to phenylalanine mutation at Y343, and lack all cytoplasmic tyrosines. Somewhat surprisingly, EPO-R HM mice are viable, demonstrating that cytoplasmic tyrosines on the EPO-R are dispensable for viability.

Further characterization by Wojchowski and colleagues revealed that the EPO-R HM mice are deficient in stress erythropoiesis [13,14]. EPO-R HM mice show delayed recovery to phenylhydrazine and 5-fluorouracil [14]. EPO-dependent activation of Oncostatin M and Pim-1 was abrogated in EPO-R HM mice [14]. While EPO-R cytoplasmic tyrosines are not required for viability, EPO-R Y343 and potentially distal regions of the cytoplasmic tail are necessary for optimal responses to erythroid stress.

ACTIVATION OF THE JAK TYROSINE KINASES

Cytokine binding results in activation of receptor-associated JAK kinases (Figure 23.1). In turn, JAK kinases (and potentially other tyrosine kinases activated downstream of JAK kinases) phosphorylate the cytoplasmic tail of cytokine

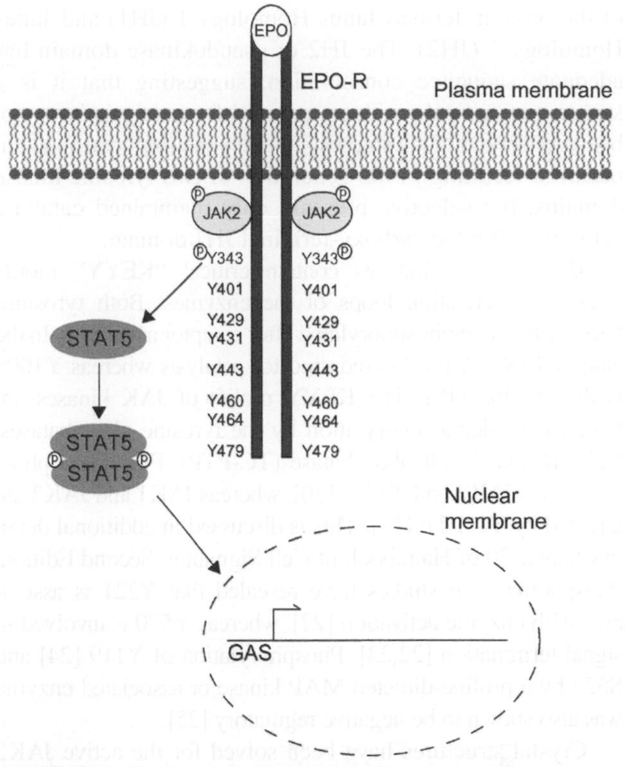

FIGURE 23.1 Cytokine signaling utilizing the erythropoietin receptor as a model.
EPO forms a high affinity receptor complex by binding to two receptor chains. This brings the associated JAK2 tyrosine kinases in close proximity, resulting in their activation. Tyrosine phosphorylation of the EPO-R cytoplasmic tail by JAK2 (or other tyrosine kinases activated downstream of JAK2) results in recruitment of SH2 domain containing proteins, including the STAT transcription factors. Once phosphorylated, STAT5 dissociates from the EPO-R, dimerizes, and shuttles to the nucleus, where it regulates transcription from gene containing GAS elements. These include Cis, Socs1, Socs3, Osm, and Pim1. Additional details can be found in the text.

receptors. There is specificity in JAK kinase-cytokine receptor interaction, believed to be imparted by selective Box-1 and Box-2 sequences expressed on the cytokine receptor. The pattern of cytokine-mediated JAK kinase activation is illustrated in Table 23.1. Additional details regarding JAK tyrosine kinases can be gleaned from Ihle and Gilliland DG, 2007 [15].

JAK kinases have four domains. A FERM domain is located at the amino terminus, and this region is responsible for receptor association. In addition, it has been illustrated that the FERM domain allows for proper receptor trafficking in the case of the EPO-R [16] and interferon-α receptor (IFNAR1) [17], although this assertion has also been recently been debated [18]. Domain alignment programs suggest that a SH2 domain is located carboxy-terminal of the FERM domain. Whether this is a *bona fide* SH2 domain has yet to be definitively proven, and phosphopeptides that are capable of binding to this region have not been identified. JAK kinases are unique because of the two kinase domains found at the carboxy-terminus of the protein, termed Janus Homology 1 (JH1) and Janus Homology 2 (JH2). The JH2 or pseudokinase domain has adequate sequence conservation, suggesting that it is a kinase, yet critical residues required for catalysis have not been conserved. This suggests that a gene duplication event occurred resulting in the generation of two tyrosine kinase domains, but selective pressure only maintained catalytic activity within the carboxy-terminal JH1 domain.

JAK tyrosine kinases contain critical "KEYY" motifs within the activation loops of the enzymes. Both tyrosines become tyrosine phosphorylated after receptor activation. In the case of JAK2, Y1007 is required for catalysis whereas Y1008 is dispensable [19]. The KEYY motifs of JAK kinases are targeted for dephosphorylation by the tyrosine phosphatases, PTP-1B and T cell phosphatase (TC-PTP). PTP-1B dephosphorylates JAK2 and TYK2 [20], whereas JAK1 and JAK3 are targeted by TC-PTP [21]. This is discussed in additional detail in Chapter 70 of Handbook of Cell Signaling, Second Edition. Phosphomapping studies have revealed that Y221 is associated with enzyme activation [22], whereas Y570 is involved in signal termination [22,23]. Phosphorylation of Y119 [24] and S523 by a proline-directed MAP kinase or associated enzyme was also shown to be negative regulatory [25].

Crystal structures have been solved for the active JAK2 kinase domain bound to JAK inhibitor-1 [26] and JAK3 JH1 domain bound to AFN941, a staurosporine derivative [27]. The JAK2 JH1 domain has features of typical tyrosine kinase domain, with a small N-terminal lobe and a large C-terminal lobe. JAK tyrosine kinases have an insertion loop (located between amino acids 1056 and 1078 in JAK2). Residues involved in coordination of the activation loop and pTyr 1007 and pTyr 1008 were identified. These data confirmed that the KEYY motif is solvent exposed and available to bind negative regulators such as SOCS-1 and PTP-1B. Superposition of the JAK2 and JAK3 structures revealed close similarity between the N-terminal lobes of both proteins. However, many differences were noted in the C domain, including the hinge region, the glycine loop, the activation loop, and the JAK insertion loop. This was due to a 5° rotation between the two structures, which may be accounted for by the differences in the binding of each inhibitor. While these structures are informative in eludicating JAK kinase domain regulation, the availability of structural information regarding a JAK JH2 and JH1 domain is critical to our understanding of JAK function. Activating mutations have been described in the JH2 domain in human diseases and in the laboratory (discussed below). How these mutations affect the catalytic activity of the JH1 domain is one of the most interesting questions currently unanswered in the field.

JAK1

Gene targeting experiments have been completed for all JAK kinases (summarized in Table 23.2). JAK1 is activated by many cytokines, including Type 1 and 2 interferons as well as the IL-2, IL-3, and IL-6 families of cytokines. JAK1-deficent mice are perinatal lethal and suffer from a plethora of defects, including defective nitric oxide production in response to IFNα or IFNγ, decreased IL-7 signaling, and lowered Leukemia Inhibitory Factor (LIF)-dependent neuronal survival [28].

TABLE 23.2 Phenotype of JAK-deficient mice

Gene	Phenotype
Tyk2	Viable
	Defective Th1 differentiation
	Defects in IFNα/β, IFNγ and IL-12
JAK1	Perinatal lethal
	Defects in IFNα/β, IFNγg
	Defects in IL-2 and IL-6
JAK2	Embryonic lethal
	Dies at E12.5 due to fetal anemia
	Phenotype similar to EPO and EPO-R knockouts
JAK3	Viable
	Autosomal SCID
	Phenotype similar to gc deficiency

JAK2

Generation of JAK2-deficient mice illustrated that this enzyme is a critical proximal component of the EPO signaling pathway [29,30]. As discussed above, EPO and EPO-R mice die at E13.5. JAK2-deficient mice die at E12.5, with a slightly more adverse phenotype, likely due to the fact that additional cytokines, including thrombopoietin (TPO), also activate JAK2. Mice deficient in JAK2 have a profound block in the production of Burst Forming Unit-Erythroid (BFU-E) and Colony Forming Unit-Erythroid (CFU-E) colonies that mimics the erythropoietic defect seen in EPO- and EPO-R-deficient mice. In addition, $JAK2^{-/-}$ mice have defective IFNγ dependent induction of target genes and antiviral responses. These genetic data support a linear pathway, since a single ligand (EPO) binds to a single receptor (EPO-R) that utilizes JAK2 as a critical proximal signaling enzyme.

JAK3

JAK3 is the only Janus kinase with a restricted tissue expression, as it is primarily found in hematopoietic tissues. JAK3 specifically associates with the γc chain of the IL-2 family of cytokines. Three groups have generated mice lacking JAK3, which demonstrates its importance in B cell maturation and T cell function [31–33]. These mice phenocopy the IL-2 R γc-deficient mice [34–36], suggesting that mutations in JAK3 that disrupt JAK3 association with the IL-2 R γc or that abrogate JAK3 kinase activity could give rise to a SCID phenotype (see below).

TYK2

Tyk2 is activated by several cytokines, including IFNα, IL-10, and IL-12. Tyk2-deficient mice show a profound reduction in IL-12 dependent signaling. IL-12 dependent STAT4 phosphorylation and IFNγ secretion was reduced in Tyk2-deficient splenocytes. IL-12 mediates T helper type 1 (TH1) differentiation through activation of STAT4. TYK2-deficient T cells display defective TH1 differentiation [37].

Previous studies illustrated that IFNα activates JAK1 and TYK2. In contrast to JAK1-deficient mice, TYK2$^{-/-}$ mice have normal IFNα signaling, suggesting that IFNα responses are mediated through JAK1 only. Similarly, cell-line studies suggest that IL-6 activates JAK1, JAK2, and TYK2. IL-6 dependent mitogenesis and STAT3 tyrosine phosphorylation is normal in TYK2-deficient thymocytes. Several IL-10 dependent functions, including STAT3 tyrosine phosphorylation, T cell proliferation, induction of MHC Class II antigen, and inhibition of LPS and IFNγ-induced TNF production, are conserved in cells isolated from TYK2-null mice.

RECRUITMENT AND ACTIVATION OF STAT TRANSCRIPTION FACTORS

Once receptor-JAK kinase complexes are activated, JAK kinases tyrosine phosphorylate the cytoplasmic tail of cytokine receptors. This serves to recruit SH2 domain containing proteins, including the Signal Transducers and Activator of Transcription (STAT) family. Tyrosine phosphorylation of STAT proteins results in release from the cytokine receptor, and two STAT proteins associate in a reciprocal SH2-mediated dimer. They are rapidly shuttled to the nucleus, where they regulate transcription of target genes that contain an interferon-γ Activated Sequence (GAS) (Figure 23.1).

There are seven mammalian STAT proteins. Conserved domains in the STAT proteins include an amino terminal oligomerization domain, a DNA binding domain, an SH2 domain, and carboxy-terminal sites of tyrosine and serine phosphorylation. The pattern of STAT activation downstream of selected cytokines is shown in Table 23.1. The specificity of cytokine-dependent STAT activation is mediated by tyrosine phosphorylation of the specific SH2 binding motif on the cytoplasmic tail of the cytokine receptor. For example, STAT5 is commonly activated by IL-2, IL-3, and EPO. Receptors for each cytokine express one or more STAT5 binding motifs (pTyr-X-X-Leu/Ile/Val). Additional details regarding STAT activation can be gleaned from the Chapter 70 of Handbook of Cell Signaling, Second Edition, as well as the excellent review by Levy and Darnell, 2002 [38].

Gene targeting experiments have been completed for all seven STAT genes, and conditional targeting phenotypes have been reported for several STAT3 and STAT5a/b knockout mice. The phenotypes of STAT knockout mice as well as the hematopoietic-targeted STAT3 and STAT5 mice will be discussed below (summarized in Table 23.3).

STAT1

STAT1-deficient mice are born in normal Mendelian ratios and are healthy unless they encounter a viral pathogen, as these animals are unable to mount a response to IFNα or IFNγ [39,40]. Exposure of STAT1-null mice to vesicular stomatitis virus (VSV) results in a defective response to IFNα, culminating in death of the mice [39]. STAT1-deficient mice are also unable to survive an immune challenge to *Listeria monocytogenes*, an intracellular pathogen that relies on effective IFNγ signaling [39]. Normal numbers of granulocytes, B lymphocytes, and monocytes are found in the fetal livers of STAT1$^{-/-}$ mice; however, the induction of interferon-stimulated genes in response to Type 1 or Type 2 IFN is impaired [39].

STAT1 is also required downstream of fibroblast growth factor (FGF) for chondrocyte proliferation and bone development [41]. However, no FGF-dependent increases in chondrocyte growth are observed in chondrocytes isolated from STAT1-deficient embryos. No increases in DNA synthesis are observed in metatarsal cartilage cultures derived from STAT1-null embryos. Furthermore, it is unclear whether there are unique transcriptional targets of STAT1 required for proper bone formation.

In addition, STAT1 was shown to be required for maintenance of stress erythropoiesis [42]. STAT1-deficient mice have increased splenic BFU-E and CFU-E, and STAT1-deficient CFU-E are hypersensitive to EPO. STAT1-deficient bone-marrow derived erythroblasts are more prone to apoptosis, and there is a global elevation of tyrosine phosphorylated proteins downstream of EPO, including JAK2, STAT5, Akt, and Erk.

STAT2

IFNα uniquely activates the ISGF3 complex, resulting in tyrosine phosphorylation of STAT1 and STAT2. Therefore, it was expected that STAT2$^{-/-}$ mice would be defective in mediating signaling responses to IFNα, but that IFNγ signaling would be normal. STAT2-null mice are healthy, except that they display a profound inability

TABLE 23.3 Phenotypes of STAT knockout mice

Knocked out gene	Phenotype
STAT1$^{-/-}$	Viable
	No IFN response
	Decreased bone marrow derieved erythroid progenitors
	Compensatory increase in splenic erythroid progenitors
STAT2$^{-/-}$	Viable
	Defective IFN response
STAT3$^{-/-}$	Embryonic lethal
	Extra-embryonic endoderm defects
Lck-Cre/STAT3$^{fl/-}$	Defective IL-6 and IL-2 induced T cell proliferation
	Enhanced T cell apoptosis
MMTV-Cre/STAT3$^{fl/fl}$	Enhanced Treg development
	Defective Th17 cell development
CD19-Cre/STAT3$^{fl/fl}$	Defect in T-dependent IgG plasma cell development and antibody production
LysM-Cre/STAT3$^{fl/fl}$	Defective IL-10 induced suppression of inflammatory cytokine production
	Th1 polarized immune response
Mx-Cre/STAT3$^{fl/fl}$	Enhanced granulopoiesis
	Defective negative regulation of G-CSF stimulation
STAT4$^{-/-}$	Viable
	No Th1 differentiation
STAT5a$^{-/-}$	Defective leptin mediated mammary cell differentiation
STAT5b$^{-/-}$	Impaired growth
STAT5a/b^{deltaN}	Viable
	Minor hematopoietic defects
	Hypomorph: a partially functional N-terminal truncated STAT5 is produced
STAT5a/b$^{-/-}$	Perinatal lethality
	E18.5 fetuses and neonates severely anemic
CD4-Cre/STAT5$^{fl/fl}$	Impaired Treg development
	Decreased number of T cells
	Enhanced expression of memory markers
CD19-Cre/STAT5$^{fl/fl}$	Defective early B cell development
	Impaired IL-7 induced B cell expansion
Tie2-Cre/STAT5$^{fl/fl}$	Anemic
	Defective erythroid development
	Impaired expression of transferrin receptor 1 (Tfr1)
STAT6$^{-/-}$	Viable
	No Th2 differentiation

to mount a productive immune response to viral exposure [43]. Surprisingly, no response to VSV viral challenge was observed in response to IFNγ. In addition, IFNγ dependent induction of ISRE-dependent genes was suppressed in STAT2$^{-/-}$ fibroblasts. These observations could be accounted for by a loss of STAT1 expression observed in the STAT2-null mice. Interestingly, tissue-specific differences in IFNα dependent gene expression was observed. IFNα dependent MHC Class I expression was STAT2 dependent in fibroblasts, whereas in macrophages MHC Class I induction was independent of STAT2. This indicates that unique responses are observed in regulating viral immunity that appear to be tissue-dependent.

STAT3

Generation of STAT3-null mice revealed that STAT3 plays a critical role in gastrulation, as these embryos show visible defects at E6.5 and are completely resorbed by E7.5 [44]. STAT3 appears to be dispensable during the pre-implantation period, as trophoblast giant cell formation is comparable between wild-type and STAT3-deficient blastocysts. In addition, egg cylinder formation is normal in STAT3$^{-/-}$ embryos, suggesting that early post-implantation is normal in this model. It is hypothesized that STAT3-deficient embryos are impaired in Leukemia Inhibitory Factor signaling, resulting in defective visceral endoderm formation. The visceral endoderm covers the egg cylinder, and is responsible for the regulation of metabolic exchange with the maternal circulation.

Conditional knockouts of STAT3 were generated to probe the requirement of STAT3 in defined tissues. Due to space limitations, this discussion will be restricted to characterization of floxed alleles within the hematopoietic compartment.

Lck-Cre/STAT3$^{flox/-}$ mice were generated in order to test involvement of STAT3 in T cell function. STAT3 is not involved in the development of T cells [45]; however, it is required for IL-2- and IL-6-induced T cell proliferation [45,46]. IL-2-mediated activation of STAT3 leads to the expression of IL-2 receptor α chain [46], and IL-6 stimulates T cell proliferation in a STAT3-dependent manner via prevention of apoptosis [31].

Regulatory T cells play a critical role in the regulation of immune response. T regulatory (Treg) cells and the developmentally related T helper 17 (Th17) cells are defined by the expression of the transcription factor Foxp3 and the production of IL-17, respectively. Both STAT3 and STAT5 play an important role in maintaining the developmental balance between Treg and Th17 cells. Studies using the MMTV-Cre/STAT3$^{flox/flox}$ mouse model have shown that STAT3 mediates the IL-6-induced repression of Foxp3, inhibiting the development of Treg cells [47]. Furthermore IL-6 induced activation of STAT3 induces the production of IL-17 by CD4^{+} T cells, thereby inducing the development of Th17 cells [48].

Interestingly, involvement of STAT3 in T cell biology extends to its expression in the thymic epithelium. The thymic epithelia of Keratin 5 (K5)-Cre/STAT3$^{flox/-}$ mice are devoid of STAT3, resulting in thymic hypoplasia in adult mice and enhanced susceptibility of the thymus to apoptosis inducing agents [49].

The highly selective role of STAT3 in B cell biology was determined via phenotypic analysis of CD19-Cre/STAT3$^{flox/flox}$ mice. STAT3-deficient B cells are unperturbed in their development and steady state maintenance, and are only deficient in T-dependent IgG plasma cell development and antibody production [50].

LysM-Cre/STAT3$^{flox/-}$ mice were generated to examine the role of STAT3 in macrophages and neutrophils. LysM-Cre/STAT3$^{flox/-}$ mice treated with LPS have elevated serum levels of TNF-α, IL-10, IL-6, IFNγ, and IL-1β in comparison to their wild-type counterparts [51]. STAT3-deficient macrophages produce significantly more IL-12 in response to LPS than do wild-type macrophages, which leads to a Th1 polarized immune response in LysM-Cre/STAT3$^{flox/-}$ mice [37]. The LysM-Cre/STAT3$^{flox/-}$ mouse model also elucidated the requirement of STAT3 for the IL-10 induced suppression of inflammatory cytokine production by macrophages and neutrophils [37].

Mx1-Cre/STAT3$^{flox/flox}$ mice were used to define the role of STAT3 in granulopoiesis. STAT3 deficiency in this model system resulted in an enhanced granulopoiesis [52]. STAT3 was found to be required for the induction of SOCS3 expression, and therefore a negative feedback loop, in response to G-CSF [38].

STAT4

Knockout mice for STAT4 revealed a novel role for this STAT protein in T cell differentiation [53]. STAT4$^{-/-}$ mice are viable, and have a normal distribution of B and T cell markers. T and natural killer (NK) cells isolated from STAT4-deficient mice failed to proliferate in response to IL-12, and STAT4-deficient T cells failed to secrete IFNγ. STAT4-null NK cells also were defective in mediating IL-12 dependent cytotoxicity. Most importantly, STAT4-deficient T cells displayed defective T_{H1} differentiation. IL-12 is a critical regulator of T_{H1} differentiation. When pretreated with IL-12 and activated by anti-CD3, STAT4-deficient T cells produce significantly less IFNγ than wild-type lymphocytes. In contrast, STAT4-deficient T cells cultured in IL-4 and IFNγ prior to T cell receptor activation produce elevated levels of the T_{H2} cytokines IL-4, IL-5, and IL-10. Therefore, STAT4 is required for T_{H1} differentiation, and its absence results in a default to T_{H2} differentiation.

STAT5

There are two highly related STAT5 genes, STAT5A and STAT5B, which are both found on mouse Chromosome 11.

Initial targeting experiments generated knockout mice that targeted Exon 1 and Exons 1 and 2. Because of space limitations, we will only discuss the double STAT5A/B knockout mice [54]. The reader is encouraged to examine the papers by Liu *et al.*, 1997 [55] and Udy *et al.*, 1997 [56] for discussion of STAT5A and STAT5B $^{-/-}$ mice, respectively.

STAT5A/B is commonly regulated downstream of multiple cytokines (Table 23.1). Consequently, this mouse model has multiple abnormalities [54]. The STAT5A/B$^{-/-}$ are smaller in stature. This phenotype is not entirely penetrant in all animals, and is growth hormone-dependent. Growth hormone regulates the Cytochrome P450 class of isozymes in a STAT5-dependent manner. Some of these enzymes are regulated in a gender-specific fashion. The GH-dependent induction of some of these enzymes is misregulated in STAT5A/B$^{-/-}$ mice. The transcription of several prolactin-dependent milk synthesis genes, including β-casein, whey acidic protein, and α-lactalbumin, are STAT5 dependent. STAT5A/B deficient females are unable to nurse their pups; however, the progeny can be fostered on surrogate mice. In addition, there is a peripheral T cell defect in STAT5A/B mice that is due to defective IL-2 signaling. Subsequent studies revealed that STAT5A/B deficient mice have a fetal anemia [57]. It now is realized that both STAT5A and STAT5B mice were hypomorphic, as truncated STAT5A and STAT5B proteins are generated. Hennighausen and colleagues have generated floxed alleles of both STAT5A and STAT5B. Again because of space limitations, we will only discuss those conditional knockouts within the hematopoietic lineage.

Studies utilizing CD4-Cre/STAT5$^{flox/flox}$ mice have demonstrated that IL-2 mediated activation of STAT5 leads to the recruitment of STAT5 to GAS sites in the Foxp3 promoter region, and induction of Foxp3 expression [47]. Furthermore IL-2 dependent inhibition of IL-17 production by CD4+ T cells is dependent on STAT5 activation [9]. Therefore, STAT5 plays an important role in IL-2 mediated effects on the development of Treg and Th17 cells. This may explain why CD4-Cre/STAT5$^{flox/flox}$ mice not only have decreased numbers of T cells, but also have enhanced expression of memory T cell markers [58].

STAT5-deficient B cells from CD19-Cre/STAT5$^{flox/-}$ mice are defective in early B cell development, but maturation and response is unaffected [59]. This defect in early B cell development in the absence of STAT5 is due to the requirement of STAT5 for IL-7 mediated B cell expansion [59].

Tie2-Cre/STAT5$^{flox/flox}$ mice, whose hematopoietic progenitors are devoid of STAT5, were used to determine the role of STAT5 in erythropoiesis. Two studies have shown that the fetal erythroid anemia observed in these mice was due to an inability to regulate transferrin receptor 1 [60,61] and iron regulatory protein-2 [61]. Therefore, the absence of STAT5 causes anemia in Tie2-Cre/STAT5$^{flox/flox}$ mice.

STAT6

Since STAT6 is required downstream of IL-4, a cytokine required for TH2 differentiation, it was predicted that STAT6 mice would have defective TH2 differentiation. STAT6-deficient mice are viable and lymphocytes isolated from these mice are defective in IL-4 dependent responses, including proliferation and distal gene regulation [62–64]. IL-4 dependent B cell immunoglobulin class switching was also altered. STAT6 is required for Cε gene transcription and IgE class switching, but negatively regulates IgG class switching [63]. T_{H2} development was compromised in STAT6$^{-/-}$ spleen cells [63]. Interestingly, the block in T_{H2} differentiation was more pronounced in the STAT6$^{-/-}$ mice when compared to IL-4 deficient mice. This is due to the fact that STAT6 integrates signals from both IL-4 and IL-13.

PARTICIPATION OF THE PHOSHATIDYLINOSITOL 3′ KINASE PATHWAY IN CELL SURVIVAL SIGNALING

Cytokines promote cell survival through the phosphatidylinositol 3′ kinase (PI3 kinase) pathway. In the case of the EPO-R, PI3 kinase can be recruited directly to the receptor through Y479, or indirectly through several adaptor proteins, including Cbl, Irs-2, Gab1, and/or Gab2. PI3 kinase couples to activation of Akt. One of the critical targets important in erythropoiesis downstream of Akt is the Foxo3a transcription factor [65–67]. Specifically, Foxo3a is a critical determinant in the regulation of oxidative stress, in part through transcriptional regulation of p21$^{CIP1/WAF1}$ [67].

Gene profiling studies have been completed that examine Foxo3a target genes in the context of EPO and stem cell factor stimulations. Microarray experiments were completed utilizing an inducible constitutively active Foxo3a, and compared with stimulations performed with EPO, SCF, and dexamethasone. Bakker and colleagues described five distinct clusters of gene regulation [66]. Interestingly, Foxo3a and STAT5 cooperated in EPO-dependent regulation of Cited2, whereas repression of Btg1 was mediated by Foxo3a and Creb. This study demonstrates that cytokine-dependent gene regulation requires the integrated control of overlapping transcriptional complexes.

ERK, JNK AND P38 ARE ALL ACTIVATED DOWNSTREAM OF CYTOKINE RECEPTOR ENGAGEMENT

Most cytokines activate all members of the MAP kinase family, including Erk and Jun kinase (Jnk)/SAP kinase (SAPK). Erk is activated downstream of Ras in the canonical Ras/Raf/MEK/Erk kinase cascade. Upstream of Ras, Grb2-Sos can be recruited directly to Grb2 binding sites on cytokine receptors, or indirectly through a plethora of adaptor proteins, including SHIP, Shc, or Shp2. It is believed that Erk activation is associated with mitogenesis, and in the case of the EPO-R is critical downstream of JAK2 activation in EPO-R HM knock-in mice.

The role of SAP K and p38 in cytokine signaling is unclear. Both of these enzymes were originally associated with integrating stress responses; however, it is clear that their role is more subtle. A distal region of the EPO-R links to SAPK activation, whereas p38 links to a proximal region of the EPO-R and does not require tyrosine residues [68]. p38 is also differentially regulated through differentiation, as p38α is expressed at high levels throughout erythroid differentiation whereas p38δ is upregulated at late stages in human erythroid cells [69]. p38$\alpha^{-/-}$ mice die during embryogenesis as p38α stabilizes the EPO transcript [70].

NEGATIVE REGULATION

Regulation of cytokine receptor signaling can be attributed in part to receptor degradation. Type I cytokine receptors, such as those for EPO, thrombopoietin (TPO), and prolactin (PRL), are commonly degraded in the lysosome [71,72] and/or the proteasome [73,74]. Receptor degradation is strongly increased by ligand stimulation, as a mechanism to "turn off" signaling: degradation inhibits effector protein recruitment to the receptor and thus attenuates phosphorylation and potentiation of downstream signaling cascades. In turn, the correct regulation of receptor degradation is critical for proliferation and survival of the cells in which these receptors are expressed.

Ligand-induced degradation of cytokine receptors frequently results from polyubiquitination. Ubiquitination is a posttranslational modification that occurs through a sequential series of covalent interactions with ubiquitin-interacting enzymes, such that the carboxy terminus of an ubiquitin (Ub) moiety forms an isopeptide bond with the lysine residue of the target protein. The ubiquitination process begins with thiol-ester linkage formation between the carboxy terminus of Ub (Glycine 76) and the active site cysteine of the ubiquitin-activating enzyme (E1). Ub is then transferred to an ubiquitin-conjugating enzyme (E2), where it forms another thiol-ester linkage with the active Cys of the E2. E3 proteins interact with the ubiquitin-bound E2s to facilitate the transfer of Ub to the target protein, and are primarily responsible for target protein specificity. Studies have shown that multiple E3 ubiquitin ligases can be responsible for the ubiquitination of a single receptor; for example, the EPO receptor can be ubiquitinated by the Skp1–Cullin1–F-box complex E3 ligase, $SCF^{\beta Trcp}$ [75], as well as the receptor-associated E3 ubiquitin ligase $p33^{RUL}$ [76]. Proteolysis of receptors can be triggered by protein polyubiquitination and subsequent ATP-dependent proteasomal degradation [77–79], or mono-ubiquitination and subsequent targeting to the lysosomal compartment [80].

It is increasingly apparent that the topology of the polyubiquitin chain is important for signaling attenuation through receptor degradation. Ubiquitin is a highly conserved, 76 amino acid protein which contains 7 lysine residues. Each of these lysine residues has the potential to covalently conjugate to the C-terminal glycine of another ubiquitin moiety, thereby allowing for diverse and specific polyubiquitin chain configurations. Data suggest that K29, K48, and K63 of ubiquitin comprise the majority of ubiquitin–ubiquitin bonds: internalization and lysosomal degradation is thought to result from K63 ubiquitin linkages [72], whereas K48 linkages generally target proteins to the proteasome. However, IFNAR1 requires both K48- and K63-mediated ubiquitin linkages for efficient internalization and lysosomal degradation [81], illustrating that the differences between ubiquitin linkages cannot be categorized easily. Since the same E3 ubiquitin ligase can generate unique polyubiquitination topologies, depending on the target protein [74,81], and each ubiquitin linkage may predispose that protein to a unique fate, the research in this area is far from complete.

Recruitment of E3 ubiquitin ligases to cytokine receptors may depend upon a consensus binding motif, such as the canonical DSG sequence and following hydrophobic threonine residue that recruits $SCF^{\beta Trcp}$ to the EPO-R [75], and phosphorylation of this motif. Recruitment of proteins involved in the proteasomal or lysosomal trafficking of ubiquitinated proteins further depends on signal sequences: lysosomal targeting of the TPO receptor, c-mpl, is dependent upon the adaptor protein, AP2, which is recruited to c-mpl by the signal sequence YRRL [71]. Phosphorylation of a signal sequence by ligand-dependent receptor stimulation is strongly tied to protein–protein interactions. For example, the EPO-R requires JAK2 activation for EPO-R–$SCF^{\beta Trcp}$ interaction. In this case, JAK2 likely activates a serine/threonine kinase that phosphorylates the $SCF^{\beta Trcp}$ binding motif on EPO-R.

In vivo models that abrogate receptor degradation display increases in receptor stability, signaling, cell growth, and proliferation. This effect has been shown in cells expressing mutant receptors which no longer bind their respective E3 ubiquitin ligase, such as PRL-R(S349A) and EPO-R(S462A), to $SCF^{\beta Trcp}$ [75,82]. Given that aberrant control of cell signaling and proliferation often leads to oncogenesis, mouse modeling studies will be vital to our understanding of the link between receptor downregulation and disease.

DEVELOPMENTAL REGULATION OF THE CYTOKINE SIGNALING PATHWAY

Considerable evidence has shown that the cytokine signaling pathway is intact in lower organisms, particularly in *Drosophila* (summarized in Figure 23.2) The entire

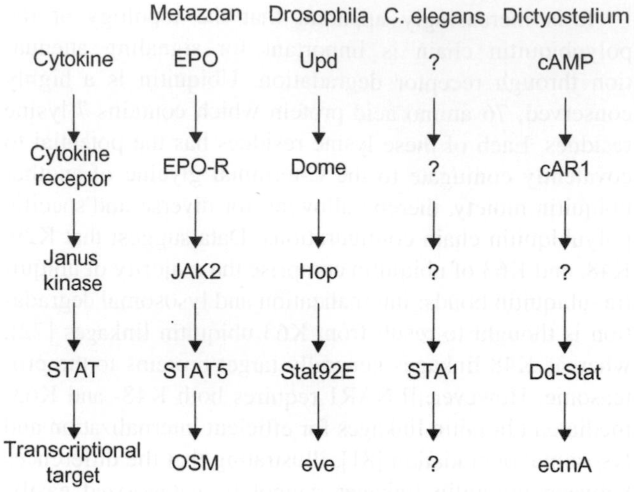

FIGURE 23.2 Developmental regulation of the cytokine signaling pathway.
Some components of the cytokine signaling pathway are conserved in *D. melanogaster, D. discoideum* and *C. elegans*. Additional details are presented within the text.

pathway is conserved in *Drosophila*, and the JAK-STAT pathway has been shown to be important in stem cell self-renewal in the fly.

The *Drosophila* ligand and cognate receptor are termed Unpaired [83] and Domeless [84], respectively. The *Drosophila* JAK kinase is called Hopscotch [85], and its activation is coupled to Stat92E activation [86,87]. Several target genes have been identified downstream of *Stat92E*, including *even-skipped (eve)*.

The JAK-STAT pathway plays a critical role in specifying male germline stem cell self-renewal. Unpaired is expressed in hub cells and is secreted in the vicinity of germline stem cells, resulting in JAK-STAT activation [88,89]. Cell division places daughter cells in distal proximity from the hub cells, resulting in their differentiation, whereas the stronger local signal preserves stem cell identity in the germline stem cells. There is no requirement of JAK-STAT signaling in female germline stem cells [90].

The first component of the JAK-STAT signaling cascade to be discovered in *Drosophila* was the JAK kinase ortholog termed Hopscotch (Hop). Overexpression of the wild-type Hop mimicked the *Tumorless-lethal (Tum-L)* phenotype originally described by Hanratty and colleagues [91]. Interestingly, *Tum-L* corresponds to a single point mutation in Hop, and is a dominant gain-of-function allele [92,93]. A second Hop mutation was identified, consisting of a highly conserved Glu within the Hop JH2 domain [94]. The significance of this mutation was underscored with the identification of activating mutations within the JH2 domain of JAK2 in myeloproliferative diseases.

Dictyostelium also has aspects of the cytokine signaling pathway conserved; however, upstream elements appear to uniquely couple to the *Dictyostelium* STAT protein. *Dictyostelium* exists in a vegetative or resting state, triggered by nutrient availability. Under conditions of nutrient deprivation, cells aggregate into a fruiting body consisting of spore and stalk cells. The *Dictyostelium* cells respond to pulsatile cAMP signaling binding to a cAMP receptor [95] and coupling to STATa expression in the tip of the slug. cAMP treatment results in tyrosine phosphorylation of STATa and nuclear localization [95]. STATa negatively regulates the expression of *ecmB* and positively regulates the expression of *CudA* in the slug tip [96].

C. elegans also expresses a developmentally regulated STAT protein, termed STA-1. However, STA-1 appears to be an orphan STAT, as homologous cytokine receptor or Janus kinases have yet to be identified in *C. elegans*. Genetic experiments revealed a requirement for components of the TGFβ signaling pathways, including DAF-1 (TGFβ receptor), and DAF-8 (Smad), in coupling to STA-1-dependent dauer formation [97]. Interestingly, STA-1 lacks the amino-terminal oligomerization domain found on STAT proteins in higher organisms [98]. STA-1 was shown not to be required for viability [98]. STA-1 could be tyrosine phosphorylated by TPR-MET when expressed in 293 cells, and displayed binding to the GAS element from the *ICAM-1* promoter [98]. Future studies will address the relevant upstream signaling pathways that couple to STA-1 activation.

INVOLVEMENT OF THE CYTOKINE SIGNALING PATHWAY IN HUMAN DISEASE

Since cytokines regulate many of the steps in proliferation, differentiation, and survival of hematopoietic cells, it is reasonable to assume that the cytokine signaling pathway may be a target for gain or loss-of-function mutations. The discussion of this section is limited to mutations within the JAK tyrosine kinases.

The first evidence that the JAK-STAT pathway played a role in disease emerged from mutagenic screens in *Drosophila*. Two temperature-sensitive mutations in *Hopscotch* were identified, termed *Tumorless-lethal,* in which there was a hyperproliferation of the *Drosophila* blood cells termed plasmatocytes. One mutation, G341E, fell within a region that is poorly conserved amongst higher Janus kinase orthologs [92,93]. However, Hop E695K fell within a highly conserved residue in all mammalian JAKs [94], and portended later discoveries in myeloproliferative diseases, as discussed below.

Subsequent research demonstrated that JAK2 is a target of chromosomal translocation. Fusion of the ets transcription factor TEL to various regions of JAK2 encompassing the JH1 domain is a rare event in human leukemia [99,100]. Animal modeling utilizing retroviral transduction bone-marrow transplantation revealed that TEL-JAK2 generates a mixed myelo-and lymphoproliferative disorder [101], and that STAT5 is required for the myeloid arm of the disease [102].

The noted hematologist William Dameshek predicted in 1951 that the myeloproliferative disorders termed polycythemia vera, essential thrombocythemia, and idiopathic myelofibrosis were interrelated [103]. This assertion proved to be particularly prescient, as five groups uncovered a single mutation, V617F, in the JH2 domain of JAK2 in 2005 [104–108]. Subsequent studies have shown additional mutations in JAK2 that are found in myeloproliferative disease [109]. Other reports have shown rare mutations in other regions of JAK2 in human disease (summarized in Table 23.4). Murine bone-marrow transplantation experiments revealed that expression of JAK2 V617F generates polycythemia [110–113], characterized by increased hematocrit and hemoglobin, reticulocytosis, splenomegaly, low plasma erythropoietin (Epo), and Epo-independent erythroid colonies in C57Bl/6 mice, augmented by the development of neutrophilia and leukocytosis when the experiment was performed in the Balb/C genetic background [113]. Conditional, floxed expression of the JAK2 V617F allele generated a disease with essential thrombocythemia when crossed with Vav-Cre reporter mice, whereas generation of Mx1-Cre x JAK2 V617F$^{fl/fl}$ mice revealed polycythemia. These specific diseases were ascribed to the ratio of JAK2 V617F protein expression to that of wild-type JAK2. JAK2 V617F protein expression is considerably lower that of wild-type JAK2 when under the transcriptional control of the Vav1 promoter, while JAK2 V617F and wild-type JAK2 protein levels are equivalent in the MX1-Cre model [114].

Approximately 10 percent of Down Syndrome patients acquire a mutation in the GATA-1 transcription factor. This mutation leads to a transient myeloproliferative disorder that resolves in the majority of patients. However, 20 percent of these patients develop acute myeloid leukemia (AML). An activating mutation was reported in JAK2 (T875N) in the CHRF-288-11 acute megakaryoblastic leukemia cell line [115]. Acute lymphoid leukemia (ALL) is also observed in Down Syndrome patients and, unlike AML associated with Down Syndrome, outcome is poor. Two recent reports describe JAK2 R683 mutations in 18 percent of Down Syndrome ALL cases [116,117].

High-throughput genomic sequencing efforts have identified JAK1 mutations in several diseases, most notably acute myeloid leukemia [118–120]. The mutations are found throughout JAK1, but appear to cluster in the JH2 and JH1 domains. Interestingly, a JAK1 V658F mutation (synonymous with JAK2 V617F) was identified [119]. JAK1 mutations are associated with shorter disease-free and overall survival in AML [120]. JAK1 mutations have also been isolated from non-small cell lung cancer and breast cancer patients [119] (Table 23.4).

Mutations in JAK3 can either be gain-of-function or loss-of-function. Sequencing of acute megakaryoblastic leukemia (AMKL) patient specimens and cell lines has revealed A572V and V722I mutations in the JH2 domain of JAK3 [121]. These mutations lead to enhanced JAK3 tyrosine kinase activity. The predominant genetic alteration in JAK3 involves loss-of-function mutations that give rise to severe combined immunodeficiency (SCID). Patients with deletion mutations of the IL-2 receptor γc display SCIDs and JAK3 mutations within the FERM domain (block γc binding to JAK3) or within the JH1 domain (kinase-inactive) [122–125]. These are summarized in Table 23.4.

A 22-year-old patient with consanguineous parents was identified with autosomal recessive hyper-IgE syndrome, characterized by recurrent skin abscesses, pneumonia, and elevated levels of serum IgE [126]. The patient was also susceptible to infection mediated by viruses, fungi, and mycobacteria. Despite normal numbers of natural killer, B and T cells, defects in multiple cytokine signaling pathways were observed. A deletion in Tyk2 was mapped at nucleotide 550, resulting in premature truncation at amino acid 90. Some unique features were observed when the patient's cells were compared to murine Tyk2-deficient cells [37], including decreased cell surface expression of IFNAR1 and IRNAR2. Similarities between the two models suggest that the Tyk2 FERM domain plays a role in receptor trafficking, as has been suggested in cell-line studies.

CONCLUDING REMARKS

There has been remarkable progress made in the understanding of cytokine signaling pathways. From the discovery of the ligands and their cognate receptors, through isolation of components of the JAK-STAT signaling pathway, a great deal is understood regarding the specificity of cytokine signaling. An active area of investigation is to understand mechanisms of negative regulation of cytokine signaling pathways, and how intracellular protein translocation affects receptor turnover, trafficking, and stability. It is evident that the STAT transcription factors do not function in isolation, but are part of an intricate regulatory network. Utilization of unbiased high throughput approaches, such as ChIP on CHIP, will aid in our understanding of cytokine-dependent gene regulation. A crystal structure of the JH2 and JH1 domains of JAK2 would facilitate greatly in understanding the mechanism underlying constitutive activation mediated by JAK2 V617F and related mutations found in myeloproliferative disease. Future studies will validate whether inhibition of JAK2 signaling is an effective means of controlling human myeloproliferative diseases.

ACKNOWLEDGEMENTS

Research in the Barber laboratory has been supported by Canadian Institutes of Health Research, National Cancer Institute of Canada, Leukemia and Lymphoma Society of

TABLE 23.4 Janus kinase mutations in disease

Janus kinase	Mutation	Domain	Disease	Reference
JAK1	L204M	FERM	B-ALL	118
	T478S	SH2	AML	120
	S512L	SH2	T-ALL	118
	V623A	JH2	AML	120
	A634D	JH2	B-ALL, T-ALL	118
	H647Y	JH2	Breast cancer	119
	L653F	JH2	AML	120
	V658F	JH2	ALL	119
	R724H	JH2	B-ALL, T-ALL	118
	L783F	JH2	ALL	119
	T872M	JH1	NSCLC	119
	R879S	JH1	T-ALL	118
	R879C	JH1	T-ALL	118
	R879H	JH1	T-ALL	118
JAK2	H538Q	Juxta JH2	MPD	109
	K539L	Juxta JH2	MPD	109
	F537-K539 deletion/L ins.	Juxta JH2	MPD	109
	N542-E543 deletion	Juxta JH2	MPD	109
	L611S	JH2	ALL	127
	V617F	JH2	MPD	104–108
	R683G	JH2	DS-ALL	116, 117
	R683S	JH2	DS-ALL	116
	R683K	JH2	DS-ALL	116
	T875N	JH1	AMKL	115
JAK3	Y100C	FERM	SCIDS	122, 125
	F408 termination	SH2	SCIDS	123
	R445 ter/C579R	SH2/JH2	SCIDS	124
	E481G/K482 ter	Inter SH2-JH2	SCIDS	128
	E481V/P517 ter	Inter SH2-JH2	SCIDS	128
	C565 termination	JH2	SCIDS	123
	L586-M592 deletion	JH2	SCIDS	128
	A572V	JH2	AMKL	121
	V715I	JH2	Breast cancer	119
	V722I	JH2	AMKL	121
	K876 termination	JH1	SCIDS	122
Tyk2	I90X	FERM	HIES	126

Canada, Leukemia and Lymphoma Society and Cancer Research Society. Mojib Javed was supported by a CIHR Master's Award.

REFERENCES

1. Richmond TD, Chohan M, Barber DL. Turning cells red: signal transduction mediated by erythropoietin. *Trends Cell Biol* 2005;**15**:146–55.
2. Livnah O, Stura EA, Middleton SA, Johnson DL, Jolliffe LK, Wilson IA. Crystallographic evidence for preformed dimers of erythropoietin receptor before ligand activation. *Science* 1999;**283**:987–90.
3. Livnah O, Johnson DL, Stura EA, et al. An antagonist peptide–EPO receptor complex suggests that receptor dimerization is not sufficient for activation. *Nature Struct Biol* 1998;**5**:993–1004.
4. Livnah O, Stura EA, Johnson DL, et al. Functional mimicry of a protein hormone by a peptide agonist: the EPO receptor complex at 2.8 A. *Science* 1996;**273**:464–71.
5. Syed RS, Reid SW, Li C, et al. Efficiency of signalling through cytokine receptors depends critically on receptor orientation. *Nature* 1998;**395**:511–16.
6. Wilson IA, Jolliffe LK. The structure, organization, activation and plasticity of the erythropoietin receptor. *Curr Opin Struct Biol* 1999;**9**:696–704.
7. Jiang G, Hunter T. Receptor signaling: when dimerization is not enough. *Curr Biol* 1999;**9**:R568–71.
8. Wu H, Liu X, Jaenisch R, Lodish HF. Generation of committed BFU-E and CFU-E progenitors does not require erythropoietin or the erythropoietin receptor. *Cell* 1995;**83**:59–68.
9. Kieran MW, Perkins AC, Orkin SH, Zon LI. Thrombopoietin rescues *in vitro* erythroid colony formation from mouse embryos lacking the erythropoietin receptor. *Proc Natl Acad Sci USA* 1996;**93**:9126–31.
10. Lin C-S, Lim S-K, D'Agati V, Costantini F. Differential effect of an erythropoietin receptor gene disruption on primitive and definitve erythropoiesis. *Genes Dev* 1996;**10**:154–64.
11. D'Andrea AD, Yoshimura A, Youssoufian H, Zon LI, Koo J-W, Lodish HF. The cytoplasmic region of the erythropoietin receptor contains nonoverlapping positive and negative growth-regulatory domains. *Mol Cell Biol* 1991;**11**:1980–7.
12. Zang H, Sato K, Nakajima H, McKay C, Ney PA, Ihle JN. The distal region and receptor tyrosines of the Epo receptor and non-essential for in vivo erythropoiesis. *EMBO J.* 2001;**20**:3156–66.
13. Li K, Menon MP, Karur VG, Hegde S, Wojchowski DM. Attenuated signaling by a phosphotyrosine-null Epo receptor form in primary erythroid progenitor cells. *Blood* 2003;**102**:3147–53.
14. Menon MP, Karur V, Bogacheva O, Bogachev O, Cuetara B, Wojchowski DM. Signals for stress erythropoiesis are integrated via an erythropoietin receptor–phosphotyrosine-343–Stat5 axis. *J Clin Invest* 2006;**116**:683–94.
15. Ihle JN, Gilliland DG. Jak2: normal function and role in hematopoietic disorders. *Curr Opin Genet Dev* 2007;**17**:8–14.
16. Huang LJ, Constantinescu SN, Lodish HF. The N-terminal domain of Janus kinase 2 is required for Golgi processing and cell surface expression of erythropoietin receptor. *Mol Cell* 2001;**8**:1327–38.
17. Gauzzi MC, Barbieri G, Richter MF, et al. The amino-terminal region of Tyk2 sustains the level of interferon alpha receptor 1, a component of the interferon alpha/beta receptor. *Proc Natl Acad Sci USA* 1997;**94**:11,839–11,844.
18. Pelletier S, Gingras S, Funakoshi-Tago M, Howell S, Ihle JN. Two domains of the erythropoietin receptor are sufficient for Jak2 binding/activation and function. *Mol Cell Biol* 2006;**26**:8527–38.
19. Feng J, Witthuhn BA, Matsuda T, Kohlhuber F, Kerr IM, Ihle JN. Activation of Jak2 catalytic activity requires phosphorylation of Y1007 in the kinase activation loop. *Mol Cell Biol* 1997;**17**:2497–501.
20. Myers MP, Andersen JN, Cheng A, et al. TYK2 and JAK2 are substrates of protein-tyrosine phosphatase 1B. *J Biol Chem* 2001;**276**:47,771–47,774.
21. Simoncic PD, Lee-Loy A, Barber DL, Tremblay ML, McGlade CJ. The T cell protein tyrosine phosphatase is a negative regulator of janus family kinases 1 and 3. *Curr Biol* 2002;**12**:446–53.
22. Argetsinger LS, Kouadio JL, Steen H, Stensballe A, Jensen ON, Carter-Su C. Autophosphorylation of JAK2 on tyrosines 221 and 570 regulates its activity. *Mol Cell Biol* 2004;**24**:4955–67.
23. Feener Jr. EP, Rosario F, Dunn SL, Stancheva Z, Myers MG Tyrosine phosphorylation of Jak2 in the JH2 domain inhibits cytokine signaling. *Mol Cell Biol* 2004;**24**:4968–78.
24. Funakoshi-Tago M, Pelletier S, Matsuda T, Parganas E, Ihle JN. Receptor specific downregulation of cytokine signaling by autophosphorylation in the FERM domain of Jak2. *Embo J* 2006;**25**:4763–72.
25. Mazurkiewicz-Munoz AM, Argetsinger LS, Kouadio JL, et al. Phosphorylation of JAK2 at serine 523: a negative regulator of JAK2 that is stimulated by growth hormone and epidermal growth factor. *Mol Cell Biol* 2006;**26**:4052–62.
26. Lucet IS, Fantino E, Styles M, et al. The structural basis of Janus kinase 2 inhibition by a potent and specific pan-Janus kinase inhibitor. *Blood* 2006;**107**:176–83.
27. Boggon TJ, Li Y, Manley PW, Eck MJ. Crystal structure of the Jak3 kinase domain in complex with a staurosporine analog. *Blood* 2005;**106**:996–1002.
28. Rodig SJ, Meraz MA, White JM, et al. Disruption of the Jak1 gene demonstrates obligatory and nonredundant roles of the Jaks in cytokine-induced biologic responses. *Cell* 1998;**93**:373–83.
29. Parganas E, Wang D, Stravopodis D, et al. Jak2 is essential for signaling through a variety of cytokine receptors. *Cell* 1998;**93**:385–95.
30. Neubauer H, Cumano A, Muller M, Wu H, Huffstadt U, Pfeffer K. Jak2 deficiency defines an essential developmental checkpoint in definitive hematopoiesis. *Cell* 1998;**93**:397–409.
31. Nosaka T, van Deursen JM, Tripp RA, et al. Defective lymphoid development in mice lacking Jak3. *Science* 1995;**270**:800–2.
32. Park SY, Saijo K, Takahashi T, et al. Developmental defects of lymphoid cells in Jak3 kinase-deficient mice. *Immunity* 1995;**3**:771–82.
33. Thomis DC, Gurniak CB, Tivol E, Sharpe AH, Berg LJ. Defects in B lymphocyte maturation and T lymphocyte activation in mice lacking Jak3. *Science* 1995;**270**:794–7.
34. Ohbo K, Suda T, Hashiyama M, et al. Modulation of hematopoiesis in mice with a truncated mutant of the interleukin-2 receptor gamma chain. *Blood* 1996;**87**:956–67.
35. DiSanto JP, Muller W, Guy-Grand D, Fischer A, Rajewsky K. Lymphoid development in mice with a targeted deletion of the interleukin 2 receptor gamma chain. *Proc Natl Acad Sci USA* 1995;**92**:377–81.
36. Cao X, Shores EW, Hu-Li J, et al. Defective lymphoid development in mice lacking expression of the common cytokine receptor gamma chain. *Immunity* 1995;**2**:233–8.
37. Shimoda K, Kato K, Aoki K, et al. Tyk2 plays a restricted role in IFN alpha signaling, although it is required for IL-12-mediated T cell function. *Immunity* 2000;**13**:561–71.
38. Levy Jr DE, Darnell JE. Stats: transcriptional control and biological impact. *Nature Rev Mol Cell Biol* 2002;**3**:651–62.
39. Durbin JE, Hackenmiller R, Simon MC, Levy DE. Targeted disruption of the mouse Stat1 gene results in compromised innate immunity to viral disease. *Cell* 1996;**84**:443–50.

40. Meraz MA, White JM, Sheehan KC, et al. Targeted disruption of the Stat1 gene in mice reveals unexpected physiologic specificity in the JAK-STAT signaling pathway. *Cell* 1996;**84**:431–42.
41. Sahni M, Amrosetti D-C, Mansukhani A, Gertner R, Levy D, Basilico C. FGF signaling inhibits chondrocyte proliferation and regulates bone development through the STAT1 pathway. *Genes Dev* 1999;**13**:1361–6.
42. Halupa A, Bailey ML, Huang K, Iscove NN, Levy DE, Barber DL. A novel role for STAT1 in regulating murine erythropoiesis: deletion of STAT1 results in overall reduction of erythroid progenitors and alters their distribution. *Blood* 2005;**105**:552–61.
43. Park C, Li S, Cha E, Schindler C. Immune response in Stat2 knockout mice. *Immunity* 2000;**13**:795–804.
44. Takeda K, Noguchi K, Shi W, et al. Targeted disruption of the mouse Stat3 gene leads to early embryonic lethality. *Proc Natl Acad Sci USA* 1997;**94**:3801–4.
45. Takeda K, Kaisho T, Yoshida N, Takeda J, Kishimoto T, Akira S. Stat3 activation is responsible for IL-6-dependent T cell proliferation through preventing apoptosis: generation and characterization of T cell-specific Stat3-deficient mice. *J Immunol* 1998;**161**:4652–60.
46. Akaishi H, Takeda K, Kaisho T, et al. Defective IL-2-mediated IL-2 receptor alpha chain expression in Stat3-deficient T lymphocytes. *Int Immunol* 1998;**10**:1747–51.
47. Yao Z, Kanno Y, Kerenyi M, et al. Nonredundant roles for Stat5a/b in directly regulating Foxp3. *Blood* 2007;**109**:4368–75.
48. Laurence A, Tato CM, Davidson TS, et al. Interleukin-2 signaling via STAT5 constrains T helper 17 cell generation. *Immunity* 2007;**26**:371–81.
49. Sano S, Takahama Y, Sugawara T, et al. Stat3 in thymic epithelial cells is essential for postnatal maintenance of thymic architecture and thymocyte survival. *Immunity* 2001;**15**:261–73.
50. Fornek JL, Tygrett LT, Waldschmidt TJ, Poli V, Rickert RC, Kansas GS. Critical role for Stat3 in T-dependent terminal differentiation of IgG B cells. *Blood* 2006;**107**:1085–91.
51. Takeda K, Clausen BE, Kaisho T, et al. Enhanced Th1 activity and development of chronic enterocolitis in mice devoid of Stat3 in macrophages and neutrophils. *Immunity* 1999;**10**:39–49.
52. Lee CK, Raz R, Gimeno R, et al. STAT3 is a negative regulator of granulopoiesis but is not required for G-CSF-dependent differentiation. *Immunity* 2002;**17**:63–72.
53. Kaplan MH, Sun YL, Hoey T, Grusby MJ. Impaired IL-12 responses and enhanced development of Th2 cells in Stat4-deficient mice. *Nature* 1996;**382**:174–7.
54. Teglund S, McKay C, Schuetz E, et al. Stat5a and Stat5b proteins have essential and nonessential, or redundant, roles in cytokine responses. *Cell* 1998;**93**:841–50.
55. Liu X, Robinson GW, Wagner KU, Garrett L, Wynshaw-Boris A, Hennighausen L. Stat5a is mandatory for adult mammary gland development and lactogenesis. *Genes Dev* 1997;**11**:179–86.
56. Udy GB, Towers RP, Snell RG, et al. Requirement of STAT5b for sexual dimorphism of body growth rates and liver gene expression. *Proc Natl Acad Sci USA* 1997;**94**:7239–44.
57. Socolovsky M, Fallon AEJ, Wang S, Brugnara C, Lodish HF. Fetal anemia and apoptosis of red cell progenitors in Stat5a$^{-/-}$5b$^{-/-}$ mice: a direct role for Stat5 in Bcl-XL induction.. *Cell* 1999;**98**:181–91.
58. Yao Z, Cui Y, Watford WT, et al. Stat5a/b are essential for normal lymphoid development and differentiation. *Proc Natl Acad Sci USA* 2006;**103**:1000–5.
59. Dai X, Chen Y, Di L, et al. Stat5 is essential for early B cell development but not for B cell maturation and function. *J Immunol* 2007;**179**:1068–79.
60. Zhu BM, McLaughlin SK, Na R, et al. Hematopoietic-specific Stat5-null mice display microcytic hypochromic anemia associated with reduced transferrin receptor gene expression. *Blood* 2008;**112**:2071–80.
61. Kerenyi MA, Grebien F, Gehart H, et al. Stat5 regulates cellular iron uptake of erythroid cells via IRP-2 and TfR-1. *Blood* 2008;**112**:3878–88.
62. Kaplan MH, Schindler U, Smiley ST, Grusby MJ. Stat6 is required for mediating responses to IL-4 and for development of Th2 cells. *Immunity* 1996;**4**:313–19.
63. Shimoda K, van Deursen J, Sangster MY, et al. Lack of IL-4-induced Th2 response and IgE class switching in mice with disrupted Stat6 gene. *Nature* 1996;**380**:630–3.
64. Takeda K, Tanaka T, Shi W, et al. Essential role of Stat6 in IL-4 signalling. *Nature* 1996;**380**:627–30.
65. Bakker WJ, Blazquez-Domingo M, Kolbus A, et al. FoxO3a regulates erythroid differentiation and induces BTG1, an activator of protein arginine methyl transferase 1. *J Cell Biol* 2004;**164**:175–84.
66. Bakker WJ, van Dijk TB, Parren-van Amelsvoort M, et al. Differential regulation of Foxo3a target genes in erythropoiesis. *Mol Cell Biol* 2007;**27**:3839–54.
67. Marinkovic D, Zhang X, Yalcin S, et al. Foxo3 is required for the regulation of oxidative stress in erythropoiesis. *J Clin Invest* 2007;**117**:2133–44.
68. Haq R, Halupa A, Beattie BK, Mason JM, Zanke BW, Barber DL. Regulation of erythropoietin-induced STAT serine phosphorylation by distinct mitogen-activated protein kinases. *J Biol Chem* 2002;**277**:17,359–17,366,.
69. Uddin S, Ah-Kang J, Ulaszek J, Mahmud D, Wickrema A. Differentiation stage-specific activation of p38 mitogen-activated protein kinase isoforms in primary human erythroid cells. *Proc Natl Acad Sci USA* 2004;**101**:147–52.
70. Tamura K, Sudo T, Senftleben U, Dadak AM, Johnson R, Karin M. Requirement for p38alpha in erythropoietin expression: a role for stress kinases in erythropoiesis. *Cell* 2000;**102**:221–31.
71. Hitchcock IS, Chen MM, King JR, Kaushansky K. YRRL motifs in the cytoplasmic domain of the thrombopoietin receptor regulate receptor internalization and degradation. *Blood* 2008;**112**:2222–31.
72. Varghese B, Barriere H, Carbone CJ, et al. Polyubiquitination of prolactin receptor stimulates its internalization, postinternalization sorting, and degradation via the lysosomal pathway. *Mol Cell Biol* 2008;**28**:5275–87.
73. Verdier F, Chretien S, Muller O, et al. Proteasomes regulate erythropoietin receptor and signal transducer and activator of transcription 5 (STAT5) activation, Possible involvement of the ubiquitinated Cis proteins. *J Biol Chem* 1998;**273**.
74. Walrafen P, Verdier F, Kadri Z, Chretien S, Lacombe C, Mayeux P. Both proteasomes and lysosomes degrade the activated erythropoietin receptor. *Blood* 2005;**105**:600–8.
75. Meyer L, Deau B, Forejtnikova H, et al. Beta-Trcp mediates ubiquitination and degradation of the erythropoietin receptor and controls cell proliferation. *Blood* 2007;**109**:5215–22.
76. Friedman AD, Nimbalkar D, Quelle FW. Erythropoietin receptors associate with a ubiquitin ligase, p33RUL, and require its activity for erythropoietin-induced proliferation. *J Biol Chem* 2003;**278**.
77. Chau V, Tobias JW, Bachmair A, et al. A multiubiquitin chain is confined to specific lysine in a targeted short-lived protein. *Science* 1989;**243**:1576–83.
78. Hough R, Pratt G, Rechsteiner M. Ubiquitin-lysozyme conjugates. Identification and characterization of an ATP-dependent protease from rabbit reticulocyte lysates. *J Biol Chem* 1986;**261**:2400–8.

79. Hough R, Pratt G, Rechsteiner M. Purification of two high molecular weight proteases from rabbit reticulocyte lysate. *J Biol Chem* 1987;**262**:8303–13.
80. Hicke L. Protein regulation by monoubiquitin. *Nature Rev Mol Cell Biol* 2001;**2**:195–201.
81. Kumar KG, Barriere H, Carbone CJ, et al. Site-specific ubiquitination exposes a linear motif to promote interferon-alpha receptor endocytosis. *J Cell Biol* 2007;**179**:935–50.
82. Li Y, Kumar KG, Tang W, Spiegelman VS, Fuchs SY. Negative regulation of prolactin receptor stability and signaling mediated by SCF(beta-TrCP) E3 ubiquitin ligase. *Mol Cell Biol* 2004;**24**:4038–48.
83. Harrison DA, McCoon PE, Binari R, Gilman M, Perrimon N. Drosophila unpaired encodes a secreted protein that activates the JAK signaling pathway. *Genes Dev* 1998;**12**:3252–63.
84. Brown S, Hu N, Hombria JC. Identification of the first invertebrate interleukin JAK/STAT receptor, the Drosophila gene domeless. *Curr Biol* 2001;**11**:1700–5.
85. Binari R, Perrimon N. Stripe-specific regulation of pair-rule genes by hopscotch, a putative Jak family tyrosine kinase in Drosophila. *Genes Dev* 1994;**8**:300–12.
86. Yan Jr. R, Luo H, Darnell JE, Dearolf CR A JAK-STAT pathway regulates wing vein formation in Drosophila. *Proc Natl Acad Sci USA* 1996;**93**:5842–7.
87. Hou XS, Melnick MB, Perrimon N. Marelle acts downstream of the Drosophila HOP/JAK kinase and encodes a protein similar to the mammalian STATs. *Cell* 1996;**84**:411–19.
88. Kiger AA, Jones DL, Schulz C, Rogers MB, Fuller MT. Stem cell self-renewal specified by JAK-STAT activation in response to a support cell cue. *Science* 2001;**294**:2542–5.
89. Tulina N, Matunis E. Control of stem cell self-renewal in Drosophila spermatogenesis by JAK-STAT signaling. *Science* 2001;**294**:2546–9.
90. Decotto E, Spradling AC. The Drosophila ovarian and testis stem cell niches: similar somatic stem cells and signals. *Dev Cell* 2005;**9**:501–10.
91. Hanratty WP, Ryerse JS. A genetic melanotic neoplasm of Drosophila melanogaster. *Dev Biol* 1981;**83**:238–49.
92. Luo H, Hanratty WP, Dearolf CR. An amino acid substitution in the Drosophila hopTum-l Jak kinase causes leukemia-like hematopoietic defects. *EMBO J* 1995;**14**:1412–20.
93. Harrison DA, Binari R, Nahreini TS, Gilman M, Perrimon N. Activation of a Drosophila Janus kinase (JAK) causes hemato-poietic neoplasia and developmental defects. *EMBO J* 1995;**14**:2857–65.
94. Luo H, Rose P, Barber D, et al. Mutation in the Jak kinase JH2 domain hyperactivates Drosophila and mammalian Jak-Stat pathways. *Mol. Cell. Biol.* 1997;**17**:1562–71.
95. Araki T, Gamper M, Early A, et al. Developmentally and spatially regulated activation of a Dictyostelium STAT protein by a serpentine receptor. *EMBO J* 1998;**17**:4018–28.
96. Fukuzawa M, Williams JG. Analysis of the promoter of the cudA gene reveals novel mechanisms of Dictyostelium cell type differentiation. *Development* 2000;**127**:2705–13.
97. Wang Y, Levy DE. C. elegans STAT cooperates with DAF-7/TGF-beta signaling to repress dauer formation. *Curr Biol* 2006;**16**:89–94.
98. Wang Y, Levy DE. *C. elegans* STAT: evolution of a regulatory switch. *FASEB J* 2006;**20**:1641–52.
99. Lacronique V, Boureux A, Valle VD, et al. A TEL-JAK2 fusion protein with constitutive kinase activity in human leukemia. *Science* 1997;**278**:1309–12.
100. Peeters P, Raynaud SD, Cools J, et al. Fusion of TEL, the ETS-variant gene 6 (ETV6), to the receptor-associated kinase JAK2 as a result of t(9;12) in a lymphoid and t(9;15;12) in a myeloid leukemia. *Blood* 1997;**90**:2535–40.
101. Schwaller J, Frantsve J, Aster J, et al. Transformation of hematopoietic cell lines to growth-factor independence and induction of a fatal myelo- and lymphoproliferative disease in mice by retrovirally transduced TEL/JAK2 fusion genes. *EMBO J* 1998;**17**:5321–33.
102. Schwaller J, Parganas E, Wang D, et al. Stat5 is essential for the myelo- and lymphoproliferative disease induced by TEL/JAK2. *Mol Cell* 2000;**6**:693–704.
103. Dameshek W. Some speculations on the myeloproliferative syndromes. *Blood* 1951;**6**:372–5.
104. Levine RL, Wadleigh M, Cools J, et al. Activating mutation in the tyrosine kinase JAK2 in polycythemia vera, essential thrombocythemia, and myeloid metaplasia with myelofibrosis. *Cancer Cell* 2005;**7**:387–97.
105. Zhao R, Xing S, Li Z, et al. Identification of an acquired JAK2 mutation in polycythemia vera. *J Biol Chem* 2005;**280**:22,788–22,792.
106. Baxter EJ, Scott LM, Campbell PJ, et al. Acquired mutation of the tyrosine kinase JAK2 in human myeloproliferative disorders. *Lancet* 2005;**365**:1054–61.
107. James C, Ugo V, Le Couedic JP, et al. A unique clonal JAK2 mutation leading to constitutive signalling causes polycythaemia vera. *Nature* 2005;**434**:1144–8.
108. Kralovics R, Passamonti F, Buser AS, et al. A gain-of-function mutation of JAK2 in myeloproliferative disorders. *N Engl J Med* 2005;**352**:1779–90.
109. Scott LM, Tong W, Levine RL, et al. JAK2 exon 12 mutations in polycythemia vera and idiopathic erythrocytosis. *N Engl J Med* 2007;**356**:459–68.
110. Bumm TG, Elsea C, Corbin AS, et al. Characterization of murine JAK2V617F-positive myeloproliferative disease. *Cancer Res* 2006;**66**:11156–65.
111. Wernig G, Mercher T, Okabe R, Levine RL, Lee BH, Gilliland DG. Expression of Jak2V617F causes a polycythemia vera-like disease with associated myelofibrosis in a murine bone marrow transplant model. *Blood* 2006;**107**:4274–81.
112. Lacout C, Pisani DF, Tulliez M, Gachelin FM, Vainchenker W, Villeval JL. JAK2V617F expression in murine hematopoietic cells leads to MPD mimicking human PV with secondary myelofibrosis. *Blood* 2006;**108**:1652–60.
113. Zaleskas VM, Krause DS, Lazarides K, et al. Molecular pathogenesis and therapy of polycythemia induced in mice by JAK2 V617F. *PLoS ONE* 2006;**1**:e18.
114. Tiedt R, Hao-Shen H, Sobas MA, et al. Ratio of mutant JAK2-V617F to wild-type Jak2 determines the MPD phenotypes in transgenic mice. *Blood* 2008;**111**:3931–40.
115. Mercher T, Wernig G, Moore SA, et al. JAK2T875N is a novel activating mutation that results in myeloproliferative disease with features of megakaryoblastic leukemia in a murine bone marrow transplantation model. *Blood* 2006;**108**:2770–9.
116. Bercovich D, Ganmore I, Scott LM, et al. Mutations of JAK2 in acute lymphoblastic leukaemias associated with Down's syndrome. *Lancet* 2008;**372**:1484–92.
117. Kearney L, Castro DGD, Yeung J, et al. Specific JAK2 mutation (JAK2R683) and multiple gene deletions in Down syndrome acute lymphoblastic leukaemia. *Blood* 2009;**113**:646–8.
118. Flex E, Petrangeli V, Stella L, et al. Somatically acquired JAK1 mutations in adult acute lymphoblastic leukemia. *J Exp Med* 2008;**205**: 751–8.

119. Jeong EG, Kim MS, Nam HK, et al. Somatic mutations of JAK1 and JAK3 in acute leukemias and solid cancers. *Clin Cancer Res* 2008;**14**:3716–21.
120. Xiang Z, Zhao Y, Mitaksov V, et al. Identification of somatic JAK1 mutations in patients with acute myeloid leukemia. *Blood* 2008;**111**:4809–12.
121. Walters DK, Mercher T, Gu TL, et al. Activating alleles of JAK3 in acute megakaryoblastic leukemia. *Cancer Cell* 2006;**10**:65–75.
122. Macchi P, Villa A, Gillani S, et al. Mutations of Jak-3 gene in patients with autosomal severe combined immune deficiency (SCID). *Nature* 1995;**377**:65–8.
123. Russell SM, Tayebi N, Nakajima H, et al. Mutation of Jak3 in a patient with SCID: essential role of Jak3 in lymphoid development. *Science* 1995;**270**:797–800.
124. Schumacher RF, Mella P, Badolato R, et al. Complete genomic organization of the human JAK3 gene and mutation analysis in severe combined immunodeficiency by single-strand conformation polymorphism. *Hum Genet* 2000;**106**:73–9.
125. Zhou YJ, Chen M, Cusack NA, et al. Unexpected effects of FERM domain mutations on catalytic activity of Jak3: structural implication for Janus kinases. *Mol Cell* 2001;**8**:959–69.
126. Minegishi Y, Saito M, Morio T, et al. Human tyrosine kinase 2 deficiency reveals its requisite roles in multiple cytokine signals involved in innate and acquired immunity. *Immunity* 2006;**25**:745–55.
127. Kratz CP, Böll S, Kontny U, et al. Mutational screen reveals a novel JAK2 mutation, L611S, in a child with acute lymphoblastic leukemia. *Leukemia* 2006;**20**:381–3.
128. Candotti F, Oakes SA, Johnston JA, et al. Structural and functional basis for JAK3-deficient severe combined immunodeficiency. *Blood* 1997;**90**:3996–4003.

Chapter 24

Growth Hormone and Prolactin Family of Hormones and Receptors: The Structural Basis for Receptor Activation and Regulation

Anthony A. Kossiakoff[1] and Charles V. Clevenger[2]

[1]*Department of Biochemistry and Molecular Biology, University of Chicago, Gordon Center for Integrative Sciences, Chicago, Illinois*

[2]*Department of Pathology, Northwestern University Medical School, Robert H. Lurie Comprehensive Cancer Center, Chicago, Illinois*

INTRODUCTION

Within the cytokine superfamily, the members of the growth hormone (GH)/prolactin (PRL) family of hormones and receptors are arguably the most extensively studied systems focused on structure–function issues and molecular recognition. These studies and those of related cytokine systems have been instrumental in defining modes of hormone action and regulation. A common misconception about these systems is that because the previous studies have produced such a breadth of important results, all the most critical issues have been resolved. The fact is that these studies have really just laid the foundations for a new generation of investigations that will produce a further level of insight about the subtleties under which molecular recognition processes drive biological function. The goal of this review is to provide a background of our current understanding and suggest areas of future investigation.

THE GROWTH HORMONE FAMILY OF HORMONES AND RECEPTORS

Growth hormone (GH), placental lactogen (PL), and prolactin (PRL) regulate an extensive variety of important physiological functions. While GH biology generally centers on the regulation and differentiation of muscle, cartilage, and bone cells, it is the PRL hormones and receptors that display a much broader spectrum of activities, ranging from their well-known effects in mammalian reproductive biology to osmoregulation in fishes and nesting behavior in birds [1]. Within the cytokine superfamily, the GH/PRL endocrine family of hormones and receptors is arguably the most extensively studied system focused on structure–function issues and molecular recognition [2–8]. These studies and those of related cytokine systems have been instrumental in defining modes of hormone action and regulation [9–13]. The structure-based mechanisms by which these systems activate are similar; however, the molecular strategies that are employed are complex and hardly predictable.

The activities of GH, PL, and PRL and their homologs are triggered by hormone-induced homodimerization of their cognate receptors, which produce subsequent signals through a series of phosphorylation events in the Janus kinase (JAK) signal transducer and activator of transcription (STAT) signaling pathway [14] (Figure 24.1). The receptors belong to the hematopoietic receptor superfamily, and have a three-domain organization [9, 15]. It is the cytoplasmic domains of the aggregated receptor complex that bind one or several JAK tyrosine kinases, which then transphosphorylate elements on themselves, the receptors, and associated transcription factors belonging to the STAT family [14].

TRIGGERING GH AND PRL RECEPTOR ACTIVATION: REVISION TO THE DOGMA

There is an extensive literature describing the molecular basis for cytokine receptor activation. The long-held notion was that for cytokine receptors, the signaling cascade is triggered by a simple concentration effect phenomenon [10, 16].

Handbook of Cell Signaling, Three-Volume Set 2 ed.

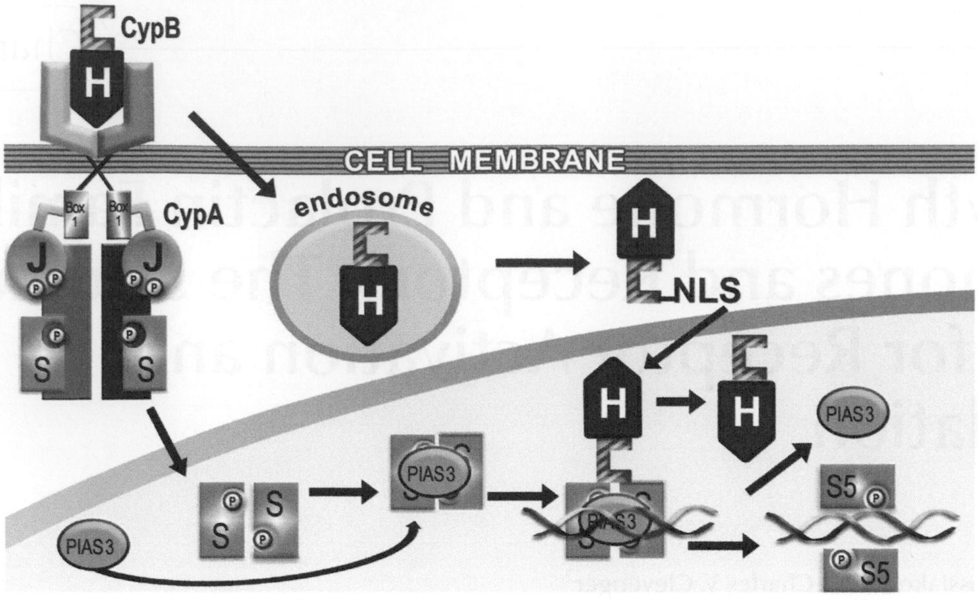

FIGURE 24.1 Mechanism of hormone-induced receptor activation.
The hormone (H) binds to a receptor homodimer. The homodimer undergoes a conformational change that allows JAK2 (J) and STAT5 (S) to bind to the cytoplasmic portion of the receptor. Phosphorylation of the receptor, JAK2, and STAT5 are triggered by conformational changes induced by CypA in the cytoplasm. The signaling complex and STAT5 dimer are endocytosed into the cytoplasm, where the hormone–CypB complex is subsequently translocated to the nucleus utilizing the nuclear localization sequence (NLS) on the N-terminus of CypB. Stat DNA binding is inhibited by PIAS3. The CypB–hormone complex binds to the STAT5–PIAS3 DNA complex and knocks off PIAS3, activating STAT5 transcription.

The role of the hormone-induced receptor homodimerization event was proposed just to bring the relevant molecules together, ensuring proximity – inherent flexibility and dynamics would then take care of the rest. However, recently a "pre-association" phenomenon, where the receptors cluster as inactive homodimers, was established for the hGH-R and hPRL-R systems [17–19]. A similar situation had previously been established for the EPO-R system [20–23]. The hormone binding acts in some way as an activation switch, but exactly how this event transforms the receptor homodimer into its signaling competent form remains to be determined.

In the hGH-R case, it was also determined that dimerization is mediated through interaction between the transmembrane (TM) α-helices [19]. Interestingly, in the case of hPRL-R, while the TM interactions appear to be important, the data suggest that additional contacts between the intracellular domains (ICD) might occur [18]. Viewed in the broader context, the generality of the receptor pre-association model is supported by observations from a number of related and distantly related receptor superfamilies. This suggests that receptors, organized in inactive oligomeric state, are probably a ubiquitous feature in eukaryotic signaling mechanics [24–27].

AN UNANTICIPATED ROLE FOR CYTOKINE HORMONES AS TRANSCRIPTIONAL ENHANCERS

The mechanistic pathway leading to GH and PRL signaling is pictured in Figure 24.1. The long-held assumption of a second messenger model for cytokine signaling limited the role of the hormone to the initial triggering event. Thus, after the initial signaling event, all downstream actions were hormone independent [16]. However, recent findings show that this model is incorrect, and that the hormones actually have crucial roles in other aspects of the signal transduction pathway. In particular, it has been found that the hormones function as direct transcriptional activators in the nucleus [28, 29].

To perform this function requires that the hormone be efficiently transported from the extracellular space to the nucleus through a process called "protein retrotransport," whereby both the hormone and receptor are conveyed in a complex that is transported across the endoplasmic reticulum (ER) and nuclear membranes by a set of chaperone proteins [30, 31]. PRL and GH are chaperoned to the nucleus by CypB, a cyclophilin prolyl-peptide isomerase (PPI), which not only serves as a transporter but also has important enzymatic functions. Establishing the functions of PPIs in the downstream signaling process is mainly based on the characterization of the PRL system, but it is expected that they are equally applicable to GH.

CypB – a Chaperone and Activator

CypB initially resides in the extracellular compartment, but performs its enzymatic function in the nucleus. CypB "hitchhikes" into the cell by binding to the 1:2 PRL–PRL-R (or GH) cell-surface complex, which is endocytosed during receptor down regulation. Once in the cell, the PRL–CypB molecules undergo a remarkable role reversal. CypB assumes the role of the active transporter by acting as the vehicle via its nuclear localization sequence (NLS) to transport PRL into the nucleus [31–33]. In the nucleus, PRL–CypB forms a direct association with the inactive STAT5–PIAS3 complex, wherein CypB isomerizes a proline bond that forces the dissociation of PIAS3 from STAT5, thereby activating STAT5-mediated gene expression (Figure 24.1) [32]. It is not known where the proline switch is located; whether it is in the PIAS3 inhibitor itself, or works indirectly by isomerizing a peptidyl proline bond in STAT5. It noteworthy that in the absence of PRL, CypB does not bind STAT5 [32]. This suggests that PRL acts as a protein scaffold to enhance the PPI activity of CypB for specific transcription factors; however, whether this scaffold functions by providing a highly structured template or plays a less defined structural role is unknown.

CypA – the PPI Catalyzed Activation Switch

Cyclophilin A (CypA) is a second related PPI that has been identified to play a key role in the receptor activation process. CypA is a cytoplasmic PPI that binds simultaneously to both JAK2 and the PRL-R ICD domain [29, 34]. This set of interactions links the kinase to the receptor, directing the PPI activity to a proline in the PPVP sequence. The catalyzed isomerization presumably serves as a conformational switch to position the JAK2 kinases and STAT5 molecules for productive cross-phosphorylation. Eliminating binding of CypA by modifying the principal proline in the so-called X-Box motif of hPRL-R inhibits signaling. By functional analogy, it is expected that GH activation would similarly utilize CypA, but this has not been verified experimentally.

STRUCTURAL BASIS FOR RECEPTOR HOMODIMERIZATION

Tertiary structure plays a role in how the GH and PRL hormones regulate receptor activation. The hormones in this family are long-chain four α-helix bundle proteins [10, 11]. A notable feature of their tertiary structure is that it contains no symmetry that might support equivalent binding environments for the receptors. How the two receptors bind to the asymmetric hormone was first revealed from the crystal structure of human growth hormone bound to extracellular domain (ECD) of its receptor (hGH-R) [35]. The structure showed that the two ECDs binding to sites 1 and 2, respectively, use essentially the same set of residues to bind to two sites on opposite faces of the hormone [35] (Figure 24.2). An identical model is seen in a prolactin hormone–receptor complex [36]. This binding is characterized by extraordinary local and global plasticity at the binding

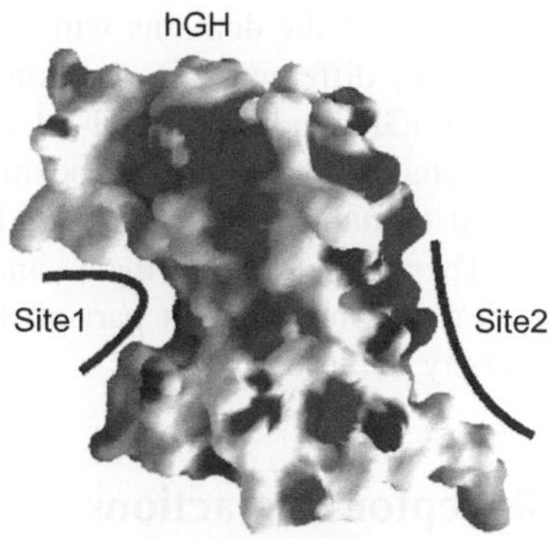

FIGURE 24.2 Molecular surface of hGH showing the different topographies of the site 1 and site 2 binding interfaces. In addition, the sites possess quite different electrostatic properties.

surfaces. The two binding sites have distinctly different topographies and electrostatic character, leading to different affinities for the receptor ECDs (Figure 24.2).

The high-affinity site, site 1, is always occupied first by ECD1 (designated as R1) [10, 11, 37]. This sequence of events is required because productive binding of ECD2 (R2) at site 2 of the hormone needs additional contacts to a patch of the C-terminal domain of R1. The binding of R2 is the programmed regulatory step for triggering biological action, and it involves a set of highly tuned interactions among binding interfaces in two spatially distinct binding sites. The energetic relationships between the R1–R2 contacts and the hormone–R2 site-2 interactions are known to be important (see below).

Hormone-Receptor Binding Sites

Binding site 1 of the hormones in the GH-PRL family is formed by residues that are exposed on helix 4 of the helix bundle, together with residues on the connecting loop between helices 1 and 2 [35, 36, 38, 39]. The total surface area buried on the hormone in the hGH–hGH-R1 and hGH–hPRL-R1 complexes is about 1200 Å^2. A similar picture has been observed for the hPRL–PRL-R1 interface [39]. The site-2 binding epitope involves residues in helices 1 and 3. In contrast to the concave surface of the hormones at site 1, binding site 2 of the hormone is a relatively flat surface and involves a somewhat smaller interface (~800 Å^2).

The receptor ECDs use essentially the same set of residues to bind to the two distinctly different site-1 and site-2 interfaces on the opposite faces of the hormones [35, 36]. The binding surfaces of the receptors are formed by six closely spaced surface loops (L1–L6) that extend from the β-sheet core in a manner somewhat similar to antigen binding loops in antibodies. Three loops reside in the N-terminal domain (L1–L3); two others in the C-terminal domain (L5–L6). Binding loop L4 serves as the five residue linker between the domains. The conformation of L4 plays a key role in orientation of the domains with respect to one another. For instance, differences in the conformation of L3 in complexes of hGH binding to hGH-R1 and hPRL-R1 create significant changes in the global positioning of the N- and C-terminal domains of the receptors bound at the site [35, 36, 38]. These differences in the N- and C- terminal domain orientation are an important part of the molecular recognition diversity in these systems.

Receptor–Receptor Interactions

A conserved structural element of the ligand-induced homodimerization of prolactin and growth hormone receptors is a set of extensive contacts between their C-terminal domains [16] (Figure 24.3). This receptor–receptor interface was described in detail for the hGH : hGH-R 1 : 2 ternary complex [35], and modeled for the hGH–hPRL-R ternary complex. Although the topology of the C-terminal domains of the rPRL-Rs is virtually identical to that of the C-terminal domain of the hGH-Rs, the receptor–receptor interfaces in these two complexes show a marked variation in their orientation and electrostatic character, and different portions of the receptors are involved in the interaction [36]. The surface area buried in the interaction between the rPRL-Rs is smaller than that buried between the hGH-Rs; the former being ~370 Å^2 compared to ~470 Å^2.

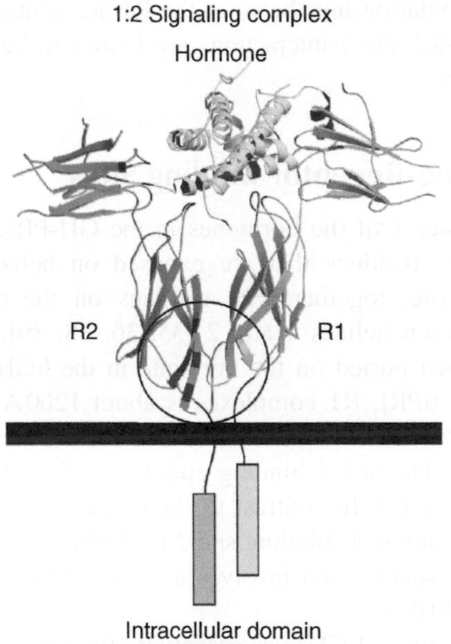

FIGURE 24.3 The GH/PRL receptors are three-domain single-pass transmembrane receptors containing an extracellular domain (ECD), a transmembrane section of about 25 amino acids, and a cytoplasmic domain that forms the binding site for the tyrosine kinase activities. The ECD consists of two fibronectin type III domains (FNIII) connected by a short linker. The hormones are four helix bundle proteins. Hormone binding consists of two steps. First the hormone binds to a high affinity site (site 1) to form a stable 1 : 1 intermediate. The hormone then binds to the second receptor through its site 2. This second binding step aligns the C-terminal domains of the receptors to form the important receptor–receptor interface.

Hormone-Receptor Binding Energetics

The high-affinity site-1 hGH–hGH-R1 interface is arguably the most highly studied protein–protein interaction using site-directed and saturation mutagenesis approaches [3, 5–7, 40]. Evaluation of the energetics of site 1 binding has led to a number of important insights into the relationships between binding and specificity determinants and protein–protein interactions in general. In particular, the seminal work of Wells and colleagues [3, 5] demonstrated that the hGH–hGH-R1 protein–protein interactions are characterized by binding "hotspots" that focus binding energies within a cluster of relatively few residues. Thirty residues of hGH make contact in the site-1 interface with hGH-R1. However, it was shown that about 85 percent of the binding energy was developed through only eight of the residues, and there were no apparent distinguishing characteristics between the interaction that are energetically important from those of the other side-chains that were energetically null [5, 40]. This very efficient use of the binding interface characterized by the concentration of the binding determinants within a relatively small area of the contact surface allows for separate adjacent areas to be used for specificity determinants without compromising the binding.

It has been determined that a similar type of hotspot also exists for the hGH–hPRL-R1 site 1 interaction and that the interface encompasses the same binding footprint as the hGH–hGH-R1 counterpart [38, 40]. The two systems differ, however, in how the energy is distributed within the footprint. For instance, residues like E174 and R167 make little energetic contribution to binding in the hGH–hGH-R1 case, but play extremely important roles in the hGH–hPRL-R1 contact surface (~ 800-fold difference in binding between the two systems when alanine is substituted at each of these sites). Interestingly, R167 makes a salt-bridge to an acidic side chain in both receptors (to E127 in hGH-R and D124 in hPRL-R), yet Ala-scan mutagenesis indicates that only in the case of hGH–hPRL-R does this salt-bridge have a positive effect on binding [40]. This is an example of how the energetics of interactions is very context-dependent in protein–protein interfaces [11].

The binding properties of R2 are intrinsically different from those for R1, because they involve the combined effects of two spatially distinct binding surfaces: binding site 2 on the hormone and a contact with the C-terminal domain of the bound R1 (Figure 24.3). Neither of these surfaces alone supports binding without the interaction of the other. In glaring contrast to the extensive characterization of the energetics of site-1 binding, there is considerably less biochemical and biophysical information for site 2 in GH–PRL systems. A study by Cunningham *et al.* [37]

showed that site 2 did not contain a well defined hotspot, as was the case for site 1. In contrast, recent data show that the receptor–receptor interface contains a hotspot in the hGH–hGH-R system, and probably plays a more important role than the hormone site 2-receptor contact in stabilizing tertiary forms of the complex [2, 41]. It was further determined that the site 2 and the receptor–receptor interactions act in an additive fashion.

Site1 and Site2 are Structurally and Functionally Coupled

It has been determined that there is a set of mutations in the site 1 of hGH that can trigger significant changes in site 2 by some form of allosteric coupling [8]. This coupling completely changes the stereochemical relationships in the hGH–hGHR site-2 interaction without affecting the overall binding efficiency of the complex.

The process whereby new binding surfaces are synthesized indirectly through molecule effects has been termed "functional cooperativity" [8]. In this mechanism it is not the mutation(s) in one site that alone affect the other site. A set of concerted changes is also involved among both the hormone and the domains of the receptor ECDs.

The finding of strong cross-molecular interaction induced during receptor dimerization establishes a new molecular recognition paradigm and opens up fundamental new areas of investigation relating to the mechanisms of biological regulation by protein–protein associations. However, it remains an open question as to how general this is, and whether evolution actually uses this strategy to influence the receptor signaling of GH/PRL systems in biologically important ways.

HORMONE SPECIFICITY AND CROSS-REACTIVITY DETERMINES PHYSIOLOGICAL ROLES

Binding to the two structurally distinct sites on the hormone, while using the same binding determinants, requires the receptor binding surfaces to undergo significant local conformational change [35, 36]. The structural requirement is further expanded by specificity factors [38]. The biology of PRL and GH is integrated on many levels [42]. However, over the 400 million years since they diverged from a common gene parent, evolution has built in different regulating components distinguishing them [43, 44]. In primates, the growth hormone receptor (GH-R) is activated solely by homodimerization through its cognate hormone. However, prolactin biology works through regulated cross-reactivity; most PRL-R receptors are programmed to bind both prolactin (PRL) and growth hormone (GH) [45].

This pattern of specificity and cross-reactivity involves some rather significant molecular recognition challenges since GHs and PRLs have little (~25 percent) sequence conservation even among the residues involved in receptor binding [36, 38]. The structure of hGH bound to the prolactin receptor (hPRL-R) showed that in these systems local conformational flexibility of the receptor binding loops, together with rigid-body movements of the receptor domains, facilitates the creation of specific, but different, interactions with the same binding site. The effects of conformational change on altering specificity were also observed in protein engineering studies that "converted" the binding site 1 of two PRL-R specific hormones, hPRL and hPL, into hGH [4, 46, 47]. This could be accomplished by substituting the hGH sequence at only five to six places in their sequence. Surprisingly, several of these positions map outside the site-1 hormone–receptor interface. Presumably, they must act as indirect specificity determinants by inducing conformational changes that subtly reorganize the contact residues into productive binding interactions. The implications of this finding are considerable, and may open up totally new ways to look at how specificity and cross-reactivity is developed in cytokine systems.

CONCLUDING REMARKS

Although the various structures of hormone–receptor complexes provide information that encompasses both the versatility and specificity components inherent in the recognition system that regulates endocrine biology, a general understanding of this process at the molecular level remains challenging to construct, even combining it with the extensive mutational database available to us. One is struck by the extraordinary adaptability of these molecules to synthesize competent binding epitopes to a wide range of large target surfaces. It appears that the binding sites for cytokine receptors using the FNIII scaffold can adapt to binding cytokines via the hormone's 1–4 helix interface (long-chain cytokines like GH and PRL) or, in the case of the short-chain cytokine motif like IL-4, through the 1–3 interface. These interactions can have quite different affinities, and might involve single or multiple binding hotspots.

In the case of GH/PRL, the nature of the adjustments required to form the optimum set of interactions between the hormones of each of its two receptors suggests that recognition and binding of the two protein surfaces is directed by an induced-fit mechanism. Distinct from the process of molecular recognition associated with the antibody–antigen paradigm, where binding is developed mainly through sequence diversity of the antibody complementarity–determining loops, the cytokine receptors can use essentially a constant set of residues to bind surfaces that are diverse both in sequence and in conformation. This is accomplished by employing conformational diversity, both local and global, and is the unifying hallmark of these systems.

REFERENCES

1. DeVlaming V. Actions of prolactin among the vertebrates. In: Barrington EJW, editor. *Hormones and evolution*. New York: Academic Press; 1979. p. 561–642.
2. Bernat B, Pal G, Sun M, Kossiakoff AA. Determination of the energetics governing the regulatory step in growth hormone-induced receptor homodimerization. *Proc Natl Acad Sci USA* 2003;**100**(3):952–7.
3. Clackson T, Wells JA. A hot spot of binding energy in a hormone–receptor interface. *Science* 1995;**267**(5196):383–6.
4. Cunningham BC, Henner DJ, Wells JA. Engineering human prolactin to bind to the human growth hormone receptor. *Science* 1990;**247**(4949 Pt 1):1461–5.
5. Cunningham BC, Wells JA. High-resolution epitope mapping of hGH-receptor interactions by alanine-scanning mutagenesis. *Science* 1989;**244**:1081–5.
6. Kouadio JL, Horn JR, Pal G, Kossiakoff AA. Shotgun alanine scanning shows that growth hormone can bind productively to its receptor through a drastically minimized interface. *J Biol Chem* 2005;**280**(27):25,524–32.
7. Pal G, Kouadio JL, Artis DR, Kossiakoff AA, Sidhu SS. Comprehensive and quantitative mapping of energy landscapes for protein–protein interactions by rapid combinatorial scanning. *J Biol Chem* 2006.
8. Walsh ST, Sylvester JE, Kossiakoff AA. The high- and low-affinity receptor binding sites of growth hormone are allosterically coupled. *Proc Natl Acad Sci USA* 2004;**101**(49):17,078–83.
9. Bazan JF. Structural design and molecular evolution of a cytokine receptor superfamily. *Proc Natl Acad Sci USA* 1990;**87**:6934–8.
10. Kossiakoff AA, De Vos AM. Structural basis for cytokine hormone-receptor recognition and receptor activation. *Adv Protein Chem* 1998;**52**:67–108.
11. Kossiakoff AA, Somers W, Ultsch M, Andow K, Muller Y, De Vos AM. Comparison of the intermediate complexes of human growth hormone bound to the human growth hormone and prolactin receptors. *Protein Sci* 1994;**3**:1697–705.
12. Sprang SR, Bazan JF. Cytokine structural taxonomy and mechanisms of receptor engagement. *Curr Opin Struct Biol* 1993;**3**:815–27.
13. Wells JA, de Vos AM. Hematopoietic receptor complexes. *Annu Rev Biochem* 1996;**65**:609–34.
14. Ihle JN, Witthuhn BA, Quelle FW, Yamamoto K, Thierfelder WE, Kreider B, Silvennoinen O. Signaling by the cytokine receptor superfamily: JAKs and STATs. *Trends Biochem Sci* 1994;**19**:222–7.
15. Cosman D, Lyman SD, Idzerda RL, Reckmann MP, Park LS, Goodwin RG, March CJ. A new cytokine receptor superfamily. *Trends Biochem Sci* 1990;**15**:265–70.
16. Kossiakoff AA. The structural basis for biological signaling, regulation, and specificity in the growth hormone–prolactin system of hormones and receptors. *Adv Protein Chem* 2004;**68**:147–69.
17. Brown RJ, Adams JJ, Pelekanos RA, Wan Y, McKinstry WJ, Palethorpe K, Seeber RM, Monks TA, Eidne KA, Parker MW, Waters MJ. Model for growth hormone receptor activation based on subunit rotation within a receptor dimer. *Nature Struct Mol Biol* 2005;**12**(9):814–21.
18. Gadd SL, Clevenger CV. Ligand-independent dimerization of the human prolactin receptor isoforms: functional implications. *Mol Endocrinol* 2006;**20**(11):2734–6.
19. Waters MJ, Hoang HN, Fairlie DP, Pelekanos RA, Brown RJ. New insights into growth hormone action. *J Mol Endocrinol* 2006;**36**(1):1–7.
20. Constantinescu SN, Keren T, Socolovsky M, Nam H, Henis YI, Lodish HF. Ligand-independent oligomerization of cell-surface erythropoietin receptor is mediated by the transmembrane domain. *Proc Natl Acad Sci USA* 2001;**98**(8):4379–84.
21. Constantinescu SN, et al. The erythropoietin receptor cytosolic juxtamembrane domain contains an essential, precisely oriented, hydrophobic motif. *Mol Cell* 2001;**7**(2):377–85.
22. Remy I, Wilson IA, Michnick SW. Erythropoietin receptor activation by a ligand-induced conformation change. *Science* 1999; **283**(5404):990–3.
23. Livnah O, Stura EA, Middleton SA, Johnson DL, Jolliffe LK, Wilson IA. Crystallographic evidence for preformed dimers of erythropoietin receptor before ligand activation. *Science* 1999;**283**(5404):987–90.
24. Chan FK, Chun HJ, Zheng L, Siegel RM, Bui KL, Lenardo MJ. A domain in TNF receptors that mediates ligand-independent receptor assembly and signaling. *Science* 2000;**288**(5475):2351–4.
25. Kim SH, Wang W, Kim KK. Dynamic and clustering model of bacterial chemotaxis receptors: structural basis for signaling and high sensitivity. *Proc Natl Acad Sci USA* 2002;**99**(18):11,611–15.
26. Papoff G, Hausler P, Eramo A, Pagano MG, Di Leve G, Signore A, Ruberti G. Identification and characterization of a ligand-independent oligomerization domain in the extracellular region of the CD95 death receptor. *J Biol Chem* 1999;**274**(53):38,241–50.
27. Siegel RM, Frederiksen JK, Zacharias DA, Chan FK, Johnson M, Lynch D, Tsien RY, Lenardo MJ. Fas preassociation required for apoptosis signaling and dominant inhibition by pathogenic mutations. *Science* 2000;**288**(5475):2354–7.
28. Clevenger CV. Nuclear localization and function of polypeptide ligands and their receptors: a new paradigm for hormone specificity within the mammary gland? *Breast Cancer Res* 2003;**5**(4):181–7.
29. Clevenger CV. Roles and regulation of stat family transcription factors in human breast cancer. *Am J Pathol* 2004;**165**(5):1449–60.
30. Rycyzyn MA, Clevenger CV. Role of cyclophilins in somatolactogenic action. *Ann NY Acad Sci* 2000;**917**:514–21.
31. Rycyzyn MA, Clevenger CV. The intranuclear prolactin/cyclophilin B complex as a transcriptional inducer. *Proc Natl Acad Sci USA* 2002;**99**(10):6790–5.
32. Rycyzyn MA, Reilly SC, O'Malley K, Clevenger CV. Role of cyclophilin B in prolactin signal transduction and nuclear retrotranslocation. *Mol Endocrinol* 2000;**14**(8):1175–86.
33. Syed F, Rycyzyn MA, Westgate L, Clevenger CV. A novel and functional interaction between cyclophilin A and prolactin receptor. *Endocrine* 2003;**20**(1–2):83–90.
34. Miller SL, DeMaria JE, Freier DO, Riegel AM, Clevenger CV. Novel association of Vav2 and Nek3 modulates signaling through the human prolactin receptor. *Mol Endocrinol* 2005;**19**(4):939–49.
35. De Vos AM, Ultsch M, Kossiakoff AA. Human growth hormone and extracellular domain of its receptor: crystal structure of the complex. *Science* 1992;**255**:306–12.
36. Elkins PA, Christinger HW, Sandowski Y, Sakal E, Gertler A, de Vos AM, Kossiakoff AA. Ternary complex between placental lactogen and the extracellular domain of the prolactin receptor. *Nature Struct Biol* 2000;**7**(9):808–15.
37. Cunningham BC, Ultsch M, De Vos AM, Mulkerrin MG, Clauser KR, Wells JA. Dimerization of the extracellular domain of the human growth hormone receptor by a single hormone molecule. *Science* 1991;**254**:821–5.
38. Somers W, Ultsch M, De Vos AM, Kossiakoff AA. The X-ray structure of the growth hormone–prolactin receptor complex. *Nature* 1994;**372**:478–81.

39. Svensson LA, Bondensgaard K, Norskov-Lauritsen L, Christensen L, Becker P, Andersen MD, Maltesen MJ, Rand KD, Breinholt J. Crystal structure of a prolactin receptor antagonist bound to the extracellular domain of the prolactin receptor. *J Biol Chem* 2008;**283**(27):19,085–94.
40. Cunningham BC, Wells JA. Comparison of a structural and a functional epitope. *J Mol Biol* 1993;**234**:554–63.
41. Walsh ST, Jevitts LM, Sylvester JE, Kossiakoff AA. Site-2 binding energetics of the regulatory step of growth hormone-induced receptor homodimerization. *Protein Sci* 2003;**12**(9):1960–70.
42. Goffin V, Shiverick KT, Kelly PA, Martial JA. Sequence–function relationships within the expanding family of prolactin, growth hormone, placental lactogen, and related proteins in mammals. *Endocrine Rev* 1996;**17**:385–410.
43. Gertler A, Grosclaude J, Strasburger CJ, Nir S, Djiane J. Real-time kinetic measurements of the interactions between lactogenic hormones and prolactin-receptor extracellular domains from several species support the model of hormone-induced transient receptor dimerization. *J Biol Chem* 1996;**271**(40):24,482–91.
44. Nicoll CS, Mayer GL, Russel SM. Structural features of prolactins and growth hormones that can be related to their biological properties. *Endocrine Rev* 1986;**7**:169–203.
45. Kelly PA, Djiane J, Banville D, Ali S, Edery M, Rozakis M. The growth hormone/prolactin receptor gene family. *Oxford Surv Eukaryotic Genes* 1991;**7**:29–50.
46. Cunningham BC, Wells JA. Rational design of receptor-specific variants of human growth hormone. *Proc Natl Acad Sci USA* 1991;**88**:3407–11.
47. Lowman HB, Cunningham BC, Wells JA. Mutational analysis and protein engineering of receptor-binding determinants in human placental lactogen. *J Biol Chem* 1991;**266**:10,982–8.

Chapter 25

Erythropoietin Receptor as a Paradigm for Cytokine Signaling

Deborah J. Stauber, Minmin Yu, and Ian A. Wilson
Department of Molecular Biology, and The Skaggs Institute for Chemical Biology, The Scripps Research Institute, La Jolla, California

INTRODUCTION

The mechanism of how cytokines and growth factors elicit a signaling response via cell surface receptors remains a major question within the field of cell signaling. It was previously thought that dimerization of the extracellular domains was sufficient to elicit activation of a signaling pathway, and, indeed, a variety of studies have shown that a number of different strategies of dimerization can cause signal activation. However, studies on the erythropoietin receptor (EPOR) have suggested that dimerization alone is not sufficient. Rather, subtle differences in ligand binding to receptor extracellular domains, as shown by various structural and biochemical data, can result in structural deviations that can modulate the signal response.

Signaling of erythropoietin (EPO) through the EPOR promotes the proliferation and differentiation of erythroid progenitor cells and is thus crucial to normal red blood cell development [1, 2]. The EPOR is a member of the cytokine receptor superfamily [3], which includes receptors for other long-chain cytokines, such as growth hormone (GH), thrombopoeitin (TPO), and granulocyte cell signaling factor (GCSF), as well as short-chain cytokines, such as interleukins (ILs) 2, 3, and 4. These receptors are single-transmembrane (TM)-spanning proteins that bind their corresponding ligands in their extracellular (EC) domains. A common motif among these receptors is the cytokine homology domain (CHD), which consists of two seven-stranded β-sandwich motifs connected by a proline linker. Signature characteristics within the CHD include inter-strand disulfide bonds within the N-terminal domain and a WSXWS-conserved motif located in the C-terminal domain. This WSXWS sequence seems to be essential for productive ligand binding and the resulting activation of the EPOR [4, 5], although it is not clear from the structural data how the WSXWS motif carries out any binding or signaling role. The cytokine receptors are coupled to members of the Janus kinase (JAK) family (non-receptor tyrosine kinases) at a proline-rich sequence in their cytosolic (CT) domain called Box 1. Agonist ligand binding in the EC domain of the receptor leads to a conformational change and reorganization that is permissive for autophosphorylation and activation of the associated JAK, resulting in phosphorylation of the CT domains of the receptors. These phosphorylation events trigger a signaling cascade via the signal transducers and activators of transcription (STATs) that ultimately leads to protein expression and cell proliferation [6]. In the case of EPOR, the associated JAK2 phophorylates many of the eight CT tyrosines, which serve as docking sites for STAT5. STAT5, after being activated by JAK2, travels to the nucleus, where it promotes genes that lead to the proliferation and survival of erythroid progenitor cells.

Erythropoietin, similar to GH [7, 8], binds to its receptor in a stoichiometry of 1:2. As EPO itself is not a symmetric molecule, two different binding interfaces exist on the cytokine surface which are each capable of interacting with the EPOR. The EPO interaction sites have been named site 1 and site 2, for which the binding affinities are 1 nM and 1 μM, respectively [9].

STRUCTURAL STUDIES ON EPOR

Studies on the binding of different ligand molecules to the EPOR have shown that dimerization of the extracellular domains itself is not sufficient for a biological response [10]. Furthermore, structural and biochemical data surprisingly revealed that the EPO receptor exists as a preformed dimer on the cell surface, and that a conformational reorganization of the receptor as a result of ligand binding is necessary

Handbook of Cell Signaling, Three-Volume Set 2 ed.

to elicit the signal transduction cascade [11, 12]. The biological appeal of a preformed dimerized receptor can be understood when one considers the low cell-surface density of the EPOR (< 1000 receptors, or 1 μM) *in vivo* [9, 12–14]. Clustering of the receptors allows EPO to act efficiently at the cell surface by binding to the high-affinity (nanomolar) site 1 immediately followed by interaction with site 2, despite its low micromolar binding affinity. In the absence of this clustering, monomeric receptor–EPO interactions would be prevalent. As a result, the efficiency of the biological response to EPO would be compromised.

The extensive amount of crystal structure data available on a variety of agonist and antagonist EPOR complexes has enabled a comparison and analysis of different activation states of the EPOR. Four EPOR structures, determined by X-ray crystallography (depicted in Figure 25.1), allow for the exploration of the dimerization interfaces involved in various signaling states of the EPOR [10, 15–17]. These structures allow us to examine how differences in the nature of the bound ligand lead to changes in the receptor structure and assembly on complex formation.

EMP1, an Agonist, Bound to EPO Receptor

The crystal structure of an EPO mimetic peptide, EMP1, was determined at 2.8-Å resolution in complex with the extracellular EPO-binding protein domain (EBP) of EPOR [15]. EMP1 is a member of a family of peptide mimetics selected by the phage display method that bind specifically and with high affinity (100–200 nM) to the EPOR and have agonist activity [18]. Co-crystals of the EMP1–EBP complex showed that EMP1 binds to two EBP molecules as a peptide dimer, resulting in a 2:2 ligand-to-receptor stoichiometry and generating an almost twofold symmetric assembly of the EMP1–EBP complex [15]. In this arrangement (Figure 25.1a), the monomers of the EBP dimer are rotated approximately 1808 from each other, and the two domains of each receptor molecule, D1 and D2, are oriented with an elbow angle of approximately 90°. The EPOR dimerization interface in the complex consists almost entirely of interactions with residues from the peptide EMP1 dimer with minimal contact occurring between the receptors themselves.

EMP33, an Antagonist, Bound to EPO Receptor

Studies on the EMP family of mimetics showed the importance of peptide Tyr4 in the peptide-mediated receptor dimerization [19]. A substitution of 3,5-dibromotryosine yielded a biologically inactive peptide, EMP33. Surprisingly, this antagonist was capable of dimerizing the EPO receptor. The crystal structure of EMP33 at 2.7-Å resolution bound to the EBP confirmed that dimerization is indeed not sufficient for signaling [10]. The structure shows an asymmetrically dimerized receptor (Figure 25.1b) in which the D1 domain of each monomer is related by a 165° rotation (in contrast to the two-fold (180°) of the EMP1–EBP dimer). Thus, only 15° of rotation distinguishes an active dimer from that of a complex that is incapable of JAK2 activation.

EPO Bound to Its Receptor

Crystal structures of a mutant [10, 20] of the natural highly potent cytokine EPO with EBP were determined at 1.9- and 2.8-Å resolutions [16]. This EPO–EBP complex binds in a 1:2 stoichiometry, similar to that of the GH–growth hormone receptor (GHR) complexes [8], using two different binding surfaces of the EPO molecule to mediate interactions with each of the EBP monomers (Figure 25.1c). These two surfaces, denoted site 1 and site 2, bind the receptor with high (K_d=1 nM) and low (K_d=1 μM) binding affinities, respectively [9]. The D1 domain of each monomer is rotated approximately 120° relative to the other. In this complex, the orientations of the D2 domains are approximately in the same plane perpendicular to the membrane, in contrast to the 45° angle of the EMP1–EBP complex. This allows for the C termini of the receptor in the EPO–EBP structure to be in close proximity. Furthermore, this complex confirms the large amount of flexibility associated with receptor assembly, which may explain how a covalently-linked EPO dimer is still capable of binding and activating the EPO receptor, presumably in a 2:2 complex [21, 22].

The Unliganded EPO Receptor

An unexpected result surfaced when the crystal structure [17] of the unliganded EPO receptor was determined at 2.4 Å. The asymmetric unit contains an EPOR dimer, such that the dimerization interface is located near the ligand binding site (Figure 25.1d). In this unliganded form, the D2 domains are oriented such that their C termini are rotated 135° away from each other [17] and at an angle to the plane of the membrane, placing their expected membrane insertion points approximately 73 Å apart. The equivalent distances of the D2 C termini of the EMP1–EBP and the EPO–EBP dimers are 39 Å and 34 Å, respectively. These values are consistent with the notion that the inter-dimer distance of the unliganded EPO receptor would be too far for the intracellularly associated JAK kinases to interact, thus causing the unbound receptor to be inactive and locked into an off state.

Hotspot in EPOR for Ligand Binding

In each of the previous structures examined (Figures 25.1a–d), the same binding surface of the EPO receptor is used to interact with the multiple ligands (Figure 25.2). Variation of this binding surface is introduced by only small conformational

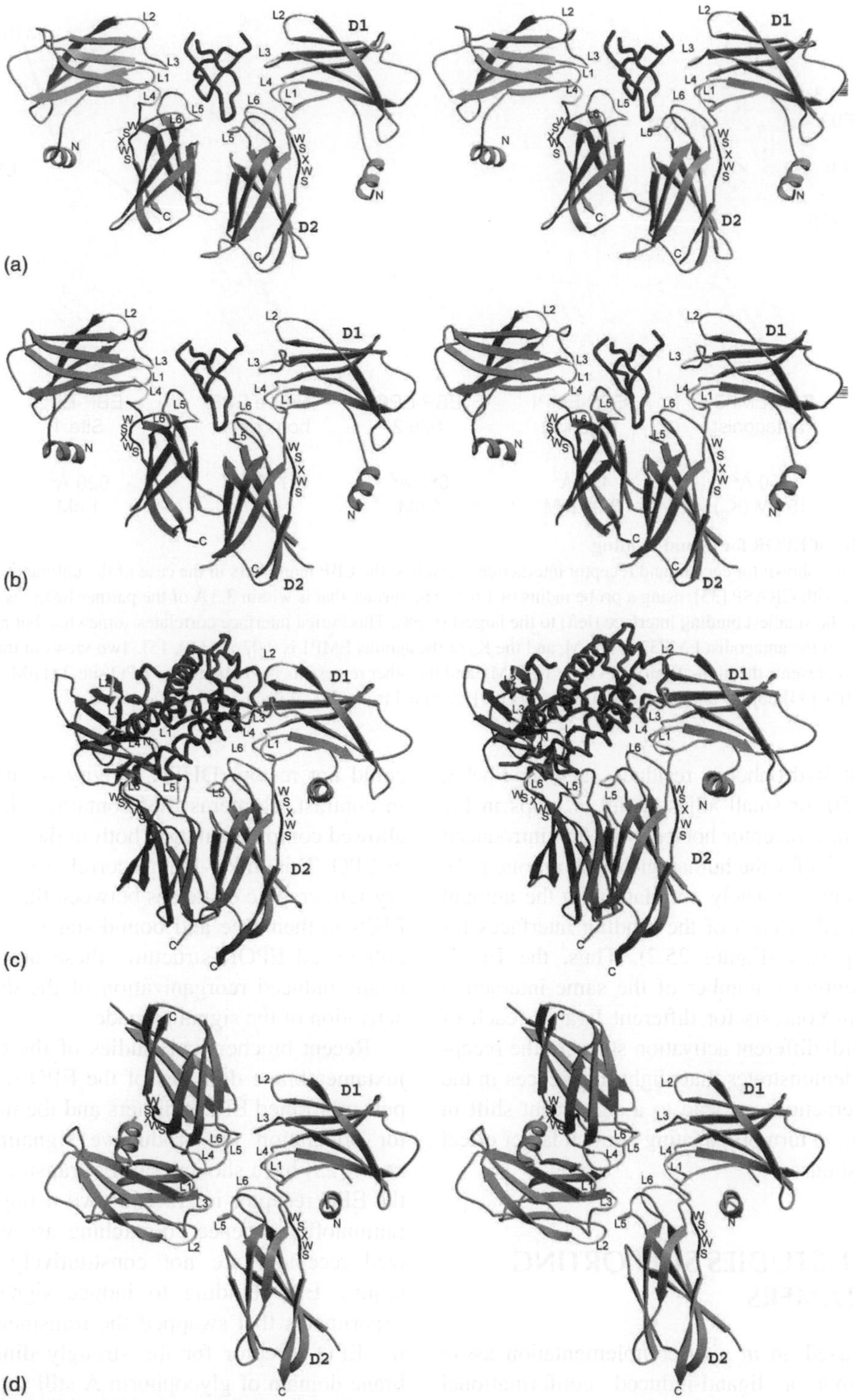

FIGURE 25.1 Stereo views of the various crystal structures of the EPOR.
In each complex, the receptors are presented as ribbons and the ligands are depicted as tubes. The ligand binding loop tips of the receptor are yellow and labeled L1–L6; the conserved WSXWS signature sequence of the cytokine receptor family is black (color shown in online version). The orientation shown for each structure corresponds to superimposing the β-strands of each receptor domain 1 (D1) onto the equivalent D1 domain of the EMP1–EBP complex. (a) Weak agonist EMP1–EBP complex (EBP, cyan; EMP1, red; PDB code 1ebp) [15]. (b) Antagonist EMP33–EBP complex (EBP, salmon red; EMP33, red; PDB code 1eba) [10]. (c) Strong agonist EPO–EBP complex (receptor site 1, green; receptor site 2, purple; EPO, red; PDB code 1cn4) [16]. (d) Self-dimer EBP–EBP native complex (EBP, orange [33]; PDB code 1ern) [17]. Molecules were made using MOLSCRIPT [34] and rendered in Raster3D [33]. Adapted from [23] (Wilson and Jolliffe, 1999).

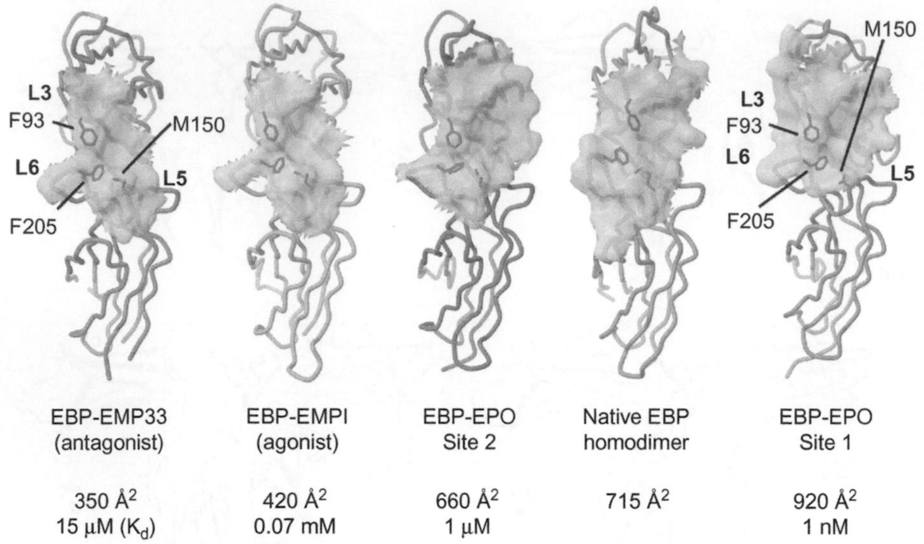

FIGURE 25.2 Plasticity of EPOR for ligand binding.
EPOR binding surfaces are shown for each ligand receptor interaction, as well as the EBP monomers in the case of the unliganded structure. Molecular surfaces were calculated with GRASP [35], using a probe radius of 1.6 Å. The surface that is within 3.3 Å of the partner ligand is shown. The representation is shown from the smallest binding interface (left) to the largest (right). This buried interface correlates somewhat, but not exactly, with the approximate K_d. The K_d of the antagonist EMP33 is 15 μM, and the K_d of the agonist EMP1 is 0.07 μM [10, 15]. Two views of the EPO–EBP structure are shown; one surface represents the high-affinity EPO site 1 (1 nM), and the other represents the low-affinity EPO site 2 (1 μM) [9]. Surfaces were converted to MOLSCRIPT [34] objects and rendered in Raster3D [33]. Adapted from [23] (Wilson and Jolliffe, 1999).

changes in hotspot hydrophobic residues such as Phe93, Phe205, and Met150, or small adjustments of loops in D1 [23]. This concept of a receptor hotspot was first introduced by Clackson and Wells for the human growth hormone [24]. Differences in K_d approximately correlate with the amount of surface area buried in each of the binding interfaces for the liganded complexes (Figure 25.2). Thus, the EPOR is capable of presenting a number of the same interaction surfaces in different contexts for different ligands, each of which correlates with different activation states of the receptor. This example demonstrates that slight differences in the contact surface interaction can lead to a significant shift in receptor orientation, in turn propagating a much larger effect on the signaling response.

BIOCHEMICAL STUDIES SUPPORTING PREFORMED DIMERS

Remy *et al.* [11] used an *in vivo* complementation assay to demonstrate that a ligand-induced conformational change of the EPOR dimer is required for activation. In these studies, chimeras that contained the extracellular and transmembrane domains of EPOR were fused to two complementary fragments of murine dihydrofolate reductase (DHFR) through flexible linkers of different lengths. Cells transfected with these chimeras express the receptors at the cell surface. In this experiment, DHFR activity is restored if the two complementary fragments are brought into close proximity. Chimeras containing a short 5-residue linker could not restore DHFR activity in the absence of EPO. In contrast, chimeras that contained the 30-residue linker allowed complementation both in the presence and absence of EPO. This linker-length correlation agrees well with the crystallographic distances between the C termini of the two EBPs in their free and bound states. Thus, taken with the unliganded EPOR structure, these observations suggest a ligand-induced reorganization of the dimer that results in activation of the signal cascade.

Recent biochemical studies of the transmembrane and juxtamembrane domains of the EPO receptor further support preformed EPOR dimers and the importance of receptor orientation for productive signaling. Constantinescu *et al.* [25] have shown that the transmembrane domains of the EPO receptor interact in vivo using antibody-mediated immunofluorescence co-patching assays. These oligomerized receptors are not constitutively active, but rather require EPO binding to induce signaling. Furthermore, experiments that swapped the transmembrane domains of the EPO receptor for the strongly dimerizing transmembrane domain of glycophorin A still showed a dependence on EPO for JAK2 activation.

Further studies demonstrate the importance of the relative orientation of a region in the cytosolic juxtamembrane domains of the EPOR in signaling [26]. Alanine scanning has indicated that three residues, Leu253, Ile257, and Trp258 (LIW), in this region were necessary for EPO-induced phosphorylation, but not for binding of JAK2. Accordingly, these residues are likely to be involved in a switch mechanism propagated from EPO binding to the EC that positions JAK2

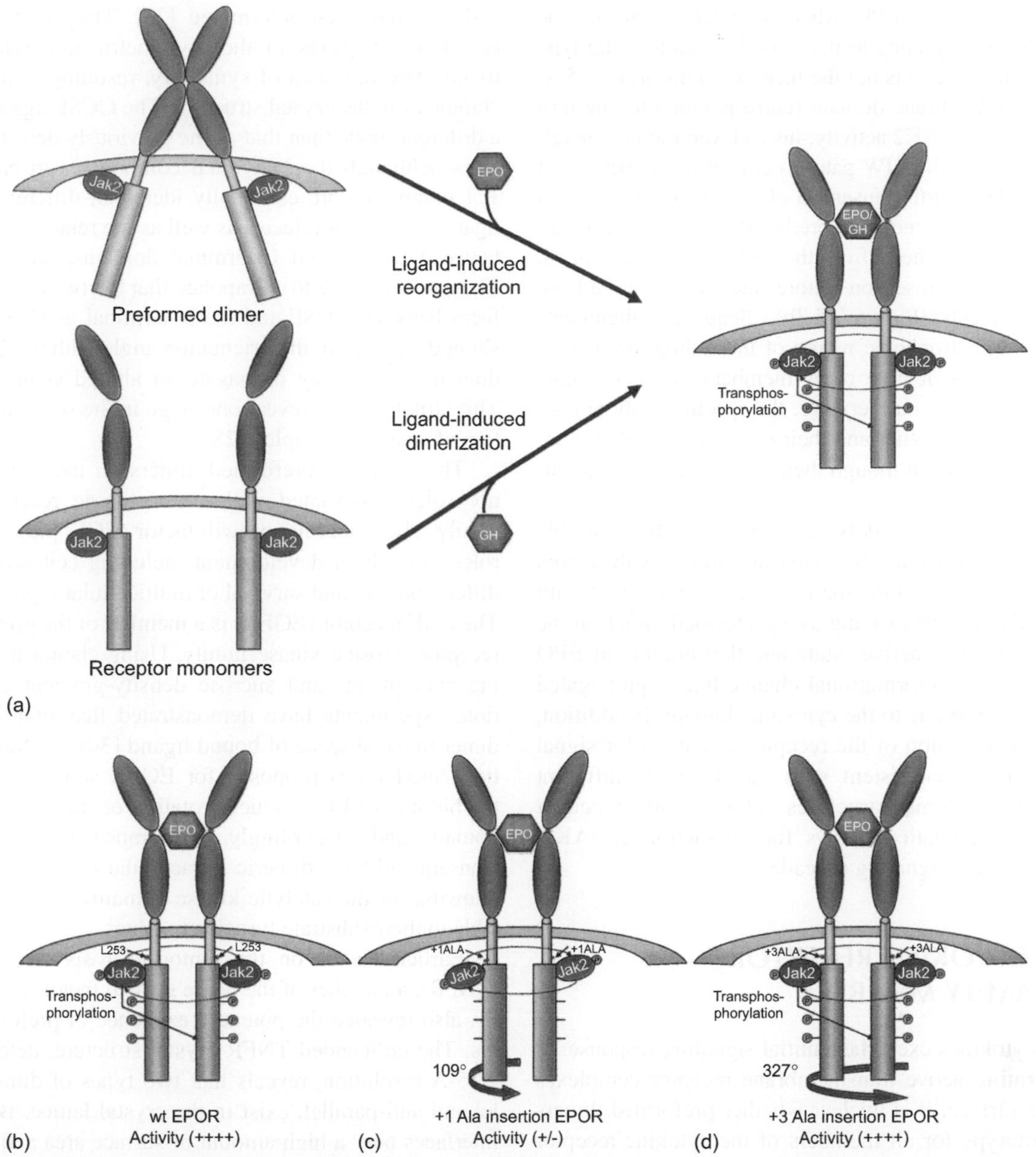

FIGURE 25.3 A model of EPO-induced dimerization and activation of the EPO receptor, JAK2 transphosphorylation, and tyrosine phosphorylation of the EPO cytosolic (CT) domain.
(a) The schematic depicts contrasting views of ligand-induced signal activation. The top panel depicts a preformed dimer whereby the binding of cytokine (e.g., EPO) induces a structural reorganization, leading to active signaling. The bottom panel (b, c, d) depicts receptor monomers on the cell surface in which the binding of cytokine (e.g., GH) leads to dimerization, resulting in active signaling. EPO receptors are dimerized by EPO. The transmembrane (TM) domain is shown as a yellow cylinder (color shown in online version) and continues as a rigid helix into the CT domain through residue W258. L253, I257, and W258 comprise a hydrophobic patch expected to be at one face of this α-helix. L253 is shown in this helix. Upon activation (b), JAK2 transphosphorylates its partner JAK2 molecule, which in turn phosphorylates the partner EPOR CT domain tyrosine residues. Phosphorylation sites are marked as red circles. Insertion of one alanine residue prior to L253 (c) rotates the putative transmembrane α-helix by 1098, causing a change in the "register" of the hydrophobic patch. In this case, EPO is capable of inducing JAK2 phosphorylation; however, tyrosine phosphorylation of the EPO CT domains does not occur. Insertion of three alanine residues prior to L253 (d) restores wild-type activity to the EPOR. A three-alanine insertion restores the "register" of the hydrophobic patch, as the helix is rotated by 3278 [26].

correctly for appropriate activation (Figure 25.3a). These three residues are likely to occur on the same face of the protein surface, as the secondary structure analysis predicts an α-helix continuing from the TM through this region. Experiments in which additional alanine residues were inserted into this juxtamembrane region were performed in order to assess the affect of changing the relative "register" of these regions and, hence the effect transmitted to the intermolecular domains (Figs. 3b–d). Each alanine insertion rotates the register of the predicted α-helix by 1098. A single insertion greatly diminishes signaling by EPO, whereas a three-residue alanine insertion restores signaling to the same level as the wild-type protein. Furthermore, it is not the increase in distance (~5 Å) from the transmembrane domain (corresponding to one turn of helix) that alters JAK2 activity; instead, variation in the relative orientation of the LIW patch is crucial for proper signal response to EPO. Further insertion of two alanine residues in the TM region confirmed the predicted secondary structure of a continuous rigid helix from the TM through the Trp258, as a two amino-acid insertion restores the register of the LIW hydrophobic patch (Figure 25.3b). Sequence alignments reveal that the hydrophobic nature of these three residues is conserved throughout nine other members of the cytokine receptor superfamily. Furthermore, the spacing of these residues relative to each other and their distance from Box 1 are strictly conserved, even though their distance from the membrane varies [26].

Two alternative models can explain the possible mechanisms of activation by cytokines and growth factors (Figure 25.3a). For EPOR, the evidence is consistent with the unliganded EPOR existing as a preformed dimer on the cell surface in an "inactive" state and that binding of EPO in vivo causes a conformational change that is propagated through the membrane to the cytosolic domain. In addition, the relative orientation of the receptor is critical for signal generation, again consistent with the distinctly different agonist and antagonist structures. This ligand-dependent structural reorganization allows for interaction of JAK2, thereby eliciting a signaling cascade.

OTHER CYTOKINE RECEPTOR SUPERFAMILY MEMBERS

While all cytokines exert their initial signaling response by way of forming active transmembrane receptor complexes on the cell surface, it is unclear whether preformed dimers are the prototype for all members of the cytokine receptor superfamily. The crystal structures of GH–GHR [8] and the ovine placental lactogen (oPL) bound to the extracellular domain of the prolactin receptor (PRLR) [27] reveal a 1:2 stoichiometry similar to that of the EPO–EBP structure. To date, no structures are available for the extracellular domains of these receptors in their unliganded form. Studies of GH mutants infer that GH binds to a monomeric GHR and then forms a stable complex with a second GHR [7] (Figure 25.3a); however, the possibility that GH binds to a preformed GHR dimer and alters its conformation is also not inconsistent with these data. The nonsymmetrical nature of the PRLR receptors in the dimerized complex reveals a large amount of flexibility between the N- and C-terminal receptor domains, suggesting a possibility of domain orientation change in response to binding.

The crystal structure of GCSF bound to the cytokine receptor homologous region of the GCSF receptor, denoted gs-CRH, has been determined [28]. This structure shows two 1:1 complexes in the asymmetric unit related by a pseudo two-fold axis of symmetry, resulting in a 2:2 stoichiometry in the crystal structure. The GCSF ligand binds in a different mode than that of the previously described structures. Although the individual components in each of the 1:1 complexes are essentially identical, differences in the ligand receptor interfaces, as well as the relative orientations between the N- and C-terminal domains, are substantial. Thus, it is possible to extrapolate that the two different interfaces between GCSF and the N-terminal gs-CRH induce a slight difference in the orientation angle with the C-terminal domain, which may propagate an altered signal response. This structure may reveal one stage in the sequential formation of an active complex [28].

The notion of preformed dimers on the cell surface is not solely associated with the cytokine receptor superfamily. The epidermal growth factor (EGF) plays important roles throughout development including cell proliferation, differentiation, and survival of multicellular organisms [29]. The EGF receptor (EGFR) is a member of the growth factor receptor tyrosine kinase family. Using chemical crosslinking experiments and sucrose density-gradient centrifugation, experiments have demonstrated that EGFR forms a dimer in the absence of bound ligand [30]. A "flexible rotation" model was proposed for EGFR activation in which the binding of EGF induces rotation of the juxtamembrane domain and, accordingly, the transmembrane domain. Consequently, the dimeric intracellular domains dissociate, allowing for the catalytic kinase domains to become accessible to their substrate tyrosine residues.

Earlier studies on the tumor necrosis factor receptor (TNFR), a member of the nerve growth factor receptor family, also revealed the potential existence of preformed dimers. The unliganded TNFR crystal structure, determined at 2.25-Å resolution, reveals that two types of dimers, parallel and anti-parallel, exist in the crystal lattice. Both dimer interfaces bury a high amount of surface area and are stabilized by large numbers of van der Waals and ionic interactions and, therefore, are thought to be biologically relevant [31]. These studies propose a model in which the TNFR, in its unliganded form, exists as the anti-parallel dimer with the cytoplasmic domains of each receptor in the pair separated by over 100 Å. Although the ligand-bound TNFR structure shows a 3:3 stoichiometry [32], in light of the EPOR studies discussed here, the unliganded TNFR dimer may indeed be biologically relevant.

CONCLUSIONS

Members of the cytokine receptor superfamily undergo major conformational reorganizations in order to respond to ligand binding and elicit appropriate signal activation. In each member of the family, strict regulation is important for specific response to the signaling molecule, and aberrant signaling can have considerable medical consequences. Studies on EPOR have led to a greater appreciation of the subtleties of receptor activation. The variety of ligands capable of binding to a single interface of the receptor is a consequence of slight variations of loops and side chains on the receptor surface. Furthermore, each of these bound ligands is capable of eliciting a different signal response by propagating slight changes in the orientations of the cytoplasmic domains of the receptors through the membrane. The knowledge gained from the structural data permits the further design of peptide mimetics and small molecules. These future advances not only may help to further our understanding of the intricacies of the cytokine receptor signaling but also may have direct applications for clinical use.

ACKNOWLEDGEMENTS

The authors are supported by NIH grants to IAW and a Damon Runyon Fellowship DRG-1668 to DJS from the Damon Runyon Cancer Research Foundation. This is manuscript #15244-MB from The Scripps Research Institute.

REFERENCES

1. Krantz SB. Erythropoietin. *Blood* 1991;**77**:419–34.
2. Goldsmith M, Mikami A, You Y, Liu K, Thomas L, Pharr P, Longmore G. Absence of cytokine receptor-dependent specificity in red blood cell differentiation in vivo. *Proc Natl Acad Sci USA* 1998;**95**:7006–11.
3. Bazan JF. Structural design and molecular evolution of a cytokine receptor superfamily. *Proc Natl Acad Sci USA* 1990;**87**:6934–8.
4. Yoshimura A, Zimmers T, Neumann D, Longmore G, Yoshimura Y, Lodish HF. Mutations in the Trp-Ser-X-Trp-Ser motif of the erythropoietin receptor abolish processing, ligand binding, and activation of the receptor. *J Biol Chem* 1992;**267**:11,619–25.
5. Quelle DE, Quelle FW, Wojchowski DM. Mutations in the WSAWSE and cytosolic domains of the erythropoietin receptor affect signal transduction and ligand binding and internalization. *Mol Cell Biol* 1992;**12**:4553–61.
6. Darnell J, Kerr I, Stark G. Jak-STAT pathways and transcriptional activation in response to IFNs and other extracellular signaling proteins. *Science* 1994;**264**:1415–20.
7. Cunningham BC, Ultsch M, deVos AM, Mulkerrin MG, Clauser KR, Wells JA. Dimerization of the extracellular domain of the human growth hormone receptor by a single hormone molecule. *Science* 1991;**254**:821–5.
8. deVos A, Ultsch M, Kossiakoff A. Human growth hormone and extracellular domain of its receptor: crystal structure of the complex. *Science* 1992;**255**:306–12.
9. Philo J, Aoki K, Arakawa T, Narhi L, Wen J. Dimerization of the extracellular domain of the erthropoietin (EPO) receptor by EPO: one high-affinity and one low-affinity interaction. *Biochemistry* 1996;**35**:1681–91.
10. Livnah O, Johnson DL, Stura EA, Farrell FX, Barbone FP, You Y, Liu KD, Goldsmith MA, He W, Krause C, Petska S, Jolliffe LK, Wilson IA. An antagonist peptide–EPO receptor complex: receptor dimerization is not sufficient for activation. *Nat Struct Biol* 1998;**5**:993–1004.
11. Remy I, Wilson IA, Michnick SW. Erythropoietin receptor activation by a ligand-induced conformation change. *Science* 1999;**283**:990–3.
12. Broudy VC, Lin N, Brice M, Nakamoto B, Papayannopoulou T. Erythropoietin receptor characteristics on primary human erythroid cells. *Blood* 1991;**77**:2583–90.
13. Sawada K, Krantz SB, Kans JS, Dessypris EN, Sawyer S, Glick AD, Civin CI. Purification of human erythroid colony-forming units and demonstration of specific binding of erythropoietin. *J Clin Invest* 1987;**80**:357–66.
14. Sawada K, Krantz SB, Sawyer ST, Civin CI. Quantitation of specific binding of erythropoietin to human erythroid colony-forming cells. *J Cell Physiol* 1988;**137**:337–45.
15. Livnah O, Stura EA, Johnson DL, Middleton SA, Mulcahy LS, Wrighton NC, Dower WJ, Jolliffe LK, Wilson IA. Functional mimicry of a protein hormone by a peptide agonist: the EPO receptor complex at 2.8 Å. *Science* 1996;**273**:464–71.
16. Syed RS, Reid SW, Li C, Cheetham JC, Aoki KH, Liu B, Zhan H, Osslund TD, Chirino AJ, Zhang J, Finer-Moore J, Elliot S, Sitney K, Katz BA, Matthews DJ, Wendoloski JJ, Egrie J, Stroud RM. Efficiency of signalling through cytokine receptors depends critically on receptor orientation. *Nature* 1998;**395**:511–16.
17. Livnah O, Stura EA, Middleton SA, Johnson DL, Jolliffe LK, Wilson IA. Crystallographic evidence for preformed dimers of erythropoietin receptor before ligand activation. *Science* 1999;**283**:987–90.
18. Wrighton NC, Farrell FX, Chang R, Kashyap AK, Barbone FP, Mulcahy LS, Johnson DL, Barrett RW, Jolliffe LK, Dower WJ. Small peptides as potent mimetics of the protein hormone erythropoietin. *Science* 1996;**273**:458–63.
19. Johnson DL, Farrell FX, Barbone FP, McMahon FJ, Tullai J, Hoey K, Livnah O, Wrighton NC, Middleton SA, Loughney DA, Stura EA, Dower WJ, Mulcahy LS, Wilson IA, Jolliffe LK. Identification of a 13 amino acid peptide mimetic of erythropoietin and description of amino acids critical for the mimetic activity of EMP1. *Biochemistry* 1998;**37**:3699–710.
20. Zhan H, Liu B, Red S, Aoki K, Li C, Syed R, Karkaria C, Koe G, Sitney K, Hayenga K, Mistry F, Cheetham LS, Egrie J, Giebel L, Stroud R. Engineering a soluble extracellular erythropoietin receptor (EPObp) in *Pichia pastoris* to eliminate microheterogeneity, and its complex with erythropoietin. *Protein Eng* 1999;**12**:505–13.
21. Qui H, Belanger A, Yoon H, Bunn H. Homodimerization restores biological activity to an inactive erythropoietin mutant. *J Biochem* 1998;**273**:11,173–6.
22. Sytkowski AJ, Lunn ED, Risinger MA, Davis KL. An erythropoietin fusion protein comprised of identical repeating domains exhibits enhanced biological properties. *J Biochem* 1999;**274**:24,773–8.
23. Wilson IA, Jolliffe LK. The structure, organization, activation and plasticity of the erythropoietin receptor. *Curr Opin Struct Biol* 1999;**9**:696–704.
24. Clackson T, Wells JA. A hot spot of binding energy in a hormone–receptor interface. *Science* 1995;**267**:383–6.
25. Constantinescu SN, Keren T, Socolovsky M, Nam H, Henis YI, Lodish HF. Ligand-independent oligomerization of the cell-surface erythropoietin receptor is mediated by the transmembrane domain. *Proc Natl Acad Sci USA* 2001;**98**:4379–84.
26. Constantinescu SN, Huang LJ, Nam H, Lodish HF. The erythropoietin receptor cytosolic juxtamembrane domain contains an essential, precisely oriented, hydrophobic motif. *Mol Cell* 2001;**7**:377–85.
27. Elkins PA, Christinger HW, Sandowski Y, Sakal E, Gertler A, deVos AM, Kossiakoff AA. Ternary complex between placental lactogen and the extracellular domain of the prolactin receptor. *Nat Struct Biol* 2000;**7**:808–14.

28. Aritomi M, Kunishima N, Okamoto T, Kuroki R, Ota Y, Morikawa K. Atomic structure of the GCSF–receptor complex showing a new cytokine–receptor recognition scheme. *Nature* 1999;**401**:713–17.
29. Moghal N, Sternberg PW. Multiple positive and negative regulators of signaling by the EGF-receptor. *Curr Opin Cell Biol* 1999;**11**:190–6.
30. Moriki T, Maruyama H, Maruyama IN. Activation of preformed EGF receptor dimers by ligand-induced rotation of the transmembrane domain. *J Mol Biol* 2001;**311**:1011–26.
31. Naismith JH, Devine TQ, Brandhuber BJ, Sprang SR. Crystallographic evidence for dimerization of unliganded tumor necrosis factor receptor. *J Biol Chem* 1995;**270**:13,303–7.
32. Banner DW, D'Arcy A, Janes W, Gentz R, Schoenfeld H, Broger C, Loetscher H, Lesslauer W. Crystal structure of the soluble human 55-kD TNF receptor–human TNF beta complex: implications for TNF receptor activation. *Cell* 1993;**73**:431–45.
33. Merritt EA, Bacon DJ. Raster3D photorealistic molecular graphics. *Methods Enzymol* 1997;**277**:505–24.
34. Kraulis PJ. MOLSCRIPT: a program to produce both detailed and schematic plots of protein structures. *J Appl Crystallogr* 1991; **24**:946–50.
35. Nicholls A, Bharadwaj R, Honig B. GRASP: graphical representation and analysis of surface properties. *Biophys J* 1993;**64**:166–7.

Chapter 26

Structure of IFNγ and its Receptors

Mark R. Walter

Department of Microbiology and Center for Macromolecular Crystallography, University of Alabama at Birmingham, Birmingham, Alabama

Interferon-γ (IFNγ) is a pleotropic cytokine that induces antiviral, antiproliferative, and immunomodulatory effects on numerous target cells [1]. These diverse biological activities are initiated by IFNγ-mediated aggregation of at least two different cell surface receptors: IFNγR1 and IFNγR2. X-ray crystallographic studies of IFNγ and its receptors have been undertaken to delineate the molecular architecture of the receptor complexes and to understand the detailed recognition mechanisms that are ultimately responsible for IFNγ biological responses. Here, these structures are summarized in the context of current biochemical and bioactivity data.

Natural forms of human IFNγ are comprised of two 143-amino-acid peptide chains that are posttranslationally modified to contain an N-terminal pyroglutamic acid residue, N-linked glycosylation at two positions, and a heterogeneous C terminus containing the positively charged sequence KTGKRKR (residues 125–131). The crystal structure of human IFNγ has revealed the tight association of two peptide chains (comprised of six α-helices, labeled A to F, from the N to C terminus) into a remarkable intertwined helix topology to form a symmetric dimer (Figure 26.1) [2]. As a result, the two-fold related domains of IFNγ are formed from the first four helices of one chain (A–D) and the last two helices (E′ and F′) from the other. Despite almost no sequence identity, the identical intertwined topology is also observed in the crystal structure of IL-10 [3]. The α-helices that form each domain are 9 to 21 residues long and are essentially linear (with the exception of helix F, which displays an ~50° bend). The helices are connected by short loops of 3 to 5 residues, except for the 13-residue AB loop that encircles helix F. C-terminal residues 124 to 143 extend away from the core of the molecule and are presumed to be flexible.

IFNγR1 and IFNγR2 are both type I membrane proteins that contain extracellular and cytoplasmic domains connected by a hydrophobic membrane spanning helix. The extracellular domain of IFNγR1 binds IFNγ with high affinity (~1 nM), while IFNγR2 exhibits essentially no affinity for IFNγ. Coexpression of IFNγR1 and IFNγR2 on cells results in a fourfold increase in affinity for IFNγ compared to cells expressing IFNγR1 alone, suggesting that the IFNγR2 binding site is formed from residues on IFNγ and IFNγR1 as presented in the IFNγ/IFNγR1 complex. IFNγ-induced formation of the biologically active complex of IFNγ, IFNγR1, and IFNγR2 activates the Janus kinases (JAK1 and JAK2) that are associated with the cytoplasmic domains. JAK-dependent phosphorylation of the intracellular domain of IFNγR1 results in the recruitment of the nuclear transcription factor STAT1 and subsequent expression of IFNγ-inducible genes [1].

At this time, structural information is only available for the extracellular domain of IFNγR1 (sIFNγR1) as it exists in complex with IFNγ (Figure 26.1) [4–6]. Based on this work, sIFNγR1 is comprised of two fibronectin type III domains (FnIII). The FnIII modules consist of a sandwich of two antiparallel β-sheets made up of seven β-strands: A, B, E, G, F, C, and C′. The N-terminal domain (D1) and the membrane-proximal or C-terminal domain (D2) are oriented at ~120° to one another. The IFNγ binding site is located at the crevice between D1 and D2. Receptor residues that contact IFN-γ are located on five different segments (labeled L2 to L6) that correspond to CC′ and EF loops of D1, the domain linker, and BC and FG loops in D2.

In agreement with solution and cell-surface binding studies, the structure of the IFNγ/IFNγR1 complex revealed the symmetric binding of two IFNγR1s to one IFNγ dimer (Figure 26.1). The two receptors in the complex do not interact with one another and are separated by about 100 Å at the putative position of the cell membrane. The large distance between the IFNγR1s is consistent with the 1:2 complex being an intermediate that is dependent on IFNγR2 binding

Handbook of Cell Signaling, Three-Volume Set 2 ed.

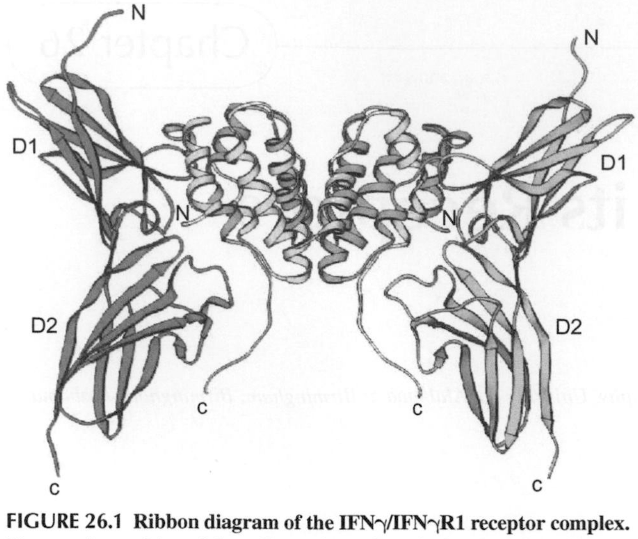

FIGURE 26.1 **Ribbon diagram of the IFNγ/IFNγR1 receptor complex.** The putative position of the cell membrane is at the bottom of the figure. The N and C termini of each molecule are labeled, as well as the D1 and D2 domains of the receptor.

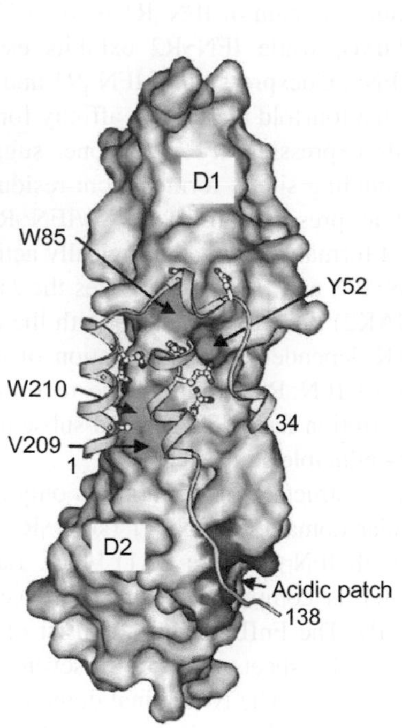

FIGURE 26.2 **The IFNγ/IFNγR1 interface.** IFNγR1 is shown as a molecular surface while IFNγ segments (residues 1–34 and 108–138) that bind IFNγR1 are represented by a ribbon. Several IFNγR1 residues that participating in important interactions in the interface are labeled by arrows. The side-chains of IFNγ residues Val-5, Glu-9, Arg-12, Ser-20, Asp-24, His-111, Glu-112, and Gln-115 are shown. The acidic patch located on IFNγR1 is labeled, and the predicted interaction with a modeled C terminus of IFNγ is shown.

and JAK2 recruitment to initiate the phosphorylation cascade. The two-fold symmetry of the IFNγ/IFNγR1 complex suggests that it contains two binding sites for IFNγR2.

The IFNγR1 binding site is comprised of IFNγ residues 1 to 34 from one chain and residues 108 to 123 on

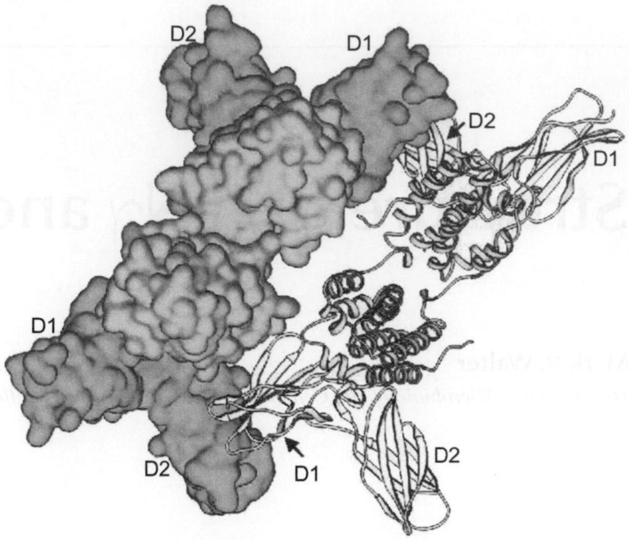

FIGURE 26.3 **Crystal structure of the 2:4 IL-10/IL-10R1 complex [7].** The view is looking down the two-fold axis of IL-10. The 1 : 2 IL-10/IL-10R1 complex represented by molecular surfaces is very similar to the IFNγ/IFNγR1 complex shown in Figure 26.1. The predicted positions of the IL-10R2s are shown as ribbons. The D1 and D2 domains for each receptor are labeled. This structure provides a possible model for the biologically active.

the other (Figure 26.2). These residues form a continuous binding site that includes helix A, the AB loop, and helix B from one chain and helix F from the other. More recently, a 2-Å structure of a mutant IFNγ/IFNγR1 complex has revealed additional details of the binding site, including the participation of five ordered waters in the interface and the reassignment of IFNγR1 Val-206 to an unfavorable phi-psi value to optimize its interactions with IFNγ [6]. In all free and bound structures reported to date, the C terminus of IFNγ that is important for binding and biological activity has not been observed; however, an acidic patch was identified on the receptor that may provide a "Velcro-like" interaction site for the basic C terminus of IFNγ [4].

Limited sequence identity confirms that IFNγR2 is structurally similar to IFNγR1; however, the inter-domain angle, receptor binding loops, and binding site cannot be accurately predicted. Currently, the most intriguing model for IFNγR2 binding has been proposed from the analysis of the crystal structure of the IL-10/IL-10R1 complex [7]. In solution and the crystals, IL-10 and soluble IL-10R1 form a complex consisting of 2 IL-10 dimers and 4 IL-10R1s (Figure 26.3). Structure and sequence comparisons of IL-10 and IFNγ receptors suggest that high-affinity (IL-10R1 and IFNγR1) and low-affinity (IL-10R2 and IFNγR2) receptors may share a common binding site on their respective cytokines. The 2:4 IL-10 receptor complex suggests a model for how the low-affinity IFNγR2 may simultaneously interact with IFNγ and IFNγR1. This structural model is supported by limited homolog scanning on IFNγR1 and radiation inactivation data showing that IFNγ biological activity requires two IFNγ dimers [1]. Confirmation of this model will require the structure determination of the IFNγ/IFNγR1/IFNγR2 complex.

REFERENCES

1. Bach EA, Aguet M, Schreiber RD. The IFN-gamma receptor: a paradigm for cytokine receptor signaling. *Annu Rev Immunol* 1997;**15**:563–91.
2. Ealick SE, Cook WJ, Vijay-Kumar S, Carson M, Nagabhushan TL, Trotta PP, Bugg CE. Three-dimensional structure of recombinant human interferon-gamma. *Science* 1991;**252**:698–702.
3. Walter MR. Structural biology of cytokines, their receptors, and signaling complexes. In: Blalock JE, editor. *Chemical immunology: neuroimmunoendrocrinology*. Basel: Karger; 1997. p. 76–98.
4. Walter MR, Windsor WT, Nagabhushan TL, Lundell DJ, Lunn CA, Zauodny PJ, Narula SK. Crystal structure of a complex between interferon-gamma and its soluble high-affinity receptor. *Nature* 1995;**376**:230–5.
5. Thiel DJ, le Du MH, Walter RL, D'Arcy A, Chene C, Fountoulakis M, Garotta G, Winkler FK, Ealick SE. Observation of an unexpected third receptor molecule in the crystal structure of human interferon-gamma receptor complex. *Struct Fold Des* 2000;**8**:927–36.
6. Randal M, Kossiakoff AA. The structure and activity of a monomeric interferon-gamma: alpha-chain receptor signaling complex. *Structure* 2001;**9**:155–63.
7. Josephson K, Logsdon NJ, Walter MR. Crystal structure of the IL-10/IL-10R1 complex reveals a shared receptor binding site. *Immunity* 2001;**14**:35–46.

Part 3

G Protein-Coupled Receptors

Chapter 27

Structures of Heterotrimeric G Proteins and their Complexes

Stephen R. Sprang
Center for Biomolecular Structure and Dynamics, University of Montana, Missoula, Montana

INTRODUCTION

The alpha subunits of heterotrimeric G proteins (Gα) belong to the superfamily of intracellular GTP hydrolases that use the energy derived from the binding of guanosine triphosphate (GTP) to effect signal transduction. GTP stabilizes an activated state of the Gα protein that is able to bind and regulate certain molecules, called effectors, in the cell. This capability is diminished or lost when the Gα protein hydrolyzes GTP. It is regained when a new molecule of GTP is bound, a process that is catalyzed by ligand-activated, seven-transmembrane helix, G protein-coupled receptors (GPCR). Molecules, such as GPCRs, that catalyze nucleotide exchange on G proteins are called Guanine nucleotide Exchange Factors (GEFs). Intracellular targets of G protein regulation comprise a small group of second-message generating molecules such as potassium and calcium ion channels, phospholipase C isoforms (PLC), adenylyl cyclases (AC), cyclic GMP phosphodiesterase – or regulators of other signaling pathways, such as p63RhoGEF and p115RhoGEF subfamilies of Rho GEFs. The sequence of GTP binding, hydrolysis, product release, and reformation of the G protein–GTP complex constitutes a signaling cycle. Steps within this cycle are subject to regulation that shapes the temporal characteristics of the signal, from ligand:receptor recognition to G protein:effector interaction. The three-dimensional structures of many of the components of this cycle have been described in several functionally relevant states (Table 27.1). These structures provide insight into the molecular mechanics of G protein-mediated signal transduction. Here, we describe the three-dimensional structures of G proteins and the molecular processes that constitute the signaling pathway. This area of research has been extensively reviewed [1–3], and the reader is directed to the primary literature for details.

Heterotrimeric G proteins comprise two functional components: Gα subunits are GTP binding proteins and, when bound to GTP, preferentially interact with effectors. Dimers composed of tightly bound β (Gβ) and γ (Gγ) chains constitute the second functional unit. G dimers act both as inhibitors of nucleotide release from G and as regulators of effector proteins, either independently or coordinately with G. Importantly, Gβγ are obligate co-substrates in the interaction of G-protein heterotrimers with GPCRs.

G SUBUNITS

Gα subunits are members of the Ras superfamily, which also includes translation elongation factors and the components of the signal recognition apparatus. In mammals, the family of Gα isoforms is encoded by 16 genes; these can be sorted into 4 closely-related homology groups or classes named for representative members of each class: $G\alpha_s$, $G\alpha_i$, $G\alpha_q$, and G_{12} (Figure 27.1) [4]. Variants of $G\alpha_s$ and $G\alpha_o$ are generated by alternative mRNA splicing. Pairwise sequence identity between members of the family ranges from 35 to 93 percent (reviewed in [5]). Effector recognition by Gα isoforms is to some degree class-specific (Figure 27.1).

Crystal structures have been determined for at least one representative of each class of Gα subunits (Table 27.1), in some cases in both GTP-bound (with the use of non-hydrolyzible analogs) and GDP-bound states. Structures of several Gα proteins have been determined bound to the complex of GDP, tetrafluroaluminate (AlF), and magnesium ion, which together mimic the transition state for GTP hydrolysis in which the phosphate is pentacoordinate.

Ras superfamily proteins are built upon a scaffold of six parallel strands, layered on each side by a set of five helices (Figure 27.2). Unique to the Gα family is an α-helical

Handbook of Cell Signaling, Three-Volume Set 2 ed.

TABLE 27.1 Selected structures of heterotrimeric G proteins and their complexes

Gα•nucleotide complexes	Nucleotide	PDB code
$G\alpha_{t/i1}$[1], $G\alpha_{i1}$, $G\alpha_s$	Mg^{2+}•GTPγS[2]	1TND, 1GIA, 1AZT
$G\alpha_{i1}$	Mg^{2+}•GppNHp[3]	1CIP
$G\alpha_{i1}$, $G\alpha_t$, $G\alpha_{12}$	Mg^{2+}•GDP•AlF_4^-	1GFI, 1TAD, 1ZCA
$G\alpha_{i1}$(G203A)	GDP•Pi	1GIT
$G\alpha_{i1}$	Mg^{2+}•GDP•SO_4^{2-}	1BOF
$G\alpha_t$	Ca^{2+}•GDP	1TAG
$G\alpha_{i1}$, $G\alpha_{13}$	GDP	1GDD, 1ZCB
Gα:GAP		
$G\alpha_{i1}$:RGS4, $G\alpha_t$:RGS9 $G\alpha_o$:RGS16[4]	Mg^{2+}•GDP•AlF_4^-	1AGR, 1FQK, 3C7K
$G\alpha_{13/i1}$:p115RhoGEF[5]	Mg^{2+}•GDP•AlF_4^-	1SHZ
Gα:effector/regulator		
$G\alpha_s$:AC[6]	Mg^{2+}•GTPγS	1AZS, 1CJU
$G\alpha_q$:p63RhoGEF:RhoA	Mg^{2+}•GDP•AlF_4^-	2RGN
$G\alpha_i$:GoLoco[7]	GDP	1KJY
Gα:effector:GAP		
$G\alpha_t$•RGS9•PDEγ[8]	Mg^{2+}•GDP•AlF_4^-	1FQJ
$G\alpha_{13/i1}$:p115RhoGEF	Mg^{2+}•GDP•AlF_4^-	1SHZ
Gβγ regulator/effector		
$G\beta_i$:$G\gamma_1$		1TBG
$G\alpha_{i1}$:$G\beta_1$:$G\gamma_2$, $G\alpha_t$:$G\beta_1$:$G\gamma_1$	GDP	1GG2, 1GOT
$G\beta_1$:$G\gamma_2$:GRK-2[9]		1OMW
$G\beta_1$:$G\gamma_1$Phosducin		1AOR,1B9X
$G\beta_5$:RGS9		2PBI
Gα effector:Gβγ		
$G\alpha_q$:GRK-2:$G\beta_1$:$G\gamma_2$	Mg^{2+}•GDP•AlF_4^-	2BCJ

[1]*a chimera of $G\alpha_{i1}$ and bovine $G\alpha_t$: the corresponding nomenclature is used to refer to other Gα chimeras.*
[2]*GTPγS: guanosine 5′-[γ-thio]triphosphate.*
[3]*GppNHp:guanosine-5′-(βγ-methylene)triphosphate.*
[4]*A structural survey of Gα:RGS complexes not included here is presented in: Soundararajan M, et al. (2008) Proc Natl Acad Sci USA **105:** 6457–6462.*
[5]*GAP function provided by peptide segment N-terminal to RGS domain, which itself binds in the mode of an effector (see text).*
[6]*AC: a complex between the C1 domain of adenylyl cyclase type V and the C2 domain of adenylyl cyclase type II. These domains comprise the catalytic unit. A soluble forskolin derivative is bound at the regulatory site of AC. In 1AZS, the domains adopt the "open" conformation. The 1CJU complex contains the ATP analog β-L, 2′,5′, dideoxy adenosine triphosphate and two magnesium ions. The AC domains adopt a "closed" conformation.*
[7]*GoLoco motif peptide from RGS14.*
[8]*PDEγ:cyclic GMP phosphodiesterase γ subunit.*
[9]*GRK-2: G protein Receptor Kinase-2.*

bundle domain inserted into the loop between the first and second strands of the Ras-like domain. Gα subunits are modified by N-terminal myristoylation [6] ($G\alpha_t$) and thioester-linked palmitoylation ($G\alpha_s$, $G\alpha_q$, $G\alpha_{13}$) or both ($G\alpha_i$, $G\alpha_o$, $G\alpha_z$) [7, 8]. The latter confers plasma membrane localization upon $G\alpha_s$ and $G\alpha_q$, but may be reversed upon activation [9]. Myristoylation is required in some cases for activity – for example, efficient inhibition of adenylyl cyclase by $G\alpha_{i1}$ [10] or sensitivity towards regulatory proteins [11].

GTP is bound between the helical and Ras-like domains, but interacts primarily with conserved sequence motifs within the Ras domain [12–16] (Figure 27.3). The P-loop, which enfolds the alpha and beta phosphates of the nucleotide, contains a ..$G^A/_T$GESGKST.. sequence characteristic of Walker A motifs [17] that is permissive for the tight turn required to encompass the phosphates. The lysine residue is a critical βγ-phosphate ligand and the serine residue that follows coordinates the catalytic Mg^{2+}

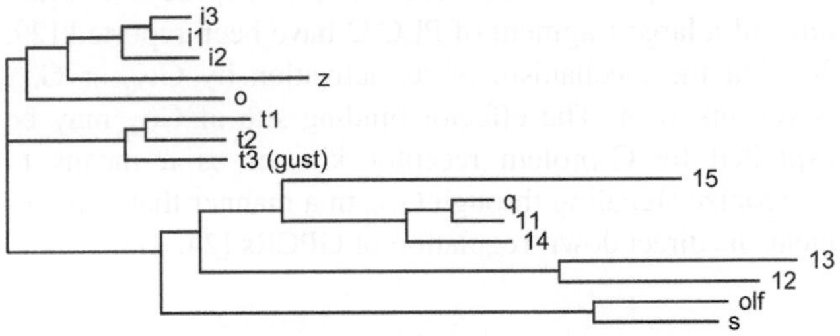

FIGURE 27.1 A phylogenetic tree, of the family of human Gα subunits, based on amino acid sequence alignment carried out using T-COFFEE [110] using the BLOSUM scoring matrix and default parameters, and computed using CLUSTALW-2 [111]. Line distances are proportional to degree of sequence divergence.

co-factor. The connector leading from the helical domain to the Ras-like domain is called Switch I, and is the functional analog to the similarly named region in ras proteins [18]. The arginine residue (178 in $G\alpha_{i1}$) within this sequence (..RVXTTG..) stabilizes the transition state for GTP hydrolysis [12, 19], in analogy with the "arginine finger" provided by GTPase activating proteins of Ras family G proteins [18]. The succeeding threonine is a second Mg^{2+}-coordinating ligand (Figure 27.3). The γ phosphate group of GTP is cradled by a tight turn (..DVGGQ..) that initiates Switch II, an irregular and conformationally mobile helix [2]. The glutamine residue in this motif (204 in $G\alpha_{i1}$) plays a critical catalytic role in GTP hydrolysis, possibly by virtue of its interaction with the water nucleophile. However, in the structures of Gα subunits bound to slowly hydrolyzable GTP analogs, the catalytic glutamine, and in $G\alpha_{i1}$ the catalytic arginine as well, are either poorly ordered or adopt conformations in which they would be incapable of providing catalytic assistance (Figure 27.3).

A key role of GTP, in conjunction with Mg^{2+}, is to maintain the activated conformation of Gα by virtue of a network of hydrogen bonds and oxygen–metal interactions that link Mg^{2+}•GTP with the P-loop, switch I and switch II. Structural studies of $G\alpha_{i1}$ and $G\alpha_t$ show that, in the GDP state, switch II is either wholly disordered, or adopts a conformation that is not conducive to effector binding [13, 15, 20]. The well-ordered state induced by GTP also promotes a set of ionic contacts between switch II and the β4–α3 loop called switch III. Upon GTP hydrolysis, the network of interactions between the three switch regions is altered or lost. The purine ring of the guanine nucleotide is cradled by two conserved loops, β5–αG and β6–α5 (Figure 27.2). The aspartate residue within the first sequence (..FLNKKD..) confers specificity towards guanine nucleotides, and mutation

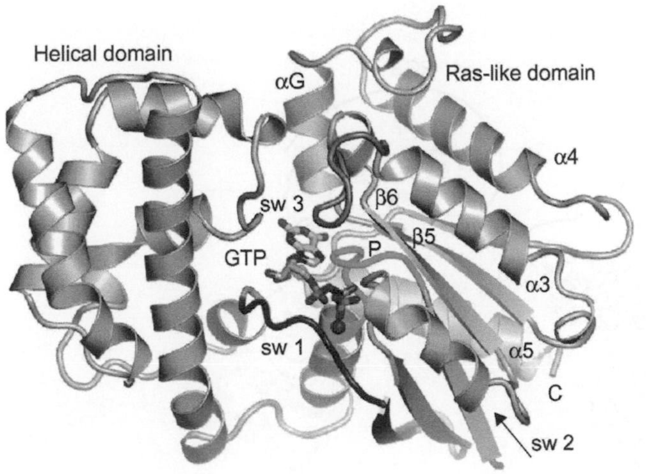

FIGURE 27.2 Schematic of the GppNHp•Mg^{2+} complex of $G\alpha_{i1}$ (Protein databank accession code 1GFI) with secondary structure elements (arrows, strands; coils, helices). Switch regions are labeled and shaded. The GTP analog is shown as a shaded ball and stick model. Figures, unless otherwise noted drawn using PyMol (http://pymol.sourceforge.net).

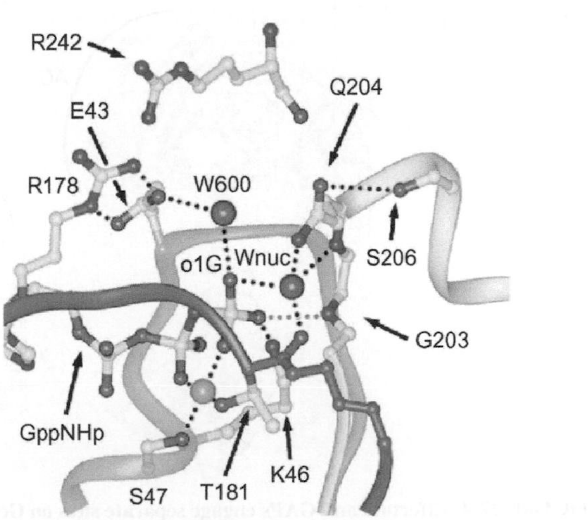

FIGURE 27.3 Catalytic site of the $G\alpha_{i1}$•Mg^{2+}•GppNHp complex, showing the amino acid side chains involved in catalysis and Mg^{2+} binding. Note that the catalytic glutamine (Q 204) and arginine (R178) adopt conformations that do not permit their involvement in catalysis. Adapted from [112] (Coleman and Sprang, 1999).

of this residue can alter the nucleotide base specificity of Gα [21]. The second loop acts in a supporting role.

G-EFFECTOR INTERACTIONS

A series of crystal structures determined over the past 10 years have defined the structural basis for Gα:effector recognition (Table 27.1). These structures demonstrate that all Gα effectors, despite their considerable structural diversity, bind at the at the cleft between switch II and the α3–β5 loop of Gα (Figure 27.4) [2]. Interactions at that locus are dependent on the "activated" state of Gα that is stabilized by GTP. Certain effectors recognize additional Gα surfaces outside of this core binding site. Notably, the Plextrin Homology (PH) and Dibble Homology (DH) domains of p63RhoGEF occupy an extensive surface of G_q that encompasses the β2–β3, α2–β4, and β4–β5 loops of $G\alpha_q$, in addition to the "core" effector binding domain [22].

The effector specificity of Gα proteins is conferred by both amino acid sequence and conformation of the effector binding segments. For example, $G\alpha_{i1}$ non-competitively inhibits the $G\alpha_s$-stimulated activity of AC isoforms I and V, possibly by interacting at a site in the C1 domain that is symmetry-related to the C2 domain binding site of $G\alpha_s$ [23]. The ability of $G\alpha_{i1}$ to discriminate its own from the $G\alpha_s$ binding site is unlikely to be due to amino acid differences between the two Gα proteins in the α2 and α3–β5 loops, since all but two amino acids in this region are conserved. Specificity may stem from differences in the spacing and orientation of the α3–β5 loop and the buttressing α4–β6 loop [16] and, importantly, by the participation of structural elements outside of the core-effector binding region [22]. Individual G proteins are also adaptive structures able to accommodate structurally diverse effectors, Thus, conformational adjustments within the effector binding segments allow $G\alpha_q$ to bind both p63 RhoGEF and G-protein receptor kinase-2 (GRK2) [22, 24]. Similarly, work from our laboratories has shown that conformational changes in α3-β5 allow $G\alpha_{13}$ to bind the disparate surfaces of p115RhoGEF and PDZ-RhoGEF [25].

The mechanisms by which Gα subunits regulate effectors are as diverse as the effectors themselves. Structural and biochemical evidence suggest that $G\alpha_s$ stimulates AC, and $G\alpha_i$ inhibits $G\alpha_s$-activated AC, through an allosteric mechanism, possibly by controlling the relative orientation of the two homologous domains that form the AC active site [23, 26]. $G\alpha_q$ and $G\alpha_{13}$ likewise appear to be allosteric activators of p63RhoGEF and RGSRhoGEFs, respectively, although by different mechanisms [22, 27]. $G\alpha_q$ may relieve an autoinhibitory inter-domain interaction in p63RhoGEF, while $G\alpha_{12/13}$ promotes the activity of RGS-RhoGEFs, by a mechanism that remains to be discovered [1]. $G\alpha_t$ (transducin) acts more passively in its activation of GMP phosphodiesterase by sequestration of the inhibitory subunit of cyclic GMP phosphodiesterase (PDEγ) [28]. Crystal structures of a large fragment of PLCβ2 have been reported [29, 30], but the mechanism of its activation by $G\alpha_q$ or Gβγ is yet unknown. The effector binding site of $G\alpha_q$ may be exploited by G-protein receptor kinases as a means to antagonize signaling through $G\alpha_q$ in a manner that complements its direct down-regulation of GPCRs [24, 31].

GTP HYDROLYSIS BY G AND ITS REGULATION BY GAPs

G proteins hydrolyze GTP with a slow catalytic rate of approximately $0.05\,s^{-1}$ at physiological temperature.

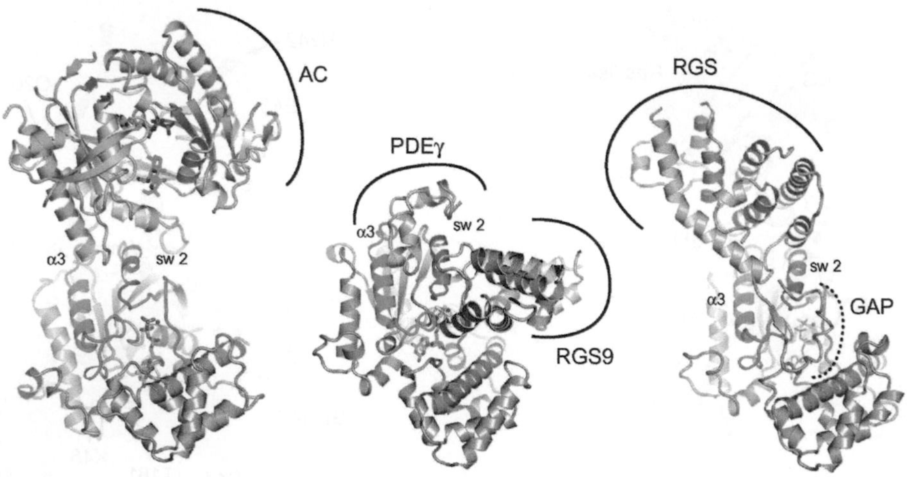

FIGURE 27.4 Effectors and GAPs engage separate sites on Gα in effector or effector–GAP complexes. Gα subunits are depicted in light gray, with switch 2 and α3, which together form a common effector binding site, labeled and shaded dark. Effector and GAP domains are labeled and shaded: *left*, $G\alpha_s$•Mg^{2+}•GTPS bound to the catalytic domains of adenylyl cyclase (1CJU); *middle*, $G\alpha_t$•GDP•Mg^{2+}•AlF_4^- bound to the RGS domain of RGS9 and the effector GMP phosphodiesterase γ(1FQJ); *right*, the rgRGS domain of p115RhoGEF bound to $G\alpha_{13/i1}$• GDP•Mg^{2+}•AlF_4^-; the peptide sequence preceding the RGS domain interacts with the GAP binding site of Gα, while the RGS domain occupies the effector binding site (1SHZ).

Since Gα•GDP binds effectors with less affinity than Gα•Mg^{2+}•GTP, the rate of hydrolysis determines the lifetime of the signal, as measured by the output of activated effector. The origin of the kinetic barrier may be deduced from the structure of the complex between Gα bound to GDP and magnesium fluoroaluminate (Mg^{2+}•AlF_4^{-1}) [12]. Fluroaluminates (AlF_3 or AlF_4^- and their hydrates) mimic the γ phosphate of GTP [32], and in the presence of GDP, promote the catalytically activated state of Gα [33]. Structural studies of $G\alpha_{i1}$ and $G\alpha_t$ GDP•Mg•AlF_4^- complexes demonstrate that AlF_4^{-1} forms a hexacoordinate complex with a phosphate oxygen of GDP and a water molecule (the presumptive nucleophile) as axial ligands, thereby approximating the pentacoordinate transition state for phosphorolysis [12, 19]. The structures show that, relative to the ground state (Figure 27.2) the switch I arginine and switch II glutamine must be substantially reoriented in order to stabilize the transition state. It is therefore possible that a conformational rearrangement within the active site contributes substantially to the overall kinetic barrier to GTP hydrolysis. Nevertheless, the chemical steps in GTP hydrolysis may, as indicated by kinetic isotope effects in Ras, be ultimately rate-limiting [34].

Many Gα-regulated physiological signaling mechanisms, for example those which regulate photoreception and potassium channel activity, require that Gα proteins be deactivated in time scales ~10^3 faster than would be possible through their intrinsic GTPase activities [35–37]. In these systems, GTPase Activating Proteins, or GAP domains within Gα effectors or binding proteins stimulate GTPase activity. In their interactions with Gα, GAP domains appear to function synergistically with effectors and GPCRs to control the spatial and temporal specificity of steady state signaling. If the Gα:effector complex is deactivated rapidly after its formation, then Gα, Gβγ and receptor will not have had the opportunity to diffuse away from each other on the plasma membrane, allowing them to rapidly reassemble for a subsequent round of Gα activation and effector regulation [38–40]. Mathematical modeling of GTP turnover by G_q in the presence of receptor and GAP provides evidence of direct synergistic interaction among these components [41].

A subset of Gα GAPs contain RGS (Regulators of G protein Signaling) domains which are characterized by an ~120 amino acid homology region termed the RGS-box [36]. RGS proteins show varying degrees of specificity towards their Gα substrates, and most Gα subunits are known to be subject to the action of one or more RGS proteins. Biochemical and crystallographic analysis indicate that RGS domains stabilize a conformation of Gα that promotes binding of GDP•Mg^{2+}•AlF_4^{-1} [42–47]. In their interactions with $G\alpha_i$-class proteins, RGS domains interact with switch I and the N-terminal half of switch II; these are adjacent to but do not overlap effector contact surfaces [16] (Figure 27.5). Therefore, it is possible for RGS domains to interact directly with effector:Gα complexes, as is illustrated in the complex of $G\alpha_t$ with the subunit of GMP phosphodiesterase and the RGS domain of RGS9 [28] (Figure 27.4). In this instance, PDEγ binding to $G\alpha_t$ is positively cooperative with that of RGS9 [48], whereas it inhibits binding of RGS7 [49]. RGS domains achieve specificity by discriminative interactions within both the switch contact regions and the helical domain [50, 51]. In contrast to certain Ras-family GAPs, RGS proteins do not supply catalytic residues to the catalytic site of Gα but, rather, stabilize the catalytically competent conformation [47]. Not all RGS homology domains that bind to Gα proteins are GAPs. The RGS-homology domain of GRK-2 binds to the effector-recognition site G-α_q [24]. Members of the RGS-RhoGEF family of effectors of $G\alpha_{12/13}$ [52], contain domains with remote sequence, and strong structural homology to RGS domains [53, 54]. These bind $G\alpha_{13}$ in the manner of effectors, rather than RGS-GAPs [25, 27] (Figure 27.4).

Certain Gα effectors contain domains that exhibit GAP activity and therefore promote rapid signal termination. This seemingly self-defeating property may, in fact, promote the high steady-state signaling activity afforded by synergistic interactions among signaling components described above. A C-terminal dimerization and binding region within phospholipase C isoforms comprises part of a domain that expresses GAP activity towards $G\alpha_{q/11}$ [55, 56] This domain was shown to have an α-helical structure and, by virtue of homologous contacts, to provide a mechanism

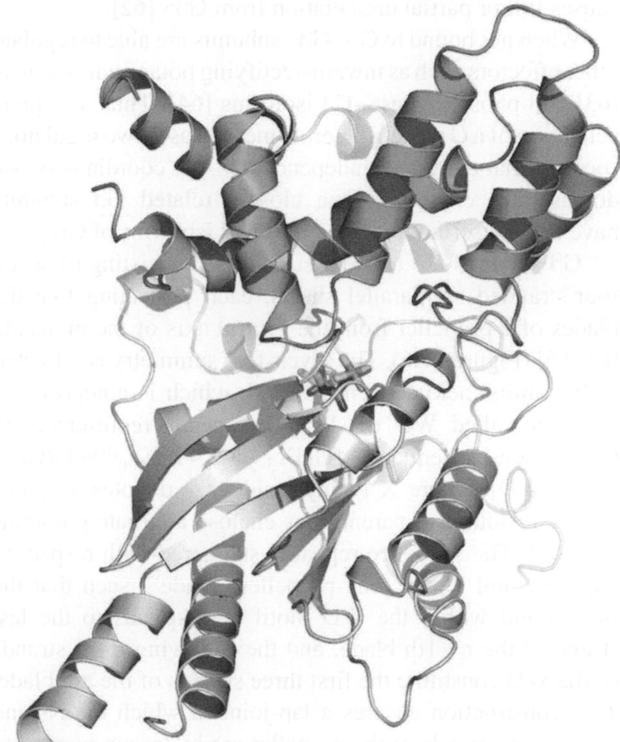

FIGURE 27.5 The complex of the RGS domain from RGS4 bound to the $G\alpha_{i1}$•GDP•$Mg^{2+}$$AlF_4^-$ (1AGR); $G\alpha_{i1}$ is light gray and the RGS domain is shaded.

for PLC dimerization [57]. A structural model for complex was proposed, and found to be consistent with mutagenesis data [58]. RGS-RhoGEF proteins possess a short, highly acidic amino acid sequence located just N-terminal to their RGS-like domains that, in p115RhoGEF and its homolog LARG, confers GAP activity [52] (Figure 27.4). Although this sequence has no similarity in structure or sequence to Gα binding elements of RGS GAPs, it forms interactions with $G\alpha_{13}$•AlF that mimic those by which RGS GAPs stabilize the catalytically-competent state of Gα subunits [27]. The corresponding sequence of PDZ-RhoGEF does not possess GAP activity, but, together with its RGS domain, appears to stabilize the activated conformation of $G\alpha_{13}$, even in the GDP-bound state [25]. Thus it is possible that regulatory domains of Gα protein effectors can prolong, as well as curtail, the signaling state of their Gα activators.

Gβγ DIMERS

Upon GTP hydrolysis, the affinity of $G\alpha_s$ for AC is reduced about 10-fold [59]. The most potent factor in signal termination may be the high affinity of Gβγ for Gα•GDP. Gβγ binds to the effector binding surface of Gα, but requires a conformation of switch II that cannot be attained in the GTP-bound state [60,61]. This non-signaling state of Gα is stable in the presence of GDP and exhibits high affinity for Gβγ. Receptor-catalyzed exchange of GDP for GTP also causes full or partial dissociation from Gβγ [62].

When not bound to Gα, Gβγ subunits are able to regulate other effectors such as inward-rectifying potassium channels [63] and phospholipase -Cβ isoforms [64]. Thus, receptor-activation of a G protein heterotrimer releases two regulatory species that can act independently or coordinately on downstream effectors. Five closely related Gβ subunits have been described, together with 12 isoforms of Gα.

Gβ subunits are toroidal structures, consisting of seven four-stranded antiparallel sheets, each projecting like the blades of a propeller from the central axis of the molecule [61, 65] (Figure 27.6). The seven-fold symmetry is reflected in the amino acid sequence of Gβ, which is composed of seven so-called WD (or WD40) repeats, represented by the consensus sequence $[GHX_{3-5}\Phi_2X\Phi X\Phi_{5-6}\Phi S/T)(G/A)X_3DX_4WD]$, where X is any residue, Φ denotes a hydrophobic residue and parentheses enclose alternate possibilities [66]. The sequence repeat is staggered with respect to the structural repeat (one propeller "blade") such that the first strand within the WD motif corresponds to the last strand of the n−1th blade, and the following three strands of the WD constitute the first three strands of the nth blade. This construction ensures a lap-joint in which the N- and C-terminal strands of the propeller are hydrogen-bonded to each other. The first of the two Asp residues in the motif is invariant, and participates in a hydrogen-bonded network with the His, Ser/Thr, and Trp residues in most of the blades. The first ~40 residues of Gβ, preceding the seven-bladed propeller, are folded into an helix. The Gγ subunit is an extended molecule consisting of three α-helical segments that do not contact each other. G subunits are farnesylated ($G\gamma_1$, $G\gamma_{11}$) or geranylgeranylated (all others) at their C-termini, thereby tethering them to the plasma membrane and promoting high-affinity interactions between Gβγ and Gα subunits and with effectors (see references in [66]).The N-terminal helix of Gγ forms a parallel coiled-coil with that of Gβ. The second and third helices lie over the surface of the Gβ torus [60, 67]. The Gγ binding surface is highly conserved among the five isoforms of Gγ. The limited selectivity between Gβ and Gγ isoforms seems to involve the interaction of hydrophobic residues in helix 2 of Gγ with its binding surface on Gβ [67].

Gα interacts with Gβ at two distinct and separate surfaces, both of which are required for high-affinity binding. The N-terminal helix of Gα contacts the side of the Gβ torus at blade 1. Switch I and switch II of Gα contact the surface of Gβ opposite that to which Gγ is bound (Figure 27.7). All of the residues of Gβ that contact Gα are conserved among Gβ isoforms. Although the Gβ-binding residues within the Ras domain of Gα are well conserved, the N-terminal helix of Gα is more variable, and this might confer some degree of conformational specificity to Gα–Gβ interactions. A series of mutagenesis studies have demonstrated that several effectors of Gβγ, including phosducin, PLC2, β-adrenergic receptor kinase (GRK), type II adenylyl cyclase, G-protein regulated inward rectifying potassium channels (GIRK), and the calcium channel 1B subunit all contact Gβγ at the same molecular surface to which Gα subunits bind [68]. Crystal structures of Gβγ bound to phosducin [24, 69], RGS9 [70], and GRK-2 bear this out (Figure 27.7). A family

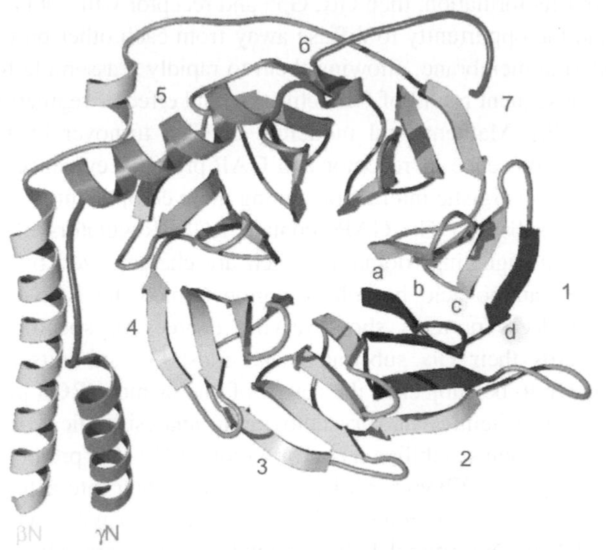

FIGURE 27.6 The complex of $G\beta_1\gamma_2$•$G\beta_1$ is light gray and the $G\gamma_2$ subunit is shaded. The four-stranded antiparallel "blades" that comprise the propeller fold are numbered. Individual strands in one repeat are lettered a–d in order of sequence. The amino termini of both subunits are labeled (1GG2).

of peptides identified by screening methods was found to target a common Gβγ "hotspot" for effector binding [71]. Remarkably, small molecules directed against different subsites within this surface exhibit pathway-specific inhibition of Gβγ regulation [72].

$G\beta_5$ is the most divergent among the five isoforms of Gβ. It has recently been shown to interact most strongly with the G-gamma-like domains (GGL) present in a variety of proteins, most notably the members of the RGS7 family [73,74]. The structure of the RGS9 : Gβγ complex has been determined, showing the GGL domain bound to Gβγ in mode analogous to Gγ [70] (Figure 27.7). The RGS9 :Gγ complex appears poised to allow binding of Gα•GDP, at the canonical binding surface of Gβ, and simultaneously, of an activated Gα subunit at the unhindered RGS domain. Thus, the complex could participate in reciprocal activation and deactivation of competing Gα-mediated pathways [75].

RECEPTOR-INDEPENDENT REGULATORS OF G PROTEIN ACTIVATION

Genetic complementation studies in yeast and biochemical analysis have led to the discovery of 25- to 30-residue G Protein Regulatory (GPR) [76, 77] or GoLoco [78] motifs that, like Gβγ, inhibit dissociation of GDP from $G\alpha_{i1}$ and $G\alpha_o$. GPR/GoLoco sequences bind only to the GDP-bound Gα proteins, and also inhibit binding of Gβ [79–82]. AGS3 is a representative member of a family of homologous proteins [83] that contain multiple GPR/GoLoco repeats, each of which is capable of engaging a single GDP-bound Gα subunit [84]. The GoLoco/GPR repeat from RGS14 adopts an extended conformation when bound to $G\alpha_{i1}$•GDP, forming contacts with switch II and crossing the gap between the Ras-like and helical domains. A conserved arginine residue from GoLoco engages the GDP β-phosphate, both stabilizing the nucleotide and at the same time, partially occupying the Mg^{2+} binding site [85]. GPR-containing molecules provide a mechanism for receptor-independent activation of Gβγ-signaling pathways, while inhibiting reactivation of $G\alpha_{i1}$. Other molecules, such as "Resistance to Inhibitors of Cholinesterase-8" (RIC-8), which catalyze exchange of GDP for GTP on Gi and Gq-class Gα proteins [86], appear to collaborate with AGS3-homologs to regulate non-membrane-localized processes such as microtubule force generation during cell division [87, 88]. Biochemical [11, 89] and structural investigations of such proteins and their complexes with Gα have only recently begun.

Gα–GPCR INTERACTIONS

Much recent progress has been made in delineating the three-dimensional structures of GPCRs. Crystal structures of inverse agonist-bound 2 adrenergic receptors [90–93] build on the pioneering structural studies of resting-state rhodopsin [94]. New structures of partially activated rhodopsin, and particularly of opsin, provide insight into the nature of the R* state of the receptor that catalyzes nucleotide exchange upon G protein heterotrimers [95, 96]. However, the nature of the heterotrimer : receptor interaction is known only from extensive mutagenesis and crosslinking studies [97–100], particularly of the rhodopsin : $G\alpha_t$ interface. These studies point to the α4–β6 loop and the C-terminal helix, α5, of $G\alpha_t$ as receptor binding regions. Mutations of residues located at the inward face of the α5 helix of $G\alpha_t$ increase

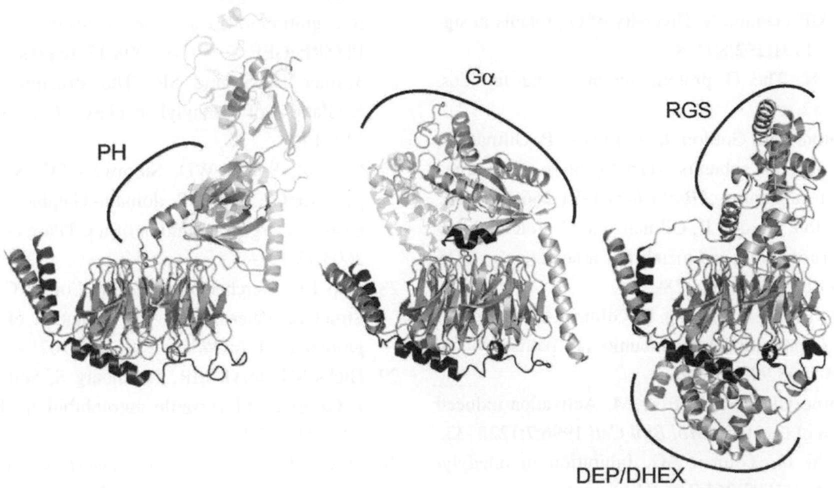

FIGURE 27.7 Gβγ:complexes: effectors and regulators occupy the same "face" of the Gβ torus. Gβγ binding partners are rendered in light gray, Gβ in medium-gray, Gγ in dark gray. Effector domains are labeled; *left*, the PH domain of GRK-2 contacts $G\beta_1\gamma_2$(1OMW); *middle*, the $G\alpha_{i1}$•GDP:$G\beta_1\gamma_2$ complex, with switch II of the G protein rendered in dark gray (1GG2); the corresponding $G\alpha_t$•GDP:$G\beta_1\gamma_1$ complex is similar in structure (1GOT); *right*, the RGS9 complex with $G\beta_5$(2PBI); the RGS domain engages the "Gα binding" surface of the Gβ torus. The GGL domain binds $G\beta_5$ in the manner of a Gγ subunit; the surface of the Gβ torus that is distal to RGS interacts with the DEP/DEHX domains.

receptor-independent rates of nucleotide release [101]. Mutation of a conserved alanine residue within the purine-contacting α4–β6 loop preceding α5 increases the intrinsic nucleotide exchange rate, and also reduces the thermostability of $G\alpha_{i1}$ [102, 103]. Some residues that affect receptor coupling are located in the Gα–Gβ interface, but distant from the putative receptor binding surface, suggesting that Gβγ plays a direct role in GPCR-mediated nucleotide exchange [104, 105]. Recent experiments with spin-labeled Gα subunits provide evidence of an order-to-disorder transition in the N-terminus of Gα in the heterotrimer upon receptor activation [106], and receptor-induced changes that propagate from the loop at the nucleotide binding site, to switch I and adjoining residues [107, 108]. NMR studies indicate that light-activated rhodopsin induces a conformationally dynamic state in $G\alpha_t$ that accompanies release of GDP from the heterotrimeric G protein [109]. If so, the crystal structure of a receptor : G protein heterotrimeric complex may remain an elusive target. Crystal structures of receptors bound to heterotrimeric G proteins, in a spectrum of functional states, will eventually be determined, and will provide some, but probably not complete, insight into receptor function. Equally important is the need to understand the organization and dynamic behavior of G protein signaling complexes at the cell membrane.

REFERENCES

1. Sternweis PC, Carter AM, Chen Z, Danesh SM, Hsiung YF, Singer WD. Regulation of Rho guanine nucleotide exchange factors by G proteins. *Adv Protein Chem* 2007;**74**:189–228.
2. Sprang SR, Chen Z, Du X. Structural basis of effector regulation and signal termination in heterotrimeric Galpha proteins. *Adv Protein Chem* 2007;**74**:1–65.
3. Oldham WM, Hamm HE. How do receptors activate G proteins. *Adv Protein Chem* 2007;**74**:67–93.
4. Simon MI, Strathmann MP, Gautam N. Diversity of G proteins in signal transduction. *Science* 1991;**252**:802–8.
5. Downes GB, Gautam N. The G protein subunit gene families. *Genomics* 1999;**62**:544–52.
6. Linder M, Pang I, Duronio R, Gordon J, Sternweis P, Gilman A. Lipid modifications of G protein subunits. Myristoylation of Go alpha increases its affinity for beta gamma. *J Biol Chem* 1991;**266**:4654–9.
7. Mumby SM, Heukeroth RO, Gordon JI, Gilman AG. G-protein alpha-subunit expression, myristoylation, and membrane association in COS cells. *Proc Natl Acad Sci USA* 1990;**87**:728–32.
8. Linder M, Middleton P, Hepler J, Taussig R, Gilman A, Mumby S. Lipid modifications of G proteins: alpha subunits are palmitoylated. *Proc Natl Acad Sci USA* 1993;**90**:3675–9.
9. Wedegaertner PB, Bourne HR, von Zastrow M. Activation-induced subcellular redistribution of Gs alpha. *Mol Biol Cell* 1996;**7**:1225–33.
10. Taussig R, Iniguez-Lluhi JA, Gilman AG. Inhibition of adenylyl cyclase by Gi alpha. *Science* 1993;**261**:218–21.
11. Tall GG, Gilman AG. Resistance to inhibitors of cholinesterase 8A catalyzes release of Galphai-GTP and nuclear mitotic apparatus protein (NuMA) from NuMA/LGN/Galphai-GDP complexes. *Proc Natl Acad Sci USA* 2005;**102**:16584–9.
12. Coleman DE, Berghuis AM, Lee E, Linder ME, Gilman AG, Sprang SR. Structures of active conformations of $G_i\alpha_1$ and the mechanism of GTP hydrolysis. *Science* 1994;**265**:1405–12.
13. Mixon MB, Lee E, Coleman DE, Berghuis AM, Gilman AG, Sprang SR. Tertiary and quaternary structural changes in $G_i\alpha_1$ induced by GTP hydrolysis. *Science* 1995;**270**:954–60.
14. Noel JP, Hamm HE, Sigler PB. The 2.2-Å crystal structure of transducin-alpha complexed with GTPγS. *Nature* 1993;**366**:654–63.
15. Lambright DG, Noel JP, Hamm HE, Sigler PB. Structural determinants for activation of the alpha-subunit of a heterotrimeric G protein. *Nature* 1994;**369**:621–8.
16. Sunahara RK, Tesmer JJG, Gilman AG, Sprang SR. Crystal structure of the adenylyl cyclase activator Gsα. *Science* 1997;**278**:1943–7.
17. Walker JE, Saraste M, Runswick MJ, Gay NJ. Distantly related sequences in the α- and β-subunits of ATP synthase, myosin, kinases and other ATP-requiring enzymes and a common nucleotide binding fold. *EMBO J* 1982;**1**:945–51.
18. Vetter IR, Wittinghofer A. The guanine nucleotide-binding switch in three dimensions. *Science* 2001;**294**:1299–304.
19. Sondek J, Lambright DG, Noel JP, Hamm HE, Sigler PB. GTPase mechanism of G proteins from the 1.7-Å crystal structure of transducin α-GDP-AlF_4^-. *Nature* 1994;**372**:276–9.
20. Kreutz B, Yau DM, Nance MR, Tanabe S, Tesmer JJ, Kozasa T. A new approach to producing functional G alpha subunits yields the activated and deactivated structures of G alpha(12/13) proteins. *Biochemistry* 2006;**45**:167–74.
21. Yu B, Simon MI. Interaction of the xanthine nucleotide binding Goalpha mutant with G protein-coupled receptors. *J Biol Chem* 1998;**273**:30,183–8.
22. Lutz S, Shankaranarayanan A, Coco C, et al. Structure of Galphaq–p63RhoGEF–RhoA complex reveals a pathway for the activation of RhoA by GPCRs. *Science* 2007;**318**:1923–7.
23. Dessauer CW, Tesmer JJ, Sprang SR, Gilman AG. Identification of a $G_i\alpha$ binding site on type V adenylyl cyclase. *J Biol Chem* 1998;**273**:25,831–9.
24. Tesmer VM, Kawano T, Shankaranarayanan A, Kozasa T, Tesmer JJ. Snapshot of activated G proteins at the membrane: the Galphaq–GRK2–Gbetagamma complex. *Science* 2005;**310**:1686–90.
25. Chen Z, Singer WD, Danesh SM, Sternweis PC, Sprang SR. Recognition of the activated states of Gα13 by the rgRGS domain of PDZRhoGEF. *Structure* 2008;**12**. in press.
26. Tesmer JJ, Sprang SR. The structure, catalytic mechanism and regulation of adenylyl cyclase. *Curr Opin Struct Biol* 1998;**8**:713–19.
27. Chen Z, Singer WD, Sternweis PC, Sprang SR. Structure of the p115RhoGEF rgRGS domain–Galpha13/i1 chimera complex suggests convergent evolution of a GTPase activator. *Nat Struct Mol Biol* 2005;**12**:191–7.
28. Slep KC, Kercher MA, He W, Cowan CW, Wensel TG, Sigler PB. Structural determinants for regulation of phosphodiesterase by a G protein at 2.0 Å. *Nature* 2001;**409**:1071–7.
29. Hicks SN, Jezyk MR, Gershburg S, Seifert JP, Harden TK, Sondek J. General and versatile autoinhibition of PLC isozymes. *Mol Cell* 2008;**31**:383–94.
30. Jezyk MR, Snyder JT, Gershberg S, Worthylake DK, Harden TK, Sondek J. Crystal structure of Rac1 bound to its effector phospholipase C-beta2. *Nat Struct Mol Biol* 2006;**13**:1135–40.
31. Carman CV, Parent JL, Day PW, et al. Selective regulation of Galpha(q/11) by an RGS domain in the G protein-coupled receptor kinase, GRK2. *J Biol Chem* 1999;**274**:34,483–92.

32. Antonny B, Chabre M. Characterization of the aluminum and beryllium fluoride species which activate transducin. Analysis of the binding and dissociation kinetics. *J Biol Chem* 1992;**267**:6710–18.
33. Sternweis PC, Gilman AG. Aluminum: a requirement for activation of the regulatory component of adenylate cyclase by fluoride. *Proc Natl Acad Sci USA* 1982;**79**:4888–91.
34. Du X, Black GE, Lecchi P, Abramson FP, Sprang SR. Kinetic isotope effects in Ras-catalyzed GTP hydrolysis: Evidence for a loose transition state. *Proc Natl Acad Sci USA* 2004;**101**:8858–63.
35. Arshavsky Jr VY, Pugh EN. Lifetime regulation of G protein–effector complex: emerging importance of RGS proteins. *Neuron* 1998; **20**:11–14.
36. Ross EM, Wilkie TM. GTPase-activating proteins for heterotrimeric G proteins: regulators of G protein signaling (RGS) and RGS-like proteins. *Annu Rev Biochem* 2000;**69**:795–827.
37. Zerangue N, Jan LY. G-protein signaling: fine-tuning signaling kinetics. *Curr Biol* 1998;**8**:R313–16.
38. Mukhopadhyay S, Ross EM. Rapid GTP binding and hydrolysis by G(q) promoted by receptor and GTPase-activating proteins. *Proc Natl Acad Sci USA* 1999;**96**:9539–44.
39. Biddlecome GH, Berstein G, Ross EM. Regulation of phospholipase C-β1 by G_q and m1 muscarinic cholinergic receptor Steady-state balance of receptor-mediated activation and GTPase-activating protein-promoted deactivation. *J Biol Chem* 1996;**271**:7999–8007.
40. Lan KL, Zhong H, Nanamori M, Neubig RR. Rapid kinetics of regulator of G-protein signaling (RGS)-mediated Galpha i and Galpha o deactivation. Galpha specificity of RGS4 and RGS7. *J Biol Chem* 2000;**275**:33,497–503.
41. Turcotte M, Tang W, Ross EM. Coordinate regulation of G protein signaling via dynamic interactions of receptor and GAP. *PLoS Comput Biol* 2008;**4**:e1000148.
42. Watson N, Linder ME, Druey KM, Kehrl JH, Blumer KJ. RGS family members: GTPase-activating proteins for heterotrimeric G- protein α-subunits. *Nature* 1996;**383**:172–5.
43. Berman DM, Wilkie TM, Gilman AG. GAIP and RGS4 are GTPase-activating proteins (GAPs) for the G_i subfamily of G protein α subunits. *Cell* 1996;**86**:445–52.
44. Chen C-K, Wieland T, Simon MI. RGS-r, a retinal specific RGS protein, binds an intermediate conformation of transducin and enhances recycling. *Proc Natl Acad Sci USA* 1996;**93**:12,885–9.
45. Srinivasa SP, Watson N, Overton MC, Blumer KJ. Mechanism of RGS4, a GTPase-activating protein for G protein α subunits. *J Biol Chem* 1998;**273**:1529–33.
46. Druey KM, Kehrl JH. Inhibition of regulator of G protein signaling function by two mutant RGS4 proteins. *Proc Natl Acad Sci USA* 1997;**94**:12851–6.
47. Tesmer JJG, Berman DM, Gilman AG, Sprang SR. Structure of RGS4 bound to AlF_4^--activated $G_i\alpha_1$:Stabilization of the transition state for GTP hydrolysis. *Cell* 1997;**89**:251–61.
48. He W, Cowan CW, Wensel TG. RGS9, a GTPase accelerator for phototransduction. *Neuron* 1998;**20**:95–102.
49. Sowa ME, He W, Slep KC, Kercher MA, Lichtarge O, Wensel TG. Prediction and confirmation of a site critical for effector regulation of RGS domain activity. *Nat Struct Biol* 2001;**8**:234–7.
50. Slep KC, Kercher MA, Wieland T, Chen CK, Simon MI, Sigler PB. Molecular architecture of Galphao and the structural basis for RGS16-mediated deactivation. *Proc Natl Acad Sci USA* 2008;**105**:6243–8.
51. Soundararajan M, Willard FS, Kimple AJ, et al. Structural diversity in the RGS domain and its interaction with heterotrimeric G protein alpha-subunits. *Proc Natl Acad Sci USA* 2008;**105**:6457–62.
52. Kozasa T, Jiang X, Hart MJ, et al. p115 RhoGEF, a GTPase activating protein for $G\alpha_{12}$ and $G\alpha_{13}$. *Science* 1998;**280**:2109–11.
53. Chen Z, Wells CD, Sternweis PC, Sprang SR. Structure of the rgRGS domain of p115RhoGEF. *Nat Struct Biol* 2001;**8**:805–9.
54. Garrard SM, Longenecker KL, Lewis ME, Sheffield PJ, Derewenda ZS. Expression, purification, and crystallization of the RGS-like domain from the Rho nucleotide exchange factor, PDZ-RhoGEF, using the surface entropy reduction approach. *Protein Expr Purif* 2001;**21**:412–16.
55. Berstein G, Blank JL, Jhon D-Y, Exton JH, Rhee SG, Ross EM. Phospholipase C-β1 is a GTPase-activating protein for $G_{q/11}$, its physiologic regulator. *Cell* 1992;**70**:411–18.
56. Lee S, Shin S, Hepler J, Gilman A, Rhee S. Activation of phospholipase C-beta 2 mutants by G protein alpha q and beta gamma subunits. *J Biol Chem* 1993;**268**:25,952–7.
57. Singer AU, Waldo GL, Harden TK, Sondek J. A unique fold of phospholipase C-beta mediates dimerization and interaction with G alpha q. *Nat Struct Biol* 2002;**9**:32–6.
58. Ilkaeva O, Kinch LN, Paulssen RH, Ross EM. Mutations in the carboxyl-terminal domain of phospholipase C-beta 1 delineate the dimer interface and a potential Galphaq interaction site. *J Biol Chem* 2002;**277**:4294–300.
59. Sunahara RK, Dessauer CW, Whisnant RE, Kleuss C, Gilman AG. Interaction of $G_s\alpha$ with the cytosolic domains of mammalian adenylyl cyclase. *J Biol Chem* 1997;**272**:22,265–71.
60. Lambright DG, Sondek J, Bohm A, Skiba NP, Hamm H, abd Sigler PB. The 2.0-Å crystal structure of a heterotrimeric G protein. *Nature* 1996;**379**:311–19.
61. Wall MA, Coleman DE, Lee E, et al. The structure of the G protein heterotrimer $G_i\alpha_1\beta_1\gamma_2$. *Cell* 1995;**80**:1047–58.
62. Bunemann M, Frank M, Lohse MJ. Gi protein activation in intact cells involves subunit rearrangement rather than dissociation. *Proc Natl Acad Sci USA* 2003;**100**:16,077–82.
63. Clapham DE, abd Neer EJ. G protein beta gamma subunits. *Annu Rev Pharmacol Toxicol* 1997;**37**:167–203.
64. Singer WD, Brown HA, abd Sternweis PC. Regulation of eukaryotic phosphatidylinositol-specific phospholipase C and phospholipase D. *Annu Rev Biochem* 1997;**66**:475–509.
65. Sondek J, Bohm A, Lambright DG, Hamm HE, Sigler PB. Crystal structure of a G-protein $\beta\gamma$ dimer at 2.1-Å resolution. *Nature* 1996;**379**:369–74.
66. Sprang SR. G protein mechanisms: insights from structural analysis. *Annu Rev Biochem* 1997;**66**:639–78.
67. Wall MA, Posner BA, Sprang SR. Structural basis of activity and subunit recognition in G protein heterotrimers. *Structure* 1998;**6**:1169–83.
68. Ford CE, Skiba NP, Bae H, et al. Molecular basis for interactions of G protein betagamma subunits with effectors. *Science* 1998;**280**: 1271–4.
69. Gaudet R, Bohm A, Sigler P. Crystal structure at 2.4 angstroms resolution of the complex of transducin betagamma and its regulator, phosducin. *Cell* 1996;**87**:577–88.
70. Cheever ML, Snyder JT, Gershburg S, Siderovski DP, Harden TK, Sondek J. Crystal structure of the multifunctional Gbeta5–RGS9 complex. *Nat Struct Mol Biol* 2008;**15**:155–62.
71. Davis TL, Bonacci TM, Sprang SR, Smrcka AV. Structural and molecular characterization of a preferred protein interaction surface on G protein betagamma subunits. *Biochemistry* 2005;**44**: 10,593–604.
72. Bonacci TM, Mathews JL, Yuan C, et al. Differential targeting of Gbetagamma-subunit signaling with small molecules. *Science* 2006;**312**:443–6.
73. Sondek J, Siderovski DP. Ggamma-like (GGL) domains: new frontiers in G-protein signaling and beta-propeller scaffolding. *Biochem Pharmacol* 2001;**61**:1329–37.

74. Snow BE, Krumins AM, Brothers GM, et al. A G protein gamma subunit-like domain shared between RGS11 and other RGS proteins specifies binding to Gbeta5 subunits. *Proc Natl Acad Sci USA* 1998;**95**:13,307–12.
75. Hajdu-Cronin YM, Chen WJ, Patikoglou G, Koelle MR, Sternberg PW. Antagonism between G(o)alpha and G(q)alpha in Caenorhabditis elegans: the RGS protein EAT-16 is necessary for G(o)alpha signaling and regulates G(q)alpha activity. *Genes Dev* 1999;**13**:1780–93.
76. Takesono A, Cismowski MJ, Ribas C, et al. Receptor-independent activators of heterotrimeric G-protein signaling pathways. *J Biol Chem* 1999;**274**:33,202–5.
77. Peterson III YK, Bernard ML, Ma H, Hazard S, Graber SG, Lanier SM. Stabilization of the GDP-bound conformation of Gialpha by a peptide derived from the G-protein regulatory motif of AGS3. *J Biol Chem* 2000;**275**:33,193–6.
78. Siderovski DP, Diverse-Pierluissi M, De Vries L. The GoLoco motif: a Galphai/o binding motif and potential guanine–nucleotide exchange factor. *Trends Biochem Sci* 1999;**24**:340–1.
79. Bernard ML, Peterson YK, Chung P, Jourdan J, Lanier SM. Selective interaction of AGS3 with G-proteins and the influence of AGS3 on the activation state of G-proteins. *J Biol Chem* 2001;**276**:1585–93.
80. Natochin M, Lester B, Peterson YK, Bernard ML, Lanier SM, Artemyev NO. AGS3 inhibits GDP dissociation from galpha subunits of the Gi family and rhodopsin-dependent activation of transducin. *J Biol Chem* 2000;**275**:40,981–5.
81. De Vries L, Zheng B, Fischer T, Elenko E, Farquhar MG. The regulator of G protein signaling family. *Annu Rev Pharmacol Toxicol* 2000;**40**:235–71.
82. De Vries L, Fischer T, Tronchere H, et al. Activator of G protein signaling 3 is a guanine dissociation inhibitor for Galpha i subunits. *Proc Natl Acad Sci USA* 2000;**97**:14364–9.
83. Blumer JB, Smrcka AV, Lanier SM. Mechanistic pathways and biological roles for receptor-independent activators of G-protein signaling. *Pharmacol Ther* 2007;**113**:488–506.
84. Adhikari A, Sprang SR. Thermodynamic characterization of the binding of activator of G protein signaling 3 (AGS3) and peptides derived from AGS3 with $G\alpha_{i1}$. *J Biol Chem* 2003;**278**:51,825–32.
85. Kimple RJ, Kimple ME, Betts L, Sondek J, Siderovski DP. Structural determinants for GoLoco-induced inhibition of nucleotide release by Galpha subunits. *Nature* 2002;**416**:878–81.
86. Tall GG, Gilman AG. Purification and functional analysis of Ric-8A: a guanine nucleotide exchange factor for G-protein alpha subunits. *Methods Enzymol* 2004;**390**:377–88.
87. Siderovski DP, Willard FS. The GAPs, GEFs, and GDIs of heterotrimeric G-protein alpha subunits. *Intl J Biol Sci* 2005;**1**:51–66.
88. Willard FS, Kimple RJ, Siderovski DP. Return of the GDI: the GoLoco motif in cell division. *Annu Rev Biochem* 2004;**73**: 925–51.
89. Thomas CJ, Tall GG, Adhikari A, Sprang SR. RIC-8A catalyzes guanine nucleotide exchange on Galphai1 bound to the GPR/GoLoco exchange inhibitor AGS3. *J Biol Chem* 2008;**12**. in press.
90. Rasmussen SG, Choi HJ, Rosenbaum DM, et al. Crystal structure of the human beta2 adrenergic G-protein-coupled receptor. *Nature* 2007;**450**:383–7.
91. Rosenbaum DM, Cherezov V, Hanson MA, et al. GPCR engineering yields high-resolution structural insights into beta2-adrenergic receptor function. *Science* 2007;**318**:1266–73.
92. Cherezov V, Rosenbaum DM, Hanson MA, et al. High-resolution crystal structure of an engineered human beta2-adrenergic G protein-coupled receptor. *Science* 2007;**318**:1258–65.
93. Warne T, Serrano-Vega MJ, Baker JG, et al. Structure of a beta1-adrenergic G-protein-coupled receptor. *Nature* 2008;**454**:486–91.
94. Palczewski K, Kumasaka T, Hori T, et al. Crystal structure of rhodopsin: A G protein-coupled receptor. *Science* 2000;**289**:739–45.
95. Salom D, Lodowski DT, Stenkamp RE, et al. Crystal structure of a photoactivated deprotonated intermediate of rhodopsin. *Proc Natl Acad Sci USA* 2006;**103**:16,123–8.
96. Park JH, Scheerer P, Hofmann KP, Choe HW, Ernst OP. Crystal structure of the ligand-free G-protein-coupled receptor opsin. *Nature* 2008;**454**:183–7.
97. Hamm HE. The many faces of G protein signaling. *J Biol Chem* 1998;**273**:669–72.
98. Onrust R, Herzmark P, Chi P, et al. Receptor and betagamma binding sites in the alpha subunit of the retinal G protein transducin. *Science* 1997;**275**:381–4.
99. Cai K, Itoh Y, Khorana HG. Mapping of contact sites in complex formation between transducin and light-activated rhodopsin by covalent crosslinking: use of a photoactivatable reagent. *Proc Natl Acad Sci USA* 2001;**98**:4877–82.
100. Itoh Y, Cai K, Khorana HG. Mapping of contact sites in complex formation between light-activated rhodopsin and transducin by covalent crosslinking: use of a chemically preactivated reagent. *Proc Natl Acad Sci USA* 2001;**98**:4883–7.
101. Marin EP, Krishna AG, Sakmar TP. Rapid activation of transducin by mutations distant from the nucleotide-binding site: evidence for a mechanistic model of receptor-catalyzed nucleotide exchange by G proteins. *J Biol Chem* 2001;**276**:27,400–5.
102. Iiri T, Herzmark P, Nakamoto JM, van Dop C, Bourne HR. Rapid GDP release from Gs alpha in patients with gain and loss of endocrine function. *Nature* 1994;**371**:164–8.
103. Posner BA, Mixon MB, Wall MA, Sprang SR, Gilman AG. The A326S mutant of $G_i\alpha_1$ as an approximation of the receptor-bound state. *J Biol Chem* 1998;**273**:21,752–8.
104. Rondard P, Iiri T, Srinivasan S, Meng E, Fujita T, Bourne HR. Mutant G protein alpha subunit activated by Gbeta gamma: a model for receptor activation. *Proc Natl Acad Sci USA* 2001;**98**: 6150–5.
105. Cherfils J, Chardin P. GEFs: structural basis for their activation of small GTP-binding proteins. *Trends Biochem Sci* 1999;**24**:306–11.
106. Medkova M, Preininger AM, Yu NJ, Hubbell WL, Hamm HE. Conformational changes in the amino-terminal helix of the G protein alpha(i1) following dissociation from Gbetagamma subunit and activation. *Biochemistry* 2002;**41**:9962–72.
107. Oldham WM, Van Eps N, Preininger AM, Hubbell WL, Hamm HE. Mechanism of the receptor-catalyzed activation of heterotrimeric G proteins. *Nat Struct Mol Biol* 2006;**13**:772–7.
108. Oldham WM, Van Eps N, Preininger AM, Hubbell WL, Hamm HE. Mapping allosteric connections from the receptor to the nucleotide-binding pocket of heterotrimeric G proteins. *Proc Natl Acad Sci USA* 2007;**104**:7927–32.
109. Abdulaev NG, Ngo T, Ramon E, Brabazon DM, Marino JP, Ridge KD. The receptor-bound "empty pocket" state of the heterotrimeric G-protein alpha-subunit is conformationally dynamic. *Biochemistry* 2006;**45**:12,986–97.
110. Notredame C, Higgins DG, Heringa J. T-Coffee: a novel method for fast and accurate multiple sequence alignment. *J Mol Biol* 2000;**302**:205–17.
111. Larkins MA, Blackshields G, Brown NP, Chenna R, McGettigan PA, McWilliam H, et al. Cluster W and Cluster X version 2.0. *Bioinformatics* 2008;**23**:2947–8.
112. Coleman DE, Sprang SR. Structure of $G_i\alpha_1$•GppNHp, autoinhibition in a $G\alpha$ protein-substrate complex. *J Biol Chem* 1999;**274**: 16,669–72.

Chapter 28

G Protein-Coupled Receptor Structures

Veli-Pekka Jaakola and Raymond C. Stevens
Department of Molecular Biology, The Scripps Research Institute, La Jolla, California

INTRODUCTION

Guanine nucleotide protein- (G protein-) coupled receptors, or GPCRs, form the largest, most ubiquitous and most variable superfamily of integral transmembrane proteins. GPCRs are heptahelical transmembrane (7-TM) receptors that initiate cellular responses from extracellular signals/stimulus by activating membrane-associated heterotrimeric guanine nucleotide binding regulatory proteins or G proteins [1, 2]. It is estimated that the GPCRs for endogenous ligands consist of approximately 400 receptors in the human; potentially more than 1000 when receptors of those ligands not yet identified are included [3–5]. About 40 percent of all prescription pharmaceuticals on the market function by using these classes of receptors [3, 4].

Consequently, near-atomic resolution structures are essential to understand ligand-GPCR : G protein signal transduction and intracellular communications in detail. Crystal structures of individual GPCRs would allow rational design of drugs to treat various human maladies [4]. Due to the many bottlenecks associated mainly with recombinant receptor production, purification, crystallization, and data collection, progress has been slow when compared to other membrane proteins [6, 7] (http://blanco.biomol.uci.edu/Membrane_Proteins_xtal.html). Until the year 2007, structural understanding of GPCRs was limited to bovine/frog visual rhodopsin (a true member of the GPCR family) and Archaeal bacteriorhodopsin (7-TM protein) systems, both abundant from their native sources in the retinal membrane of the eye and purple membranes, respectively [2, 8–11]. In 2007–2008, however, the crystallographic structures of three unique GPCRs, namely human β_2-adrenergic receptor (β_2-AR) [12–15], turkey β_1-adrenergic receptor (β_1-AR) [16], and human adenosine A2A receptor [17], that bind diffusible antagonists or receptor blockers (such as small molecules and neurotransmitters) were obtained. More recently, the crystallographic structure of a rhodopsin with G protein peptide bound in an active conformation was also solved [18]. Altogether, as of November 2008, the Protein Data Bank (PDB; http://www.rcsb.org/pdb/) had more than 15 bacterial and mammalian entries for crystallographic and nuclear magnetic resonance spectroscopic (NMR) structures of GPCRs (Table 28.1), and was beginning to provide real insight into the molecular structural understanding of signal transduction across the cell membrane.

CLASSIFICATION

The GPCRs share a common structural motif of seven transmembrane helices, connected by three cytosolic loops/domains, three extracellular loops/domains, with an extracellular amino terminal end/domain and a cytosolic carboxyl terminal end/domain (Figure 28.1; http://www.gpcr.org/). This basic motif was first visualized by vitrified cryo-electron microscopy reconstitution of visual rhodopsin (and bacteriorhodopsin) using two-dimensional crystals, and later verified by protein crystallography using three-dimensional crystals, and by other biophysical/biochemical studies [2, 19–21]. Functionally, GPCRs are responsible for the communication of events from the extracellular environment to the inside of the cell by activating the internal membrane-associated heterotrimeric G proteins, although more recent results indicate G protein independent signaling; therefore, a more general term, heptahelical proteins (or 7-TM), has been proposed [2, 6]. The extracellular and 7-TM core regions are involved in ligand binding, whereas the intracellular regions are involved in signaling (Figure 28.1).

Based on primary amino acid sequence comparison, the GPCRs can be classified into six different receptor classes (1–5 or A–F; see also, www.gpcr.org/7tm/ [3]). Class 1 (also referred as the rhodopsin-like receptor family) is the largest and most extensively characterized. It is determined by the presence of 20 conserved residues in the 7-TM core. Class 2 has a large amino terminal domain that contains six highly conserved cysteine residues, in addition to the 20 conserved residues found in the 7-TM core. The large amino terminus is shown to participate in ligand binding.

Handbook of Cell Signaling, Three-Volume Set 2 ed.

TABLE 28.1 Overview of 7-TM, GPCR and GPCR domain structures deposited in the Protein Data Bank

Receptor	Resolution	PDB ID#
G protein-coupled receptors		
Bovine rhodopsin		
Rod outer segment	2.8 Å	1F88
Rod outer segment	2.6 Å	1L9H
Rod outer segment	2.65 Å	1GZM
Rod outer segment	2.2 Å	1U19
Recombinant (N2C/D282C)	3.4 Å	2J4Y
Photoactivated, rod outer segment	4.15 Å	2I37
(Rhod)opsin in ligand-free state (opsin, rod outer segment)	2.9 Å	3CAP
(Rhod)opsin (opsin (R*) – Gα transducin peptide complex, bovine rod outer segment)	3.2 Å	3DQB
Squid rhodopsin		
Gq protein type coupled	2.5 Å	2Z73
Gq protein type coupled (shows intracellularly extended cytoplasmic region)	3.7 Å	2ZIY
Non-rhodopsin		
Recombinant Turkey β1 Adrenergic Receptor with (Several Ala/Leu mutations, C-terminal deletion, Antagonist Cyanopindolol Bound)	2.7 Å	2VT4
Recombinant Human β2 Adrenergic Receptor Antibody complex, C-terminal deletion	3.4/3.7 Å	2R4R, 2R4S
Recombinant Human β2 Adrenergic Receptor T4 lysozyme replaces third intracellular loop, C-terminal deletion, partial inverse agonist carazolol bound	2.4 Å	2RH1
Recombinant Human β2 Adrenergic Receptor E122W mutation, T4 lysozyme replaces third intracellular loop, C-terminal deletion, FDA approved timolol bound, confirms cholesterol binding region	2.8 Å	3D4S
Recombinant human adenosine A2A receptor (T4 lysozyme replaces third intracellular loop, C-terminal deletion, antagonist ZM241385 bound)	2.6 Å	3EML
Other heptahelical proteins		
H. salinarum Bacteriorhodopsin		
Two-dimensional crystals (electron diffraction)	3.5 Å	2BRD
Two-dimensional crystals (electron diffraction)	3.0 Å	1AT9
Three-dimensional crystals (first *in cubo* structure)	2.35 Å	1AP9
Three-dimensional crystals (lipids)	1.9 Å	1QHJ
K intermediate	2.1 Å	1QKO, 1KQP
Three-dimensional crystals (lipids)	2.3 Å	1BRX
Three-dimensional crystals (high resolution)	1.55 Å	1C3W

(Continued)

TABLE 28.1 (Continued)

Receptor	Resolution	PDB ID#
D96N mutant in M bR state	1.80/2.00 Å	1C8S,1C8R
Three-dimensional crystals (Highest resolution *in surfo* Structure)	2.9 Å	1BRR
H. salinarum Halorhodopsin		
Three-dimensional crystals	1.8 Å	1E12
Sensory Rhodopsin II		
N. pharaonis	2.4 Å	1JGJ
N. pharaonis	2.1 Å	1H68
Anabaena (Nostoc) sp. PCC7120	2.0 Å	1XIO
Sensory Rhodopsin II with Transducer		
N. pharaonis	1.93 Å	1H2S
Archaerhdopsins		
Halorubrum sp. aus -1 Archaerhodopsin 1	3.4 Å	1UAZ
Halorubrum sp. aus -2 Archaerhodopsin2	2.5 Å	1VGO
Halorubrum sp. aus-2 Archaerhodopsin-2 (Crystallized with the Carotenoid Bacterioruberin)	2.10 Å	2EI4
Non-membrane associated GPCR domains		
SDF1 in complex with the CXCR4 N-terminus	NMR	2K03
Rat extracellular region of the Group II metabotropic glutamate receptor complexed with L-glutamate	2.35 Å	2E4U
Human follicle stimulating hormone complexed with N terminal domain of receptor	2.92 Å	1XWD
Mouse cysteine-rich domain of frizzled receptor	1.35 Å	1IJY
Drosophila ectodomain of Methuselah receptor	2.30 Å	1FJR
Mouse N-terminal domain of a type B1 GPCR in complex with peptide ligand	NMR	2JND

Table is adapted from information provided at http://blanco.biomol.uci.edu/Membrane_Proteins_xtal.html and http://www.rcsb.org/pdb/

Class 3 (the Ca^{2+} sensing receptor family and the metabotropic neurotransmitter) is characterized by a very long amino terminal domain that contains nearly 20 conserved cysteine residues, in addition to those conserved residues localized in the 7-TM core. Class 4 contains yeast pheromone receptors. Class 5 includes the frizzled/smoothed receptors of fly embryonic development. Finally, there is also an additional sixth GPCR class, the so-called cyclic adenosine monophosphate receptors, which have been found in the slime mold *Dictyostelium discoideum.*

BASIC CONCEPT OF GPCR; HETEROTRIMERIC G PROTEINS; THE VAST COMPLEXITY OF GPCR SIGNALING

The basic signaling concept includes a ligand, a GPCR, a heterotrimeric G protein complex, and a G protein regulated secondary effector(s) (for a recent review, see [22]). In the non-activated or resting state of the G protein signaling cycle, the heterotrimeric G protein complex ($\alpha\beta\gamma$ subunits) is associated with a guanosine diphosphate

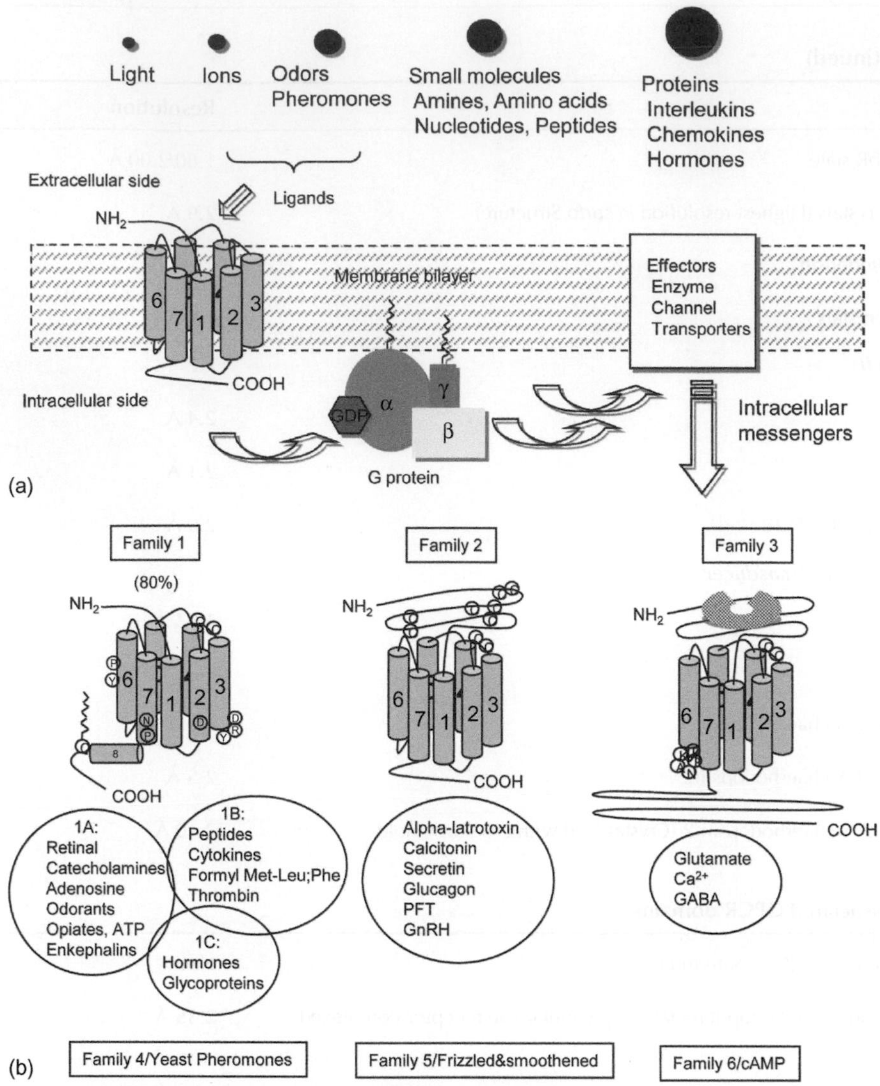

FIGURE 28.1 The classification and binding models of GPCRs (http://www.gpcr.org/).
(a) Schematic presentation of basic components for GPCR signaling machinery (ligand, receptor, membrane barrier, G protein, effectors). The ligand size for receptors can vary from light to nearly megaDalton protein complexes. (b) GPCRs have three main families based on their amino acid sequences. Receptors from different families have very low sequence similarity. Family 1 contains the majority of GPCRs, including receptors for odorants.

molecule bond. In this state, the heterotrimeric G protein can be recognized by an appropriate activated receptor (this can also be spontaneous, non-ligand associated), which interacts with the heterotrimeric G protein. The receptor:heterotrimeric G protein interaction results in the exchange of guanosine diphosphate for guanosine triphosphate in the G protein α subunit. Guanosine triphosphate binding to the G protein α subunit induces a conformational change, which causes dissociation of the G protein α subunit and the G protein βγ complex from each other and from the heptahelical receptor. The guanosine triphosphate-bound G protein α subunit, as well as the G protein βγ dimer, is now able to interact with effector proteins such as channels, transporters, and enzymes. The slow inherent guanosine triphosphate hydrolysis activity in the G protein α subunit terminates the G protein signaling. The resultant G protein α:guanosine diphosphate complex is inactive, and, by reassociating with the βγ heterodimer complex, it inactivates the βγ complex. The vast complexity of receptor signaling has been further underlined by results that some receptors undergo spontaneous activation in the absence of agonist [23]. There are about 20 different G subunits characterized so far; αβγ isoforms can form different combinations that have different selectivity for a particular receptor. Also, one receptor type can activate more than one class of G proteins, depending on the cellular environment and number of G proteins/receptors in the membranes or surrounding microdomains (receposomes).

G protein signaling is a transient and non-constant process; high activation of a receptor reduces its capability to be stimulated in the future (this is called desensitization), while low activation increases its capability (sensitization). Also, agonist-induced desensitization has been classified into agonist-specific (homologous) or agonist-non-specific (heterologous) desensitization. After agonist binding, desensitization is controlled by receptor phosphorylation. This happens by second-messenger kinases or by a distinct family of GPCR kinases (for reviews, see [22, 24]). Second-messenger kinases, such as protein kinase A (PKA) and protein kinase C (PKC), directly uncouple receptors from their G proteins. The second general mechanism for regulating receptor occurs *via* the GPCR kinases (GRKs): β-arrestin mediated system; the agonist-occupied or activated receptor is phosphorylated by G protein regulators (GPRs). GPR phosphorylation causes binding of β-arrestin to the receptor, which inhibits binding between the receptor and the G protein.

MODELS FOR RECEPTOR ACTIVATION

The most widely used model for ligand-dependent receptor activation has been described by the so-called extended ternary complex, and its extension, the cubic ternary complex (for a recent review, see [2, 6, 21, 22]). These two-state models assume that, in the absence of a ligand, there is equilibrium between two functionally and structurally different states: the inactive (low affinity; R) and the active (high affinity; R*). The ability of ligands to induce signaling responses (the efficacy of the ligand) varies; a full agonist binds and stabilizes the active state (R*), which has high coupling efficiency for a specific G protein (R*G). A partial agonist binds and stabilizes the active state less efficiently, and causes only partial G protein activation. A neutral antagonist recognizes all receptor forms equally, and has no effect on the equilibrium between (R) and (R*) states. An inverse agonist binds and stabilizes the inactive state (R), thus decreasing the basal agonist-independent G protein activity. The extended ternary complex model can describe these different events. The cubic ternary complex model includes the possibility that the inactive receptor forms (R and AR) interact with G protein, which is not included in the extended ternary complex model. The initial models were based on studies using radioligand binding effectors (G protein, guanyl nucleotides, ligands, and allosteric modulators) combined with mutagenesis. All of these change the binding properties, and thereby the binding curve shape (cooperatively, Hill number) of the receptor, which can be explicated by the models.

A few years ago, a sequential binding model was described for some extensively studied receptors [23]. These sequential binding models were based on intermolecular fluorescent probes or inherent tryptophan fluorescence monitoring of local conformational changes within purified receptor. In the model, agonist binding results in receptor activation through a series of isolated intermediate conformational states ($R + A \leftrightarrow AR1 \leftrightarrow ARn \leftrightarrow AR^*$) that can be associated with different structural origins of the ligands.

STRUCTURES OF EXTRACELLULAR DOMAINS OF GPCRs

The amino terminal domains have the largest sequence variability, and therefore most probable structural variability, within GPCRs [1, 3]. The amino terminus forms the primary ligand binding site for some classes of GPCRs. Consequently, various structural studies have been carried out to characterize the recombinantly expressed large amino terminus without the 7-TM core domain, using crystallography as well as NMR (Figure 28.2). The follicle stimulating hormone (FSH) receptor amino terminal domain FSH–glycoprotein complex forms a tubular-like structure where FSH–glycoprotein is tightly bound around it [25]. It has been suggested that this might be a conserved binding motif among this class of GPCRs, since a similar motif is also found in the thyroid-stimulating hormone (TSH) receptor antibody complex. Most of the class 3 GPCRs have a so-called *Venus Flytrap* ligand binding domain with a cysteine-rich linker domain. The structures of the Venus Flytrap domain have been determined for a few metabotropic glutamate receptors, or mGluRs, consisting of two separate lobes with a ligand binding cavity located in the gap between lobes [26]. The cysteine-rich ligand binding domain structures of corticotrobin-releasing factor and fly's Methuselah might form prototypes of peptide-ligand binding domains among class 2 GPCRs [27]. The binding domains consist mainly of beta-sheets, and are stabilized by several Cys–Cys bridges. Class 5 GPCRs comprise the frizzled/smoothed receptors, the crystallographic structure of mouse frizzled ligand binding domain has been determined [28]. The cysteine-rich structure is mainly alpha-helical. Finally, the NMR structure of chemokine CXCR4 receptor amino terminal domain [29]: SDF-1 ligand complex was solved, providing insight into chemokine recognition.

Although various amino terminal structures, with and without the ligand bound, have been determined, very little is known about the structural origin of signal transduction to the 7-TM. Based on mutagenesis studies (with and without the amino terminal domain), 7-TMs fold correctly and are localized to the plasma membrane. In the case where small molecules or allosteric molecules exist, receptors are able to signal when agonized and/or allosterically modulated. This might suggest that the activation mechanism is conserved among GPCRs, and activation or inactivation is probably due to a conserved conformational change in the 7-TM core.

STRUCTURES PROBING THE INACTIVE STATE(S): LIGAND ENTRY, BINDING, AND MODES FOR ACTIVITY BLOCKING

The overall fold in the 7-TM core found in the various rhodopsin, human/turkey β-ARs and human adenosine

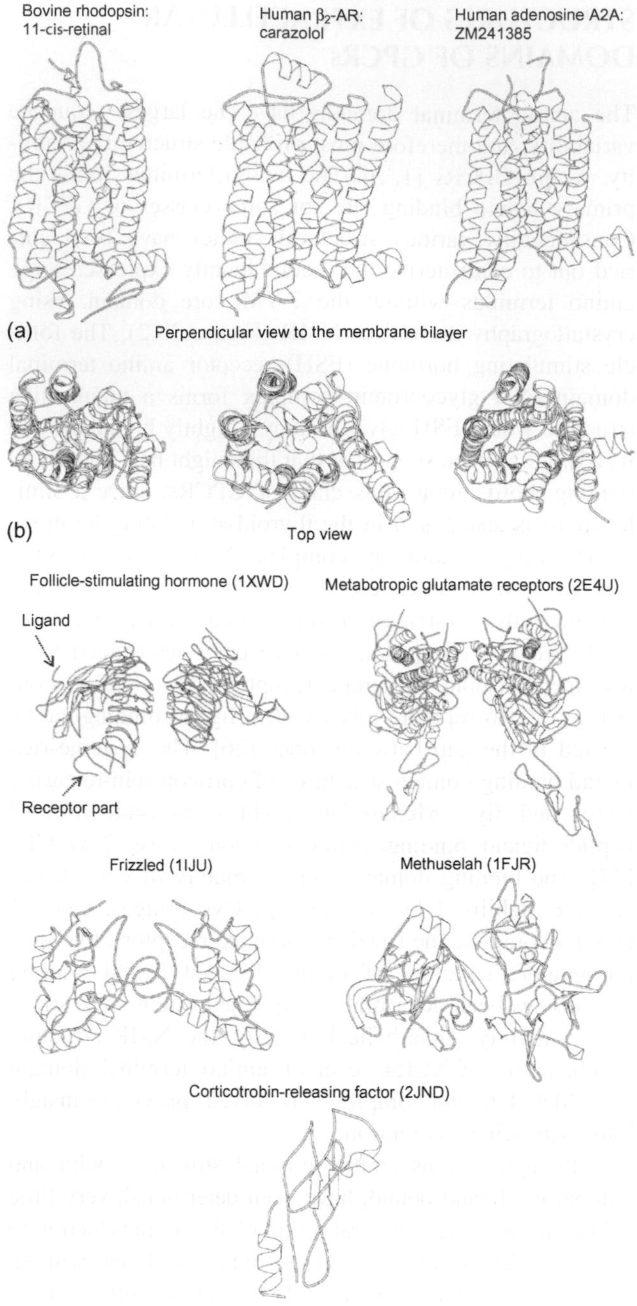

FIGURE 28.2 Structures of known GPCRs and GPCR domains. (a) Representative structures of 7-TM core GPCRs: bovine rhodopsin : 11-*cis*-retinal complex (1U19), human β_2-AR : carazolol complex (2HR1), and human adenosine A2A : ZM241385 (3EML). Side (a) and top (b) views are given. Note the open entry to the binding cavity in 2HR1 and 3EML structures when compared to 1U19. The ligand position is very similar between 1U19 and 2HR1, but differs from 3EML. (c) Representative set of GPCR domain structures solved by X-ray and NMR. Note that the majority of the domains (with ligands) form dimers in the crystals. It is not known if these are physiologically relevant, or artifacts from crystallization contacts. Interestingly, a majority of the domains are cysteine-rich, which are forming intramolecular Cys–Cys bridges.

A2A receptor structures is very similar, with a root mean square deviation value *ca.* 1–2.5 Å between these structures (Figure 28.2) [7, 30]. Functionally, the rhodopsin structure with 11-*cis*-retinal bound [8], β_2-AR-lysozyme T4 (T4) structures with timolol/carazolol bound [12, 13, 15], β_2-AR-antibody fragment structure with carazolol bound (not visible in electron density) [14], turkey β_1-AR (various residues mutated to alanine or leucine to promote stability/restrict conformational flexibility) structure with cyanopindolol bound [16], and the human adenosine A2A-T4 structure with ZM241385 bound [17], represent inactive (R) ligand blocked states. The representative ligand binding pocket for a small molecule in these receptors, which are all class 1 GPCRs, lies within the 7-TM core. Just a few years ago, nearly all atomic resolution information was based on visual rhodopsin structure(s). The human β_2-AR structures that appeared first provided new information about the diffusible ligand binding [12, 13, 15]. First, the hairpin structure of the rhodopsin extracellular domain that is formed mainly of two beta-strands of the extracellular (EC)-2 loop is not present in other 7-TM structures. In β-ARs, the EC-2 loop forms an alpha helix followed by a short beta-strand that is constrained by two Cys–Cys bridges. This arrangement forms an open cavity in the ligand binding pocket, allowing diffusible ligands to enter. Another interesting feature in this arrangement for diffusible ligands is the electrostatic distribution on the surface of β-ARs that is favorable for charged amine-energic ligands. Also, there are coordinated sodium ion/water ions that lie near an alpha-helical region and which could be related to allosteric modulation [8, 10].

Interestingly, the open access of diffusible ligands is organized differently in the human adenosine A2A structure [17]. The EC-2 of adenosine A2A is mainly a random coil, containing only a very short alpha helix element that forms a critical aromatic phi stacking interaction to bound ligand. The coil structure is spatially constrained, and keeps the binding cavity blocked by four Cys–Cys bridges; three of them are unique for the adenosine A2A structure. There are short missing regions that were not elucidated in the final model due to a very weak electron density; however, in the trace-modeled version this region does not block the binding cavity (Jaakola *et al.*, personal communication).

It is interesting that in the unligated rhodopsin and unligated/activated (rhod)opsin G protein peptide complex structures, discussed in more detail below, the hairpin beta-strand structures blocking the binding cavity are still present [18, 31]. How is the ultimately covalently bound 11-*trans*-retinal transported/diffused into the binding cavity?

A second major finding in the new structures is the binding models of ligands. The ligand binding location and position between 11-*cis*-retinal and beta-blockers (timolol, carazolol, and cyanopindolol) in respective structures is quite similar [8, 12, 13, 15, 16]. In this sense, some of the three-dimensional models based on rhodopsin homology modeling, and used to understand structure–activity relationships and rational model-based drug-design of new beta-blockers, were relatively valid [7, 10, 30]. The β_2-AR structures shed a new light on a role for EC-2 in ligand binding interactions, selectivity, and kinetics. The timolol, carazolol, and cyanopindolol represent one similar class of beta-blockers; it will be interesting to see how the other larger beta-blockers with long hydrophobic tails, or linkers, will accumulate within the same binding cavity. These drugs might be useful in the treatment of, for example, obesity disorders, or for other non-traditional beta-blocker therapies [3].

The big surprise in the human adenosine A2A structure was the location and position of the antagonist binding cavity [17]. The ZM241385 is a typical high-affinity adenosine A2A selective antagonist with a characteristic adenosine-blocker ring system. Based on various homology models and supporting mutagenesis data, it was predicted that the location and position of the drug would be similar to retinal or beta-blockers. However, based on the structure, ZM241385 binds in an extended orientation perpendicular to the bilayer lipid plane and co-linear with helixes 5, 6, and 7. The drug makes several unpredicted interactions both with EC-2 and EC-3 that form subtype selectivity of the ZM241385. Notably, almost all mutagenesis data could be re-rationalized and explained based on the structure. The binding cavity forms slight differences in the helical position and orientation relatively to the rhodopsin and β-AR structures. It will be interesting to see how the diverse adenosine-blockers from ZM241385 will bind to the receptors. Nevertheless, the adenosine A2A ZM241385 structure creates a refreshed foundation for the rational drug design of adenosine-blockers that could be useful for various neurological disorders such as Parkinson's and Huntington's diseases [3].

As discussed above, classical drugs can be categorized as agonists (activate), inverse agonists (inactivate), or neutral antagonists (block), based on the physiological responses they engender [6, 21]. Depending on the ligand, the receptor can "dial" almost any response/conformation, from fully activated to fully inactivated; therefore, ligands are subcategorized in terms as neutral, weak, partial, and full. Inverse agonism was discovered in the context of recombinant systems (which typically have very high expression when compared to native cells) and constitutive activity mutants. The difference between true neutral antagonism and inverse agonism is related to the study system, and has been somewhat difficult to classify. Now, with different antagonist/inverse antagonist bound or unligated structures, we can begin to understand antagonism/inverse antagonism events in near-atomic terms.

Functional studies combined with mutagenesis, fluorescent probes, and biophysical spectroscopic studies have suggested that there are two main intermolecular constraints controlling the receptor activation mechanism: the so-called "toggle" switch and "ionic lock" [2, 10]. In the "toggle" switch, a cluster of aromatic residues in helix 6 (with especially conserved tryptophan residue in the middle of helix 6) changes side-chain rotamers that alter the configuration of the conserved proline residue of helix 6 related kink, and causes movement of the cytoplasmic end of helices 5, 6, and 3 during the activation process. This allows the opening of the cavity for G protein binding and activity. The direct and indirect interaction with these aromatic residues varies between the ligand and the "toggle" switch, and might be related to basal activity of the receptor. The visual rhodopsin: retinal system has virtually no activity (high number of contacts with "toggle" residues), while the adenosine A2A:ZM241385 and β_1-AR:cyanopindolol has relatively low basal or spontaneous activity (a medium number of contacts) and β_2-AR : carazolol/timolol has a relatively high basal activity that can be suppressed by these beta-blockers (only a few contacts) (Figure 28.3).

Another triggering site that has been proposed and plays a critical role in activation is the "ionic lock;" this is interactions between conversed residues in the cytoplasmic end of helices 3 and 6 (D/E–RY motif). In the rhodopsin structure, the "ionic lock" is intact [2, 10]. In the adenosine A2A and β_1-AR structure, the "ionic lock" is broken; however, the DRY motif of helix 3 forms similar polar interactions between cytosolic loop 2 (CL-2) in the both structures (Figure 28.3). Interestingly, in the β_2-AR structures there are no polar interactions with CL-2, even though the tyrosine residue is present in the same position, leaving the CL-2 more open in the β_2-AR structure when compared to adenosine A2A or β_1-AR (Figure 28.3). Prior to gaining this structural information, based on mutagenesis experiments, it was suggested that CL-2 works as a control switch for G protein activation using a helical elements in the CL-2; indeed, the structural data now support this assumption. Based on the basal activity profiles of rhodopsin, adenosine A2A, and β-ARs, the DRY motif and CL-2 interactions may have direct implications in spontaneous G protein activation.

STRUCTURE OF ACTIVE STATE(S)

Our structural understanding of activated states in GPCR is very limited [2, 10]. Only very recently, the first convincing structure of visual (rho)dopsin bound to a short peptide mimicking G protein tranducin carboxyl terminus was reported at modest resolution [18]. The most notable difference between activated (rhod)opsin and inactivated rhodopsin is a large tilt movement of helix 6, as well as a small movement of helix 7, forming the binding cavity for the G protein peptide on the intracellular side. It is noteworthy that these motions match very well those described in various biophysical studies using spectroscopic techniques, such as spin labeling and NMR [21, 32]. It is interesting

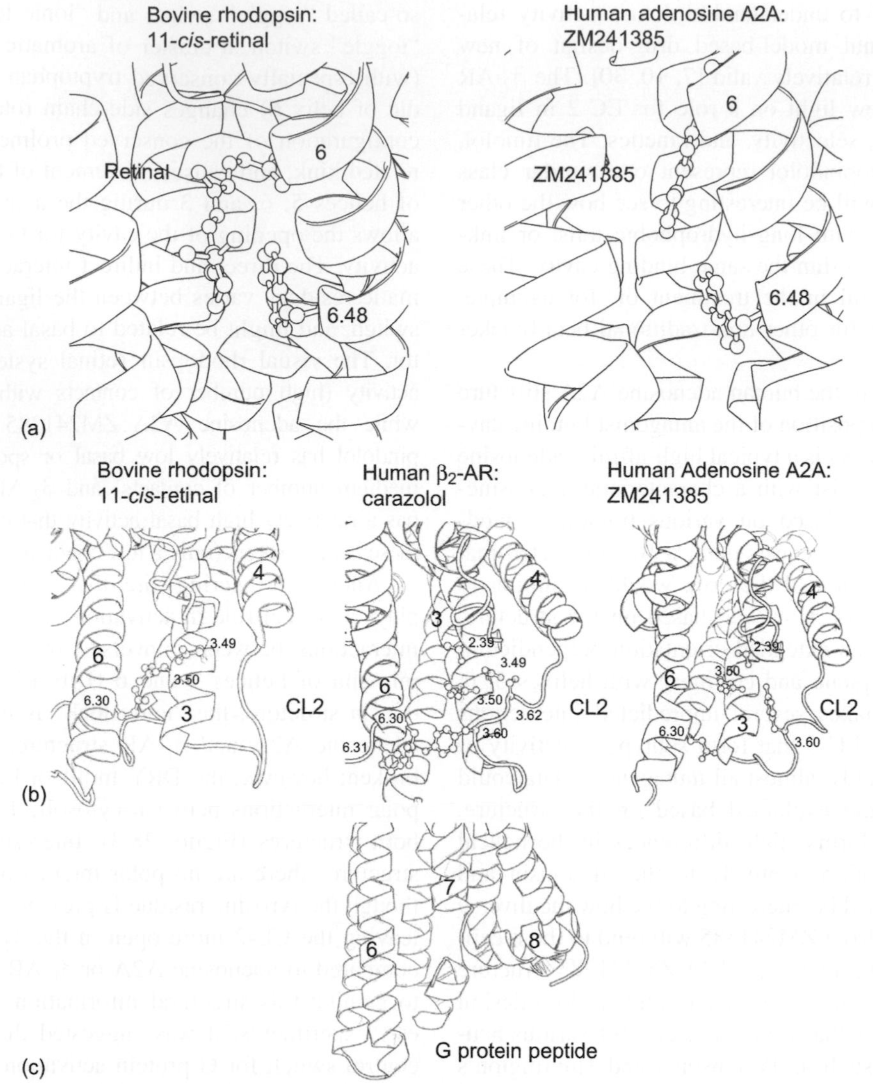

FIGURE 28.3 Some key features keeping GPCRs in inactive or active state(s).
(a) Comparison of the "toggle" switch residue of Trp 6.48 (Ballesteros–Weinstein numbering system) between bovine rhodopsin : 11-cis-retinal (1U19) and human adenosine A2A (3EML). Besides the different ligand chemistry and overall orientation of the ligands, the ligands have similar effects with the Trp 6.48 rotamer, constraining its rotamer position that has been proposed to keep the receptor in an inactive conformation. (b) Comparison of "ionic lock" and cytosolic loop 2 structures between bovine rhodopsin : 11-*cis*-retinal complex (1U19), human β_2-AR : carazolol complex (2HR1), and human adenosine A2A : ZM241385 (3EML). Bovine rhodopsin (left) has a canonical "ionic lock" between the D/E–RY motif of helix 3 and helix 6. These interactions are probably required to keep the receptor totally silent in the absence of light. This will reduce the "noise" in the visual signaling system. The "ionic lock" is broken in the β-ARs (middle) and adenosine receptor (right) systems; however, other stabilizing interactions occur in the same regions. Human adenosine A2A and turkey β_1-AR (not shown) have a helical cytosolic loop 2, and residue 3.50 (Arg102) of helix 3 may have a role in shifting the *p*Ka of the adjacent 3.49 (Asp101) resulting in hydrogen bonding interactions between helix 2 and cytosolic loop 2. Turkey β_1AR participates in similar interactions to A2A-T4L-ΔC, without the hydrogen bond to helix 2. β_2-AR does not have a helical portion in cytosolic loop 2, and has different set of interactions. These arrangements might explain the relatively high basal activity of β_2-AR, and the relatively low basal activities of adenosine A2A and β1-ARs. The residue numbering is given via the Ballesteros–Weinstein numbering system. (c) The cytosolic view of activated bovine (rhod)opsin with G protein transducin peptide bound. The structure represents the first glance of a receptor in active conformation. When compared to inactive state structures or the unligated (rhod)opsin structure, the helix 6 is tilted and forms an open cavity for G protein peptide.

that the activating covalently-bound ligand, all-*trans*-retinal, is not present in the structure [18].

The structure of (rho)opsin : G protein peptide is limited currently at 3.2-Å resolution, and thus does not allow detection of any ordered water molecules. A chain of conserved intrahelical water molecules (five to nine) connecting the ligand binding cavity via a so-called NP–XX–Y motif in helix 7 was found in the high-resolution structures of bovine/squid rhodopsins and β_2-AR–T4 lysozyme [11–13, 33]. Some of the known constitutively active and uncoupling mutants of β-ARs interact with these water molecules, and it was proposed that the water chain might take part in the activating mechanism [13, 33] It will be interesting to see, from studying active-state high-resolution structures, whether this is the case or not.

Above are some limited examples of the complexity of GPCR signaling, and descriptions of what we at first glance have noticed in these fascinating new structures. The challenging goals for the future will be obtaining the crystal structures of activated state(s) between agonist(s), receptor(s), and G protein(s). Prior to that, the crystallization of other GPCRs, β-ARs, and adenosine receptors bound to other ligands, peptides, or allosteric modulators will be extremely beneficial, both for basic and applied science (in better drug design, for instance). A crystal structure represents only a static picture of one conformation in a limited timeframe. With only a few structures available, we cannot clearly see what the limitations of crystallography are for understanding these classes of dynamic receptors; nevertheless, a new page has been turned in the GPCR story.

ACKNOWLEDGEMENTS

This work was supported in part by the NIH Roadmap Initiative grant P50 GM073197 (JCIMPT) for technology development and the Protein Structure Initiative grant U54 GM074961 (ATCG3D) for GPCR processing. The authors thank the GPCR team, including Vadim Cherezov, Mike Hanson, Chris Roth, Ellen Chien, Mark Griffith, Jeffrey Velasquez, Tam Trinh, and Kirk Allin, as well as Angela Walker for assistance with manuscript preparation and Raymond Benoit for laboratory management.

REFERENCES

1. Pierce KL, Premont RT, Lefkowitz RJ. Seven-transmembrane receptors. *Nature Rev* 2002;**3**:639–50.
2. Park PS, Lodowski DT, Palczewski K. Activation of G-protein-coupled receptors: beyond two-state models and tertiary conformational changes. *Annu Rev Pharmacol Toxicol* 2008;**48**:107–41.
3. Foord SM, Bonner TI, Neubig RR, Rosser EM, Pin JP, Davenport AP, Spedding M, Harmar AJInternational Union of Pharmacology. XLVI. G protein-coupled receptor list. *Pharmacol Rev* 2005;**57**:279–88.
4. Tyndall JD, Sandilya R. GPCR agonists and antagonists in the clinic. *Med Chem (United Arab Emirates)* 2005;**1**:405–21.
5. Vassilatis DK, Hohmann JG, Zeng H, Li F, Ranchalis JE, Mortrud MT, Brown A, Rodriguez SS, Weller JR, Wright AC, Bergmann JE, Gaitanaris GA. The G protein-coupled receptor repertoires of human and mouse. *Proc Natl Acad Sci USA* 2003;**100**:4903–8.
6. Kobilka BK, Deupi X. Conformational complexity of G-protein-coupled receptors. *Trends Pharmacol Sci* 2007;**28**:397–406.
7. Lefkowitz RJ, Sun JP, Shukla AK. A crystal clear view of the beta2-adrenergic receptor. *Nature Biotechnol* 2008;**26**:189–91.
8. Palczewski K, Kumasaka T, Hori T, Behnke CA, Motoshima H, Fox BA, Le Trong I, Teller DC, Okada T, Stenkamp RE, Yamamoto M, Miyano M. Crystal structure of rhodopsin: a G protein-coupled receptor. *Science* 2000;**289**:739–45.
9. Pebay-Peyroula E, Rummel G, Rosenbusch JP, Landau EM. X-ray structure of bacteriorhodopsin at 2.5 angstroms from microcrystals grown in lipidic cubic phases. *Science* 1997;**277**:1676–81.
10. Audet M, Bouvier M. Insights into signaling from the beta2-adrenergic receptor structure. *Nature Chem Biol* 2008;**4**:397–403.
11. Okada T, Sugihara M, Bondar AN, Elstner M, Entel P, Buss V. The retinal conformation and its environment in rhodopsin in light of a new 2.2 A crystal structure. *J Mol Biol* 2004;**342**:571–83.
12. Cherezov V, Rosenbaum DM, Hanson MA, Rasmussen SG, Thian FS, Kobilka TS, Choi HJ, Kuhn P, Weis WI, Kobilka BK, Stevens RC. High-resolution crystal structure of an engineered human beta2-adrenergic G protein-coupled receptor. *Science* 2007;**318**:1258–65.
13. Rosenbaum DM, Cherezov V, Hanson MA, Rasmussen SG, Thian FS, Kobilka TS, Choi HJ, Yao XJ, Weis WI, Stevens RC, Kobilka BK. GPCR engineering yields high-resolution structural insights into beta2-adrenergic receptor function. *Science* 2007;**318**:1266–73.
14. Rasmussen SG, Choi HJ, Rosenbaum DM, Kobilka TS, Thian FS, Edwards PC, Burghammer M, Ratnala VR, Sanishvili R, Fischetti RF, Schertler GF, Weis WI, Kobilka BK. Crystal structure of the human beta2 adrenergic G-protein-coupled receptor. *Nature* 2007;**450**:383–7.
15. Hanson MA, Cherezov V, Griffith MT, Roth CB, Jaakola VP, Chien EY, Velasquez J, Kuhn P, Stevens RC. A specific cholesterol binding site is established by the 2.8 A structure of the human beta2-adrenergic receptor. *Structure* 2008;**16**:897–905.
16. Warne T, Serrano-Vega MJ, Baker JG, Moukhametzianov R, Edwards PC, Henderson R, Leslie AG, Tate CG, Schertler GF. Structure of a beta1-adrenergic G-protein-coupled receptor. *Nature* 2008;**454**:486–91.
17. Jaakola VP, Griffith MT, Hanson MA, Cherezov V, Chien EY, Lane JR, IJzerman AP, Stevens RC. The 2.6 angstrom crystal structure of a human A2A adenosine receptor bound to an antagonist. *Science* 2008. in press.
18. Scheerer P, Park JH, Hildebrand PW, Kim YJ, Krauss N, Choe HW, Hofmann KP, Ernst OP. Crystal structure of opsin in its G-protein-interacting conformation. *Nature* 2008;**455**:497–502.
19. Henderson R, Unwin PN. Three-dimensional model of purple membrane obtained by electron microscopy. *Nature* 1975;**257**:28–32.
20. Baldwin JM. The probable arrangement of the helices in G protein-coupled receptors. *EMBO J* 1993;**12**:1693–703.
21. Schwartz TW, Frimurer TM, Holst B, Rosenkilde MM, Elling CE. Molecular mechanism of 7-TM receptor activation–a global toggle switch model. *Annu Rev Pharmacol Toxicol* 2006;**46**:481–519.
22. Oldham WM, Hamm HE. Heterotrimeric G-protein activation by G protein-coupled receptors. *Nature Rev* 2008;**9**:60–71.
23. Swaminath G, Xiang Y, Lee TW, Steenhuis J, Parnot C, Kobilka BK. Sequential binding of agonists to the beta2 adrenoceptor. Kinetic evidence for intermediate conformational states. *J Biol Chem* 2004;**279**:686–91.
24. Gainetdinov RR, Premont RT, Bohn LM, Lefkowitz RJ, Caron MG. Desensitization of G-protein-coupled receptors and neuronal functions. *Annu Rev Neurosci* 2004;**27**:107–44.
25. Fan QR, Hendrickson WA. Structure of human follicle-stimulating hormone in complex with its receptor. *Nature* 2005;**433**:269–77.
26. Muto T, Tsuchiya D, Morikawa K, Jingami H. Structures of the extracellular regions of the group II/III metabotropic glutamate receptors. *Proc Natl Acad Sci USA* 2007;**104**:3759–64.
27. West Jr AP, Llamas LL, Snow PM, Benzer S, Bjorkman PJ. Crystal structure of the ectodomain of Methuselah, a Drosophila G protein-coupled receptor associated with extended lifespan. *Proc Natl Acad Sci USA* 2001;**98**:3744–9.

28. Dann CE, Hsieh JC, Rattner A, Sharma D, Nathans J, Leahy DJ. Insights into Wnt binding and signalling from the structures of two Frizzled cysteine-rich domains. *Nature* 2001;**412**:86–90.
29. Veldkamp III CT, Seibert C, Peterson FC, De la Cruz NB, Haugner JC, Basnet H, Sakmar TP, Volkman BF. Structural basis of CXCR4 sulfotyrosine recognition by the chemokine SDF-1/CXCL12. *Sci Signal* 2008;**1**:ra4.
30. Shukla AK, Sun JP, Lefkowitz RJ. Crystallizing thinking about the beta2-adrenergic receptor. *Mol Pharmacol* 2008;**73**:1333–8.
31. Park JH, Scheerer P, Hofmann KP, Choe HW, Ernst OP. Crystal structure of the ligand-free G-protein-coupled receptor opsin. *Nature* 2008;**454**:183–7.
32. Altenbach C, Kusnetzow AK, Ernst OP, Hofmann KP, Hubbell WL. High-resolution distance mapping in rhodopsin reveals the pattern of helix movement due to activation. *Proc Natl Acad Sci USA* 2008;**105**:7439–44.
33. Murakami M, Kouyama T. Crystal structure of squid rhodopsin. *Nature* 2008;**453**:363–7.

Chapter 29

Heterotrimeric G-Protein Signaling at Atomic Resolution

David G. Lambright

Program in Molecular Medicine and Department of Biochemistry and Molecular Pharmacology, University of Massachusetts Medical School, Worcester, Massachusetts

INTRODUCTION

Heterotrimeric G proteins (subunits Gαβγ) mediate a variety of physiological responses, including sensory perception, hormone action, polarization, chemotaxis, and growth control [1–3]. In the conventional paradigm for G-protein signaling, ligand-bound (or light-activated) heptahelical receptors catalyze release of GDP from the Gα subunit, resulting in a complex between the receptor and the nucleotide-free Gαβγ heterotrimer. Association of GTP with Gα triggers release from the receptor and dissociation of Gα-GTP and Gβγ. Depending on the particular signaling pathway, either Gα-GTP or the released Gβγ subunits interact with downstream effectors until the Gα subunit is deactivated by GTP hydrolysis. Although the Gα subunit possesses an intrinsic GTP hydrolytic activity, regulator of G-protein signaling (RGS) domains, present in a variety of modular proteins, accelerate the rate of GTP hydrolysis [4].

Crystallographic studies of two different G-protein signaling pathways, involving the visual G protein transducin (G_t) as well as the hormone activated G proteins that stimulate (G_s) and inhibit (G_i) adenylyl cyclase (AC), reveal a highly conserved structural basis for heterotrimer assembly, activation by nucleotide exchange, and deactivation by GTP hydrolysis. The various structures explain a wealth of biochemical and cell biological data accumulated over the years, and provide a springboard for mutational analyses aimed at dissecting structure–function relationships for the myriad diverse biological responses mediated by G proteins. The most salient observations are highlighted below. Interested readers are encouraged to consult the cited references for in-depth discussion of particular structures.

ARCHITECTURE AND SWITCHING MECHANISM OF THE Gα SUBUNITS

As illustrated in Figure 29.1a, the Gα subunits share a conserved architecture with a Ras-like domain, consisting of a core six-stranded β sheet (β1–β6) surrounded by five helices (α1–α5), and a helical domain comprising a long helix (αA) enveloped on three sides by five shorter helices (αB–αF) [5]. Compared with monomeric GTPases, the helical domain of the Gα subunits represents an insertion within the α1β2 loop (known as the "effector binding loop" in Ras). Consequently, the Ras-like and helical domains are connected by two extended strands, one joining α1 of the Ras-like domain to αA of the helical domain and the other traversing from αF of the helical domain to β2 of the Ras-like domain. In all of the Gα-GDP and Gα-GTP structures determined to date, the αDαE loop of the helical domain engages the Ras-like domain, thereby capping the nucleotide binding site [5–10]. In contrast, the ribose and phosphate moieties of the nucleotide are partially exposed in monomeric GTPases, particularly in the GDP-bound conformation in which the α1β2 loop is highly flexible. Consistent with the structural observations, the Gα subunits bind GDP with high affinity in the absence of Mg^{2+} whereas monomeric GTPases require Mg^{2+} for high-affinity binding, exhibiting rapid rates of nucleotide release in the presence of EDTA [11]. As an obvious consequence of the conserved domain architecture in the Gα subunits, intrinsic as well as receptor-catalyzed GDP release necessarily involves a transition to an "open" intermediate in which the helical domain has rotated away from the Ras-like domain.

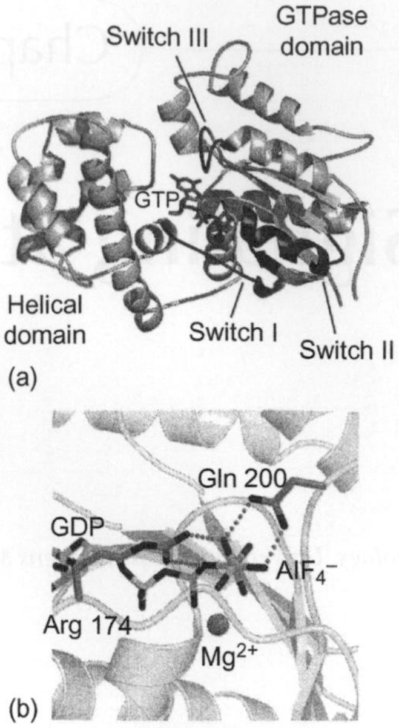

FIGURE 29.1 Active and transition states of a Gα subunit. (a) Structure of the active form of $G_t\alpha$ bound to GTPγS, a poorly hydrolyzable GTP analog. The conformational switch regions, deduced by comparison with the inactive (GDP-bound) form, are highlighted in dark gray. The Mg^{2+} ion is represented as a dark-gray sphere. (b) Structure of AlF_4^- activated $G_t\alpha$GDP.

Structural changes between the GDP-bound (inactive) and GTP-bound (active) conformations are localized to three non-consecutive conformational switch regions corresponding to the first linker strand (switch I), a region extending from the C-terminus of β3 to the C-terminus of the α2β4 loop (switch II), and the β4α3 loop (switch III) [6, 9]. In the GDP-bound structures, the switch regions are either disordered [9] or adopt a relaxed conformation stabilized by crystal contacts [6]. In the GTP-bound form, the γ-phosphate is detected by direct hydrogen bonding interactions with the side-chain hydroxyl of the invariant threonine residue in the RxxxT motif of switch I and the backbone NH group of the invariant glycine residue in the DxxGQ motif of switch II. The conformation of switch III is indirectly coupled to the nucleotide state by a hydrogen-bonding interaction between the side-chain carboxylate of an invariant glutamate residue in switch III and a main chain NH group in switch II. The active conformation is further stabilized through Mg^{2+} coordination by the side-chain hydroxyl of the invariant threonine residue in switch I, and through an extensive hydrophobic/ionic interface between the α2 helix of switch II and the α3-helix. Finally, the zippering of the β2 and β3 strands, resulting in two additional hydrogen bonds in the GTP-bound form, suggests cooperativity in the switching mechanism.

INSIGHT INTO THE GTP HYDROLYTIC MECHANISM FROM AN UNEXPECTED TRANSITION STATE MIMIC

Aluminum fluoride (AlF_4^-) binds to Gα-GDP near the binding site for the γ-phosphate of GTP, inducing a conformational change that results in artificial activation of Gα-GDP and dissociation from Gβγ. Crystal structures of the GDP•AlF_4^--bound forms of $G_i\alpha$ [7] and $G_t\alpha$ [8] revealed an unexpected finding. Although the conformation of all three switch regions closely resembles that of the GTP-bound form, AlF_4^- does not mimic the tetrahedral geometry of the γ-phosphate but rather adopts an octahedral geometry with four fluoride ligands arranged in an equatorial plane and two axial ligands consisting of an oxygen from the β-phosphate of GDP and a water molecule (Figure 29.1b). The bound aluminum fluoride interacts with and orders the side-chains of two critical residues previously implicated in GTP hydrolysis, namely the arginine of the switch-I RxxxT motif and the glutamine of the switch-II DxxGQ motif. These observations led to the hypothesis that AlF_4^- activates Gα-GDP by approximating the expected stereochemistry of the pentavalent intermediate for GTP hydrolysis. This notion is strongly supported by the remarkable observation that AlF_4^- also binds to and stabilizes complexes of both Ras and Rho GTPases bound to their respective GAPs [12, 13].

Gβγ WITH AND WITHOUT Gα

Parallel crystallographic studies revealed the stunningly beautiful structure of the Gβγ heterodimer alone [14] and in complex with Gα-GDP [15, 16]. The seven distinctive WD repeats of Gβ fold into a seven-bladed β propeller in which each blade consists of a four-stranded antiparallel β-sheet (Figure 29.2a). It is interesting that the WD repeats do not coincide precisely with the individual blades of the propeller, but rather each repeat begins with the outer strand of one blade and extends through the inner three strands of the next blade. Gγ possesses an N-terminal helix, which forms a parallel coiled-coil with the N-terminal helix of Gβ, followed by an internal helix and a region of coil that extends across the bottom of the Gβ propeller. Although Gγ has well-defined secondary structure, it is devoid of intramolecular interactions characteristic of tertiary structure and thus could not adopt a properly folded structure in the absence of Gβ. Indeed, roughly half of the residues in Gγ are buried in an extensive hydrophobic interface with Gβ, a finding that explains the unusually high stability of the Gβγ heterodimer.

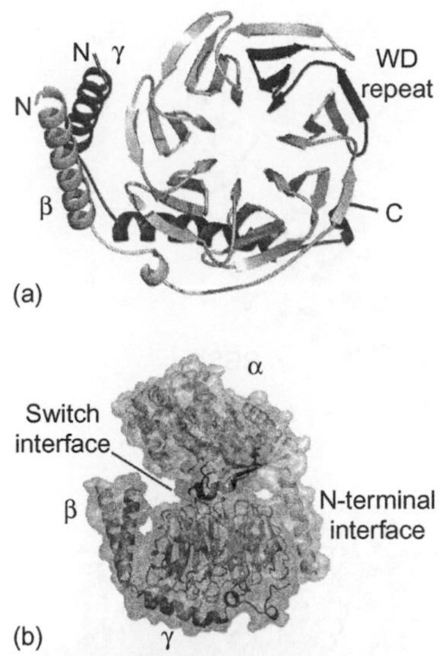

FIGURE 29.2 Structure of a heterotrimeric G protein.
(a) Ribbon rendering of $G_t\beta\gamma$ with the γ subunit and one of the WD repeats in the β subunit shown in dark gray. (b) Heterotrimeric complex of a $G_t\alpha/G_i\alpha$ chimera and the unprenylated form of $G_t\beta\gamma$.

The interaction between Gα-GDP and Gβγ occurs at two distinct interfaces (Figure 29.2b). The most extensive interface involves the switch-I and -II regions of Gα, which contact residues from the loops and turns at the top of the Gβ propeller. The second interface forms between the N-terminal helix of Gα and the side of Gβ. In the complex with Gβγ, the switch regions of α adopt a well-ordered conformation that is incompatible with the active conformation of Gα-GTP. In contrast, it appears that the interaction with the N-terminal helix of Gα would not be directly influenced by the state of the bound nucleotide, consistent with a residual low-affinity interaction between Gα-GTP and the released Gβγ subunits.

PHOSDUCIN AND Gβγ

The first insight into how phosducin engages Gβγ and promotes membrane dissociation came from the crystal structure of a phosducin complex with an unprenylated form of retinal Gβγ [17]. Phosducin contains two domains, a small N-terminal helical domain composed primarily of hydrophilic residues and a C-terminal domain with a thioredoxin-like fold (Figure 29.3a). The N-terminal domain interacts with the top of the Gβ propeller, overlapping extensively with the epitope for interaction with the switch regions of Gα, whereas the thioredoxin domain contacts the side of Gβ at a site distinct from the N-terminal epitope for Gα. These observations explain why

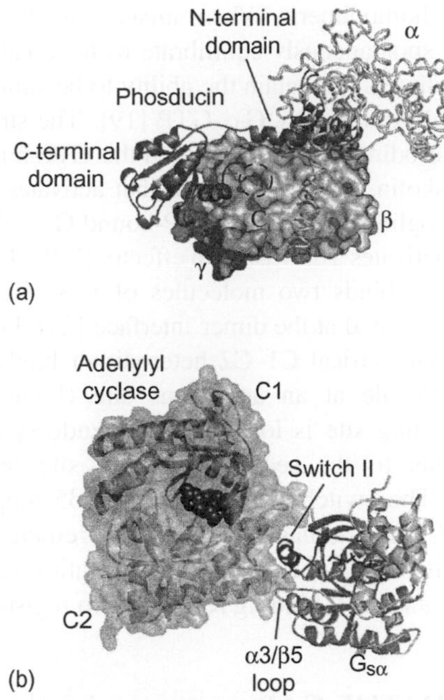

FIGURE 29.3 Regulatory and effector complexes with phosducin and adenylyl cyclase.
(a) Structure of the unprenylated $G_t\beta\gamma$ (light surface) in complex with retinal phosducin (dark ribbon). For comparison, the $G_t\alpha/G_i\alpha$ chimera from Figure 29.2b is overlaid as a light coil. A dashed circle denotes the location of the proposed binding cavity for the farnesyl moiety. (b) Structure of an active adenylyl cyclase C1–C2 heterodimer (ribbons and semitransparent surface) bound to forskolin (dark spheres) and $G_s\alpha$-GTPγS (ribbons). The active site is located to the left of the forskolin-binding site. Note that the interaction epitope of $G_s\alpha$ comprises the Switch II region and the α3β5 loop.

the interaction of Gβγ with phosducin and Gα is mutually exclusive. Furthermore, electrostatic calculations indicate that the presence of phosducin's thioredoxin-like domain introduces a substantial negative electrostatic potential near the prenylation site at the C-terminus of Gγ, thereby destabilizing the association with acidic membranes. Finally, the interaction with phosducin's N-terminal domain perturbs the conformation of three loops at the top of the Gβ propeller. A subsequent structure of phosducin bound to farnesylated Gβγ suggested that the conformational changes in Gβ open a pocket of appropriate dimensions to accommodate the hydrophobic farnesyl group of Gγ [18].

$G_S\alpha$ AND ADENYLYL CYCLASE

Adenylyl cyclase (AC) consists of two hexahelical transmembrane domains, each followed by a similar cytoplasmic domain referred to as C1 and C2. Expressed independently, the isolated C1 or C2 domains form soluble but inactive homodimers. When mixed, the C1 and C2 homodimers spontaneously equilibrate to form catalytically active heterodimers that retain the ability to be stimulated by $G_s\alpha$-GTP and inhibited by $G_i\alpha$-GTP [19]. The structure of a C1–C2 heterodimer in complex with the GTP-bound form $G_s\alpha$ and forskolin, a plant terpenoid that activates AC, provided the first glimpse of how a GTP-bound Gα subunit recognizes and activates a downstream effector [20]. The inactive C2 homodimer binds two molecules of forskolin at symmetrical sites located at the dimer interface [21]. In contrast, the pseudo-symmetrical C1–C2 heterodimer binds a single forskolin molecule at an analogous site (Figure 29.3b). The ATP binding site is located at a pseudo-symmetrical site analogous to the second forskolin site in the C2 homodimer. The switch II region and α3β5 loop of $G_s\alpha$ contact the C1–C2 domains at a location remote from the active site, thereby inducing a domain rotation that brings key catalytic and ATP-binding residues into register.

FILLING IN THE GAP

RGS domains present in a variety of modular proteins accelerate GTP hydrolysis for Gα subunits [4]. The underlying structural basis was established by the crystal structure of the helical RGS domain of RGS4 in complex with $G_i\alpha$-GDP and AlF_4^- [22]. In contrast to GAPs (GTPase-activating proteins) for Ras and Rho GTPases, which supply an "arginine finger" analogous to the catalytic arginine in switch I of Gα, RGS proteins promote GTP hydrolysis by engaging the switch-I and -II regions so as to reorient the catalytic arginine and glutamine residues of $G_i\alpha$ to stabilize the pentavalent intermediate (Figure 29.4a).

VISUAL FIDELITY

Structures of the GTPγS-bound and AlF_4^--activated forms of $G_t\alpha$ in complex with RGS9 and/or the inhibitory subunit of the retinal phosphodiesterase (PDEγ) provided further insight into the cooperative mechanism of effector recognition and RGS stimulation of GTP hydrolysis in the visual system [23]. PDEγ forms a predominately hydrophobic interface with residues in the switch II/α3 cleft of $G_t\alpha$, consistent with mutational data. This interaction sequesters C-terminal residues of PDEγ implicated in PDEαβ inhibition. RGS9 engages the switch-I and -II regions of $G_t\alpha$ in a manner analogous to that observed for RGS4 and $G_i\alpha$. As shown in Figure 29.4b, a small interface between PDEγ and a unique loop of RGS9, near the critical aspargine residue involved in positioning the catalytic glutamine of $G_t\alpha$, couples the maximal GAP activity of RGS9 to the interaction of $G_t\alpha$ with PDEγ, thereby enhancing the fidelity of visual signal transduction.

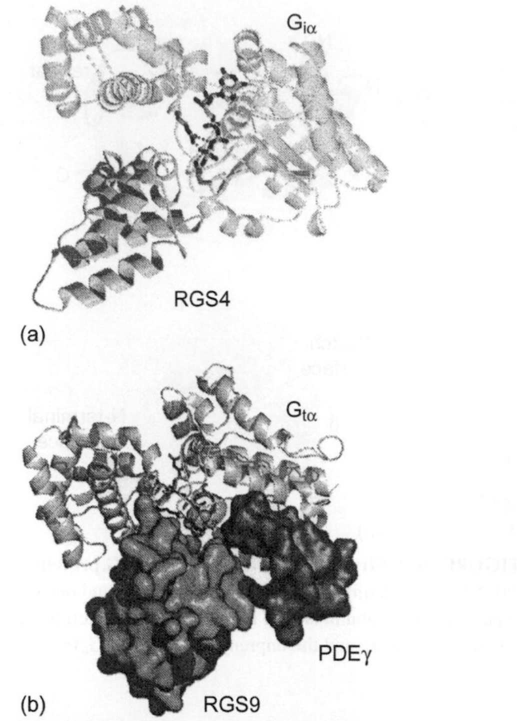

FIGURE 29.4 RGS and effector complexes with Gα subunits. (a) Structure of RGS4 (dark ribbon) bound to AlF_4^--activated $G_i\alpha$GDP (light ribbon). Also shown are the conserved arginine and glutatmine residues in the switch regions as well as GDP-AlF_4^-. (B) Structure of AlF_4^--activated $G_t\alpha$GDP (ribbons) bound to PDEγ (dark surface) and RGS9 (light surface). AlF_4^- is depicted as light spheres, whereas GDP as well as the conserved arginine and threonine residues are shown as bonded cylinders.

WHAT STRUCTURES MAY FOLLOW

Clearly the most important unresolved structural question is how an activated receptor engages a heterotrimeric G protein so as to catalyze nucleotide exchange on the Gα subunit. The resolution of this question requires the crystal structure of a complex between a ligand-bound or light-activated receptor and the nucleotide-free form of a G-protein heterotrimer. Only then can we claim to have glimpsed the conversion of extracellular signals into intracellular second messengers at atomic resolution.

REFERENCES

1. Conklin BR, Bourne HR. Structural elements of Gα subunits that interact with Gβγ, receptors, and effectors. *Cell* 1993;**73**:631–41.
2. Neer EJ. Heterotrimeric G proteins: organizers of transmembrane signals. *Cell* 1995;**80**:249–57.
3. Hepler JR, Gilman AG. G proteins. *Trends Biochem Sci* 1992;**17**:383–7.
4. Ross EM, Wilke TM. GTPase-activating proteins for heterotrimeric G proteins: regulators of G protein signaling (RGS) and RGS-like proteins. *Annu Rev Biochem* 2000;**69**:795–827.

5. Noel JP, Hamm HE, Sigler PB. The 2.2-Å crystal structure of transducin-alpha complexed with GTPγS. *Nature* 1993;**366**:654–63.
6. Lambright DG, Noel JP, Hamm HE, Sigler PB. Structural determinants for activation of a G-protein α subunit. *Nature* 1994;**369**:621–8.
7. Coleman DE, Berghuis AM, Lee E, Linder ME, Gilman AG, Sprang SR. Structures of active conformations of $G_i\alpha 1$ and the mechanism of GTP hydrolysis. *Science* 1994;**265**:1405–12.
8. Sondek J, Lambright DG, Noel JP, Hamm HE, Sigler PB. GTPase mechanism of G proteins from the 1.7-Å crystal structure of transducin α-GDP–AlF_4^-. *Nature* 1994;**372**:276–9.
9. Mixon MB, Lee E, Coleman DE, Berghuis AM, Gilman AG, Sprang SR. Tertiary and quaternary structural changes in G_i alpha 1 induced by GTP hydrolysis. *Science* 1995;**270**:954–60.
10. Sunahara RK, Tesmer JJ, Gilman AG, Sprang SR. Crystal structure of the adenylyl cyclase activator $G_{s\alpha}$. *Science* 1997;**278**:1943–7.
11. Sprang SR. G protein mechanisms: insights from structure analysis. *Annu Rev Biochem* 1997;**66**:639–78.
12. Scheffzek K, Ahmadian MR, Kabsch W, Wiesmüller L, Lautwein A, Schmitz F, Wittinghofer A. The Ras–RasGAP complex: structural basis for GTPase activation and its loss in oncogenic Ras mutants. *Science* 1997;**277**:333–8.
13. Rittinger K, Walker PA, Eccleston JF, Smerdon SJ, Gamblin SJ. Structure at 1.65 Å of RhoA and its GTPase-activating protein in complex with a transition-state analogue. *Nature* 1997;**389**:758–62.
14. Sondek J, Bohm A, Lambright DG, Hamm HE, Sigler PB. Crystal structure of a G-protein βγ dimer at 2.1-Å resolution. *Nature* 1996;**379**:369–74.
15. Wall MA, Coleman DE, Lee E, Iniguez-Lluhi JA, Posner BA, Gilman AG, Sprang SR. The structure of the G protein heterotrimer $G_i\alpha 1\beta 1\gamma 2$. *Cell* 1995;**83**:1047–58.
16. Lambright DG, Skiba N, Hamm HE, Sigler PB. The 2.0-Å structure of a heterotrimeric G-protein. *Nature* 1996;**379**:311–16.
17. Gaudet R, Bohm A, Sigler PB. Crystal structure at 2.4-Å resolution of the complex of transducin βγ and its regulator, phosducin. *Cell* 1996;**87**:577–88.
18. Loew A, Ho YK, Blundell T, Bax B. Phosducin induces a structural change in transducin beta gamma. *Structure* 1998;**6**:1007–19.
19. Tang WJ, Gilman AG. Construction of a soluble adenylyl cyclase activated by $G_s\alpha$ and forskolin. *Science* 1995;**268**:1769–72.
20. Tesmer JJ, Sunahara RK, Gilman AG, Sprang SR. Crystal structure of the catalytic domains of adenylyl cyclase in a complex with $G_s\alpha$ GTPγS. *Science* 1997;**278**:1907–16.
21. Zhang G, Liu Y, Ruoho AE, Hurley JH. Structure of the adenylyl cyclase catalytic core. *Nature* 1997;**386**:247–53.
22. Tesmer JJ, Berman DM, Gilman AG, Sprang SR. Structure of RGS4 bound to AlF4-activated $G_i\alpha 1$: stabilization of the transition state for GTP hydrolysis. *Cell* 1997;**89**:251–61.
23. Slep KC, Kercher MA, He W, Cowan CW, Wensel TG, Sigler PB. Structural determinants for regulation of phosphodiesterase by a G protein at 2.0 Å. *Nature* 2001;**409**:1071–7.

Chapter 30

Structure and Function of G-Protein-Coupled Receptors: Lessons from Recent Crystal Structures

Thomas P. Sakmar
Laboratory of Molecular Biology and Biochemistry, The Rockefeller University, New York, New York

INTRODUCTION

The crystal structure of the rod cell visual pigment, rhodopsin, was reported in 2000, and allowed the critical evaluation of a decade of earlier structure–activity relationship (SAR) studies. Recent reports of crystal structures of additional G-protein-coupled receptors (GPCRs) and of a ligand-free opsin now provide an opportunity to compare and contrast modes of ligand binding and G-protein activation among the entire superfamily of heptahelical receptors. For example, crystal structures of engineered human β_2 adrenergic receptors (ARs) in complex with an inverse agonist ligand, carazolol, provide three-dimensional snapshots of the disposition of its seven transmembrane helices and the ligand binding site. As expected, β_2AR shares substantial structural similarities with rhodopsin. However, although carazolol and the 11-*cis*-retinylidene chromophore of rhodopsin, which acts as an inverse agonist, are situated in the same general binding pocket, the second extracellular (E2) loop structures are quite distinct. E2 in rhodopsin shows β-sheet structure and forms part of the chromophore binding site. In the β_2AR, E2 is α-helical and seems to be distinct from the receptor's active site, allowing a potential entry pathway for diffusible ligands. The structure of an engineered human A2A adenosine receptor in complex with the subtype specific antagonist ZM241385 provides a number of additional surprises. The ligand binds to the A2A adenosine receptor in an extended conformation essentially perpendicular to the plane of the membrane. Stuctures of ligand-free opsin and of opsin in complex with a peptide corresponding to the carboxyl-terminal tail of its cognate G-protein α subunit, which probably represent the active-state structure of the receptor, strongly support the "helix movement model" of receptor activation. Compared with the structure of rhodopsin, the opsin structures show prominent structural changes in the highly conserved E(D)RY and NpxxY(x)5,6F regions, and in transmembrane helices 5, 6 and 7 (H5–H7). Remarkably H6 is tilted outward by 6–7 Å in the opsin structure. These new structures, together with extensive SAR data from earlier studies, provide insight about structural determinants of ligand specificity, and suggest how the binding of agonist ligands might cause the structural changes that activate GPCRs.

RECENT ADVANCES IN STRUCTURAL STUDIES OF G-PROTEIN-COUPLED RECEPTORS

GPCR signaling complexes are allosteric machines. Agonist receptor ligands outside of the cell induce guanine-nucleotide exchange on a heterotrimeric guanine-nucleotide binding regulatory protein (G protein) inside of the cell where the ligand binding site on the receptor and the nucleotide binding site and the G protein are in the order of 8–10 nm or more apart. Additional non-canonical signaling pathways facilitate cross-talk between linear GPCR-signaling pathways and receptor tyrosine kinase (RTK)-mediated signaling pathways. Receptor phosphorylation by receptor-specific kinases and the binding of various cellular adaptor proteins also regulates receptor desensitization, internalization, sequestration, and recycling. Receptor oligomerization, or in some cases hetero-oligomerization, is thought to modulate receptor cell-surface expression, ligand binding affinity, and downstream signaling specificity for at least some classes of GPCRs. Heptahelical receptors are also arguably the most important single class of pharmaceutical drug targets in the human genome. According to Overington *et al.* [1], of the

Handbook of Cell Signaling, Three-Volume Set 2 ed.

266 human targets for approved drugs, a remarkable 27 percent correspond to rhodopsin-like, or Family A, GPCRs. GPCRs will remain important drug targets in the foreseeable future, especially since well over 100 of the estimated 726 heptahelicals encoded by the human genome remain "orphan" receptors and most of these are expressed in the central nervous system. Orphan receptors are expressed but have no known endogenous ligand, although such ligands are presumed to exist.

The concepts of affinity and efficacy are used to describe the pharmacology of a drug (ligand). Affinity refers to the ability of the drug to bind to its molecular target, and efficacy refers to the ability of a drug to induce a biological response in its molecular target. Partial agonists exhibit less efficacy compared with (full) agonists. Neutral antagonists do not change the basal activity of the receptor (zero efficacy), and inverse agonists are able to stabilize the inactive state of the receptor. They essentially exhibit inversion of efficacy. Initially, in classical work, medicinal chemistry and pharmacology studies generated a vast inventory of SAR data. Receptor subtypes were defined by specific ligands (mostly antagonists) and validated in tissue physiology models, if available. With the cloning and heterologous expression of receptors (and chimeric/mutant receptors), a second generation of experiments led to mapping of ligand binding sites on receptors representing members of the main pharmacological targets (muscarinic, α and β adrenergic, dopamine, and serotonin receptors). However, only relatively few subtypes were analyzed in this way. The third generation of experiments was focused on mutagenesis-based structure determination to map the interactions between the highly conserved residues on TM helices, which, together with the bacteriorhodopsin structure and the available three-dimensional low-resolution cryo-electron microscopy projection structures of rhodopsin, resulted in the first models of the TM domains of a larger series of homologous receptors. The fourth generation of experiments focused on the dynamics of receptor activation, culminating in the "helix movement model" of receptor activation and some level of understanding of the structural basis for constitutive receptor activity and constitutively-activating mutations (CAMs) [2–6]. Today, with a new generation of receptor structures, we are finally able to compare and contrast the high-resolution structure of rhodopsin with that of other GPCRs on the background of deduced primary structures of a substantial number of related receptors from various genomes.

In a landmark report in 2000, a high-resolution crystal structure for rhodopsin, from bovine rod cells, was reported [7]. In the past few years, additional structures of rhodopsin and of thermally trapped early photoproducts of rhodopsin with all-*trans*-retinylidene chromophore have also been reported [8–10]. Most recently, high-resolution crystal structures of a number of additional GPCRs have become available: engineered expressed monoaminergic human β_2 adrenergic receptors (ARs) [11–13], engineered expressed monoaminergic turkey β_1AR [14], engineered expressed human A2A adenosine receptor [15], squid rhodopsin [16–17], and ligand-free opsins [18–19]. Table 30.1 shows the complete list of structures currently available.

TABLE 30.1 GPCR crystal structures

Receptor structure	PDB Code	Resolution (Å)	Reference
Bovine rhodopsin	1F88	2.8	Palczewski *et al.* [7]
Bovine rhodopsin	1HZX	2.8	Teller *et al.* [37]
Bovine rhodopsin	1L9H	2.6	Okada *et al.* [38]
Bovine rhodopsin	1U19	2.2	Okada *et al.* [39]
Bovine rhodopsin	1GZM	2.65	Li *et al.* [40]
Bovine rhodopsin photoproduct	2I35-37	4.15	Salom *et al.* [41]
Bovine bathorhodopsin	2G87	2.6	Nakamichi and Okada [8]
Bovine lumirhodopsin	2HPY	2.8	Nakamichi and Okada [9]
Bovine isorhodopsin	2PED	2.95	Nakamichi *et al.* [10]
Expressed engineered bovine rhodopsin	2J4Y	3.4	Standfuss *et al.* [42]
Expressed engineered human β_2 adrenergic receptor	2RH1	2.4	Cherezov *et al.* [11]
Expressed engineered human β_2 adrenergic receptor	2R4R/S	3.4/3.7	Rasmussen *et al.* [13]
Squid rhodopsin	2Z73	2.5	Murakami and Kouyama [16]
Squid rhodopsin	2ZIY	3.7	Shimamura *et al.* [17]
Expressed engineered human β_2 adrenergic receptor	3D4S	2.8	Hanson *et al.* [43]
Expressed engineered turkey β_1 adrenergic receptor	2VT4	2.7	Warne *et al.* [14]
Bovine opsin	3CAP	2.9	Park *et al.* [18]
Bovine opsin	3DQB	3.2	Scheerer *et al.* [19]
Expressed engineered human A2A adenosine receptor	3EML	2.6	Jaakola *et al.* [15]

CRYSTAL STRUCTURES OF HUMAN β_2AR

The crystal structures of the β_2AR provide a excellent data set to carry out a detailed analysis of ligand binding principles in Family A GPCRs, as has been recently reviewed [20]. The crystal structure of an engineered human β_2AR consists of a receptor construct modified at the N-terminus by addition of a hemagglutinin (HA) signal sequence followed by a FLAG epitope [13]. The receptor was truncated after position 365 so that it lacks 48 amino acid residues at its C-terminus. A glycosylation site was also removed by site-directed mutagenesis (N187E). A second receptor construct was also crystallized that contained a Tobacco Etch Virus (TEV) protease cleavage site so that the receptor lacked 24 amino acids from the N-terminus after proteolysis. The receptor clones were expressed in *Sf9* insect cells, solubilized in dodecylmaltoside detergent, and purified by successive antibody affinity and ligand (alprenolol) affinity chromatography. N-linked glycosylation was removed by glycosidase treatment and the FLAG epitope was removed where possible with the AcTEV protease, an enhanced form of TEV protease. The purified protein was relatively stable, but only about one-half of the purified receptors appeared to be functional and capable of ligand binding.

A monoclonal antibody was developed in parallel that binds to the C3 loop of the native receptor, but not to denatured receptor protein. Fab fragments were prepared and reacted with the purified receptor, which was further stabilized by carazolol, an inverse agonist ligand with pM affinity. The resulting complex was purified by size-exclusion chromatography, and mixed with bicelles composed of the phospholipid DMPC and the detergent CHAPSO suitable for crystallization trials. Long, thin, plate-like crystals were cryoprotected, and an entire data set was collected from a single crystal. Although the initial resolution was about 3.0 Å, the diffraction was somewhat anisotropic and the resolution of the final model was 3.4 Å in the plane of the putative membrane, and 3.7 Å perpendicular to the plane of the membrane. The final structures (one for each of the two constructs; PDB : 2R4R and PDB : 2R4S) were determined using molecular replacement using the immunoglobulin-domain search models for the Fab.

Several regions of the receptor are either unresolved in the crystal structure or obscured by the Fab fragment. In particular, the C-terminal tail region is missing from the engineered-receptor construct, and the C3 loop is bound to the Fab fragment. However, the binding of the Fab apparently did not affect agonist or antagonist binding, and did not prevent the conformational changes concomitant with receptor activation. The ligand binding site can be identified as an extended flat structure near to the extracellular receptor surface, but the carazolol itself is not resolved and the active site of the receptor is not seen in detail. The Fab itself and the TM helices near to the cytoplasmic surface of the receptor are the best-resolved regions of the structure.

In simultaneous work, Cherezov *et al.* [11] engineered a β_2AR with the same modifications in the N- and C-terminal tails and the same N187E substitution. However, instead of using an Fab fragment to stabilize the receptor, they created a fusion protein by inserting a synthetic gene encoding a slightly modified version of the enzyme T4 lysozyme between Ile233(5.72) and Arg260(6.22). The β_2AR–T4 lysozyme fusion construct was also expressed in *Sf9* insect cells and purified by a method very similar to that of Rasmussen *et al.* Final diffraction data from more than 40 crystals were considered to obtain a full dataset at complete 2.4-Å resolution. Initial phases of the construct were obtained by molecular replacement using both T4 lysozyme and a polyamine model of rhodopsin as search models. Most likely due to the use of the LCP method, the quality of the structure reported by Cherezov *et al.* [11] (PDB : 2RH1) is superior to that reported by Rasmussen *et al.* [13].

Excellent quality electron density is observed for residues 29 to 342 of the 365 amino-acid full-length construct. Two disulfide bonds, Cys106–Cys191 and Cys184–Cys190, are observed, and a palmitic acid linked to Cys341 is clearly visible in the Fo–Fc omit maps. Importantly, the active site contains a well-resolved carazolol ligand. The borders of all of the TM helices and loops can be defined from the structure, except, of course, where the T4 lysozyme was introduced in the C3 loop. Interestingly, a short helical segment near to the middle of the E2 loop is seen, which is not present in rhodopsin and was also not predicted by automated secondary structure prediction algorithms. Interactions between residues of the T4 lysozyme and receptor components are discussed in some detail in Cherezov *et al.* [11]. Basically, there are minimal intermolecular interactions between the T4 lysozyme component of one molecule and the receptor component of a neighboring molecule in the LCP crystal lattice, and it is argued that these interactions cause no particular structural perturbation.

It is informative to compare and contrast the structures of rhodopsin and the β_2AR. The N-terminus of the β_2AR is disordered, whereas that of rhodopsin is resolved clearly and interacts extensively with extracellular domains and forms a small four-strand β-sheet in concert with the E2 loop, which essentially blocks access to the retinal binding site [21]. In the β_2AR a short helical segment on the E2 loop, which includes a disulfide-bonded Cys, sits well above the carazolol binding site. Carazolol and retinal are situated in similar binding sites in the core of the helical bundle, but near to the extracellular surface. The TM helical segments of rhodopsin and β_2AR superimpose nearly precisely, except for TM helix 1. The TM helix 1 of β_2AR, despite containing a Pro not found in rhodopsin, is straight, and angled away from the axis of the receptor by about 18°.

The functional properties of the expressed β_2AR–T4 lysozyme fusion construct were studied in detail [12]. In summary, the construct bound the antagonist [^{3}H]-dihydroalprenolol (DHA) and the inverse agonist ICI-118,551 with the same affinities as the native expressed receptor. However, the affinities of the engineered construct for agonists (isoproterenol, epinephrine and formoterol) and partial agonist (salbutamol) were two- to three-fold higher than those of the native receptor. A shift in agonist affinity is associated with constitutive receptor activity, and it is possible that the engineered receptor conformation, at least in membranes, is partially active. The engineered receptor fusion could not bind to heterotrimeric G proteins. However, studies of a receptor construct containing a fluorescent probe linked to Cys265 at the cytoplasmic end of TM helix 6 were consistent with a partial active conformational equilibrium. Finally, when compared with the structure of the β_2AR solved in complex with the Fab fragment, one prominent difference is the altered packing of Phe264 at the cytoplasmic end of TM helix 6 in the fusion receptor. In the β_2AR–Fab complex, the interactions between Phe264 and residues in TM helices 5 and 6, and the C2 loop, may be important in maintaining β_2AR in its basal "off" state [12].

UNDERSTANDING LIGAND BINDING SPECIFICITY IN GPCRs

The central dogma in GPCR molecular pharmacology is that receptor activation and desensitization are both mediated via strictly stimulus-dependent interaction of the receptor with another protein (heterotrimeric G protein, G-protein-coupled receptor kinases, and β-arrestins) [22]. The ternary complex of agonist, receptor, and G protein is believed to trigger allosterically nucleotide exchange on the G protein. The consequence is that agonist binding to the receptor should exhibit allosteric modulation by the G protein. In the absence of the G protein, the binding affinity for agonists is typically reduced [23]. The question is whether or not the agonist–receptor complex can formally switch to the active conformation without the G protein. Receptor activation is accompanied by movement of cytoplasmic end of TM helix 6, as demonstrated by restraints imposed by engineered metal ion binding sites [24] and disulfide bridges [25], and by interaction of spin labels on double cysteine mutants [25,26].

In rhodopsin, photon capture isomerizes the inverse agonist form of the covalently-bound 11-*cis*-retinylidene ligand to the all-*trans* agonist form. Chromophore isomerization results in large conformational changes, as demonstrated by several biophysical methods [27]. It should be noted that rhodopsin also functions as a receptor for diffusible ligands, such as all-*trans*-retinal [28], even though its standard function is mediated by the covalently-bound ligand, which allows detection of photons. Moreover, in a mutant receptor K296G, an artificial amine ligand binds and activates as a strictly diffusible ligand [29].

How can the β_2AR crystal structure be used to gain a better understanding of ligand–receptor interactions and receptor activation? One approach is to employ advanced computational molecular dynamics (MD) simulations. For example, a recent MD simulation placed the receptor bound to carazolol in a typical bilayer membrane environment, and then addressed the question of whether or not the agonist ligand, adrenaline, would itself be stable in fundamentally the same receptor structural conformation as when it binds carazolol [20]. The nanosecond timescale of these MD simulations was certainly not sufficient to capture the slow transition to the active receptor conformation. However, a principal component analysis of the movements of the backbone Cα atoms demonstrated a global change of the receptor structure in its transition from the carazolol- to the adrenaline-bound form. The change is consistent with the elongated structure of the antagonist pushing the extracellular ends of TM helices 2 and 6 outward into the bilayer, and of TM helix 7 toward water. In fluorescently labeled β_2AR (tetramethylrhodamine-5-maleimide labeled at Cys265(6.27), or monobromobimane labeled at Cys271(6.33)) in detergent, noradrenaline induces a change in fluorescence with biphasic kinetics of 2.8-s and 70-s half-life times of the fast and slow phases, respectively [30–32]. However, the activation kinetics in cyan-fluorescent protein/yellow-fluorescent protein (CFP/YFP) fusion constructs of α_2AR *in vivo* with comparable ligand concentrations is significantly faster. Noradrenaline induces a conformational change within about 40 ms [33], but the time traces appear to contain a second slow phase in the seconds timescale. Possible reasons for this difference in activation rates compared to β_2AR have been discussed; for example, the presence of the cell membrane and G proteins *in vivo* [34].

STRUCTURAL BASIS OF THE ACTIVE STATE

Do any of the new crystal structures correspond to that of a constitutively active basal state? A constitutively active receptor is expected to have higher agonist binding affinities in the absence of G proteins. The β_2AR–T4L fusion protein, compared with β_2AR, exhibits higher affinities for agonists, such as (−)-isoproterenol and (−)-epinephrine [12]. The experiments were performed in Sf9 membranes in the absence of GTP. In Sf9 membranes β_2AR does not contain sufficient G-protein concentrations to exhibit a GTP-sensitive high-affinity receptor form [23]. The higher agonist affinity is consistent with constitutive activity of the fusion protein. Moreover, the ligand-dependent changes in fluorescence of bimane-labeled Cys265 indicate an altered basal state [12]. Interestingly, the crystal structure shows that Glu268(6.30) interacts with T4L instead of Arg131(3.50) in the conserved E(D)RY motif at the cytoplasmic end of TM helix 3 [12]. Arg131(3.50) interacts

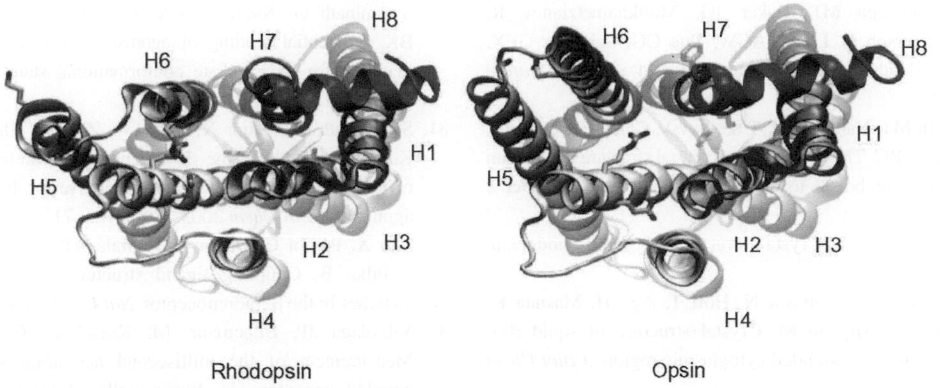

FIGURE 30.1 Comparison of structures of the structures of rhodopsin and "active" opsin.
Backbone ribbon depictions are presented for rhodopsin (PDB:1GZM) (left) and opsin (PDB:3CAP) (right). The view is from the cytoplasmic surface side of the receptor. Helices are progressively shaded from light gray (H1) to dark gray (cytoplasmic amphipathic helix 8). The 11-*cis*-retinylidene chromophore is removed from the rhodopsin structure for clarity. Several key structural changes are noted. The cytoplasmic surface is generally more opened in the opsin structure. H6 moves outward and rotates, affecting the orientation of "ionic lock" residues. The distance between Lys296 on H7 and Glu113 on H3 increases, and the distance between Lys296 and Glu181 on the E3 loop decreases. The structure of the E3 loop changes dramatically and small pores appear between H5 and H6 and H7 and H1 (these changes are not well seen in this depiction, however).

with Asp130(3.29) and a sulfate ion, and in the MD simulations there is evidence for local density of chloride ions. It has been shown that the E268A(6.30) substitution in a β_2AR–Gα_s fusion protein results in higher agonist affinity for the low affinity state [35]. Based on homology with rhodopsin, the "ionic lock" formed by Asp130(3.49), Arg131(3.50), and Glu268(6.30) is thought to stabilize the inactive receptor conformation. Both Shukla *et al.* [36] and Rosenbaum *et al.* [12] concluded that their crystal structures may in fact represent an "active-like state" or a "partial constitutively active" receptor, respectively.

The structure of ligand-free opsin, however, most certainly represents an active-state structure (Figure 30.1). Compared with rhodopsin, prominent structural changes are noted in the highly conserved E(D)RY and NpxxY(x)5,6F regions, and in H5–H7. Furthermore, H6 is tilted out by a remarkable 6–7P Å. A follow-up study reported the structure of the complex of opsin with an 11-amino acid residue peptide from the alpha-subunit of the rod cell G protein, transducin. This peptide was known to bind to light-activated rhodopsin and to stabilize the active state of rhodopsin called metarhodopsin II. The bound peptide makes extensive contacts with opsin along the inner surface of H5 and H6, which induce a helical structure in the peptide not found in solution. If confirmed to be the active-state structure, the "helix movement model" of receptor activation would be validated. The challenge now remains to obtain the structure of the complex between an agonist-bound GPCR and a G protein.

REFERENCES

1. Overington JP, Al-Lazikani B, Hopkins AL. How many drug targets are there? *Nature Rev Drug Discov* 2006;**5**:993–6.
2. Han M, Smith SO, Sakmar TP. Constitutive activation of opsin by mutation of methionine 257 on transmembrane helix 6. *Biochemistry* 1998;**37**:8253–61.
3. Robinson PR, Cohen GB, Zhukovsky EA, Oprian DD. Constitutively active mutants of rhodopsin. *Neuron* 1992;**9**:719–25.
4. Kjelsberg MA, Cotecchia S, Ostrowski J, Caron MG, Lefkowitz RJ. Constitutive activation of the α1B-adrenergic receptor by all amino acid substitutions at a single site. Evidence for a region which constrains receptor activation. *J Biol Chem* 1992;**267**:1430–3.
5. Cotecchia S. Constitutive activity and inverse agonism at the α1adrenoceptors. *Biochem Pharmacol* 2007;**73**:1076–83.
6. Parnot C, Miserey-Lenkei S, Bardin S, Corvol P, Clauser E. Lessons from constitutively active mutants of G protein-coupled receptors. *Trends Endocrinol Metab* 2002;**13**:336–43.
7. Palczewski K, Kumasaka T, Hori T, Behnke CA, Motoshima H, Fox BA, Le Trong I, Teller DC, Okada T, Stenkamp RE, Yamamoto M, Miyano M. Crystal structure of rhodopsin: A G protein-coupled receptor. *Science* 2000;**289**:739–45.
8. Nakamichi H, Okada T. Crystallographic analysis of primary visual photochemistry. *Angew Chem Intl Ed Engl* 2006;**45**:4270–3.
9. Nakamichi H, Okada T. Local peptide movement in the photoreaction intermediate of rhodopsin. *Proc Natl Acad Sci USA* 2006; **103**:12729–34.
10. Nakamichi H, Buss V, Okada T. Photoisomerization mechanism of rhodopsin and 9-cis-rhodopsin revealed by x-ray crystallography. *Biophys J* 2007;**92**:L106–8.
11. Cherezov V, Rosenbaum DM, Hanson MA, Rasmussen SG, Thian FS, Kobilka TS, Choi HJ, Kuhn P, Weis WI, Kobilka BK, Stevens RC. High-resolution crystal structure of an engineered human β2-adrenergic G protein-coupled receptor. *Science* 2007;**318**:1258–65.
12. Rosenbaum DM, Cherezov V, Hanson MA, Rasmussen SG, Thian FS, Kobilka TS, Choi HJ, Yao XJ, Weis WI, Stevens RC, Kobilka BK. GPCR engineering yields high-resolution structural insights into β2-adrenergic receptor function. *Science* 2007;**318**:1266–73.
13. Rasmussen SG, Choi HJ, Rosenbaum DM, Kobilka TS, Thian FS, Edwards PC, Burghammer M, Ratnala VR, Sanishvili R, Fischetti RF, Schertler GF, Weis WI, Kobilka BK. Crystal structure of the human β2 adrenergic G-protein-coupled receptor. *Nature* 2007;**450**:383–7.

14. Warne A, Serrano-Vega MJ, Baker JG, Moukhametzianov R, Edwards PC, Henderson R, Leslie AGW, Tate CG, Schertler GFX. Structure of the β1-adrenergic G protein-coupled receptor. *Nature* 2008;**454**:486–92.
15. Jaakola VP, Griffith MT, Hanson MA, Cherezov V, Chien EY, Lane JR, Ijzerman AP, Stevens RC. The 2.6-Angstrom crystal structure of a human A2A adenosine receptor bound to an antagonist. *Science* 2008;**Oct 2** [Epub ahead of print].
16. Murakami M, Kouyama T. Crystal structure of squid rhodopsin. *Nature* 2008;**453**:363–7.
17. Shimamura T, Hiraki K, Takahashi N, Hori T, Ago H, Masuda K, Takio K, Ishiguro M, Miyano M. Crystal structure of squid rhodopsin with intracellularly extended cytoplasmic region. *J Biol Chem* 2008;**283**:17753–6.
18. Park JH, Scheerer P, Hofmann KP, Choe H-W, Ernst OP. Crystal structure of the ligand-free G-protein-coupled receptor opsin. *Nature* 2008;**454**:183–7.
19. Scheerer P, Park JH, Hildebrand PW, Kim YJ, Krauss N, Choe H-W, Hofmann KP, Ernst OP. Crystal structure of opsin in its G-protein-interacting conformation. *Nature* 2008;**455**:497–503.
20. Huber T, Menon S, Sakmar TP. Structural basis for ligand binding and specificity in adrenergic receptors: Implications for GPCR-targeted drug discovery. *Biochemistry* 2008. in press.
21. Menon ST, Han M, Sakmar TP. Rhodopsin: structural basis of molecular physiology. *Physiol Rev* 2001;**81**:1659–88.
22. Lefkowitz RJ. Seven transmembrane receptors: something old, something new. *Acta Physiol (Oxf)* 2007;**190**:9–19.
23. Seifert R, Lee TW, Lam VT, Kobilka BK. Reconstitution of β2-adrenoceptor–GTP binding–protein interaction in Sf9 cells – high coupling efficiency in a β2-adrenoceptor-Gsα fusion protein. *Eur J Biochem* 1998;**255**:369–82.
24. Sheikh SP, Zvyaga TA, Lichtarge O, Sakmar TP, Bourne HR. Rhodopsin activation blocked by metal-ion binding sites linking transmembrane helices C and F. *Nature* 1996;**383**:347–50.
25. Farrens DL, Altenbach C, Yang K, Hubbell WL, Khorana HG. Requirement of rigid-body motion of transmembrane helices for light activation of rhodopsin. *Science* 1996;**274**:768–70.
26. Altenbach C, Kusnetzow AK, Ernst O, Hofmann KP, Hubbell WL. High resolution distance mapping in rhodopsin reveals the pattern of helix movement due to activation. *Proc Natl Acad Sci USA* 2008;**105**:7439–44.
27. Shieh T, Han M, Sakmar TP, Smith SO. The steric trigger in rhodopsin activation. *J Mol Biol* 1997;**269**:373–84.
28. Cohen GB, Oprian DD, Robinson PR. Mechanism of activation and inactivation of opsin – role of Glu113 and Lys296. *Biochemistry* 1992;**31**:12592–601.
29. Zhukovsky EA, Robinson PR, Oprian DD. Transducin activation by rhodopsin without a covalent bond to the 11-cis-retinal chromophore. *Science* 1991;**251**:558–60.
30. Swaminath G, Xiang Y, Lee TW, Steenhuis J, Parnot C, Kobilka BK. Sequential binding of agonists to the β2 adrenoceptor. Kinetic evidence for intermediate conformational states. *J Biol Chem* 2004; **279**:686–91.
31. Swaminath G, Deupi X, Lee TW, Zhu W, Thian FS, Kobilka TS, Kobilka B. Probing the β2 adrenoceptor binding site with catechol reveals differences in binding and activation by agonists and partial agonists. *J Biol Chem* 2005;**280**:22165–71.
32. Yao X, Parnot C, Deupi X, Ratnala VR, Swaminath G, Farrens D, Kobilka B. Coupling ligand structure to specific conformational switches in the β2-adrenoceptor. *Nat Chem Biol* 2006;**2**:417–22.
33. Vilardaga JP, Bunemann M, Krasel C, Castro M, Lohse MJ. Measurement of the millisecond activation switch of G-protein-coupled receptors in living cells. *Nature Biotechnol* 2003;**21**: 807–12.
34. Lohse MJ, Hoffmann C, Nikolaev VO, Vilardaga JP, Bunemann M. Kinetic analysis of G-protein-coupled receptor signaling using fluorescence resonance energy transfer in living cells. *Adv Protein Chem* 2007;**74**:167–88.
35. Ghanouni P, Schambye H, Seifert R, Lee TW, Rasmussen SG, Gether U, Kobilka BK. The effect of pH on β2 adrenoceptor function. Evidence for protonation—dependent activation. *J Biol Chem* 2000;**275**:3121–7.
36. Shukla AK, Sun JP, Lefkowitz RJ. Crystallizing thinking about the β2-adrenergic receptor. *Mol Pharmacol* 2008;**73**:1333–8.
37. Teller DC, Okada T, Behnke CA, Palczewski K, Stenkamp RE. Advances in determination of a high-resolution three-dimensional structure of rhodopsin, a model of G-protein-coupled-receptors (GPCRs). *Biochemistry* 2001;**40**:7761–72.
38. Okada T, Fujioshi Y, Silow M, Navarro J, Landau EM, Schichida Y. Functional role of internal water molecules in rhodopsin revealed by X-ray crystallography. *Proc Natl Acad Sci USA* 2002;**99**:5982–7.
39. Okada T, Sugihara M, Bondar AN, Elstner M, Entel P, Buss V. The retinal conformation and its environment in rhodopsin in light of a new 2.2-Å crystal structure. *J Mol Biol* 2004;**342**:571–83.
40. Li J, Edwards PC, Burghammer M, Villa C, Schertler GF. Structure of bovine rhodopsin in a trigonal crystal form. *J Mol Biol* 2004; **343**:1409–38.
41. Salom D, Lodowski DT, Stenkamp RE, Trong IL, Golczak M, Jastrzebska B, Harris T, Ballesteros JA, Palczewski K. Crystal structure of a photoactivated deprotonated intermediate of rhodopsin. *Proc Natl Acad Sci USA* 2006;**103**:16123–8.
42. Standfuss J, Xie G, Edwards PC, Burghammer M, Oprian DD, Schertler GF. Crystal structure of a thermally stable rhodopsin mutant. *J Mol Biol* 2007;**372**:1179–88.
43. Hanson MA, Cherezov V, Griffith MT, Roth CB, Jaakola VP, Chien EY, Velasquez J, Kuhn P, Stevens RC. A specific cholesterol binding site is established by the 2.8 A structure of the human β2-adrenergic receptor. *Structure* 2008;**16**:897–905.

Chapter 31

Chemokines and Chemokine Receptors: Structure and Function

Carol J. Raport[1] and Patrick W. Gray[2]
[1]*ICOS Corporation, Bothell, Washington*
[2]*Macrogenics, Inc., Seattle, Washington*

INTRODUCTION

The name chemokine is derived from "chemotactic cytokine," and the hallmark activity of chemokines is chemotaxis, the ability to induce directed cell movement. Chemokines are encoded by a large gene family with at least 45 members. The receptors for chemokines also belong to a gene family with at least 18 members, and all are G-protein-coupled receptors. The sequence similarities found in the chemokine gene family are reflected in their similar three-dimensional structures; however, chemokines display a diverse range of activities. Chemokines were originally identified as potent leukocyte attractants involved in inflammatory disease. More recently, they have been found to play critical roles in the natural development and regulation of the immune system. In addition, chemokines and their receptors have been utilized by pathogens to subvert the host immune system. This review will explore the many functions of this important family of immune modulators.

CHEMOKINE STRUCTURE AND FUNCTION

Although their discovery has been relatively recent (mostly within the past 15 years), chemokines and their receptors have rapidly become appreciated for their impact on health and disease. They are involved in a broad variety of natural biological processes, including development, inflammation, immunity, and angiogenesis. In addition, these sequences have been corrupted by pathogens to subvert the innate and adaptive immune responses. This chapter provides a brief introduction to chemokine structure and activities. More detailed reviews can be found in the reference section [1–4].

Over 45 human chemokines have been characterized. The first chemokines to be identified were associated with inflammatory disease, but their discrete biologic activities were not known. Once interleukin-8 (IL-8), monocyte chemoattractant protein-1 (MCP-1), and others were shown to attract leukocytes, other proteins with similar structure were also identified with leukocyte chemotactic activity [5–8]. The earliest identified chemokines were isolated by standard protein purification techniques, including heparin affinity chromatography. Others were soon identified as induced sequences in cDNA libraries that encoded similar protein structures [9–11]. Most of the newer chemokines were found in expressed sequence tag (EST) cDNA libraries by sequence similarity, while others have been discovered in the course of sequencing the human genome. Because many laboratories and many methods have been responsible for chemokine discovery, the original chemokine nomenclature is confusing. Recently a more comprehensive nomenclature has been developed [12]. This is presented in Table 31.1, along with the receptors and cell types with which they interact.

Chemokines are 8- to 10-kDa proteins that share significant homology in their amino acid sequences. Chemokines share between 20 and 80 percent identity and are found as gene families in all species of vertebrates. The four families of chemokines have been distinguished on the basis of the relative position of their cysteine residues. The α and β chemokines contain four cysteines (sometimes six) and are the largest families. One amino acid separates the first two cysteine residues (cysteine-X amino acid-cysteine, or CXC) in the α chemokines. In the β chemokines, the first two cysteine residues are adjacent to each other (cysteine–cysteine, or CC). The other two families of chemokines contain a single member each: lymphotactin, with only two cysteines [13], and fractalkine, in which the first two cysteine residues are separated by three amino acids (CXXXC) and the chemokine domain is at the amino terminus of a membrane-bound glycoprotein (fractalkine is also unusual in its size, 95 kDa) [14].

Handbook of Cell Signaling, Three-Volume Set 2 ed.

TABLE 31.1 Chemokine receptors and ligands

Receptor	Expression	Ligands
CXCR1	Neutrophils	CXCL8 (IL-8)
CXCR2	Neutrophils	CXCL8 (IL-8), CXCL1–3 (gro-α/β/γ), CXCL5 (ENA-78), CXCL6 (GCP-2), CXCL7 (NAP-2)
CXCR3	Activated T cells (Th1)	CXCL9 (mig), CXCL10 (IP-10), CXCL11 (ITAC)
CXCR4	T cells and other leukocytes	CXCL12 (SDF-1)
CXCR5	B cells	CXCL13 (BLC, BCA-1)
CXCR6	Activated T cells	CXCL16
CCR1	Monocytes, activated T cells	CCL3 (MIP-1α), CCL5 (RANTES), CCL7 (MCP-3), CCL 14–16 (HCC 1, 2, 4), CCL23 (MPIF-1)
CCR2	Monocytes, activated T cells	CCL2 (MCP-1), CCL8 (MCP-2), CCL7 (MCP-3), CCL13 (MCP-4)
CCR3	Eosinophils, basophils	CCL11 (Eotaxin), CCL24 (Eotaxin-2), CCL26 (Eotaxin-3), CCL5 (RANTES)
CCR4	T cells (Th2)	CCL22 (MDC), CCL17 (TARC)
CCR5	Macrophages, Th1 cells	CCL3 (MIP-1α) , CCL4 (MIP-1β) , CCL5 (RANTES)
CCR6	Activated T cells, dendritic cells	CCL20 (LARC, MIP-3a)
CCR7	Naïve lymphocytes, mature dendritic cells	CCL19 (ELC, MIP-3b), CCL21 (SLC, 6Ckine)
CCR8	T cells (Th2)	CCL1 (I-309)
CCR9	Gut homing α4β7+ T cells	CCL25 (TECK)
CCR10	Skin homing CLA+ T cells	CCL27(CTACK), CCL28 (MEC)
CX_3CR1	Monocytes, microglia, T cells	CX3CL1 (Fractalkine)
XCR1	Lymphocytes	XCL1 (Lymphotactin)

The three-dimensional structures of many chemokines have been determined by either X-ray crystallography or nuclear magnetic resonance (NMR) (reviewed in Clore and Gronenborn [15] and Rojo *et al.* [16]). Because of their multiple cysteine residues, the structures are confined by disulfide bridges. The first amino terminal cysteine forms a disulfide bond with the third, and the second cysteine with the fourth. Because of the disulfide constraints and relatively high sequence similarity, different chemokines share quite similar structural features. As also shown by structural studies, chemokines are isolated as dimers; however, dimerization does not appear to be necessary for receptor binding and may occur most frequently at the high concentrations used for the structural studies.

The structures of chemokines are critical for function. Alteration of a single residue, especially near the amino terminus, can greatly affect activity. For example, amino-terminal modification of RANTES results in a potent receptor antagonist [17]. Several chemokines undergo natural amino-terminal proteolytic processing after secretion which can alter their activity. Platelet basic protein is inactive until processed at its amino-terminal end by monocyte proteases to form neutrophil-activating peptide 2, a potent neutrophil chemoattractant [18]. The protease CD26 removes two amino acids from macrophage-derived chemokine (MDC) [19]; this destroys its activity on CCR4 but enhances its binding to CCR5, enabling it to inhibit HIV entry. HCC-1 is found in serum at relatively high concentrations, but it is inactive until its amino-terminal end is cleaved off [20]. Limited proteolysis may be a general mechanism that allows local factors to regulate chemokine activity.

The α chemokines (with 16 human members thus far) can be divided into two functional groups. The CXC chemokines that contain the sequence glutamic acid–leucine–arginine (ELR) preceding the CXC sequence are chemotactic for neutrophils [21]. Such chemokines play an important role in acute inflammatory diseases. The other (non-ELR) group of CXC chemokines tend to act on lymphocytes. For example, IP-10 and MIG (monokine induced by interferon-γ) attract activated T cells [22], and stromal-cell-derived factor 1 (SDF-1) acts on resting lymphocytes [23]; SDF-1 also plays a critical role in cardiac and neuronal development [24, 25].

There are at least 28 human β chemokines. In general, these CC chemokines attract monocytes, eosinophils, basophils, and lymphocytes with variable selectivity, but they do not attract neutrophils. The four monocyte chemoa tractant proteins and eotaxin form a subfamily for which the members are approximately 65 percent identical to each other [26]. As with the CXC family, the amino-terminal amino acids preceding the CC residues of β chemokines are critical for biologic activity and leukocyte selectivity [27].

Chemokines can be divided into two general classes based on whether they are induced by pro-inflammatory cytokines or are constitutively expressed [4]. The induced chemokines are upregulated very quickly at sites of infection or trauma and control the recruitment of leukocytes to the affected area. The constitutive chemokines are generally more involved in controlling migration of leukocytes through various tissues. This allows for naïve T and B cells to encounter antigen in secondary lymphoid organs and results in their subsequent activation and differentiation. In addition, the migration of T cells in the thymus is regulated by chemokines as thymocytes proceed through development [28].

CHEMOKINE RECEPTORS

Chemokines induce cell migration and activation through interactions with a family of cell-surface receptors containing seven transmembrane regions [2]. Typical of all heptahelical receptors, ligand binding induces a cascade of intracellular events mediated by activation of G proteins. Early studies showed that chemokine responses were sensitive to inhibition by pertussis toxin, confirming coupling of these receptors through the G_i family of G proteins [29]. A key result of chemokine binding is the activation of phospholipase C, producing inositol 1,4,5-trisphosphate (IP_3) and diacylglycerol [30]. Elevation of these products leads to release of Ca^{2+} from intracellular stores and activation of protein kinase C. Chemokine receptor engagement can also lead to stimulation of other intracellular enzymes such as phosphatidylinositol 3-kinase [31, 32], mitogen-activated protein kinase, and the ras family of GTP binding proteins [33, 34, 35]. These signaling events are key mediators of the ultimate cellular responses to chemokines, most notably cell migration. Different chemokines may participate in alternative downstream signaling pathways. For example, chemokines that have acute inflammatory properties ultimately activate nuclear factor κB (NFκB) and modulate transcription of inflammatory cytokines. Other chemokines are involved in alternative signaling cascades responsible for cellular differentiation or angiogenesis [36].

The first chemokine receptors identified were the IL-8 receptors in 1991 [37, 38]. Since then, a total of 18 chemokine receptors have been identified. They are divided into several classes, depending on the type of chemokine they bind (see Table 31.1). Six CXC receptors (CXCRs) that interact with one or more of the CXC chemokines have been identified. In addition, ten CCRs bind only CC chemokines. There are also single members in the CX3CR and XCR classes. Most of the CXCRs and CCRs interact with multiple chemokines, resulting in considerable apparent redundancy of chemokine function [1, 2].

Chemokine receptors contain conserved motifs found in all members of the family of chemoattractant receptors. Certain regions in the extracellular domains have been implicated in chemokine ligand binding. The amino terminus of the receptors was found to be necessary and to confer specificity for binding to their cognate ligands. However, a region of the third extracellular loop, extending into the transmembrane domain, was also determined to be involved in high-affinity ligand binding and receptor signaling [39, 40]. Most chemokine receptors have two disulfide bonds in their extracellular domains that are necessary for chemokine binding, while the majority of GPCRs have a single disulfide bridge [41]. Regions in the third intracellular loop and carboxy terminus have been shown to interact with G proteins, similar to other G-protein-coupled receptors [42]. The carboxy terminus is also rich in serine and threonine residues, which are thought to be phosphorylated, leading to interaction with arrestin to turn off the receptor signal [43]. Posttranslational modifications of the chemokine receptors include glycosylation (residues in the amino terminus and third extracellular domain) and sulfation (tyrosine residues in the amino terminus), which are necessary for high-affinity interactions with chemokine ligands [44].

Each chemokine receptor is expressed on a subset of cells and confers responsiveness of those cells to particular chemokines. Some receptors are restricted to a particular leukocyte. For example, CXCR1 is found almost exclusively on neutrophils [38]. Others are more broadly expressed, such as CXCR4. Other chemokine receptors are expressed only by a subset of cells in a certain activation state. CXCR3 for example is expressed only on activated T cells of the Th1 subtype. T cells especially seem to regulate their expression of many chemokine receptors in response to activating cytokines and other external stimuli [22]. Receptor regulation with cell activation allows for a selective amplified response to a particular antigen.

In addition to their role in cell migration, chemokine receptors are utilized by various pathogens to gain access to host cells [45]. The most striking example is HIV, which can interact with CXCR4 and CCR5, along with CD4, to infect T cells and macrophages (reviewed in Clapham and McKnight [46]). This discovery, made in the mid-1990s, has inspired great efforts to produce inhibitors of the HIV/receptor interaction as a treatment for AIDS.

While chemokines are necessary for coordinating leukocyte defense against external invaders, inflammation often occurs inappropriately and can lead to a variety of diseases [47]. Overexpression of chemokines and chemokine receptors has been reported in many conditions, leading to leukocyte accumulations in tissues. These include rheumatoid arthritis, multiple sclerosis, psoriasis, ulcerative colitis, asthma, and arteriosclerosis. In animal models for many of these diseases, chemokine inhibitors (generally blocking antibodies) have been found to prevent development of inflammatory lesions. Chemokines and their receptors have also been implicated in carcinogenesis [48].

Small molecule inhibitors of chemokine/receptor interactions are being developed for treatment of many human inflammatory diseases [49, 50]. GPCRs have historically been very good targets for drug development and hold promise for success in the chemokine area. Some chemokine inhibitors that have reached human clinical trials target CCR1, CCR5, and CXCR4. Others that are in preclinical development include inhibitors of CCR2, CCR3, CCR4, CCR6, CXCR2, and CXCR3. By targeting specific receptors, the hope is that only subsets of leukocytes involved in disease will be affected while general immune functions can still occur. The future of chemokine inhibitors looks bright, and we should soon have clinical data to confirm the potential for utilizing these inhibitors for treating disease.

REFERENCES

1. Luster AD. Chemokines – chemotactic cytokines that mediate inflammation. *New Engl J Med* 1998;**338**:436–45.
2. Murphy PM. The molecular biology of leukocyte chemoattractant receptors. *Annu Rev Immunol* 1994;**12**:593–633.
3. Baggiolini M, Dewald B, Moser B. Human chemokines: an update. *Annu Rev Immunol* 1997;**15**:675–705.
4. Yoshie O, Imai T, Nomiyama H. Chemokines in immunity. *Adv Immunol* 2001;**78**:57–110.
5. Yoshimura T, Matsushima K, Tanaka S, Robinson EA, Appella E, Oppenheim JJ, Leonard EJ. Purification of a human monocyte-derived neutrophil chemotactic factor that has peptide sequence similarity to other host defense cytokines. *Proc Natl Acad Sci USA* 1987;**84**:9233–7.
6. Wolpe SD, Davatelis G, Sherry B, Beutler B, Hesse DG, Nguyen HT, Moldawer LL, Nathan CF, Lowry SF, Cerami A. Macrophages secrete a novel heparin-binding protein with inflammatory and neutrophil chemokinetic properties. *J Exp Med* 1988;**167**:570–81.
7. Yoshimura T, Robinson EA, Tanaka S, Appella E, Kuratsu J, Leonard EJ. Purification and amino acid analysis of two human glioma-derived monocyte chemoattractants. *J Exp Med* 1989;**169**:1449–59.
8. Matsushima K, Larsen CG, DuBois GC, Oppenheim JJ. Purification and characterization of a novel monocyte chemotactic and activating factor produced by a human myelomonocytic cell line. *J Exp Med* 1989;**169**:1485–90.
9. Hieshima K, Imai T, Opdenakker G, Van Damme J, Kusuda J, Tei H, Sakaki Y, Takatsuki K, Miura R, Yoshie O, Nomiyama H. Molecular cloning of a novel human CC chemokine liver and activation-regulated chemokine (LARC) expressed in liver. Chemotactic activity for lymphocytes and gene localization on chromosome 2. *J Biol Chem* 1997;**272**:5846–53.
10. Nagira M, Imai T, Hieshima K, Kusuda J, Ridanpaa M, Takagi S, Nishimura M, Kakizaki M, Nomiyama H, Yoshie O. Molecular cloning of a novel human CC chemokine secondary lymphoid-tissue chemokine that is a potent chemoattractant for lymphocytes and mapped to chromosome 9p13. *J Biol Chem* 1997;**272**:19, 518–24.
11. Hromas R, Gray PW, Chantry D, Godiska R, Krathwohl M, Fife K, Bell GI, Takeda J, Aronica S, Gordon M, Cooper S, Broxmeyer HE, Klemsz MJ. Cloning and characterization of exodus, a novel beta-chemokine. *Blood* 1997;**89**:3315–22.
12. Zlotnik A, Yoshie O. Chemokines: a new classification system and their role in immunity. *Immunity* 2000;**12**:121–7.
13. Kelner GS, Kennedy J, Bacon KB, Kleyensteuber S, Largaespada DA, Jenkins NA, Copeland NG, Bazan JF, Moore KW, Schall TJ, Zlotnik A. Lymphotactin: a cytokine that represents a new class of chemokine. *Science* 1994;**266**:1395–9.
14. Bazan JF, Bacon KB, Hardiman G, Wang W, Soo K, Rossi D, Greaves DR, Zlotnik A, Schall TJ. A new class of membrane-bound chemokine with a CX3C motif. *Nature* 1997;**385**:640–4.
15. Clore GM, Gronenborn AM. Three-dimensional structures of alpha and beta chemokines. *FASEB J* 1995;**9**:57–62.
16. Rojo D, Suetomi K, Navarro J. Structural biology of chemokine receptors. *Biol Res* 1999;**32**:263–72.
17. Proudfoot AEI, Buser R, Borlat F, Alouani S, Soler D, Offord RE, Schroder J-M, Power CA, Wells TNC. Amino-terminally modified RANTES analogues demonstrate differential effects on RANTES receptors. *J Biol Chem* 1999;**274**:32478–85.
18. Walz A, Dewald B, von Tscharner V, Baggiolini M. Effects of the neutrophil-activating peptide NAP-2, platelet basic protein, connective tissue-activating peptide III and platelet factor 4 on human neutrophils. *J Exp Med* 1989;**170**:1745–50.
19. Proost P, Struyf S, Schols D, Opdenakker G, Sozzani S, Allavena P, Mantovani A, Augustyns K, Bal G, Haemers A, Lambeir AM, Scharpe S, Van Damme J, De Meester I. Truncation of macrophage-derived chemokine by CD26/dipeptidyl-peptidase IV beyond its predicted cleavage site affects chemotactic activity and CC chemokine receptor 4 interaction. *J Biol Chem* 1999;**274**:3988–93.
20. Detheux M, Standker L, Vakili J, Munch J, Forssmann U, Adermann K, Pohlmann S, Vassart G, Kirchhoff F, Parmentier M, Forssmann WG. Natural proteolytic processing of hemofiltrate CC chemokine 1 generates a potent CC chemokine receptor (CCR)1 and CCR5 agonist with anti-HIV properties. *J Exp Med* 2000;**192**:1501–8.
21. Clark-Lewis I, Schumacher C, Baggiolini M, Moser B. Structure–activity relationships of interleukin-8 determined using chemically synthesized analogs: critical role of NH2-terminal residues and evidence for uncoupling of neutrophil chemotaxis, exocytosis, and receptor binding activities. *J Biol Chem* 1991;**266**:23128–34.
22. Loetscher M, Gerber B, Loetscher P, Jones SA, Piali L, Clark-Lewis I, Baggiolini M, Moser B. Chemokine receptor specific for IP10 and mig: structure, function, and expression in activated T-lymphocytes. *J Exp Med* 1996;**184**:963–9.
23. Bleul CC, Fuhlbrigge C, Casasnovas JM, Aiuti A, Springer TA. A highly efficacious lymphocyte chemoattractant, stromal cell-derived factor 1 (SDF-1). *J Exp Med* 1996;**184**:1101–9.

24. Zou YR, Kottmann AH, Kuroda M, Taniuchi I, Littman DR. Function of the chemokine receptor CXCR4 in haematopoiesis and in cerebellar development. *Nature* 1998;**393**:595–9.
25. Nagasawa T, Hirota S, Tachibana K, Takakura N, Nishikawa S, Kitamura Y, Yoshida N, Kikutani H, Kishimoto T. Defects of B-cell lymphopoiesis and bone-marrow myelopoiesis in mice lacking the CXC chemokine PBSF/SDF-1. *Nature* 1996;**382**:635–8.
26. Luster AD, Rothenberg ME. Role of the monocyte chemoattractant protein and eotaxin subfamily of chemokines in allergic inflammation. *J Leukoc Biol* 1997;**62**:620–33.
27. Gong J-H, Clark-Lewis I. Antagonists of monocyte chemoattractant protein 1 identified by modification of functionally critical NH2-terminal residues. *J Exp Med* 1995;**181**:631–40.
28. Mantovani A, Gray PW, Van Damme J, Sozzani S. Macrophage-derived chemokine (MDC). *J Leukoc Biol* 2000;**68**:400–4.
29. Thelen M, Peveri P, Kernen P, von Tscharner V, Walz A, Baggiolini M. Mechanism of neutrophil activation by NAF, a novel monocyte-derived peptide agonist. *FASEB J* 1988;**2**:2702–6.
30. Bokoch GM. Chemoattractant signaling and leukocyte activation. *Blood* 1995;**86**:1649–60.
31. Turner SJ, Domin J, Waterfield MD, Ward SG, Westwick J. The CC chemokine monocyte chemotactic peptide-1 activates both the class I p85/p110 phosphatidylinositol 3-kinase and the class II PI3K-C2alpha. *J Biol Chem* 1998;**273**:25987–95.
32. Hirsch E, Katanaev VL, Garlanda C, Azzolino O, Pirola L, Silengo L, Sozzani S, Mantovani A, Altruda F, Wymann MP. Central role for G protein-coupled phosphoinositide 3-kinase in inflammation. *Science* 2000;**287**:1049–53.
33. Knall C, Young S, Nick JA, Buhl AM, Worthen GS, Johnson GL. Interleukin-8 regulation of the Ras/Raf/mitogen-activated protein kinase pathway in human neutrophils. *J Biol Chem* 1996;**271**:2832–8.
34. Laudanna C, Campbell JJ, Butcher EC. Role of Rho in chemoattractant-activated leukocyte adhesion through integrins. *Science* 1996;**271**:981–3.
35. Wang D, Yang W, Du J, Devalaraja MN, Liang P, Matsumoto K, Tsubakimoto K, Endo T, Richmond A. MGSA/GRO-mediated melanocyte transformation involves induction of Ras expression. *Oncogene* 2000;**19**:4647–59.
36. Muller G, Hopken UE, Stein H, Lipp M. Systemic immunoregulatory and pathogenic functions of homeostatic chemokine receptors. *J Leukoc Biol* 2002;**72**:1–8.
37. Murphy PM, Tiffany HL. Cloning of complementary DNA encoding a functional human interleukin-8 receptor. *Science* 1991;**253**:1280–3.
38. Holmes WE, Lee J, Kuang W-J, Rice GC, Wood WI. Structure and functional expression of a human interleukin-8 receptor. *Science* 1991;**253**:1278–80.
39. Wells TNC, Power CA, Lusti-Narasimhan M, Hoogewerf AJ, Cooke RM, Chung C, Peitsch MC, Proudfoot AEI. Selectivity and antagonism of chemokine receptors. *J Leukoc Biol* 1996;**59**:53–60.
40. Siciliano SJ, Rollins TE, DeMartino J, Konteatis Z, Malkowitz L, Van Riper G, Bondy S, Rosen H, Springer MS. Two-site binding of C5a by its receptor: an alternative binding paradigm for G protein-coupled receptors. *Proc Natl Acad Sci USA* 1994;**91**:1214–8.
41. Blanpain C, Lee B, Vakili J, Doranz BJ, Govaerts C, Migeotte I, Sharron M, Dupriez V, Vassart G, Doms RW, Parmentier M. Extracellular cysteines of CCR5 are required for chemokine binding, but dispensable for HIV-1 coreceptor activity. *J Biol Chem* 1999;**274**:18902–8.
42. Gosling J, Monteclaro FS, Atchison RE, Arai H, Tsou CL, Goldsmith MA, Charo IF. Molecular uncoupling of C-C chemokine receptor 5-induced chemotaxis and signal transduction from HIV-1 coreceptor activity. *Proc Natl Acad Sci USA* 1997;**94**:5061–6.
43. Franci C, Gosling J, Tsou CL, Coughlin SR, Charo IF. Phosphorylation by a G protein-coupled kinase inhibits signaling and promotes internalization of the monocyte chemoattractant protein-1 receptor Critical role of carboxyl-tail serines/threonines in receptor function. *J Immunol* 1996;**157**:5606–12.
44. Bannert N, Craig S, Farzan M, Sogah D, Santo NV, Choe H, Sodroski J. Sialylated O-glycans and sulfated tyrosines in the NH2-terminal domain of CC chemokine receptor 5 contribute to high affinity binding of chemokines. *J Exp Med* 2001;**194**:1661–73.
45. Seet BT, McFadden G. Viral chemokine-binding proteins. *J Leukoc Biol* 2002;**72**:24–34.
46. Clapham PR, McKnight A. HIV-1 receptors and cell tropism. *Br Med Bull* 2001;**58**:43–59.
47. Gerard C, Rollins BJ. Chemokines and disease. *Nature Immunol* 2001;**2**:108–15.
48. Muller A, Homey B, Soto H, Ge N, Catron D, Buchanan ME, McClanahan T, Murphy E, Yuan W, Wagner SN, Barrera JL, Mohar A, Verastegui E, Zlotnik A. Involvement of chemokine receptors in breast cancer metastasis. *Nature* 2001;**410**:50–6.
49. Proudfoot AE. Chemokine receptors: multifaceted therapeutic targets. *Nature Rev Immunol* 2002;**2**:106–15.
50. Schwarz MK, Wells TNC. New therapeutics that modulate chemokine networks. *Nature Rev Drug Discov* 2002;**1**:347–58.

The β_2 Adrenergic Receptor as a Model for G-Protein-Coupled Receptor Structure and Activation by Diffusible Hormones

Daniel M. Rosenbaum, Søren G.F. Rasmussen, and Brian K. Kobilka

Department of Molecular and Cellular Physiology, Stanford University School of Medicine, Palo Alto, California

INTRODUCTION

G-protein-coupled receptors (GPCRs) represent the largest family of membrane proteins in the human genome, with members grouped into five classes based on sequence and functional similarity [1]. Structurally, all GPCRs are characterized by the presence of seven membrane-spanning α-helical segments. These receptors couple the binding of agonists to heterotrimeric G-protein activation [2], thereby playing central roles in cellular responses to hormones, neurotransmitters, and other molecules involved in a wide variety of physiological processes. Many GPCR signaling systems are complex and tightly regulated, involving multiple G-protein subtypes and desensitization pathways [3].

The β_2 adrenergic receptor (β_2AR) is among the most extensively characterized family A GPCRs. The β_2AR is activated by epinephrine and norepinephrine, and plays key regulatory roles in cardiovascular and pulmonary physiology through mediation of sympathetic neurotransmission [4]. Due to the importance of the β_2AR and other adrenergic receptors in cardiovascular and pulmonary diseases, many drugs have been developed that target these GPCRs. Such compounds have been instrumental for studying how the β_2AR functions at a molecular level. We begin this chapter by reviewing the biochemical properties of the β_2AR, in particular the evidence for multiple conformational states of the receptor. Next we describe various features of the recently published β_2AR crystal structures. Finally, we discuss potential molecular mechanisms for β_2AR activation, based on the structures as well as the existing biochemical and biophysical data on β_2AR conformational changes. The combination of high-resolution structure and *in vitro* biophysical data makes the β_2AR unique among GPCRs recognizing diffusible ligands.

A MODEL SYSTEM FOR GPCRS RECOGNIZING DIFFUSIBLE LIGANDS

The β_2AR was the second GPCR to be cloned (after rhodopsin), and the first receptor for diffusible hormones whose primary sequence was known [4,5]. The cloning of the β_2AR quickly led to the isolation of many related receptor genes, and established the structural conservation among GPCRs of seven hydrophobic transmembrane (TM) domains connected by short extracellular (ECL) and intracellular (ICL) loops. Following the initial cloning of the receptor, the molecular determinants of agonist and antagonist binding were studied through the pharmacological characterization of receptor mutants [6]. Specifically, it was discovered that agonists such as epinephrine and norepinephrine, as well as antagonists, bind at a site buried between the transmembrane helical domains. Polar amino acids from TM3 and TM7 were shown to be critical for recognition of the common alkylamino functional group of most adrenergic ligands [7–9], while a cluster of serines on TM5 were shown to interact with the catechol hydroxyls of agonists or the heterocyclic ring substituents of certain antagonists [10, 11]. Thus the general position of the hormone binding site within the β_2AR was determined, although many of the details would only be known after the structure was solved. Further experiments using chimeras between the β_2AR and the related α_{2A}AR established that the region between the cytoplasmic ends of TM5 and TM6 is critical for G-protein interaction [12]. In addition, mutants distant from the ligand binding site were identified that exhibit altered signaling behavior: Uncoupling mutants (UCMs) resulted in reduced agonist-stimulated signaling without the loss of expression or ligand binding [13]; conversely, constitutively active mutants (CAMs) resulted in elevated basal (agonist-independent)

Handbook of Cell Signaling, Three-Volume Set 2 ed.

activity, while generally leading to reduced expression and increased agonist affinity [14]. The majority of these mutations are presumed to affect intra-molecular interactions within the protein and thereby alter the energetic landscape of receptor conformations.

CONFORMATIONAL STATES ON THE PATHWAY TO ACTIVATION

From studies on the ligand binding properties and efficacy profiles of different GPCRs, it has been appreciated that activation of such receptors cannot be modeled as a simple two-state transition between inactive and active conformations [15]. The ability to purify the β_2AR to homogeneity and site-selectively incorporate fluorescent probes has enabled *in vitro* experimental evidence of multiple conformational states to be obtained for this receptor. In experiments utilizing a tetramethylrhodamine dye covalently attached at Cys265$^{6.27}$ (at the cytoplasmic end of TM6) as a probe, fluorescence intensity changes stimulated by full agonists could only be effectively fit by a two-component exponential function (Figure 32.1) [16, 17]. Note that Ballesteros–Weinstein numbering [18], allowing comparison of residues between Class A GPCRs, is indicated in superscript to identify amino acid positions. Partial agonists for the receptor, such as dopamine, can only stimulate the rapid component of this fluorescence response. Even catechol, a small fragment of epinephrine that is known to interact with TM5, can induce the conformational change being reported by the rapid component; this response is not blocked by a saturating concentration of non-catechol agonists [19]. These studies suggest a model in which different functional groups on an agonist sequentially engage different chemical groups on the receptor, triggering distinct conformational changes that lead to intermediates along the pathway to full activation. Crucially, the ability of different ligands to promote these conformational changes correlates with their known efficacies. Using a different biophysical probe – quenching between a bimane fluorescent probe incorporated by cysteine modification at position 271$^{6.33}$ (on TM6) and a tryptophan placed at position 135$^{3.54}$ (cytoplasmic end of TM3) – it was possible to monitor disruption of the so-called ionic lock: a highly conserved interaction between the conserved E/DRY motif on TM3 and a glutamate at the cytoplasmic end of TM6 that is thought to stabilize the inactive state of family A GPCRs [20]. In contrast to the studies mentioned above, both partial and full agonists were observed to maximally stimulate disruption of the ionic lock, indicating that this conformational change is necessary, but not sufficient, for full activation of the receptor [21]. These studies and others have led to a complex

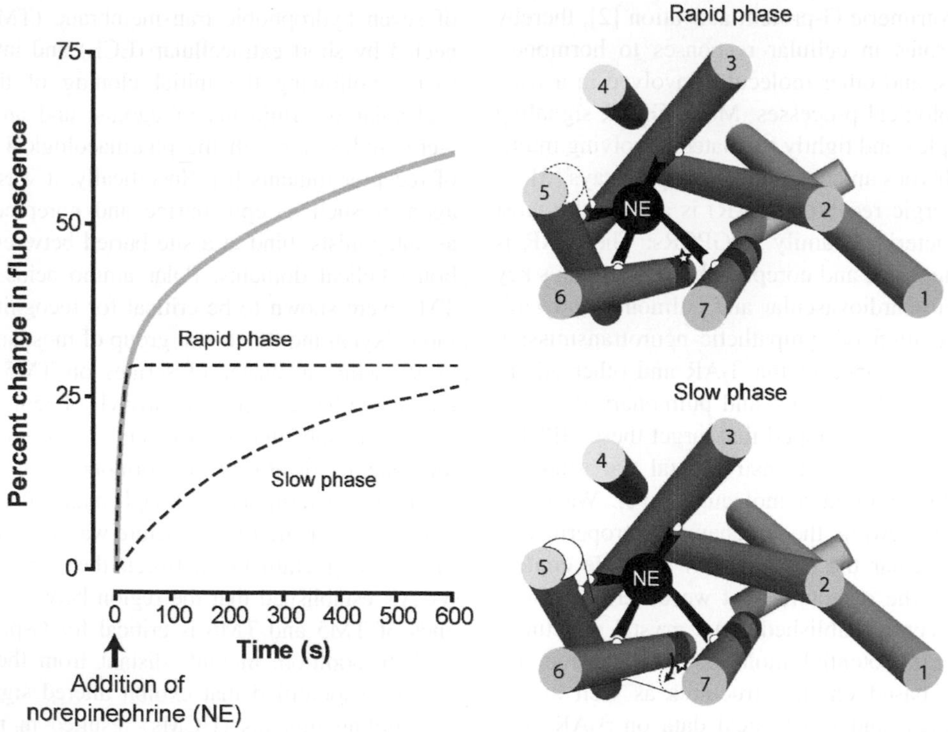

FIGURE 32.1 In biophysical studies of the β_2AR, catecholamine agonist-induced changes in the fluorescence of a covalently attached probe (indicated by the black star) at the cytoplasmic end of TM6) could only be fit by a two-component exponential function (adapted from [19]). We hypothesize that the rapid phase of this conformational change is due to engagement of residues on TMs 5 and 6 by the catechol ring of agonists. The slow phase, which is only stimulated by full agonists, reflects formation of additional contacts between the agonist and TMs 3 and 7, reordering of the binding pocket, and coupled changes in the positions of TMs 5 and 6 at the cytoplasmic surface.

picture of β_2AR activation, in which different agonists can promote different combinations of molecular switches that together specify the ligand's efficacy [22].

CRYSTAL STRUCTURES OF THE HUMAN β_2AR

The crystal structures of the β_2AR bound to the inverse agonist carazolol represent the first high-resolution picture of a GPCR recognizing a diffusible ligand (Figure 32.2a), and provide an initial framework for interpreting biochemical and biophysical data on this receptor. Major technical developments were required to obtain these structures: high-level overexpression in a recombinant host (Sf9 insect cells); use of functional (ligand-affinity) purification [23]; use of the stabilizing high-affinity inverse agonist ligand carazolol [24]; selection of conformation-specific antibodies against the folded state of the protein [25]; engineering of functional receptor–T4 lysozyme fusions [26]; application of bicelle and lipidic cubic phase crystallization methods [27]; and use of microdiffraction synchrotron beamlines. Two parallel approaches resulted in complementary structures:

1. A complex between the wild-type β_2AR and an antibody Fab fragment recognizing the ICL3 region was crystallized from DMPC bicelles [28]. While the extracellular and ligand binding regions of the β_2AR in these crystals were poorly ordered, this structure represents a more native conformation at the intracellular surface of the receptor.
2. A fusion protein in which the ICL3 of the β_2AR is replaced by the soluble protein T4 lysozyme was crystallized from a cholesterol-doped monoolein cubic phase [26, 29]; this structure was solved at higher resolution, and fully elucidates the conformations of the extracellular loops and the interactions with the ligand carazolol. The very close agreement between these two structures reinforces the validity of the observed protein conformation.

COMPARISON TO THE STRUCTURE OF RHODOPSIN

The β_2AR structures represent the second high-resolution picture of a GPCR obtained after rhodopsin [30]. Given that both receptors were co-crystallized with bound inverse agonists (11-cis retinal for rhodopsin, carazolol for the β_2AR), we can now assess the structural conservation between Class A (rhodopsin family) receptors in light of the hypothesized common mechanisms of agonist-induced activation. Overall, the structural superposition of β_2AR and rhodopsin TM domains produces an rmsd of 2–3 Å, depending on the regions specified. While this degree of overlap indicates that the two proteins share a highly similar overall architecture, the divergence is still high enough to signify important differences in helical packing.

As might be expected from the functional differences between receptors, the major divergence between rhodopsin and the β_2AR lies in the extracellular loops and ligand binding region. The ECL2 of rhodopsin forms a short β-sheet that caps the covalently bound 11-cis-retinal, shielding the chromophore from bulk solvent and preventing Schiff base hydrolysis. In contrast, the ECL2 of the β_2AR contains a short α-helix that is stabilized by intra- and inter-loop disulfide bonds. Furthermore, the glycosylated N-terminus of rhodopsin adopts a structured conformation at the extracellular apex of the protein, while the analogous N-terminal region of the β_2AR is disordered. The resulting more open and solvent-exposed extracellular surface (Figure 32.2b) of the β_2AR relative to rhodopsin helps to explain its lower thermal stability, while presumably being required to accommodate the diffusible catecholamine neurotransmitters.

The sites of carazolol and 11-cis-retinal binding are partly overlapping in superpositions of the β_2AR and rhodopsin structures. However, the position of carazolol within the β_2AR is slightly more extracellular than 11-cis-retinal in rhodopsin. This difference in positions of the ligands results in a significant difference in inverse agonist interaction with the residue Trp286$^{6.48}$, which is hypothesized to undergo a key rotamer conformational transition in GPCR activation – the "rotamer toggle switch" [31]. While the ionone ring of retinal makes direct contact with Trp286$^{6.48}$ in rhodopsin, carazolol in the β_2AR packs against a set of aromatic residues that shield Trp286$^{6.48}$ from the binding site. The less direct coupling between inverse agonist binding and the inactive conformation of the rotamer toggle switch in the β_2AR may explain some of the elevated basal activity of this receptor relative to rhodopsin. The amino acids from the β_2AR that are observed making direct contacts with the carazolol ligand are in excellent agreement with the biochemical literature. Asp113$^{3.32}$ on TM3 and Asn312$^{7.39}$ on TM7 form hydrogen bonds with the alkylamine and hydroxyl groups of the ligand, while Ser204$^{5.43}$ on TM5 forms an expected hydrogen bond with the nitrogen of the carbazole heterocycle [11]. Furthermore, several of the hydrophobic and aromatic interactions between the β_2AR and carazolol have biochemical precedent either for the β_2AR or for closely related GPCRs. Although the identities of many of these interacting amino acids were predicted, the three-dimensional conformations of these residues at the binding site, and their precise interactions with a bound ligand could only by unveiled through structure determination (Figure 32.2c).

The polypeptide backbone conformation at the intracellular surface is highly similar between rhodopsin and the β_2AR. However, one significant difference is observed in the "ionic lock" between the highly conserved E/DRY

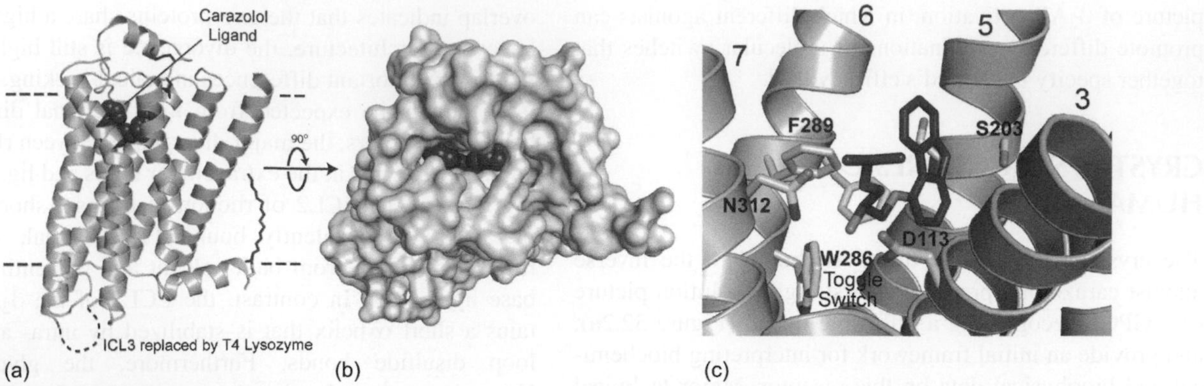

FIGURE 32.2 Structural features of the β₂AR.
(a) Overview of the receptor structure from the plane of the membrane, with the transmembrane helices shown as a light-gray cartoon, the disulfide bonds restraining ECL2 as dark sticks, and the carazolol inverse agonist as dark spheres. (b) Extracellular view of the receptor showing solvent accessibility of the ligand binding pocket and complementary packing with carazolol. The receptor is shown in light-gray surface representation, with carazolol as dark spheres. (c) Illustration of several key interactions at the ligand binding site. Receptor residues are represented as light-gray sticks, with carazolol shown as dark sticks. Figures were derived from the structure of the βAR–T4 Lysozyme fusion protein [26, 29], and made using Pymol.

motif on TM3 and Glu268$^{6.30}$ on TM6. In rhodopsin, the equivalent amino acids form a network of polar interactions bridging the two TM helices, which is thought to stabilize the inactive state conformation (Figure 32.3c) [32]. In contrast, for both β_2AR structures solved, the non-covalent polar interactions of the ionic lock are broken (Figure 32.3b). The lack of an ionic lock in the β_2AR crystal structures can be interpreted in two ways: either this interaction does not exist in the carazolol-bound receptor, or else the interaction is weak enough that it can be easily overcome energetically by crystal packing interactions. Either way, this observation is compatible with the elevated constitutive activity of the β_2AR relative to rhodopsin.

MECHANISM OF AGONIST-INDUCED ACTIVATION

The fundamental question of mechanism with respect to the β_2AR remains: how does binding of an agonist and the resulting changes in interactions at the ligand binding pocket lead to conformational changes that are propagated from the extracellular portion of the molecule to the cytoplasmic surface involved in G-protein binding? The recent structure of opsin at low pH, which is thought to resemble the activated state of rhodopsin, sheds light on the changes in transmembrane helix packing that are to be expected as a consequence of agonist binding [33]. In this structure, the cytoplasmic end of TM6 is shifted more than 6 Å outwards from the center of the bundle relative to its position in the inactive state, and at the same time moves closer to TM5 (Figure 32.3d). This rigid-body movement of TM6, which has been experimentally observed in studies of both rhodopsin [34] and the β_2AR [21], is accompanied by changes in several key interactions. Most importantly, the ionic lock is completely broken and new interactions are formed between previously associated residues (Arg135$^{3.50}$ of the ERY motif and Glu247$^{6.30}$ on TM6) and different amino acids on TM5. Additionally, Tyr306$^{7.53}$ of the highly conserved NPXXY motif (on TM7) undergoes a conformational change and inserts into space occupied by TM6 in dark state rhodopsin, thereby stabilizing the active conformation. The end result of the changes in going from inactive rhodopsin to "active-state" opsin is the creation of a small cavity between TMs 3, 5, and 6, in which the G protein can bind. The structure of opsin bound to a C-terminal peptide of transducin has recently demonstrated that this cleft on the receptor does indeed provide the interaction surface for the most crucial binding epitope of the G protein [35].

Based on the conserved three-dimensional structure and G-protein signaling mechanism between Class A GPCRs, it is reasonable to hypothesize that β_2AR activation will be accompanied by similar changes in transmembrane helix packing to those observed in the opsin structures. In fact, biophysical studies of the β_2AR are in good agreement with such a mechanism [21]. This still leaves the question, how does agonist binding far from the cytoplasmic surface lead to the expected packing rearrangements? In the structure of β_2AR bound to carazolol, it is apparent that the inactive state conformation cannot allow for simultaneous contacts between agonist-binding amino acids and both the catechol and chiral hydroxyl groups of catecholamines. This incompatibility between the carazolol-bound β_2AR structure and agonist binding is analogous to the fact that the retinal binding pocket in the dark state of rhodopsin cannot accommodate the photon-activated all-trans conformation of the chromophore.

Thus conformational changes at the ligand binding site are likely to accompany agonist binding in the β_2AR. One hypothesized change is movement of the upper region

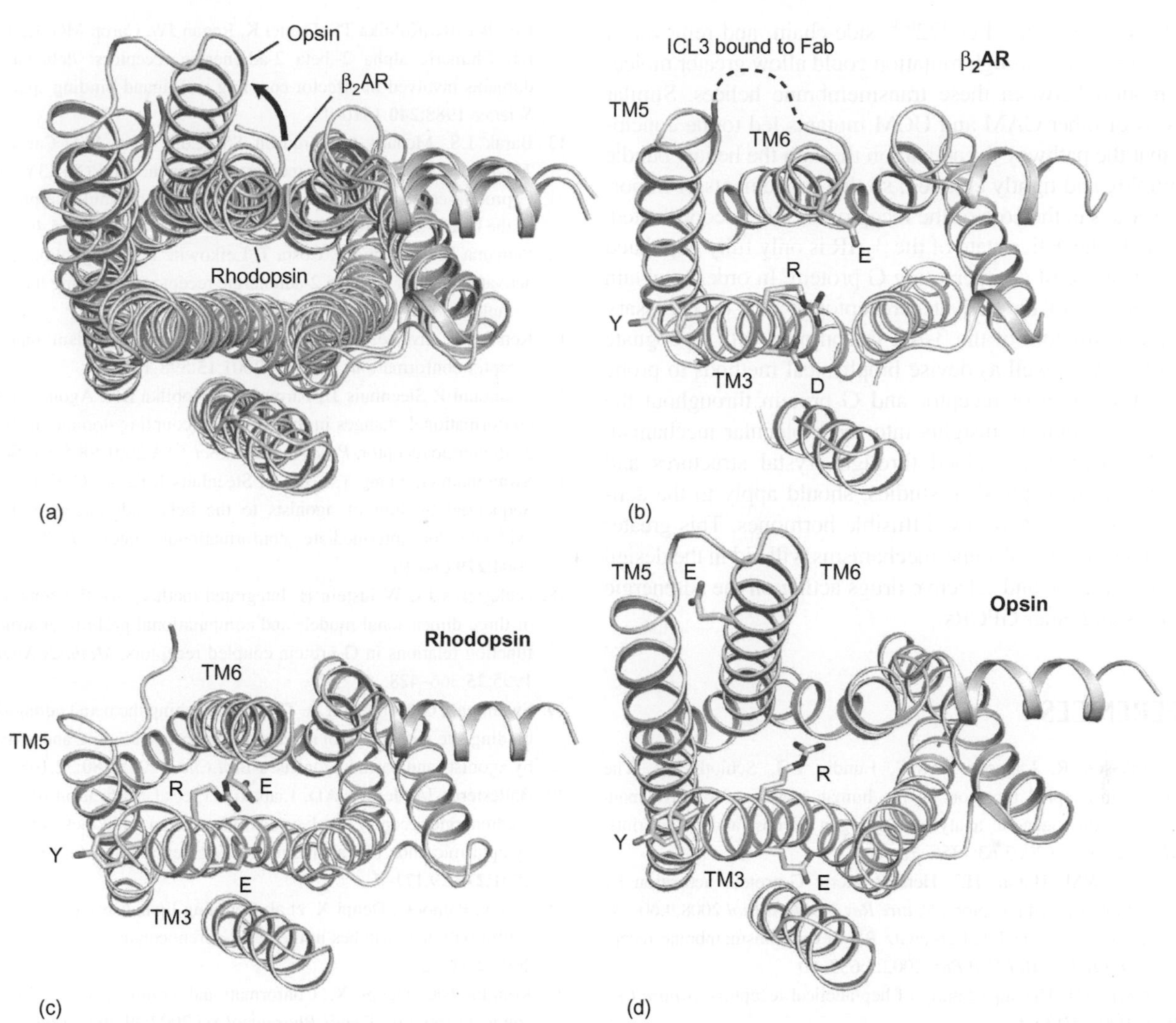

FIGURE 32.3 Intracellular surface of the β_2AR, in comparison with rhodopsin and opsin. (a) Rhodopsin and opsin were each superimposed onto the β_2AR using Cαs of structurally homologous regions of TMs 1–5 and 7 for least-squares fitting. ECL2, ECL3, and the C-termini were removed for clarity. (b), β_2AR alone, derived from the Fab complex structure [28]. The residues of the conserved DRY motif on TM3 and Glu268$^{6.30}$ from TM6, proposed to participate in the "ionic lock," are shown as light-gray sticks. (c) Rhodopsin alone [30], with the homologous ionic lock residues shown. (d) Opsin alone [33], showing the significant rigid-body movement of the cytoplasmic portion of TM6 relative to the other structures.

of TM5, containing Serines 203$^{5.42}$, 204$^{5.43}$, and 207$^{5.46}$, closer to TM3 (e.g. Asp113$^{3.32}$) and TM7 (e.g. Asn312$^{7.39}$). Simultaneous engagement of the ligand by TM5-catechol hydrogen-bonding and TM3/7-alkylamine polar contacts could facilitate changes in packing of nearby amino acids such as Phe289$^{6.51}$ and Phe290$^{6.52}$. These residues pack between carazolol and Trp268$^{6.48}$ in the inverse agonist-bound crystal structure (Figure 32.2c). In this manner, binding of an agonist could be coupled to movements of the rotamer toggle switch. It was previously noted that the compounds dopamine and catechol, which lack chemical substituents of epinephrine but retain the ability to interact with the serines on TM5, were distinct from full agonists in biophysical studies: the rapid component of stimulated fluorescence changes in a covalently-attached tetramethylrhodamine probe was still present, while the slow component was absent. Assuming the rapid component of the fluorescence change is due to engagement of residues on TM5 and TM6 by the catechol ring, the inability of these compounds to induce the slow component could reflect their inability to reorder and compress the binding pocket and thereby couple binding to changes at the cytoplasmic surface (Figure 32.1).

Future structures and biophysical studies are needed to elucidate the multiple changes in molecular interactions that allow energetic coupling between the binding site and the packing of helices at the receptor's cytoplasmic surface. Some clues come from inspecting the positions of the CAM and UCM mutants in the structure of carazolol-bound β_2AR [26]. For example, mutation of Leu272$^{6.34}$ in TM6 to Cys has been demonstrated to elevate the basal activity of the receptor [36], as well as destabilize the receptor biochemically [37]. Tight packing of amino acids from TMs 3, 5, and 6 constrain the movement of the $Leu272^{6.34}$ side-chain, and removal of these constraints through mutation could allow greater molecular motion between these transmembrane helices. Similar analysis of other CAM and UCM mutants led to the conclusion that the pathway for activation through the helical bundle is complex and tightly coupled, so that movements of important residues in the core of the receptor are likely coordinated. Ultimately the active state of the β_2AR is only fully populated in the presence of an interacting G protein. In order to obtain a true picture of the activated receptor, it will be necessary to solve a structure of the β_2AR in complex with its cognate Gαs protein, as well as devise biophysical methods to probe the conformation of receptor and G protein throughout the process of coupling. Insights into the molecular mechanism of β_2AR activation, gained through crystal structures and solution-based biophysical studies, should apply to the activation of other GPCRs by diffusible hormones. This greater understanding of molecular mechanisms will aid in the design of more effective and selective drugs acting on the adrenergic receptors and other GPCRs.

REFERENCES

1. Fredriksson R, Lagerstrom MC, Lundin LG, Schioth HB. The G-protein-coupled receptors in the human genome form five main families. Phylogenetic analysis, paralogon groups, and fingerprints. *Mol Pharmacol* 2003;**63**:1256–72.
2. Oldham WM, Hamm HE. Heterotrimeric G-protein activation by G-protein-coupled receptors. *Nature Rev Mol Cell Biol* 2008;**9**:60–71.
3. Pierce KL, Premont RT, Lefkowitz RJ. Seven-transmembrane receptors. *Nature Rev Mol Cell Biol* 2002;**3**:639–50.
4. Lefkowitz RJ. The superfamily of heptahelical receptors. *Nature Cell Biol* 2000;**2**:E133–6.
5. Dixon RA, Kobilka BK, Strader DJ, et al. Cloning of the gene and cDNA for mammalian beta-adrenergic receptor and homology with rhodopsin. *Nature* 1986;**321**:75–9.
6. Tota MR, Candelore MR, Dixon RA, Strader CD. Biophysical and genetic analysis of the ligand-binding site of the beta-adrenoceptor. *Trends Pharmacol Sci* 1991;**12**:4–6.
7. Strader CD, Sigal IS, Register RB, Candelore MR, Rands E, Dixon RA. Identification of residues required for ligand binding to the beta—adrenergic receptor. *Proc Natl Acad Sci USA* 1987;**84**:4384–8.
8. Strader CD, Sigal IS, Candelore MR, Rands E, Hill WS, Dixon RA. Conserved aspartic acid residues 79 and 113 of the beta-adrenergic receptor have different roles in receptor function. *J Biol Chem* 1988;**263**:10,267–71.
9. Suryanarayana S, Kobilka BK. Amino acid substitutions at position 312 in the seventh hydrophobic segment of the beta 2-adrenergic receptor modify ligand-binding specificity. *Mol Pharmacol* 1993;**44**:111–14.
10. Strader CD, Candelore MR, Hill WS, Sigal IS, Dixon RA. Identification of two serine residues involved in agonist activation of the beta-adrenergic receptor. *J Biol Chem* 1989;**264**:13,572–578.
11. Liapakis G, Ballesteros JA, Papachristou S, Chan WC, Chen X, Javitch JA. The forgotten serine. A critical role for Ser-2035.42 in ligand binding to and activation of the beta 2-adrenergic receptor. *J Biol Chem* 2000;**275**:37,779–788.
12. Kobilka BK, Kobilka TS, Daniel K, Regan JW, Caron MG, Lefkowitz RJ. Chimeric alpha 2-,beta 2-adrenergic receptors: delineation of domains involved in effector coupling and ligand binding specificity. *Science* 1988;**240**:1310–16.
13. Barak LS, Menard L, Ferguson SS, Colapietro AM, Caron MG. The conserved seven-transmembrane sequence NP(X)2,3Y of the G-protein-coupled receptor superfamily regulates multiple properties of the beta 2-adrenergic receptor. *Biochemistry* 1995;**34**:15,407–14.
14. Samama P, Cotecchia S, Costa T, Lefkowitz RJ. A mutation-induced activated state of the beta 2-adrenergic receptor. Extending the ternary complex model. *J Biol Chem* 1993;**268**:4625–36.
15. Kenakin T. Inverse, protean, and ligand-selective agonism: matters of receptor conformation. *FASEB J* 2001;**15**:598–611.
16. Ghanouni P, Steenhuis JJ, Farrens DL, Kobilka BK. Agonist-induced conformational changes in the G-protein-coupling domain of the beta 2 adrenergic receptor. *Proc Natl Acad Sci USA* 2001;**98**:5997–6002.
17. Swaminath G, Xiang Y, Lee TW, Steenhuis J, Parnot C, Kobilka BK. Sequential binding of agonists to the beta2 adrenoceptor. Kinetic evidence for intermediate conformational states. *J Biol Chem* 2004;**279**:686–91.
18. Ballesteros JA, Weinstein H. Integrated methods for the construction of three-dimensional models and computational probing of structure-function relations in G-protein coupled receptors. *Methods Neurosci.* 1995;**25**:366–428.
19. Swaminath G, Deupi X, Lee TW, et al. Probing the beta2 adrenoceptor binding site with catechol reveals differences in binding and activation by agonists and partial agonists. *J Biol Chem* 2005;**280**:22,165–71.
20. Ballesteros JA, Jensen AD, Liapakis G, et al. Activation of the beta 2-adrenergic receptor involves disruption of an ionic lock between the cytoplasmic ends of transmembrane segments 3 and 6. *J Biol Chem* 2001;**276**:29,171–7.
21. Yao X, Parnot C, Deupi X, et al. Coupling ligand structure to specific conformational switches in the beta2-adrenoceptor. *Nature Chem Biol* 2006;**2**:417–22.
22. Kobilka BK, Deupi X. Conformational complexity of G-protein-coupled receptors. *Trends Pharmacol Sci* 2007;**28**:397–406.
23. Kobilka BK. Amino and carboxyl terminal modifications to facilitate the production and purification of a G-protein-coupled receptor. *Analyt Biochem* 1995;**231**:269–71.
24. Tota MR, Strader CD. Characterization of the binding domain of the beta-adrenergic receptor with the fluorescent antagonist carazolol. Evidence for a buried ligand binding site. *J Biol Chem* 1990;**265**:16,891–897.
25. Day PW, Rasmussen SG, Parnot C, et al. A monoclonal antibody for G-protein-coupled receptor crystallography. *Nature Methods* 2007;**4**:927–9.
26. Rosenbaum DM, Cherezov V, Hanson MA, et al. GPCR engineering yields high-resolution structural insights into beta2-adrenergic receptor function. *Science* 2007;**318**:1266–73.
27. Caffrey M. Membrane protein crystallization. *J Struct Biol* 2003;**142**:108–32.
28. Rasmussen SG, Choi H-J, Rosenbaum DM, et al. Crystal structure of the human beta2 adrenergic G-protein-coupled receptor. *Nature* 2007;**450**:383–7.
29. Cherezov V, Rosenbaum DM, Hanson MA, et al. High-resolution crystal structure of an engineered human beta2-adrenergic G-protein-coupled receptor. *Science* 2007;**318**:1258–65.
30. Okada T, Sugihara M, Bondar AN, Elstner M, Entel P, Buss V. The retinal conformation and its environment in rhodopsin in light of a new 2.2-Å crystal structure. *J Mol Biol* 2004;**342**:571–83.

31. Shi L, Liapakis G, Xu R, Guarnieri F, Ballesteros JA, Javitch JA. Beta2 adrenergic receptor activation. Modulation of the proline kink in transmembrane 6 by a rotamer toggle switch. *J Biol Chem* 2002;**277**:40,989–96.
32. Vogel R, Mahalingam M, Ludeke S, Huber T, Siebert F, Sakmar TP. Functional role of the "ionic lock"– an interhelical hydrogen-bond network in family A heptahelical receptors. *J Mol Biol* 2008;**380**:648–55.
33. Park JH, Scheerer P, Hofmann KP, Choe HW, Ernst OP. Crystal structure of the ligand-free G-protein-coupled receptor opsin. *Nature* 2008;**454**:183–7.
34. Farrens DL, Altenbach C, Yang K, Hubbell WL, Khorana HG. Requirement of rigid-body motion of transmembrane helices for light activation of rhodopsin. *Science* 1996;**274**:768–70.
35. Scheerer P, Park JH, Hildebrand PW, et al. Crystal structure of opsin in its G-protein-interacting conformation. *Nature* 2008;**455**:497–502.
36. Jensen AD, Guarnieri F, Rasmussen SG, Asmar F, Ballesteros JA, Gether U. Agonist-induced conformational changes at the cytoplasmic side of transmembrane segment 6 in the beta 2 adrenergic receptor mapped by site-selective fluorescent labeling. *J Biol Chem* 2001;**276**:9279–90.
37. Gether U, Ballesteros JA, Seifert R, Sanders-Bush E, Weinstein H, Kobilka BK. Structural instability of a constitutively active G-protein-coupled receptor. Agonist-independent activation due to conformational flexibility. *J Biol Chem* 1997;**272**:2587–90.

Chapter 33

Agonist-Induced Desensitization and Endocytosis of G-protein-Coupled Receptors

Michael Tanowitz and Mark von Zastrow

Departments of Psychiatry and Cellular and Molecular Pharmacology, University of California, San Francisco

Multiple mechanisms contribute to the physiological regulation of G-protein-coupled receptors (GPCRs). Early studies delineated the existence of distinct functional processes of receptor regulation in natively expressing cells and tissues. Some of the biochemical mechanisms underlying these regulatory processes are now understood, and there is accelerating progress toward elucidating their physiological significance. We will begin this chapter by reviewing basic processes of GPCR regulation, as defined operationally in early studies, then review current information regarding biochemical and cell biological mechanisms that underlie them. In doing so, we will focus primarily on well-characterized mechanisms that appear to be relevant to the regulation of many GPCRs. Finally, we will briefly mention recent insights to previously unanticipated features of GPCR regulation and signaling, which may be of interest for future therapeutic exploration.

GENERAL PROCESSES OF GPCR REGULATION

Desensitization and Resensitization: Rapid Regulation of the Functional Activity of Receptors

Many GPCRs are regulated very rapidly after agonist-induced activation, a process that has been characterized in considerable detail in studies of the β_2 adrenergic receptor (β_2AR) [1–3]. Upon binding of agonist, β_2AR promotes guanine nucleotide exchange on its cognate heterotrimeric G protein (G_s), leading to activation of adenylyl cyclase. Receptor-mediated signaling by this mechanism occurs within seconds after the initial addition of agonist to cells or tissues. Within several minutes, the ability of receptors to mediate agonist-induced signaling diminishes. This process of *rapid desensitization* has been studied extensively in cultured cells, and typically is manifest as a 10- to 100-fold right-shift of the dose response curve for adenylyl cyclase activation observed when agonist re-challenge is carried out immediately after washout. In most cell models agonist responsiveness of the β_2AR recovers quite rapidly, so that the dose–response relationship observed upon agonist re-challenge returns to baseline within several minutes. This recovery of signaling potential from the desensitized state is thought to be important physiologically for mediating sustained (rather than transient) signaling responses in the continued presence of agonist. Rapid desensitization and resensitization of β_2AR-mediated signaling can occur without significant effects on other signaling pathways, and without any detectable change in the total number of receptors present in cells or tissues, suggesting that these processes reflect primarily changes in the *functional activity* of receptors.

Sequestration: Rapid Regulation of the Subcellular Localization of Receptors

Agonists can also cause a pronounced decrease in the number of receptors present in the plasma membrane, which is typically observed over a similarly rapid time-course to that of functional desensitization. This process, often called *sequestration*, was defined originally by pharmacological studies investigating the number of β_2AR ligand binding sites accessible to membrane-impermeant radioligands in intact cells [4]. In general, sequestration occurs without any change in the total number of receptors present in cells or tissues, as detected using membrane-permeant radioligands

Handbook of Cell Signaling, Three-Volume Set 2 ed.

or by carrying out binding assays on disrupted membrane preparations, suggesting that this process reflects primarily a change in the subcellular *localization* of GPCRs.

Downregulation and Upregulation: Slower Modulation of the Total Number of Receptors Present in Cells or Tissues

The process of *downregulation* is characterized by a decrease in total number of receptor sites present in cells or tissues, generally determined by radioligand binding assay conducted using membrane-permeant compounds or on disrupted membrane preparations. Downregulation, indicated by reduced B_{max} calculated from saturation binding isotherms, generally occurs over a period of hours after prolonged agonist activation of receptors or activation of downstream signaling pathways; this process can be distinguished from changes in agonist binding affinity (K_d) that often are associated with rapid desensitization [5, 6]. It is thought that downregulation functions to reduce cell or tissue responsiveness under conditions of prolonged agonist activation. Recovery of receptor number (and signaling responsiveness) after downregulation also occurs relatively slowly, and typically requires biosynthesis of new receptor protein. Some ligands (usually antagonists) can induce an opposite process of *upregulation*, which refers to a gradual increase in the B_{max} detected by radioligand binding [7]. Therefore the processes of downregulation and upregulation are thought to reflect primarily a change in the total *number* of GPCRs present in cells or tissues.

MECHANISMS OF GPCR DESENSITIZATION AND ENDOCYTOSIS

Functional Uncoupling of GPCRs from Heterotrimeric G Proteins Mediated by Receptor Phosphorylation

Pioneering studies of rhodopsin (a light-activated GPCR) and several ligand-activated GPCRs (such as β_2AR), established a conserved mechanism that regulates the functional activity of many GPCRs in all animal species examined [1–3,8]. This mechanism involves the phosphorylation of receptors by a specific family of G-protein-coupled receptor kinases (GRKs) followed by the interaction of phosphorylated receptors with cytoplasmic accessory proteins called arrestins. These events effectively prevent receptor interaction with heterotrimeric G proteins, disrupting the pathway of GPCR-mediated signal transduction at the earliest stage (Figure 33.1a and b).

Biochemical analysis of isolated rod outer segment preparations identified a protein called rhodopsin kinase (or GRK1), which inhibited the ability of light-activated rhodopsin to stimulate transducin (the heterotrimeric G protein that couples to vertebrate rhodopsin). Light-activated rhodopsin is a good substrate for phosphorylation by rhodopsin kinase, whereas rhodopsin that has not been activated by light is a poor substrate [9]. The signaling activity of rhodopsin was not completely abrogated by GRK1-mediated phosphorylation, however. A second protein, *visual arrestin*, was identified from cytoplasmic fractions of rod cells according to its ability to completely inhibit, or arrest, activation of transducin by phosphorylated rhodopsin [10].

Studies using functional reconstitution of β_2AR-mediated activation of adenylyl cyclase revealed a role of phosphorylation in mediating rapid desensitization of this ligand-activated GPCR [11]. Biochemical purification of the cytoplasmic activity responsible for this regulation identified a protein called the β adrenergic receptor kinase (BARK, or GRK2). This enzyme was shown to preferentially phosphorylate agonist-occupied receptors [12], and is closely similar in structure to rhodopsin kinase [13]. Biochemical reconstitution studies indicated that increasingly purified fractions of the kinase exhibited reduced ability to attenuate β_2AR-mediated signal transduction in a reconstituted membrane preparation. Further analysis identified a distinct protein component that restored functional desensitization when added back to a highly purified fraction of the kinase [14, 15]. This protein cofactor turned out to be a protein similar to visual arrestin, and was therefore named "non-visual" arrestin or *beta-arrestin* (βArr). cDNA cloning has identified a family of arrestins involved in regulating the function of phosphorylated GPCRs [3].

Arrestin interaction with GPCRs such as β_2AR is promoted both by agonist-induced phosphorylation of receptors and by agonist binding to the receptor itself [16]. This dual control assures that functional uncoupling occurs specifically at agonist-activated receptors. Other receptors that are not activated, including co-expressed GPCRs that recognize other ligands and are potentially desensitized by the same mechanism, are largely unaffected. Indeed, GRK-mediated phosphorylation and subsequent binding of arrestins is considered to be a paradigm for *homologous* desensitization, a form of desensitization that is specific only to the activated GPCR at hand and is not influenced by activation of other receptors in the same cell [12].

Desensitization of GPCRs by Other Kinases: Example of A Mechanism Mediating Heterologous Desensitization

Kinases other than GRKs, such as the so-called second messenger-regulated kinases, can also function in GPCR desensitization. β_2AR, for example, can be phosphorylated by cyclic AMP-dependent protein kinase (PKA). PKA-mediated phosphorylation of a single residue located in the third intracellular loop of β_2AR is sufficient to reduce receptor coupling to G_s, thereby attenuating receptor-mediated

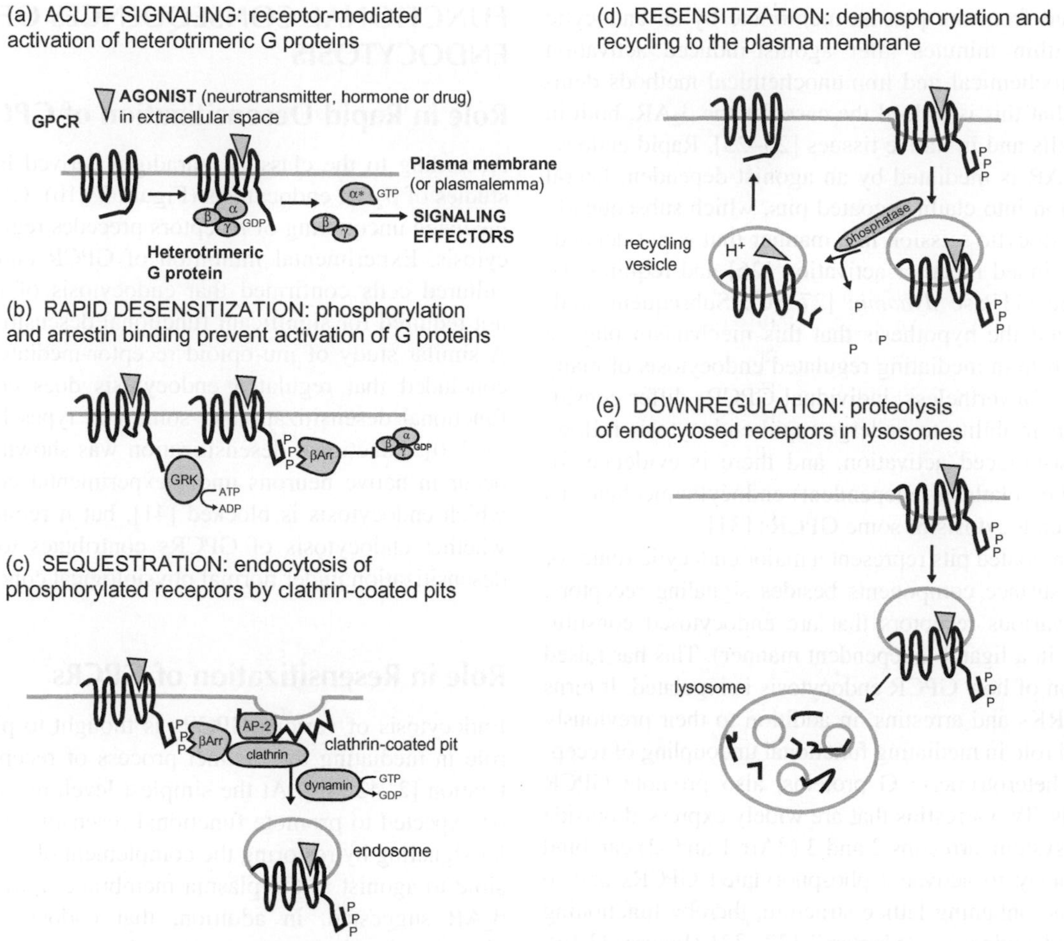

FIGURE 33.1 Major mechanisms of GPCR desensitization and endocytosis.
(a) Classical pathway of GPCR signaling via receptor-mediated activation of heterotrimeric G proteins. (b) Rapid desensitization (functional uncoupling) of GPCRs mediated by GRKs and arrestins. (c) Role of GRKs and non-visual (beta-) arrestins (βArr) in promoting endocytosis of GPCRs via clathrin-coated pits. (d) GPCR resensitization by dephosphorylation and recycling. (e) Downregulation of GPCRs by endocytic trafficking to lysosomes.

signaling [17–19]. Phosphorylation of this residue is thought to impair receptor–G-protein coupling directly, without requiring any known protein cofactor such as an arrestin. PKA is activated by cyclic AMP, a signaling intermediate produced as a result of β_2AR activation, so PKA-mediated phosphorylation of β_2AR is generally viewed as a negative feedback mechanism contributing to cellular homeostasis. PKA-mediated phosphorylation of β_2AR differs fundamentally from GRK-mediated phosphorylation because PKA can efficiently phosphorylate inactive β_2ARs, and PKA activity can be stimulated by any receptor (including but not limited to β_2AR) that promotes adenylyl cyclase activation. For these reasons, phosphorylation by second messenger-regulated kinases is considered a paradigm for *heterologous* desensitization; that is, a form of desensitization of one type of GPCR that can be induced by activation of another (heterologous) receptor. Heterologous desensitization, in contrast to homologous desensitization that is restricted to activated GPCRs, may function in mediating signal integration across distinct GPCRs and diverse signaling mechanisms.

Additional proteins contribute to functional attenuation of GPCR signaling, but do so by acting on downstream substrates distinct from the GPCR itself. Regulators of G-protein signaling (RGS proteins) bind preferentially to the GTP-activated form of the G-protein α subunit and accelerate its intrinsic GTPase activity, thereby helping to terminate the G-protein-mediated signal [20]. Arrestins, in addition to mediating direct actions on phosphorylated GPCRs, can recruit specific phosphodiesterases that accelerate degradation of cAMP produced by receptor-mediated signaling [21]. Thus, signaling through GPCRs can be terminated by GRK/arrestin-dependent uncoupling of the receptor, RGS protein-mediated inactivation of heterotrimeric G proteins, and by arrestin-promoted degradation of second messengers.

Agonist-induced Endocytosis of GPCRs

Pharmacological studies of the process of agonist-induced sequestration led to the hypothesis that certain GPCRs are removed from the plasma membrane by an endocytic process within minutes after agonist-induced activation [4, 22]. Biochemical and immunochemical methods demonstrated that this is indeed the case for the β_2AR, both in cultured cells and in native tissues [23–25]. Rapid endocytosis of β_2AR is mediated by an agonist-dependent lateral redistribution into clathrin-coated pits, which subsequently undergo endocytic scission in a manner that is not dependent on continued receptor activation [26] and requires the cytoplasmic GTPase *dynamin* [27–30]. Subsequent studies supported the hypothesis that this mechanism plays a conserved role in mediating regulated endocytosis of many GPCRs [3]. Nevertheless, individual GPCRs differ considerably in their ability to undergo rapid endocytosis following agonist-induced activation, and there is evidence for alternate (i.e., clathrin-independent) endocytic mechanisms mediating endocytosis of some GPCRs [31].

Clathrin-coated pits represent a major endocytic route for many cell-surface components besides signaling receptors, including various receptors that are endocytosed constitutively (i.e., in a ligand-independent manner). This has raised the question of how GPCR endocytosis is regulated. It turns out that GRKs and arrestins, in addition to their previously established role in mediating functional uncoupling of receptors from heterotrimeric G proteins, also promote GPCR endocytosis. Two arrestins that are widely expressed outside the visual system, arrestins 2 and 3 (βArr-1 and -2) can bind simultaneously to activated phosphorylated GPCRs and to the clathrin-containing lattice structure, thereby functioning as regulated endocytic "adapters" [32, 33] (Figure 33.1c). Arrestins regulate endocytosis of many GPCRs, although there is also evidence for arrestin-independent mechanism(s) mediating concentration of some GPCRs in clathrin-coated pits [34]. Such diversity in mechanism and regulation of GPCR endocytosis, although still poorly understood, may have important effects on GPCR regulation and signaling specificity in native tissues [35].

Although regulated endocytosis promoted by arrestins is generally thought to be highly selective for agonist-activated GPCRs, analogous to the selectivity observed in homologous desensitization of GPCR signaling, this is not always the case. Of particular interest, certain GPCRs shown to exist in hetero-oligomeric complexes when co-expressed in cultured cells can undergo rapid endocytosis in the absence of direct activation, provided that another GPCR in the oligomeric complex is activated. Conversely, there are examples in which agonist-induced endocytosis of one GPCR is inhibited by oligomer formation with a non-internalizing GPCR [36, 37]. Oligomerization can also affect the membrane trafficking of GPCRs at later stages of trafficking, thereby determining the functional consequences of agonist-induced endocytosis [38]. While much remains to be learned about the occurrence and significance of these phenomena in native tissues, there is considerable interest currently in their potential implications for GPCR-directed pharmacotherapy.

FUNCTIONAL CONSEQUENCES OF GPCR ENDOCYTOSIS

Role in Rapid Desensitization of GPCRs

According to the classical paradigm derived largely from studies of β_2AR endocytosis (Figure 33.1b), GRK/arrestin-mediated uncoupling of receptors precedes regulated endocytosis. Experimental inhibition of GPCR endocytosis in cultured cells confirmed that endocytosis of receptors is not required for significant functional desensitization [39]. A similar study of mu-opioid receptor-mediated signaling concluded that regulated endocytosis does contribute to functional desensitization in some cell types but not others [40]. Functional desensitization was shown recently to occur in native neurons under experimental conditions in which endocytosis is blocked [41], but it remains unclear whether endocytosis of GPCRs contributes to functional desensitization under normal physiological conditions.

Role in Resensitization of GPCRs

Endocytosis of various GPCRs is thought to play a major role in mediating the distinct process of receptor resensitization [2, 42, 43]. At the simplest level, recycling would be expected to promote functional resensitization of cellular signaling by restoring the complement of GPCRs accessible to agonist at the plasma membrane. Early studies of β_2AR suggested, in addition, that endocytosis delivers phosphorylated receptors to an endosome-associated phosphatase that acts on receptors previously phosphorylated (hence "desensitized") at the cell surface [39, 44]. The degree to which this model (Figure 33.1d) can be applied as a general paradigm remains unclear, as there is evidence that dephosphorylation and functional resensitization of some GPCRs can occur efficiently under conditions in which rapid endocytosis is prevented [41, 45].

Role in Mediating Proteolytic Downregulation of GPCRs

Endocytosis is also thought to play an important role in mediating downregulation of many GPCRs by promoting their proteolytic degradation. The most extensively studied pathway mediating proteolytic downregulation of GPCRs involves endocytosis of receptors followed by membrane trafficking to lysosomes (Figure 33.1e). Additional proteolytic machinery, such as proteasomes or membrane-associated endoproteases, may also contribute to downregulation of certain GPCRs (46). GPCRs can be targeted to lysosomes after initial endocytosis by clathrin-coated pits, but there is also some evidence for a distinct membrane pathway to lysosomes involving alternate endocytic mechanism(s) [6, 46]. To a first approximation, however, it is thought that the functional consequences of GPCR endocytosis are determined by a molecular "sorting" operation, occurring in the endosome membrane, which dictates whether internalized receptors recycle to the plasma membrane (resensitization) or traffic to lysosomes (downregulation). Distinct GPCRs differ significantly in their sorting between these divergent membrane pathways even when co-expressed and activated in the same cells [47,48]. Recent studies have described a highly conserved endosome-associated mechanism that is thought to function in sorting many integral membrane proteins, including GPCRs, to lysosomes [49]. A number of GPCRs are sorted by this mechanism following covalent modification by ubiquitin; other GPCRs can be efficiently sorted to lysosomes in the absence of ubiquitination via distinct, non-covalent protein interactions with the cytoplasmic surface of the internalized receptor [50]. Moreover, a number of GPCRs contain additional sequences that promote [51,52] or prevent [53] recycling of receptors to the plasma membrane. It is likely that there exist multiple biochemical mechanisms that distinguish the post-endocytic sorting of specific GPCRs, and which play a critical role in determining the precise functional consequences of agonist-induced endocytosis under physiological conditions [54].

Role in Controlling the Specificity of Signal Transduction

Endocytosis of GPCRs may also control the specificity with which receptors signal to particular downstream effectors, such as mitogen-activated protein (MAP) kinase modules. A number of mechanisms for redirecting GPCR signaling to MAP kinases have been proposed, generally involving the formation on endosome membranes of a protein complex including both internalized GPCRs and signal-transducing kinases (such as c-Src) recruited from the cytoplasm [55,56] or receptor tyrosine kinases (such as epidermal growth factor receptors) co-endocytosed from the plasma membrane [57]. For many GPCR-mediated signaling events, non-visual arrestins appear to function as molecular scaffolds that organize signaling complexes in association with endocytosed or activated GPCRs. An early example demonstrated formation of complexes between the β_2AR, arrestin 2 (βArr-1), and c-Src following receptor activation [55]. Subsequent studies have found that arrestins can also directly recruit and direct MAP kinase phosphorylation cascades via arrestin-mediated scaffolding [56,58]. The characteristics that distinguish arrestin-mediated from G-protein-mediated MAPK activation entail both temporal differences and changes in subcellular localization. For example, G-protein-dependent phosphorylation of ERK1/2 peaks within several minutes after GPCR activation, whereas arrestin-mediated ERK phosphorylation can persist for a much longer time period. Prolonged signaling to ERK is particularly notable in studies of GPCRs such as the angiotensin 1A receptor, which remain associated with arrestins in the endosome membrane for an unusually prolonged time period, suggesting arrestin-mediated signaling to ERK occurs from endosomes [59]. Phospho-ERK produced by the G-protein-dependent mechanism can be observed in the nucleus (as is observed for phospho-ERK produced by receptor tyrosine kinase activation), while phospho-ERK produced by the arrestin-dependent mechanism is restricted from the nucleus and can be localized to endosomes [60]. While much remains to be learned, the present data support the hypothesis that arrestins can serve as multifunctional signaling scaffolds both at the plasma membrane and on endosome membranes [61].

REFERENCES

1. Lefkowitz RJ, Pitcher J, Krueger K, Daaka Y. Mechanisms of beta-adrenergic receptor desensitization and resensitization. *Adv Pharmacol* 1998;**42**:416–20.
2. Ferguson SS, Zhang J, Barak LS, Caron MG. Molecular mechanisms of G-protein-coupled receptor desensitization and resensitization. *Life Sci* 1998;**62**:1561–5.
3. Carman CV, Benovic JL. G-protein-coupled receptors: turn-ons and turn-offs. *Curr Opin Neurobiol* 1998;**8**:335–44.
4. Staehelin M, Simons P. Rapid and reversible disappearance of beta-adrenergic cell surface receptors. *EMBO J* 1982;**1**:187–90.
5. Clark RB. Receptor desensitization. *Adv Cyclic Nucl Prot Phos Res* 1986;**20**:151–209.
6. Koenig JA, Edwardson JM. Endocytosis and recycling of G-protein-coupled receptors. *Trends Pharmacol Sci* 1997;**18**:276–87.
7. Doss RC, Perkins JP, Harden TK. Recovery of beta-adrenergic receptors following log term exposure of astrocytoma cells to catecholamine: role of protein synthesis. *J Biol Chem* 1981;**256**:12,281–6.
8. Krupnick JG, Benovic JL. The role of receptor kinases and arrestins in G-protein-coupled receptor regulation. *Annu Rev Pharmacol Toxicol* 1998;**38**:289–319.
9. McDowell JH, Kuhn H. Light-induced phosphorylation of rhodopsin in cattle photoreceptor membranes: substrate activation and inactivation. *Biochemistry* 1977;**16**:4054–60.
10. Bennett N, Sitaramayya A. Inactivation of photoexcited rhodopsin in retinal rods: the roles of rhodopsin kinase and 48-kDa protein (arrestin). *Biochemistry* 1988;**27**:1710–15.
11. Sibley DR, Strasser RH, Caron MG, Lefkowitz RJ. Homologous desensitization of adenylate cyclase is associated with phosphorylation of the beta-adrenergic receptor. *J Biol Chem* 1985;**260**:3883–6.
12. Benovic JL, Strasser RH, Caron MG, Lefkowitz RJ. Beta-adrenergic receptor kinase: identification of a novel protein kinase that phosphorylates the agonist-occupied form of the receptor. *Proc Natl Acad Sci USA* 1986;**83**:2797–801.
13. Benovic JL, Stone WC, Huebner K, Croce C, Caron MG, Lefkowitz RJ. cDNA cloning and chromosomal localization of the human beta-adrenergic receptor kinase. *FEBS Letts* 1991;**283**:122–6.
14. Lohse MJ, Benovic JL, Codina J, Caron MG, Lefkowitz RJ. beta-Arrestin: a protein that regulates beta-adrenergic receptor function. *Science* 1990;**248**:1547–50.
15. Benovic JL, Kuhn H, Weyand I, Codina J, Caron MG, Lefkowitz RJ. Functional desensitization of the isolated beta-adrenergic receptor by the beta-adrenergic receptor kinase: potential role of an analog of

the retinal protein arrestin (48-kDa protein). *Proc Natl Acad Sci USA* 1987;**84**:8879–82.

16. Gurevich VV, Benovic JL. Mechanism of phosphorylation-recognition by visual arrestin and the transition of arrestin into a high affinity binding state. *Mol Pharmacol* 1997;**51**:161–9.
17. Hausdorff WP, Lohse MJ, Bouvier M, Liggett SB, Caron MG, Lefkowitz RJ. Two kinases mediate agonist-dependent phosphorylation and desensitization of the beta 2-adrenergic receptor. *Symp Soc Exp Biol* 1990;**44**:225–40.
18. Bouvier M, Hausdorff WP, De Blasi A, et al. Removal of phosphorylation sites from the beta 2-adrenergic receptor delays onset of agonist-promoted desensitization. *Nature* 1988;**333**:370–3.
19. Benovic JL, Bouvier M, Caron MG, Lefkowitz RJ. Regulation of adenylyl cyclase-coupled beta-adrenergic receptors. *Annu Rev Cell Biol* 1988;**4**:405–28.
20. De Vries L, Zheng B, Fischer T, Elenko E, Farquhar MG. The regulator of G-protein signaling family. *Annu Rev Pharmacol Toxicol* 2000;**40**:235–71.
21. Perry SJ, Baillie GS, Kohout TA, et al. Targeting of cyclic AMP degradation to beta 2-adrenergic receptors by beta-arrestins. *Science* 2002; **298**:834–6.
22. Toews ML, Perkins JP. Agonist-induced changes in beta-adrenergic receptors on intact cells. *J Biol Chem* 1984;**259**:2227–35.
23. von Zastrow M, Kobilka BK. Ligand-regulated internalization and recycling of human beta 2-adrenergic receptors between the plasma membrane and endosomes containing transferrin receptors. *J Biol Chem* 1992;**267**:3530–8.
24. Kurz JB, Perkins JP. Isoproterenol-initiated beta-adrenergic receptor diacytosis in cultured cells. *Mol Pharmacol* 1992;**41**:375–81.
25. Keith DE, Anton B, Murray SR, et al. mu-Opioid receptor internalization: opiate drugs have differential effects on a conserved endocytic mechanism in vitro and in the mammalian brain. *Mol Pharmacol* 1998;**53**:377–84.
26. von Zastrow M, Kobilka BK. Antagonist-dependent and -independent steps in the mechanism of adrenergic receptor internalization. *J Biol Chem* 1994;**269**:18,448–52.
27. van der Bliek AM, Redelmeier TE, Damke H, Tisdale EJ, Meyerowitz EM, Schmid SL. Mutations in human dynamin block an intermediate stage in coated vesicle formation. *J Cell Biol* 1993;**122**:553–63.
28. Herskovits JS, Burgess CC, Obar RA, Vallee RB. Effects of mutant rat dynamin on endocytosis. *J Cell Biol* 1993;**122**:565–78.
29. Zhang J, Ferguson S, Barak LS, Menard L, Caron MG. Dynamin and beta-arrestin reveal distinct mechanisms for G-protein-coupled receptor internalization. *J Biol Chem* 1996;**271**:18,302–5.
30. Cao TC, Mays RW, von Zastrow M. Regulated endocytosis of G-protein-coupled receptors by a biochemically and functionally distinct subpopulation of clathrin-coated pits. *J Biol Chem* 1998;**273**:24,592–2.
31. Tsao PI, von Zastrow M. Diversity and specificity in the regulated endocytic membrane trafficking of G-protein-coupled receptors. *Pharmacol Ther* 2001;**89**:139–47.
32. Goodman OJ, Krupnick JG, Santini F, et al. Beta-arrestin acts as a clathrin adaptor in endocytosis of the beta2-adrenergic receptor. *Nature* 1996;**383**:447–50.
33. Laporte SA, Oakley RH, Holt JA, Barak LS, Caron MG. The interaction of beta-arrestin with the AP-2 adaptor is required for the clustering of beta 2-adrenergic receptor into clathrin-coated pits. *J Biol Chem* 2000;**275**:23,120–6.
34. Lee KB, Pals RR, Benovic JL, Hosey MM. Arrestin-independent internalization of the m1, m3, and m4 subtypes of muscarinic cholinergic receptors. *J Biol Chem* 1998;**273**:12,967–72.
35. Roettger BF, Rentsch RU, Pinon D, et al. Dual pathways of internalization of the cholecystokinin receptor. *J Cell Biol* 1995;**128**:1029–41.
36. Devi LA. Heterodimerization of G-protein-coupled receptors: pharmacology, signaling and trafficking. *Trends Pharmacol Sci* 2001;**22**:532–7.
37. Xu J, He J, Castleberry AM, Balasubramanian S, Lau AG, Hall RA. Heterodimerization of alpha 2A- and beta 1-adrenergic receptors. *J Biol Chem* 2003;**278**:10,770–7.
38. Cao TT, Brelot A, von Zastrow M. The composition of the beta-2 adrenergic receptor oligomer affects its membrane trafficking after ligand-induced endocytosis. *Mol Pharmacol* 2005;**67**:288–97.
39. Pippig S, Andexinger S, Lohse MJ. Sequestration and recycling of beta 2–adrenergic receptors permit receptor resensitization. *Mol Pharmacol* 1995;**47**:666–76.
40. Pak Y, Kouvelas A, Scheideler MA, Rasmussen J, O'Dowd BF, George SR. Agonist-induced functional desensitization of the mu-opioid receptor is mediated by loss of membrane receptors rather than uncoupling from G protein. *Mol Pharmacol* 1996;**50**:1214–22.
41. Arttamangkul S, Torrecilla M, Kobayashi K, Okano H, Williams JT. Separation of mu-opioid receptor desensitization and internalization: endogenous receptors in primary neuronal cultures. *J Neurosci* 2006;**26**:4118–25.
42. Pippig S, Andexinger S, Daniel K, et al. Overexpression of beta-arrestin and beta-adrenergic receptor kinase augment desensitization of beta 2-adrenergic receptors. *J Biol Chem* 1993;**268**:3201–8.
43. Yu SS, Lefkowitz RJ, Hausdorff WP. Beta-adrenergic receptor sequestration. A potential mechanism of receptor resensitization. *J Biol Chem* 1993;**268**:337–41.
44. Pitcher JA, Payne ES, Csortos C, DePaoli RA, Lefkowitz RJ. The G-protein-coupled receptor phosphatase: a protein phosphatase type 2A with a distinct subcellular distribution and substrate specificity. *Proc Natl Acad Sci USA* 1995;**92**:8343–7.
45. Gardner B, Liu ZF, Jiang D, Sibley DR. The role of phosphorylation/dephosphorylation in agonist-induced desensitization of D1 dopamine receptor function: evidence for a novel pathway for receptor dephosphorylation. *Mol Pharmacol* 2001;**59**:310–21.
46. Tsao P, Cao T, von Zastrow M. Role of endocytosis in mediating downregulation of G-protein-coupled receptors. *Trends Pharmacol Sci* 2001;**22**:91–6.
47. Gagnon AW, Kallal L, Benovic JL. Role of clathrin-mediated endocytosis in agonist-induced down-regulation of the beta2-adrenergic receptor. *J Biol Chem* 1998;**273**:6976–81.
48. Tsao PI, von Zastrow M. Type-specific sorting of G-protein-coupled receptors after endocytosis. *J Biol Chem* 2000;**275**:11,130–40.
49. Saksena S, Sun J, Chu T, Emr SD. ESCRTing proteins in the endocytic pathway. *Trends Biochem Sci* 2007;**32**:561–73.
50. Marchese A, Paing MM, Temple BR, Trejo J. G-protein-coupled receptor sorting to endosomes and lysosomes. *Annu Rev Pharmacol Toxicol* 2008;**48**:601–29.
51. Cao TT, Deacon HW, Reczek D, Bretscher A, von Zastrow M. A kinase-regulated PDZ-domain interaction controls endocytic sorting of the beta2-adrenergic receptor. *Nature* 1999;**401**:286–90.
52. Tanowitz M, von Zastrow M. A novel endocytic recycling signal that distinguishes the membrane trafficking of naturally occurring opioid receptors. *J Biol Chem* 2003;**278**:45,978–86.
53. Innamorati G, Sadeghi HM, Tran NT, Birnbaumer M. A serine cluster prevents recycling of the V2 vasopressin receptor. *Proc Natl Acad Sci USA* 1998;**95**:2222–6.
54. Hanyaloglu AC, von Zastrow M. Regulation of GPCRs by membrane trafficking and its potential implications. *Annu Rev Pharmacol Toxicol* 2008;**48**:537–68.
55. Luttrell LM, Ferguson SS, Daaka Y, et al. Beta-arrestin-dependent formation of beta2 adrenergic receptor-Src protein kinase complexes. *Science* 1999;**283**:655–61.

56. DeFea KA, Zalevsky J, Thoma MS, Dery O, Mullins RD, Bunnett NW. beta-arrestin-dependent endocytosis of proteinase-activated receptor 2 is required for intracellular targeting of activated ERK1/2. *J Cell Biol* 2000;**148**:1267–81.
57. Maudsley S, Pierce KL, Zamah AM, et al. The beta(2)-adrenergic receptor mediates extracellular signal-regulated kinase activation via assembly of a multi-receptor complex with the epidermal growth factor receptor. *J Biol Chem* 2000;**275**:9572–80.
58. Luttrell LM, Roudabush FL, Choy EW, et al. Activation and targeting of extracellular signal-regulated kinases by beta-arrestin scaffolds. *Proc Natl Acad Sci USA* 2001;**98**:2449–54.
59. Ahn S, Shenoy SK, Wei H, Lefkowitz RJ. Differential kinetic and spatial patterns of beta-arrestin and G-protein-mediated ERK activation by the angiotensin II receptor. *J Biol Chem* 2004;**279**: 35,518–25.
60. Tohgo A, Pierce KL, Choy EW, Lefkowitz RJ, Luttrell LM. Beta-arrestin scaffolding of the ERK cascade enhances cytosolic ERK activity but inhibits ERK-mediated transcription following angiotensin AT1a receptor stimulation. *J Biol Chem* 2002;**277**: 9429–36.
61. DeWire SM, Ahn S, Lefkowitz RJ, Shenoy SK. Beta-arrestins and cell signaling. *Annu Rev Physiol* 2007;**69**:483–510.

Chapter 34

Functional Role(s) of Dimeric Complexes Formed from G-Protein-Coupled Receptors

Raphael Rozenfeld and Lakshmi A. Devi

Department of Pharmacology and Systems Therapeutics, Mount Sinai School of Medicine, New York, New York

INTRODUCTION

The classic notion that G-protein-coupled receptors (GPCR) function as monomers and the recent observation that GPCRs exist and function as homo- and/or hetero-oligomeric complexes have raised the question of whether receptors have to form dimers to function or if monomers are functional. Research by several laboratories has provided evidence that support both hypotheses; that individual protomers are able to function (i.e., couple to G proteins) [1, 2] and that GPCRs form physiologically relevant oligomers [3]. Studies using reconstituted systems provide convincing evidence indicating that a single GPCR molecule is able to activate a G protein [1, 2]. For example, insertion of a single rhodopsin molecule (monomer) into lipid bilayers of high-density lipoprotein particles appears to activate transducin (a G protein) with the same efficacy as two rhodopsin molecules inserted into the same particle [1]. This suggests that either one rhodopsin molecule is accessible for G-protein coupling when two receptor molecules are present, or that dimerization does not enhance, but in fact reduces, the efficiency of G-protein coupling. Interestingly, an increasing number of studies show that GPCR oligomers exist under physiological conditions, and that dimerization/oligomerization appears to be necessary for receptor biosynthesis/folding to ensure proper receptor maturation, exit from the endoplasmic reticulum (ER), and transport to the plasma membrane [4–6]. Receptor oligomerization is thus likely to be necessary for the physiological function of GPCRs. Therefore, it is important to make a distinction between the intrinsic properties of GPCRs determined with purified systems, and their properties in a physiological context. Furthermore, the idea (proposed from the results obtained with reconstituted systems) that one receptor monomer coupling to one G protein is the most efficient model for a functional GPCR [1, 2] is also challenged by an observation from a physiologically relevant system in human monocytes from atherosclerotic patients, where a covalently linked GPCR homodimer (due to crosslinking between individual receptors) displays enhanced signaling [7]. This example, in which an obligatory dimer displays enhanced signaling, contradicts the model in which dimerization reduces the efficiency of G-protein coupling. However, these opposing models raise the exciting possibility that since a receptor monomer is able to function in isolation, association with other receptors could affect its activity. In support of this, findings from emerging studies with a variety of GPCRs indicate that GPCR hetero-oligomerization alters receptor function by forming new signaling units with distinct properties, which are involved in specific pathophysiological mechanisms. In this review, we describe evidence for receptor dimerization/oligomerization and how this leads to the modulation of the properties of individual receptors.

HISTORICAL PERSPECTIVE

Indirect Evidence

Although the notion of GPCR dimerization has only recently become accepted, studies carried out in the past 30 years had provided evidence, albeit indirect, that GPCRs exist and function as complexes (rather than as individual monomeric units). Cooperativity between binding sites of a variety of GPCRs (including β-adrenergic receptor (βAR) or muscarinic receptor) was observed in endogenous tissues [8–13]. To explain this and other similar observations, the "receptor mosaic hypothesis" was proposed that stated that the formation of receptor complexes (called mosaics) at synapses could represent a molecular adaptation to various patterns of transmitters released by the source neuron

Handbook of Cell Signaling, Three-Volume Set 2 ed.

[14]. The functional unit within the "mosaic" was thought to be a receptor dimer (or an oligomer consisting of a dimer of dimers). Biochemical studies further supported the idea that GPCRs associate to form functional complexes consisting of dimers; gel-filtration analysis [15], radiation inactivation analysis [16–18], and affinity labeling studies [19, 20] showed that the size of the functional unit was greater than expected, suggesting that the receptors exist as higher molecular weight complexes.

Functional Complementation

Functional complementation analysis has been applied to demonstrate direct association of proteins to form functional receptors. This analysis explores the formation of a functional unit by co-expression of two non-functional protein partners. Such an approach was first applied to GPCRs by Maggio and co-workers [21], who reconstituted an adrenergic and a muscarinic receptor binding site by co-expression of two complementary chimeras of m3 muscarinic and α_2 adrenergic receptors (α_2AR), which did not contain binding sites for their respective ligands when they were individually expressed [21]. These studies indicated the possibility of a physical association between m3 muscarinic and α_2AR leading to the formation of the ligand binding pocket for both receptors. In another study, Monnot and co-workers [22] generated two different mutants of the AT1 receptor that did not bind to angiotensin. When co-expressed, the ability to bind the ligand was restored, suggesting that an association between the two AT1 protomers leads to the reconstitution of a functional ligand binding site [22]. More recently, using a heterodimeric complex of leukotriene B4 receptors in which each of the protomers could be selectively activated by a structurally different ligand, it was established that agonist binding to one of the receptor protomers leads to conformational changes on the other protomer. This is in agreement with the model of a cross-conformational change transmitted between the two protomers within a receptor dimer [23]. Taken together, these studies support the notion that GPCRs exist as functional dimers.

Biochemical/Biophysical Evidence

The existence of receptor complexes in recombinant systems (and, in a few cases, in tissues) has been demonstrated by a variety of techniques:

1. Crosslinking of proteins using a variety of agents, followed by Western blot analysis to detect receptor complexes as a higher molecular weight form [24–26]
2. Co-immunoprecipitation of differentially epitope-tagged receptors followed by Western blot analysis to indicate an interaction between the two tagged receptors [26]
3. Bioluminescence resonance energy transfer (BRET), a proximity-based biophysical method that examines receptor–receptor associations under physiological conditions, in living cells [27, 28]
4. MALDI mass spectrometry of reconstituted receptors with G proteins, which has revealed the presence of a pentameric unit consisting of two GPCR monomers and one heterotrimeric G protein [29].

The results from these studies using distinct biochemical and biophysical techniques are in agreement with the formation of GPCR dimers.

Physical Evidence

In a handful of cases, direct evidence for receptor dimerization has come from techniques that allow the visualization of GPCR complexes, such as atomic force microscopy (AFM) or X-ray crystallography. Fotiadis and co-workers used atomic force microscopy (AFM) to demonstrate that in isolated murine disc membranes, rhodopsin is organized in linear arrays of receptor dimers [30]. The same paracrystalline organization could also be observed in the case of opsin isolated from photoreceptors of Rpe65−/− mutant mice, which do not produce the chromophore 11-cis-retinal [31], suggesting that the dimeric arrangement of the receptors is independent of their activation state. The dimeric nature of a class C GPCR, metabotropic glutamate receptor (mGluR), was confirmed by structural data obtained from the resolution of the crystal structure of the extracellular domain of these receptors. This established the key contribution of the N-terminal extracellular domain of these receptors to dimerization [32, 33]. Recently, the crystal structure of the β_2AR has been solved. To obtain crystals, Kobilka and co-workers stabilized the receptor using a Fab antibody fragment that recognizes a conformational epitope at both ends of the third intracellular loop [34] or by replacement of the third intracellular loop with T4 lysozyme [35]. This resulted in the successful generation of crystals that diffracted at 3.4 Å/3.7 Å [34] and 2.4 Å [35] respectively. Although the crystals of the β_2AR–Fab complex showed no inter-receptor contacts, in the β_2AR–T4 chimera crystals, lipid-mediated contacts between helixes 1 and 8 suggested the possibility of formation of receptor complexes. It remains to be seen if intact receptors (lacking replacements in the third intracellular loop) are able to form dimers/oligomers. Taken together, the information accumulated using these different techniques strongly supports the notion that GPCRs exist in oligomeric complexes. In this review, we focus on GPCR heterodimers and summarize the current knowledge of how heterodimerization affects receptor function, and its consequences on pathophysiological processes. Since the majority of the existing techniques do not allow us to make a distinction between dimers and oligomers, and for the sake of simplicity, we refer to heteromeric (dimeric or oligomeric) units as *heterodimers* and the event as *heterodimerization*.

HETERODIMERIZATION ALTERS RECEPTOR FUNCTION

A number of studies have examined the effect of association between two distinct receptors on their function. Most of these studies were carried out in recombinant or isolated cell-culture systems that allow close examination of receptor pharmacology and signaling, but do not necessarily represent physiological phenomena. Nevertheless, these studies are of scientific interest since they examine the influence of heterodimerization on receptor function at the molecular level. The first important concept about GPCR heterodimerization is that heterodimers represent a functional unit, distinct from their cognate protomers. This is supported by the observation that receptor heterodimers exhibit distinct binding [26, 36–38], signaling [36, 39–42], trafficking (Figure 34.1a), and subcellular localization [43, 44] as compared to their cognate protomers. The second concept is that GPCR heterodimerization allows for a large degree of plasticity in the receptors' ability to respond to ligands. The heterodimer, in addition to being able to bind individual ligands at either protomer, can bind to heterodimer-specific ligands that only bind to the receptors when in a heterodimeric complex [37, 45, 46]. More importantly, this leads to changes in the extent of receptor coupling and signaling [39, 45] (Figure 34.1b). Hence, GPCR heterodimerization allows for a new level of complexity in the regulation of the properties and thus functions of GPCRs.

Modulation of Receptor Pharmacology

A number of studies have described modulation of the binding properties of GPCRs by heterodimerization [26, 36–38]. This is thought to be the consequence of the formation of altered binding pockets. For example, a decrease in the affinity and potency of selective agonists for either κ or δ opioid receptors (KOR; DOR, respectively) was observed in the case of KOR–DOR heterodimers [26], suggesting that heterodimerization leads to conformational changes of the receptors and that some ligands may preferentially

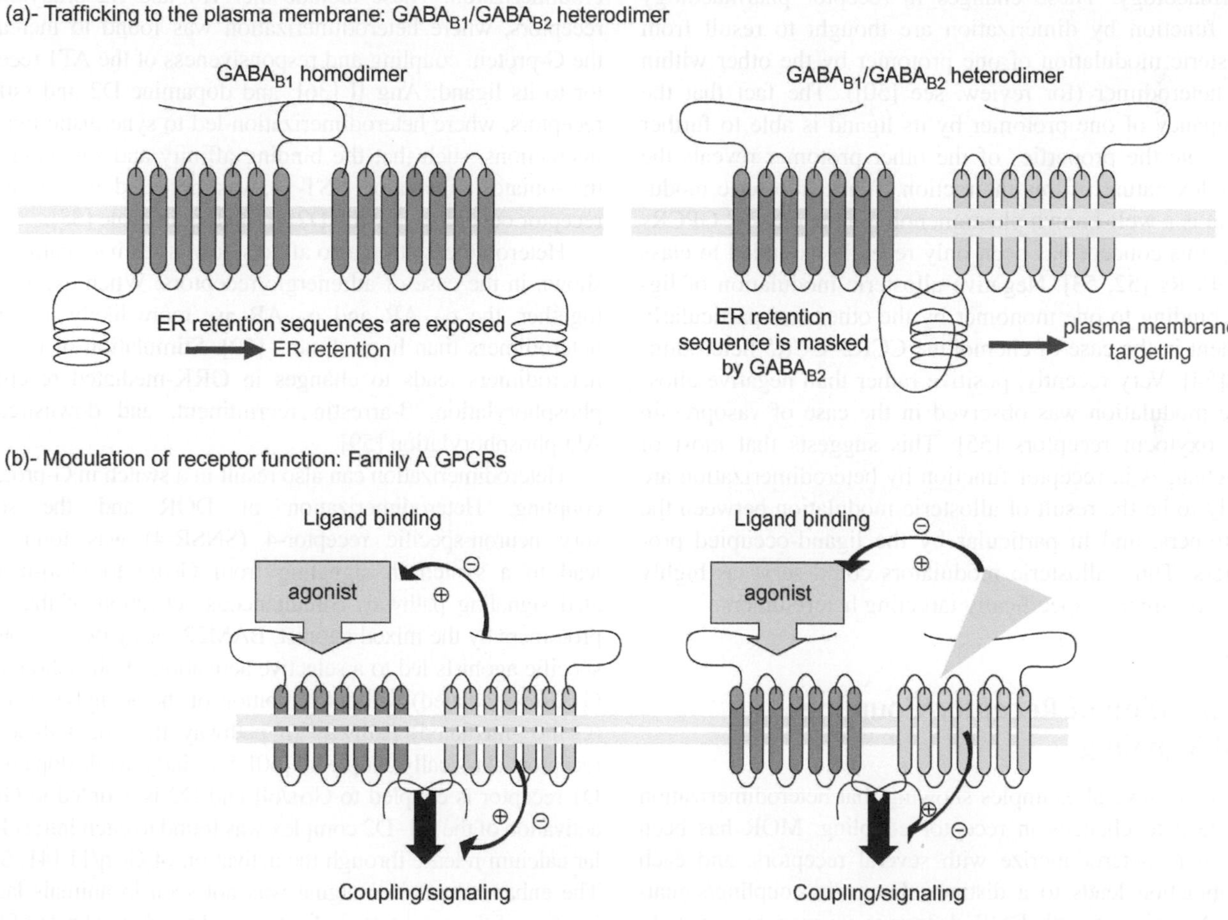

FIGURE 34.1 Regulation of GPCR properties by heterodimerization.
(a) The $GABA_{B1}$ contains an ER retention sequence in its C-terminal domain. This causes the retention of this receptor in the ER and prevents its traffic to the plasma membrane. Dimerization with $GABA_{B2}$ receptor masks the ER retention sequence and allows plasma membrane trafficking and functional activity of the $GABA_{B1}/GABA_{B2}$ heterodimer. (b) Examples of modulation of receptor properties by heterodimerization: within a GPCR heterodimer, a receptor molecule can allosterically modulate ligand binding, coupling, and signaling of the associated receptor molecule; receptor occupancy can further modulate the binding, coupling and signaling of the associated receptor molecule.

bind to receptors in their heterodimeric conformation. This notion was further supported by the identification of a specific ligand for the KOR–DOR opioid receptor heterodimer, 6′-guanidinonaltrindole (6′-GNTI) [46]. This ligand does not bind to KOR or DOR, but only to the KOR–DOR complex.

Studies have also shown that heterodimerization between μ opioid receptor (MOR) and DOR leads to a reduction in affinity and potency for highly selective agonists [47]. This decrease could be reversed by co-treatment with ligands to the individual receptor partners [37, 45]. A similar observation was made with the somatostatin sst2A/sst3 receptor pair, where heterodimerization was found to reduce the affinity and the intrinsic activity of a sst3A-selective ligand [48]. In the case of adrenergic receptors, a recent study reported that heterodimerization of β_1AR and β_2AR leads to a marked increase in the binding affinity for β_2AR ligands [49]. Taken together, these studies indicate that heterodimerization does not produce a general effect on receptor pharmacology, but that each receptor pair represents a unique moiety exhibiting alteration in receptor pharmacology. These changes in receptor pharmacology and function by dimerization are thought to result from allosteric modulation of one protomer by the other within the heterodimer (for review, see [50]). The fact that the occupancy of one protomer by its ligand is able to further modulate the properties of the other protomer reveals the complex nature of this interaction. While allosteric modulation is a well-known phenomenon among class C GPCRs [51], this concept has been only recently extended to class A GPCRs [52, 53]. Negative allosteric modulation of ligand binding to one monomer by the other was particularly evident in the case of chemokine CCR2–CCR5 heterodimers [54]. Very recently, positive rather than negative allosteric modulation was observed in the case of vasopressin and oxytocin receptors [55]. This suggests that most of the changes in receptor function by heterodimerization are likely to be the result of allosteric modulation between the protomers, and in particular by the ligand-occupied protomers. Thus, allosteric modulators could serve as highly selective ligands specifically targeting heterodimers.

Modulation of Receptor Coupling and Signaling

There are several examples showing that heterodimerization can lead to changes in receptor coupling. MOR has been shown to heterodimerize with several receptors, and each combination leads to a distinct change in coupling/signaling. Association with DOR decreases its responsiveness to selective agonists (37), association with α_2AR increases its responsiveness [56], and association with CB_1 cannabinoid receptors (CB_1R) does not affect its responsiveness [57]. However, the mechanism by which this modulation occurs and by which heterodimerization affects G-protein coupling is poorly understood. Recent biophysical approaches using fluorescence resonance energy transfer FRET microscopy have allowed the examination of the detailed molecular events underlying the functional interaction between at least one heterodimeric pair, namely MOR and α_{2A}AR [58]. In this case, MOR and α_{2A}AR were found to communicate with each other through a cross-conformational switch that permits direct inhibition of one receptor by the other, and this exhibited subsecond kinetics. Whereas previous studies have shown that morphine binding to MOR triggers a conformational change in the norepinephrine-bound α_{2A}AR that inhibits its signaling to Gα(i) and the downstream MAP kinase cascade [56], Vilardaga and co-workers, using FRET, demonstrated a conformational change that propagates from one receptor to the other, causing the rapid inactivation of the second receptor with subsecond kinetics [58]. Thus, one could speculate that heterodimerization induces a change in receptor conformation that is responsible for the new properties of the receptors in the heterodimer. There are other examples of modulation of receptor coupling and signaling by heterodimerizaton. These include the AT1 and B2 bradykinin receptors, where heterodimerization was found to increase the G-protein coupling and responsiveness of the AT1 receptor to its ligand; Ang II [36]; and dopamine D2 and sstR5 receptors, where heterodimerization led to synergistic ligand interactions, such that the binding affinity and signaling by the somatostatin analog, SST-14, was increased as a result of receptor occupancy by the D2 agonist, quinpirole [38].

Heterodimerization also affects downstream signaling, as shown in the case of adrenergic receptors. When expressed together, the α_{2A}AR and α_{2C}AR are more likely to form heterodimers than homodimers [59]. Stimulation of α_{2a}AR heterodimers leads to changes in GRK-mediated receptor phosphorylation, β-arrestin recruitment, and downstream Akt phosphorylation [59].

Heterodimerization can also result in a switch in G-protein coupling. Heterodimerization of DOR and the sensory neuron-specific receptor-4 (SNSR-4) was found to lead to a switch in signaling from Gαi/o to Gαq-mediated signaling pathway. Simultaneous activation of the two protomers by the mixed agonist, BAM22, or by two receptor-specific agonists led to a selective activation of phospholipase C (Gαq-mediated) without inhibition of the adenylyl cyclase (Gαi/o -mediated) pathway (a pathway that the individual receptors classically couple to) [60]. Similarly, while dopamine D1 receptor is coupled to Gαs/olf and D2 is coupled to Gαi, activation of the D1–D2 complex was found to elicit intracellular calcium release through the activation of Gαq/11 [41, 61]. The enhanced calcium signal was not seen in animals lacking one of the receptors, indicating a physiological role of the D1–D2 complex in the central nervous system [41]. Finally, in the case of the MOR–DOR heterodimer, studies have shown that the heterodimer selectively couples to Gαz (whereas individual receptors couple to Gαi)[40] or to β-arrestin

[39]. The recruitment of β-arrestin and signaling by the β-arrestin-mediated pathway can be reversed by a combination of MOR and DOR ligands, since this treatment dissociates β-arrestin from the receptor complex and allows the restoration of a G-protein-mediated receptor signaling [39]. This was also demonstrated in dorsal root ganglia neurons [62], suggesting that similar β-arrestin-mediated coupling of the MOR–DOR heterodimer occurs *in vivo*. The fact that β-arrestin knockout mice exhibit aberrant responses to morphine [63] and do not develop morphine tolerance [64] suggests a potential role for MOR–DOR heterodimers in opiate tolerance.

In short, these studies show that GPCR heterodimerization leads to both quantitative and qualitative changes in coupling, from modulating the strength of an agonist-mediated response, to switching to different G-protein coupling and/or associated signaling complexes. Furthermore, some of these changes have also been observed in a physiological context, and are likely to extend the repertoire of GPCR-mediated regulation *in vivo*.

Modulation of Receptor Maturation, Trafficking, and Localization

Maturation

Accumulating evidence indicates that the majority of GPCRs form constitutive dimers in the endoplasmic reticulum (ER), and that dimerization is necessary for GPCR trafficking to the cell surface [4–6]. The best documented example is that of γ-aminobutyric acid B ($GABA_B$) receptors that consist of $GABA_{B1}$ and $GABA_{B2}$ proteins, which themselves are not functional when either is expressed alone. As protomers, $GABA_{B1}$ are retained in the ER, unable to go to the cell surface, whereas $GABA_{B2}$ are expressed at the cell surface, couple to G proteins, but do not bind the ligand GABA. Heterodimerization leads to the masking of an ER retention sequence in $GABA_{B1}$ by $GABA_{B2}$ [65–68], allowing the heterodimers to be efficiently expressed at the cell surface. Furthermore, studies have shown that, within the heterodimer, $GABA_{B1}$ binds to the ligand while $GABA_{B2}$ couples to the G protein and elicits signal transduction, demonstrating that $GABA_{B1}$–$GABA_{B2}$ is an obligate heterodimer and that heterodimerization is required for functional activity. Similarly, heterodimerization of α_{1D}AR with either α_{1B}AR or β_2AR is necessary for efficient cell surface expression and receptor activity [69]. Finally, in the case of the taste receptors, heterodimerization is required not only for efficient cell surface expression but also for agonist selectivity. Dimerization of T1R1 with T1R2 leads to receptors that bind to sweet-tasting molecules, whereas dimerization with T1R3 leads to receptors that bind to "umami" tasting molecules–i.e., L-amino acids [70, 71]. Taken together, these and other emerging studies with other Family A and Family C GPCRs propose a role for heterodimerization in GPCR folding/maturation, exit from the ER, and trafficking to the cell surface.

Endocytosis

Heterodimerization has also been shown to modulate receptor endocytosis. An elegant study of this phenomenon was carried out with the V1a and V2 vasopressin receptors [44]. V1a and V2 receptors can form heterodimers that are co-internalized, and their post-endocytic fate depends on the activated protomer: stimulation of V1a leads to recycling of the heterodimer while stimulation of V2 induces a stable interaction of the receptor complex with β-arrestin, preventing recycling of the receptors [44]. In the case of MOR-neurokinin-1 receptor (NK1R) heterodimers, the NK1R agonist and/or the MOR agonist promoted internalization of both MOR and NK1R receptors into the same endosomal compartment as a stable complex with β−arrestin [72]. Therefore, association of MOR and NK1R affects the endocytic properties of MOR, which on its own does not interact stably with β-arrestin. Finally, heterodimerization of the orexin receptor with the CB_1R receptor influences their trafficking properties [43]. Interestingly, treatment with the CB_1R antagonist, SR141716A, induced the cell surface localization of the endocytosed CB_1R-orexin receptor heterodimer [43], suggesting that receptor localization, and most likely function, can be modulated by antagonists of the heterodimeric partner. Taken together, these examples show that heterodimerization can change the trafficking properties of GPCRs, and that a ligand for one protomer can affect the localization and desensitization of the partner GPCR provided that the two receptors are in an interacting heteromeric complex.

Degradation

Little is known about the effects of heterodimerization on receptor degradation. It should be noted that heterodimer formation and trafficking to the cell surface has been shown to be a regulated event in the case of MOR–DOR heterodimers. These receptors, when in an interacting complex, are degraded to a greater extent than MOR or DOR homodimers [73]. Binding of RTP4, a chaperone involved in sorting of taste receptors [74], leads to a an increase in the level of MOR–DOR heterodimer and enhanced cell surface expression [73]. Since the MOR–DOR heterodimer exhibits distinct ligand-binding and signaling properties (as described earlier), modulation of the relative levels of heterodimers at the cell surface is likely significantly to impact the response to opiates *in vivo*.

RECEPTOR HETERODIMERIZATION IN PHYSIOLOGY AND PATHOLOGY

Growing evidence indicates that heterodimerization leads to the formation of new complexes with altered function in specific tissues, and especially in pathological contexts

(Table 34.1). In some cases, these new functional complexes could underlie the molecular mechanism of the diseases and thereby represent novel drug targets.

AT1R–B2R in Pregnancy-Induced Pre-Eclampsia:

It has been shown that AT1R–B2R heterodimers exist *in vivo* in platelets and omental vessels of pregnant women [75]. Blood pressure and vascular tone are regulated by a balance between the vasopressor effects of circulating Ang II (AT1R agonist) and the vasodepressor effects of bradykinin (B2R agonist). It appears that during pregnancy there is an increase in the levels of the AT1R–B2R heterodimers, resulting in an enhanced efficacy and potency of Ang II; this, in turn, pushes the balance towards a vasoconstrictive response [75]. Hence, heterodimerization of AT1R with B2R plays a major role in Ang II-induced hypersensitivity in pre-eclampsia-induced hypertension [75]. More recently, it was shown that the AT1R–B2R heterodimer also contributes to Ang II hyper-responsiveness in mesangial cells in experimental hypertension, such as in spontaneously hypertensive rats [76]. In these cases, receptor heterodimerization provides a new level of fine tuning for the regulation of blood pressure.

β_2AR–AT1R and β_2AR–β_1AR in Cardiomyocytes

In mouse cardiomyocytes, βARs form complexes with AT1R, and within this complex the βAR response is sensitive to AT1R antagonists [77]. Blocking AT1R results in a significant reduction in the maximal response to catecholamine-induced elevation in heart rate [77].Cardiac contractility is also regulated by β_2AR–β_1AR heterodimers. This receptor pair displays an increased binding affinity for subtype-specific ligands, leading to an increase in the efficacy of βAR agonist to elicit myocyte contraction and cAMP formation. These results indicate that heterodimerization of βAR subtypes results in an increased sensitization to agonist stimulation. Furthermore, since dimerization with β_1AR suppresses the spontaneous activity of β_2ARs in cardiac myocytes, it is thought that heterodimerization leads to the formation of a novel population of βARs with distinct functional and pharmacological properties, which demonstrate enhanced signaling efficiency in response to agonist stimulation. This is thought to increase the efficiency of β-adrenergic modulation of cardiac contractility [49]. These two examples demonstrate how GPCR heterodimerization contributes to the fine tuning of a fundamental physiological process such as heart muscle function.

β_2AR-Prostaglandin Receptor (EP1R) in Asthma

β_2AR and EP1R have been shown to interact in airway smooth muscle cells [78]. The receptor complex exhibits unique signaling properties: EP1 acts as a modulator of β_2AR signaling, by altering the extent of coupling of β_2AR to Gs, thereby reducing the bronchodilatory potential of the β_2AR agonist [78]. This regulation of bronchodilatation by

TABLE 34.1 Heterodimers involved in pathophysiological regulations

Receptor pair	Localization	Modulation	Physiological effect	Pathophysiology	References
AT1R–B2R	Platelets; omental vessels	Increased AT1 agonist efficacy and potency	Increase in blood pressure	Pregnancy-related hypertension	[75]
AT1R–β2AR	Cardiomyocytes	Cross-antagonism	Regulation of adrenergic response	Regulation of heart rate	[77]
β2AR–β1AR	Cardiomyocytes	Increased binding affinity for subtype-specific ligands	Increase cAMP formation; increase efficacy of myocyte contraction	Regulation of cardiomyocyte contractility	[49]
β2AR–EP1	Airway smooth muscles	Reduced coupling of β2AR	Decreased adrenergic response	Asthma	[78]
5HT2AR–mGluR2	Cortex	Modulation of G-protein coupling of 5HT2A-R	Mediation of hallucinogenic effects of drugs	Schizophrenia	[42, 79]
DOR–KOR	Spinal cord	Agonist affinity and potency	Enhanced spinal analgesia	Pain	[26, 46]
MOR–DOR	Spinal cord/SNC	Modulation of receptor binding, coupling and signaling	Decreased responsiveness to MOR ligands	Pain; tolerance to opiates	[45, 81]

receptor heterodimerization could offer new therapeutic strategies for the treatment of asthma.

5HT2AR–mGluR2 in Schizophrenia

In a recent study, serotonin receptor 5-HT2AR and the metabotropic glutamate receptor mGluR2 were shown to form a complex in the cerebral cortex that exhibited a distinct pharmacology as compared to that of individual receptors [42]. This study suggests that abnormal dimerization between these receptors may underlie some of the symptoms of schizophrenia. To address this issue, the authors used radioligand binding to show that expression levels of 5-HT2AR and mGluR2/3 receptors were increased and decreased, respectively, in the dorsolateral prefrontal cortex from untreated subjects with schizophrenia. In addition, the authors showed that the level of mGluR2 mRNA, but not mGluR3, was decreased in the same CNS regions in subjects with schizophrenia, suggesting that the interaction between 5-HT2AR and mGluR2 to be specific. The possibility that changes in the relative levels of homodimers to that of heterodimers could play a role in schizophrenia is an exciting new direction that warrants the development of 5HT2AR–mGluR2 specific ligands as potential drugs for the treatment of this disorder, especially since this receptor complex is responsible for mediating the hallucinogenic effect of drugs such as psilocybin and lysergic acid diethylamide [42, 79].

DOR–KOR in Spinal Analgesia

Opioid receptor heterodimers appear to play a role in the modulation of analgesic responses. The existance of the DOR–KOR heterodimer *in vivo* has been demonstrated indirectly by the binding of DOR–KOR specific ligands in select regions of the nervous system, such as the spinal cord [80]. In agreement with this, DOR–KOR heterodimers have been shown to be involved in spinal analgesia. The heterodimer specific agonist 6′-GNTI elicits analgesia when administered directly into the spinal cord, but has no effect when administered directly to the brain [46]. This supports the notion that the heterodimers are likely to represent a tissue-specific target. Moreover, this spinal-selective analgesic effect was blocked by a bivalent DOR–KOR selective receptor antagonist, confirming the involvment of the DOR–KOR heterodimer in modulation of pain/analgesia and supporting the possibility that this heterodimer represents a functional target for analgesia *in vivo*.

MOR–DOR in Opiate Analgesia and Tolerance

MOR–DOR complexes have been isolated from spinal cord neurons [45], indicating that the receptors heterodimers exist *in vivo*. Furthermore, in agreement with the previously described pharmacological properties of the heterodimer, where MOR responsiveness is increased by the occupancy of the DOR binding site, recent studies have shown that a DOR antagonist enhances morphine-induced analgesia [45] and decreases development of tolerance to morphine [81]. This unique property of the heterodimer opens an avenue for improving pharmacological intervention by targeting both receptors for the treatment of pain.

These studies showcase the critical role of GPCR heterodimers in physiological processes and underscore the importance of these complexes as drug targets, especially in disease states, under conditions where one of the partners could be up- or downregulated, leading to changes in the relative ratio of heterodimers to homodimers.

CONCLUSION

The evidence that GPCRs form heterodimers *in vivo*, albeit not as numerous as the *in vitro* data, demonstrates convincing cases where receptor complexes not only exist but also play a distinct role in pathophysiological processes. Hence, a role for GPCR heterodimers has been shown for fundamental physiological mechanisms such as regulation of heart rate, pain, and analgesia, as well as regulation of blood pressure. They represent a new level of fine tuning for the regulation of receptor function that is tissue specific, receptor specific and ligand specific.

Changes in receptor expression during pathophysiologies can alter heterodimer-mediated regulation and contribute to the molecular mechanisms of these diseases. This suggests that GPCR heterodimers represent new drug targets, with high tissue specificity, especially in pathophysiological contexts.

ACKNOWLEDGEMENTS

We thank Dr Ivone Gomes and Ittai Bushlin for critical reading of the manuscript. The work was supported by NIH grants (DA008863, DA019521 and NS053751 to LAD).

REFERENCES

1. Bayburt TH, Leitz AJ, Xie G, Oprian DD, Sligar SG. Transducin activation by nanoscale lipid bilayers containing one and two rhodopsins. *J Biol Chem* 2007;**282**(20):14,875–81.
2. Whorton MR, Jastrzebska B, Park PS, et al. Efficient coupling of transducin to monomeric rhodopsin in a phospholipid bilayer. *J Biol Chem* 2008;**283**(7):4387–94.
3. Guo W, Urizar E, Kralikova M, et al. Dopamine D2 receptors form higher order oligomers at physiological expression levels. *EMBO J* 2008;**27**(17):2293–304.

4. Bulenger S, Marullo S, Bouvier M. Emerging role of homo- and heterodimerization in G-protein-coupled receptor biosynthesis and maturation. *Trends Pharmacol Sci* 2005;**26**(3):131–7.
5. Terrillon S, Durroux T, Mouillac B, et al. Oxytocin and vasopressin V1a and V2 receptors form constitutive homo- and heterodimers during biosynthesis. *Mol Endocrinol* 2003;**17**(4):677–91.
6. Salahpour A, Angers S, Mercier JF, Lagace M, Marullo S, Bouvier M. Homodimerization of the beta2-adrenergic receptor as a prerequisite for cell surface targeting. *J Biol Chem* 2004;**279**(32):33,390–7.
7. AbdAlla S, Lother H, Langer A, el Faramawy Y, Quitterer U. Factor XIIIA transglutaminase crosslinks AT1 receptor dimers of monocytes at the onset of atherosclerosis. *Cell* 2004;**119**(3):343–54.
8. Limbird LE, Meyts PD, Lefkowitz RJ. Beta-adrenergic receptors: evidence for negative cooperativity. *Biochem Biophys Res Commun* 1975;**64**(4):1160–8.
9. Limbird LE, Lefkowitz RJ. Negative cooperativity among beta-adrenergic receptors in frog erythrocyte membranes. *J Biol Chem* 1976;**251**(16):5007–14.
10. Hirschberg BT, Schimerlik MI. A kinetic model for oxotremorine M binding to recombinant porcine m2 muscarinic receptors expressed in Chinese hamster ovary cells. *J Biol Chem* 1994;**269**(42):26,127–35.
11. Wreggett KA, Wells JW. Cooperativity manifest in the binding properties of purified cardiac muscarinic receptors. *J Biol Chem* 1995;**270**(38):22,488–99.
12. Mattera R, Pitts BJ, Entman ML, Birnbaumer L. Guanine nucleotide regulation of a mammalian myocardial muscarinic receptor system. Evidence for homo- and heterotropic cooperativity in ligand binding analyzed by computer-assisted curve fitting. *J Biol Chem* 1985;**260**(12):7410–21.
13. Potter LT, Ballesteros LA, Bichajian LH, et al. Evidence of paired M2 muscarinic receptors. *Mol Pharmacol* 1991;**39**(2):211–21.
14. Agnati LF, Fuxe K, Zoli M, Rondanini C, Ogren SO. New vistas on synaptic plasticity: the receptor mosaic hypothesis of the engram. *Med Biol* 1982;**60**(4):183–90.
15. Fraser CM, Venter JC. The size of the mammalian lung beta 2-adrenergic receptor as determined by target size analysis and immunoaffinity chromatography. *Biochem Biophys Res Commun* 1982;**109**(1):21–9.
16. Venter JC. Muscarinic cholinergic receptor structure. Receptor size, membrane orientation, and absence of major phylogenetic structural diversity. *J Biol Chem* 1983;**258**(8):4842–8.
17. Venter JC, Horne P, Eddy B, Greguski R, Fraser CM. Alpha 1-adrenergic receptor structure. *Mol Pharmacol* 1984;**26**(2):196–205.
18. Lilly L, Fraser CM, Jung CY, Seem an P, Venter JC. Molecular size of the canine and human brain D2 dopamine receptor as determined by radiation inactivation. *Mol Pharmacol* 1983;**24**(1):10–14.
19. Herberg JT, Codina J, Rich KA, Rojas FJ, Iyengar R. The hepatic glucagon receptor. Solubilization, characterization, and development of an affinity adsorption assay for the soluble receptor. *J Biol Chem* 1984;**259**(14):9285–94.
20. Frame LT, Yeung SM, Venter JC, Cooper DM. Target size of the adenosine Ri receptor. *Biochem J* 1986;**235**(2):621–4.
21. Maggio R, Vogel Z, Wess J. Coexpression studies with mutant muscarinic/adrenergic receptors provide evidence for intermolecular "cross-talk" between G-protein-linked receptors. *Proc Natl Acad Sci USA* 1993;**90**(7):3103–7.
22. Monnot C, Bihoreau C, Conchon S, Curnow KM, Corvol P, Clauser E. Polar residues in the transmembrane domains of the type 1 angiotensin II receptor are required for binding and coupling. Reconstitution of the binding site by co-expression of two deficient mutants. *J Biol Chem* 1996;**271**(3):1507–13.
23. Damian M, Martin A, Mesnier D, Pin JP, Baneres JL. Asymmetric conformational changes in a GPCR dimer controlled by G-proteins. *EMBO J* 2006;**25**(24):5693–702.
24. Guo W, Urizar E, Kralikova M, Mobarec JC, Shi L, Filizola M, Javitch JA. Dopamine D2 receptors form higher order oligomers at physiological expression levels. *EMBO J* 2008;**27**:2293–304.
25. Cvejic S, Devi LA. Dimerization of the delta opioid receptor: implication for a role in receptor internalization. *J Biol Chem* 1997;**272**(43):26,959–64.
26. Jordan BA, Devi LA. G-protein-coupled receptor heterodimerization modulates receptor function. *Nature* 1999;**399**(6737):697–700.
27. Gomes I, Filipovska J, Jordan BA, Devi LA. Oligomerization of opioid receptors. *Methods* 2002;**27**(4):358–65.
28. Angers S, Salahpour A, Joly E, et al. Detection of beta 2-adrenergic receptor dimerization in living cells using bioluminescence resonance energy transfer (BRET). *Proc Natl Acad Sci USA* 2000;**97**(7):3684–9.
29. Baneres JL, Parello J. Structure-based analysis of GPCR function: evidence for a novel pentameric assembly between the dimeric leukotriene B4 receptor BLT1 and the G-protein. *J Mol Biol* 2003;**329**(4):815–29.
30. Fotiadis D, Liang Y, Filipek S, Saperstein DA, Engel A, Palczewski K. Atomic-force microscopy: rhodopsin dimers in native disc membranes. *Nature* 2003;**421**(6919):127–8.
31. Liang Y, Fotiadis D, Filipek S, Saperstein DA, Palczewski K, Engel A. Organization of the G-protein-coupled receptors rhodopsin and opsin in native membranes. *J Biol Chem* 2003;**278**(24):21,655–62.
32. Tsuchiya D, Kunishima N, Kamiya N, Jingami H, Morikawa K. Structural views of the ligand-binding cores of a metabotropic glutamate receptor complexed with an antagonist and both glutamate and Gd^{3+}. *Proc Natl Acad Sci USA* 2002;**99**(5):2660–5.
33. Muto T, Tsuchiya D, Morikawa K, Jingami H. Structures of the extracellular regions of the group II/III metabotropic glutamate receptors. *Proc Natl Acad Sci USA* 2007;**104**(10):3759–64.
34. Rasmussen SG, Choi HJ, Rosenbaum DM, et al. Crystal structure of the human beta2 adrenergic G-protein-coupled receptor. *Nature* 2007;**450**(7168):383–7.
35. Rosenbaum DM, Cherezov V, Hanson MA, et al. GPCR engineering yields high-resolution structural insights into beta2-adrenergic receptor function. *Science* 2007;**318**(5854):1266–73.
36. AbdAlla S, Lother H, Quitterer U. AT1-receptor heterodimers show enhanced G-protein activation and altered receptor sequestration. *Nature* 2000;**407**(6800):94–8.
37. Gomes I, Jordan BA, Gupta A, Trapaidze N, Nagy V, Devi LA. Heterodimerization of mu and delta opioid receptors: a role in opiate synergy. *J Neurosci* 2000;**20**(22):RC110.
38. Rocheville M, Lange DC, Kumar U, Patel SC, Patel RC, Patel YC. Receptors for dopamine and somatostatin: formation of hetero-oligomers with enhanced functional activity. *Science* 2000;**288**(5463):154–7.
39. Rozenfeld R, Devi LA. Receptor heterodimerization leads to a switch in signaling: beta-arrestin2-mediated ERK activation by mu-delta opioid receptor heterodimers. *FASEB J* 2007;**21**(10):2455–65.
40. Fan T, Varghese G, Nguyen T, Tse R, O'Dowd BF, George SR. A role for the distal carboxyl tails in generating the novel pharmacology and G-protein activation profile of mu and delta opioid receptor hetero-oligomers. *J Biol Chem* 2005;**280**(46):38,478–88.
41. Rashid AJ, So CH, Kong MM, et al. D1–D2 dopamine receptor heterooligomers with unique pharmacology are coupled to rapid activation of Gq/11 in the striatum. *Proc Natl Acad Sci USA* 2007;**104**(2):654–9.
42. Gonzalez-Maeso J, Ang RL, Yuen T, et al. Identification of a serotonin/glutamate receptor complex implicated in psychosis. *Nature* 2008;**452**(7183):93–7.

43. Ellis J, Pediani JD, Canals M, Milasta S, Milligan G. Orexin-1 receptor-cannabinoid CB1 receptor heterodimerization results in both ligand-dependent and -independent coordinated alterations of receptor localization and function. *J Biol Chem* 2006;**281**(50):38,812–24.
44. Terrillon S, Barberis C, Bouvier M. Heterodimerization of V1a and V2 vasopressin receptors determines the interaction with beta-arrestin and their trafficking patterns. *Proc Natl Acad Sci USA* 2004;**101**(6):1548–53.
45. Gomes I, Gupta A, Filipovska J, Szeto HH, Pintar JE, Devi LA. A role for heterodimerization of mu and delta opiate receptors in enhancing morphine analgesia. *Proc Natl Acad Sci USA* 2004;**101**(14):5135–9.
46. Waldhoer M, Fong J, Jones RM, et al. A heterodimer-selective agonist shows in vivo relevance of G-protein-coupled receptor dimers. *Proc Natl Acad Sci USA* 2005;**102**(25):9050–5.
47. George SR, Fan T, Xie Z, et al. Oligomerization of mu- and delta-opioid receptors. Generation of novel functional properties. *J Biol Chem* 2000;**275**(34):26,128–35.
48. Pfeiffer M, Koch T, Schroder H, et al. Homo- and heterodimerization of somatostatin receptor subtypes. Inactivation of sst(3) receptor function by heterodimerization with sst(2A). *J Biol Chem* 2001;**276**(17):14,027–36.
49. Zhu WZ, Chakir K, Zhang S, et al. Heterodimerization of beta1- and beta2-adrenergic receptor subtypes optimizes beta-adrenergic modulation of cardiac contractility. *Circ Res* 2005;**97**(3):244–51.
50. Rozenfeld R, Décaillot F, IJzerman AP, Devi LA. Heterodimers of G-protein-coupled receptors as novel and distinct drug targets. *Drug Disc Today: Ther Strat* 2006;**3**(4):437–43.
51. Pin JP, Kniazeff J, Liu J, et al. Allosteric functioning of dimeric class C G-protein-coupled receptors. *FEBS J* 2005;**272**(12):2947–55.
52. Armstrong D, Strange PG. Dopamine D2 receptor dimer formation: evidence from ligand binding. *J Biol Chem* 2001;**276**(25):22,621–9.
53. Urizar E, Montanelli L, Loy T, et al. Glycoprotein hormone receptors: link between receptor homodimerization and negative cooperativity. *EMBO J* 2005;**24**(11):1954–64.
54. Springael JY, Le Minh PN, Urizar E, Costagliola S, Vassart G, Parmentier M. Allosteric modulation of binding properties between units of chemokine receptor homo- and hetero-oligomers. *Mol Pharmacol* 2006;**69**(5):1652–61.
55. Albizu L, Balestre MN, Breton C, et al. Probing the existence of G-protein-coupled receptor dimers by positive and negative ligand-dependent cooperative binding. *Mol Pharmacol* 2006;**70**(5):1783–91.
56. Jordan BA, Gomes I, Rios C, Filipovska J, Devi LA. Functional interactions between mu opioid and alpha 2A-adrenergic receptors. *Mol Pharmacol* 2003;**64**(6):1317–24.
57. Rios C, Gomes I, Devi LA. mu opioid and CB1 cannabinoid receptor interactions: reciprocal inhibition of receptor signaling and neuritogenesis. *Br J Pharmacol* 2006;**148**(4):387–95.
58. Vilardaga JP, Nikolaev VO, Lorenz K, Ferrandon S, Zhuang Z, Lohse MJ. Conformational cross-talk between alpha2A-adrenergic and mu-opioid receptors controls cell signaling. *Nature Chem Biol* 2008; **4**(2):126–31.
59. Small KM, Schwarb MR, Glinka C, et al. Alpha2A- and alpha2C-adrenergic receptors form homo- and heterodimers: the heterodimeric state impairs agonist-promoted GRK phosphorylation and beta-arrestin recruitment. *Biochemistry* 2006;**45**(15):4760–7.
60. Breit A, Gagnidze K, Devi LA, Lagace M, Bouvier M. Simultaneous activation of the delta opioid receptor (deltaOR)/sensory neuron-specific receptor-4 (SNSR-4) hetero-oligomer by the mixed bivalent agonist bovine adrenal medulla peptide 22 activates SNSR-4 but inhibits deltaOR signaling. *Mol Pharmacol* 2006;**70**(2):686–96.
61. So CH, Verma V, O'Dowd BF, George SR. Desensitization of the dopamine D1 and D2 receptor hetero-oligomer mediated calcium signal by agonist occupancy of either receptor. *Mol Pharmacol* 2007;**72**(2):450–62.
62. Rozenfeld R, Abul-Husn NS, Gomez I, Devi LA. An emerging role for the delta opioid receptor in the regulation of mu opioid receptor function. *Sci World J* 2007;**7**:64–73.
63. Bohn LM, Lefkowitz RJ, Gainetdinov RR, Peppel K, Caron MG, Lin FT. Enhanced morphine analgesia in mice lacking beta-arrestin 2. *Science* 1999;**286**(5449):2495–8.
64. Bohn LM, Gainetdinov RR, Lin FT, Lefkowitz RJ, Caron MG. Muopioid receptor desensitization by beta-arrestin-2 determines morphine tolerance but not dependence. *Nature* 2000;**408**(6813): 720–3.
65. Jones KA, Borowsky B, Tamm JA, et al. GABA(B) receptors function as a heteromeric assembly of the subunits GABA(B)R1 and GABA(B)R2. *Nature* 1998;**396**(6712):674–9.
66. Kaupmann K, Malitschek B, Schuler V, et al. GABA(B)-receptor subtypes assemble into functional heteromeric complexes. *Nature* 1998;**396**(6712):683–7.
67. White JH, Wise A, Main MJ, et al. Heterodimerization is required for the formation of a functional GABA(B) receptor. *Nature* 1998;**396**(6712):679–82.
68. Kuner R, Kohr G, Grunewald S, Eisenhardt G, Bach A, Kornau HC. Role of heteromer formation in GABAB receptor function. *Science* 1999;**283**(5398):74–7.
69. Uberti MA, Hague C, Oller H, Minneman KP, Hall RA. Heterodimerization with beta2-adrenergic receptors promotes surface expression and functional activity of alpha1D-adrenergic receptors. *J Pharmacol Exp Ther* 2005;**313**(1):16–23.
70. Nelson G, Hoon MA, Chandrashekar J, Zhang Y, Ryba NJ, Zuker CS. Mammalian sweet taste receptors. *Cell* 2001;**106**(3):381–90.
71. Nelson G, Chandrashekar J, Hoon MA, et al. An amino-acid taste receptor. *Nature* 2002;**416**(6877):199–202.
72. Pfeiffer M, Kirscht S, Stumm R, et al. Heterodimerization of substance P and mu-opioid receptors regulates receptor trafficking and resensitization. *J Biol Chem* 2003;**278**(51):51,630–7.
73. Behrens M, Bartelt J, Reichling C, Winnig M, Kuhn C, Meyerhof W. Members of RTP and REEP gene families influence functional bitter taste receptor expression. *J Biol Chem* 2006;**281**(29):20,650–9.
74. Decaillot FM, Rozenfeld R, Gupta A, Devi LA. Cell surface targeting of mu-delta opioid receptor heterodimers by RTP4. *Proc Natl Acad Sci USA* 2008;**105**(41):16,045–50.
75. AbdAlla S, Lother H, el Massiery A, Quitterer U. Increased AT(1) receptor heterodimers in preeclampsia mediate enhanced angiotensin II responsiveness. *Nature Med* 2001;**7**(9):1003–9.
76. AbdAlla S, Abdel-Baset A, Lother H, el Massiery A, Quitterer U. Mesangial AT1/B2 receptor heterodimers contribute to angiotensin II hyperresponsiveness in experimental hypertension. *J Mol Neurosci* 2005;**26**(2–3):185–92.
77. Barki-Harrington L, Luttrell LM, Rockman HA. Dual inhibition of beta-adrenergic and angiotensin II receptors by a single antagonist: a functional role for receptor-receptor interaction in vivo. *Circulation* 2003;**108**(13):1611–18.
78. McGraw DW, Mihlbachler KA, Schwarb MR, et al. Airway smooth muscle prostaglandin-EP1 receptors directly modulate beta2-adrenergic receptors within a unique heterodimeric complex. *J Clin Invest* 2006;**116**(5):1400–9.
79. Sealfon SC, Gonzalez-Maeso J. Receptor pair for schizophrenia. *Pediatr Res* 2008;**64**(1):1.

80. Bhushan RG, Sharma SK, Xie Z, Daniels DJ, Portoghese PS. A bivalent ligand (KDN-21) reveals spinal delta and kappa opioid receptors are organized as heterodimers that give rise to delta(1) and kappa(2) phenotypes. Selective targeting of delta-kappa heterodimers. *J Med Chem* 2004;**47**(12):2969–72.

81. Abul-Husn NS, Sutak M, Milne B, Jhamandas K. Augmentation of spinal morphine analgesia and inhibition of tolerance by low doses of mu- and delta-opioid receptor antagonists. *Br J Pharmacol* 2007;**151**(6):877–87.

Part 4

TGFβ Receptors

Chapter 35

Receptor–Ligand Recognition in the TGFβ Superfamily as Suggested by Crystal Structures of their Ectodomain Complexes

Matthias K. Dreyer
Sanofi-Aventis Deutschland GmbH, Structural Biology, Industriepark Höchst, Frankfurt, Germany

INTRODUCTION

The transforming growth factor β (TGFβ[1]) superfamily is a large group of soluble, dimeric proteins that initiate and control proliferation and differentiation of many cell types in animals, thus playing important roles in embryonal development and adult tissue homeostasis. Malfunctioning of this tightly controlled signaling system leads to developmental disorders and to severe defects in organ function, and is the cause of several diseases, including various types of cancer [1, 2].

Signaling by these factors requires binding of the ligand to two homologous but functionally distinct types of transmembrane serine/threonine receptor kinases, designated type I and type II. Upon ligand-complex formation, the constitutively active type II receptor kinases activate the type I receptor kinases through phosphorylation in a conserved juxtamembrane region (GS box). Activated type I receptors are then able to phosphorylate downstream targets of the signaling pathways, which primarily include Smad proteins, Tak1/Tab1, and possibly some others [1, 3–5].

The large number of ligands in the TGFβ superfamily is opposed to a comparatively small number of receptors and, in contrast to other highly specific receptor–ligand systems, the members of the TGFβ superfamily display a certain promiscuity: several ligands are able to recognize different receptors, and most of the receptors can bind different ligands.

With a focus on BMP2 and TGFβ, this chapter will summarize the current knowledge of the initiating events in signal transduction within the TGFβ superfamily based on crystal and solution structures of ligands, receptor ectodomains and receptor–ligand complexes.

LIGAND STRUCTURES

Currently, around 40 different ligands within the TGFβ superfamily have been described. Based on their sequence homology, which varies between ~25 and more than 90 percent, they can be categorized in the subfamilies of the name-giving *TGFβs*; the bone morphogenetic proteins (BMPs), which form the largest subgroup; the growth and differentiation factors (GDFs); the activins/inhibins; and a few other more distant members [6]. The three-dimensional structures of the unbound ligands TGFβ1 [7], TGFβ2 [8, 9], TGFβ3 [10], BMP7 [11], GDNF [12], BMP2 [13], ActA [14], BMP6, and BMP3 [15] show that their monomeric chains share a common fold, which is characterized by an extended, partly interrupted four stranded β-sheet and one α-helix. Significant differences are mainly observed in loops and at the N-termini. This monomer fold is stabilized by a network of disulfide bridges forming the so-called cystine knot, a motif that is also found in a number of other growth factors [16, 17]. The overall fold of a monomer has been described as a curled hand, in which the extended β-sheets represent the "fingers" and the α-helix the "wrist" [8, 18] region of the hand. The functionally competent molecule, however, is a dimer – usually, but not necessarily, a homodimer – with an antiparallel palm-to-palm orientation of the subunits, formed by a covalent connection of two monomers through a disulfide bridge. The inherent flexibility of the dimers, and hence the angles between the

Handbook of Cell Signaling, Three-Volume Set 2 ed.

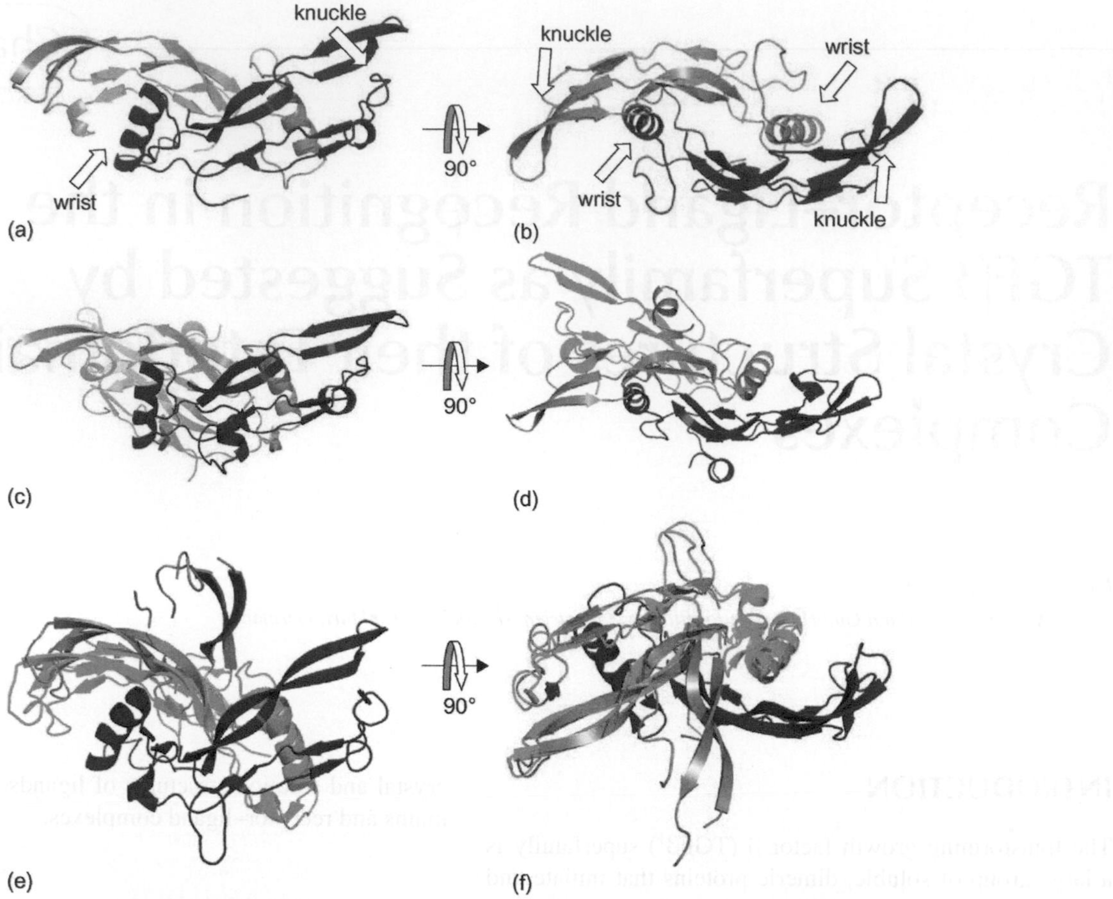

FIGURE 35.1 Structures and flexibility of the ligands.
(a) Free BMP2 [13; *3bmp*] with arrows pointing to the type I receptor interface (wrist epitope) and the type II receptor interface (knuckle epitope). The membrane-facing side is on the bottom. The two monomer chains are colored light and dark gray. (b) Top view onto BMP2 along the two-fold symmetry axis of the dimer obtained from (a) by a rotation of ~90° about the horizontal axis. All four epitopes of the dimer are indicated. (c) TGFβ3 structures: dark-gray ribbons (I, II) show free TGFβ3 [10; *1tgj*] with the front monomer chain aligned to the dark-gray monomer chain in BMP2 (a); the light-gray ribbon represents one monomer of receptor bound TGFβ3 [40; *1ktz*], the second monomer is aligned to and superimposes well with chain I but is not shown for clarity. The arrow indicates the relative domain movement upon binding. (d) Same as (c), but rotated by 90° about the horizontal axis to better visualize the relative domain movement upon receptor binding. (e) and (f) ActA structures in two orientations: black ribbons (I, II) show free ActA [14; *2arv*]; the light-gray chain is from the ActA:Follistatin complex, the medium-gray chain from receptor bound ActA [43; *2b0u* and *1nyu*], the respective second chains were aligned with monomer I of free ActA, but are not shown for clarity. ActA displays the largest observed subunit movement.

long axes of the monomers, can vary substantially between the dimers of different subfamilies (Figure 35.1).

The palm-to-palm orientation of the monomer hands distinguishes the TGFβ superfamily from the VEGF and NGF families, whose monomer structures are similar but which dimerize in different manners (see [19], and corresponding chapters in this *Handbook*).

RECEPTOR STRUCTURES

The receptor chains are composed of a small extracellular (ec) ligand binding domain, comprising around 100–140 amino acid residues, one short transmembrane fragment, and a cytoplasmic part, that contains the catalytic kinase domain. Recently, the crystal structures of unbound ectodomains of the type II receptors ActRIIA$_{ec}$[20], TGFβRII$_{ec}$ [21], and BMPRII$_{ec}$ [22] were reported. In addition, further receptor structures were elucidated in complex with ligands (see below), including the current only type I receptor BMPRIA$_{ec}$ [23]. They all fold into similar single, compact domains with a rigid central core formed by an antiparallel β-sheet whose overall architecture is maintained by five to six disulfide bridges. Their pattern is conserved largely within and partially between the two receptor classes (Figure 35.2). This fold, which has previously been described for a number of snake venom toxins that don't have any reported functional connection with the receptors, can be compared to an open left hand, and was named the "three finger toxin fold" [24].

Structural variations of the receptor chain folds occur mainly at the edges of the sheet and in loop regions, and can have important consequences for the functionality. While

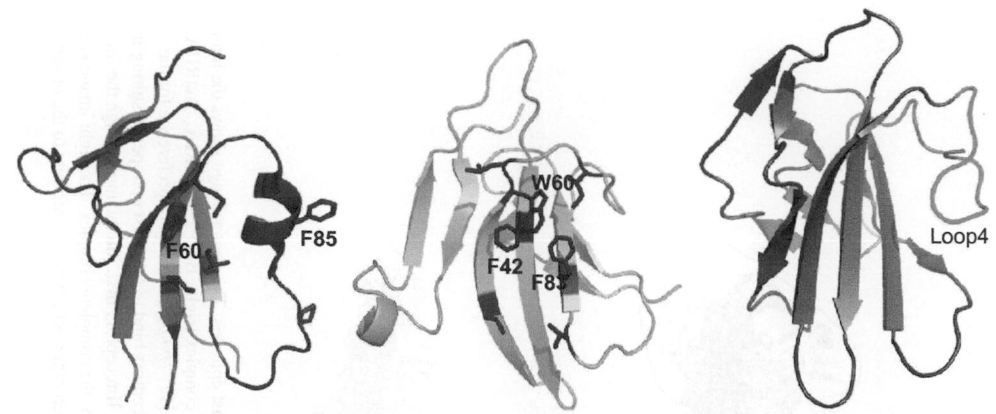

FIGURE 35.2 Structures of the receptor ectodomains.
(a) $BMPRIA_{ec}$ as in the complex with BMP2 [23; *1es7*]; the view is onto the ligand binding epitope, i.e., the "palm" side of the receptor hand; residues interacting with BMP2 are highlighted in black, those forming the hydrophobic interface are shown as sticks. (b) Free $ActRIIA_{ec}$ [20; *1bte*], in the same orientation as (a), residues forming the hydrophobic interface are highlighed as black sticks. (c) Free $TGF\beta RII_{ec}$ [21; *1m9z*] in the same orientation as (a); loop4 blocks access to the central β-sheet; in contrast to $BMPRIA_{ec}$ and $ActRIIA_{ec}$, the ligand binding epitope is formed by residues of β-strands 1, 2, and 4 (black) and is located at the edge of the central β-sheet.

in BMPRIA, BMPRII, ActRIIA, and ActRIIB structurally equivalent residues on the concave central β-sheet, i.e. the "palm" side of the receptor hand, form the ligand binding epitope, in TGFβRII a loop packs against this surface and the ligand binding epitope is located at a completely different part of the molecule (Figure 35.2c). Despite the similarity of their overall folds, there is no report of cross-specificity between type I and type II receptors.

RECEPTOR–LIGAND COMPLEXES

Although BMPs and TGFβs are very similar in structure, there are significant differences in the molecular mechanisms of recognition and signal activation. In TGFβ signaling, complex formation is initiated by high-affinity interactions between the ligand and the type II receptor TGFβRII. This primary complex subsequently recruits the low-affinity type I receptor (TGFβRI), which is not able to bind to the ligands in the absence of TGFβRII [25, 26]. The sequence of events is likely to be reversed in the case of some BMPs, as exemplified by studies with BMP2: BMP2 has higher affinity for the type I receptors BMPRIA and BMPRIB, but lower affinity for BMPRII and ActRII. Affinity for type II receptors is slightly increased in the presence of type I receptors, but, even in the absence of the latter, BMP2 is able to interact with the type II chains [18, 27, 28] and the cooperative effect observed with TGFβ is less pronounced. In addition, a population of BMP receptors exists in preformed complexes composed of BMPRII and BMPRIA or BMPRIB prior to ligand binding [29]. Binding of BMP2 to either such *preformed* receptor complexes or in a sequential way first to the high-affinity BMPRI receptors and then to BMPRII, yielding *ligand-induced* receptor complexes, results in induction of different signaling pathways [30].

BMPs and Activins Bind the Receptors on their Knuckle and Wrist Epitopes

Comprehensive mutational studies on BMP2 and similar studies with ActA mapped two symmetrical pairs of spatially separated and functionally independent epitopes on the surfaces of the ligand dimers for interactions with the two receptor classes [18, 31–33]. Based on their locations on the BMP2 hand, the type I receptor binding site was called the "wrist epitope" and the type II receptor site the "knuckle epitope" (Figure 35.1a). This epitope assignment was fully compatible with the first crystal structure of a tetrameric[2] complex between BMP2 and a pair of its high-affinity receptor chains $BMPRIA_{ec}$ [23]. In the complex, the two-fold rotational symmetry of the ligand is retained, with the symmetry axis normal to the plane of the membrane (Figure 35.3a). Each of the two receptor molecules binds with its concave "palm" side to the "wrist" of one of the two BMP2 monomer hands. An extended groove on the receptor surface harbors ligand residues of the so-called pre-helix loop and the α-helix (residues Phe49–Val63 in BMP2). The binding site on BMP2 extends to the inner side of the "fingers" of the adjacent BMP2 subunit, and thus covers the complete "finger-helix cavity" [13]. Central to the contacts in this second part of the binding epitope is a hydrophobic pocket on the ligand surface, which is found in all currently known ligand structures of the superfamily [23]. The counterpart to this pocket is a large hydrophobic residue (Phe85 in $BMPRIA_{ec}$) that is functionally conserved within the class of type I receptors (except for ALK-1). This residue is located in helix α1, a secondary structure element which is not present in the type II receptors ActRII and TGFβRII. Based on the high degree of conservation of this knob-into-hole like motif, it was deduced that this is a key interaction between the ligand and the type I receptors.

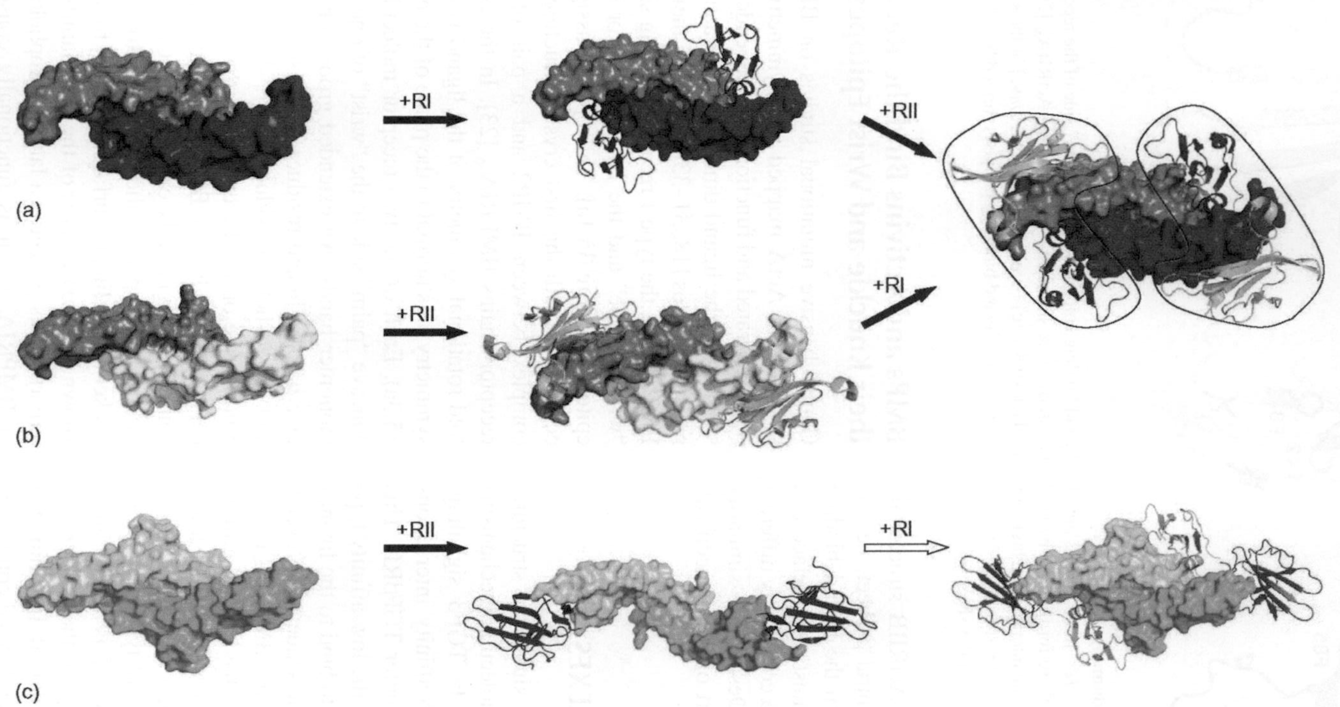

FIGURE 35.3 Different scenarios of complex formation.
(a) For BMP2 (left image, [10, *1tgj*]) or BMP4, the type I receptors are the high affinity receptors and therefore likely to bind first, occupying the ligand's wrist epitopes. The center image shows the intermediate binary complex BMP2:BMPRIA$_{ec}$ [23; *1es7*]. The lower-affinity type II receptors then bind to the knuckle epitopes, yielding the complete signaling complex (right; BMP2: BMPRIA$_{ec}$:ActRIIA; [39; *2goo*]). Pairs of receptors that are close in space to each other that are expected to interact with their cytoplasmic domains are encircled. (b) BMP7 (left; [11; *1bmp*]) has a higher preference for the type II receptors and likely forms a binary complex as exemplified by the BMP7:ActRIIA$_{ec}$ complex (center; [36; *1lx5*]). In the subsequent step, the type I receptors bind to the wrist epitopes, resulting in the same overall arrangement for the signaling complex as with BMP2. (c) In a more stringent and cooperative sequence, TGFβs bind their high-affinity type II receptors first, interacting through the fingertip epitopes as shown in the TGFβ3:TGFβRII$_{ec}$ structure (center; [40; *1ktz*]). Only then can the low-affinity type I receptor bind to the wrist epitope to form the complete complex. The right image is a theoretical model of such a complex, based on a superposition with the BMP2:BMPRIA$_{ec}$ structure. Here, the type I and type II receptors possibly contact each other, which may contribute to the observed cooperativity.

While it is generally accepted that the type I receptor binding site is common to all ligands in the TGFβ superfamily, this does not hold true for the type II receptor sites.

Type II Receptors Bind to the Knuckle Epitope in BMPs and Activins

The complex structures of activin type II receptors with either ActA [34, 35] or BMP7 [36] as ligands provided the first proof for the location of the type II receptor epitope on the "knuckles" of the ligand fingers – i.e. on the outer surface of the extended β-sheets. $ActRII_{ec}$ uses its concave palm side with a central cluster of three hydrophobic residues (Tyr60, Trp78, and Phe101) to interact with the convex "knuckle" epitope. This observation is in full agreement with the mutagenesis experiments, which showed that alanine substitutions of these residues strongly decreased the ability of ActRII to bind activin [37]. The hydrophobic center of the interface is surrounded by several polar interactions. Based on an analysis of sequence alignments in conjunction with affinity measurements of native and mutant proteins, it is proposed that pairwise ligand–receptor interactions in this area of the interface fine-tune the specificity within the BMP/activin subfamily [34, 36].

Activin is Highly Flexible

The ActRII–ActA complexes unexpectedly also revealed the inherent flexibility of the activin ligand: in one of the analyzed crystals the activin adopts a rather folded-up conformation in which the bound receptor chains are in direct contact which each other [34], while in the second crystal the activin protomer chains are spread apart [35] in a similar fashion as in free ActA [14] or BMP7 and BMP2 (Figure 35.1e, f), and the receptor chains are not in mutual contact. In all ActA protomers the interfaces for the type II receptors are virtually identical and seem to be rather independent of the relative subunit orientation, but significant differences were observed at the presumed interfaces for the type I receptor. In the folded-up conformation this interface is completely disordered and could not bind a type I receptor, while in the spread-out conformation the type I receptor interface is at least partly restored. Elegant surface plasmon resonance experiments in conjunction with the observed structural variance indicate that it is the binding of two high-affinity membrane-embedded type II receptor chains that is required to capture a spread-out activin conformation with an intact type I receptor interface that is then capable of binding the low-affinity type I receptor in order to form the complete hexameric complex. A full discussion of all factors determining binding affinity also includes thermodynamic considerations of ligand–receptor interaction [38].

The Complete Assembly

Very recently, the hexameric structure of BMP2:$BMPRIA_{ec}$:$ActRII_{ec}$ was solved [39], the first (and hitherto the only) complete ligand–receptor complex in the TGFβ superfamily, representing a milestone and a completion of the structural jigsaw puzzle work of receptor and ligand complexes. The overall arrangement, relative orientations, and conformations of subunits within the complex are very similar to those observed in the binary structures of BMP2:$BMPRIA_{ec}$ and BMP7:$ActRII_{ec}$, and in free BMP2. In contrast to ActA, BMP2 is apparently a much more rigid ligand, and does not experience significant conformational changes upon binding to the receptors[3].

As previously predicted based on the binary complexes [36,38], the receptor binding epitopes on the surface of BMP2 do not overlap and the receptor chains do not contact each other within the amino acid segments visible in the crystal structure. The rigidity of the ligand and the spatial separation of receptor epitopes may explain why only little cooperativity is observed upon binding of the two receptor types to BMP2 (Figure 35.3).

In the hexameric arrangement, one pair of type I and II receptor chains is in close spatial proximity (33 Å distance between the visible C-termini, bound to wrist and knuckle epitope of different BMP2 monomers) while the second pair is much farther apart (76 Å, wrist and knuckle epitopes of the same monomer). Although not all residues of the receptor ectodomains can be seen in the structure and not much is known about the interaction in the cytoplasmic domains, it is likely that intracellular transphosphorylation takes place within a close pair of receptor chains.

TGFβ-RII Binds to the Fingertips of TGFβ3

The general picture of the complete receptor–ligand arrangement, based on the BMP and activin complex structures with well-separated type I and type II binding epitopes, that was assumed to be valid for the whole TGFβ superfamily, does not hold true for the complex formation with TGFβ itself, and had to be revised with the publication of the crystal structure of TGFβ3 in complex with $TGF\beta RII_{ec}$ [40]. Surprisingly, the receptor does not bind to a region on the ligand corresponding to the knuckle epitope on BMP2. Instead, it interacts with a hydrophobic cleft in the "fingertips" of the ligand, and is in close proximity to the expected type I receptor binding site. Although $TGF\beta RII_{ec}$ shares the overall fold topology of $ActRII_{ec}$ and $BMPRIA_{ec}$, a few structural modifications in loop regions result in a completely different binding mode to the ligand. Most prominent among these is an insertion of seven residues in loop 4 of TGFβRII that packs against the central β-sheet in such a way that the residues that correspond to those that are crucial to binding the ligand in $ActRII_{ec}$ as well as in $BMPRIA_{ec}$ are no longer accessible. The ligand

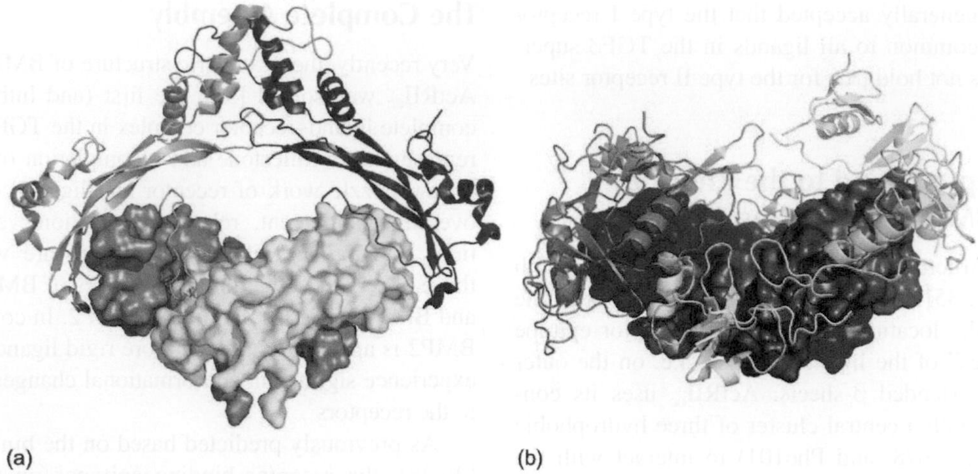

FIGURE 35.4 Inhibition by extracellular antagonists.
(a) Complex of BMP7 and Noggin [42; *1m4u*]. BMP7 is shown as a surface model in a similar orientation as BMP2 in Figure 35.1a. The two noggin chains are shown as light and dark ribbons, forming a molecular clamp that lies across the BMP7 dimer axis. (b) Complex between ActA (light- and dark-surface model) and follistatin (light- and dark-gray ribbons) that wraps around the equator of the ActA dimer. In both complexes, all four receptor binding epitopes are blocked by the inhibitors.

binding residues in TGFβRII$_{ec}$ are located in β-strands 1 and 4, whereas the equivalent amino acids in BMPRIA$_{ec}$ do not contact the ligand.

Based on these findings, a model for the hexameric signaling complex of TGFβs was suggested that combines the evidence from the known crystal structures of ligand–receptor complexes [40]. In this model, the type I receptor binds to TGFβ in a way described by the BMP2–BMPRIA$_{ec}$ complex, as this is accepted to be common within the superfamily, and the type II receptor binds as found in the TGFβ3–TGFβRII$_{ec}$ complex (Figure 35.3c). As a consequence, both receptors would be in direct contact with each other, which may account for the cooperativity found for TGFβ receptor-complex formation.

Signaling can be Regulated by Soluble Extracellular Mediators

Signaling in the TGFβ superfamily is negatively and positively regulated on various levels and by various mechanisms, such as inhibitory smad proteins, expression of the pseudo-receptor BAMBI or of co-receptors, and ligand sequestration through soluble receptor ectodomains, or through binding of other antagonistic proteins to the ligands, such as noggin, follistatin, chordin, or DAN (reviewed in [41]). The crystal structures of a BMP7:Noggin complex [42] and of the ActA:Follistatin complex [43] have been solved, and show that both antagonists apply similar but structurally distinct strategies by blocking the receptor binding sites of the ligands. Noggin, a BMP-specific antagonist consisting of two identical, disulfide-linked subunits, bends over the complete BMP7 dimer like a huge clamp (Figure 35.4a). It makes extensive interactions with both knuckle epitopes of BMP7, and both Noggin N-termini meander into the finger-helix cavity at the wrist epitopes, where they mimic the hydrophobic knob-into-hole motif of type I receptors. Hence, all four receptor sites are blocked by one Noggin dimer. In the follistatin complex with activin A, two antagonist monomers wrap around the activin dimer in such a way that one follistatin molecule blocks one distant pair of type I and type II epitopes (Figure 35.4b). This leaves the possibility that asymmetric complexes with one follistatin molecule and one type I receptor may also form, as observed with BMP4 and BMPRIA$_{ec}$ [44]. Such a complex would, however, not be able to produce a signal, because a signaling-competent complex requires a nearby receptor pair.

CONCLUDING REMARKS

The structural understanding of signaling in the TGFβ superfamily has seen tremendous progress in the past few years. Various ligand–receptor and ligand–inhibitor complexes have contributed to the current image of different interaction patterns in the BMP/activin *vs* the TGFβ subfamily, and document how conformational flexibility and sometimes subtle amino acid exchanges regulate the function and specificity in this system. Future studies will, however, be necessary to assess whether these structural principles will be valid for the complete TGFβ superfamily, as some more distant members may display yet other surprising properties.

NOTES

1. Abbreviations: TGF, transforming growth factor; BMP, bone morphogenetic protein; Act, activin; TGFβRI, TGFβ receptor type I (accordingly: TGFβRII,

BMPRIA, BMPRIB, BMPRII, ActRII); ec, extracellular, or ectodomain.

2. Due to the covalent connection of the two protomers, the formally correct chemical description for the BMP2 ligand would be as a single molecule. However, here it is regarded as a "dimer," as it better emphasizes the origin and symmetric nature of the ligands. Consequently, the various described complexes are called "tetrameric" and "hexameric" (instead of "trimeric" and "pentameric," as found in some papers).
3. It can, however, not fully be excluded that BMP2 is also flexible and that any other than the spread-out conformation have not yet been captured in crystals.

REFERENCES

1. Massague J, Blain SW, Lo RS. TGFβ signaling in growth control, cancer, and heritable disorders. *Cell* 2000;**103**:295–309.
2. Akhurst RJ, Derynck R. TGF-β signaling in cancer – a double-edged sword. *Trends Cell Biol* 2001;**11**:S44–S51.
3. Kawabata M, Imamura T, Miyazono K. Signal transduction by bone morphogenetic proteins. *Cytokine Growth Factor Rev* 1998;**9**:49–61.
4. Piek E, Heldin C-H, ten Dijke P. Specificity, diversity, and regulation in TGF-β superfamily signaling. *FASEB J* 1999;**13**:2105–24.
5. Nohe A, Keating E, Knaus P, Petersen NO. Signal transduction of bone morphogenetic protein receptors. *Cell Signal* 2004;**16**:291–9.
6. Massagué J. TGF-β signal transduction. *Annu Rev Biochem* 1998;**67**:753–91.
7. Hinck AP, Archer SJ, Qian SW, Roberts AB, Sporn MB, Weatherbee JA, Tsang ML-S, Lucas R, Zhang B-L, Wenker J, Torchia DA. Transforming growth factor β1: three-dimensional structure in solution and comparison with the X-ray structure of transforming growth factor β2. *Biochemistry* 1996;**35**:8517–34.
8. Daopin S, Piez KA, Ogawa Y, Davies DR. Crystal structure of transforming growth factor-β2: an unusual fold for the superfamily. *Science* 1992;**257**:369–73.
9. Schlunegger MP, Grütter MG. An unusual feature revealed by the crystal structure at 2.2-Å resolution of human transforming growth factor-β2. *Nature* 1992;**358**:430–4.
10. Mittl PR, Priestle JP, Cox DA, McMaster G, Cerletti N, Grütter MG. The crystal structure of TGF-β3 and comparison to TGF-β2: implications for receptor binding. *Protein Sci* 1996;**5**:1261–71.
11. Griffith DL, Keck PC, Sampath TK, Rueger DC, Carlson WD. Three-dimensional structure of recombinant humanosteogenic protein 1: structural paradigm for the transforming growth factor β superfamily. *Proc Natl Acad Sci USA* 1996;**93**:878–83.
12. Eigenbrot C, Gerber N. X-ray structure of glial cell-derived neurotrophic factor at 1.9-Å resolution and implications for receptor binding. *Nat Struct Biol* 1997;**4**:435–8.
13. Scheufler C, Sebald W, Hülsmeyer M. Crystal structure of human bone morphogenetic protein-2 at 2.7-Å resolution. *J Mol Biol* 1999;**287**:103–15.
14. Harrington AE, Morris-Triggs SA, Ruotolo BT, Robinson CV, Ohnuma S, Hyvönen M. Structural basis for the inhibition of activin signaling by follistatin. *EMBO J* 2006;**25**:1035–45.
15. Allendorph GP, Isaacs MJ, Kawakami Y, Izpisua Belmonte JC, Choe S. BMP-3 and BMP-6 structures illuminate the nature of binding specificity with receptors. *Biochemistry* 2007;**46**:12,238–12,247.
16. McDonald NQ, Hendrickson WA. A structural superfamily of growth factors containing a cystine knot motif. *Cell* 1993;**73**:421–4.
17. Murray-Rust J, McDonald NQ, Blundell TJ, Hosang M, Oefner C, Winkler F, Bradshaw RA. Topological similarities in TGF-β2, PDGF-BB and NGF define a superfamily of polypeptide growth factors. *Curr Biol* 1993;**1**:153–9.
18. Kirsch T, Nickel J, Sebald W. BMP-2 antagonists emerge from alterations in the low-affinity binding epitope for receptor BMPRII. *EMBO J* 2000;**13**:3314–24.
19. Wiesmann C, de Vos AM. Variations on ligand–receptor complexes. *Nat Struct Biol* 2000;**7**:440–2.
20. Greenwald J, Fischer WH, Vale WW, Choe S. Three-finger toxin fold for the extracellular ligand-binding domain of the type II activin receptor serine kinase. *Nat Struct Biol* 1999;**6**:18–22.
21. Boesen CC, Radaev S, Motyka SA, Patamawenu A, Sun PD. The 1.1-Å crystal structure of human TGF-β type II receptor ligand binding domain. *Structure* 2002;**10**:913–19.
22. Mace PD, Cutfield JF, Cutfield SM. High-resolution structures of the bone morphogenetic protein type II receptor in two crystal forms: implications for ligand binding. *Biochem Biophys Res Comm* 2006;**351**:831–8.
23. Gray PC, Greenwald J, Blount AL, Kunitake KS, Donaldson CJ, Choe S, Vale W. Identification of a binding site on the type II activin receptor for activin and inhibin. *J Biol Chem* 2000;**275**:3206–12.
24. Kirsch T, Sebald W, Dreyer MK. Crystal structure of the BMP-2-BRIA ectodomain complex. *Nat Struct Biol* 2000;**7**:492–6.
25. Rees B, Bilwes A. Three-dimensional structures of neurotoxins and cardiotoxins. *Chem Res Toxicol* 1993;**6**:385–406.
26. Wrana JL, Attisano L, Wieser R, Ventura F, Massagué J. Mechanism of activation of the TGF-β receptor. *Nature* 1994;**370**:341–7.
27. Wrana JL, Attisano L, Carcamo J, Zentella A, Doody J, Laiho M, Wang XF, Massagué J. TGFβ signals through a heteromeric protein kinase receptor complex. *Cell* 1992;**71**:1003–14.
28. Nohno T, Ishikawa T, Saito T, Hosokawa K, Noji S, Wolsing DH, Rosenbaum JS. Identification of a human type II receptor for bone morphopogenetic protein-4 that forms differential heteromeric complexes with bone morphogentic protein type I receptors. *J Biol Chem* 1995;**270**:22, 522–22, 526.
29. Liu F, Ventura F, Doody J, Massagué J. Human type II receptor for bone morphogenetic proteins (BMPs): extension of the two-kinase receptor model the BMPs. *J Mol Cell Biol* 1995;**15**:3479–86.
30. Gilboa L, Nohe A, Geissendörfer T, Sebald W, Henis YI, Knaus P. Bone morphogenetic protein receptor complexes on the surface of live cells: a new oligomerization mode for serine/threonine kinase receptors. *Mol Cell Biol* 2000;**11**:1023–35.
31. Nohe A, Hassel S, Ehrlich M, Neubauer F, Sebald W, Henis YI, Knaus P. The mode of bone morphogenetic protein (BMP) receptor oligomerization determines different BMP-2 signaling pathways. *J Biol Chem* 2002;**277**:5330–8.
32. Wuytens G, et al. Identification of two amino acids in activin A that are important for biological activity and binding to the activin type II receptors. *J Biol Chem* 1999;**274**:9821–7.
33. Fischer WH, Park M, Donaldson C, Wiater E, Vaughan J, Bilezikjian LM, Vale W. Residues in the C-terminal region of activin A determine specificity for follistatin and type II receptor binding (2003). *J Endocrinol* 2003;**176**:61–8.
34. Harrison CA, Gray PC, Fischer WH, Donaldson C, Choe S, Vale W. An activin mutant with disrupted ALK4 binding blocks signaling via type II receptors. *J Biol Chem* 2004;**279**:28,036–28,044.

35. Thompson TB, Woodruff TK, Jardetzky TS. Structures of an ActRIIB: activin complex reveal a novel bindingmode for TGF-β ligand:receptor interactions. *EMBO J* 2003;**22**:1555–66.
36. Greenwald J, Vega ME, Allendorph GP, Fischer WH, Vale W, Choe S. A flexible activin explains the membrane-dependent cooperative assembly of TGF-β family receptors. *Mol Cell* 2004;**15**: 485–9.
37. Greenwald J, Groppe J, Gray P, Wiater E, Kwiatkowski W, Vale W, Choe S. The BMP7/ActRII extracellular domain complex provides new insights into the cooperative nature of receptor assembly. *Mol Cell* 2003;**11**:605–17.
38. Sebald W, Mueller TD. The interaction of BMP-7 and ActRII implicates a new mode of receptor assembly. *Trends Biochem Sci* 2003;**28**:518–21.
39. Allendorph GP, Vale WV, Choe S. Structure of the ternary signaling complex of a TGF-β superfamily member. *Proc Natl Acad Sci* 2006;**103**:7643–8.
40. Hart PJ, Deep S, Taylor ZS, Hinck CS, Hinck AP. Crystal structure of the human TβR2 ectodomain–TGF-β3 complex. *Nat Struct Biol* 2002;**9**:203–8.
41. Balemans W, Van Hul W. Extracellular regulation of BMP signaling. *Dev Biol* 2002;**250**:231–50.
42. Groppe J, Greenwald J, Wiater E, Rodriguez-Leon J, Economides AN, Kwiatkowski W, Affolter M, Vale WW, Izpisua Belmonte JC, Choe S. Structural basis of BMP signaling inhibition by the cystine knot protein Noggin. *Nature* 2002;**420**:636–42.
43. Thompson TB, Lerch TF, Cook RW, Woodruff TK, Jardetzky TS. The structure of the follistatin:activin complex reveals antagonsim of both type I and type II receptor binding. *Dev Cell* 2005;**9**:535–43.
44. Iemura S, Yamamoto TS, Takagi C, Uchiyama H, Natsume T, Shimasaki S, Sugino H, Ueno N. Direct binding of follistatin to a complex of bone-morphogenetic protein and its receptor inhibits ventral and epidermal cell fates in early Xenopus embryo. *Proc Natl Acad Sci* 1998;**95**:9337–42.

Chapter 36

TGFβ Signal Transduction

Cristoforo Silvestri[1,5], Rohit Bose[2,4], Liliana Attisano,[1,3,5], and Jeffrey L. Wrana[2,4]
[1]*Institute of Medical Sciences*
[2]*Department of Molecular Genetics*
[3]*Department of Biochemistry, University of Toronto, Toronto, Ontario, Canada*
[4]*Program in Molecular Biology and Cancer, Samuel Lunenfeld Research Institute, Mt Sinai Hospital, Toronto, Ontario, Canada*
[5]*Donnelly Centre for Cellular and Biomolecular Research, University of Toronto, Toronto, Ontario, Canada*

INTRODUCTION

Transforming Growth Factor β (TGFβ) superfamily members are expressed and function ubiquitously throughout early development and the lifetime of higher animals from worms to humans [1–10]. This large family of extracellular signaling molecules regulates numerous biological processes, including, but not limited to, the regulation of cellular growth, differentiation, motility, and apoptosis. TGFβ ligands transduce their signals into cells via transmembrane serine/threonine kinase receptors. In general, the formation of a heteromeric complex of ligand and receptors induces the phosphorylation of the receptor-regulated Smads (R-Smads) within the cytoplasm. This phosphorylation results in R-Smad activation, freeing the R-Smad from the receptors and allowing for the formation of a heteromeric complex with the common Smad, Smad4. These active R-Smad/Smad4 complexes accumulate in the nucleus, and function in conjunction with a host of nuclear co-factors to regulate the transcription of target genes. Moreover, Smad-independent signaling events that contribute to cellular responses and morphological changes such as the control of cell polarity have been elucidated. Every step of these processes is subject to regulatory events, which refine the signaling cascade, allowing for specific cellular responses to a TGFβ signal.

TGFβ LIGANDS

TGFβ is the prototypical member of a large family of related growth factor ligands comprised of at least 30 genes in mammals [2–7]. Multiple members have also been identified in frogs, fish, flies, and worms. The large number of related ligands likely reflects the need to fine-tune signal activation through various regulatory mechanisms in order to properly specify the myriad of biological processes influenced by TGFβs. In general, the TGFβs can be subdivided into two groups; TGFβs/Activins/Nodals, and Bone Morphogenetic Proteins (BMPs)/Growth Differentiation Factors (GDFs)/ Mullerian Inhibitory Substance (MIS) (Figure 36.1). This grouping is not based solely on structural differences but, more importantly, on the ability of ligands to activate one of two distinct groups of Smad signaling proteins (see [11], and below). Most TGFβ family members exist as homodimers that have been described as having a butterfly shape, though heterodimers have been identified [6]. They are initially translated with a large amino-terminal prodomain, which is divergent in sequence between factors, and a more highly conserved carboxy-terminal end, which forms the active ligand [1]. The individual, mature monomers of TGFβ family members are dominated by a structure termed the "cysteine knot." This structure results from the formation of three disulfide bonds between six highly conserved cysteines within the carboxy-terminal domain, which act to interlock several anti-parallel β strands into a defined structure, while a seventh cysteine conserved in most family members mediates dimerization.

Ligands are processed intracellularly by proteases that cleave the prodomain from the bioactive carboxy-terminal end [1, 12]. The prodomain, also known as the latency associated propeptide, remains non-covalently linked to the ligand during secretion, and acts to keep the complex inactive. The latency-associated peptide interacts with Latent TGFβ Binding Proteins (LTBPs), resulting in the formation of a

Handbook of Cell Signaling, Three-Volume Set 2 ed.

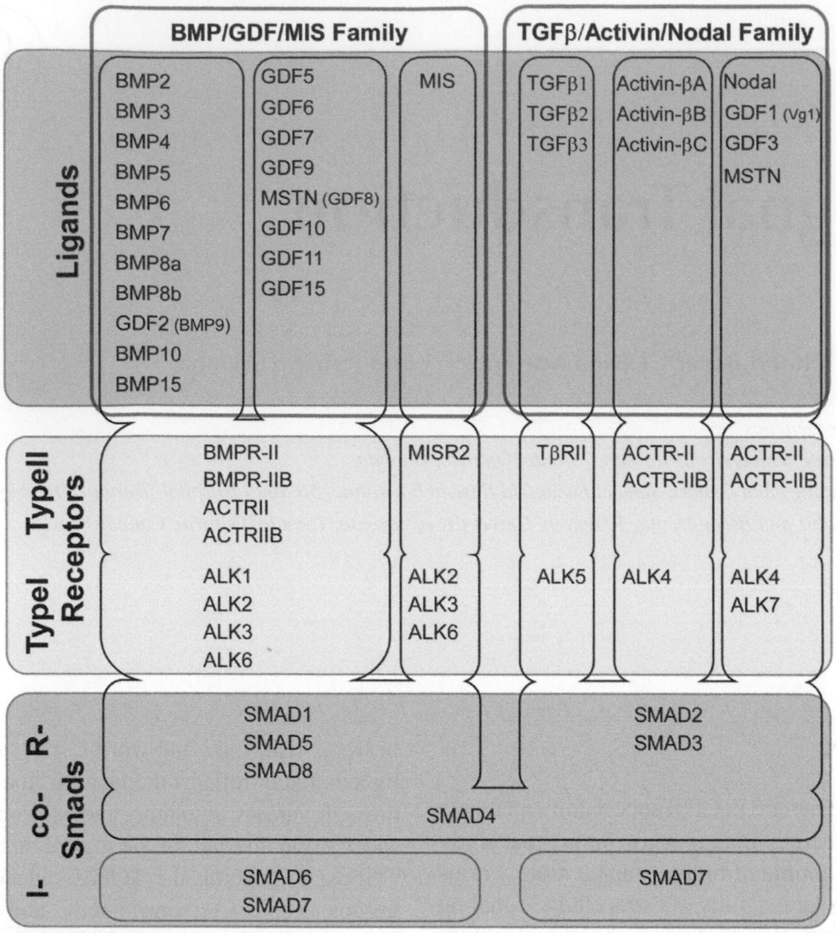

FIGURE 36.1 TGFβ superfamily members.
List of human TGFβ superfamily pathways members with ligands subdivided into the BMP/GDF/MIS and TGF/Activin/Nodal subfamilies. Receptors (Type II and Type I) and SMADS (receptor regulated; R, common; co-, inhibitory; I) are shared between many ligands, as indicated by the bubbles.

large latent complex (Figure 36.2). LTBPs regulate ligand activity by targeting them to the ECM. This sequesters the complexes, limiting their access to receptors, and allows for a store to build up which is readily available upon demand. In some cases, proteolytic cleavage by metalloproteases such as BMP1 is required to free the active ligand dimers that go on to signal via their cognate receptors [13].

Many "ligand traps" have been identified as inhibitors of TGFβ signaling through their ability to bind to various ligands and thus regulate ligand–receptor interactions [12, 14]. These include the Cerberus/DAN family, which inhibits BMP and Nodal signals, and the Noggins, which inhibit BMP and other signals, both of which are characterized by cysteine knot structures. Their vital importance is evident in the host of biological processes that these molecules influence [14].

RECEPTORS

TGFβ family members transduce signals across the plasma membrane via the formation of an active heteromeric complex of type II and type I receptors [2, 3, 6, 7, 15]. Together they are unique, as they are the only identified class of transmembrane serine/threonine kinases. Despite the fact that a rather large number of TGFβ ligands exist, relatively few receptors (five type II and seven type I receptors) initiate the intracellular signaling cascades. Moreover, TGFβ ligands are promiscuous, signaling through diverse combinations of type II and type I receptors. Within the heterotrimeric ligand/receptor complex, the type I receptor determines which TGFβ intracellular signaling pathway is activated (Figure 36.1). In general, the TGFβ/Activin/Nodal subfamily of ligands functions via ALK4, 5, and 7 type I receptors to activate Smad2 and Smad3, while the BMP/GDF/MIS ligands work through ALK1, 2, 3, and 6 to activate Smad1, 5, and 8 [7, 15]. It should be noted that these distinctions are not absolute. Multiple GDFs have been found to signal through ALK4, 5, and 7 to activate Smad2 and 3 [16, 17]. Further, TGFβ can form a complex with ALK5 as well as ALK1 in endothelial cells to activate Smad2 and 3, as well as Smad1, 5, and 8 [18]. It is this use of different combinations of receptors, along with the regulation of ligand expression and the actions of

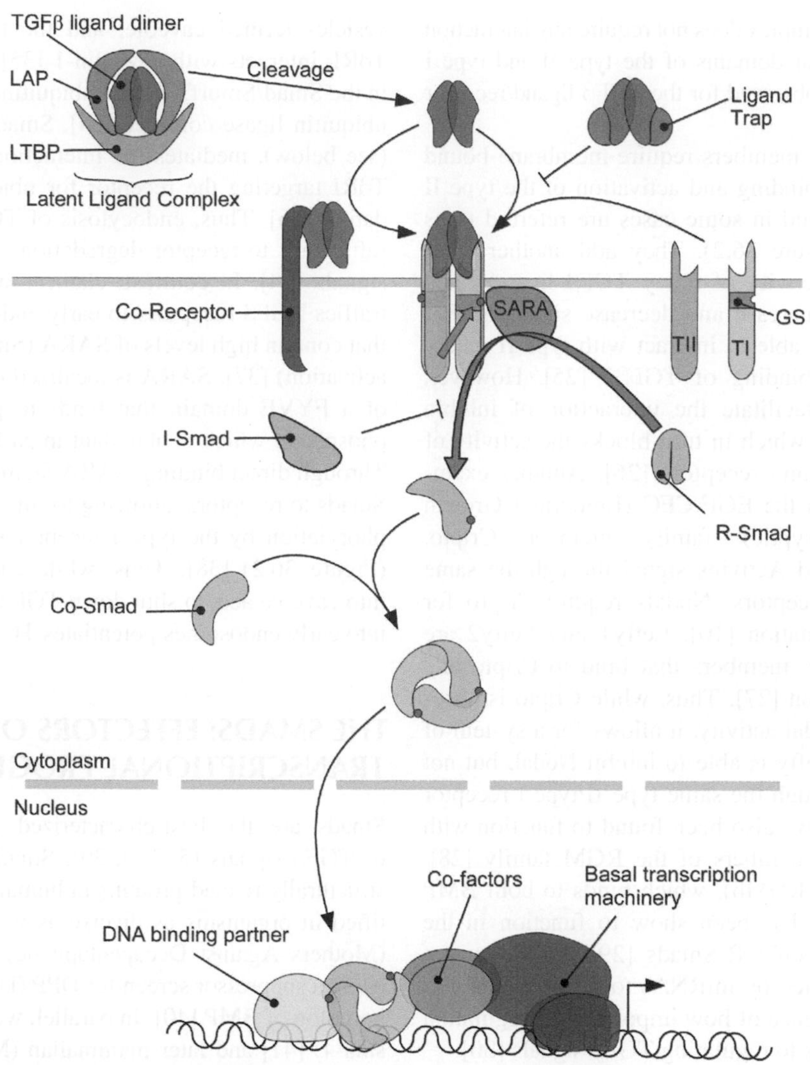

FIGURE 36.2 General TGFβ superfamily signaling pathway.
A generalized schematic of the TGFβ signaling pathway. Dimeric ligands are held in an inactive latent complex until processing frees the active dimers. Ligand traps inhibit signal transcduction by blocking ligand–receptor interactions, which bring together the Type II (TII) and Type I (TI) receptors. The constitutively active Type II receptor then transphophorylates (arrows) the Type I receptor in the juxtamembrane glysine-serine domain (GS), thereby activating it. Membrane bound SARA facilitates the interaction of R-Smads with the Type I receptor. Phosphorylation of the R-Smad carboxy-terminal SSXS motif releases R-Smads from the receptor complex, allowing for the formation of tri-heteromeric Smad complexes typically composed of two R-Smads and one Co-Smad. Active Smad complexes then translocate to the nucleus, where they interact with sequence specific DNA-binding partners and Smad binding elements as well as recruit co-factors to regulate transcriptional activation. Dots indicate phosphorylated peptide residues.

extracellular binding proteins, that allows for the diversity of biological responses elicited by TGFβs.

ACTIVATION AND REGULATION OF RECEPTORS

In the absence of an interacting ligand, type II and type I TGFβ receptors exist as homodimers [3, 6, 7, 15]. Both receptor types contain divergent, cysteine-rich amino-terminal extracellular domains that interact with the ligands, single-pass transmembrane domains, and relatively well conserved serine/threonine kinase domains located in the intracellular carboxy-terminal end. The divergence between the two TGFβ subfamilies appears to also extend to how the different subgroups interact with extracellular domains of their cognate receptors. In general, TGFβs and Activins bind to type II receptors with high affinities in the absence of any type I receptors [19–22]. The type I receptors are then recruited to the ligand/receptor complex via their interaction with a binding surface that is created at the ligand/type II receptor interface [23]. In contrast, BMPs generally have little affinity for their type II receptors in the absence of type I receptors. Thus, BMPs bind to their type I receptors initially, which then allows the type II receptors to join into a high-affinity heteromeric complex that is able to activate Smad signaling [24].

The formation of this complex does not require any interaction between the extracellular domains of the type II and type I receptors, unlike those observed for the TGFβ ligand/receptor interactions [22, 23].

Some TGFβ family members require membrane bound co-receptors to allow binding and activation of the type II and type I receptors, and in some cases are referred to as type III receptors (Figure 36.2). They add another layer of regulation to the activity of many TGFβ ligands, and can function to both increase and decrease signaling. For instance, betaglycan is able to interact with type II receptors to facilitate the binding of TGFβ2 [25]. However, Betaglycan can also facilitate the interaction of inhibin with type II receptors, which in turn blocks the activity of Activin through the same receptors [26]. Another example of a co-receptor is the EGF-CFC (Epidermal Growth Factor-Cripto-FRL1-Cryptic) family member Cripto. While both Nodals and Activins signal through the same type II and type I receptors, Nodals require Cripto for receptor complex formation [10]. Lefty1 and Lefty2 are divergent TGFβ family members that bind to Cripto and inhibit Nodal interaction [27]. Thus, while Cripto is absolutely required for Nodal activity, it allows for a system of regulation in which Lefty is able to inhibit Nodal, but not Activin signaling, through the same type II/type I receptor combination. BMPs have also been found to function with co-receptors, such as members of the RGM family [28]. For instance, Dragon (RGMb), which binds to both BMP receptors and ligands, has been show to function in the activation of BMP-specific R-Smads [29]. Evidence that Lefty levels are regulated by miRNAs to refine Nodal signals gives further evidence of how important the regulation of receptor activation is to regulating TGFβ signals [30].

While both type II and type I receptors contain intracellular kinase domains, those of the type II receptors are constitutively active, while those of the type I receptors require the formation of a ligand/receptor complex for activation [19]. All type I receptors have a signature juxtamembrane glycine/serine-rich domain, termed the GS domain, upstream of their kinases. Upon receptor complex formation, several serine residues within the GS domain are targeted for phosphorylation by the kinase activity of the type II receptor (Figure 36.2) [19, 31]. This phosphorylation event activates the type I receptor kinase, which in turn interacts with and phosphorylates a specific subset of Smads, which are intracellular transcriptional effectors of TGFβ signaling. Thus, through phosphorylation, the GS domain acts as a switch to activate the kinase of the type I receptor.

Internalization of receptors by various endocytic pathways serves to regulate cell signaling by managing receptor activity, recycling and degradation [32]. TGFβ receptors are internalized by lipid-raft as well as clathrin-mediated endocytotic mechanisms [33, 34]. Lipid-raft mediated endocytosis trafficks TGFβ receptors into caveolin-1 positive vesicles termed caveole, and the type I TGFβ receptor, TβRI, interacts with caveolin-1 [35]. Caveole are enriched in the Smad/Smurf2 (Smad ubiquitination regulatory factor) ubiquitin ligase complex [34]. Smad7, an inhibitory Smad (see below), mediates the interaction between Smurf2 and TβRI targeting the receptor for ubiquitin-mediated degradation [36]. Thus, endocytosis of TGFβ receptors by lipid rafts leads to receptor degradation and inhibition of TGFβ signals [34]. In contrast, clathrin-mediated internalization traffics TGFβ receptors to early endosomes, compartments that contain high levels of SARA (Smad anchor for receptor activation) [37]. SARA is localized to endosomes by virtue of a FYVE domain that binds to phosphatidylinositol-3-phosphate, which is abundant in early endosomes [32, 38]. Through direct binding, SARA facilitates the recruitment of Smads to receptors, allowing for more efficient Smad phosphorylation by the type I receptor serine/threonine kinase (Figure 36.2) [38]. Thus, while endocytosis of receptors into caveole acts to shut down TGFβ signaling, endocytosis into early endosomes potentiates TGFβ signaling [34].

THE SMADS: EFFECTORS OF TGFβ FAMILY TRANSCRIPTIONAL PROGRAMS

Smads are the best-characterized intracellular mediators of TGFβ signals [5–7, 9, 39]. Smads are a group of eight structurally related proteins in humans with homologs identified in organisms as diverse as worms to humans. MAD (Mothers Against Decapentaplegic) was identified in flies using a suppressor screen for DPP (Decapentaplegic), the fly homolog of BMP [40]. In parallel, worm (sma-2, sma-3, and sma-4) [41] and later mammalian (MADR; MAD Related) homologs were identified. The term "Smad" (a conjunction of sma and MAD) was adopted as a unified nomenclature for the vertebrate genes of this family [42].

There are three classes of Smad proteins (Figure 36.1) [5–7, 9, 39]. Receptor regulated Smads (R-Smads) are the only members of the family that are directly activated by the kinase activity of the TGFβ type I receptors (Figure 36.2). R-Smads can be further subdivided into two groups based on the receptors that generally activate them. Smad2 and Smad3 are activated by TGFβs/Activins/Nodals, while Smad1, Smad5, and Smad8 are activated by BMPs/GDFs/MIS. The second class of Smads is comprised of only one member, Smad4. Smad4 is able to participate in signaling cascades downstream of all TGFβ family members through its interactions with the R-Smads, and is therefore often referred to as the common or co-Smad. The third class of Smads consists of the inhibitory Smads (I-Smads), Smad6 and Smad7. As their name implies, these Smads serve as negative regulators of TGFβ family member activity.

Smad proteins have two conserved domains, an amino terminal MAD homology domain (MH1) and a carboxy terminal MAD homology domain (MH2), joined together by a poorly conserved linker region (Figure 36.3). While the MH2 domain is well conserved among all Smads, the amino terminal domain of I-Smads only weakly resembles the MH1 domains of the R-Smads and Smad4 [43]. The domains of the Smads mediate different functions through their unique structures and interactions with numerous proteins. The MH1 domain is responsible for nuclear import, DNA binding, and interactions with transcription factors, while the MH2

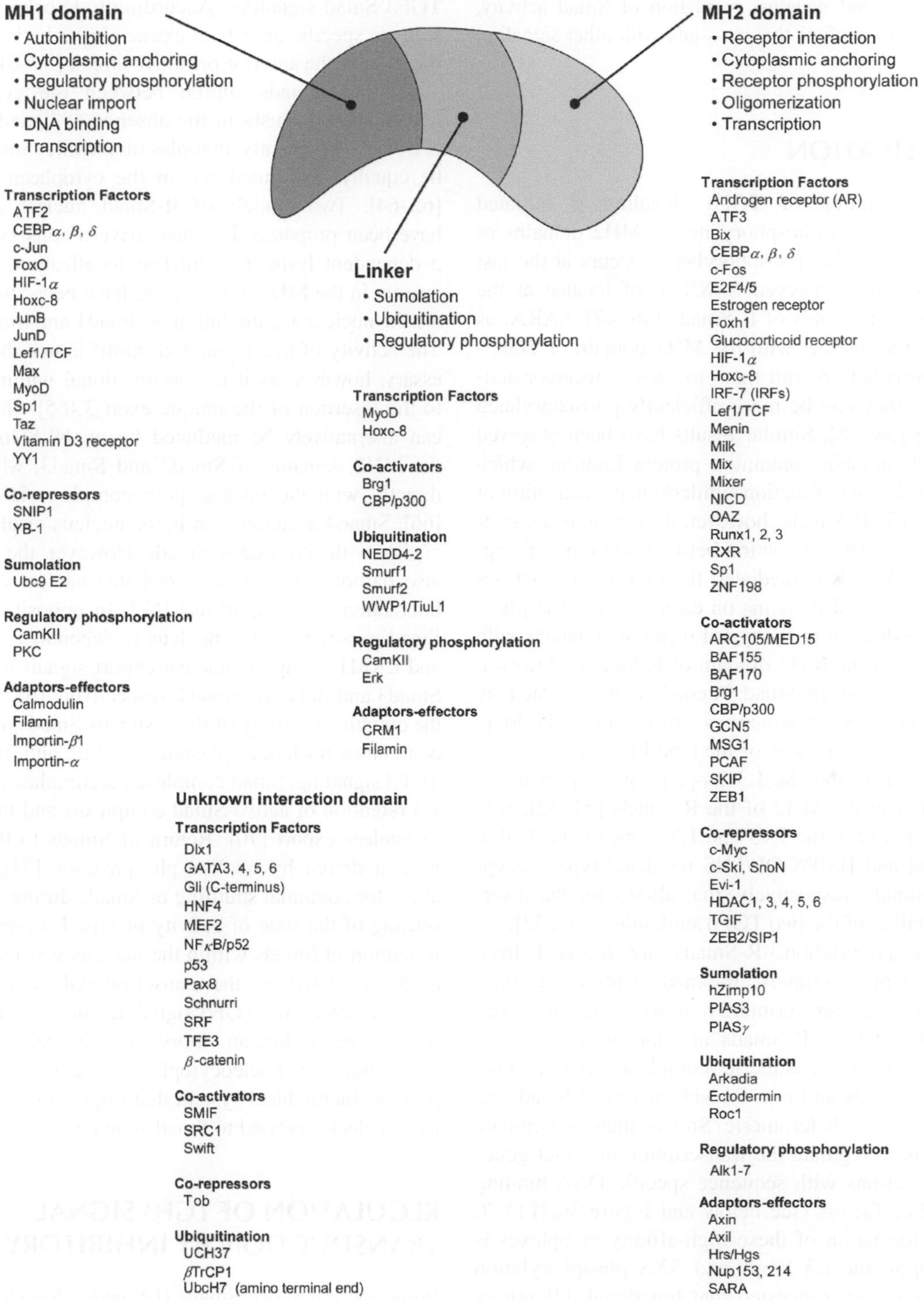

FIGURE 36.3 Simple schematic of Smad domains and Smad-interacting proteins. Smad MAD Homology domains (MH1, MH2) as well as the linker have unique biological actions and interact with a plethora of proteins which regulate domain function. Listed interactors include regulators of transcription, nuclear shuttling, and protein stability, and those that result in protein modifications such as phosphorylation, sumoylation, and ubiquitination.

mediates interactions with receptors, Smad oligomerization, and interactions with transcription factors, co-activators, and co-repressors [5–7, 9, 39]. The linker region between the MH1 and MH2 is divergent in sequence among Smads, and contains multiple phosphorylation sites for numerous kinases acting downstream of multiple pathways [9]. These sites allow for both positive and negative regulation of Smad activity, and are utilized to mediate the cross-talk with other signaling pathways.

SMAD ACTIVATION

Intracellular activation of TGFβ signaling is initiated when type I receptors phosphorylate the MH2 domains of R-Smads [44, 45]. This phosphorylation occurs at the last two serines of the conserved SSXS motif located at the extreme carboxy-terminus of R-Smads [46, 47]. SARA, as well as Hgs/Hrs, interact with the MH2 domain of TGFβ-activated R-Smads to recruit them to sites of receptor activation so that they can be more efficiently phosphorylated (Figure 36.2) [38, 48]. Similar results have been observed for the FYVE domain containing protein Endofin, which has also been shown to function similarly in the activation of the BMP specific R-Smads; however, these results are contentious [49, 50]. The interaction between the type I receptors and the R-Smads is mediated through the interactions of at least two sets of domains on each protein. The phosphorylated GS domain of the type I receptors interacts with the basic patch in the MH2 domain of R-Smads. However, specificity of receptor/R-Smad interaction is governed by the interaction between structures termed the L45 loop, located in the kinase domain of the type I receptor between β strands 4 and 5, and the L3 loop, located upstream of the basic patch in the MH2 of the R-Smads [51, 52]. It is the variation between the L45 and L3 loops of the TGFβ/Activin/Nodal and BMP/GDF/MIS regulated type I receptors and R-Smads, respectively, that allows for the divergence in signaling of the two TGFβ subfamilies [51, 52].

Upon phosphorylation, R-Smads are released from the active receptor complex. R-Smads with unphosphorylated SXS sites are primarily monomeric; however, upon phosphorylation, R-Smads are able to form active heterotrimeric or heterodimeric complexes that include either two R-Smads and one Smad4, or one R-Smad and one Smad4. These heteromeric Smads then accumulate in the nucleus to regulate the transcription of target genes through interactions with sequence specific DNA binding partners and co-factors (see below and Figure 36.2) [5–7, 9, 39]. The formation of these high-affinity complexes is dependent upon the L3 loops and SXS phosphorylation [53–55]. It has been suggested that functional differences may exist between heterotrimeric and heterodimeric complexes, and that their formation may be dependent on target gene promoter context [56]. All transcriptional regulatory complexes analyzed to date contain Smad4, however, in cells lacking Smad4, TGFβ is still able to regulate a subset of its transcriptional targets and in the fly, Medea (Smad4) is not required for all of the biological activities of DPP [57, 58]. Genetic analysis in the mouse suggests that Smad4 is required in cells that require the highest level of TGFβ Smad signaling. Accordingly, knocking out Smad4 leads to specific defects in extraembryonic tissue and specification of the anterior primitive streak [59–61].

Though Smads shuttle between the cytoplasm and nucleus continuously in the absence of ligand stimulation, R-Smads are mainly cytoplasmic, while Smad4 tends to be equally distributed within the cytoplasm and nucleus [62–64]. Two models of R-Smad nuclear accumulation have been proposed. R-Smads have a conserved importin-β dependent lysine-rich nuclear localization signal (NLS) present in the MH1 domain, and have been shown to mediate the nuclear accumulation of Smad1 and Smad3 [64, 65]. The activity of this lysine-rich motif is not absolutely necessary, however, as it is non-functional within Smad2 due to the insertion of the unique exon 3 [65]. Nuclear import can alternatively be mediated by an NLS located within the MH2 domains of Smad2 and Smad3, which interacts directly with the nuclear pore complex via nucleoporins [66]. Smad4 accumulation in the nucleus results from association with activated R-Smads. However, the Smad4 MH1 also harbors an NLS that mediates nuclear entry through interaction with importin-β [67]. In opposition to import, Smad export from the nucleus is dependent on exportin 4 and CRM1 acting via nuclear export signals in the MH2 of Smad3 and linker of Smad4, respectively [68, 69]. Through the combined activity of these signals Smads thus engage in continuous nucleocytoplasmic shuttling and, in response to TGFβ signaling, Smad complexes accumulate in the nucleus via retention of active Smad complexes and thus inhibition of nuclear export [70]. Return of Smads to the cytoplasm is then driven by nuclear phosphatases [71], which thus allow for continual shuttling of Smads during signaling and sensing of the state of activity of type I receptors [62, 72]. Retention of Smads within the nucleus was recently shown to be regulated by the transcriptional co-activator TAZ, and is essential for TGFβ signaling and the maintenance of pluripotency in human embryonic stem cells [73]. As TAZ itself undergoes nucleocytoplasmic shuttling, this creates a possible hierarchical system that might control the sensitivity of cells to respond to Smad signaling.

REGULATION OF TGFβ SIGNAL TRANSDUCTION BY INHIBITORY SMADS

While the inhibitory Smads (I-Smads), Smad6 and Smad7, share structural similarities with the R-Smads, they function to negatively regulate TGFβ signaling [43]. Moreover, I-Smads are induced by TGFβ superfamily members, and thus function within a negative feedback loop. Smad6 appears to preferentially inhibit BMP-type signals, while Smad7 negatively regulates both TGFβ and BMP signaling with similar efficiency (Figure 36.1) [74].

I-Smads inhibit TGFβ signaling by targeting different points within the signaling cascade. At the level of the TGFβ receptors I-Smads act through at least three mechanisms, all of which lead to reduced levels of active R-Smad/Smad4 complexes. First, I-Smad interaction with activated type I receptors physically blocks R-Smad recruitment and phosphorylation [75–77]. Second, Smad7 targets the serine/threonine phosphatase, PP1, to active receptor complexes, mediating inactivation of the type I receptors via dephosphorylation, presumably, of the type II receptor kinase target sites in the GS domain [78]. Finally, Smad7 induces TGFβ receptor degradation by recruiting Smurf E3 ubiquitin ligases to receptor complexes [36, 79, 80]. Similarly, at the level of Smads, Smad7 can induce Smad4 degradation by mediating Smad4/ubiquitin ligase interactions [81]. Further, Smad6 has been reported to interact with activated Smad1, competitively inhibiting the formation of active Smad1/Smad4 complexes [82]. I-Smads inhibit TGFβ-like signals within the nucleus as well; by interacting with Smad DNA-binding partners or binding to Smad-responsive elements directly, inhibiting the formation of Smad-dependent transcription complexes and thus TGFβ−mediated gene activation [83, 84].

SMADS ARE DNA-BINDING PROTEINS

Once in the nucleus, Smad complexes bind to DNA on Smad binding elements (SBEs) and interact with a host of sequence specific transcription factors and general coregulators to modulate gene expression [5–7, 9, 39]. Smad DNA-binding is independent of interactions with other transcription factors, and is mediated by a conserved β-hairpin loop located within the MH1 domain, which interacts with residues in the major groove of an SBE [85]. Smad2 is unable to bind to DNA directly, due to the insertion of exon 3 immediately upstream of the β-hairpin loop; however, a splice variant of Smad2 that lacks exon 3 is able to bind DNA similar to Smad3 and Smad4 [85, 86]. Activated full-length Smad2 can be localized to DNA enhancer elements through inclusion in active Smad complexes, and interactions with sequence specific transcription factors.

Use of binding site selection revealed that the optimal binding site for Smad3 and Smad4 is the palindromic sequence GTCTAGACA [87]. However, Smad3 and Smad4 have relaxed binding specificity and also bind to CG-rich sequences [88]. BMP-activated Smads also interact with GC-rich motifs [89]; however, Smad5 binds to the optimal GTCTAGACA site determined for Smad3 and Smad4 as well [87, 90]. While Smads bind to DNA in a sequence-specific manner, the interaction has a low affinity; thus it is believed that Smads are targeted to enhancer elements through interactions with a large variety of sequence specific transcription factors to form higher-order complexes with higher affinity for regulatory elements (Figure 36.3). In fact, no endogenous TGFβ response elements containing a single SBE have been observed, and *in vitro* the optimal SBE must be concatemerized in order to respond to activated Smads complexes [87]. It is interactions at the DNA level with independently regulated transcription factors that dictate much of the cross-talk between TGFβ and other signaling pathways.

SMADS COOPERATE WITH DNA-BINDING PARTNERS

The first Smad DNA-binding partner was identified through the study of an activin-responsive element (ARE) that mediated the immediate early activation of *Mix.*2 expression in the *Xenopus* embryo [91]. Activin treatment of embryos resulted in the formation of an ARE-binding protein complex termed ARF (activin-responsive factor) [91]. The ARF was found to be composed of Smad2, Smad4, and the novel protein FAST1 (forkhead activin signal transducer 1) [91, 92], which is now called FoxH1. Shortly thereafter, FoxH1-related genes were identified in mouse, human, and zebrafish (*schmalspur*/*sur* mutant locus) ([88, 93–96]). Analysis of FoxH1 activity downstream of TGFβ signaling using a mouse *Gsc* enhancer element [88], together with studies on frog FoxH1, yielded a model in which FoxH1 forms a higher-order DNA-binding complex with an R-Smad and Smad4 in response to TGFβ signals [88, 97]. FoxH1-dependent, TGFβ-mediated transcriptional activation required the formation of this protein/DNA complex that was also reliant upon the presence of both Smad and FoxH1 DNA-binding sites within the *Gsc* enhancer element [88]. Thus, Smads regulate target response elements via composite elements in which a high-affinity DNA-binding Smad partner provides specificity for Smad recruitment to the element, and Smad binding to the DNA at adjacent sites stabilizes formation of a higher-order transcriptional regulatory complex. By defining the rules governing functional FoxH1-Smad composite elements, efficient prediction of TGFβ-FoxH1 target elements can be achieved; this led to the discovery of a key role for FoxH1 in initiating retinoic acid signaling in the forebrain [98]. Moreover, studies on the role of FoxH1 in the heart revealed that FoxH1 mediates TGFβ-dependent activation of target elements in conjunction with the heart-specific transcription factor Nkx2-5, and is required for formation of the secondary heart field in the mouse [99]. FoxH1 also interacts with the homeobox protein Gsc, to restrict Mixl1 expression in gastrulating mouse embryos to the posterior primitive streak [100]. Notably, while FoxH1 is expressed during early development, by later stages and in the adult [95] FoxH1 is not expressed and thus FoxH1 target genes remain silent in response to Smad signaling [101]. Therefore, FoxH1 confers key developmental and cell-type-specific responses to TGFβ signaling during early vertebrate development. The importance of this pathway in humans has recently been revealed by the discovery of mutations in FoxH1 in patients with congenital heart disease and holoprosencephaly [102].

The biochemical and genetic analyses of FoxH1 have established important characteristics of TGFβ-Smad signaling, in particular highlighting the requirement for DNA-binding Smad partners that in turn provide tremendous flexibility and cell type-specificity to TGF signaling. Accordingly, since the discovery of FoxH1, a plethora of Smad DNA-binding partners have been identified (summarized in Figure 36.3). These belong to a wide range of families, including the bHLH, bZIP, homeodomain, nuclear receptor, Zn finger, and forkhead families [39]. It is the interaction of Smads with this large array of DNA-binding partners on the regulatory elements of target genes that confers the diversity of transcriptional responses to TGF family members in a wide range of cell types.

SMADS INTERACT WITH TRANSCRIPTION CO-ACTIVATORS AND REPRESSORS

The first indication that Smads could function as transcriptional activators came with the observation that heterologous fusion of the Smad MH2 domain to GAL4 was capable of inducing expression of a Gal4 binding site-containing reporter construct [103]. Moreover, the importance of Smad interaction with CBP/p300 for TGFβ-mediated transcription is well documented [9]. CBP/p300 co-activates transcription downstream of TGFβ and other signals mainly through the permissive modification of chromosome structure via the acetylation of histones, which creates a microenvironment that enhances transcription. Interestingly, Smads themselves are also targets of the acetyltransferase activity of CBP/p300, which results in the enhanced DNA-binding affinity of the MH1 domains and more efficient gene activation [104]. Further, activated Smads interact with numerous co-activators via their MH2 domains, including (but not limited to) MSG1, Arc105, Zeb1, and Smif [5]. These interactions increase transcriptional activation of target genes by promoting Smad/CBP/p300 interaction or the recruitment of the RNA Polymerase II complex to promoter elements.

Smads also regulate transcription through interactions with a number of transcriptional co-repressors (Figure 36.3) that silence transcription by multiple mechanisms. Competitive interactions of c-Ski, TGIF, YB-1, and SNIP1 with Smads are able to disrupt R-Smad/Smad4 and or Smad/CBP/p300 complex formation, thus inhibiting TGFβ-mediated gene activation [5–7, 9, 39]. Additionally, co-repressors such as c-Ski, SnoN, Evi-1, TGIF, Zeb2, and Gsc oppose the histone acetyltransferase activity of CBP/p300 by recruiting histone deacetylases (HDACs) into transcriptional complexes, resulting in a restrictive chromosome structure that inhibits transcription [5–7, 9, 39]. Finally, by binding to Smads, c-Myc inhibits the functional cooperation of Smads with Sp1, therefore inhibiting transcription; though the exact mechanism involved is not yet known, it may involve histone deacetylation [105].

NON-SMAD SIGNALING PATHWAYS

A tremendous amount of effort has focused on understanding how Smads are regulated and mediate transcriptional responses to TGFβ family members. However, quite a bit less is understood about how TGFβ regulates non-Smad pathways. Earlier studies on TGFβ signaling have suggested that TGFβ regulates various MAPK cascades [106], in particular the ERK, JNK, and p38 pathways, which may be important in mediating epithelial to mesenchymal transition and apoptotic responses. Studies with mutated variants of the TGFβ type I receptor that are incapable of activating Smads further suggests that these pathways can be regulated independently of Smad activation. Furthermore, a complex comprised of the kinase TAK1, its activating partners TAB1 and TAB2, and the apoptotic inhibitor Xiap have been linked to Ser/Thr kinase receptors, and TAK1 can regulate fate choice during early development [107], as well as the development of the vascular system in the mouse [108].

Although most TGFβ family member receptors possess a type II receptor characterized by a short tail sequence downstream of the kinase, the BMP type II receptor, BMPRII possesses a long extension that is also found in orthologous receptors in *Drosophila* and *C. elegans*, suggesting it has an important cell signaling function. Indeed, the tail of BMPRII possesses binding sites for LIM kinase (LIMK), a key regulator of actin cytoskeletal dynamics [109, 110]. Binding of LIMK to BMPRII stimulates LIMK catalytic activity and, upon BMP ligand treatment, synergizes with Cdc42 to stimulate phosphorylation of the LIMK substrate, cofilin. In cortical neurons, LIMK interaction with BMPRII plays a key role in BMP-dependent dendritogenesis [110].

Considerable attention has been directed at TGFβ for its role in promoting tumor progression in late-stage cancers, in particular, through regulation of epithelial to mesenchymal transitions [4, 8, 111]. Epithelial-to-mesenchymal transition (EMT) is a process by which epithelial cells acquire the characteristics of mesenchymal or fibroblast-type cells. EMT is an essential process during embryogenesis; however, similar processes occur during the progression of carcinoma to an invasiveness/metastatic phenotype, as well as in fibrosis. For example, in mouse models of renal fibrosis

[112] and TGFβ1-induced lung interstitial disease [113], lineage analysis and cell labeling techniques revealed that epithelial cells are a major source of the pathogenic fibroblasts. In cancer, EMT underlies the transition of a malignant but non-invasive epithelial tumor (carcinoma *in situ*) to a highly motile, spindle-shaped cell that can detach from surrounding cells; all changes that occur as carcinomas invade through the basement membrane and intravasate during the early stages of metastasis [114]. Imaging of mouse carcinoma models has confirmed that individual cells spread from primary tumors, as predicted in a EMT-type progression model of cancer [115].

There are three major components to EMT. Cells lose their epithelial-specific junctions, which include tight junctions and "classic" zona adherens composed of E-cadherin [116]. The cells also undergo cytoskeletal modification [114]. Specifically, cortical actin redistributes to form contractile stress fibers attached to focal adhesions, thereby promoting rapid cell migration [117]. These junctional and cytoskeletal changes occur quite quickly; microscopic changes are observed a few hours after addition of ligand [114]. In the third component of EMT, the gene expression program of the cells changes from an epithelial to a fibroblast-type program, thereby ensuring that the acquisition of fibroblastoid changes is not a temporary one. This is achieved through upregulation of transcription factors such as Snail and Twist, which repress epithelial-specific genes, and induce expression of fibroblast-type genes such as vimentin [118, 119]. In a number of cell lines, such as Normal Murine Mammary Gland cells (NMuMG), TGFβ is a key regulator of EMT [120], and in breast cancer models cooperates with oncogenic Ras mutations to stimulate metastatic progression by inducing EMT [121]. In terms of gene expression associated with EMT, activation of the Smad pathway leads to an increase in the expression of the EMT-related transcription factors Snail and Twist [122, 123]. However, recent studies have begun to uncover key roles for non-Smad signaling pathways in controlling the morphological transformation associated with EMT–specifically, the loss of apical polarity and tight junctions along with the rearrangement of the actin cytoskeleton associated with mesenchymal cell types.

Polarity in most, if not all, cell types is controlled by proteins that engage in multiple macromolecular complexes [124]. In epithelial cells, acquisition and maintenance of apical-basal polarity are tightly regulated by these polarity complexes. Recently, interactions between TGFβ receptors and the junctional components occludin [125] and cadherins [126] have been characterized, as well as interactions with the polarity regulator Par6 [127]. While interactions with occludin and cadherins function to maintain TGFβ receptors in the junctional region of polarized cells, the association of the TGFβ type I receptor with Par6 plays a key role in scaffolding Par6 into the activated heteromeric TGFβ receptor complex. This serves to bring Par6 to the TGFβ type II receptor kinase domain, which then phosphorylates Par6 on a serine residue near the carboxy-terminus. This leads to recruitment of the Smurf ubiquitin ligases to Par6, and subsequent localized targeting of the GTPase RhoA for degradation [127]. This is essential to allow remodeling of the cortical actin ring during EMT. Accordingly, blocking the TGFβ-Par6-Smurf pathway by a variety of means can prevent dissolution of tight junctions and EMT, despite the presence of normal Smad signaling. Indeed, analysis of this pathway during heart development has revealed a key role for the TGFβ-Par6 pathway during valve morphogenesis [128], a process that is critically dependent on EMT. Interestingly, TGFβ has also been shown to activate RhoA during EMT events [129], although the mechanism linking TGFβ receptors to RhoA activation has yet to be delineated. Nevertheless, these findings suggest that spatiotemporal control of receptor partners, such as Par6 and putative RhoA activators, may allow TGFβ to dynamically control actin cytoskeleton dynamics during the complex process of cell morphogenesis. How EMT and in particular these non-Smad pathways contribute to cancer progression to an invasive/metastatic phenotype is an area of intense investigation.

In summary much has been learned about how the TGFβ family regulates development and homeostasis, and how aberrant TGFβ signaling contributes to a wide array of diseases. However, much has yet to be learned, particularly in defining how TGFβ and its downstream signaling pathways are embedded in much larger morphogen super-networks that act to provide key contextual framework in which extracellular cues are translated into complex cell fate choices and behaviors.

REFERENCES

1. Massagué J. The transforming growth factor-β family. *Annu Rev Cell Biol* 1990;**6**:597–641.
2. Attisano L, Wrana JL. Signal transduction by the TGF-β superfamily. *Science* 2002;**296**:1646–7.
3. Derynck R, Miyazono KE. *The TGF-beta Family*. Cold Spring Harbor: Cold Spring Harbor Laboratory Press; 2008.
4. Massague J. TGFβ in cancer. *Cell* 2008;**134**:215–30.
5. ten Dijke P, Heldin CHE. Smad signal transduction. Smads in proliferation, differentiation and disease. In: Ridley A, Frampton J, editors. *Proteins and Cell Regulation*. Amsterdam: Springer; 2006:472.
6. Schmierer B, Hill CS. TGFβ-SMAD signal transduction: molecular specificity and functional flexibility. *Nat Rev Mol Cell Biol* 2007;**8**:970–82.
7. Feng XH, Derynck R. Specificity and versatility in TGF-β signaling through Smads. *Annu Rev Cell Dev Biol* 2005;**21**:659–93.
8. Bierie B, Moses HL. Tumour microenvironment: TGFβ: the molecular Jekyll and Hyde of cancer. *Nat Rev Cancer* 2006;**6**:506–20.
9. Ross S, Hill CS. How the Smads regulate transcription. *Intl J Biochem Cell Biol* 2008;**40**:383–408.
10. Schier AF. Nodal signaling in vertebrate development. *Annu Rev Cell Dev Biol* 2003;**19**:589–621.

11. Miyazawa K, Shinozaki M, Hara T, Furuya T, Miyazono K. Two major Smad pathways in TGF-β superfamily signalling. *Genes Cells* 2002;**7**:1191–204.
12. Annes JP, Munger JS, Rifkin DB. Making sense of latent TGFβ activation. *J Cell Sci* 2003;**116**:217–24.
13. Ge G, Greenspan DS. BMP1 controls TGFβ1 activation via cleavage of latent TGFβ-binding protein. *J Cell Biol* 2006;**175**:111–20.
14. Balemans W, Van Hul W. Extracellular regulation of BMP signaling in vertebrates: a cocktail of modulators. *Dev Biol* 2002;**250**:231–50.
15. de Caestecker M. The transforming growth factor-β superfamily of receptors. *Cytokine Growth Factor Rev* 2004;**15**:1–11.
16. Rebbapragada A, Benchabane H, Wrana JL, Celeste AJ, Attisano L. Myostatin signals through a transforming growth factor β-like signaling pathway to block adipogenesis. *Mol Cell Biol* 2003;**23**:7230–42.
17. Andersson O, Reissmann E, Ibanez CF. Growth differentiation factor 11 signals through the transforming growth factor-β receptor ALK5 to regionalize the anterior-posterior axis. *EMBO Rep* 2006;**7**:831–7.
18. Goumans MJ, Valdimarsdottir G, Itoh S, et al. Activin receptor-like kinase (ALK)1 is an antagonistic mediator of lateral TGFβ/ALK5 signaling. *Mol Cell* 2003;**12**:817–28.
19. Wrana JL, Attisano L, Wieser R, Ventura F, Massagué J. Mechanism of activation of the TGF-β receptor. *Nature* 1994;**370**:341–7.
20. Boesen CC, Radaev S, Motyka SA, Patamawenu A, Sun PD. The 1.1-Å crystal structure of human TGF-β type II receptor ligand binding domain.. *Structure* 2002;**10**:913–19.
21. Hart PJ, Deep S, Taylor AB, Shu Z, Hinck CS, Hinck AP. Crystal structure of the human TβR2 ectodomain–TGFβ3 complex. *Nat Struct Biol* 2002;**9**:203–8.
22. Greenwald J, Vega ME, Allendorph GP, Fischer WH, Vale W, Choe S. A flexible activin explains the membrane-dependent cooperative assembly of TGF-β family receptors. *Mol Cell* 2004;**15**:485–9.
23. Groppe J, Hinck CS, Samavarchi-Tehrani P, et al. Cooperative assembly of TGF-β superfamily signaling complexes is mediated by two disparate mechanisms and distinct modes of receptor binding. *Mol Cell* 2008;**29**:157–68.
24. Keller S, Nickel J, Zhang JL, Sebald W, Mueller TD. Molecular recognition of BMP-2 and BMP receptor IA. *Nat Struct Mol Biol* 2004;**11**:481–8.
25. López-Casillas F, Wrana JL, Massagué J. Betaglycan presents ligand to the TGF-β signaling receptor. *Cell* 1993;**73**:1435–44.
26. Lewis KA, Gray PC, Blount AL, et al. Betaglycan binds inhibin and can mediate functional antagonism of activin signalling. *Nature* 2000;**404**:411–14.
27. Cheng SK, Olale F, Brivanlou AH, Schier AF. Lefty blocks a subset of TGFβ signals by antagonizing EGF–CFC co-receptors. *PLoS Biol* 2004;**2**:E30.
28. Matsunaga E, Chedotal A. Repulsive guidance molecule/neogenin: a novel ligand–receptor system playing multiple roles in neural development. *Dev Growth Diff* 2004;**46**:481–6.
29. Samad TA, Rebbapragada A, Bell E, et al. DRAGON, a bone morphogenetic protein co-receptor. *J Biol Chem* 2005;**280**:14122–9.
30. Choi WY, Giraldez AJ, Schier AF. Target protectors reveal dampening and balancing of Nodal agonist and antagonist by miR-430. *Science* 2007;**318**:271–4.
31. Attisano L, Wrana JL, Montalvo E, Massagué J. Activation of signalling by the activin receptor complex. *Mol Cell Biol* 1996;**16**:1066–73.
32. Le Roy C, Wrana JL. Clathrin- and non-clathrin-mediated endocytic regulation of cell signalling. *Nat Rev Mol Cell Biol* 2005;**6**:112–26.
33. Anders RA, Doré JJE, Arline SL, Garamszegi N, Leof EB. Differential requirement for type I and type II transforming growth factor-β receptor kinase activity in ligand-mediated receptor endocytosis. *J Biol Chem* 1998;**273**:23,118–125.
34. Di Guglielmo GM, Le Roy C, Goodfellow AF, Wrana JL. Distinct endocytic pathways regulate TGF-β receptor signalling and turnover. *Nat Cell Biol* 2003;**5**:410–21.
35. Ranzani B, Zhang XL, Bizer M, von Gersdorff G, Bottinger E, Lisanti MP. Caveolin-1 regulates transforming growth factor (TGF)-β/SMAD signaling through an interaction with the TGF-β type I receptor. *J Biol Chem* 2001;**276**:6727–38.
36. Kavsak P, Rasmussen RK, Causing CG, et al. Smad7 binds to Smurf2 to form an E3 ubiquitin ligase that targets TGFβ receptor for degradation. *Mol Cell* 2000;**6**:1365–75.
37. Hayes S, Chawla A, Corvera S. TGF β receptor internalization into EEA1-enriched early endosomes: role in signaling to Smad2. *J Cell Biol* 2002;**158**:1239–49.
38. Tsukazaki T, Chiang TA, Davison AF, Attisano L, Wrana JL. SARA, a FYVE domain protein that recruits Smad2 to the TGF-β receptor. *Cell* 1998;**95**:779–91.
39. Lin X, Chen Y-G, Feng X-H. Transcriptional control via Smads. In: Derynck R, Miyazono K, editors. *The TGF-beta Family*. Cold Spring Harbor: Cold Spring Harbor Laboratory Press; 2008.
40. Sekelsky JJ, Newfeld SJ, Raftery LA, Chartoff EH, Gelbart WM. Genetic characterization and cloning of Mothers against dpp, a gene required for decapentaplegic function in *Drosophila melanogaster*. *Genetics* 1995;**139**:1347–58.
41. Savage C, Das P, Finelli A, et al. The *C. elegans sma-2, sma-3* and *sma-4* genes define a novel conserved family of TGF-β pathway components. *Proc Natl Acad Sci USA* 1996;**93**:790–4.
42. Derynck R, Gelbart WM, Harland RM, et al. Nomenclature: vertebrate mediators of TGFβ family signals. *Cell* 1996;**87**:173.
43. Nakao A. Inhibitory Smads: mechanisms of action and roles in human diseases. In: ten Dijke P, Heldin CH, editors. *Smad Signal Transduction*. Amsterdam: The Netherlands; 2006.
44. Macías-Silva M, Abdollah S, Hoodless PA, Pirone R, Attisano L, Wrana JL. MADR2 is a substrate of the TGFβ receptor and its phosphorylation is required for nuclear accumulation and signalling. *Cell* 1996;**87**:1215–24.
45. Kretzschmar M, Liu F, Hata A, Doody J, Massagué J. The TGF-β family mediator Smad1 is phosphorylated directly and activated functionally by the BMP receptor kinase. *Genes Dev* 1997;**11**:984–95.
46. Abdollah S, Macias-Silva M, Tsukazaki T, Hayashi H, Attisano L, Wrana JL. TβRI phosphorylation of Smad2 on Ser465 and Ser467 is required for Smad2–Smad4 complex formation and signaling. *J Biol Chem* 1997;**272**:27,678–685.
47. Souchelnytskyi S, Tamaki K, Engström U, Wernstedt C, ten Dijke P, Heldin C-H. Phosphorylation of Ser^{465} and Ser^{467} in the C terminus of Smad2 mediates interaction with Smad4 and is required for Transforming Growth Factor-β signaling. *J Biol Chem* 1997;**272**:28,107–115.
48. Miura S, Takeshita T, Asao H, et al. Hgs (Hrs), a FYVE domain protein, is involved in Smad signalling through cooperation with SARA. *Mol Cell Biol* 2000;**20**:9346–55.
49. Shi W, Chang C, Nie S, Xie S, Wan M, Cao X. Endofin acts as a Smad anchor for receptor activation in BMP signaling. *J Cell Sci* 2007;**120**:1216–24.
50. Chen YG, Wang Z, Ma J, Zhang L, Lu Z. Endofin, a FYVE domain protein, interacts with Smad4 and facilitates transforming growth factor-β signaling. *J Biol Chem* 2007;**282**:9688–95.
51. Chen Y-G, Hata A, Lo RS, et al. Determinants of specificity in TGF-β signal transduction. *Genes Dev* 1998;**12**:2144–52.

52. Feng X-H, Derynck R. A kinase subdomain of transforming growth factor-β (TGF-β) type 1 receptor determines the TGF-β intracellular signaling specificity. *EMBO J* 1997;**16**:3912–23.
53. Chacko BM, Qin BY, Tiwari A, et al. Structural basis of heteromeric smad protein assembly in TGF-β signaling. *Mol Cell* 2004;**15**:813–23.
54. Jayaraman L, Massagué J. Distinct oligomeric states of Smad proteins in the Transforming Growth Factor-β pathway. *J Biol Chem* 2000;**275**:40710–17.
55. Wu J-W, Hu M, Chai J, et al. Crystal structure of a phosphorylated Smad2: recognition of phosphoserine by the MH2 domain and insights on Smad function in TGFb signaling. *Mol Cell* 2001;**8**:1277–89.
56. Inman GJ, Hill CS. Stoichiometry of active smad-transcription factor complexes on DNA. *J Biol Chem* 2002;**277**:51,008–016.
57. Sirard C, Kim S, Mirtsos C, et al. Targeted disruption in murine cells reveals variable requirement for Smad4 in transforming growth factor β-related signaling. *J Biol Chem* 2000;**275**:2063–70.
58. Wisotzkey RG, Mehra A, Sutherland DJ, et al. *Medea* is a *Drosophila Smad 4* homolog that is differentially required to potentiate DPP responses. *Development* 1998;**125**:1433–45.
59. Sirard C, de la Pompa JL, Elia A, et al. The tumor suppressor gene Smad4/Dpc4 is required for gastrulation and later for anterior development of the mouse embryo. *Genes Dev* 1998;**12**:107–19.
60. Yang X, Li C, Xu X, Deng C. The tumor suppressor SMAD4/DPC4 is essential for epiblast proliferation and mesoderm induction in mice. *Proc Natl Acad Sci USA* 1998;**95**:3667–72.
61. Chu GC, Dunn NR, Anderson DC, Oxburgh L, Robertson EJ. Differential requirements for Smad4 in TGFβ-dependent patterning of the early mouse embryo. *Development* 2004;**131**:3501–12.
62. Inman GJ, Nicolas FJ, Hill CS. Nucleocytoplasmic shuttling of Smads 2, 3, and 4 permits sensing of TGF-β receptor activity. *Mol Cell* 2002;**10**:283–94.
63. Watanabe M, Masuyama N, Fukuda M, Nishida E. Regulation of intracellular dynamics of Smad4 by its leucine-rich nuclear export signal. *EMBO Rep* 2000;**1**:176–82.
64. Xiao Z, Watson N, Rodriguez C, Lodish HF. Nucleocytoplasmic shuttling of Smad1 conferred by its nuclear localization and nuclear export signals. *J Biol Chem* 2001;**276**:39,404–410.
65. Kurisaki A, Kose Y, Yoneda Y, Heldin C-H, Moustakas A. Transforming Growth Factor-β induces nuclear import of Smad3 in an importin-β and Ran-dependent manner. *Mol Biol Cell* 2001;**12**:1079–91.
66. Xu L, Kang Y, Col S, Massague J. Smad2 nucleocytoplasmic shuttling by nucleoporins CAN/Nup214 and Nup153 feeds TGFβ signaling complexes in the cytoplasm and nucleus. *Mol Cell* 2002;**10**:271–82.
67. Xiao Z, Latek R, Lodish HF. An extended bipartite nuclear localization signal in Smad4 is required for its nuclear import and transcriptional activity. *Oncogene* 2003;**22**:1057–69.
68. Pierreux CE, Nicolas FJ, Hill CS. Transforming growth factor β-independent shuttling of Smad4 between the cytoplasm and nucleus. *Mol Cell Biol* 2000;**20**:9041–54.
69. Kurisaki A, Kurisaki K, Kowanetz M, et al. The mechanism of nuclear export of Smad3 involves exportin 4 and Ran. *Mol Cell Biol* 2006;**26**:1318–32.
70. Schmierer B, Hill CS. Kinetic analysis of Smad nucleocytoplasmic shuttling reveals a mechanism for transforming growth factor β-dependent nuclear accumulation of Smads. *Mol Cell Biol* 2005;**25**:9845–58.
71. Lin X, Duan X, Liang YY, et al. PPM1A functions as a Smad phosphatase to terminate TGFβ signaling. *Cell* 2006;**125**:915–28.
72. Schmierer B, Tournier AL, Bates PA, Hill CS. Mathematical modeling identifies Smad nucleocytoplasmic shuttling as a dynamic signal-interpreting system. *Proc Natl Acad Sci USA* 2008;**105**:6608–13.
73. Varelas X, Sakuma R, Samavarchi-Tehrani P, et al. TAZ controls Smad nucleocytoplasmic shuttling and regulates human embryonic stem-cell self-renewal. *Nat Cell Biol* 2008;**10**:837–48.
74. Miyazono K. Positive and negative regulation of TGF-β signaling. *J Cell Sci* 2000;**113**(Pt 7):1101–9.
75. Nakao A, Afrakhte M, Morén A, et al. Identification of Smad7, a TGFβ-inducible antagonist of TGF-β signalling. *Nature* 1997;**389**:631–5.
76. Imamura T, Takase M, Nishihara A, et al. Smad6 inhibits signalling by the TGF-β superfamily. *Nature* 1997;**389**:622–6.
77. Hayashi H, Abdollah S, Qiu Y, et al. The MAD-related protein Smad7 associates with the TGFβ receptor and functions as an antagonist of TGFβ signaling. *Cell* 1997;**89**:1165–73.
78. Shi W, Sun C, He B, et al. GADD34-PP1c recruited by Smad7 dephosphorylates TGFβ type I receptor. *J Cell Biol* 2004;**164**:291–300.
79. Ebisawa T, Fukuchi M, Murakami G, et al. Smurf1 interacts with transforming growth factor-β type I receptor through Smad7 and induces receptor degradation. *J Biol Chem* 2001;**276**:12477–80.
80. Ogunjimi AA, Briant DJ, Pece-Barbara N, et al. Regulation of Smurf2 ubiquitin ligase activity by anchoring the E2 to the HECT domain. *Mol Cell* 2005;**19**:297–308.
81. Moren A, Imamura T, Miyazono K, Heldin CH, Moustakas A. Degradation of the tumor suppressor Smad4 by WW and HECT domain ubiquitin ligases. *J Biol Chem* 2005;**280**:22,115–123.
82. Hata A, Lagna G, Massague J, Hemmati-Brivanlou A. Smad6 inhibits BMP/Smad1 signaling by specifically competing with the Smad4 tumor suppressor. *Genes Dev* 1998;**12**:186–97.
83. Zhang S, Fei T, Zhang L, et al. Smad7 antagonizes transforming growth factor β signaling in the nucleus by interfering with functional Smad-DNA complex formation. *Mol Cell Biol* 2007;**27**:4488–99.
84. Bai S, Shi X, Yang X, Cao X. Smad6 as a transcriptional corepressor. *J Biol Chem* 2000;**275**:8267–70.
85. Shi Y, Wang YF, Jayaraman L, Yang H, Massagué J, Pavletich NP. Crystal structure of a Smad MH1 domain bound to DNA: insights on DNA binding in TGF-β signaling. *Cell* 1998;**94**:585–94.
86. Yagi K, Goto D, Hamamoto T, Takenoshita S, Kato M, Miyazono K. Alternatively Spliced Variant of Smad2 Lacking Exon3. *J Biol Chem* 1999;**274**:703–9.
87. Zawel L, Dai JL, Buckhaults P, et al. Human Smad3 and Smad4 are sequence-specific transcription activators. *Mol Cell* 1998;**1**:611–17.
88. Labbe E, Silvestri C, Hoodless PA, Wrana JL, Attisano L. Smad2 and Smad3 positively and negatively regulate TGF β-dependent transcription through the forkhead DNA-binding protein FAST2. *Mol Cell* 1998;**2**:109–20.
89. Kim J, Johnson K, Chen HJ, Carroll S, Laughon A. *Drosophila* Mad binds to DNA and directly mediates activation of *vestigial* by decapentaplegic. *Nature* 1997;**388**:304–8.
90. Li W, Chen F, Nagarajan RP, Liu X, Chen Y. Characterization of the DNA-binding property of Smad5. *Biochem Biophys Res Commun* 2001;**286**:1163–9.
91. Huang H-C, Murtaugh LC, Vize PD, Whitman M. Identification of a potential regulator of early transcriptional responses to mesoderm inducer in the frog embryo. *EMBO J* 1995;**14**:5965–73.
92. Chen X, Rubock MJ, Whitman M. A transcriptional partner for MAD proteins in TGF-β signalling. *Nature* 1996;**383**:691–6.
93. Liu B, Dou CL, Prabhu L, Lai E. FAST-2 is a mammalian winged-helix protein which mediates transforming growth factor β signals. *Mol Cell Biol* 1999;**19**:424–30.

94. Zhou S, Zawel L, Lengauer C, Kinzler KW, Vogelstein B. Characterization of human FAST-1, a TGFβ and activin signal transducer. *Mol Cell* 1998;**2**:121–7.
95. Weisberg E, Winnier GE, Chen X, Farnsworth CL, Hogan BL, Whitman M. A mouse homologue of FAST-1 transduces TGFβ superfamily signals and is expressed during early embryogenesis. *Mech Dev* 1998;**79**:17–27.
96. Pogoda HM, Solnica-Krezel L, Driever W, Meyer D. The zebrafish forkhead transcription factor FoxH1/Fast1 is a modulator of nodal signaling required for organizer formation. *Curr Biol* 2000;**10**:1041–9.
97. Chen X, Weisberg E, Fridmacher V, Watanabe M, Naco G, Whitman M. Smad4 and FAST-1 in the assembly of activin-responsive factor. *Nature* 1997;**389**:85–9.
98. Silvestri C, Narimatsu M, von Both I, et al. Genome-wide identification of Smad/Foxh1 targets reveals a role for Foxh1 in retinoic acid regulation and forebrain development. *Dev Cell* 2008;**14**:411–23.
99. von Both I, Silvestri C, Erdemir T, et al. Foxh1 is essential for development of the anterior heart field. *Dev Cell* 2004;**7**:331–45.
100. Izzi L, Silvestri C, von Both I, et al. Foxh1 recruits Gsc to negatively regulate Mixl1 expression during early mouse development. *EMBO J* 2007;**26**:3132–43.
101. Attisano L, Silvestri C, Izzi L, Labbé E. The transcriptional role of Smads and FAST (FoxH1) in TGFβ and activin signalling. *Mol Cell Endocrinol* 2001;**180**:3–11.
102. Roessler E, Ouspenskaia MV, Karkera JD, et al. Reduced NODAL signaling strength via mutation of several pathway members including FOXH1 is linked to human heart defects and holoprosencephaly. *Am J Hum Genet* 2008;**83**:18–29.
103. Liu F, Hata A, Baker J, et al. A human Mad protein acting as a BMP-regulated transcriptional activator. *Nature* 1996;**381**:620–3.
104. Simonsson M, Kanduri M, Gronroos E, Heldin CH, Ericsson J. The DNA binding activities of Smad2 and Smad3 are regulated by coactivator-mediated acetylation. *J Biol Chem* 2006;**281**:39,870–880.
105. Feng XH, Liang YY, Liang M, Zhai W, Lin X. Direct interaction of c-Myc with Smad2 and Smad3 to inhibit TGF-β-mediated induction of the CDK inhibitor p15(Ink4B). *Mol Cell* 2002;**9**:133–43.
106. Zhang Y. Non-Smad TGF-β Signaling Pathways. In: Derynck R, Miyazono K, editors. *The TGF-beta Family*. Cold Spring Harbor: Cold Spring Harbor Laboratory Press; 2008.
107. Ohkawara B, Shirakabe K, Hyodo-Miura J, et al. Role of the TAK1–NLK–STAT3 pathway in TGF-β-mediated mesoderm induction. *Genes Dev* 2004;**18**:381–6.
108. Jadrich JL, O'Connor MB, Coucouvanis E. The TGF β activated kinase TAK1 regulates vascular development in vivo. *Development* 2006;**133**:1529–41.
109. Foletta VC, Lim MA, Soosairajah J, et al. Direct signaling by the BMP type II receptor via the cytoskeletal regulator LIMK1. *J Cell Biol* 2003;**162**:1089–98.
110. Lee-Hoeflich ST, Causing CG, Podkowa M, Zhao X, Wrana JL, Attisano L. Activation of LIMK1 by binding to the BMP receptor, BMPRII, regulates BMP-dependent dendritogenesis. *EMBO J* 2004;**23**:4792–801.
111. Derynck R, Akhurst RJ. Differentiation plasticity regulated by TGF-β family proteins in development and disease. *Nat Cell Biol* 2007;9:1000–4.
112. Iwano M, Plieth D, Danoff TM, Xue C, Okada H, Neilson EG. Evidence that fibroblasts derive from epithelium during tissue fibrosis. *J Clin Invest* 2002;**110**:341–50.
113. Kim KK, Kugler MC, Wolters PJ, et al. Alveolar epithelial cell mesenchymal transition develops in vivo during pulmonary fibrosis and is regulated by the extracellular matrix. *Proc Natl Acad Sci USA* 2006;**103**:13,180–185.
114. Thiery JP. Epithelial-mesenchymal transitions in tumour progression. *Nature Rev Cancer* 2002;**2**:442–54.
115. Condeelis J, Segall JE. Intravital imaging of cell movement in tumours. *Nature Rev Cancer* 2003;**3**:921–30.
116. Thiery JP, Sleeman JP. Complex networks orchestrate epithelial-mesenchymal transitions. *Nat Rev Mol Cell Biol* 2006;**7**:131–42.
117. Pellegrin S, Mellor H. Actin stress fibres. *J Cell Sci* 2007;**120**:3491–9.
118. Nakajima Y, Yamagishi T, Hokari S, Nakamura H. Mechanisms involved in valvuloseptal endocardial cushion formation in early cardiogenesis: roles of transforming growth factor (TGF)-β and bone morphogenetic protein (BMP). *Anat Rec* 2000;**258**:119–27.
119. Yang J, Mani SA, Donaher JL, et al. Twist, a master regulator of morphogenesis, plays an essential role in tumor metastasis. *Cell* 2004;**117**:927–39.
120. Miettinen PJ, Ebner R, Lopez AR, Derynck R. TGF-β induced transdifferentiation of mammary epithelial cells to mesenchymal cells: involvement of type I receptors. *J Cell Biol* 1994;**127**:2021–36.
121. Oft M, Peli J, Rudaz C, Schwarz H, Beug H, Reichmann E. TGF-β1 and Ha-Ras collaborate in modulating the phenotypic plasticity and invasiveness of epithelial tumor cells. *Genes Dev* 1996;**10**:2462–77.
122. Peinado H, Quintanilla M, Cano A. Transforming growth factor β-1 induces snail transcription factor in epithelial cell lines: mechanisms for epithelial mesenchymal transitions. *J Biol Chem* 2003;**278**:21,113–123.
123. Dong YF, Soung do Y, Chang Y, et al. Transforming growth factor-β and Wnt signals regulate chondrocyte differentiation through Twist1 in a stage-specific manner. *Mol Endocrinol* 2007;**21**:2805–20.
124. Harder JL, Margolis B. SnapShot: tight and adherens junction signaling. *Cell* 2008;**133**(1118):e1111–2.
125. Barrios-Rodiles M, Brown KR, Ozdamar B, et al. High-throughput mapping of a dynamic signaling network in mammalian cells. *Science* 2005;**307**:1621–5.
126. Rudini N, Felici A, Giampietro C, et al. VE-cadherin is a critical endothelial regulator of TGF-β signalling. *Embo J* 2008;**27**:993–1004.
127. Ozdamar B, Bose R, Barrios-Rodiles M, Wang HR, Zhang Y, Wrana JL. Regulation of the polarity protein Par6 by TGFβ receptors controls epithelial cell plasticity. *Science* 2005;**307**:1603–9.
128. Townsend TA, Wrana JL, Davis GE, Barnett JV. Transforming growth factor-β-stimulated endocardial cell transformation is dependent on Par6c regulation of RhoA. *J Biol Chem* 2008;**283**:13834–41.
129. Bhowmick NA, Ghiassim M, Bakin A, et al. Transforming growth factor-β1 mediates epithelial to mesenchymal transdifferentiation through a RhoA-dependent mechanism. *Mol Biol Cell* 2001;**12**:27–36.

Chapter 37

The Smads

Malcolm Whitman

Department of Developmental Biology, Harvard School of Dental Medicine, Boston, Massachusetts

HISTORY AND CATEGORIZATION: R-SMADS, CO-SMADS, AND I-SMADS

The founding member of the Smad family, MAD, was identified in *Drosophila* as a modifier of the effects of dpp, a *Drosophila* BMP homolog [1]. A related set of genes, Sma1, 2, 3, were identified as required for signaling by a TGFβ homolog in *C. elegans* [2], and both homology based searches and functional screens in vertebrates subsequently identified vertebrate MAD/Sma homologs (reviewed in [3]), defining a broadly conserved family, renamed Smads [4], as transducers of TGFβ signals. In both vertebrates and invertebrates, the Smads fall into three functional categories (reviewed in [5]: (1) the receptor regulated (R)-Smads, which act downstream of specific subsets of TGFβ superfamily ligands and are directly phosphorylated by Type I receptors; (2) co-Smads, which are not pathway specific, but that interact with R-Smads in response to ligand stimulation and are required for R-Smad signaling; (3) inhibitory (I)-Smads, which are often induced as early responses to TGFβ stimuli and that appear to act primarily as feedback inhibitors of signaling. A total of eight Smads have been identified in vertebrates (reviewed in [5]): Smads 1, 2, 3, 5, and 8 are R-Smads, Smad4 (and the closely related Smad4β, to date identified only in frogs) are co-Smads, and Smads 6 and 7 are I-Smads. All the Smads share two distinct regions of homology, an N-terminal MH1 domain and a C-terminal MH2 domain, separated by a linker that varies in size among the different Smad family members. The MH2 of Smad4 crystallizes as a trimer, and biochemical data indicate that Smads can homo-oligomerize via the MH2 domain *in vivo* in the presence or absence of ligand stimulation (reviewed in [6]). R-Smads and Smad4 also hetero-oligomerize in response to ligand stimulation and R-Smad phosphorylation, and the composition and relative proportion of homo-oligomeric and hetero-oligomeric Smad complexes is influenced by the transcription factors with which the Smad complexes interact [7].

SMAD REGULATION BY RECEPTORS AND NUCLEOCYTOPLASMIC SHUTTLING

Following the binding of TGFβ ligand to specific Type I and Type II transmembrane receptors, signaling is initiated when the Type I receptor phosphorylates an R-Smad at a conserved C-terminal SSXS motif (reviewed in [3]). Smad2 and Smad3 are activated downstream of activin, TGFβ, and nodal family ligands, while Smad1, Smad5, and Smad8 are activated primarily downstream of BMP ligands. A number of cytoplasmic/membrane associated proteins have been identified that both serve to retain unphosphorylated R-Smads in the cytoplasm and to facilitate R-Smad receptor interaction [8–10]. The best characterized of these is SARA a FYVE domain containing protein that participates in the functional coupling of Smad2 and Smad3 with their upstream Type I receptors [8]. A FYVE domain protein that may link Smad1 to its activating receptor has also been identified [11, 12]. While these linking proteins are likely to have an important role in modulating signaling through R-Smads, whether they have an essential role in this process remains to be demonstrated.

Following phosphorylation by the Type I receptor, R-Smads heterodimerize with the co-Smad Smad4 [13, 14]. R-Smad translocation is not dependent on Smad4, however, and nuclear accumulation of both R-Smads and Smad4 are controlled by a complex set of nuclear import and export signals [15–18]. The nuclear accumulation of the Smads is associated with an active nucleocytoplasmic shuttling process, providing a mechanism by which the state of receptor activation is continuously monitored by R-Smads as they move between the cytoplasm and the nucleus [15, 19–21].

TRANSCRIPTIONAL REGULATION BY SMADS

Following translocation to the nucleus, the R-Smad/Smad4 complex can interact with DNA by two distinct mechanisms: (1)

direct, site specific binding of DNA by the MH1 domain [22], and (2) interaction with additional site specific transcription factors [23, 24]. The Smad DNA binding domains preferentially recognize the motif GTCT, although binding to other target sites, particularly GC rich sequences, has also been reported (reviewed in [6]). While tandem repeats of the GTCT motif are sufficient for TGFβ regulation of synthetic reporter constructs, specific transcriptional regulation by Smads is thought to be directed in large part by the association of Smad complexes with additional site specific DNA binding factors. The first Smad associated DNA binding protein to be identified was FAST-1 (now FoxH1) [25, 26], a transcription factor that mediates signaling by nodals (a subset of TGFβ superfamily ligands) in the early vertebrate embryo [27]. FoxH1 associates with Smad2 or Smad3 and Smad4 in response to receptor activation and Smad phosphorylation via the direct interaction between a modular, conserved Smad interaction domain in the C-terminus of FoxH1 and the MH2 domain of Smad2/3 [26, 28]. Enhancer binding by the resulting complex is dependent on the forkhead DNA binding domain of FoxH1, but this binding is significantly enhanced by the DNA binding activity of the Smad MH1 domain [29]. Adjacent FoxH1 and Smad consensus binding sites can be found in several enhancers of FoxH1 regulated genes, indicating that specific transcriptional regulation by the FoxH1/Smad complex results from synergistic activity of FoxH1 and Smad DNA binding activities [29]. Interactions of R-Smad complexes with a wide variety of transcription factors have now been reported [23, 24, 30]; many of these are cell type specific, and some interact preferentially with Smad2/Smad3 [26, 31] while others interact with Smad1/5/8 [32], providing an explanation for both ligand specific and cell type specific transcriptional responses to TGFβ superfamily ligands.

Once bound to DNA, complexes of Smads and other transcription factors can regulate both transcriptional activation and repression. R-Smads and Smad4 have been shown to interact with a variety of transcriptional coactivators, including histone acetyltransferases (HATs) p300 [33–36], ARC105 [37], P/CAF [38], SWIFT, and SMIF [39, 40]. In addition to binding coactivators, Smads may activate transcription by displacing or sequestering transcriptional inhibitors, as in the case of the displacement of the transcriptional repressor Brinker from target promoters by the binding of the *Drosophila* Smad *Mad* to an overlapping DNA target site [41] reviewed in [23].

Transcriptional repression by Smads, like transcriptional activation, typically involves complex formation with both additional site specific DNA binding proteins and with coregulators. In the case of TGFβ repression of the c-Myc promoter, the corepressor p107 interacts with both Smad3 and the transcription factor E2F4/5 to form a repressor complex that binds to a transcription inhibitory element (TIE) in the c-Myc promoter [42]. R-Smads can also interact cooperatively with histone deacetylases (HDACs) and the transcription factors Runx2 or Nkx3.2 to form gene specific transcription inhibitory complexes [43, 44]. In these examples, receptor mediated activation of R-Smads leads to the formation of gene specific inhibitory complexes. In a number of additional examples, corepressors such as TGIF, Sno/Ski, and Evi-1, interact with R-Smads to inhibit Smad transcriptional activation, Smad DNA binding, or both [45–49]. In these cases, the result is thought to be repression of ligand mediated transcriptional activation, rather then ligand directed transcriptional repression.

While early studies of Smads indicated that signaling by R-Smads is dependent on association with Smad4, it is by now clear that Smad4 is not universally required for R-Smad signaling. Comparison of R-Smad loss of function phenotypes with Smad4 loss of function phenotypes in both mice and flies suggests that R-Smad function is not fully dependent on Smad4 [50–52]. In addition, several other cofactors for R-Smad transcriptional regulation that appear to physically and functionally replace Smad4 have been identified. The RBCC domain protein TIF1γ can displace Smad4 from activated Smad2/3, and the resulting R-Smad/TIF1gamma complex participates in TGFβ regulated hematopoietic differentiation [53]. In keratinocytes, Smad2/3 binds to IκB kinase alpha to form a complex that regulates transcription of the c-Myc antagonist MAD1, independent of Smad4 [54]. TIF1γ and IκB kinase share no evident sequence or structural similarity with Smad4 or with each other, indicating that a structural diverse range of factors can, in specific cases, functionally replace Smad4 in the regulation of transcription by R-Smads.

In addition to acting in the regulation of transcriptional responses downstream of TGFβ superfamily ligands, several other functionally significant interactions have been described for the Smads. Both Smad3 and Smad4 can interact directly with the canonical wnt signaling components LEF-1 and β-catenin, providing a mechanism for the synergistic regulation of transcription by wnt and TGFβ superfamily signals [55–57]. In addition to this pathway synergy, however, Smad4 has been shown to modulate transcriptional regulation by Wnt signaling even in the absence of upstream TGFβ superfamily receptor activation [56]. This observation suggests that Smad4 may function, in at least some cases, as a component of the Wnt signaling pathway in a capacity distinct from its established role in TGFβ superfamily signaling. Recent observations that critical regulators of β-catenin stability in the canonical Wnt pathway, axin, and GSK3 also regulate the stability of Smad3, further highlights the intertwining roles of components of the TGFβ and Wnt signaling pathways [58]. Smad3 also interacts with Akt, independently both of receptor activation of that Smad and of the kinase activity of Akt. This interaction appears to sequester Smad3 in the cytoplasm, and can modulate the responsiveness of cells to TGFβ signals [59, 60]. Since Smad3 activation is, under some circumstances, pro-apoptotic, while Akt is broadly anti-apoptotic, this interaction has been proposed to be a point of signal dependent regulation of apoptosis [59]. While it has not been established that this interaction is limiting for Smad3 activity under physiological circumstances *in vivo*, it provides a potential mechanism for cross-regulation of TGFβ signaling and a variety of Akt linked signaling pathways.

Recently, a novel function for R-Smads in the regulation of micro-RNA (miRNA) maturation has been identified [61]. Both TGFβ and BMP regulated R-Smads interact directly with the DROSHA miRNA processing complex in vascular smooth muscle cells to regulate cell behavior. This association mediates the TGFβ/BMP triggered maturation of the miRNA miR-21, which in turn modifies expression of PDCD4, a regulator of smooth muscle contractile genes. This novel posttranscriptional mechanism of R-Smad signaling provides an intriguing new facet to Smad function that generates a new family of questions about how this mechanism is integrated with Smad regulation of transcription.

DOWNREGULATION AND CROSS-REGULATION OF SMADS

A variety of mechanisms have now been identified by which Smad signaling can be downregulated, including dephosphorylation by phosphatases, specific protein degradative pathways, kinase mediated inhibition by other signaling pathways, and the induction of I-Smads [23]. Since the original demonstration that R-Smads are activated by C-terminal phosphorylation, there has been an expectation that phosphatases are responsible for terminating signaling activity by removing these activating phosphorylations. Only relatively recently, however, have the phosphatases responsible for these dephosphorylations been identified. The proteins PPM1A and SCP1-3 have been identified as phosphatases that dephosphorylate and inactivate the signaling function of R-Smads [62, 63], but whether these are the only phosphatases with this activity, and whether these phosphatases are themselves subject to regulation by extracellular signals, remains to be determined.

A family of ubiquitin E3 ligases called Smurfs were first characterized as regulators of the specific degradation of R-Smads [64]. Ubiquitin mediated degradation of R-Smads may be either ligand independent or stimulated by TGFβ ligands [64–68]. Smurf regulated ubiquitin dependent degradation of the R-Smads can be modulated by phosphorylation in the Smad linker region, which can be catalyzed by both MAP kinase family kinases and GSK3 [69]. This mechanism for R-Smad degradation provides a mechanism for the cross-regulation of Smads by a wide variety of signals that activate ERKs, stress activated kinases, or GSK3 [69, 70]. Smad linker phosphorylation can also control the nuclear localization of R-Smads, providing a second layer of regulation via these cross-regulatory mechanisms. Several other ubiquitin ligases can target Smads for degradation, including the SCF β–TrCP1 complex, NEDD4–2, and the R-Smad coregulator TIF1gamma/ectodermin, and may serve to link the stability of Smads to additional regulatory pathways [71–73].

The inhibitory Smads (I-Smads), vertebrate Smad6 and Smad7 and *Drosophila* Dad, were originally identified as members of the Smad family that inhibit, rather than transduce, TGFβ family signals [74]. I-Smads are often induced as feedback inhibitors of TGFβ superfamily signals, but can be induced by a variety of other signals as well. I-Smads can inhibit R-Smad signaling by several distinct mechanisms: (1) interfering with the interaction of R-Smads with Type I receptors; (2) promoting the degradation of Type I receptors (3) preventing the association of R-Smads with Smad4, and (4) directly interacting with transcription factors and recruiting corepressors to inhibit transcription (reviewed in [74]). In addition to these activities, Smad7 expression has also been shown to induce the activation of several MAP kinase family kinases [75]. How the activation of protein kinases by Smad7 occurs, and whether this activation is strictly linked to TGFβ signaling or may rather follow from any stimulus that increases Smad7 expression, is not known.

FUNCTION OF SMADS *IN VIVO*: GAIN OF FUNCTION AND LOSS OF FUNCTION EXPERIMENTS

Both loss of function and gain of function experiments *in vivo* have confirmed a crucial role for Smads in TGFβ signaling during a variety of biological processes (reviewed in [27, 76–79]. In early studies of Smad function, the ectopic expression of Smad1 and Smad2 in frog embryos recapitulated the developmental effects of BMPs and nodals, respectively [80–82], providing the first distinction in signaling function between different Smads. Subsequent genetic analysis in flies, zebrafish, and mice has confirmed an essential role for Smads in a wide variety of developmental and physiological processes [83]. A full consideration of the genetic analysis of Smad function is well beyond the scope of this chapter, and the reader is directed to several recent reviews [24, 79].

Study of the functional roles of the Smads *in vivo* is complicated both by partial functional redundancy among Smads and the embryonic lethality that results from the genetic inactivation of Smads 1, 2, 4, and 5 [52, 84–86]. Smads can have essential roles in both embryonic and extra-embryonic tissues, and both chimeric mice and conditional inactivation of alleles has been necessary to sort out specific functional roles of different Smads. While Smad2/3 signaling is essential for early embryonic patterning, Smad3 appears to be dispensable for this process, as Smad3$^{-/-}$ embryos develop to term normally, while Smad2$^{-/-}$ embryos fail to develop past gastrulation [85, 86]. Detailed analysis of Smad2 and Smad3 interaction in both embryos and somatic cells demonstrates, however, that each Smad makes distinctive contributions to signaling by TGFβ superfamily ligands [85]. Additional analysis of conditional and compound Smad knockouts will ultimately be necessary to sort out the distinctive role of each Smad in specific physiological processes *in vivo*. Antibodies directed against the C-terminally phosphorylated, activated forms of the R-Smads provide an additional tool for investigating exactly where and when Smads function *in vivo* [87].

While Smads have emerged as key mediators of TGFβ signals, a growing body of data indicate that TGFβs utilize Smad independent as well as Smad dependent signaling pathways in the regulation of multiple biological processes [reviewed in [88, 89], and the relative contribution of each pathway to physiological regulation *in vivo* remains poorly understood. The identification and characterization of the Smads has provided a coherent framework for understanding how specificity is generated in the wide range of biological actions of TGFβs. Future work will flesh out the full range of transcriptional targets for Smad interaction, the mechanisms and significance of Smad interaction with other signaling pathways, and the relationship between Smad dependent and Smad independent signals in mediating TGFβ action.

ACKNOWLEDGEMENTS

Due to space constraints, reviews have been extensively cited at the expense of primary references. The author apologizes to workers whose work is not directly cited due to these constraints. The author is supported by grants from the NICHD.

REFERENCES

1. Sekelsky JJ, Newfeld SJ, Raftery LA, et al. Genetic characterization and cloning of mothers against dpp, a gene required for decapentaplegic function in drosophila melanogaster. *Genetics* 1995;**139**:1347–58.
2. Savage C, Das P, Finelli AL, et al. Caenorhabditis elegans genes sma2, sma-3, and sma-4 define a conserved family of transforming growth factor beta pathway components. *Proc Nat Acad Sci USA* 1996;**93**:790–4.
3. Massague J. TGF-beta signal transduction. *Annu Rev Biochem* 1998;**67**:753–91.
4. Derynck R, Gelbart WM, Harland RM, et al. Nomenclature: vertebrate mediators of TGF-beta family signals. *Cell* 1996;**87**:173.
5. Attisano L, Tuen Lee-Hoeflich S. The Smads. *Genome Biol* 2001;**2**:1. reviews, 3010.1–3010.8.
6. Shi Y. Structural insights on Smad function in TGFbeta signaling. *BioEssays* 2001;**23**:223–32.
7. Inman GJ, Hill CS. Stoichiometry of active smad-transcription factor complexes on DNA. *J Biol Chem* 2002;**277**:51,008–51,016.
8. Tsukazaki T, Chiang TA, Davison AF, et al. SARA, a FYVE domain protein that recruits Smad2 to the TGFbeta receptor. *Cell* 1998;**95**:779–91.
9. Hocevar BA, Smine A, Xu XX, et al. The adaptor molecule Disabled-2 links the transforming growth factor beta receptors to the Smad pathway. *EMBO J* 2001;**20**:2789–801.
10. Yamakawa N, Tsuchida K, Sugino H. The rasGAP-binding protein, Dok-1, mediates activin signaling via serine/threonine kinase receptors. *EMBO J* 2002;**21**:1684–94.
11. Shi W, Chang C, Nie S, et al. Endofin acts as a Smad anchor for receptor activation in BMP signaling. *J Cell Sci* 2007;**120**:1216–24.
12. Chen YG, Wang Z, Ma J, et al. Endofin, a FYVE domain protein, interacts with Smad4 and facilitates transforming growth factor-beta signaling. *J Biol Chem* 2007;**282**:9688–95.
13. Lagna G, Hata A, Hemmati-Brivanlou A, et al. Partnership between DPC4 and SMAD proteins in TGF-beta signalling pathways. *Nature* 1996;**383**:832–6.
14. Zhang Y, Feng X, We R, et al. Receptor-associated Mad homologues synergize as effectors of the TGF-beta response. *Nature* 1996;**383**:168–72.
15. Pierreux CE, Nicolas FJ, Hill CS. Transforming growth factor beta-independent shuttling of Smad4 between the cytoplasm and nucleus. *Mol Cell Biol* 2000;**20**:9041–54.
16. Xiao Z, Liu X, Henis YI, et al. A distinct nuclear localization signal in the N terminus of Smad 3 determines its ligand-induced nuclear translocation. *Proc Natl Acad Sci U S A* 2000;**97**:7853–8.
17. Xu L, Chen YG, Massague J. The nuclear import function of Smad2 is masked by SARA and unmasked by TGFbeta-dependent phosphorylation. *Nat Cell Biol* 2000;**2**:559–62.
18. Xu L, Kang Y, Col S, et al. Smad2 nucleocytoplasmic shuttling by nucleoporins CAN/Nup214 and Nup153 feeds TGFbeta signaling complexes in the cytoplasm and nucleus. *Mol Cell* 2002;**10**:271–82.
19. Nicolas FJ, De Bosscher K, Schmierer B, et al. Analysis of Smad nucleocytoplasmic shuttling in living cells. *J Cell Sci* 2004;**117**:4113–25.
20. Schmierer B, Hill CS. Kinetic analysis of Smad nucleocytoplasmic shuttling reveals a mechanism for transforming growth factor beta-dependent nuclear accumulation of Smads. *Mol Cell Biol* 2005;**25**:9845–58.
21. Inman GJ, Nicolas FJ, Hill CS. Nucleocytoplasmic shuttling of Smads 2, 3, and 4 permits sensing of TGF-beta receptor activity. *Mol Cell* 2002;**10**:283–94.
22. Kim J, Johnson K, Chen H, et al. Drosophila Mad binds to DNA and directly mediates activation of vestigial by Decapentaplegic. *Nature* 1997;**388**:304–8.
23. Massague J, Seoane J, Wotton D. Smad transcription factors. *Genes Dev* 2005;**19**:2783–810.
24. Derynck R, Miyazono K. *The TGFβ family cold spring harbor monograph series*. Cold Spring Harbor, NY: Cold Spring Harbor Press; 2008.
25. Chen X, Rubock MJ, Whitman M. A transcriptional partner for Mad proteins in TGF-β signalling. *Nature* 1996;**383**:691–6.
26. Chen X, Weisberg E, Fridmacher V, et al. Smad4 and FAST-1 in the assembly of activin-response factor. *Nature* 1997;**389**:85–9.
27. Whitman M. Nodal signaling in early vertebrate embryos: themes and variations. *Dev Cell* 2001;**1**:605–17.
28. Germain S, Howell M, Esslemont GM, et al. Homeodomain and winged-helix transcription factors recruit activated Smads to distinct promoter elements via a common Smad interaction motif. *Genes Dev* 2000;**14**:435–51.
29. Yeo C-Y, Chen X, Whitman M. The role of FAST-1 and Smads in transcriptional regulation by activin during early *Xenopus* embryogenesis. *J Biol Chem* 1999;**274**:26,584–26,590.
30. Ross S, Hill CS. How the Smads regulate transcription. *Int J Biochem Cell Biol* 2008;**40**:383–408.
31. Germain S, Howell M, Esslemont GM, et al. Homeodomain and winged-helix transcription factors recruit activated Smads to distinct promoter elements via a common Smad interaction motif. *Genes Dev* 2000;**14**:435–51.

32. Hata A, Seoane J, Lagna G, et al. OAZ uses distinct DNA- and protein-binding zinc fingers in separate BMP-Smad and Olf signaling pathways. *Cell* 2000;**100**:229–40.
33. Pouponnot C, Jayaraman L, Massagué J. Physical and functional interaction of SMADs and p300/CBP. *J. Biol. Chem* 1998;**273**:22,865–22,868.
34. Feng XH, Zhang Y, Wu RY, et al. The tumor suppressor Smad4/DPC4 and transcriptional adaptor CBP/p300 are coactivators for smad3 in TGF-beta-induced transcriptional activation. *Genes Dev* 1998;**12**:2153–63.
35. Janknecht R, Wells NJ, Hunter T. TGF-beta-stimulated cooperation of smad proteins with the coactivators CBP/p300. *Genes Dev* 1998;**12**:2114–19.
36. de Caestecker MP, Yahata T, Wang D, et al. The Smad4 activation domain (SAD) is a proline-rich, p300-dependent transcriptional activation domain. *J Biol Chem* 2000;**275**:2115–22.
37. Kato Y, Habas R, Katsuyama Y, et al. A component of the ARC/Mediator complex required for TGF beta/Nodal signalling. *Nature* 2002;**418**:641–6.
38. Itoh S, Ericsson J, Nishikawa J, et al. The transcriptional co-activator P/CAF potentiates TGF-beta/Smad signaling. *Nucleic Acids Res* 2000;**28**:4291–8.
39. Shimizu K, Bourillot PY, Nielsen SJ, et al. Swift is a novel BRCT domain coactivator of Smad2 in transforming growth factor beta signaling. *Mol Cell Biol* 2001;**21**:3901–12.
40. Bai RY, Koester C, Ouyang T, et al. SMIF, a Smad4-interacting protein that functions as a co-activator in TGFbeta signalling. *Nat Cell Biol* 2002;**4**:181–90.
41. Saller E, Bienz M. Direct competition between Brinker and Drosophila Mad in Dpp target gene transcription. *EMBO Rep* 2001;**2**:298–305.
42. Chen CR, Kang Y, Siegel PM, et al. E2F4/5 and p107 as Smad cofactors linking the TGFbeta receptor to c-myc repression. *Cell* 2002;**110**:19–32.
43. Kim DW, Lassar AB. Smad-dependent recruitment of a histone deacetylase/Sin3A complex modulates the bone morphogenetic protein-dependent transcriptional repressor activity of Nkx3.2. *Mol Cell Biol* 2003;**23**:8704–17.
44. Kang JS, Alliston T, Delston R, et al. Repression of Runx2 function by TGF-beta through recruitment of class II histone deacetylases by Smad3. *EMBO J* 2005;**24**:2543–55.
45. Wotton D, Massague J. Smad transcriptional corepressors in TGF beta family signaling. *Curr Top Microbiol Immunol* 2001;**254**:145–64.
46. Wotton D, Lo RS, Swaby LA, et al. Multiple modes of repression by the Smad transcriptional corepressor TGIF. *J Biol Chem* 1999;**274**:37,105–37,110.
47. Alliston T, Ko TC, Cao Y, et al. Repression of bone morphogenetic protein and activin-inducible transcription by Evi-1. *J Biol Chem* 2005;**280**:24,227–24,237,.
48. Izutsu K, Kurokawa M, Imai Y, et al. The corepressor CtBP interacts with Evi-1 to repress transforming growth factor beta signaling. *Blood* 2001;**97**:2815–22.
49. Stroschein SL, Wang W, Zhou S, et al. Negative feedback regulation of TGF-beta signaling by the SnoN oncoprotein. *Science* 1999;**286**:771–4.
50. Wisotzkey R, Mehra A, Sutherland D, et al. Medea is a *Drosophila* Smad4 homolog that is differentially required to potentiate DPP responses. *Development* 1998;**125**:1433–45.
51. Sirard C, Kim S, Mirtsos C, et al. Targeted disruption in murine cells reveals variable requirement for Smad4 in transforming growth factor beta-related signaling. *J Biol Chem* 2000;**275**:2063–70.
52. Sirard C, de la Pompa JL, Elia A, et al. The tumor suppressor gene Smad4/Dpc4 is required for gastrulation and later for anterior development of the mouse embryo. *Genes Dev* 1998;**12**:107–19.
53. He W, Dorn DC, Erdjument-Bromage H, et al. Hematopoiesis controlled by distinct TIF1gamma and Smad4 branches of the TGFbeta pathway. *Cell* 2006;**125**:929–41.
54. Descargues P, Sil AK, Sano Y, et al. IKKalpha is a critical coregulator of a Smad4-independent TGFbeta-Smad2/3 signaling pathway that controls keratinocyte differentiation. *Proc Natl Acad Sci U S A* 2008;**105**:2487–92.
55. Labbe E, Letamendia A, Attisano L. Association of Smads with lymphoid enhancer binding factor 1/T cell- specific factor mediates cooperative signaling by the transforming growth factor-beta and wnt pathways. *Proc Natl Acad Sci U S A* 2000;**97**:8358–63.
56. Hussein SM, Duff EK, Sirard C. Smad4 and beta-catenin co-activators functionally interact with lymphoid-enhancing factor to regulate graded expression of Msx2. *J Biol Chem* 2003;**278**:48,805–48,814.
57. Nishita M, Hashimoto MK, Ogata S, et al. Interaction between Wnt and TGF-beta signalling pathways during formation of Spemann's organizer. *Nature* 2000;**403**:781–5.
58. Guo X, Ramirez A, Waddell DS, et al. Axin and GSK3- control Smad3 protein stability and modulate TGF- signaling. *Genes Dev* 2008;**22**:106–20.
59. Conery AR, Cao Y, Thompson EA, et al. Akt interacts directly with Smad3 to regulate the sensitivity to TGF-beta induced apoptosis. *Nat Cell Biol* 2004;**6**:366–72.
60. Remy I, Montmarquette A, Michnick SW. PKB/Akt modulates TGF-beta signalling through a direct interaction with Smad3. *Nat Cell Biol* 2004;**6**:358–65.
61. Davis BN, Hilyard AC, Lagna G, et al. SMAD proteins control DROSHA-mediated microRNA maturation. *Nature* 2008;**454**:56–61.
62. Lin X, Duan X, Liang YY, et al. PPM1A Functions as a Smad Phosphatase to Terminate TGFbeta Signaling. *Cell* 2006;**125**:915–28.
63. Sapkota G, Knockaert M, Alarcon C, et al. Dephosphorylation of the linker regions of Smad1 and Smad2/3 by small C-terminal domain phosphatases has distinct outcomes for bone morphogenetic protein and transforming growth factor-beta pathways. *J Biol Chem* 2006;**281**:40,412–40,419.
64. Zhu H, Kavsak P, Abdollah S, et al. A SMAD ubiquitin ligase targets the BMP pathway and affects embryonic pattern formation. *Nature* 1999;**400**:687–93.
65. Kavsak P, Rasmussen RK, Causing CG, et al. Smad7 binds to Smurf2 to form an E3 ubiquitin ligase that targets the TGF beta receptor for degradation. *Mol Cell* 2000;**6**:1365–75.
66. Podos SD, Hanson KK, Wang YC, et al. The DSmurf ubiquitin-protein ligase restricts BMP signaling spatially and temporally during Drosophila embryogenesis. *Dev Cell* 2001;**1**:567–78.
67. Zhang Y, Chang C, Gehling DJ, et al. Regulation of Smad degradation and activity by Smurf2, an E3 ubiquitin ligase. *Proc Natl Acad Sci U S A* 2001;**98**:974–9.
68. Lo RS, Massagué J. Ubiquitin-dependent degradation of TGF-beta-activated Smad2. *Nat Cell Biolog* 1999;**1**:472–8.
69. Fuentealba LC, Eivers E, Ikeda A, et al. Integrating patterning signals: Wnt/GSK3 regulates the duration of the BMP/Smad1 signal. *Cell* 2007;**131**:980–93.
70. Sapkota G, Alarcon C, Spagnoli FM, et al. Balancing BMP signaling through integrated inputs into the Smad1 linker. *Mol Cell* 2007;**25**:441–54.
71. Dupont S, Zacchigna L, Cordenonsi M, et al. Germ-layer specification and control of cell growth by Ectodermin, a Smad4 ubiquitin ligase. *Cell* 2005;**121**:87–99.

72. Moren A, Imamura T, Miyazono K, et al. Degradation of the tumor suppressor Smad4 by WW and HECT domain ubiquitin ligases. *J Biol Chem* 2005;**280**:22,115–22,123.
73. Wan M, Tang Y, Tytler EM, et al. Smad4 protein stability is regulated by ubiquitin ligase SCF beta-TrCP1. *J Biol Chem* 2004;**279**:14,484–14,487.
74. Miyazono K. Regulation of TGF-β family signaling by Inhibitory Smads. In: Derynck R, Miyazono K, editors. *The TGF-β Family*. Cold Spring Harbor, NY: Cold Spring Harbor Press; 2008. p. 363–87.
75. Edlund S, Landstrom M, Heldin CH, et al. Smad7 is required for TGF-beta-induced activation of the small GTPase Cdc42. *J Cell Sci* 2004;**117**:1835–47.
76. Weinstein M, Yang X, Deng C. Functions of mammalian Smad genes as revealed by targeted gene disruption in mice. *Cytokine Growth Factor Rev* 2000;**11**:49–58.
77. Whitman M. Smads and early developmental signaling by the TGFβ superfamily. *Genes Dev* 1998;**12**:2443–53.
78. Lu CC, Brennan J, Robertson EJ. From fertilization to gastrulation: axis formation in the mouse embryo. *Curr Opin Genet Dev* 2001;**11**:384–92.
79. Chang H, Lau AL, Matzuk MM. Studying TGF-beta superfamily signaling by knockouts and knockins. *Mol Cell Endocrinol* 2001;**180**:39–46.
80. Graff JM, Bansal A, Melton DA. *Xenopus* Mad proteins transduce distinct subsets of signals for the TGFβ superfamily. *Cell* 1996;**85**:479–87.
81. Baker JC, Harland RM. A novel mesoderm inducer, Madr2, functions in the activin signal transduction pathway. *Genes Dev* 1996;**10**:1880–9.
82. Thomsen GH. Xenopus mothers against decapentaplegic is an embryonic ventralizing agent that acts downstream of the Bmp-2/4 receptor. *Development* 1996;**122**:2359–66.
83. Hild M, Dick A, Rauch GJ, et al. The *smad5* mutation *somitabun* blocks Bmp2b signaling during early dorsoventral patterning of the zebrafish embryo. *Development* 1999;**126**:2149–59.
84. Tremblay KD, Dunn NR, Robertson EJ. Mouse embryos lacking Smad1 signals display defects in extra-embryonic tissues and germ cell formation. *Development* 2001;**128**:3609–21.
85. Dunn NR, Vincent SD, Oxburgh L, et al. Combinatorial activities of Smad2 and Smad3 regulate mesoderm formation and patterning in the mouse embryo. *Development* 2004;**131**:1717–28.
86. Waldrip W, Bikoff E, Hoodless P, et al. Smad2 signaling in extraembryonic tissues determines anterior–posterior polarity of the early mouse embryo. *Cell* 1998;**92**:797–808.
87. Lee MA, Heasman J, Whitman M. Timing of endogenous activin-like signals and regional specification of the Xenopus embryo. *Development* 2001;**128**:2939–52.
88. von Bubnoff A, Cho KW. Intracellular BMP signaling regulation in vertebrates: pathway or network?. *Dev Biol* 2001;**239**:1–14.
89. Massague J. How cells read TGF-beta signals. *Nat Rev Mol Cell Biol* 2000;**1**:169–78.

Part 5

TNF Receptors

Chapter 38

Structure and Function of Tumor Necrosis Factor (TNF) at the Cell Surface

Hao Wu[1] and Sarah G. Hymowitz[2]
[1] *Department of Biochemistry, Weill Cornell Medical College, New York, New York*
[2] *Department of Structural Biology, Genentech Inc., South San Francisco, California*

INTRODUCTION

The tumor necrosis factor (TNF) superfamily (TNFSF) and the TNF receptor (TNFR) superfamily (TNFRSF) form the corresponding ligand and receptor systems that are widely distributed in different tissues and cell types. Collectively they play critical roles in numerous aspects of mammalian biology, including embryonic development, innate and adaptive immunity, and maintenance of cellular homeostasis [1–3]. Agents that manipulate the signaling of these receptors are being used or showing promise towards the treatment and prevention of many human diseases [4–6].

Historically, the phrase *tumor necrosis factor* referred to a "factor" induced by bacterial infections that caused tumor regression in anecdotal cases. As early as the late nineteenth century, attempts were made to treat many kinds of cancers by provoking acute local skin infections, sometimes with success [7]. In 1975, it was discovered that bacterial endotoxin induced the production and release of anti-tumor activity from host cells. This activity caused hemorrhagic necrosis of transplanted tumors in mice, and killed transformed cell lines [8]. Its promise as a cancer cure prompted many laboratories to search the molecular identity of TNF. This eventually led to the purification, characterization, and cloning of TNFα, and the realization that the wasting-inducing factor cachexia is identical to TNFα [9–12].

It was soon discovered that TNFα has a wide range of biological effects in host defense against pathogens. On a cellular level, it is capable of inducing cell survival, proliferation, and differentiation, as well as both apoptotic and necrotic cell death under certain conditions [13, 14]. These collections of effects are mediated by the two receptors of TNFα, TNF-R1 and TNF-R2 [15]. TNFα does not generally provoke cell killing as in its anti-tumor activity, but more often promotes gene transcription and cell activation. There are currently 18 TNFSF members and 28 TNFRSF members that comprise signaling receptors, decoy receptors, and orphan receptors (Figure 38.1, Table 38.1). Some ligands and receptors interact with more than one partner, increasing the regulatory flexibility and complexity. Over 30 years of research since the cloning of TNFα has led to the thriving field of TNFSF and TNFRSF, with an estimated number of publications of over 100,000 and a TNF congress that meets every 2 years.

STRUCTURAL FEATURES

TNFSF members are generally homotrimeric type II transmembrane proteins, many of which can be shed from the cell surface to act as soluble signaling molecules. The defining feature of this family of extracellular ligands is the trimeric *TNF homology domain* (THD), comprising of three jelly roll protomers (Figure 38.2a, Table 38.2). Each protomer is formed by two β-sheets composed of strands A′AHCF and B′BGDE. These domains are exclusively located at the C-terminal region of the protein. The family can be divided into three groups (the conventional, the EF-disulfide containing, and the divergent), based on sequence and structural features in the THD. The "conventional" TNFSF members include TNFα, LTα, LTβ, LTαβ2, Apo2L/TRAIL, TL1A, LIGHT, FasL, RANKL and CD40L. This group is well characterized functionally and structurally, with crystal structures available for most of the ligands [16]. These ligands all have relatively long loops connecting the CD, DF, and DE strands, resulting in a characteristic pyramidal shape of the trimer. All "conventional" ligands are expected to bind receptors in a similar manner, with the elongated receptors nestled in the ligand–protomer

Handbook of Cell Signaling, Three-Volume Set 2 ed.

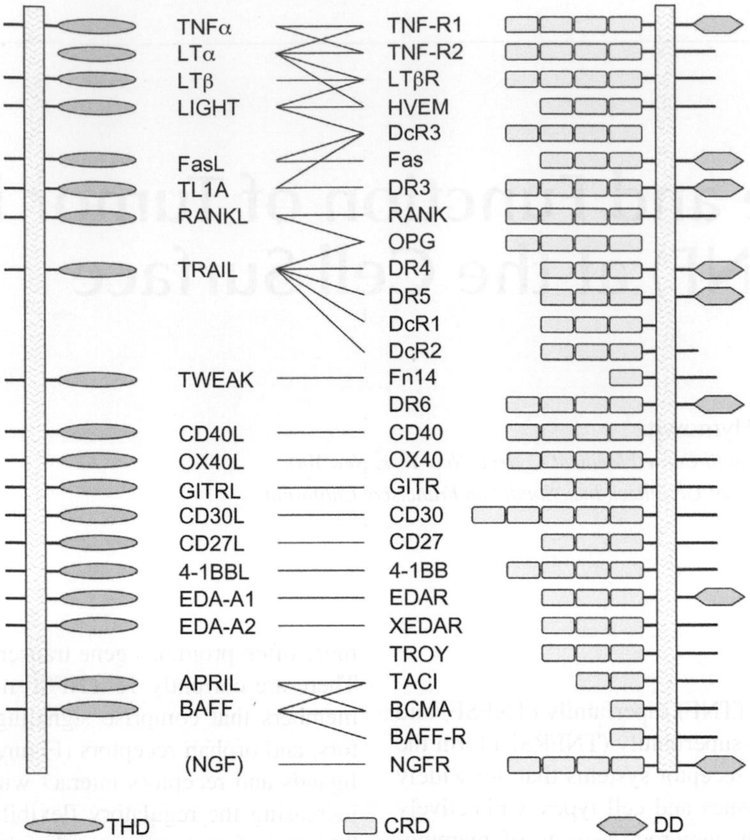

FIGURE 38.1 Ligand : receptor interaction map between the TNFSF and the TNFRSF.
THD, TNF homology domain; CRD, cysteine-rich domain; DD, death domain.

interfaces with two significant contact areas. One of the contact areas includes interactions between the receptor and conserved hydrophobic features in the ligand DE loop.

The second TNFSF subfamily, the "EF-disulfide" group, consists of APRIL, BAFF, TWEAK, and EDA, all of which possess a characteristic disulfide connecting the E and F strands. In addition, these ligands have shorter CD and EF loops, which lead to a more globular THD in contrast to the pyramidal conventional ligands. Crystal structures are available for APRIL, BAFF, and EDA ([17–21]). Receptor binding by this TNFSF group also differs from the conventional ligands, as they lack the conserved hydrophobic residues in the DE loop. Three of these ligands (APRIL, BAFF, and TWEAK) interact with small, atypical TNFRSF members (BAFF-R, TACI, BCMA, and Fn14) [16].

The third "divergent" ligand group contains the remaining members of the TNFSF (CD27L, CD30L, GITRL, 4-1BBL, and OX40L). These ligands are characterized by very divergent sequences, both from each other and from either the conventional or EF-disulfide groups. This group, despite the greater sequence divergence, is the least well studied crystallographically. Crystal structures have only been determined for the ligands OX40L and GITRL [22–24], and for the receptor–ligand pair OX40–OX40L [23]. These structures have shown evidence of greater plasticity in both the ligands and ligand–receptor complexes than had previously been appreciated. For instance, the trimer interface in both OX40L and GITRL is considerably smaller than in other ligands, leading the ligands to have an "open" appearance. Based on biophysical characterization of GITRL, the dissociation constant for trimer assembly appears to be lower than for other members of the TNFSF, with more extensive trimeric interfaces. Most surprisingly, two independent crystal structures of the extracellular domain of recombinant mouse GITRL, but not human GITRL, indicate a dimer both in the crystal and in solution [25] [24,26]. The functional consequences of this alternative dimeric packing in mouse GITRL are still being explored.

The corresponding TNFRSF members are type I transmembrane proteins that share certain structural features in their extracellular domains [3]. In contrast to the globular ligands, the typical multidomain TNFRSF members are elongated molecules composed of an extracellular domain of multiple ~40-residue pseudo repeats, typically containing six cysteines forming three disulfides (Figure 38.2b). These modules are termed *CRDs* (cysteine-rich domains), and can be further subdivided into smaller submodules based on the number of cysteines and topology of the cysteine connectivity [27]. A typical CRD is composed of A1 and B2

TABLE 38.1 The TNFSF and TNFRSF

Standardized names	Common names	Standardized names	Common names
TNFSF1	LTα, TNFβ, LT	TNFRSF1A	TNF-R1, CD120a, p55-R, TNFR60
TNFSF2	TNFα, TNF, cachectin	TNFRSF1B	TNF-R2, CD120b, p75-R, TNFR80
TNFSF3	LTβ	TNFRSF3	LTβR
TNFSF4	OX40L	TNFRSF4	OX40, CD134
TNFSF5	CD40L, CD154	TNFRSF5	CD40
TNFSF6	FasL, CD95	TNFRSF6	Fas, CD95, Apo-1
		TNFRSF6B	DcR3
TNFSF7	CD27L, CD70	TNFRSF7	CD27
TNFSF8	CD30L	TNFRSF8	CD30
TNFSF9	4-1BBL	TNFRSF9	4-1BB, CD137
TNFSF10	TRAIL, Apo-2L, TL2	TNFRSF10A	DR4, Apo-2, TRAIL-R1
		TNFRSF10B	DR5, TRAIL-R2
		TNFRSF10C	DcR1, TRAIL-R3
		TNFRSF10D	DcR2, TRAIL-R4
TNFSF11	RANKL, TRANCE, OPGL	TNFRSF11A	RANK, TRANCE-R,
		TNFRSF11B	OPG, osteoprotegerin
TNFSF12	TWEAK, Apo-3L	TNFRSF12A	Fn14, TWEAK-R
TNFSF13	APRIL	TNFRSF13B	TACI, CAML interactor
TNFSF13B	BAFF, Blys, TALL1	TNFRSF13C	BAFF-R, BR3
		TNFRSF17	BCMA, BCM
TNFSF14	LIGHT, HVEM-L, LTγ	TNFRSF14	HVEM, HveA, ATAR, LIGHT-R
TNFSF15	TL1A, VEGI	TNFRSF12	DR3, Apo-3, TRAMP
TNFSF18	GITRL, AITRL, TL6	TNFRSF18	GITR, AITR
	Eda, Ectodysplasin		EDAR
			XEDAR, EDA-A2R
	(NGF)	TNFRSF16	NGFR, p75
		TNFRSF19	TROY, TAJ, TRADD
		TNFRSF21	DR6

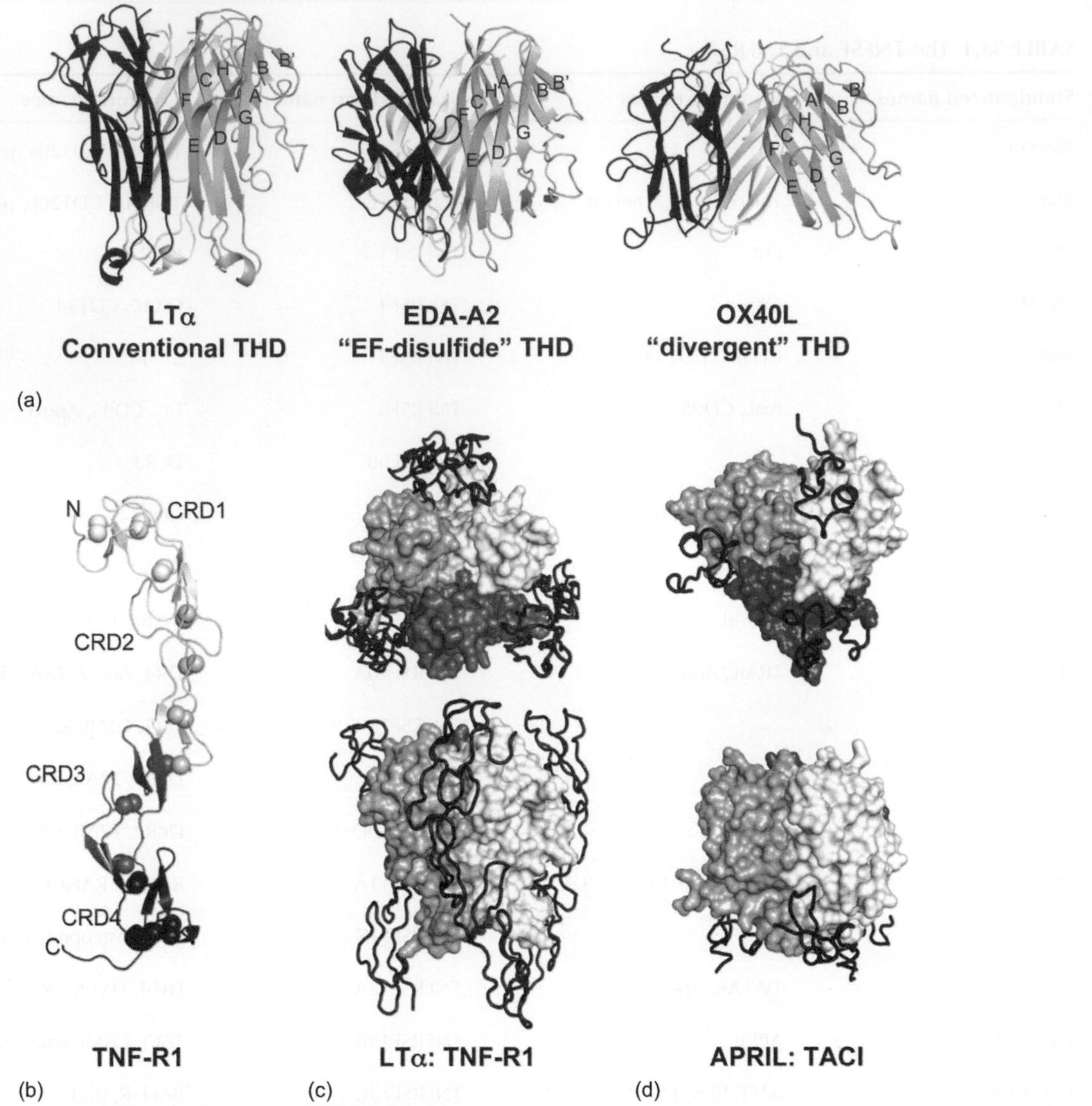

FIGURE 38.2 Structures of TNFSF and TNFRSF members and complexes.
(a) Three subtypes of TNFSF members. Ligands are oriented such that the ligand termini are on the top. Left, Cα trace of LTα trimer showing the pyramidal shape of the "conventional" family members (pdb code, 1TNR). Middle, Cα trace of the EDA-A2 trimer showing the globular shape of the "EF-disulfide" family members (pdb code, 1RJ8). Right, Cα trace of the OX40L trimer showing the open packing of the "divergent" family members (pdb code, 2HEV). (b) Cα trace of the TNF-R1 extracellular domain (pdb code, 1EXT). The sulfur atoms of the characteristic disulfide linkages are shown as spheres. The receptor is oriented such that the receptor cell membrane would be at the bottom of the page. CRDs are labeled. (c) A typical ligand : receptor complex illustrating that the receptor binds in the protomer interfaces (pdb code, 1TNR). LTα is shown as a molecular surface and the TNF-R1 extracellular domain is shown as dark gray Cα ribbon. (d) An "EF-disulfide" ligand : single-CRD receptor complex illustrating that the receptor binds in the protomer interfaces (pdb code, 1XU1). APRIL is shown as a molecular surface and the TACI extracellular domain is shown as dark gray Cα ribbon. Top, axis view highlighting the three-receptor : one-ligand trimer assembly; Bottom, side view highlighting the receptor binding sites in the ligand protomer–protomer interfaces.

tandemly linked subdomains. The A1 subdomain contains a single disulfide (the 1–2 disulfide), while the B2 subdomain contains two disulfides that are linked in a 3–5, 4–6 topology. Other subdomain variants exist, such as the A2 that contains two disulfides or the B1 that lacks one of the characteristic disulfides. This fold is preserved in viral proteins, such as CrmE, that modulate host immune systems [28].

Though less common, there are some TNFRSF members (BCMA, Fn14, BAFF-R) that only contain a single or partial CRD [16]. The BAFF and APRIL receptor TACI also belongs to this group of the TNFRSF, despite the apparent appearance of two CRDs in the TACI sequence. Biophysical, structural, and sequence analysis suggests that the two CRDs of TACI are likely the consequence of a relatively recent duplication event, and that only the second CRD is functional [29].

Despite the variety of ligand and receptor diversity within the family, to date all of the TNFSF:TNFRSF

TABLE 38.2 Crystal structures of extracellular domains of TNF family ligands and receptors

Names	Years	PDB codes	References
Ligands			
TNFα	1989, 1997, 1998	1TNF, 2TUN, 2TNF, 5TSW, 4TSV, 1A8M	99–102
TNFα: compound	2005	2AZ5	91
CD40L	1995	1ALY	103
CD40L : Fab	2001	1I9R	104
TRAIL	1999, 2000	1D2Q, 1DG6	105, 106
RANKL	2001, 2002	1JTZ, 1IQA	107, 108
BAFF	2002	1KD7, 1JH5, 1KXG	18–20
APRIL	2004	1U5X, 1U5Y, 1U5Z	21
EDA-A1	2003	1RJ8	17
EDA-A2	2003	1RJ7	17
OX40L	2006	2HEW	23
TL1A	2007	2O0O, 2RE9, 2QE3	109
hGITL	2007, 2008	3B93, 3B94, 2R32, 2R30, 2Q1M	25, 110
mGITL	2008	3B9I, 2QDN, 2Q8O	24, 26
Receptors			
TNF-R1	1995, 1996	1NCF, 1EXT	111, 112
TNF-R1 : compound	2001	1FT4	89
DR5 : Fab	2005, 2006	1ZA3, 2H9G	113, 114
HVEM : gD	2001	1JMA	115
HVEM : BTLA	2005	2AW2	116
CrmE	2007	2UWI	28
NGFR : NGF	2004	1SG1	117
NGFR : NT3	2008	3BUK	118
Ligand : receptor complexes			
LTα:TNF-R1	1993	1TNR	30
TRAIL : DR5	1999, 2000	1D0G1, 1D4V, 1DU3	31–33
BAFF : BAFF-R	2003	1OQE	35
BAFF : BCMA	2003	1OQD	35
April : BCMA	2005	1XU2	29
April : TACI	2005	1XU1	29
OX40L : OX40	2006	2HEY, 2HEV	23

H, human; m, mouse. The structures of human and mouse GITL are trimers and domain-swapped dimers, respectively
Fab, antibody Fab fragment; gD, Herpes simplex virus envelope glycoprotein D; BTLA, B and T lymphocyte attenuator, a CD28-like protein; CrmE, a Vaccinia virus-encoded tumour necrosis factor receptor; NGF, nerve growth factor; NT3, neurotrophin-3.

complexes oligomerize by binding receptor at each of the ligand–protomer interfaces to create a 3:3 heterotrimer (Figure 38.2c). For the multi-domain receptors, structures of LTα:TNF-R1 [30], TRAIL:DR5 [31–33], and OX40:OX40L [23] revealed that the ligand:receptor binding surface is formed primarily by residues from CRD2 and 3, but that additional contacts from CRD1 are also possible. The small single-domain receptors such as BCMA or BAFF-R also bind the ligand–protomer interfaces, but the interaction is focused at the end of the protomer interface away from the ligand termini [29,34,35] (Figure 38.2d). In either case, complex formation results in clustering of receptor extracellular domains in a signaling-competent manner. In the BAFF:BCMA and BAFF:BAFF-R complexes, the extracellular domain of BAFF forms a virus-like assembly that further cluster the receptors into higher order [35].

SIGNALING PATHWAYS AND REGULATION

Unlike many receptors, TNFRSF members do not possess enzymatic activity. Instead, the oligomeric complexes presumably place the intracellular regions of the receptors into proximity for recruitment of signaling proteins with enzymatic activities to amplify the signal transduction. Prior to ligand binding, at least some TNFRSF members, such as Fas, TNF-R1, TNF-R2, DR4, and CD40, appear to exist in pre-formed non-signaling oligomers through a region of the extracellular domain named the *pre-ligand-binding assembly domain* (PLAD) [36, 37]. The PLAD is likely physically separate from the ligand binding site and appears to be required for many aspects of receptor signaling, including dominant interference by receptors with pathogenic mutations at the PLAD region [37].

The signaling pathways of TNFRSF members differ depending on the domain and sequence in the intracellular region. Some of the receptors do not contain a structural module known as the death domain (DD) in their intracellular domains and are "survival" receptors, which directly recruit adaptor proteins known as the *TNF receptor associated factors* (TRAFs) [38–40]. Some examples of "survival" receptors include TNF-R2, CD40, CD30, OX40, 4-1BB, LTβR, RANK, and TACI. Seven mammalian TRAFs (TRAF1–7) have been identified so far, some of which are ubiquitin ligases [41, 42]. Among these, TRAF1, 2, 3, 5, and 6 participate in the signal transduction of the TNFRSF, leading to activation of transcription factors in the nuclear factor κ-B (NF-κB) and activator protein-1 (AP-1) family [43, 44].

Some TNFRSF members, such as Fas and TNF-R1, contain an intracellular DD and are known as death receptors [4, 45]. Fas is an effective prototypical cell-killing receptor. The intracellular DD of Fas directly recruits a DD-containing protein known as Fas-associated DD (FADD) via DD–DD interactions [46]. FADD also contains a *death-effector domain* (DED), which further recruits the DED-containing pro-caspase-8 or pro-caspase-10 to elicit caspase activation and apoptosis [47–49]. DR4 and DR5 also recruit FADD and caspase-8 or caspase-10, similar to Fas [50]. TNF-R1-like death receptors, on the other hand, possess the intrinsic capability of both cell-death and cell-survival induction. The underlying mechanism for this duality lies on the recruitment of a multifunctional protein, TNF receptor-associated DD (TRADD), via DD:DD domain interactions, by TNF-R1 [51]. TRADD recruits TRAF2 [51–53] and FADD [51, 54], leading to both survival and death signaling in a "cellular context"-dependent manner.

Many structures of TRAF proteins in complex with receptor sequences and adaptor proteins have been determined, which showed a matching trimeric symmetry of TRAFs and specificity of recognition [42, 55]. For a more detailed review of these structures, please refer to Chapter 49 of Handbook of Cell Signaling, Second Edition. Many DD and DED structures are known, including those involved in TNFRSF signaling. However, no structures of DD:DD or DED:DED complexes in the TNFRSF pathways are currently available [56]. A recent oligomeric structure of a DD:DD complex involved in caspase activation following DNA damage showed a completely asymmetric assembly mechanism that may provide a template for understanding these interactions [57].

Receptor signaling may be regulated by various means, such as by naturally occurring decoy receptors including OPG and DcR3 and by receptor shedding from the cell surface [58]. The latter may be a mechanism for terminating inflammation in a temporally controlled manner, and genetic defect in the shedding of TNF-R1 is a major cause of periodic fever syndromes [59].

BIOLOGICAL FUNCTIONS

Biological functions of the TNFSF and TNFRSF reflect their signaling capabilities, such as activation of transcription factors NFκB and AP-1 for cell survival and differentiation, and activation of caspases for cell-death induction. Although some TNFRSF members share similar intracellular signaling pathways, they can exert specific, non-redundant biological functions. First, many receptors exhibit specific tissue distribution patterns, and are induced by different developmental cues or environmental stimuli. Second, the intracellular signaling pathways of these receptors are differentially regulated, resulting in diverse cellular effects.

TNFSF and TNFRSF are major coordinators in the development of many organs, such as lymphoid organs, mammary glands and hair follicles. Secondary lymphoid organs are located at strategic sites where foreign antigens can be efficiently brought together with immune system regulatory and effector cells. The organized structure of secondary lymphoid tissues is thought to enhance sensitivity of antigen recognition, and to support proper regulation of activation of antigen-responsive lymphoid cells. LTβ, LTβR, RANKL, and

RANK are indispensable for the development of such lymphoid organs, including the spleen and lymph nodes [60–63]. Genetic studies suggest that the requirement for RANKL: RANK and for LTβ:LTβR in lymph-node formation may be sequential. CD4+ hematopoietic precursor cells emigrating into the primordial lymph node first use RANKL:RANK for survival and differentiation [63]. The ensuing expression of LTβ then allows interaction of these precursor cells with stromal connective tissue cells that express LTβR to establish spatial constraints in lymphoid organ definition, and to complete maturation of the node [60].

The RANKL:RANK system is important for terminal differentiation of mammary gland for lactation during pregnancy [64]. It is interesting that RANKL:RANK is also important for bone homeostasis (see below); coordinated activation of osteoclasts and maturation of mammary gland may be important for mobilization of minerals from the mother's bone to the newborn. The TNFRSF members EDAR, XEDAR, and TROY play roles in hair-follicle and sweat-gland development. Mice deficient in either EDA or EDAR, or humans with mutations in these proteins, have no primary hair follicles or sweat glands [65, 66]. In contrast, another TNFRSF member, p75, appears to temporally coordinate hair follicle development [67]. Interestingly, p75 is a receptor for the dimeric NGF, which is not part of the TNFSF. It plays multiple roles in neuronal development, including sensory neuron development, and p75-deficient mice exhibit decreased sensory innervation and serious cutaneous sensorineural defects [68, 69].

The TNFSF and TNFRSF are major coordinators in innate immune response and acute inflammation. Pathogen-associated molecular patterns such as lipopolysaccharides (LPS), peptidoglycan, flagellin, bacterial DNA CpG motifs, and viral RNAs activate cell surface and intracellular receptors, such as the Toll-like receptors (TLRs) and Nod-like receptors [70–72]. Cytokines such as TNFα are secreted upon pathogen recognition, which in turn lead to production and upregulation of chemokines and adhesion molecules. This is mostly mediated by the ability of TNFα to activate transcription factors in the NFκB family, which target the expression of proteins in immune and inflammatory responses. Chemokines and adhesion molecules are crucial for rapid recruitment of inflammatory cells such as granulocytes, monocytes, and lymphocytes to the site of infection. Massive reaction to pathogens can lead to septic shock, and TNFα or TNFR deficiency attenuates these events [73].

TNFSF and TNFRSF also participate in acute adaptive immune response, primarily through their effect on formation of B-cell rich germinal centers (GCs) within secondary lymphoid organs during an antigen response. TNFSF and TNFRSF members such as TNFα, LTα, LTβ, TNF-R1, and LTβR coordinate the formation of networks of follicular dendritic cells (FDC) that provide the environment of GCs [60]. FDCs are specialized mesenchymal cells that collect antigens in draining lymph nodes and interact with clonally expanding B cells in GCs. The engagement of the TNFRSF member CD40 on the B cells by CD40L on T cells is crucial for B cell somatic hypermutation and subsequent selection of high-affinity B cells for Ig class switching [74, 75]. Another TNFSF member, BAFF, on dendritic cells, interacts with its receptors on B cells to enhance B cell survival, which can augment autoimmunity under pathological conditions [3, 75]. TNFRSF members such as OX40, 4-1BB, CD27, CD30, HVEM, and GITR promote the expansion and survival of CD4+ and CD8+ T cells upon stimulation by dendritic cells that bear the corresponding ligands [76, 77].

The TNFRSF comprises various death receptors (DRs). One major function of certain DRs is to mediate the killing of virus-infected cells by CD8+ cytotoxic T cells. This effect has been well known for Fas [78]. Another function of DR-mediated death is immune homeostasis to balance recurrent lymphocyte expansion in response to antigen. This is crucial because of the limited space of lymphoid organs and toxic effects of massive lymphocyte expansion [79]. High or repeated antigen stimulation of activated T cells may induce these death molecules, and causes apoptosis in a fraction of the expanding cell population. Genetic impairment of Fas-induced apoptosis in humans or mice causes a dramatic loss of lymphocyte homeostasis and autoimmunity. While some DRs can induce caspase activation and apoptosis, some DRs, such as TNF-R1, can also mediate death of viral-infected cells that more resembles necrosis [80].

In addition to their role in lymphocyte homeostasis, the TNFSF and TNFRSF members such as RANK, RANKL, and TNFα play important roles in bone homeostasis. RANK and RANKL are important for differentiation and activation of osteoclasts from a monocyte precursor, and their absence leads to overly dense bones [62]. TNFα both synergizes with RANKL [81] and acts independently to induce osteoclast development in RANK deficient mice [82]. Bone homeostasis is not autonomous, but integrated with immune and hormonal functions. Activated T cells promote bone loss because RANKL expression is induced by antigen receptor engagement, which contributes to joint inflammation, bone and cartilage destruction, and crippling in arthritis [83]. The soluble RANKL decoy receptor OPG is induced by estrogen [84], and estrogen deficiency induces bone loss by enhancing T cell production of TNFα [85]. Therefore, the decrease in estrogen levels after menopause may explain the prevalence of osteoporosis.

THERAPEUTICS AND FUTURE EXPECTATIONS

Due to their role in a variety of important biological processes, including inflammation and apoptosis, many TNFSF and TNFRSF members have been targeted with therapeutic agents. Three anti-TNF agents targeting TNFα to suppress its inflammatory activities have been approved, including the TNF-R2–Fc fusion Enteracept (EnbrelTM), and two antibody-based agents, Infliximab (RemicadeTM) and Adalimumab (HumiraTM). These drugs have been used for treatment against autoimmune diseases such as rheumatoid arthritis, psoriasis, Crohn's disease, and ulcerative colitis. Recent biochemical characterizations have shown that the epitopes of Enbrel and Infliximab are energetically distinct, although they can compete for TNFα under some circumstances [86]. Recombinant TNFα has been approved for isolated limb perfusion therapy in melanoma [87].

TRAIL receptors DR4 and DR5 are highly expressed on many kinds of cancer cells, and recombinant human TRAIL and agonistic DR4 and DR5 antibodies are being investigated in clinical trials for their uses in cancer treatment [88]. On the other hand, inhibiting Fas with the use of the soluble Fas-Fc decoy receptor is being considered for inhibiting apoptosis in spinal cord injury [87]. Due to the relatively large interfaces, targeting the extracellular portion of these receptors and ligands with small molecules as agonists or antagonists is a challenging task. However, a variety of approaches have been tried, with some success [89–93]. There is also the possibility of inhibiting the intracellular signal transduction of TNFRSF members. One example is the use of cell-permeable TRAF6 binding peptides in downregulating RANK signaling and osteoclast differentiation in primary monocytes [94].

Interestingly, although recombinant TNFα has only been approved for limited use in melanoma, a striking recent finding showed that TNFα is a major agent of tumor-cell killing by Smac mimetics [95]. Smac is a protein that normally resides in the inter-membrane space of mitochondria and is released to cytosol during apoptosis to antagonize inhibition of apoptosis proteins (IAPs). A recent finding is that Smac mimetics activate the ubiquitin ligase activity of cIAP1 and cIAP2, leading to activation of NF-κB and induction of TNFα secretion [96–98]. The low level of TNFα in turn causes tumor-cell death in the presence of these mimetics. There are great expectations of the TNFSF and TNFRSF for their wider uses in future therapies against cancer and immune diseases.

ACKNOWLEDGEMENT

This work was funded by the National Institute of Health (AI45937, AI50872, and AI76927). We apologize to all whose work has not been appropriately reviewed or cited due to space limitations.

REFERENCES

1. Smith CA, Farrah T, Goodwin RG. The TNF receptor superfamily of cellular and viral proteins: activation, costimulation and death. *Cell* 1994;**76**:959–62.
2. Gravestein LA, Borst J. Tumor necrosis factor receptor family members in the immune system. *Sem Immunol* 1998;**10**:423–34.
3. Locksley RM, Killeen N, Lenardo MJ. The TNF and TNF receptor superfamilies: integrating mammalian biology. *Cell* 2001;**104**:487–501.
4. Ashkenazi A, Dixit VM. Death receptors: signaling and modulation. *Science* 1998;**281**:1305–8.
5. Leonen WAM. Editorial overview: CD27 and (TNFR) relatives in the immune system: their role in health and disease. *Sem Immunol* 1998;**10**:417–22.
6. Newton RC, Decicco CP. Therapeutic potential and strategies for inhibiting tumor necrosis factor-a. *J Med Chem* 1999;**42**:2295–314.
7. Coley WB. The treatment of malignant tumors by repeated inoculations of erysipelas: with a report of ten original cases. *Am J Med Sci* 1893;**105**:487–511.
8. Carswell EA, Old LJ, Kassel RL, Green S, Fiore N, Williamson B. An endotoxin-induced serum factor that causes necrosis of tumors. *Proc Natl Acad Sci USA* 1975;**72**:3666–70.
9. Pennica D, Nedwin GE, Hayflick JS, Seeburg PH, Derynck R, Palladino MA, Kohr WJ, Aggarwal BB, Goeddel DV. Human tumour necrosis factor: precursor structure, expression and homology to lymphotoxin. *Nature* 1984;**312**:724–9.
10. Wang AM, Creasey AA, Ladner MB, Lin LS, Strickler J, Van Arsdell JN, Yamamoto R, Mark DF. Molecular cloning of the complementary DNA for human tumor necrosis factor. *Science* 1985;**228**:149–54.
11. Shirai T, Yamaguchi H, Ito H, Todd CW, Wallace RB. Cloning and expression in *Escherichia coli* of the gene for human tumour necrosis factor. *Nature* 1985;**313**:803–6.
12. Beutler B, Cerami A. Cachectin and tumour necrosis factor as two sides of the same biological coin. *Nature* 1986;**320**:584–8.
13. Goeddel DV, Aggarwal BB, Gray PW, Leung DW, Nedwin GE, Palladino MA, Patton JS, Pennica D, Shepard HM, Sugarman BJ, et al. Tumor necrosis factors: gene structure and biological activities. *Cold Spring Harb Symp Quant Biol* 1986;**51**(Pt 1):597–609.
14. Fiers W. Tumor necrosis factor. Characterization at the molecular, cellular and in vivo level. *FEBS Letts* 1991;**285**:199–212.
15. Lewis M, Tartaglia LA, Lee A, Bennett GL, Rice GC, Wong GH, Chen EY, Goeddel DV. Cloning and expression of cDNAs for two distinct murine tumor necrosis factor receptors demonstrate one receptor is species specific. *Proc Natl Acad Sci USA* 1991;**88**:2830–4.
16. Bodmer JL, Schneider P, Tschopp J. The molecular architecture of the TNF superfamily. *Trends Biochem Sci* 2002;**27**:19–26.
17. Hymowitz SG, Compaan DM, Yan M, Wallweber HJ, Dixit VM, Starovasnik MA, de Vos AM. The crystal structures of EDA-A1 and EDA-A2: splice variants with distinct receptor specificity. *Structure* 2003;**11**:1513–20.
18. Karpusas M, Cachero TG, Qian F, Boriack-Sjodin A, Mullen C, Strauch K, Hsu YM, Kalled SL. Crystal structure of extracellular human BAFF, a TNF family member that stimulates B lymphocytes. *J Mol Biol* 2002;**315**:1145–54.
19. Liu Y, Xu L, Opalka N, Kappler J, Shu HB, Zhang G, et al. Crystal structure of sTALL-1 reveals a virus-like assembly of TNF family ligands. *Cell* 2002;**108**:383–94.
20. Oren DA, Li Y, Volovik Y, Morris TS, Dharia C, Das K, Galperina O, Gentz R, Arnold E. Structural basis of BLyS receptor recognition. *Nature Struct Biol* 2002;**9**:288–92.
21. Wallweber HJ, Compaan DM, Starovasnik MA, Hymowitz SG. The crystal structure of a proliferation-inducing ligand, APRIL. *J Mol Biol* 2004;**343**:283–90.

22. Chattopadhyay K, et al. Structural basis of inducible costimulator ligand costimulatory function: determination of the cell surface oligomeric state and functional mapping of the receptor binding site of the protein. *J Immunol* 2006;**177**:3920–9.
23. Compaan DM, Hymowitz SG. The crystal structure of the costimulatory OX40–OX40L complex. *Structure* 2006;**14**:1321–30.
24. Zhou Z, Tone Y, Song X, Furuuchi K, Lear JD, Waldmann H, Tone M, Greene MI, Murali R. Structural basis for ligand-mediated mouse GITR activation. *Proc Natl Acad Sci USA* 2008;**105**:641–5.
25. Chattopadhyay K, Ramagopal UA, Mukhopadhaya A, Malashkevich VN, Dilorenzo TP, Brenowitz M, Nathenson SG, Almo SC. Assembly and structural properties of glucocorticoid-induced TNF receptor ligand: implications for function. *Proc Natl Acad Sci USA* 2007;**104**:19,452–7.
26. Chattopadhyay K, Ramagopal UA, Brenowitz M, Nathenson SG, Almo SC. Evolution of GITRL immune function: murine GITRL exhibits unique structural and biochemical properties within the TNF superfamily. *Proc Natl Acad Sci USA* 2008;**105**:635–40.
27. Naismith JH, Sprang SR. Modularity in the TNF-receptor family. *Trends Biochem Sci* 1998;**23**:74–9.
28. Graham SC, Bahar MW, Abrescia NG, Smith GL, Stuart DI, Grimes JM. Structure of CrmE, a virus-encoded tumour necrosis factor receptor. *J Mol Biol* 2007;**372**:660–71.
29. Hymowitz SG, Patel DR, Wallweber HJ, Runyon S, Yan M, Yin J, Shriver SK, Gordon NC, Pan B, Skelton NJ, Kelley RF, Starovasnik MA. Structures of APRIL-receptor complexes: like BCMA, TACI employs only a single cysteine-rich domain for high affinity ligand binding. *J Biol Chem* 2005;**280**:7218–27.
30. Banner DW, D'Arcy A, Janes W, Gentz R, Schoenfeld JJ, Broger C, Loetscher H, Lesslauer W. Crystal structure of the soluble human 55-kD TNF receptor–human TNF-β complex: implications for TNF receptor activation. *Cell* 1993;**73**:431–45.
31. Cha SS, Sung BJ, Kim YA, Song YL, Kim HJ, Kim S, Lee MS, Oh BH. Crystal structure of TRAIL–DR5 complex identifies a critical role of the unique frame insertion in conferring recognition specificity. *J Biol Chem* 2000;**275**:31,171–7.
32. Hymowitz SG, Christinger HW, Fuh G, Ultsch M, O'Connell M, Kelley RF, Ashkenazi A, de Vos AM. Triggering cell death: the crystal structure of Apo2L/TRAIL in a complex with death receptor 5. *Mol Cell* 1999;**4**:563–71.
33. Mongkolsapaya J, Grimes JM, Chen N, Xu XN, Stuart DI, Jones EY, Screaton GR. Structure of the TRAIL–DR5 complex reveals mechanisms conferring specificity in apoptotic initiation. *Nature Struct Biol* 1999;**6**:1048–53.
34. Kim HM, Yu KS, Lee ME, Shin DR, Kim YS, Paik SG, Yoo OJ, Lee H, Lee JO. Crystal structure of the BAFF–BAFF-R complex and its implications for receptor activation. *Nature Struct Biol* 2003;**10**:342–8.
35. Liu Y, Hong X, Kappler J, Jiang L, Zhang R, Xu L, Pan CH, Martin WE, Murphy RC, Shu HB, Dai S, Zhang G. Ligand–receptor binding revealed by the TNF family member TALL-1. *Nature* 2003;**423**:49–56.
36. Chan FK, Chun HJ, Zheng L, Siegel RM, Bui KL, Lenardo MJ. A domain in TNF receptors that mediates ligand-independent receptor assembly and signaling. *Science* 2000;**288**:2351–4.
37. Siegel RM, Frederiksen JK, Zacharias DA, Chan FK, Johnson M, Lynch D, Tsien RY, Lenardo MJ. Fas preassociation required for apoptosis signaling and dominant inhibition by pathogenic mutations. *Science* 2000;**288**:2354–7.
38. Rothe M, Wong SC, Henzel WJ, Goeddel DV. A novel family of putative signal transducers associated with the cytoplasmic domain of the 75-kDa tumor necrosis factor receptor. *Cell* 1994;**78**:681–92.
39. Arch RH, Gedrich RW, Thompson CB. Tumor necrosis factor receptor-associated factors (TRAFs) – a family of adapter proteins that regulates life and death. *Genes Dev* 1998;**12**:2821–30.
40. Chung JY, Park YC, Ye H, Wu H. All TRAFs are not created equal: common and distinct molecular mechanisms of TRAF-mediated signal transduction. *J Cell Sci* 2002;**115**:679–88.
41. Wu H. Assembly of post-receptor signaling complexes for the tumor necrosis factor receptor superfamily. *Adv Protein Chem* 2004;**68**:225–79.
42. Chung JY, Lu M, Yin Q, Wu H. Structural revelations of TRAF2 function in TNF receptor signaling pathway. *Adv Exp Med Biol* 2007;**597**:93–113.
43. Ghosh S, Karin M. Missing pieces in the NF-κB puzzle. *Cell* 2002;**109**(Suppl.):S81–96.
44. Shaulian E, Karin M. AP-1 as a regulator of cell life and death. *Nature Cell Biol* 2002;**4**:E131–6.
45. Nagata S. Apoptosis by death factor. *Cell* 1997;**88**:355–65.
46. Chinnaiyan AM, O'Rourke K, Tewari M, Dixit VM. FADD, a novel death domain-containing protein, interacts with the death domain of Fas and initiates apoptosis. *Cell* 1995;**81**:505–12.
47. Boldin MP, Goncharov TM, Goltsev YV, Wallach D. Involvement of MACH, a novel MORT1/FADD-interacting protease, in Fas/APO-1- and TNF receptor-induced cell death. *Cell* 1996;**85**:803–15.
48. Muzio M, Chinnaiyan AM, Kischkel FC, O'Rourke K, Shevchenko A, Ni J, Scaffidi C, Bretz JD, Zhang M, Gentz R, Mann M, Krammer PH, Peter ME, Dixit VM. FLICE, a novel FADD-homologous ICE/CED-3-like protease, is recruited to the CD95 (Fas/APO-1) death-inducing signaling complex. *Cell* 1996;**85**:817–27.
49. Wang J, Chun HJ, Wong W, Spencer DM, Lenardo MJ. Caspase-10 is an initiator caspase in death receptor signaling. *Proc Natl Acad Sci USA* 2001;**98**:13,884–8.
50. LeBlanc HN, Ashkenazi A. Apo2L/TRAIL and its death and decoy receptors. *Cell Death Diff* 2003;**10**:66–75.
51. Hsu H, Shu H-B, Pan M-G, Goeddel DV. TRADD–TRAF2 and TRADD–FADD interactions define two distinct TNF receptor 1 signal transduction pathways. *Cell* 1996;**84**:299–308.
52. Yeh WC, Shahinian A, Speiser D, Kraunus J, Billia F, Wakeham A, de la Pompa JL, Ferrick D, Hum B, Iscove N, Ohashi P, Rothe M, Goeddel DV, Mak TW. Early lethality, functional NF-κB activation, and increased sensitivity to TNF-induced cell death in TRAF2-deficient mice. *Immunity* 1997;**7**:715–25.
53. Kelliher MA, Grimm S, Ishida Y, Kuo F, Stanger BZ, Leder P. The death-domain kinase RIP mediates the TNF-induced NF-kB signal. *Immunity* 1998;**8**:297–303.
54. Micheau O, Tschopp J. Induction of TNF receptor I-mediated apoptosis via two sequential signaling complexes. *Cell* 2003;**114**:181–90.
55. Chung JY, Lu M, Yin Q, Lin SC, Wu H. Molecular basis for the unique specificity of TRAF6. *Adv Exp Med Biol* 2007;**597**:122–30.
56. Park HH, Lu M, Yin Q, Lin SC, Wu H. The death domain superfamily in intracellular signaling of apoptosis and inflammation. *Annu Rev Immunology* 2007;**25**:561–86.
57. Park HH, Logette E, Rauser S, Cuenin S, Walz T, Tschopp J, Wu H. Death domain assembly mechanism revealed by crystal structure of the oligomeric PIDDosome core complex. *Cell* 2007;**128**:533–46.
58. Kepler TB, Chan C. Spatiotemporal programming of a simple inflammatory process. *Immunol Rev* 2007;**216**:153–63.

59. Stojanov S, McDermott MF. The tumour necrosis factor receptor-associated periodic syndrome: current concepts. *Expert Rev Mol Med* 2005;**7**:1–18.
60. Fu YX, Chaplin DD. Development and maturation of secondary lymphoid tissues. *Annu Rev Immunol* 1999;**17**:399–433.
61. Dougall WC, Logette E, Rauser S, Cuenin S, Walz T, Tschopp J, Wu H. RANK is essential for osteoclast and lymph node development. *Genes Dev* 1999;**13**:2412–24.
62. Kong YY, Yoshida H, Sarosi I, Tan HL, Timms E, Capparelli C, Morony S, Oliveira-dos-Santos AJ, Van G, Itie A, Khoo W, Wakeham A, Dunstan CR, Lacey DL, Mak TW, Boyle WJ, Penninger JM. OPGL is a key regulator of osteoclastogenesis, lymphocyte development and lymph-node organogenesis. *Nature* 1999;**397**:315–23.
63. Kim D, Mebius RE, MacMicking JD, Jung S, Cupedo T, Castellanos Y, Rho J, Wong BR, Josien R, Kim N, Rennert PD, Choi Y. Regulation of peripheral lymph node genesis by the tumor necrosis factor family member TRANCE. *J Exp Med* 2000;**192**:1467–78.
64. Fata JE, Kong YY, Li J, Sasaki T, Irie-Sasaki J, Moorehead RA, Elliott R, Scully S, Voura EB, Lacey DL, Boyle WJ, Khokha R, Penninger JM. The osteoclast differentiation factor osteoprotegerin-ligand is essential for mammary gland development. *Cell* 2000;**103**:41–50.
65. Headon DJ, Overbeek PA. Involvement of a novel TNF receptor homologue in hair follicle induction. *Nature Genet* 1999;**22**:370–4.
66. Monreal AW, Ferguson BM, Headon DJ, Street SL, Overbeek PA, Zonana J. Mutations in the human homologue of mouse dl cause autosomal recessive and dominant hypohidrotic ectodermal dysplasia. *Nature Genet* 1999;**22**:366–9.
67. Botchkareva NV, Botchkarev VA, Chen LH, Lindner G, Paus R. A role for p75 neurotrophin receptor in the control of hair follicle morphogenesis. *Dev Biol* 1999;**216**:135–53.
68. Lee KF, Li E, Huber LJ, Landis SC, Sharpe AH, Chao MV, Jaenisch R. Targeted mutation of the gene encoding the low affinity NGF receptor p75 leads to deficits in the peripheral sensory nervous system. *Cell* 1992;**69**:737–49.
69. Bibel M, Barde YA. Neurotrophins: key regulators of cell fate and cell shape in the vertebrate nervous system. *Genes Dev* 2000;**14**:2919–37.
70. Martinon F, Tschopp J. NLRs join TLRs as innate sensors of pathogens. *Trends Immunol* 2005;**26**:447–54.
71. Kawai T, Akira S. TLR signaling. *Cell Death Diff* 2006;**13**:816–25.
72. Seth RB, Sun L, Chen ZJ. Antiviral innate immunity pathways. *Cell Res* 2006;**16**:141–7.
73. Yeh WC, Hakem R, Woo M, Mak TW. Gene targeting in the analysis of mammalian apoptosis and TNF receptor superfamily signaling. *Immunol Rev* 1999;**169**:283–302.
74. Grewal IS, Flavell RA. CD40 and CD154 in cell-mediated immunity. *Annu Rev Immunol* 1998;**16**:111–35.
75. Cozine CL, Wolniak KL, Waldschmidt TJ. The primary germinal center response in mice. *Curr Opin Immunol* 2005;**17**:298–302.
76. So T, Lee SW, Croft M. Tumor necrosis factor/tumor necrosis factor receptor family members that positively regulate immunity. *Intl J Hematol* 2006;**83**:1–11.
77. Scheu S, Alferink J, Potzel T, Barchet W, Kalinke U, Pfeffer K. Targeted disruption of LIGHT causes defects in costimulatory T cell activation and reveals cooperation with lymphotoxin beta in mesenteric lymph node genesis. *J Exp Med* 2002;**195**:1613–24.
78. Nagata S, Golstein P. The Fas death factor. *Science* 1995;**267**:1449–56.
79. Brenner D, Krammer PH, Arnold R. Concepts of activated T cell death. *Crit Rev Oncol Hematol* 2008;**66**:52–64.
80. Li M, Beg AA. Induction of necrotic-like cell death by tumor necrosis factor alpha and caspase inhibitors: novel mechanism for killing virus-infected cells. *J Virol* 2000;**74**:7470–7.
81. Lam J, Takeshita S, Barker JE, Kanagawa O, Ross FP, Teitelbaum SL. TNF-α induces osteoclastogenesis by direct stimulation of macrophages exposed to permissive levels of RANK ligand. *J Clin Invest* 2000;**106**:1481–8.
82. Li J, Sarosi I, Yan XQ, Morony S, Capparelli C, Tan HL, McCabe S, Elliott R, Scully S, Van G, Kaufman S, Juan SC, Sun Y, Tarpley J, Martin L, Christensen K, McCabe J, Kostenuik P, Hsu H, Fletcher F, Dunstan CR, Lacey DL, Boyle WJ. RANK is the intrinsic hematopoietic cell surface receptor that controls osteoclastogenesis and regulation of bone mass and calcium metabolism. *Proc Natl Acad Sci USA* 2000;**97**:1566–71.
83. Wu H, Arron JR. TRAF6, a molecular bridge spanning adaptive immunity, innate immunity and osteoimmunology. *Bioessays* 2003;**25**:1096–105.
84. Hofbauer LC, Khosla S, Dunstan CR, Lacey DL, Spelsberg TC, Riggs BL. Estrogen stimulates gene expression and protein production of osteoprotegerin in human osteoblastic cells. *Endocrinology* 1999;**140**:4367–70.
85. Cenci S, Weitzmann MN, Roggia C, Namba N, Novack D, Woodring J, Pacifici R. Estrogen deficiency induces bone loss by enhancing T-cell production of TNF-α. *J Clin Invest* 2000;**106**:1229–37.
86. Kim MS, Lee SH, Song MY, Yoo TH, Lee BK, Kim YS. Comparative analyses of complex formation and binding sites between human tumor necrosis factor-alpha and its three antagonists elucidate their different neutralizing mechanisms. *J Mol Biol* 2007;**374**:1374–88.
87. Fischer U, Schulze-Osthoff K. Apoptosis-based therapies and drug targets. *Cell Death Diff* 2005;**12**(Suppl. 1):942–61.
88. Ashkenazi A. Targeting the extrinsic apoptosis pathway in cancer. *Cytokine Growth Factor Rev* 2008;**19**:325–31.
89. Carter PH, Scherle PA, Muckelbauer JK, Voss ME, Liu RQ, Thompson LA, Tebben AJ, Solomon KA, Lo YC, Li Z, Strzemienski P, Yang G, Falahatpisheh N, Xu M, Wu Z, Farrow NA, Ramnarayan K, Wang J, Rideout D, Yalamoori V, Domaille P, Underwood DJ, Trzaskos JM, Friedman SM, Newton RC, Decicco CP. Photochemically enhanced binding of small molecules to the tumor necrosis factor receptor-1 inhibits the binding of TNF-α. *Proc Natl Acad Sci USA* 2001;**98**:11,879–84.
90. Fournel S, Wieckowski S, Sun W, Trouche N, Dumortier H, Bianco A, Chaloin O, Habib M, Peter JC, Schneider P, Vray B, Toes RE, Offringa R, Melief CJ, Hoebeke J, Guichard G. C3-symmetric peptide scaffolds are functional mimetics of trimeric CD40L. *Nature Chem Biol* 2005;**1**:377–82.
91. He MM, Smith AS, Oslob JD, Flanagan WM, Braisted AC, Whitty A, Cancilla MT, Wang J, Lugovskoy AA, Yoburn JC, Fung AD, Farrington G, Eldredge JK, Day ES, Cruz LA, Cachero TG, Miller SK, Friedman JE, Choong IC, Cunningham BC. Small-molecule inhibition of TNF-α. *Science* 2005;**310**:1022–5.
92. Murali R, Smith AS, Oslob JD, Flanagan WM, Braisted AC, Whitty A, Cancilla MT, Wang J, Lugovskoy AA, Yoburn JC, Fung AD, Farrington G, Eldredge JK, Day ES, Cruz LA, Cachero TG, Miller SK, Friedman JE, Choong IC, Cunningham BC. Disabling TNF receptor signaling by induced conformational perturbation of tryptophan–107. *Proc Natl Acad Sci USA* 2005;**102**:10,970–5.
93. Trouche N, Wieckowski S, Sun W, Chaloin O, Hoebeke J, Fournel S, Guichard G. Small multivalent architectures mimicking homotrimers of the TNF superfamily member CD40L: delineating the relationship between structure and effector function. *J Am Chem Soc* 2007;**129**:13,480–92.
94. Ye H, Arron JR, Lamothe B, Cirilli M, Kobayashi T, Shevde NK, Segal D, Dzivenu OK, Vologodskaia M, Yim M, Du K, Singh S, Pike JW, Darnay BG, Choi Y, Wu H. Distinct molecular mechanism for initiating TRAF6 signalling. *Nature* 2002;**418**:443–7.

95. Wu H, Tschopp J, Lin SC. Smac mimetics and TNF-α: a dangerous liaison? *Cell* 2007;**131**:655–8.
96. Petersen SL, Wang L, Yalcin-Chin A, Li L, Peyton M, Minna J, Harran P, Wang X. Autocrine TNFα signaling renders human cancer cells susceptible to Smac-mimetic-induced apoptosis. *Cancer Cell* 2007;**12**:445–56.
97. Vince JE, Wong WW, Khan N, Feltham R, Chau D, Ahmed AU, Benetatos CA, Chunduru SK, Condon SM, McKinlay M, Brink R, Leverkus M, Tergaonkar V, Schneider P, Callus BA, Koentgen F, Vaux DL, Silke J. IAP antagonists target cIAP1 to induce TNFα-dependent apoptosis. *Cell* 2007;**131**:682–93.
98. Varfolomeev E, Blankenship JW, Wayson SM, Fedorova AV, Kayagaki N, Garg P, Zobel K, Dynek JN, Elliott LO, Wallwebe HJ. IAP antagonists induce autoubiquitination of c-IAPs, NF-κB activation, and TNFα-dependent apoptosis. *Cell* 2007;**131**:669–81.
99. Jones EY, Stuart DI, Walker NP. Structure of tumour necrosis factor. *Nature* 1989;**338**:225–8.
100. Eck MJ, Sprang SR. The structure of tumor necrosis factor-alpha at 2.6 Å resolution. Implications for receptor binding. *J Biol Chem* 1989;**264**:17,595–17,605,.
101. Cha SS, Kim JS, Cho HS, Shin NK, Jeong W, Shin HC, Kim YJ, Hahn JH, Oh BH. High resolution crystal structure of a human tumor necrosis factor-alpha mutant with low systemic toxicity. *J Biol Chem* 1998;**273**:2153–60.
102. Reed C, Fu ZQ, Wu J, Xue YN, Harrison RW, Chen MJ, Weber IT. Crystal structure of TNF-α mutant R31D with greater affinity for receptor R1 compared with R2. *Protein Eng* 1997;**10**:1101–7.
103. Karpusas M, Hsu YM, Wang JH, Thompson J, Lederman S, Chess L, Thomas D. 2 A crystal structure of an extracellular fragment of human CD40 ligand. *Structure* 1995;**3**:1031–9.
104. Karpusas M, Lucci J, Ferrant J, Benjamin C, Taylor FR, Strauch K, Garber E, Hsu YM. Structure of CD40 ligand in complex with the Fab fragment of a neutralizing humanized antibody. *Structure* 2001;**9**:321–9.
105. Cha SS, Kim MS, Choi YH, Sung BJ, Shin NK, Shin HC, Sung YC, Oh BH. 2.8 A resolution crystal structure of human TRAIL, a cytokine with selective antitumor activity. *Immunity* 1999;**11**:253–61.
106. Hymowitz SG, O'Connell MP, Ultsch MH, Hurst A, Totpal K, Ashkenazi A, de Vos AM, Kelley RF. A unique zinc-binding site revealed by a high-resolution X-ray structure of homotrimeric Apo2L/TRAIL. *Biochemistry* 2000;**39**:633–40.
107. Lam J, Nelson CA, Ross FP, Teitelbaum SL, Fremont DH. Crystal structure of the TRANCE/RANKL cytokine reveals determinants of receptor-ligand specificity. *J Clin Invest* 2001;**108**:971–9.
108. Ito S, Wakabayashi K, Ubukata O, Hayashi S, Okada F, Hata T. Crystal structure of the extracellular domain of mouse RANK ligand at 2.2-A resolution. *J Biol Chem* 2002;**277**:6631–6.
109. Jin T, Wakabayashi K, Ubukata O, Hayashi S, Okada F, Hata T. X-ray crystal structure of TNF ligand family member TL1A at 2.1 Å. *Biochem Biophys Res Commun* 2007;**364**:1–6.
110. Zhou Z, Wakabayashi K, Ubukata O, Hayashi S, Okada F, Hata T. Human glucocorticoid-induced TNF receptor ligand regulates its signaling activity through multiple oligomerization states. *Proc Natl Acad Sci USA* 2008;**105**:5465–70.
111. Naismith JH, Devine TQ, Brandhuber BJ, Sprang SR. Crystallographic evidence for dimerization of unliganded tumor necrosis factor receptor. *J Biol Chem* 1995;**270**:13,303–7.
112. Naismith JH, Devine TQ, Kohno T, Sprang SR. Structures of the extracellular domain of the type I tumor necrosis factor receptor. *Structure* 1996;**4**:1251–62.
113. Fellouse FA, Li B, Compaan DM, Peden AA, Hymowitz SG, Sidhu SS. Molecular recognition by a binary code. *J Mol Biol* 2005;**348**:1153–62.
114. Li B, Russell SJ, Compaan DM, Totpal K, Marsters SA, Ashkenazi A, Cochran AG, Hymowitz SG, Sidhu SS. Activation of the proapoptotic death receptor DR5 by oligomeric peptide and antibody agonists. *J Mol Biol* 2006;**361**:522–36.
115. Carfi A, Willis SH, Whitbeck JC, Krummenacher C, Cohen GH, Eisenberg RJ, Wiley DC. Herpes simplex virus glycoprotein D bound to the human receptor HveA. *Mol Cell* 2001;**8**:169–79.
116. Compaan DM, Gonzalez LC, Tom I, Loyet KM, Eaton D, Hymowitz SG. Attenuating lymphocyte activity: the crystal structure of the BTLA–HVEM complex. *J Biol Chem* 2005;**280**:39,553–61.
117. He XL, Garcia KC. Structure of nerve growth factor complexed with the shared neurotrophin receptor p75. *Science* 2004;**304**:870–5.
118. Gong Y, Cao P, Yu HJ, Jiang T. Crystal structure of the neurotrophin-3 and p75NTR symmetrical complex. *Nature* 2008;**454**:789–93.

Chapter 39

Tumor Necrosis Factor Receptor-Associated Factors in Immune Receptor Signal Transduction

Qian Yin, Su-Chang Lin, Yu-Chih Lo, Steven M. Damo, and Hao Wu
Department of Biochemistry, Weill Medical College of Cornell University, New York, New York

INTRODUCTION

It is crucial for cells to detect extracellular and intracellular stresses and act correspondingly. Sophisticated receptor families have been developed to sense both extracellular and intracellular stress signals, such as pathogen-associated molecular patterns (PAMPs). While Toll-like receptors (TLRs) sense extracellular PAMPs, the NOD-like receptors (NLRs) and the more recently identified double-strand RNA (dsRNA) receptors sense intracellular PAMPs, eliciting signal transduction for immune and inflammatory responses as part of the innate immunity against pathogens [1]. Cytokine secretion and induction of specific adaptive immunity ensue to complete the complex network of host defense. Tumor necrosis factor (TNF) receptor-associated factors (TRAFs) are signaling proteins that participate in many aspects of the signal transduction of these immune receptors, including receptors for TNF and related cytokines, the IL-1 receptor, TLRs, NLRs, dsRNA receptors, the T cell receptors (TCR), and the B cell receptors (BCR) [1, 2]. TRAF activation eventually leads to activation of transcription factors such as NFκB and AP-1 (Figure 39.1).

DISCOVERY OF TRAF PROTEINS

The first two and founding members of the TRAF family, TRAF1 and TRAF2, were discovered in 1994 via yeast two-hybrid using the cytoplasmic region of TNF-R2 [3] as the bait. From then on, five more human TRAF family members have been identified and characterized and shown to participate in signal transduction beyond the TNF receptor superfamily (Table 39.1). Unlike many cell surface receptors, TNF receptors (TNFRs) do not possess any catalytic activity in their cytoplasmic tails, and rely on TRAFs and other intracellular proteins for signal transduction. TRAFs mainly reside in cytosolic fractions, which are consistent with their functions as receptor-proximal signaling proteins. However, they can be found in membrane fractions when overexpressed. TRAF4 has been found in nucleus, but its nuclear function remains unknown.

BIOLOGICAL FUNCTIONS OF TRAF PROTEINS

Mouse genetic studies have revealed that TRAF proteins regulate innate and adaptive immunity, embryonic development, stress responses, and bone metabolism through the modulation of cell survival, proliferation, differentiation, and cell death [4, 5]. TRAF homologs have been found in lower eukaryotes, including *Danio rerio* [6], *Drosophila melanogaster* [7–9], *C. elegans* [10], and *Dictyostelium discoideum* [11]. In drosophila, TRAFs are essential for dorsoventral polarization and innate host defense [12, 13].

The biological functions of TRAF proteins appear to be mediated through activation of transcription factors such as NFκB and AP-1. In fact, almost all NFκB activation is mediated by TRAFs except that involved in DNA damage responses [14]. The interceding steps between TRAF activation and transcription factor activation involve kinase cascades such as the IκB kinases (IKKs) and mitogen-activated protein kinases (MAPKs), including c-Jun N-terminal kinase (JNK) and p38. TRAF proteins can play either a positive or a negative role in NFκB activation.

TRAF proteins can interact with each other and with numerous receptors, kinases, and signaling proteins from the immune pathways they participate in [10]. Therefore,

Handbook of Cell Signaling, Three-Volume Set 2 ed.

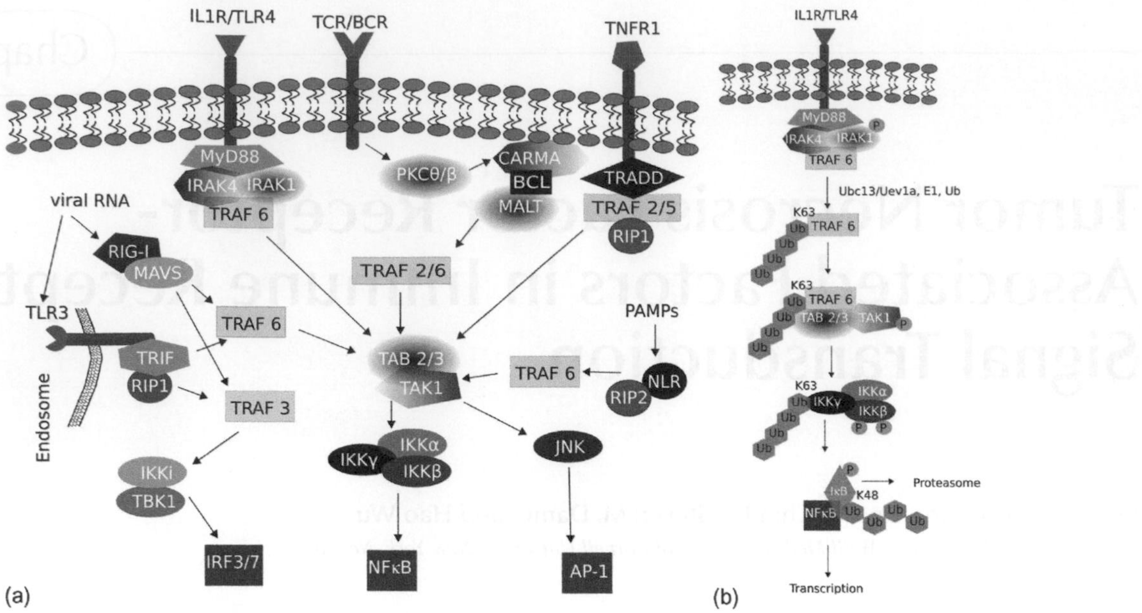

FIGURE 39.1 TRAF-mediated signaling pathways.
(a) An overall view of the pathways that TRAFs engage in. (b) TRAFs, in particular TRAF6, mediate Lys63-linked polyubiquitination and NF-κB signaling.

TABLE 39.1 Discovery of human TRAF proteins

	Identification methods	Reference
TRAF1	Yeast two-hybrid (TNF-R2)	3
TRAF2	Yeast two-hybrid (TNF-R2)	3
TRAF3	Yeast two-hybrid (CD40)	11, 56, 57
TRAF4	Screening of a cDNA library from breast cancer-derived metastatic lymph nodes	58
TRAF5	Yeast two-hybrid (CD40)	59
	Degenerate oligonucleotide PCR (conserved residues in the TRAF domain of TRAF1, 2 and 3)	60
TRAF6	EST database search (TRAF-C domain of TRAF2)	29
	Yeast two-hybrid (CD40)	61
TRAF7	EST database search	62
	(TRAF-like RING)	63
	TAP/LC-MS/MS	

In parentheses are the baits used in yeast two-hybrid screens, templates used in expressed sequence tag (EST) database search or primers used for degenerate PCR.

tight regulation of the myriad of interactions is required for precise and specific signaling. Which TRAF protein is activated and which downstream pathway is elicited upon its activation are dependent on both the nature of the stimuli and the cell types [15].

DOMAIN ORGANIZATIONS AND STRUCTURES OF TRAFs

The N-terminal part of TRAFs 2–7 consists of one really interesting new gene (RING) type zinc binding domain followed by several CCHC zinc fingers (Figure 39.2). TRAF1 lacks the most N-terminal RING domain, but retains one predicted CCHC zinc finger. With the exception of TRAF7, all TRAF proteins share the characteristic TRAF domain, or merpin and TRAF homology (MATH) domain at their C-termini, which mediates interaction with receptors and adaptor proteins (Figure 39.2). Instead of a TRAF domain, TRAF7 has seven WD40 repeats that are also well known to mediate protein–protein interactions.

The N-Terminal RING Domain and Zinc Fingers

The RING domain is a special type of zinc binding domain. The consensus RING sequence is $CX_2CX_{(9-39)}CX_{(1-3)}HX_{(2-3)}$ $C/HX_2CX_{(4-48)}CX_2C$ with eight cysteines and histidines in a "cross-brace" topology to coordinate two zinc ions [16]. RING domains are present in many E3 ubiquitin–protein ligases to mediate the interaction between E2 ubiquitin conjugating enzymes and their substrates. Zinc fingers such as the CCHC type in TRAFs, on the other hand, were first identified in transcription factors [17]. They wind themselves on one zinc ion, and often occur in tandem. Today, the zinc finger family has expanded rapidly to mediate protein–DNA, protein–RNA, protein–protein and protein–lipid interactions [18–20]. They have even been

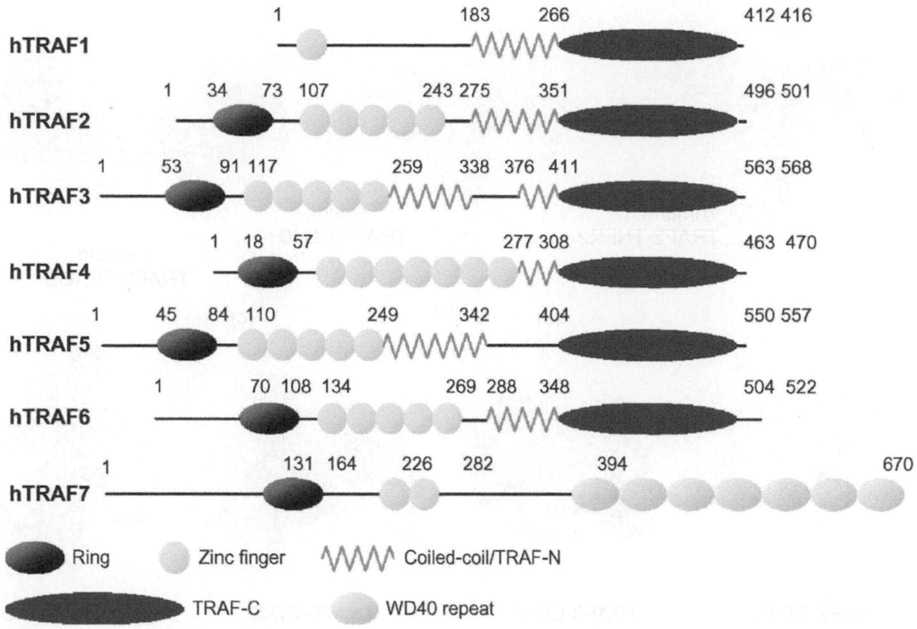

FIGURE 39.2 Domain Organizations of human TRAF1–7.

exploited by protein engineering to create artificial zinc fingers with novel binding specificity [21].

The C-Terminal TRAF Domain

The C-terminal TRAF domain can be further divided into TRAF-N and TRAF-C domains. TRAF-N domain is a coiled-coil region that mediates homo- and hetero-oligomerization among TRAF family members. TRAF-C domain is responsible for physical association with upstream receptors and adaptor proteins. Crystal structures of TRAF domains from different TRAF proteins have shown that they form mushroom-shaped trimers with the TRAF-C domain as the "head" of the mushroom and the bundled coiled-coils as the "stalk" [22–24] (Figure 39.3). Both TRAF-N and TRAF-C domains contribute to the trimer interface; however, without the coiled-coil TRAF-N domain, the TRAF-C domain alone exists as a monomer in solution [25]. Each TRAF-C domain adopts a unique β-sandwich topology composed of eight anti-parallel β strands, which is similar to the conformation observed in merpin metalloproteases and the drosophila Siah protein.

X-ray crystallographic studies have also revealed the interaction details of TRAFs with peptides derived from receptors [22–24] and with the adaptor protein TRADD (TNF receptor associated protein with a death domain) [26] (Figure 39.3). One peptide molecule binds exclusively to one TRAF protomer. The receptor-derived peptides bind to a shallow surface groove on the TRAF-C domains, but the binding details differ in different TRAF-peptide pairs. In particular, TRAF6 has a peptide binding specificity that is distinct from TRAF2 and TRAF3. TRAF trimerization enhances the otherwise weak interactions between TRAF proteins and the peptides via avidity [27]. TRAF2–TRADD interaction is quite distinctive from TRAF–peptide interactions [26]. It is much stronger and can effectively compete with TRAF2–receptor complexes, which may ensure recruitment of anti-apoptotic cIAPs to the TNFR1 signaling complex to suppress apoptosis.

THE UNIQUE TRAF6

Evolutionarily, TRAF6 is probably the most ancient mammalian TRAF and most resembles the primordial TRAF protein, while the divergence of TRAFs 1, 2, 3 and 5 from TRAF4 and among themselves is the result of recent gene duplications [28]. Consistently, the TRAF-C domain homology suggests that TRAF6 is most distant from other TRAFs [29]. Along with TRAF2, 3, and 5, TRAF6 physically associates with and mediates signaling from TNFR superfamily members, including CD40, RANK, and TACI [10]. However, TRAF6 distinguishes itself by binding to a different motif on TNF receptors [30].

Moreover, TRAF6 is the major transducer of IL-1 receptor/TLR signaling. TRAF6 was first identified as a mediator of IL-1 signaling [29]. Ligation of IL-1 to IL-1R first recruits myeloid differentiation primary response gene 88 (MyD88) via the interaction between the Toll/interleukin receptor domains (TIRs) present in both IL-1R cytoplasmic tail and MyD88 (Figure 39.1). MyD88 then recruits IL-1 receptor-associated kinase 1 (IRAK1) and its homolog IRAK4 via death domain (DD)–DD interactions.

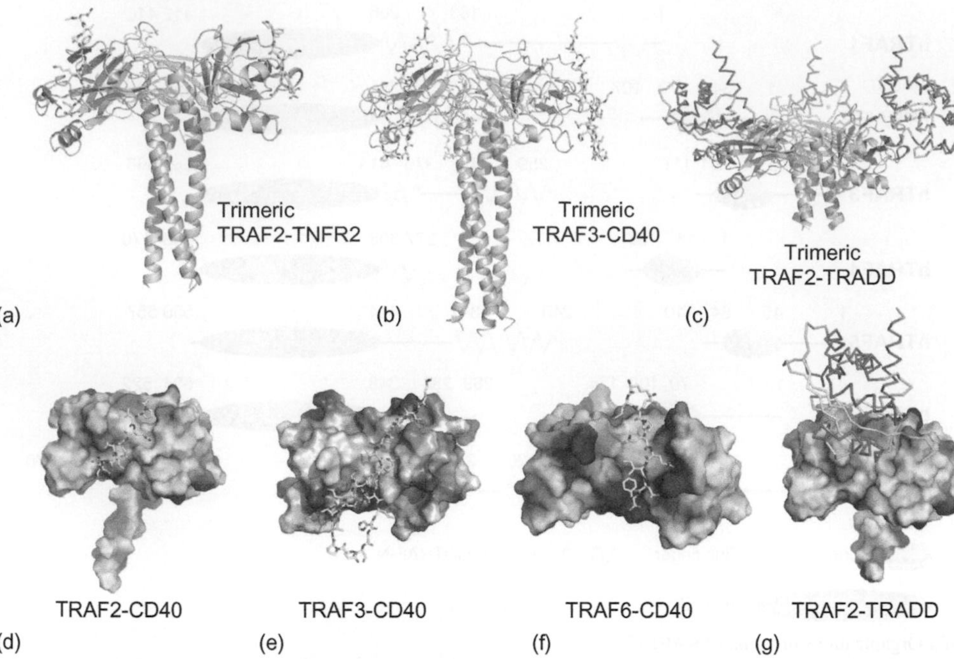

FIGURE 39.3 Structural studies of the TRAF domains and their complexes.
(a) Trimeric TRAF domain structure of TRAF2 (ribbon) in complex with the TNFR2 peptide (stick model); (b) trimeric TRAF domain structure of TRAF3 (ribbon) in complex with the CD40 peptide (stick model); (c) trimeric TRAF domain structure of TRAF2 (ribbon) in complex with the N-terminal domain of TRADD (Cα trace); (d) surface electrostatic diagram of TRAF2 shown in complex with the stick model of the TRAF2/3/5 binding peptide of CD40; (e) surface electrostatic diagram of TRAF3 shown in complex with the stick model of the TRAF2/3/5-binding peptide of CD40; (f) surface electrostatic diagram of TRAF6 shown in complex with the stick model of the TRAF6 binding peptide of CD40; (g) surface electrostatic diagram of TRAF2 shown in complex with the Cα trace of the N-terminal domain of TRADD.

The kinase activity of IRAK4 is required for IL-1 signaling, while that of IRAK1 is not required. When recruited to the receptor proximal complex, IRAK4 phosphorylates IRAK1 and TRAF6 is recruited to this complex via IRAK1 [24]. Recent studies have shown that while TRAF6 mediates NFκB and MAP kinase activation by these receptors, TRAF3 may be involved in interferon production upon TLR activation [31, 32]. For T cell receptor and B cell receptor mediated NFκB activation, TRAF6 acts downstream of the CARMA1–Bcl10–MALT1 (CBM) complex [33]. In addition, the role of TRAF6 in signaling of the dsRNA sensor RIG-I and NLRs such as Nod1 and Nod2 has been implicated.

TRAF SIGNALING AND LYS63 LINKED POLYUBIQUITINATION

Biochemical fractionation and *in vitro* reconstitution have revealed that MAP kinase and NFκB activation by TRAF6 requires a novel form of polyubiquitination [34, 35]. Ubiquitination is one of the most prevalent posttranslational modifications that is accomplished in three steps by ATP-dependent attachment of ubiquitin (Ub) via thioester bond to Ub activating enzyme (E1), transfer of Ub from E1 to the active site Cys of Ub conjugating enzyme (E2), and transfer of Ub from E2 active site to Lys residues of substrates with the aid of a Ub ligase (E3). TRAF6 is a RING-type E3 that ubiquitinates via the Lys63 linkage instead of the Lys48 linkage for proteasomal degradation.

On activation by the relevant signaling pathways upon ligand stimulation, TRAF6 promotes Lys63-linked polyubiquitination of itself and downstream signaling proteins, a process that requires a heterodimeric E2 of Ubc13 and the ubiquitin E2 variant (Uev) known as Uev1A [36]. The Lys63-linked polyubiquitin chains function as anchors to recruit the transforming growth factor (TGF)-β-activated kinase 1 (TAK1) complex consisting of TAK1, the TAK1 binding protein 1 (TAB1), and TAB2 or TAB3. Both TAB2 and TAB3 contain ubiquitin binding motifs, which mediate the interaction between the TAK1 complex and ubiquitinated TRAF6 [37]. Activated TAK1 then phosphorylates and activates IκB kinase β (IKKβ), a catalytic subunit of the IKK complex. Phosphorylation of IκB by the IKK complex leads to its ubiquitination via the Lys48 linkage, and its destruction by the 26S proteasome. Rid of its inhibitor, the previously cytosol-trapped NFκB is free to translocate to the nucleus, induce transcriptional activation of various genes, and launch a battery of immune responses to encounter the stress signals. NFκB essential modulator (NEMO) or IKKγ, the regulatory subunit of the IKK complex, is also a substrate of TRAF6 ubiquitin ligase activity. TAK1 also directly phosphorylates MAP kinases, leading to activation of AP-1 transcription factors.

For other TRAF family members, Lys63-linked polyubiquitination also appears to be the common mechanism of their downstream action. It has been shown that TNF activated endogenous TAK1, and the kinase-negative TAK1 acted as a dominant negative inhibitor against TNF induced NFκB activation [38]. More convincingly, siRNA-mediated knockdown showed that TAK1 is critical for TNF-induced activation of the NFκB pathway [39]. It was shown that TNF-induced IKK activation is mediated through the TRAF2, TRAF5, and TAK1 signaling pathway [40]. Similarly, the role of the TAB2-related protein TAB3 in TNFR signaling has been uncovered (41–43). The role of Lys63 ubiquitination in TRAF2 and TRAF5 signaling has also been inferred from the negative regulation by the Lys63 linkage specific de-ubiquitinating enzyme CYLD [44–46].

REGULATION OF TRAF SIGNALING

TRAF signaling is tightly regulated by many mechanisms; in particular, oligomerization and ubiquitination. These regulatory mechanisms ensure that TRAFs are activated upon ligand stimulation and turned off at the appropriate times. Oligomerization appears to be the common theme of TRAF6 activation in all known signaling pathways: oligomerization by TNFRs, by TIR signaling complexes, and by the CBM complex. The C-terminal TRAF domain mediates its trimerization. Forced dimerization by fused dimerizing domain also activates TRAF6 [47]. A cytosolic TRAF-interacting protein known as TIFA, with a forkhead-associated (FHA) domain, can enhance TRAF6 oligomerization and activation [48].

Several de-ubiquitinating and ubiquitinating enzymes, such as A20 and CYLD, have been shown to provide feedback inhibition of TRAF-mediated NFκB activation. A20 was originally characterized as an early response gene to TNF stimulation [49], and possesses dual ubiquitin editing functions [50]. While the N-terminal domain of A20 is a de-ubiquitinating enzyme (DUB) for Lys63-linked polyubiquitination of signaling mediators such as TRAF6 and RIP, its C-terminal domain is a ubiquitin ligase (E3) for Lys48-linked degradative polyubiquitination of the same substrates [50–54]. CYLD is Lys63-specific de-ubiquitinating enzyme, whose mutations are the underlying causes of familial cylindromatosis, with predisposition to tumors of skin appendages called cylindromas [44–46, 54, 55].

SUMMARY AND PERSPECTIVES

Since the identification of the first two TRAF family members in 1994, it has become clear that different TRAFs exhibit specific biological functions. The membrane-proximal events for initiating differential TRAF signal transduction have been relatively well established from the wealth of structural and functional studies. The biggest challenge now lies ahead – to further elucidate the molecular mechanisms of the E3 activities of the different TRAF family members. While the E3 activity of TRAF6 has been demonstrated both *in vitro* and in cells, relatively less is known regarding the E3 activity of other TRAFs and their specificity. Therefore, this aspect of TRAF signaling will remain an important focus of investigation for a wide range of biological interests.

ACKNOWLEDGEMENT

This work was supported by National Institute of Health (RO1 AI045937 to HW). SCL and YCL are postdoctoral fellows of the Cancer Research Institute.

REFERENCES

1. Meylan E, Tschopp J, Karin M. Intracellular pattern recognition receptors in the host response. *Nature* 2006;**442**:39–44.
2. Schulze-Luehrmann J, Ghosh S. Antigen-receptor signaling to nuclear factor κB. *Immunity* 2006;**25**:701–15.
3. Rothe M, Wong SC, Henzel WJ, Goeddel DV. A novel family of putative signal transducers associated with the cytoplasmic domain of the 75-kDa tumor necrosis factor receptor. *Cell* 1994;**78**:681–92.
4. Bradley JR, Pober JS. Tumor necrosis factor receptor-associated factors (TRAFs). *Oncogene* 2001;**20**:6482–91.
5. Chung JY, Park YC, Ye H, Wu H. All TRAFs are not created equal: common and distinct molecular mechanisms of TRAF-mediated signal transduction. *J Cell Sci* 2002;**115**:679–88.
6. Phelan PE, Mellon MT, Kim CH. Functional characterization of full-length TLR3, IRAK-4, and TRAF6 in zebrafish (Danio rerio). *Mol Immunol* 2005;**42**:1057–71.
7. Medzhitov R, Janeway C. Jr. Innate immune recognition: mechanisms and pathways. *Immunol Rev* 2000;**173**:89–97.
8. Liu H, Su YC, Becker E, Treisman J, Skolnik EY. A Drosophila TNF-receptor-associated factor (TRAF) binds the ste20 kinase Misshapen and activates Jun kinase. *Curr Biol* 1999;**9**:101–4.
9. Zapata JM, Matsuzawa S, Godzik A, Leo E, Wasserman SA, Reed JC. The Drosophila tumor necrosis factor receptor-associated factor-1 (DTRAF1) interacts with Pelle and regulates NFκB activity. *J Biol Chem* 2000;**275**:12,102–107.
10. Wajant H, Henkler F, Scheurich P. The TNF-receptor-associated factor family: scaffold molecules for cytokine receptors, kinases and their regulators. *Cell Signal* 2001;**13**:389–400.
11. Regnier CH, Tomasetto C, Moog-Lutz C, et al. Presence of a new conserved domain in CART1, a novel member of the tumor necrosis factor receptor-associated protein family, which is expressed in breast carcinoma. *J Biol Chem* 1995;**270**:25,715–21.
12. Preiss A, Johannes B, Nagel AC, Maier D, Peters N, Wajant H. Dynamic expression of Drosophila TRAF1 during embryogenesis and larval development. *Mech Dev* 2001;**100**:109–13.
13. Imler JL, Hoffmann JA. Toll receptors in innate immunity. *Trends Cell Biol* 2001;**11**:304–11.
14. Hayden MS, Ghosh S. Shared principles in NF-κB signaling. *Cell* 2008;**132**:344–62.
15. Bishop GA. The multifaceted roles of TRAFs in the regulation of B-cell function. *Nature Rev Immunol* 2004;**4**:775–86.

16. Freemont PS. RING for destruction?. *Curr Biol* 2000;**10**:R84–7.
17. Miller J, McLachlan AD, Klug A. Repetitive zinc-binding domains in the protein transcription factor IIIA from Xenopus oocytes. *EMBO J* 1985;**4**:1609–14.
18. Brown RS. Zinc finger proteins: getting a grip on RNA. *Curr Opin Struct Biol* 2005;**15**:94–8.
19. Hall TM. Multiple modes of RNA recognition by zinc finger proteins. *Curr Opin Struct Biol* 2005;**15**:367–73.
20. Gamsjaeger R, Liew CK, Loughlin FE, Crossley M, Mackay JP. Sticky fingers: zinc-fingers as protein-recognition motifs. *Trends Biochem Sci* 2007;**32**:63–70.
21. Giesecke AV, Fang R, Joung JK. Synthetic protein–protein interaction domains created by shuffling Cys2His2 zinc-fingers. *Mol Syst Biol* 2006;**2**:2006–11.
22. Park YC, Burkitt V, Villa AR, Tong L, Wu H. Structural basis for self-association and receptor recognition of human TRAF2. *Nature* 1999;**398**:533–8.
23. Ni CZ, Welsh K, Leo E, et al. Molecular basis for CD40 signaling mediated by TRAF3. *Proc Natl Acad Sci USA* 2000;**97**:10,395–9.
24. Ye H, Arron JR, Lamothe B, et al. Distinct molecular mechanism for initiating TRAF6 signalling. *Nature* 2002;**418**:443–7.
25. Wu H, Park YC, Ye H, Tong L. Structural studies of human TRAF2. *Cold Spring Harb Symp Quant Biol* 1999;**64**:541–9.
26. Park YC, Ye H, Hsia C, et al. A novel mechanism of TRAF signaling revealed by structural and functional analyses of the TRADD–TRAF2 interaction. *Cell* 2000;**101**:777–87.
27. Ye H, Wu H. Thermodynamic characterization of the interaction between TRAF2 and receptor peptides by isothermal titration calorimetry. *Proc Natl Acad Sci* 2000;**97**:8961–6.
28. Grech A, Quinn R, Srinivasan D, Badoux X, Brink R. Complete structural characterisation of the mammalian and Drosophila TRAF genes: implications for TRAF evolution and the role of RING finger splice variants. *Mol Immunol* 2000;**37**:721–34.
29. Cao Z, Xiong J, Takeuchi M, Kurama T, Goeddel DV. TRAF6 is a signal transducer for interleukin-1. *Nature* 1996;**383**:443–6.
30. Pullen SS, Dang TT, Crute JJ, Kehry MR. CD40 signaling through tumor necrosis factor receptor-associated factors (TRAFs). Binding site specificity and activation of downstream pathways by distinct TRAFs. *J Biol Chem* 1999;**274**:14,246–54.
31. Hacker H, Redecke V, Blagoev B, et al. Specificity in Toll-like receptor signalling through distinct effector functions of TRAF3 and TRAF6. *Nature* 2006;**439**:204–7.
32. Oganesyan G, Saha SK, Guo B, et al. Critical role of TRAF3 in the Toll-like receptor-dependent and -independent antiviral response. *Nature* 2006;**439**:208–11.
33. Sun L, Deng L, Ea CK, Xia ZP, Chen ZJ. The TRAF6 ubiquitin ligase and TAK1 kinase mediate IKK activation by BCL10 and MALT1 in T lymphocytes. *Mol Cell* 2004;**14**:289–301.
34. Deng L, Wang C, Spencer E, et al. Activation of the IκB kinase complex by TRAF6 requires a dimeric ubiquitin-conjugating enzyme complex and a unique polyubiquitin chain. *Cell* 2000;**103**: 351–61.
35. Wang C, Deng L, Hong M, Akkaraju GR, Inoue J, Chen ZJ. TAK1 is a ubiquitin-dependent kinase of MKK and IKK. *Nature* 2001;**412**: 346–51.
36. Pickart CM, Eddins MJ. Ubiquitin: structures, functions, mechanisms. *Biochim Biophys Acta* 2004;**1695**:55–72.
37. Kanayama A, Seth RB, Sun L, et al. TAB2 and TAB3 activate the NF-κB pathway through binding to polyubiquitin chains. *Mol Cell* 2004;**15**:535–48.
38. Sakurai H, Miyoshi H, Toriumi W, Sugita T. Functional interactions of transforming growth factor beta-activated kinase 1 with IκB kinases to stimulate NF-κB activation. *J Biol Chem* 1999;**274**:10,641–8.
39. Takaesu G, Surabhi RM, Park KJ, Ninomiya-Tsuji J, Matsumoto K, Gaynor RB. TAK1 is critical for IκB kinase-mediated activation of the NF-κB pathway. *J Mol Biol* 2003;**326**:105–15.
40. Sakurai H, Suzuki S, Kawasaki N, et al. Tumor necrosis factor-alpha-induced IKK phosphorylation of NF-κB p65 on serine 536 is mediated through the TRAF2, TRAF5, and TAK1 signaling pathway. *J Biol Chem* 2003;**278**:36,916–23.
41. Ishitani T, Takaesu G, Ninomiya-Tsuji J, Shibuya H, Gaynor RB, Matsumoto K. Role of the TAB2-related protein TAB3 in IL-1 and TNF signaling. *Embo J* 2003;**22**:6277–88.
42. Cheung PC, Nebreda AR, Cohen P. TAB3, a new binding partner of the protein kinase TAK1. *Biochem J* 2004;**378**:27–34.
43. Jin G, Klika A, Callahan M, et al. Identification of a human NF-κB-activating protein, TAB3. *Proc Natl Acad Sci USA* 2004;**101**:2028–33.
44. Kovalenko A, Chable-Bessia C, Cantarella G, Israel A, Wallach D, Courtois G. The tumour suppressor CYLD negatively regulates NF-κB signalling by de-ubiquitination. *Nature* 2003;**424**:801–5.
45. Brummelkamp TR, Nijman SM, Dirac AM, Bernards R. Loss of the cylindromatosis tumour suppressor inhibits apoptosis by activating NF-κB. *Nature* 2003;**424**:797–801.
46. Trompouki E, Hatzivassiliou E, Tsichritzis T, Farmer H, Ashworth A, Mosialos G. CYLD is a de-ubiquitinating enzyme that negatively regulates NF-κB activation by TNFR family members. *Nature* 2003;**424**:793–6.
47. Wang C, Deng L, Hong M, Akkaraju GR, Inoue J, Chen ZJ. TAK1 is a ubiquitin-dependent kinase of MKK and IKK. *Nature* 2001;**412**:346–51.
48. Ea CK, Sun L, Inoue J, Chen ZJ. TIFA activates IκB kinase (IKK) by promoting oligomerization and ubiquitination of TRAF6. *Proc Natl Acad Sci USA* 2004;**101**:153,18–23.
49. Opipari Jr. AW, Boguski MS, Dixit VM The A20 cDNA induced by tumor necrosis factor alpha encodes a novel type of zinc finger protein. *J Biol Chem* 1990;**265**:14,705–8.
50. Wertz IE, O'Rourke KM, Zhou H, et al. De-ubiquitination and ubiquitin ligase domains of A20 downregulate NF-κB signalling. *Nature* 2004;**430**:694–9.
51. Heyninck K, Beyaert R. The cytokine-inducible zinc finger protein A20 inhibits IL-1-induced NF-κB activation at the level of TRAF6. *FEBS Letts* 1999;**442**:147–50.
52. Boone DL, Turer EE, Lee EG, et al. The ubiquitin-modifying enzyme A20 is required for termination of Toll-like receptor responses. *Nature Immunol* 2004;**5**:1052–60.
53. Lin SC, Chung JY, Lamothe B, et al. Molecular basis for the unique de-ubiquitinating activity of the NF-κB inhibitor A20. *J Mol Biol* 2008;**376**:526–40.
54. Komander D, Barford D. Structure of the A20 OTU domain and mechanistic insights into de-ubiquitination. *Biochem J* 2008;**409**:77–85.
55. Komander D, Lord CJ, Scheel H, et al. The structure of the CYLD USP domain explains its specificity for Lys63-linked polyubiquitin and reveals a B Box module. *Mol Cell* 2008;**29**:451–64.
56. Hu HM, O'Rourke K, Boguski MS, Dixit VM. A novel RING finger protein interacts with the cytoplasmic domain of CD40. *J Biol Chem* 1994;**269**:30,069–72.
57. Cheng G, Cleary AM, Ye ZS, Hong DI, Lederman S, Baltimore D. Involvement of CRAF1, a relative of TRAF, in CD40 signaling. *Science* 1995;**267**:1494–8.

58. Sato T, Irie S, Reed JC. A novel member of the TRAF family of putative signal transducing proteins binds to the cytosolic domain of CD40. *FEBS Letts* 1995;**358**:113–18.
59. Ishida TK, Tojo T, Aoki T, et al. TRAF5, a novel tumor necrosis factor receptor-associated factor family protein, mediates CD40 signaling. *Proc Natl Acad Sci USA* 1996;**93**:9437–42.
60. Nakano H, Oshima H, Chung W, et al. TRAF5, an activator of NF-κB and putative signal transducer for the lymphotoxin-beta receptor. *J Biol Chem* 1996;**271**:14,661–4.
61. Ishida T, Mizushima S, Azuma S, et al. Identification of TRAF6, a novel tumor necrosis factor receptor-associated factor protein that mediates signaling from an amino-terminal domain of the CD40 cytoplasmic region. *J Biol Chem* 1996;**271**:28,745–8.
62. Xu LG, Li LY, Shu HB. TRAF7 potentiates MEKK3-induced AP1 and CHOP activation and induces apoptosis. *J Biol Chem* 2004;**279**:17,278–82.
63. Bouwmeester T, Bauch A, Ruffner H, et al. A physical and functional map of the human TNF-alpha/NF-κ B signal transduction pathway. *Nature Cell Biol* 2004;**6**:97–105.

Part 6

Guanylyl Cyclases

Chapter 40

Guanylyl Cyclases

Lincoln R. Potter

Department of Biochemistry, Molecular Biology and Biophysics, University of Minnesota–Twin Cities, Minneapolis, Minnesota

HISTORICAL PERSPECTIVE

Although cGMP signaling is now recognized as a critical signal transduction system, many years of painstaking research were conducted on this topic in relative obscurity before the importance of cGMP signaling was fully appreciated. Cyclic GMP was first purified and identified in rat urine in 1963 [1]. Six years later, three independent groups described an enzymatic activity that catalyzed the conversion of GTP into cGMP [2–4]. The cGMP synthesizing the activity was initially called guanyl cyclase, but it is more often referred to as guanylyl cyclase in the recent literature. The Nomenclature Committee of the International Union of Biochemistry and Molecular Biology indicates that the accepted name is guanylate cyclase (GTP pyrophosphate-lyase, EC 4.6.1.2) (http://www.chem.qmul.ac.uk/iubmb/enzyme/EC4/6/1/2.html). The reaction catalyzed is: divalent metal bound GTP→cGMP+PPi. Early studies revealed distinct soluble and particular forms of the enzyme with unique kinetic properties and tissue distributions [5, 6]. However, the true diversity of the family was not fully appreciated until the late 1980s and early 1990s, when individual cDNAs encoding the various family members were identified by molecular cloning.

OVERVIEW OF MAMMALIAN GUANYLYL CYCLASES

Humans express four soluble (α_1, α_2, β_1, and β_2) guanylyl cyclase subunits and five *bona fide* single membrane-spanning (GC-A, GC-B, GC-C, GC-E, and GC-F) forms (see Table 40.1 for ligands and "knockout" phenotypes associated with each cyclase). The soluble forms exist as heterodimers and the transmembrane members exist as homeric structures containing at least two molecules per complex. The minimal guanylyl cyclase catalytic unit appears to be a dimer [7, 8]. The soluble forms contain amino terminal heme-binding, dimerization, and carboxyl-terminal catalytic domains (see Figure 40.1 for topology of each cyclase). The membrane-spanning forms contain an extracellular ligand binding, transmembrane, kinase homology, dimerization and carboxyl-terminal catalytic domain. The latter domain is the most conserved between family members, and is homologous to the catalytic domain of adenylyl cyclase [9, 10]. No crystal or solution structure for a guanylyl cyclase has been reported. However, modeling and mutagenesis studies predict that soluble guanylyl cyclases form head-to-tail heterodimers containing one catalytic domain, and that membrane guanylyl cyclases form head-to-tail homodimers containing two catalytic domains [9, 10].

Rats and mice have 7 single membrane-spanning and 4 soluble family members (all human guanylyl cyclases plus GC-D and GC-G), whereas *Caenorhabdiis elegans* is predicted to have at least 26 membrane-spanning and 7 soluble guanylyl cyclases based on genome analysis [11]. Hence, guanylyl cyclase diversity appears to be declining as a function of evolution, which may indicate that the remaining human guanylyl cyclases have adopted the functions of the cyclases that were lost. Multimembrane-spanning guanylyl cyclases similar to adenylyl cyclases have been found in *Dictyostelium discoidium*, *Plasmodium falciparum*, *Paramecium tetraurelia*, and *Tetrahymena pyriformis*. Based on topology, these guanylyl cyclases probably evolved from adenylyl cyclase [12]. One guanylyl cyclase has been identified in cyanobacteria [13], but no guanylyl cyclase genes have been identified in the genomes of yeast, fungi, or bacteria [14].

SOLUBLE GUANYLYL CYCLASE, NITRIC OXIDE, AND NITRIC OXIDE SYNTHASE

The primary and best-studied endogenous activator of soluble guanylyl cyclase (sGC) is nitric oxide (NO), which was originally described as endothelium-derived relaxing

Handbook of Cell Signaling, Three-Volume Set 2 ed.

TABLE 40.1 Guanylyl cyclases, tissue distribution, cognate ligand, and effect of inactivation

Guanylyl cyclase	Tissue expression	Ligand	Effect of inactivation
GC-A	Lung, kidney, adrenal, vascular smooth muscle, endothelium, heart, adipose	ANP, BNP	Hypertension, cardiac hypertrophy and fibrosis
GC-B	Bone, vascular smooth muscle, lung, brain, heart, liver, uterus	CNP	Acromesomelic dysplasia type Maroteaux dwarfism
GC-C	Intestinal epithelium	Guanylin, uroguanylin and bacterial heat-stable enterotoxin	Increased colonic epithelial cell proliferation
GC-D	Olfactory bulb	?	Loss of guanylin- and uroguanylin-dependent olfactory signaling
GC-E	Retina, pineal gland	GCAPs	Cone dystrophy, Leber congenital amaurosis type 1, blindness no obvious phenotype
GC-F	Retina	GCAPs	
GC-G	Lung, intestine, skeletal muscle, testes	?	Increased deleterious responses to renal ischemia/reperfusion injury
Soluble alpha 1	Platelets, lung, brain, vasculature	NO, CO	Loss of NO-dependent platelet aggregation
Soluble alpha 1	Brain, vasculature	NO, CO	
Soluble beta 1	Platelets, lung, brain, vasculature	NO, CO	Hypertension, loss of NO-dependent platelet aggregation, sterility
Soluble beta 2	Kidney and liver	?	?

ANP, atrial natriuretic peptide; BNP, B-type natriuretic peptide; CNP, C-type natriuretic peptide; CO, carbon monoxide; GCAPs, guanylyl cyclase activator proteins; Gn, guanylin; Hb, hemoglobin; NO, nitric oxide; Sta, heat-stable enterotoxin; Uro, uroguanylin.

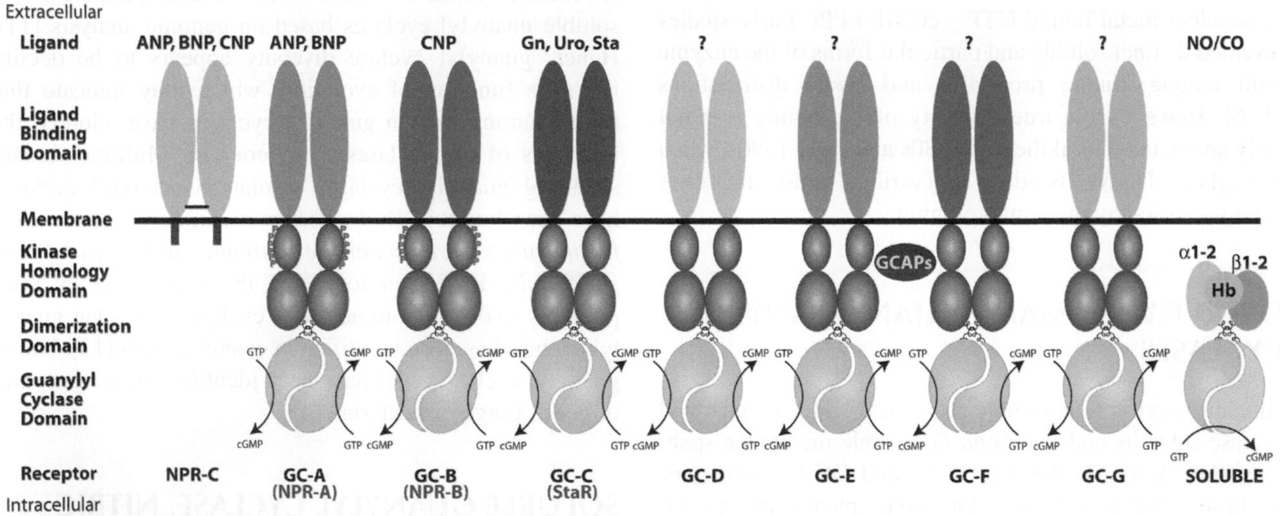

FIGURE 40.1 Schematic of mammalian guanylyl cyclases and their ligands.
There are seven transmembrane and four soluble subunits that encode guanylyl cyclases in mammals. The structure and function of each member is discussed in the text. The "P" indicates known phosphorylation sites. The black bar between NPR-C subunits indicates an intermolecular disulfide bond. Abbreviations: ANP, atrial natriuretic peptide; BNP, B-type natriuretic peptide; CNP, C-type natriuretic peptide; CO, carbon monoxide; GCAPs, guanylyl cyclase activator proteins; Gn, guanylin; Hb, hemoglobin; NO, nitric oxide; NPR-C, natriuretic peptide receptor C; Sta, heat-stable enterotoxin; Uro, uroguanylin.

factor for its potent ability to relax blood vessels in response to vasodilators like acetylcholine or bradykinin [15]. NO is synthesized by nitric oxide synthase (NOS), which catalyzes a NADPH-dependent 5-electron oxidation of L-arginine by O_2 to yield NO and citrulline. There are three genetically distinct isoforms of NOS: a constitutively expressed and calcium-activated endothelial form (eNOS or NOS3), a neuronal isozyme (nNOS or NOS1), and a cytokine- or endotoxin-induced, calcium-independent form (iNOS or NOS2) that is highly expressed in macrophages and neutrophils. Endothelium-produced NO migrates into blood and surrounding vascular smooth muscle cells to inhibit platelet aggregation and stimulate vasorelaxation, respectively. Blood pressures in mice lacking endothelial NOS are about 20 mm Hg higher than pressures in wild-type littermates [16, 17]. Neuronal NOS also causes vasodilation, and may be a retrograde messenger in long-term potentiation. Mice with disruptions in exon two of nNOS have an enlarged pyloric sphincter and increased aggressive behavior, but normal blood pressure [18, 19]. The main role of inducible NOS is to kill invading microorganisms by producing extremely high levels of NO, which combine with superoxide to form peroxynitrite. Thus, this immune function of NO is independent of guanylyl cyclase. Mice lacking iNOS have a compromised ability to defend against microorganism infection [20, 21].

Soluble GCs are found in most tissues, often in combination with one or more membrane-spanning family members. Lung, brain, kidney, and vascular tissue are particularly rich in sGC. In humans, two α and two β subunits have been identified. The best-characterized heterodimers are the α_1/β_1 and the α_2/β_1 forms. β_1 is 619 residues, and contains an evolutionarily conserved amino-terminal heme prosthetic group binding domain of about 200 residues, a dimerization region, and a carboxyl-terminal guanylyl cyclase domain of about 250 residues. The β_2 subunit has an additional 86 carboxyl-terminal amino acids compared to the β_1 isoform that contains a consensus CAAX sequence for possible isoprenylation or carboxymethylation [22]. In contrast to the other subunits, β_2 has been shown to form an active homodimer [23]. The human α_1 subunit is 717 amino acids, and is about 34 percent homologous to the β_1 subunit [24]. The α_2 subunit is 48 percent identical to the α_1 subunit at the amino acid level [25]. α_1 is the major subunit in platelets and lung, but only accounts for about half the NO-dependent guanylyl cyclase activity in the brain [26]. Although α_2 only represents 6 percent of the soluble cyclase activity in the vasculature, it is sufficient to yield maximum smooth muscle relaxation in response to NO [26].

At the amino portion of the enzyme, His-105 of the β_1 subunit is the axial ligand of the pentacoordinated reduced iron center of heme [27]. His-105 and a heme prosthetic group containing reduced iron are required for NO activation of the enzyme. NO activates sGC by binding to the sixth position of the heme ring and breaking the bond between the axial histidine and iron to form a 5-coordinated ring with NO in the fifth position. Carbon monoxide can also activate sGC by binding heme to form a 6-coordinated complex. However, since NO activates the enzyme 100- to 200-fold whereas CO only activates it about 4-fold, the physiologic significance of CO activation is unclear.

Genetic disruption of the α_1 subunit leads to the loss of NO-dependent platelet aggregation but not loss of the vasorelaxation response [26]. Disruption of the β_1 subunit [28] leads to the loss of both these responses, with blood pressures being elevated 26 mm Hg in the knockout compared to wild-type animals [28]. In addition, the homozygous knockout animals exhibited severely reduced parastalsis leading to gastrointestinal obstruction similar to that observed in animals lacking cGMP dependent protein kinase I (PKGI/cGKI) [29]. Additionally, the males are infertile [29]. Hence, sGC appears to be the sole mediator of the major cardiovascular and sexual arousal effects of NO.

GC-A/NPR-A/NPR1

In contrast to the soluble heterdimeric guanylyl cyclases that are activated by gases, transmembrane guanylyl cyclases are homodimers that are activated by peptides. The best-characterized transmembrane guanylyl cyclase is NPR-A, which is also called GC-A or NPR1. It binds and is activated by atrial natriuretic peptide (ANP) and B-type natriuretic peptide (BNP) [30]. Mature ANP is a 28 amino acid disulfide-linked peptide released from atrial granules. Atrial wall stretch due to increased intravascular volume causes the release of ANP into the circulation. BNP is found in atrial granules as well, but it is released in greater concentrations from the ventricles due to increased mass. It is not stored in granules in the ventricles, but is regulated at the level of transcription. Genetic ablation of ANP results in hypertensive mice with cardiac hypertrophy [31, 32]. Disruption of BNP increases ventricular fibrosis, but does not affect blood pressure [33].

NPR-A is highly expressed in kidney, lung, adrenal, vasculature, brain, liver, endothelial, and adipose tissues [34, 35]. It is expressed at lower levels in the heart. The extracellular domain of rat NPR-A contains three intramolecular disulfide bonds and five N-linked glycosylation sites [36]. In the basal state, NPR-A exists as a homodimer or homotetramer and ligand binding does not result in further oligomerization. Crystallographic studies indicate that ANP binds to NPR-A in a 1 : 2 stoichiometry and activates the receptor by causing an intermolecular twist with little intramolecular conformational change [37]. NPR-A is phosphorylated on four serines and two threonines within the amino-terminal portion of its kinase homology domain in the unstimulated state [38]. Phosphorylation of NPR-A is required for hormonal activation, and dephosphorylation in response to protein kinase C activation or prolonged ANP exposure inhibits NPR-A [39]. The role of ATP in the regulation of NPR-A is controversial. Some studies indicate that it is required for NPR-A activation [40, 41], whereas others indicate that it is not required for initial activation but prolongs activity by reducing the Km for GTP [42, 43]. NPR-A-null mice exhibit cardiac hypertrophy, high blood pressure, and ventricular fibrosis [44, 45].

GC-B/NPR-B/NPR2

NPR-B, which is also called GC-B or NPR2, has a similar topology to NPR-A but is activated by CNP, which exists in 22 and 53 amino acid forms [30, 46], neither of which are stored in granules. CNP does not circulate at high levels; rather, it signals in a paracrine manner. Recently, patients with translocations of portions of chromosome 2 of the CNP gene were shown to have elevated CNP levels and skeletal overgrowth associated with a Marfanoid phenotype [47, 48].

NPR-B is abundantly expressed in brain, lung, bone, heart, and ovary tissue [49–51]. It is also expressed at relatively high levels in fibroblast and vascular smooth muscle cells. Like NPR-A, it contains three intramolecular disulfide bonds and is highly glycosylated on asparagine residues. NPR-B is highly phosphorylated, and dephosphorylation is associated with receptor inhibition in response to elevations of intracellular calcium [52], activation of protein kinase C [53], or prolonged exposure to CNP [54]. CNP-dependent NPR-B cyclase activity is increased in the presence of ATP [42, 43].

Two loss-of-function mouse models exist for NPR-B. In one model, the exons that encode the carboxyl-terminal half of the extracellular domain and transmembrane segment were deleted by homologous recombination [55]. These mice are dwarfs, and the females are sterile. The heterozygous mice were significantly shorter than the wild-type animals. The other mouse model has a spontaneous mutation of an arginine for a highly conserved leucine in the catalytic domain. This so called cn/cn mouse that contains two defective alleles also displays dwarfism, although female sterility was not noted [56]. In humans, a homozygous loss of function mutation has been identified. These patients have acromesomelic dysplasia, type Maroteaux, a rare form of short-limbed dwarfism [57]. These patients are not sterile and, like the "knockout" mice, individuals with one normal and one abnormal allele display normal limb proportions but are statistically shorter that the average person from their respective populations [58]. Surprisingly, NPR-B, not NPR-A, is the primary particulate guanylyl cyclase in the failed heart [59]. Recent reports indicate the CNP signaling is downregulated in patients with aortic valve stenosis [60], and that NPR-B and PKGI are essential for sensory axon bifurcation in the spinal cord of mice [61].

NPR-C/NPR-3

The natriuretic peptide clearance receptor, which is also called NPR-C or NPR3, shares 35 percent amino acid identity with NPR-A and NPR-B in its extracellular domain, and binds all three known natriuretic peptides with similar affinities [62]. It also binds osteocrin, which acts a decoy ligand to increase CNP levels in bone tissue [63, 64]. NPR-C is expressed in most tissues, usually at levels that are significantly higher than those of NPR-A or NPR-B. NPR-C is glycosylated, and contains two sets of intramolecular disulfide bonds that are conserved in NPR-A and NPR-B. It also contains one or two intermolecular bonds, depending on the species [65]. The internalization of NPR-C is constitutive [66], and its endocytosis is mediated by a clathrin-dependent mechanism [67]. Ligand bound to NPR-C undergoes lysosomal hydrolysis followed by NPR-C recycling to the cell surface [66, 68]. Its main function is to clear natriuretic peptides from the circulation through receptor-mediated internalization and degradation, as is evidenced by loss of function mutations in mice [69, 70]. However, a number of groups have also reported that NPR-C exhibits signaling functions via heterotrimeric GTP binding proteins [71].

GC-C/STAR

Guanylyl cyclase C (GC-C) has similar topology to GC-A and GC-B, but contains about a 60 amino acid carboxyl terminal extension that renders the receptor detergent insoluble. GC-C is primarily found highly concentrated at the apical membrane of intestinal epithelial cells, and is the target for small peptides secreted from pathogenic bacteria called heat-stable enterotoxins (Sta) that contain three stabilizing disulfide bonds. Hence, it is sometimes referred to as the heat-stable enterotoxin receptor, abbreviated to StaR. Sta activation of GC-C leads to intracellular cGMP elevations, PKGII-dependent phosphorylation of the cystic fibrosis transmembrane regulator, and increased Cl^- secretion in the gut. Lumenal ion secretion increases intestinal water content, which results in diarrhea.

The first endogenous ligand of GC-C was purified from rat jejunum using a bioassay involving cGMP elevations in human colon T84 cells [72]. The mature form of the peptide, called guanylin, is a 15 amino acid peptide that contains two intramolecular disulfide bonds. A second 19 amino acid endogenous GC-C ligand with conserved disulfide bonds to guanylin was purified from opossum urine and therefore named uroguanylin [73]. Human guanylin and uroguanylin are 53 percent identical. Both peptides compete for binding with Sta to the extracellular domain of GC-C. Mice lacking functional guanylin display increased colonic epithelial cell proliferation, but no change in blood pressure or sodium excretion [74]. In contrast, mice lacking uroguanylin display decreased ability to excrete an enteral NaCl load as well as salt-independent hypertension [75]. Hence, uroguanylin may be an antihypertensive factor in mice.

Like NPR-A and NPR-B, the extracellular domain of GC-C contains many N-linked glycosylation sites [76–78] and multiple intramolecular disulfide bonds [79]. GC-C exists as a higher-ordered structure; dimer [80] or trimer forms have been reported [81]. ATP increases the ligand-dependent guanylyl cyclase activity of GC-C, possibly by stabilizing the active form of the receptor [80, 82, 83]. Unlike GC-A and GC-B, GC-C is not basally phosphorylated [80], but treatment of cells with phorbol esters increases the phosphate content and activity of GC-C, consistent with protein kinase C-dependent regulation [80, 84]. Mice lacking functional GC-C are resistant to heat-stable enterotoxin infection and have increased proliferation of intestinal epithelia cells [85–87]. Recent work indicates that GC-C is a marker [88] and possible therapeutic target for colon cancer [89–91].

GC-D

In 1995 a novel member of the single membrane-spanning guanylyl cyclase family, named GC-D, was molecularly cloned from rat olfactory epithelium [92]. GC-D was subsequently co-localized by immunofluorescence detection with phosphodiesterase 2 in mouse olfactory cilia [93], and with phosphodiesterase 2 and the selective cGMP gated nucleotide channel, CNGA3, in neurons that project to the necklace glomeruli of the olfactory bulb in rats [94]. Uniquely, GC-D positive cells completely lack essential components of the well-known cAMP olfactory signal transduction system. Recently, uroguanylin or guanylin but not Sta were shown to elicit an excitatory calcium elevation signal in GC-D positive neurons. The response to both compounds was abolished in mice lacking either GC-D or CNGA3 [95]. Whether guanylin or uroguanylin directly bind GC-D is not known. However, since GC-D positive cells do not respond to Sta, this seems unlikely [95]. GC-D is a pseudogene in humans that was likely lost some time during primate evolution [96].

GC-E/RET-GC1 AND GC-F/RET-GC2

Cyclic GMP is an essential regulator of phototransduction; a process where rods and cones in the retina convert photons into electrical signals producing vision. In the presence of light, photons bind the seven membrane-spanning protein, rhodopsin, which leads to transducin-dependent activation of phosphodiesterase 6 and reduced intracellular cGMP concentrations. As a result, cGMP-gated sodium and calcium channels close, causing hyperpolarization. In the dark, rhodopsin signaling is reduced and cGMP degradation is slowed. This, in combination with cyclase activation (see below), causes the opening of cGMP-gated cation channels, elevation of intracellular sodium and calcium concentrations, and cell depolarization, which results in photorecovery.

Initial studies in the early 1980s indicated that guanylyl cyclase activity in the retina is inhibited by calcium [97]. Later, calcium sensitivity was shown to be mediated by a factor that could be separated from the cyclase by a high salt wash [98]. Complementary DNA cloning in the 1990s revealed that cyclic GMP is synthesized in the retina by two single membrane-spanning guanylyl cyclases, GC-E and GC-F, also known as RetGC-1 and RetGC-2, respectively [99–101]. GC-E is expressed in the retina and pineal gland, whereas GC-F is only expressed in the retina [99–101]. Both retinal cyclases contain the same signature domains as GC-A, GC-B, and GC-C but, unlike these known hormone receptors, no extracellular activators of GC-E or GC-F have been identified [99–101]. Instead, small (20- to 24-kDa), soluble, intracellular calcium-binding molecules called guanylyl cyclase activator proteins (GCAPs) regulate GC-E and GC-F [102–105]. In vertebrate retinas, at least two GCAPs (GCAP1 and GCAP2) are present. Both are fatty acylated on their amino termini and contain four calcium binding EF-hand domains, although only three are functional. GCAPs are constitutively associated with GC-E, but only activate the cyclase under low calcium conditions, possibly by facilitating oligomerization of the catalytic domain. Surprisingly, when intracellular calcium levels rise, GCAPs inhibit cyclase activity.

Inactivation of GC-E in mice caused suppression of a and b waves of electroretinograms from dark-adapted null mice, and cone dystrophy, but had no effect on rods [106]. Inactivating GC-E/RetGC-1 mutations in humans cause Leber congenital amaurosis type 1, an autosomal recessive disease characterized by early onset rod and cone dystrophy and blindness [107, 108]. Inactivation of GC-F in mice causes no electroretinographic effects, but mice lacking both retinal cyclases have unstable and nonfunctional cones and rods [109]. No mutations in human GC-F/RetGC-2 have been reported. Functional inactivation of GCAP1 and GCAP2 results in mice that lack calcium-dependent retinal guanylyl cyclase regulation, and a delayed recovery of the photoresponse [110]. Expression of GCAP1 in a double GCAP knockout background restores wild-type flash responses in rods and cones, suggesting that GCAP1 is sufficient for normal retina photoregulation [111, 112]. Finally, several missense mutations within GCAP1 are associated with autosomal dominant cone dystrophy in humans [113].

GC-G

GC-G was initially identified as a unique PCR product that was amplified from a rat small intestine cDNA library with degenerate primers common to single membrane-spanning guanylyl cyclases [114]. Several additional cloning and amplification techniques based on cDNA and gene structures were required to assemble a full-length clone [115]. Comparison of the extracellular domain of GC-G indicated that it is most similar to natriuretic peptide receptors. Expression in Cos7 and 293 cells revealed elevated basal guanylyl cyclase activity that was not increased by ANP, BNP, CNP, or Sta. Northern hybridization analysis revealed high GC-G mRNA expression in lung, intestine, and skeletal muscle. Cloning of the apparent mouse homolog of GC-G revealed high expression in the testes [116]. Mice lacking functional GC-G were not histologically different from wild-type animals. Despite apparent low expression in the kidney, serum creatinine and urea levels were reduced in GC-G knockout mice compared to wild-type animals in response to renal ischemia/reperfusion injury [117]. Finally, GC-G is a psuedogene in humans [118].

ACKNOWLEDGEMENTS

The article is dedicated to the memory of Dr David L. Garbers, who made repeated seminal contributions to the guanylyl cyclase field during his distinguished career. I thank Dr Deborah Dickey for critically reading the manuscript. Research in Dr Potter's laboratory is supported by grants from the National Institutes of Health, the American Heart Association, Minnesota Medical Foundation, and a Medica funded Minnesota-Mayo Partnership Grant.

REFERENCES

1. A shman DF, Lipton R, Melicow MM, Price TD. Isolation of adenosine 3′,5′-monophosphate and guanosine 3′,5′-monophosphate from rat urine. *Biochem Biophys Res Commun* 1963;**11**:330–4.
2. Hardman JG, Sutherland EW. Guanyl cyclase, an enzyme catalyzing the formation of guanosine 3′,5′-monophosphate from guanosine trihosphate. *J Biol Chem* 1969;**244**:6363–70.
3. Schultz G, Bohme E, Munske K. Guanyl cyclase. Determination of enzyme activity. *Life Sci* 1969;**8**:1323–32.
4. White AA, Aurbach GD. Detection of guanyl cyclase in mammalian tissues. *Biochim Biophys Acta* 1969;**191**:686–97.
5. Chrisman TD, Garbers DL, Parks MA, Hardman JG. Characterization of particulate and soluble guanylate cyclases from rat lung. *J Biol Chem* 1975;**250**:374–81.
6. Kimura H, Murad F. Evidence for two different forms of guanylate cyclase in rat heart. *J Biol Chem* 1974;**249**:6910–16.
7. Liu Y, Ruoho AE, Rao VD, Hurley JH. Catalytic mechanism of the adenylyl and guanylyl cyclases: modeling and mutational analysis. *Proc Natl Acad Sci USA* 1997;**94**:13,414–13,419.
8. Thompson DK, Garbers DL. Dominant negative mutations of the guanylyl cyclase-A receptor. Extracellular domain deletion and catalytic domain point mutations. *J Biol Chem* 1995;**270**:425–30.
9. Sunahara RK, Beuve A, Tesmer JJ, Sprang SR, Garbers DL, Gilman AG. Exchange of substrate and inhibitor specificities between adenylyl and guanylyl cyclases. *J Biol Chem* 1998;**273**:16,332–16,338.
10. Tucker CL, Hurley JH, Miller TR, Hurley JB. Two amino acid substitutions convert a guanylyl cyclase, RetGC-1, into an adenylyl cyclase. *Proc Natl Acad Sci USA* 1998;**95**:5993–7.
11. Chen N, Harris TW, Antoshechkin I, et al. WormBase: a comprehensive data resource for Caenorhabditis biology and genomics. *Nucleic Acids Res* 2005;**33**:D383–9.
12. Linder JU, Schultz JE. Guanylyl cyclases in unicellular organisms. *Mol Cell Biochem* 2002;**230**:149–58.
13. Ochoa De Alda JA, Ajlani G, Houmard J. Synechocystis strain PCC 6803 cya2, a prokaryotic gene that encodes a guanylyl cyclase. *J Bacteriol* 2000;**182**:3839–42.
14. Fitzpatrick DA, O'Halloran DM, Burnell AM. Multiple lineage specific expansions within the guanylyl cyclase gene family. *BMC Evol Biol* 2006;**6**:26.
15. Furchgott RF, Zawadzki JV. The obligatory role of endothelial cells in the relaxation of arterial smooth muscle by acetylcholine. *Nature* 1980;**288**:373–6.
16. Huang PL, Huang Z, Mashimo H, et al. Hypertension in mice lacking the gene for endothelial nitric oxide synthase. *Nature* 1995;**377**:239–42.
17. Shesely EG, Maeda N, Kim HS, et al. Elevated blood pressures in mice lacking endothelial nitric oxide synthase. *Proc Natl Acad Sci USA* 1996;**93**:13,176–13,181.
18. Huang PL, Dawson TM, Bredt DS, Snyder SH, Fishman MC. Targeted disruption of the neuronal nitric oxide synthase gene. *Cell* 1993;**75**:1273–86.
19. Nelson RJ, Demas GE, Huang PL, et al. Behavioural abnormalities in male mice lacking neuronal nitric oxide synthase. *Nature* 1995;**378**:383–6.
20. MacMicking JD, Nathan C, Hom G, et al. Altered responses to bacterial infection and endotoxic shock in mice lacking inducible nitric oxide synthase. *Cell* 1995;**81**:641–50.
21. Wei XQ, Charles IG, Smith A, et al. Altered immune responses in mice lacking inducible nitric oxide synthase. *Nature* 1995;**375**:408–11.
22. Yuen PS, Potter LR, Garbers DL. A new form of guanylyl cyclase is preferentially expressed in rat kidney. *Biochemistry* 1990;**29**:10,872–10,878.
23. Koglin M, Vehse K, Budaeus L, Scholz H, Behrends S. Nitric oxide activates the beta 2 subunit of soluble guanylyl cyclase in the absence of a second subunit. *J Biol Chem* 2001;**276**:30,737–30,743.
24. Giuili G, Scholl U, Bulle F, Guellaen G. Molecular cloning of the cDNAs coding for the two subunits of soluble guanylyl cyclase from human brain. *FEBS Letts* 1992;**304**:83–8.
25. Harteneck C, Wedel B, Koesling D, Malkewitz J, Bohme E, Schultz G. Molecular cloning and expression of a new alpha-subunit of soluble guanylyl cyclase. Interchangeability of the alpha-subunits of the enzyme. *FEBS Letts* 1991;**292**:217–22.
26. Mergia E, Friebe A, Dangel O, Russwurm M, Koesling D. Spare guanylyl cyclase NO receptors ensure high NO sensitivity in the vascular system. *J Clin Invest* 2006;**116**:1731–7.
27. Wedel B, Humbert P, Harteneck C, et al. Mutation of His-105 in the beta 1 subunit yields a nitric oxide-insensitive form of soluble guanylyl cyclase. *Proc Natl Acad Sci USA* 1994;**91**:2592–6.
28. Friebe A, Mergia E, Dangel O, Lange A, Koesling D. Fatal gastrointestinal obstruction and hypertension in mice lacking nitric oxide-sensitive guanylyl cyclase. *Proc Natl Acad Sci USA* 2007;**104**:7699–704.
29. Pfeifer A, Klatt P, Massberg S, et al. Defective smooth muscle regulation in cGMP kinase I-deficient mice. *EMBO J* 1998;**17**:3045–51.

30. Suga S, Nakao K, Hosoda K, et al. Receptor selectivity of natriuretic peptide family, atrial natriuretic peptide, brain natriuretic peptide, and C-type natriuretic peptide. *Endocrinology* 1992;**130**:229–39.
31. John SW, Krege JH, Oliver PM, et al. Genetic decreases in atrial natriuretic peptide and salt-sensitive hypertension. *Science* 1995;**267**:679–81. (published erratum in Science 267:1753).
32. John SW, Veress AT, Honrath U, et al. Blood pressure and fluid-electrolyte balance in mice with reduced or absent ANP. *Am J Physiol* 1996;**271**:R109–14.
33. Tamura N, Ogawa Y, Chusho H, et al. Cardiac fibrosis in mice lacking brain natriuretic peptide. *Proc Natl Acad Sci USA* 2000;**97**:4239–44.
34. Bryan PM, Smirnov D, Smolenski A, et al. A sensitive method for determining the phosphorylation status of natriuretic peptide receptors: cGK-Ialpha does not regulate NPR-A. *Biochemistry* 2006;**45**:1295–303.
35. Potter LR, Abbey-Hosch S, Dickey DM. Natriuretic peptides, their receptors, and cyclic guanosine monophosphate-dependent signaling functions. *Endocr Rev* 2006;**27**:47–72.
36. Potter LR, Hunter T. Guanylyl cyclase-linked natriuretic peptide receptors: structure and regulation. *J Biol Chem* 2001;**276**:6057–60.
37. Ogawa H, Qiu Y, Ogata CM, Misono KS. Crystal structure of hormone-bound atrial natriuretic peptide receptor extracellular domain: rotation mechanism for transmembrane signal transduction. *J Biol Chem* 2004;**279**:28,625–28,631.
38. Potter LR, Hunter T. Phosphorylation of the kinase homology domain is essential for activation of the A-type natriuretic peptide receptor. *Mol Cell Biol* 1998;**18**:2164–72.
39. Potter LR, Garbers DL. Dephosphorylation of the guanylyl cyclase-A receptor causes desensitization. *J Biol Chem* 1992;**267**:14,531–14,534.
40. Chinkers M, Singh S, Garbers DL. Adenine nucleotides are required for activation of rat atrial natriuretic peptide receptor/guanylyl cyclase expressed in a baculovirus system. *J Biol Chem* 1991;**266**:4088–93.
41. Kurose H, Inagami T, Ui M. Participation of adenosine 5′-triphosphate in the activation of membrane-bound guanylate cyclase by the atrial natriuretic factor. *FEBS Letts* 1987;**219**:375–9.
42. Antos LK, Abbey-Hosch SE, Flora DR, Potter LR. ATP-independent activation of natriuretic peptide receptors. *J Biol Chem* 2005;**280**:26,928–26,932.
43. Antos LK, Potter LR. Adenine nucleotides decrease the apparent Km of endogenous natriuretic peptide receptors for GTP. *Am J Physiol Endocrinol Metab* 2007;**293**:E1756–63.
44. Lopez MJ, Wong SK, Kishimoto I, et al. Salt-resistant hypertension in mice lacking the guanylyl cyclase-A receptor for atrial natriuretic peptide. *Nature* 1995;**378**:65–8.
45. Oliver PM, Fox JE, Kim R, et al. Hypertension, cardiac hypertrophy, and sudden death in mice lacking natriuretic peptide receptor A. *Proc Natl Acad Sci USA* 1997;**94**:14,730–14,735.
46. Koller KJ, Lowe DG, Bennett GL, et al. Selective activation of the B natriuretic peptide receptor by C-type natriuretic peptide (CNP). *Science* 1991;**252**:120–3.
47. Bocciardi R, Giorda R, Buttgereit J, et al. Overexpression of the C-type natriuretic peptide (CNP) is associated with overgrowth and bone anomalies in an individual with balanced t(2;7) translocation. *Hum Mutat* 2007;**28**:724–31.
48. Moncla A, Missirian C, Cacciagli P, et al. A cluster of translocation breakpoints in 2q37 is associated with overexpression of NPPC in patients with a similar overgrowth phenotype. *Hum Mutat* 2007;**12**:1183–8.
49. Schulz S, Singh S, Bellet RA, et al. The primary structure of a plasma membrane guanylate cyclase demonstrates diversity within this new receptor family. *Cell* 1989;**58**:1155–62.
50. Nagase M, Katafuchi T, Hirose S, Fujita T. Tissue distribution and localization of natriuretic peptide receptor subtypes in stroke-prone spontaneously hypertensive rats. *J Hypertens* 1997;**15**:1235–43.
51. Chrisman TD, Schulz S, Potter LR, Garbers DL. Seminal plasma factors that cause large elevations in cellular cyclic GMP are C-type natriuretic peptides. *J Biol Chem* 1993;**268**:3698–703.
52. Potthast R, Abbey-Hosch SE, Antos LK, Marchant JS, Kuhn M, Potter LR. Calcium-dependent dephosphorylation mediates the hyperosmotic and lysophosphatidic acid-dependent inhibition of natriuretic peptide receptor-B/guanylyl cyclase-B. *J Biol Chem* 2004;**279**:48,513–48,519.
53. Potter LR, Hunter T. Activation of PKC stimulates the dephosphorylation of natriuretic peptide receptor-B at a single serine residue: a possible mechanism of heterologous desensitization. *J Biol Chem* 2000;**275**:31,099–31,106.
54. Potter LR. Phosphorylation-dependent regulation of the guanylyl cyclase-linked natriuretic peptide receptor B: dephosphorylation is a mechanism of desensitization. *Biochemistry* 1998;**37**:2422–9.
55. Tamura N, Doolittle LK, Hammer RE, Shelton JM, Richardson JA, Garbers DL. Critical roles of the guanylyl cyclase B receptor in endochondral ossification and development of female reproductive organs. *Proc Natl Acad Sci USA* 2004;**101**:17,300–17,305.
56. Tsuji T, Kunieda T. A loss-of-function mutation in natriuretic peptide receptor 2 (Npr2) gene is responsible for disproportionate dwarfism in cn/cn mouse. *J Biol Chem* 2005;**280**:14,288–14,292.
57. Bartels CF, Bukulmez H, Padayatti P, et al. Mutations in the transmembrane natriuretic peptide receptor NPR-B impair skeletal growth and cause acromesomelic dysplasia, type Maroteaux. *Am J Hum Genet* 2004;**75**:27–34.
58. Olney RC, Bukulmez H, Bartels CF, et al. Heterozygous mutations in natriuretic peptide receptor-B (NPR2) are associated with short stature. *J Clin Endocrinol Metab* 2006;**91**:1229–32.
59. Dickey DM, Flora DR, Bryan PM, Xu X, Chen Y, Potter LR. Differential regulation of membrane guanylyl cyclases in congestive heart failure: natriuretic peptide receptor (NPR)-B, Not NPR-A, is the predominant natriuretic peptide receptor in the failing heart. *Endocrinology* 2007;**148**:3518–22.
60. Peltonen TO, Taskinen P, Soini Y, et al. Distinct downregulation of C-type natriuretic peptide system in human aortic valve stenosis.. *Circulation* 2007;**116**:1283–9.
61. Schmidt H, Stonkute A, Juttner R, et al. The receptor guanylyl cyclase Npr2 is essential for sensory axon bifurcation within the spinal cord. *J Cell Biol* 2007;**179**:331–40.
62. Potter LR. Domain analysis of human transmembrane guanylyl cyclase receptors: implications for regulation. *Front Biosci* 2005;**10**:1205–20.
63. Moffatt P, Thomas G, Sellin K, et al. Osteocrin is a specific ligand of the natriuretic Peptide clearance receptor that modulates bone growth. *J Biol Chem* 2007;**282**:36,454–36,462.
64. Thomas G, Moffatt P, Salois P, et al. Osteocrin, a novel bone-specific secreted protein that modulates the osteoblast phenotype. *J Biol Chem* 2003;**278**:50,563–50,571.
65. Stults JT, O'Connell KL, Garcia C, et al. The disulfide linkages and glycosylation sites of the human natriuretic peptide receptor-C homodimer. *Biochemistry* 1994;**33**:11,372–11,381.
66. Nussenzveig DR, Lewicki JA, Maack T. Cellular mechanisms of the clearance function of type C receptors of atrial natriuretic factor. *J Biol Chem* 1990;**265**:20,952–20,958.
67. Cohen D, Koh GY, Nikonova LN, Porter JG, Maack T. Molecular determinants of the clearance function of type C receptors of natriuretic peptides. *J Biol Chem* 1996;**271**:9863–9.

68. Fan D, Bryan PM, Antos LK, Potthast RJ, Potter LR. Down-regulation does not mediate natriuretic peptide-dependent desensitization of natriuretic peptide receptor (NPR)-A or NPR-B: guanylyl cyclase-linked natriuretic peptide receptors do not internalize. *Mol Pharmacol* 2005;**67**:174–83.
69. Jaubert J, Jaubert F, Martin N, et al. Three new allelic mouse mutations that cause skeletal overgrowth involve the natriuretic peptide receptor C gene (Npr3). *Proc Natl Acad Sci USA* 1999;**96**:10,278–10,283.
70. Matsukawa N, Grzesik WJ, Takahashi N, et al. The natriuretic peptide clearance receptor locally modulates the physiological effects of the natriuretic peptide system. *Proc Natl Acad Sci USA* 1999;**96**:7403–8.
71. Anand-Srivastava MB, Trachte GJ. Atrial natriuretic factor receptors and signal transduction mechanisms. *Pharmacol Rev* 1993;**45**:455–97.
72. Currie MG, Fok KF, Kato J, et al. Guanylin: an endogenous activator of intestinal guanylate cyclase. *Proc Natl Acad Sci USA* 1992;**89**:947–51.
73. Hamra FK, Forte LR, Eber SL, et al. Uroguanylin: structure and activity of a second endogenous peptide that stimulates intestinal guanylate cyclase. *Proc Natl Acad Sci USA* 1993;**90**:10,464–10,468.
74. Steinbrecher KA, Wowk SA, Rudolph JA, Witte DP, Cohen MB. Targeted inactivation of the mouse guanylin gene results in altered dynamics of colonic epithelial proliferation. *Am J Pathol* 2002;**161**:2169–78.
75. Lorenz JN, Nieman M, Sabo J, et al. Uroguanylin knockout mice have increased blood pressure and impaired natriuretic response to enteral NaCl load. *J Clin Invest* 2003;**112**:1244–54.
76. Ghanekar Y, Chandrashaker A, Tatu U, Visweswariah SS. Glycosylation of the receptor guanylate cyclase C: role in ligand binding and catalytic activity. *Biochem J* 2004;**379**:653–63.
77. Hasegawa M, Hidaka Y, Wada A, Hirayama T, Shimonishi Y. The relevance of N-linked glycosylation to the binding of a ligand to guanylate cyclase C. *Eur J Biochem* 1999;**263**:338–46.
78. Vaandrager AB, Schulz S, De Jonge HR, Garbers DL. Guanylyl cyclase C is an N-linked glycoprotein receptor that accounts for multiple heat-stable enterotoxin-binding proteins in the intestine. *J Biol Chem* 1993;**268**:2174–9.
79. Hasegawa M, Matsumoto-Ishikawa Y, Hijikata A, Hidaka Y, Go M, Shimonishi Y. Disulfide linkages and a three-dimensional structure model of the extracellular ligand-binding domain of guanylyl cyclase C. *Protein J* 2005;**24**:315–25.
80. Vaandrager AB, van der Wiel E, de Jonge HR. Heat-stable enterotoxin activation of immunopurified guanylyl cyclase C. Modulation by adenine nucleotides. *J Biol Chem* 1993;**268**:19,598–19,603.
81. Vaandrager AB, van der Wiel E, Hom ML, Luthjens LH, de Jonge HR. Heat-stable enterotoxin receptor/guanylyl cyclase C is an oligomer consisting of functionally distinct subunits, which are non-covalently linked in the intestine. *J Biol Chem* 1994;**269**:16,409–16,415.
82. Bhandari R, Srinivasan N, Mahaboobi M, Ghanekar Y, Suguna K, Visweswariah SS. Functional inactivation of the human guanylyl cyclase C receptor: modeling and mutation of the protein kinase-like domain. *Biochemistry* 2001;**40**:9196–206.
83. Gazzano H, Wu HI, Waldman SA. Activation of particulate guanylate cyclase by Escherichia coli heat-stable enterotoxin is regulated by adenine nucleotides. *Infect Immun* 1991;**59**:1552–7.
84. Crane JK, Shanks KL. Phosphorylation and activation of the intestinal guanylyl cyclase receptor for Escherichia coli heat-stable toxin by protein kinase C. *Mol Cell Biochem* 1996;**165**:111–20.
85. Li P, Lin JE, Chervoneva I, Schulz S, Waldman SA, Pitari GM. Homeostatic control of the crypt-villus axis by the bacterial enterotoxin receptor guanylyl cyclase C restricts the proliferating compartment in intestine. *Am J Pathol* 2007;**171**:1847–58.
86. Mann EA, Jump ML, Wu J, Yee E, Giannella RA. Mice lacking the guanylyl cyclase C receptor are resistant to STa-induced intestinal secretion. *Biochem Biophys Res Commun* 1997;**239**:463–6.
87. Schulz S, Lopez MJ, Kuhn M, Garbers DL. Disruption of the guanylyl cyclase-C gene leads to a paradoxical phenotype of viable but heat-stable enterotoxin-resistant mice. *J Clin Invest* 1997;**100**:1590–5.
88. Carrithers SL, Parkinson SJ, Goldstein S, Park P, Robertson DC, Waldman SA. Escherichia coli heat-stable toxin receptors in human colonic tumors. *Gastroenterology* 1994;**107**:1653–61.
89. Pitari GM, Di Guglielmo MD, Park J, Schulz S, Waldman SA. Guanylyl cyclase C agonists regulate progression through the cell cycle of human colon carcinoma cells. *Proc Natl Acad Sci USA* 2001;**98**:7846–51.
90. Pitari GM, Zingman LV, Hodgson DM, et al. Bacterial enterotoxins are associated with resistance to colon cancer. *Proc Natl Acad Sci USA* 2003;**100**:2695–9.
91. Shailubhai K, Yu HH, Karunanandaa K, et al. Uroguanylin treatment suppresses polyp formation in the Apc(Min/+) mouse and induces apoptosis in human colon adenocarcinoma cells via cyclic GMP. *Cancer Res* 2000;**60**:5151–7.
92. Fulle HJ, Vassar R, Foster DC, Yang RB, Axel R, Garbers DL. A receptor guanylyl cyclase expressed specifically in olfactory sensory neurons. *Proc Natl Acad Sci USA* 1995;**92**:3571–5.
93. Juilfs DM, Fulle HJ, Zhao AZ, Houslay MD, Garbers DL, Beavo JA. A subset of olfactory neurons that selectively express cGMP-stimulated phosphodiesterase (PDE2) and guanylyl cyclase-D define a unique olfactory signal transduction pathway. *Proc Natl Acad Sci USA* 1997;**94**:3388–95.
94. Meyer MR, Angele A, Kremmer E, Kaupp UB, Muller F. A cGMP-signaling pathway in a subset of olfactory sensory neurons. *Proc Natl Acad Sci USA* 2000;**97**:10,595–10,600.
95. Leinders-Zufall T, Cockerham RE, Michalakis S, et al. Contribution of the receptor guanylyl cyclase GC-D to chemosensory function in the olfactory epithelium. *Proc Natl Acad Sci USA* 2007;**104**: 14,507–14,512.
96. Young JM, Waters H, Dong C, Fulle HJ, Liman ER. Degeneration of the olfactory guanylyl cyclase D gene during primate evolution. *PLoS ONE* 2007;**2**:e884.
97. Lolley RN, Racz E. Calcium modulation of cyclic GMP synthesis in rat visual cells. *Vision Res* 1982;**22**:1481–6.
98. Koch KW, Stryer L. Highly cooperative feedback control of retinal rod guanylate cyclase by calcium ions. *Nature* 1988;**334**:64–6.
99. Lowe DG, Dizhoor AM, Liu K, et al. Cloning and expression of a second photoreceptor-specific membrane retina guanylyl cyclase (RetGC), RetGC-2. *Proc Natl Acad Sci USA* 1995;**92**:5535–9.
100. Shyjan AW, de Sauvage FJ, Gillett NA, Goeddel DV, Lowe DG. Molecular cloning of a retina-specific membrane guanylyl cyclase.. *Neuron* 1992;**9**:727–37.
101. Yang RB, Foster DC, Garbers DL, Fulle HJ. Two membrane forms of guanylyl cyclase found in the eye. *Proc Natl Acad Sci USA* 1995;**92**:602–6.
102. Dizhoor AM, Lowe DG, Olshevskaya EV, Laura RP, Hurley JB. The human photoreceptor membrane guanylyl cyclase, RetGC, is present in outer segments and is regulated by calcium and a soluble activator. *Neuron* 1994;**12**:1345–52.
103. Gorczyca WA, Gray-Keller MP, Detwiler PB, Palczewski K. Purification and physiological evaluation of a guanylate cyclase

activating protein from retinal rods. *Proc Natl Acad Sci USA* 1994;**91**:4014–18.

104. Palczewski K, Subbaraya I, Gorczyca WA, et al. Molecular cloning and characterization of retinal photoreceptor guanylyl cyclase-activating protein. *Neuron* 1994;**13**:395–404.
105. Dizhoor AM, Olshevskaya EV, Henzel WJ, et al. Cloning, sequencing, and expression of a 24-kDa Ca(2+)-binding protein activating photoreceptor guanylyl cyclase. *J Biol Chem* 1995;**270**:25,200–25,206.
106. Yang RB, Robinson SW, Xiong WH, Yau KW, Birch DG, Garbers DL. Disruption of a retinal guanylyl cyclase gene leads to cone-specific dystrophy and paradoxical rod behavior. *J Neurosci* 1999;**19**:5889–97.
107. Tucker CL, Ramamurthy V, Pina AL, et al. Functional analyses of mutant recessive GUCY2D alleles identified in Leber congenital amaurosis patients: protein domain comparisons and dominant negative effects. *Mol Vis* 2004;**10**:297–303.
108. Hanein S, Perrault I, Gerber S, et al. Leber congenital amaurosis: comprehensive survey of the genetic heterogeneity, refinement of the clinical definition, and genotype-phenotype correlations as a strategy for molecular diagnosis. *Hum Mutat* 2004;**23**:306–17.
109. Baehr W, Karan S, Maeda T, et al. The function of guanylate cyclase 1 and guanylate cyclase 2 in rod and cone photoreceptors. *J Biol Chem* 2007;**282**:8837–47.
110. Mendez A, Burns ME, Sokal I, et al. Role of guanylate cyclase-activating proteins (GCAPs) in setting the flash sensitivity of rod photoreceptors. *Proc Natl Acad Sci USA* 2001;**98**:9948–53.
111. Pennesi ME, Howes KA, Baehr W, Wu SM. Guanylate cyclase-activating protein (GCAP) 1 rescues cone recovery kinetics in GCAP1/GCAP2 knockout mice. *Proc Natl Acad Sci USA* 2003;**100**:6783–8.
112. Howes KA, Pennesi ME, Sokal I, et al. GCAP1 rescues rod photoreceptor response in GCAP1/GCAP2 knockout mice. *EMBO J* 2002;**21**:1545–54.
113. Palczewski K, Sokal I, Baehr W. Guanylate cyclase-activating proteins: structure, function, and diversity. *Biochem Biophys Res Commun* 2004;**322**:1123–30.
114. Schulz S, Green CK, Yuen PS, Garbers DL. Guanylyl cyclase is a heat-stable enterotoxin receptor. *Cell* 1990;**63**:941–8.
115. Schulz S, Wedel BJ, Matthews A, Garbers DL. The cloning and expression of a new guanylyl cyclase orphan receptor. *J Biol Chem* 1998;**273**:1032–7.
116. Kuhn M, Ng CK, Su YH, et al. Identification of an orphan guanylate cyclase receptor selectively expressed in mouse testis. *Biochem J* 2004;**379**:385–93.
117. Lin H, Cheng CF, Hou HH, et al. Disruption of guanylyl cyclase-G protects against acute renal injury. *J Am Soc Nephrol* 2008;**19**:339–48.
118. Manning G, Whyte DB, Martinez R, Hunter T, Sudarsanam S. The protein kinase complement of the human genome. *Science* 2002;**298**:1912–34.

Section D

Other Transmembrane Signaling Proteins

Part 1

Adhesion Molecules

Chapter 41

Mechanistic Features of Cell-Surface Adhesion Receptors

Steven C. Almo[1, 2], Anne R. Bresnick[1], and Xuewu Zhang[3]

[1] *Department of Biochemistry, Albert Einstein College of Medicine, Bronx, New York, New York*

[2] *Center for Synchrotron Biosciences*

[3] *Department of Cell Biology, Albert Einstein College of Medicine, Bronx, New York, New York*

Living cells constantly interact with their environment. As a consequence, a number of sensory systems have evolved for the collection, processing, and integration of a remarkable range of environmental stimuli arising from cell–cell and cell–substrate interactions. For instance, developmental and morphological processes in higher eukaryotes rely on the orchestrated migration of cells in response to specific physical and chemical cues; T cell activation relies on the localization and compartmentalization of cell-adhesion and signaling molecules; and adherent cells must respond to a variety of intracellular and extracellular mechanical forces. All of these processes rely on the engagement of specific cell-surface receptors with the appropriate extracellular ligand to report on the immediate physical environment by transducing extracellular signals across the plasma membrane. This review examines the diversity of mechanisms thought to be involved in adhesion and signaling and highlights some of the shared principles that must be considered for all signaling pathways utilizing cell-surface receptors.

MECHANOSENSORY MECHANISMS

The ability to detect and respond to alterations in applied mechanical force is required for a number of cellular and developmental functions. This is particularly critical for adherent cells that directly contact the extracellular matrix (ECM) and are subject to considerable physical deformation. For example, sheer forces associated with blood flow are major determinants of arterial tone and vascular reorganization. At the cellular level, morphology and orientation are optimized to minimize mechanical stress and damage associated with variations in flow-related forces (see, for example, [1–3]). Similarly, fibroblasts must be highly responsive to the mechanical forces associated with alterations in the ECM (reviewed in [4]).

Considerable evidence points to focal adhesions, the sites of cell–substrate contact, as the sensors of mechanical force. Central to focal adhesion assembly and function are the integrins, a family of α–β heterodimeric transmembrane glycoproteins that provide essential adhesive functions for cell migration and the establishment and maintenance of normal tissue architecture. At least 18α and 8β chains allow for the formation of multiple integrin heterodimers that are able to display a spectrum of specificities for cell-surface adhesion molecules and for a range of ECM components, including laminin, collagen, and fibronectin. The integrin cytoplasmic domains bind a variety of scaffolding and actin regulatory proteins, which in turn recruit a large number of adaptor and signaling molecules. These physical links couple the integrins to the downstream activation of numerous signaling molecules, including MAP kinase, focal adhesion kinase, Src, and PI3-kinase (see, for example, [4, 5]). Furthermore, integrin affinity is modulated by the activation state of the particular cell in question, and this "inside-out" signaling is thought to control the tertiary and quaternary structural rearrangements required for high-affinity ligand binding. The focal adhesion may thus be viewed as a highly dynamic sensory organelle that exploits the direct linkage between the ECM and actin cytoskeleton to respond to mechanical force through a wide range of signaling pathways.

The mechanisms underlying integrin-associated signaling rely on the determinants of mechanical strain, including tension provided by cytoskeletal motor proteins, such as myosin-II, and the intrinsic mechanical properties of the underlying ECM (Figure 41.1). For example, the growth of cells on soft, or pliable, surfaces does not support integrin signaling nor the formation of focal adhesions [6], while

Handbook of Cell Signaling, Three-Volume Set 2 ed.

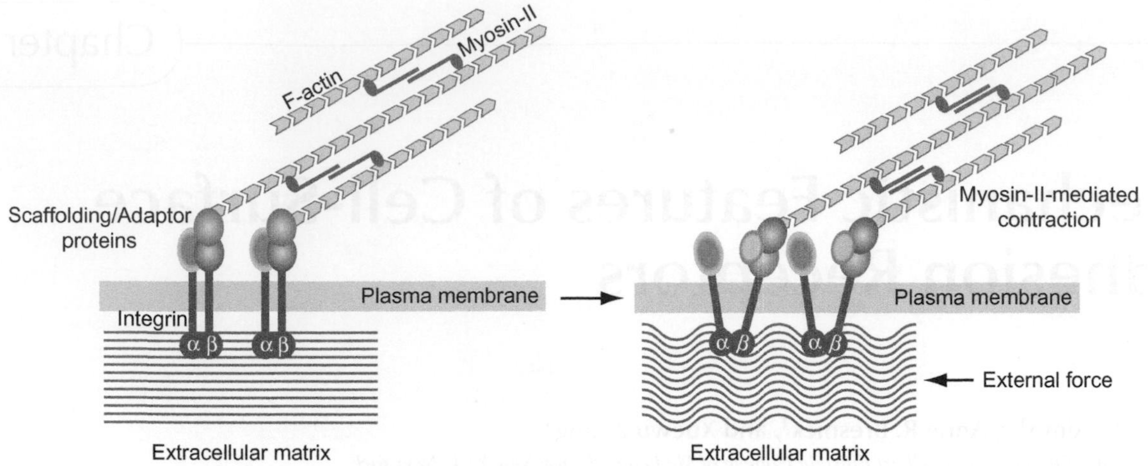

FIGURE 41.1 Model for the mechanochemical signaling mechanism of integrins at focal adhesions. The extracellular domain of integrin binds ECM components, such as collagen and fibronectin. The cytoplasmic domain of integrin contacts a series of scaffolding proteins and cytoskeletal regulatory proteins (pink ellipse), including talin and paxillin, which provides a direct physical linkage between the ECM and the actomyosin cytoskeleton. Alterations in the ECM generate tension that may result in tertiary and quaternary structural changes (illustrated here as a scissor-like motion between the α- and β-integrin chains). These structural changes are propagated to the cytoplasm, which may uncover cryptic binding and recruitment sites for additional signaling molecules (blue ellipse). The ability to couple force generation to alterations in the composition of integrin-associated focal adhesion molecules provides a direct mechanism for mechanochemical signaling. (Color shown in online version.)

"stretching" of these substrates supports both focal adhesion formation and integrin signaling [7, 8], presumably by allowing for a sufficient level of tension to be achieved. At the molecular level, mechanical force may be transduced into a cytoplasmic signal through a number of possible mechanisms. The application of force may disrupt or distort various intermolecular binding interfaces, resulting in the reorganization of focal adhesions by enhancing the entry or exit of specific signaling molecules through either free or facilitated diffusion. A related potential mechanism is the force-induced conformational reorganization of integrin-associated focal adhesion molecules, which may uncover cryptic binding and recruitment sites for additional signaling molecules. This notion is consistent with the fact that a number of focal adhesion components, including vinculin and ERM proteins, exist in multiple conformations (see [9, 10] and references therein). Of special note are a series of structural [11–15] and biochemical studies (reviewed in references [12, 13, 16]) describing the localized ligand-induced conformational rearrangements and a model for integrin activation [12, 13]. This model suggests that a large-scale conformational reorganization, including a scissor-like motion, may be required for high-affinity ligand binding. Some aspects of this conformational plasticity may also play a role in transducing mechanical force into cytoplasmic signals. These mechanisms, whether affecting the dynamic assembly/disassembly properties of the focal adhesion as a whole or directing conformational reorganization of a specific focal adhesion protein, can provide a direct linkage between cell surface-ECM adhesive interactions, focal adhesion composition, and cytoplasmic signaling. Furthermore, recent studies demonstrate a complex rela tionship between valency and geometric organization of the ligand and the strength of integrin-associated signaling [17], suggesting some mechanistic similarities with the c described below. Thus, integrin-associated signaling provides one of the clearest couplings of signaling and the adhesive properties of a receptor–ligand pair.

CELL–CELL ADHESIONS/ADHERENS JUNCTIONS

The cadherins are a family of cell-surface receptors that form calcium-dependent homophilic interactions between the surfaces of adjacent cells. These interactions result in the formation of intercellular adhesions, adherens junctions, which play essential roles in the establishment and maintenance of cell polarity and tissue architecture and in the recognition and migratory events associated with developmental and morphological processes. These adhesive interactions are supported by a catenin-mediated linkage to the underlying actin cytoskeleton, as the carboxy-terminal cytoplasmic tail of cadherin binds β-catenin, and via an interaction with α-catenin is linked to the cortical actin network (Figure 41.2) (see [18] and references therein). The importance of this cytoskeletal connection is highlighted by the observation that disruption of normal catenin function prevents the formation of mature adherens junctions and is associated with increased motility and invasiveness of tumor cells (reviewed in [19]).

β-Catenin plays a dual role in cell physiology, as in addition to being an essential structural component of the adherens junction it serves as a transcriptional activator of several

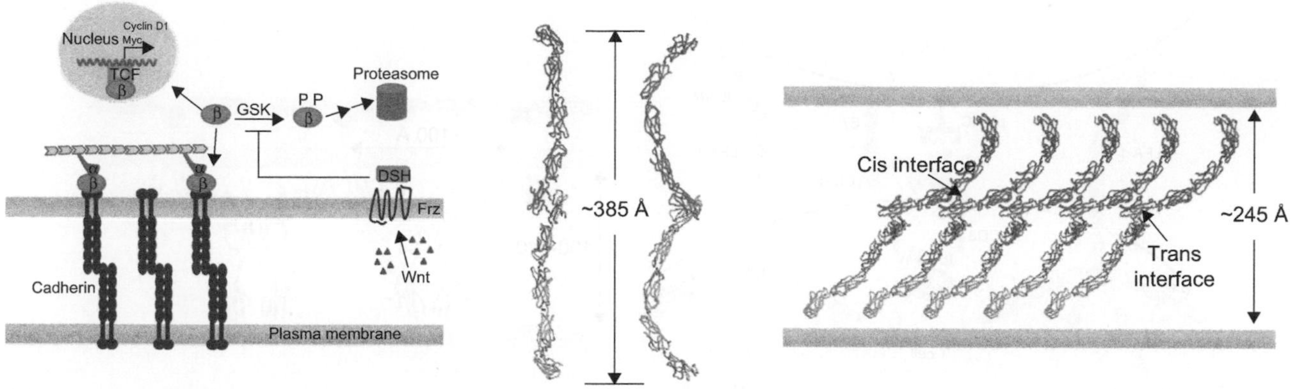

(a) (b) (c)

FIGURE 41.2 Adherens junctions and cadherin function.
(a) Schematic of adherens junction organization and associated signaling pathways. The catenins provide a direct physical linkage between the homophilic cadherin-mediated cell–cell contacts and the underlying actin cytoskeleton and support the integrity of the adherens junctions. In turn, the actin cytoskeleton provides β-catenin docking sites that serve to modulate β-catenin signaling by buffering the soluble concentration of β-catenin. (b) Structure of C-cadherin showing the arched arrangement formed by the five individual cadherin domains (EC1–EC5) and a model for the trans (cell–cell) interaction from abutting EC1 domains. Two orthogonal views are shown, with the arched nature of the structure evident in the right figure. (c) A model of the trans and cis interactions at the adherens junction based on contacts present in the C-cadherin crystal structure. In this model, the individual cadherin molecules in the adherens junction are tilted by ~45° with respect to the plasma membrane, implying an intermembrane separation of ~245 Å. Figure 41.2a is adapted from [18] (Conacci-Sorrell *et al.*, 2002).

genes involved in cellular proliferation and invasion, including Myc, cyclin D1, metalloproteinases, and fibronectin [18]. A number of regulatory mechanisms modulate β-catenin signaling. In the absence of Wnt signaling, cytoplasmically disposed soluble β-catenin is a substrate for phosphorylation by glycogen synthase phosphorylase, which serves to mark it for degradation by the 26S proteasome; however, activation of the Wnt pathway inhibits this phosphorylation and β-catenin is shunted to the nucleus, where it forms a complex with the T cell factor (TCF) to activate selected genes. The formation of normal adherens junctions appears critical for control of β-catenin signaling, as a loss of cadherin expression correlates with increased nuclear β-catenin. Thus, there appears to be a close linkage between cadherin-mediated adhesion and β-catenin-mediated signaling pathways, with the adherens junction acting as a buffer of soluble β-catenin (Figure 41.2) [20].

Structural studies have suggested several models for the homophilic adhesive interactions formed by the cadherins at adherens junctions. The recent report of the structure of the entire extracellular domain of C-cadherin by Boggon and colleagues [21] provides new insights into both the cis (intracellular) and trans (intercellular) interactions that are essential for the formation and maintenance of adherens junctions (Figure 41.2). The structure shows that the five extracellular cadherin domains (EC1–EC5) form an arched structure, and the abutment of two N-terminal EC1 domains in the crystal provides a model for the trans adhesive interaction. Additional crystal contacts suggest a model of the cis contact, and together the interactions observed in the crystalline state provide a detailed model for the periodic organization of cadherin molecules within the adherens junction. Of particular note is the suggestion that the cadherin molecules in the adherens junction are tilted by ~45° with respect to the plasma membrane, implying an intermembrane separation of ~245 Å. This feature of the model is particularly noteworthy, as there is a strong bias to view intrinsic membrane proteins as projecting perpendicular to the plane of the plasma membrane; *a priori* there is no fundamental reason for this assumption.

T CELL CO-STIMULATION

An optimal T cell response requires the integration of a number of distinct extracellular signaling and adhesive events at the T cell-antigen-presenting cell (APC) interface, which has been termed the immunological synapse. Engagement of T cell receptors (TCRs) on the surfaces of T cells with major histocompatibility complex (MHC)/peptide complexes displayed on the surfaces of APCs is essential, but not sufficient, for complete T cell activation [22]. The subsequent engagement of a series of co-stimulatory receptor–ligand pairs provides the additional signals needed for efficient T cell activation, as well as the negative signals required to attenuate the immune response (Figure 41.3) [23–25]. The most extensively characterized T cell co-stimulatory receptors are CD28 and CTLA-4, which share ~30 percent identity and bind the B7-1 and B7-2 ligands presented on APCs. Together with signaling through the TCR, the engagement of CD28 by the B7 ligands leads to optimal T cell activation [22], while the interaction of B7 with CTLA-4 provides inhibitory signals required for downregulation of the response.

Initial TCR engagement is followed by a remarkable reorganization and compartmentalization of signaling and adhesive molecules at the immunological synapse. The central zone of

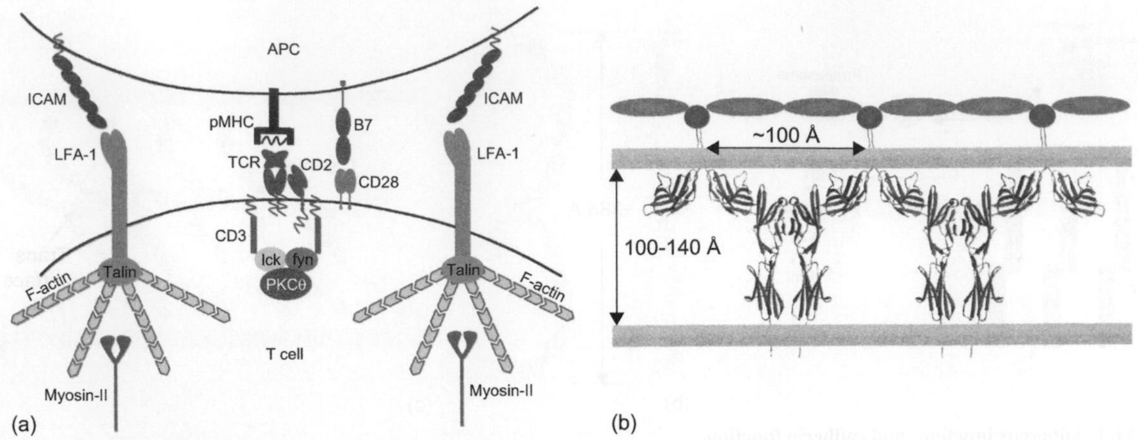

FIGURE 41.3 T cell activation and the immunological synapse.
(a) Schematic of the immunological synapse highlighting the compartmentalization of specific signaling components into discrete zones. The central zone is enriched in cell-surface signaling molecules (i.e., TCR, MHC/peptide complex, and co-stimulatory receptors and ligands) and cytoplasmically associated scaffolding and signaling proteins (i.e., Src family kinases, etc.). Surrounding this signaling complex is the peripheral zone, which is composed of large adhesion molecules and cytoplasmically associated cytoskeletal components required for the observed pattern of localization. (b) Model for the co-stimulatory signaling network at the T cell–APC interface. The disulfide-linked CTLA-4 dimers are shown in red, while the non-covalent B7-1 dimers are blue. The interactions between these two dimeric bivalent molecules in the crystal result in a periodic array of CTAL-4 and B7 homodimers with a characteristic spacing of ~100 Å. This periodicity may result in the organized recruitment of signaling molecules (pink and red) and may in some circumstances provide further adhesive interactions required for productive signaling. (Color shown in online version.)

the synapse contains the receptor–ligand pairs, including the TCR-CD3/MHC-peptide complex, CD28/B7 co-stimulatory complex, and CD2/CD58 complexes, as well as non-covalently associated intracellular signaling molecules, such as fyn, lck, and PKC-theta [26]. The central zone is bordered by the peripheral zone, which is composed of large adhesion molecules, including LFA-1 and ICAM-1, and components of the actin cytoskeleton (Figure 41.3) [26]. This organization appears to be dependent on an uncompromised actomyosin cytoskeleton, thus providing another example of the intimate involvement of the actin-based cytoskeleton in a fundamental signaling pathway. A number of potential functions have been proposed for the molecular organization in the synapse, including the polarized secretion of cytokines, TCR recycling, and the promotion of co-stimulatory receptor–ligand engagement [27, 28]. In addition, the B7 ligands appear to control APC function, as crosslinking the B7 isoforms modulates both B cell proliferation and antibody production [22, 29–31]. Thus, engagement of the co-stimulatory receptor–ligand pairs represents an outstanding example of bidirectional signaling.

Of particular note are the recent structural descriptions of the CTLA-4/B7 receptor–ligand complexes, which exhibit an alternating arrangement of bivalent CTLA-4 and B7 dimers (Figure 41.3) [32, 33]. The observation of this linear periodic array suggests a model for the organization of these cell surface molecules at the immunological synapse. Importantly, the observed spacing between the extracellular receptor domains is also imposed on any cytoplasmically associated signaling molecules, and suggests that the oligomerization of multiple (i.e., at least two) CTLA-4 dimers may be required to afford a biologically optimal organization and local concentration of intracellular signaling molecules.

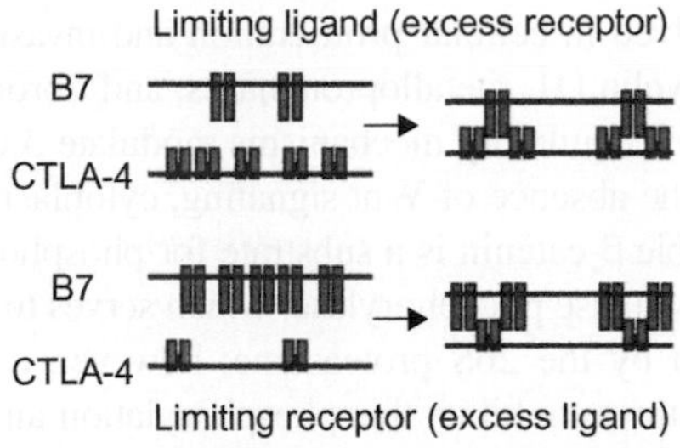

FIGURE 41.4 Effect of stoichiometry on the signaling complexes formed by multivalent receptor–ligand pairs.
(Top) Limiting ligand will favor the formation of cell-surface complexes composed of two receptor dimers (e.g., CTLA-4) linked by a single ligand dimer (e.g., B7-1). This assembly would impose a constraint between the two adjacent receptors and any associated cytoplasmic signaling molecules (i.e., ~100 Å in the case of the CTLA-4/B7 complex). (Bottom) Excess ligand would favor complexes composed of a single receptor linking two independent ligand dimers. This association would not enforce any specific spatial relationship between individual receptor molecules but would still direct the localization of the receptor and ligand to the immunological synapse and could result in a sufficiently high local concentration of individual receptor dimers to support signaling.

In considering the types of assemblies that are formed *in vivo* by multivalent receptor–ligand pairs, it is essential to bear in mind the relative concentrations of the binding partners (Figure 41.4). For example, a large excess of either receptor or ligand will favor the formation of "isolated" signaling complexes. In the case of limiting ligand, a cell-surface complex composed of two receptor dimers (e.g., CTLA-4) linked by a single ligand dimer (e.g., B7-1) would be favored, and such an assembly would impose an ~100-Å constraint between the two adjacent receptors and any associated cytoplasmic signaling molecules. In contrast, the presence of excess ligand would favor complexes composed of a single receptor linking two independent ligand dimers. This association would not enforce any specific spatial relationship between individual receptor molecules but would still direct the localization of the receptor and ligand to the immunological synapse and could result in a sufficiently high local concentration of individual CTLA-4 dimers to support signaling. Finally, equivalent amounts of receptor and ligand at a cell–cell interface would favor the formation of more extensive periodic networks. Importantly, this is a general consideration relevant to all multivalent receptor–ligand pairs.

In addition to playing a direct role in signaling, cell-surface receptor–ligand engagement constrains the approach of the adjacent plasma membranes (as in the case of the adherens junction discussed above) and may play a role in directing the organization of molecules at the cell–cell or cell–ECM interface. The maximal dimension of the CTLA-4/B7 complexes (~100-140 Å) is compatible with those of other receptor–ligand pairs present in the central zone of the synapse (i.e., MHC/TCR [34, 35] and CD2/CD58 [36]). In contrast, the adhesive complexes present in the peripheral zone (e.g., the LFA-1/ICAM-1 complex) are significantly larger in maximal extent, and this difference has led to the suggestion that the compartmentalization observed in the immunological synapse is the consequence of a mechanical sorting mechanism based on relative molecular dimension [37, 38]. While this is an appealing hypothesis, it is based on the assumption that intrinsic membrane proteins extend perpendicular to the plasma membrane and ignores the possibility that a molecule of large extent can be accommodated within the central zone by tilting with respect to the plasma membrane, as was suggested in the model of C-cadherin in the adherens junction (Figure 41.2).

While the adhesive functions of ICAM and LFA-1 are essential to synapse formation and T cell function, engagement of these molecules is also likely to play a direct signaling role in T cell activation and function. Recent studies have shown that ICAM-1 binding is associated with LFA-1 clustering, enhanced actin polymerization, and F-actin bundling within T cells [39]. Conversely, crosslinking of ICAM-1 in lymphocytes stimulates calcium signaling and PKC activity, which results in cytoskeletal rearrangements associated with migration [40]. These observations indicate a strong coupling between adhesive and signaling functions and suggest that reciprocal bidirectional signaling may be associated with ICAM/LFA-1 adhesive interactions (see, for example, [41]).

As the localization of adhesive partners at cell–cell and cell–ECM interfaces necessarily results in the localization of cytoplasmically associated species, it is relevant to ask whether situations exist in which adhesive functions are fully uncoupled from signaling events. For instance, the one-dimensional lattice observed in the CTLA-4/B7 crystal structures exhibits considerable similarities to the adhesive assembly formed by the cadherins (Figure 41.2), and on this basis it is tempting to suggest that co-stimulatory receptor–ligand engagement might also provide adhesive interactions required for efficient T cell function. Although no data bear directly on the adhesive properties of CTLA-4, recent studies indicate that CD28 does not make any significant contri butions to the adhesive properties of naïve T cells [37]. These results differ from earlier studies indicating that the CD28/B7 interaction significantly enhanced adhesion. However, these earlier studies utilized systems in which either receptor or ligand was overexpressed [42,43], again stressing the importance of accurately knowing the cell surface densities of the binding partners in order to correctly predict mechanism. These recent studies also indicated that only ~30 percent of the CD28 molecules exhibited free lateral diffusion in the plasma membrane [37], implying that only a fraction of the total population may be available to bind B7 at the immunological synapse. While no evidence supports limited diffusional freedom as a general feature of cell-surface proteins, these studies nonetheless stress the potential importance of considering the "available" receptor and ligand concentrations, as opposed to total cellular concentrations.

AXON GUIDANCE AND NEURAL DEVELOPMENT

The Eph family of receptor tyrosine kinases and their associated ephrin ligands play a central role in neural development by providing repulsive guidance cues that direct axonal targeting. Specifically, a migrating growth cone expressing a given Eph receptor will turn away from cells expressing cognate ephrin ligands, as a result of the disassembly or redistribution of filamentous actin networks at the leading edge [44]. Two classes of ephrins are defined on the basis of their mode of cell surface attachment. The ephrin A ligands utilize a glycophosphatidylinositol (GPI) linkage for cell-surface attachment and bind the EphA receptors, while ephrinB ligands are transmembrane proteins that bind EphB receptors.

Recent structural characterization of the ephrin-B2/EphB2 receptor complex provides new insights into the potential signaling mechanisms utilized (Figure 41.5) [45]. This structure provides details of the receptor–ligand binding site and of a "circular" 2:2 receptor–ligand complex that is thought to be relevant to signaling. The organization observed in the crystal structure is consistent with ligand-induced clustering of the EphB2 receptor, resulting in the trans-autophosphorylation required for activation and subsequent recruitment of signaling molecules, including src family kinases and GTP-activating proteins (GAPS) [46]. Engagement also results in clustering of the ephrin ligand, providing another example of bidirectional signaling, as the

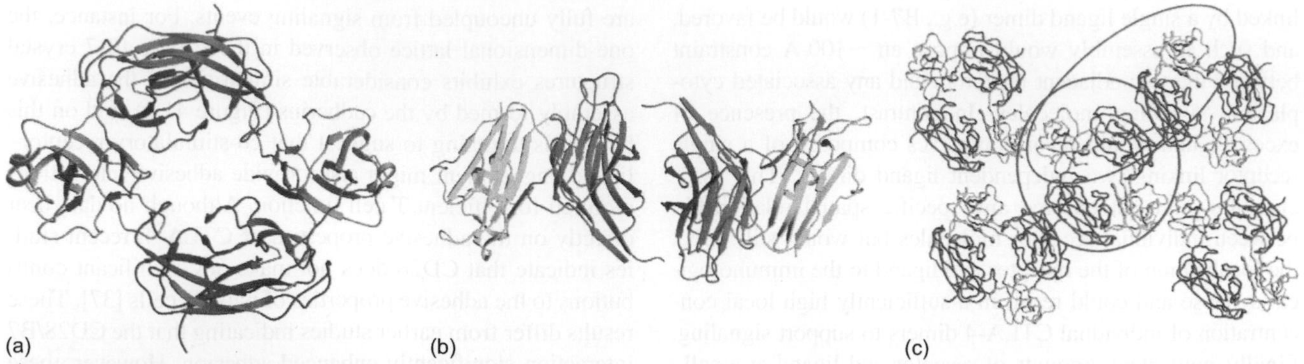

(a) (b) (c)

FIGURE 41.5 Structure of the ephrin-B2/EphB2 receptor complex.
(a) Circular tetramer formed by the interaction of two EphB2 receptors (green) with two ephrinB2 ligands (yellow) thought to represent the favored receptor–ligand organization *in vivo*. Note that each ligand contacts two receptor molecules, but there are no ligand–ligand or receptor–receptor contacts. (b) Crystal packing results in another tetramer (elliptical), in which an extensive interface is formed between two receptor molecules. The physiological relevance of this binding interaction remains to be proven but may be consistent with the propensity of Eph/ephrin molecules to form higher order oligomers. (c) A "layer" from the ephrin-B2/EphB2 receptor complex crystal structure showing the long-range, two-dimensional ordered array formed by the combination of both the circular (highlighted in red) and elliptical (highlighted in blue) tetramers. Such an organized network could potentially play roles in signaling and/or adhesion. (Color shown in online version.)

cytoplasmic domain of ephrin-B2 is required for normal angiogenesis and vascular morphogenesis [46]. Furthermore, consistent with the propensity to form higher order oligomers, the crystal structure suggests the formation of an extended two-dimensional signaling complex (supercluster) of receptors and ligands at the cell–cell interface (in contrast to the one-dimensional array proposed for the CTLA-4/B7 complexes), which might afford enhanced signaling.

The proposed long-range organization suggests that, in addition to a direct role in signaling, engagement of the Eph receptor-ligand pairs may also provide essential adhesive functions. The first evidence supporting this notion came from the observation that ~17 percent of mice defective in ephrinA5 exhibit neural tube defects, which is not consistent with the classical repulsive effects attributed to ephrin/Eph receptor function [47]. These studies also revealed that the expression of splice variants of an ephrinA5 receptor (i.e., EphA7) which lack the intracellular kinase domain support direct adhesive interactions with ephrinA5-expressing cells [48]. This provides yet another example of the close linkage between signaling and adhesive interactions.

CONCLUSIONS

As illustrated, biology depends on a vast array of information processing activities that are coordinated by diverse cell-surface adhesion receptors and their cognate ligands. Though these receptor–ligand pairs differ in chemical and structural terms, there are common principles that must be carefully considered in order to construct viable molecular and atomic mechanisms for signaling. The engagement of receptor–ligand pairs leads to an increase in their local density/concentration at cell–cell and cell–ECM interfaces, and in many cases may support a natural coupling between signaling and adhesive function. Of particular importance is the quantitative understanding of both cell-surface oligomeric state and the available concentration of receptor and ligand on their cell surfaces, as they dictate the relative stoichiometries and the type of signaling complexes that can be formed at cell-cell and cell–ECM interfaces. Finally, as a general cautionary note, while direct structural information, in the form of X-ray and nuclear magnetic resonance (NMR) structures, may provide enormous insights into function and mechanism, in the absence of confirmatory biochemical data great care should be exercised in extrapolating intermolecular contacts observed in crystal structures to physiologically relevant protein–protein interfaces.

REFERENCES

1. Girard PR, Nerem RM. Shear stress modulates endothelial cell morphology and F-actin organization through the regulation of focal adhesion-associated proteins. *J Cell Physiol* 1995;**163**:179–93.
2. Girard PR, Nerem RM. Endothelial cell signaling and cytoskeletal changes in response to shear stress. *Front Med Biol Eng* 1993;**5**:31–6.
3. Tzima E, del Pozo MA, Shattil SJ, Chien S, Schwartz MA. Activation of integrins in endothelial cells by fluid shear stress mediates Rho-dependent cytoskeletal alignment. *EMBO J* 2001;**20**:4639–47.
4. Schwartz MA, Ginsberg MH. Networks and crosstalk: integrin signalling spreads. *Nat Cell Biol* 2002;**4**:E65–8.
5. Geiger B, Bershadsky A. Exploring the neighborhood: adhesion-coupled cell mechanosensors. *Cell* 2002;**110**:139–42.
6. Pelham RJ, Jr. Wang Y. Cell locomotion and focal adhesions are regulated by substrate flexibility. *Proc Natl Acad Sci USA* 1997;**94**:13,661–5.
7. Sawada Y, Sheetz MP. Force transduction by Triton cytoskeletons. *J Cell Biol* 2002;**156**:609–15.
8. Wang HB, Dembo M, Hanks SK, Wang Y. Focal adhesion kinase is involved in mechanosensing during fibroblast migration. *Proc Natl Acad Sci USA* 2001;**98**:11,295–300.

9. Johnson RP, Craig SW. Actin activates a cryptic dimerization potential of the vinculin tail domain. *J Biol Chem* 2000;**275**:95–105.
10. Bretscher A, Edwards K, Fehon RG. ERM proteins and merlin: integrators at the cell cortex. *Natl Rev Mol Cell Biol* 2002;**3**:586–99.
11. Emsley J, Knight CG, Farndale RW, Barnes MJ, Liddington RC. Structural basis of collagen recognition by integrin $\alpha 2\beta 1$. *Cell* 2000;**101**:47–56.
12. Beglova N, Blacklow SC, Takagi J, Springer TA. Cysteine-rich module structure reveals a fulcrum for integrin rearrangement upon activation. *Natl Struct Biol* 2002;**9**:282–7.
13. Liddington RC. Will the real integrin please stand up? *Structure (Camb)* 2002;**10**:605–7.
14. Xiong JP, et al. Crystal structure of the extracellular segment of integrin $\alpha V\beta 3$ in complex with an Arg-Gly-Asp ligand. *Science* 2002;**296**:151–5.
15. Xiong JP, et al. Crystal structure of the extracellular segment of integrin $\alpha V\beta 3$. *Science* 2001;**294**:339–45.
16. Shimaoka M, Takagi J, Springer TA. Conformational regulation of integrin structure and function. *Annu Rev Biophys Biomol Struct* 2002;**31**:485–516.
17. Koo LY, Irvine DJ, Mayes AM, Lauffenburger DA, Griffith LG. Co-regulation of cell adhesion by nanoscale RGD organization and mechanical stimulus. *J Cell Sci* 2002;**115**:1423–33.
18. Conacci-Sorrell M, Zhurinsky J, Ben-Ze'ev A. The cadherin–catenin adhesion system in signaling and cancer. *J Clin Invest* 2002;**109**:987–91.
19. Okegawa T, Li Y, Pong RC, Hsieh JT. Cell adhesion proteins as tumor suppressors. *J Urol* 2002;**167**:1836–43.
20. Gottardi CJ, Wong E, Gumbiner BM. E-cadherin suppresses cellular transformation by inhibiting β-catenin signaling in an adhesion-independent manner. *J Cell Biol* 2001;**153**:1049–60.
21. Boggon TJ, et al. C-cadherin ectodomain structure and implications for cell adhesion mechanisms. *Science* 2002;**296**:1308–13.
22. Lenschow DJ, Walunas TL, Bluestone JA. CD28/B7 system of T cell costimulation. *Annu Rev Immunol* 1996;**14**:233–58.
23. Nishimura H, Nose M, Hiai H, Minato N, Honjo T. Development of lupus-like autoimmune diseases by disruption of the PD-1 gene encoding an ITIM motif-carrying immunoreceptor. *Immunity* 1999;**11**:141–51.
24. Nishimura H, et al. Autoimmune dilated cardiomyopathy in PD-1 receptor-deficient mice. *Science* 2001;**291**:319–22.
25. Greenwald RJ, Boussiotis VA, Lorsbach RB, Abbas AK, Sharpe AH. CTLA-4 regulates induction of anergy *in vivo*. *Immunity* 2001;**14**:145–55.
26. Bromley SK, et al. The immunological synapse. *Annu Rev Immunol* 2001;**19**:375–96.
27. Lee KH, et al. T cell receptor signaling precedes immunological synapse formation. *Science* 2002;**295**:1539–42.
28. van Der Merwe PA, Davis SJ. Immunology: the immunological synapse – a multitasking system. *Science* 2002;**295**:1479–80.
29. Hirokawa M, Kuroki J, Kitabayashi A, Miura AB. Transmembrane signaling through CD80 (B7-1) induces growth arrest and cell spreading of human B lymphocytes accompanied by protein tyrosine phosphorylation. *Immunol Letts* 1996;**50**:95–8.
30. Suvas S, Singh V, Sahdev S, Vohra H, Agrewala JN. Distinct role of CD80 and CD86 in the regulation of the activation of B cell and B cell lymphomas. *J Biol Chem* 2001;**28**:28.
31. Jeannin P, et al. CD86 (B7-2) on human B cells. A functional role in proliferation and selective differentiation into IgE- and IgG4-producing cells. *J Biol Chem* 1997;**272**:15,613–9.
32. Schwartz JC, Zhang X, Fedorov AA, Nathenson SG, Almo SC. Structural basis for co-stimulation by the human CTLA-4/B7-2 complex. *Nature* 2001;**410**:604–8.
33. Stamper CC, et al. Crystal structure of the B7-1/CTLA-4 complex that inhibits human immune responses. *Nature* 2001;**410**:608–11.
34. Garboczi DN, et al. Structure of the complex between human T-cell receptor, viral peptide and HLA-A2. *Nature* 1996;**384**:134–41.
35. Garcia KC, et al. An $\alpha\beta$ T cell receptor structure at 2.5 Å and its orientation in the TCR–MHC complex. *Science* 1996;**274**:209–19.
36. Wang JH, et al. Structure of a heterophilic adhesion complex between the human CD2 and CD58 (LFA-3) counterreceptors. *Cell* 1999;**97**:791–803.
37. Bromley SK, et al. The immunological synapse and CD28–CD80 interactions. *Nat Immunol* 2001;**2**:1159–66.
38. Wild MK, et al. Dependence of T cell antigen recognition on the dimensions of an accessory receptor–ligand complex. *J Exp Med* 1999;**190**:31–41.
39. Porter JC, Bracke M, Smith A, Davies D, Hogg N. Signaling through integrin LFA-1 leads to filamentous actin polymerization and remodeling, resulting in enhanced T cell adhesion. *J Immunol* 2002;**168**:6330–5.
40. Etienne-Manneville S, et al. ICAM-1-coupled cytoskeletal rearrangements and transendothelial lymphocyte migration involve intracellular calcium signaling in brain endothelial cell lines. *J Immunol* 2000;**165**:3375–83.
41. Lupher Jr. ML, et al Cellular activation of leukocyte function associated antigen-1 and its affinity are regulated at the I domain allosteric site. *J Immunol* 2001;**167**:1431–9.
42. Linsley PS, Clark EA, Ledbetter JA. T-cell antigen CD28 mediates adhesion with B cells by interacting with activation antigen B7/BB-1. *Proc Natl Acad Sci USA* 1990;**87**:5031–5.
43. Kaga S, Ragg S, Rogers KA, Ochi A. Stimulation of CD28 with B7-2 promotes focal adhesion-like cell contacts where Rho family small G proteins accumulate in T cells. *J Immunol* 1998;**160**:24–7.
44. Carter N, Nakamoto T, Hirai H, Hunter T. EphrinA1-induced cytoskeletal re-organization requires FAK and p130(cas). *Nat Cell Biol* 2002;**4**:565–73.
45. Himanen JP, et al. Crystal structure of an Eph receptor–ephrin complex. *Nature* 2001;**414**:933–8.
46. Wilkinson DG. Eph receptors and ephrins: regulators of guidance and assembly. *Intl Rev Cytol* 2000;**196**:177–244.
47. Adams RH, et al. The cytoplasmic domain of the ligand ephrinB2 is required for vascular morphogenesis but not cranial neural crest migration. *Cell* 2001;**104**:57–69.
48. Holmberg J, Clarke DL, Frisen J. Regulation of repulsion versus adhesion by different splice forms of an Eph receptor. *Nature* 2000;**408**:203–6.

Chapter 42

Structural Basis of Integrin Signaling

Robert C. Liddington
Program on Cell Adhesion, The Burnham Institute, La Jolla, California

INTRODUCTION

The integrins are a family of proteins that reside in the plasma membrane of most cells of multicellular organisms [1]. They are the primary receptors that recognize the protein components of the extracellular matrix (ECM). Binding to the ECM triggers intracellular signaling pathways that regulate adhesion, migration, growth, and survival [2]. These pathways often intersect with those generated by receptors for soluble factors [3]. However, integrins differ from "classical" signaling receptors in a number of ways. First, because the ECM is static and polyvalent, integrins cluster at the sites of attachment. Second, ligand-bound integrins form connections with the cytoskeleton that regulate cell shape and rigidity, as well as providing platforms for signaling complexes. Third, integrins can also transmit signals from the inside of the cell to the outside. Thus, integrin signaling is a bidirectional process that evolves rapidly in time and space as the cell adapts to its environment, allowing integrins to be sensors and messengers of the surroundings and shape of the cell, as well as the mechanical forces acting upon it [2].

Integrin signaling typically involves conformational changes within the integrin molecule that are propagated across the plasma membrane. Under some circumstances, lateral self-association of integrins ("clustering") is sufficient for signaling [4, 5], and a number of molecules that associate laterally with integrins have also been identified that contribute to signaling [6]. However, the focus of this section is on the conformational changes within individual molecules that control the recognition of extracellular and intracellular binding partners.

STRUCTURE

Integrins are αβ heterodimers, consisting of a head domain from which emerge two legs, one from each subunit, ending in a pair of single-pass transmembrane helices and short cytoplasmic tails (Figure 42.1). In the absence of ligand, bonds between the legs and tails are believed to hold the head in an inactive or resting conformation that has low affinity for ligand [7, 8]. During outside-in signaling, ECM binding to the head triggers conformational changes that are propagated down the legs and through the plasma membrane, leading to a reorganization of the C-terminal tails that allows them to bind intracellular proteins [3]. During inside-out signaling, cytosolic proteins bind and sequester one or both of

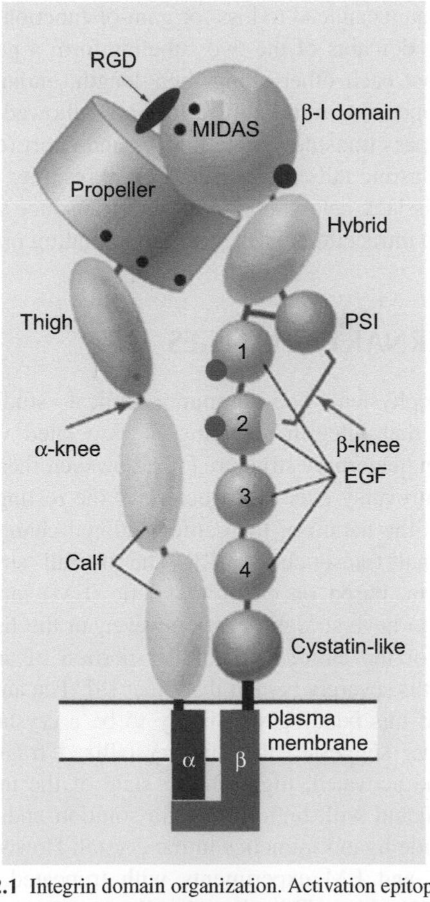

FIGURE 42.1 Integrin domain organization. Activation epitopes [13, 34] are shown as red, blue, and cyan disks (color shown in online version).

Handbook of Cell Signaling, Three-Volume Set 2 ed.

the cytoplasmic tails, triggering conformational changes in the head that lead to a high-affinity active integrin.

The integrin "head" is composed of a seven-bladed propeller from the α-subunit that makes an intimate contact with a GTP-ase-like domain of the β-subunit (called either an A or I domain by different authors, and I domain here), in a manner that strongly resembles the heterotrimeric G proteins [9]. Instead of a catalytic center, the I domain contains an invariant ligand binding site called MIDAS (metal ion-dependent adhesion site), in which a metal ion is coordinated by three loops from the I domain, and a glutamic or aspartic acid from the ligand completes an octahedral coordination sphere around the metal. Specificity is provided by ligand contacts to the surface surrounding the MIDAS, which is highly variable among integrin family members, and in some cases by additional contact to the α-subunit propeller. A helix that emanates from one of the MIDAS loops packs against the central axis of the propeller, thus providing a potential link between ligand binding and the quaternary structure. In certain integrins, an additional I domain (α-I) is inserted into the α-subunit, between two loops on the upper surface of the propeller, where it forms the major ligand binding site. Modeling studies indicate that this domain will form contacts with both the propeller and the β-I domain that regulate the conformation and ligand affinity of the domain, and indeed mutations to the outer surface of the domain can lead to loss- or gain-of-function [10]. The remaining domains of the two subunits form a pair of legs that contact each other along their length, ending at their closely apposed C termini. The legs are followed by a pair of single-pass transmembrane helices and short (except for β4) cytoplasmic tails, typically 20 to 50 residues in length. These tails lack catalytic activity and transduce signals by binding to intracellular structural and signaling proteins.

QUATERNARY CHANGES

Early biophysical and immunochemical studies demonstrated that integrin signaling is associated with large changes in quaternary structure [11]; however, there remains much controversy over the structure of the resting integrin, as well as the nature of the conformational changes underlying signal transduction [12]. The overall structure of the integrin, based on electromagnetic (EM) images, was expected to have straight legs. However, in the first crystal structure of the entire extracellular portion of an integrin (αVβ3), it is severely bent at the knees [9]. The authors proposed that this bending was likely to be a crystal artifact; they further suggested that the crystallized fragment represents the activated, high-affinity state of the integrin, as it binds ligand with high affinity in solution and is able to bind peptide ligand mimetics in the crystal. However, based on NMR and EM experiments with truncated integrins, Springer, Blacklow, Takagii and colleagues have suggested that the "knees-bent" or genuflected integrin represents the inactive conformation *in vivo*, and that a "switch-blade" opening of the integrin is associated with activation [13,14]. In a third model of quaternary changes in integrins, Hantgan and colleagues [15] have provided evidence, using EM and hydrodynamic studies of peptide-bound integrin, that the α- and β-head segments separate on activation. Such a model would extend the analogy with the heterotrimeric G proteins [16]. In the G proteins, the GTP-ase domain locks onto the propeller domain, regulating ligand binding in two ways: steric blockade of the propeller and allosteric control of the GTP-ase domain. On binding GTP, the GTP-ase domain dissociates from the propeller, enabling both domains to bind their respective ligands. Binding of an RGD-style ligand could play an role analogous to GTP. The G-protein model is also consistent with the observations of Mould and colleagues [17], who mapped two distinct binding sites, one on the propeller and the other at the MIDAS motif, for two different regions of fibronectin, separated by ~40 Å. In the crystal structure of αVβ3, the fibronectin binding sites are closer together, suggesting that the head must reorganize in order to engage both sites on fibronectin.

Curiously, none of these integrin models is consistent with the assignment of a long-range disulfide in the β-subunit [18], although Yan and Smith [19] have provided evidence that disulfide shuffling occurs in integrin αIIbβ3 and modulates activation, raising the possibility of further, thus far uncharacterized, large-scale quaternary changes underlying activation and signaling.

TERTIARY CHANGES

Crystal structures of recombinant α-I domains with and without ligand have demonstrated a dramatic conformational switch between closed and open states involving a change in the details of metal coordination at the MIDAS motif that is mechanically linked to a 10-Å downward shift of the C-terminal helix (α7) [20]. Mutational studies of the I domain in the context of the intact integrin have confirmed that these conformational changes underlie affinity control and that the conformational state of the I domain is regulated by the quaternary organization of the integrin [10, 21, 22].

The conformation of the β-I domain in the crystal structure is much more similar to the closed (i.e., inactive) conformation of the α-I domain, although Xiong and colleagues have proposed the opposite [9]. Furthermore, the same group recently soaked a short circular RGD peptide that acts as a ligand mimetic into the same crystals [23], and the conformational changes are consistent with those expected for a liganded domain within the context of a closed quaternary structure, in which tertiary changes are in the direction of those observed in the α-I domain but are frustrated by the closed quaternary structure [12]. "Indeed, three mutations that suppress activation map to the loops that link one of the MIDAS loops to the C-terminal (α7) helix [24], and it is conceivable that in the activated integrin, a large shift of the α7, comparable to that observed in the α-I domain, occurs in concert with a hinge-like motion of β-I with respect to the β-hybrid domain that may affect the interface with the propeller. The crystal structure of an authentic active integrin–ligand complex is required to test this proposal."

TAIL INTERACTIONS

Abundant biochemical and genetic data support the notion that interactions between integrin α and β cytoplasmic tails hold the resting integrin in a low-affinity conformation [7, 25, 26]. For example, a classic study by Ginsberg and colleagues [7] showed that a salt bridge between the αIIb Arg^{995} and β3 Asp^{723} was necessary and sufficient to hold the integrin in its resting state.

The NMR solution structure of the integrin αβ tail interaction has recently been published [27], and reveals a pair of interacting helices that largely confirm models previously proposed from mutational and functional studies. The cytoskeletal protein talin, which binds isolated integrin β-tails and activates platelets when overexpressed in cells [28], disrupts the αβ association detected by NMR. The crystal structure of an activating fragment talin in complex with the integrin β3 tail shows that talin engages integrin via a PhosphoTyrosine-Binding (PTB)-like domain, utilizing an NPxY motif, similar to the recognition motifs of classical PTB domains [29]. The footprint of talin on the β-chain partially overlaps that of the α-tail, suggesting how activation occurs. The model of αβ tail dissocation upon integrin activation predicts that the α-tail also becomes available to bind effector molecules involved in signal transduction. One such interaction is between the α4 tail and paxillin, which regulates cell migration [30], although no structure is available.

These data support the notion that "inside-out" signaling occurs by disruption of stabilizing bonds between the α- and β-tails. It remains to be seen how this separation is linked to conformational changes in the extracellular domains. Two simple models are supported by data. The first is a scissors model, in which the integrin pivots about some point between its legs, leading to a separation of both the head and tail domains [2, 31]; such a movement is consistent with the EM studies of Hantgan and colleagues [15]. Inside-out signaling is simple to envisage in such a model, and would simply require that a cytosolic protein bound tightly to one or both tails, pulling them apart. A second possibility is a piston model, in which one or both tails moves up and down with respect to the plasma membrane, changing the border of the transmembrane and cytoplasmic domains. Typically an R or K is positioned approximately 23 hydrophobic amino acid residues carboxy terminal to the predicted start of the transmembrane domain, followed by 4-6 hydrophobic residues [25] which also appear to be membrane-imbedded in the resting integrin [32]. It has been proposed that changes in the length or orientation of the integrin transmembrane domain could occur during physiological integrin activation. For example, the binding site of the β2 integrin regulatory protein, cytohesin-1, is in the hydrophobic membrane proximal region [33]. A final possibility is that the legs become widely separated upon activation, as seen in EM images of truncated integrins [14].

CONCLUDING REMARKS

In spite of major advances in the past 12 months, understanding the structural basis of integrin signaling is far from complete. The major missing data include the crystal structure of a true ligand bound, active integrin, and EM studies of intact integrins in different physiological ligation states showing how conformational changes in the head are transduced to and from the cytoplasmic tails.

REFERENCES

1. Hynes RO. Integrins: versatility, modulation, and signalling in cell adhesion. *Cell* 1992;**69**:11–25.
2. Schwartz MA, Schaller MD, Ginsberg MH. Integrins: emerging paradigms of signal transduction. *Annu Rev Cell Dev Biol* 1995;**11**:549–99.
3. Schwartz MA, Ginsberg MH. Networks and crosstalk: integrin signalling spreads. *Nat Cell Biol* 2002;**4**:E65–8.
4. Bazzoni G, Hemler ME. Are changes in integrin affinity and conformation overemphasized? *Trends Biochem Sci* 1998;**23**:30–4.
5. Hogg N, Leitinger B. Shape and shift changes related to the function of leukocyte integrins LFA-1 and Mac-1. *J Leukoc Biol* 2001;**69**:893–8.
6. Woods A, Couchman JR. Integrin modulation by lateral association. *J Biol Chem* 2000;**275**:24,233–6.
7. Hughes P, Diaz-Gonzalez F, Leong L, Wu C, McDonald J, Shattil SJ, Ginsberg MH. Breaking the integrin hinge. *J Biol Chem* 1996; **271**:6571–4.
8. Takagi J, Erickson HP, Springer TA. C-terminal opening mimics "inside-out" activation of integrin α5β1. *Nat Struct Biol* 2001;**8**:412–16.
9. Xiong J-P, Stehle T, Diefenbach B, Zhang R, Dunker R, Scott DL, Joachimiak A, Goodman SL, Arnaout MA. Crystal structure of the extracellular segment of integrin αVβ3. *Science* 2001;**294**:339–45.
10. Lupher MLJ, Harris EA, Beals CR, Sui LM, Liddington RC, Staunton DE. Cellular activation of leukocyte function-associated antigen-1 and its affinity are regulated at the I domain allosteric site. *J Immunol* 2001;**167**:1431–9.
11. Du X, Gu M, Weisel JW, Nagaswami C, Bennett JS, Bowditch RD, Ginsberg MH. Long range propagation of conformational changes in integrin αIIbβ3. *J Biol Chem* 1993;**268**:23,087–92.
12. Liddington RC. Will the real integrin please stand up?. *Structure* 2002;**10**:605–7.
13. Beglova N, Blacklow SC, Takagi J, Springer TA. Cysteine-rich module structure reveals a fulcrum for integrin rearrangement upon activation. *Nat Struct Biol* 2002;**9**:282–7.
14. Takagi J, Petre BM, Walz T, Springer TA. Global conformational rearrangements in integrin extracellular domains in outside-in and outside-in signaling. *Cell* 2002;**110**:599–611.

15. Hantgan RR, Paumi C, Rocco M, Weisel JW. Effects of ligand-mimetic peptides Arg-Gly-Asp-X (X=Phe, Trp, Ser) on αIIbβ3 integrin conformation and oligomerization. *Biochemistry* 1999;**38**:14,461–74.
16. Bohm A, Gaudet R, Sigler PB. Structural aspects of heterotrimeric G-protein signaling. *Curr Opin Biotechnol* 1997;**8**:480–7.
17. Mould AP, Askari JA, Aota S, Yamada KM, Irie A, Takada Y, Mardon HJ, Humphries MJ. Defining the topology of integrin α5β1-fibronectin interactions using inhibitory anti-α5 and anti-β1 monoclonal antibodies: evidence that the synergy sequence of fibronectin is recognized by the amino-terminal repeats of the α5 subunit. *J Biol Chem* 1997;**272**:17,283–92.
18. Calvete JJ, Mann K, Alvarez MV, López MM, González-Rodriguéz J. Proteolytic dissection of the isolated platelet fibrinogen receptor, integrin GPIIb/IIIa. Localization of GPIIb and GPIIIa sequences putatively involved in the subunit interface and in intrasubunit and intrachain contacts. *Biochem J* 1992;**282**:523–32.
19. Yan B, Smith JW. Mechanism of integrin activation by disulfide bond reduction. *Biochemistry* 2001;**40**:8861–7.
20. Emsley J, Knight CG, Farndale RW, Barnes MJ, Liddington RC. Structural basis of collagen recognition by integrin α2β1. *Cell* 2000; **101**:47–56.
21. Li R, Rieu P, Griffith DL, Scott D, Arnaout MA. Two functional states of the CD11b A-domain: correlations with key features of two Mn^{2+}-complexed crystal structures. *J Cell Biol* 1998;**143**:1523–34.
22. Oxvig C, Lu C, Springer TA. Conformational changes in tertiary structure near the ligand binding site of an integrin I domain. *Proc Natl Acad SciUSA* 1999;**96**:2215–20.
23. Xiong JP, Stehle T, Zhang R, Joachimiak A, Frech M, Goodman SL, Arnaout MA. Crystal structure of the extracellular segment of integrin αVβ3 in complex with an Arg-Gly-Asp ligand. *Science* 2002;**296**:151–5.
24. Baker EK, Tozer EC, Pfaff M, Shattil SJ, Loftus JC, Ginsberg MH. A genetic analysis of integrin function: Glanzmann thrombasthenia *in vitro*. *Proc Natl Acad Sci USA* 1997;**94**:1973–8.
25. Williams MJ, Hughes PE, O'Toole TE, Ginsberg MH. The inner world of cell adhesion: integrin cytoplasmic domains. *Trends Cell Biol* 1994;**4**:109–12.
26. Ginsberg MH, Yaspan B, Forsyth J, Ulmer TS, Campbell ID, Slepak M. A membrane-distal segment of the integrin αIIb cytoplasmic domain regulates integrin activation. *J Biol Chem* 2001;**276**:22,514–21.
27. Vinogradova O, Velyvis A, Velyviene A, Hu B, Haas T, Plow EF, Qin J. A structural mechanism of integrin αIIbβ3 "inside-out" activation as regulated at its cytoplasmic face. *Cell* 2002;**110**:587–97.
28. Calderwood DA, Zent R, Grant R, Rees DJG, Hynes RO, Ginsberg MH. The talin head domain binds to integrin beta subunit cytoplasmic tails and regulates integrin activation. *J Biol Chem* 1999;**274**: 28,071–4.
29. Garcia-Alvarez B, de Pereda JM, Calderwood DA, Ulmer TS, Critchley DR, Campbell ID, Ginsberg MH, Liddington RC. Structural determinants of integrin recognition by talin. *Mol Cell* 2003;**11**:49–58.
30. Liu S, Thomas SM, Woodside DG, Rose DM, Kiosses WB, Pfaff M, Ginsberg MH. Binding of paxillin to α4 integrins modifies integrin-dependent biological responses. *Nature* 1999;**402**:676–81.
31. Loftus JC, Liddington RC. New insights into integrin–ligand interaction. *J Clin Invest* 1997;**99**:2302–6.
32. Armulik A, Nilsson I, von Heijne G, Johansson S. Determination of the border between the transmembrane and cytoplasmic domains of human integrin subunits. *J Biol Chem* 1999;**274**:37,030–4.
33. Nagel W, Zeitlmann L, Schilcher P, Geiger C, Kolanus J, Kolanus W. Phosphoinositide 3-OH kinase activates the β2 integrin adhesion pathway and induces membrane recruitment of cytohesin-1. *J Biol Chem* 1998;**273**:14,853–61.
34. Mould AP, Askari JA, Barton S, Kline AD, McEwan PA, Craig SE, Humphries MJ. Integrin activation involves a conformational change in the alpha 1 helix of the beta subunit A-domain. *J Biol Chem* 2002;**1000**:1–5.

Chapter 43

Carbohydrate Recognition and Signaling

James M. Rini[1] and Hakon Leffler[2]

[1]*Departments of Molecular and Medical Genetics and Biochemistry, University of Toronto, Toronto, Ontario, Canada*

[2]*Section MIG, Department of Laboratory Medicine, University of Lund, Lund, Sweden*

INTRODUCTION

The recognition of extracellular and cell surface carbohydrates by specific carbohydrate-binding proteins, or lectins, is an important component of many biological processes. Here, we review the main principles of protein-carbohydrate recognition with particular reference to examples where structural data are available and signaling is known to be important. In conclusion, we explore the suggestion that carbohydrate-mediated interactions provide unique cell-signaling mechanisms.

BIOLOGICAL ROLES OF CARBOHYDRATE RECOGNITION

Carbohydrates, in the form of oligosaccharides or glycoconjugates, are found on the cell surfaces and extracellular proteins of virtually all living organisms. Although roles for carbohydrates in endogenous physiological interactions had long been suspected, it was not until the 1970s, with the discovery of the hepatic asialoglycoprotein receptor and Man-6-phosphate-mediated intracellular protein targeting, that firm evidence for such roles began to emerge. Since then, a number of animal lectin families have been identified [1, 2] and their functions, in processes ranging from protein folding and quality control to leukocyte homing, have been the subject of considerable study.

The cloning of glycosyltransferases and the generation of null-mutant mice have also provided further clear evidence of roles for endogenous carbohydrate recognition in the development and functioning of the immune and nervous systems [3–5]. In addition, mutations affecting the elaboration of complex carbohydrates are now known to be the basis for a growing number of human diseases collectively known as the congenital disorders of glycosylation (CDGs) [6]. The discovery that aberrant glycosylation of dystroglycan results in various forms of muscular dystrophy provides the most recent example [7–9].

Perhaps most surprising has been the finding that carbohydrates are also involved in the regulation of a number of signaling pathways. Fringe, for example, is a β1,3 N-acetylglucosaminyltransferase [10, 11] whose action modulates the interaction of the Notch receptor with its ligands, and mutations in Brainiac, a glycolipid-specific β1,3 N-acetylglucosaminyltransferase [12, 13], effect oogenesis. Genetic studies have also shown that proteoglycans/glycosaminylglycans play key roles in development, and, in Drosophila and Caenorhabditis elegans, they have been shown to be involved in regulating the fibroblast growth factor, Wnt, transforming growth factor-β, and Hedgehog signaling pathways [14].

CARBOHYDRATE STRUCTURE AND DIVERSITY

The structural diversity characteristic of the oligosaccharides found in nature stem principally from three sources: (1) a large number of monosaccharide types, (2) the multiple ways in which the monosaccharides can be linked together, and (3) the fact that oligosaccharides can be further modified chemically (e.g., sulfate, phosphate, and acetyl). The basic themes are illustrated in Figure 43.1, which shows the structure of sulfated sialy Lewis x, a tetrasaccharide important in selectin-mediated recognition. Because most monosaccharides have more than one hydroxyl group available for glycosidic bond formation, oligosaccharides, unlike their peptide counterparts, can form branched structures. The oligosaccharide structures linked to lipid or to protein, through Ser/Thr (O-linked) or Asn (N-linked), typically contain

Handbook of Cell Signaling, Three-Volume Set 2 ed.

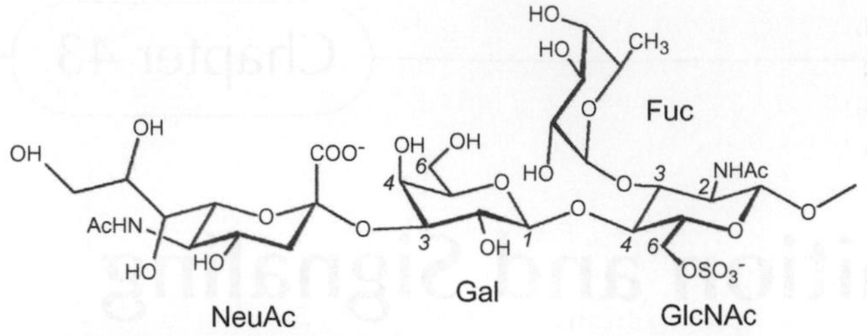

FIGURE 43.1 Structural representation of the sulfated sialyl Lewis x tetrasaccharide, NeuAcα2-3Galβ1-4[Fucα1-3](6-sulfo)GlcNAc. NeuAc, Gal, GlcNAc, and Fuc label the monosaccharide moieties N-acetylneuraminic acid, galactose, N-acetylglucosamine, and fucose, respectively.

between 1 and 20 monosaccharide moieties and may be branched or linear. The much longer linear glycosaminylglycans, either in isolation or as the oligosaccharide chains of proteoglycans, are found on cell surfaces and in the extracellular matrix.

In vivo, oligosaccharides are synthesized by glycosyltransferases, each of which typically has a unique donor, acceptor, and linkage specificity. As such, a very large number of glycosyltransferases and related enzymes are required to generate the oligosaccharide diversity seen in nature. Although the basis for this diversity is not fully understood, general themes are beginning to emerge. The so-called terminal elaborations (e.g., sialic acid, galactose, and sulfate) typical of the N-linked oligosaccharides of multicellular organisms, for example, seem to have appeared as part of the machinery required to mediate cell–cell and cell–matrix interactions [15]. In addition, it seems likely that oligosaccharide diversity has also been driven by evolutionary pressures arising from the need to differentiate self from non-self [16].

LECTINS AND CARBOHYDRATE RECOGNITION

Carbohydrate-binding proteins or lectins, like their saccharide counterparts, are also found in organisms ranging from microbes to humans [1]. The canonical carbohydrate recognition domain (CRD), characteristic of a given lectin type, can be found either in isolation or in conjunction with other protein domains, including coiled-coil domains and membrane-spanning motifs. Although many of the known CRD types are completely unrelated at the protein structural level [17–19], they can be grouped into two broad classes [20]. The type I CRDs are typified by the bacterial carbohydrate transporters and are characterized by deep carbohydrate-binding sites that essentially envelop their small saccharide ligands. In type II CRDs, the carbohydrate-binding sites are more shallow in nature and the saccharide remains relatively exposed to solvent, even when bound to the CRD. As a result, the dissociation constants (K_d) for small mono- or disaccharides can approach 0.1 μM for the type I CRDs, while the type II CRDs tend to bind small saccharides with K_d in the range of 0.1–1.0 mM.

Despite their relatively weak affinities for small saccharides, type II CRDs often show a strict mono- or disaccharide binding specificity. From a structural standpoint, this is achieved by a complementarity of fit between the CRD and the saccharide moiety which includes both hydrogen bond and van der Waals interactions. The structural and thermodynamic basis for this specificity has, in fact, been well studied and reviewed in detail elsewhere [17–23].

Given that the type II CRDs bind small saccharides relatively weakly, most of these lectin types have employed multivalency as a means of conferring additional affinity and specificity on their binding interactions with larger oligosaccharides [17]. In addition to the monosaccharide in the primary site, the CRD may possess subsites for interaction with other monosaccharides of the oligosaccharide. Alternately, many lectins cluster their CRDs as a means of making multivalent interactions with larger oligosaccharides or other extended structures such as cell surfaces. Members of the C-type lectin family, for example, are known to form monomers, trimers, tetramers, pentamers, and hexamers, as well as higher order oligomers, and in some cases a single polypeptide chain will possess more than one canonical CRD.

CARBOHYDRATE-MEDIATED SIGNALING

Lectins as Receptors

Most of the current evidence for the biological roles of complex carbohydrates comes from systems where they act as ligands for membrane-bound receptors that are lectins. Typically, these receptors have one or more extracellular CRDs, a single transmembrane-spanning region, and a relatively short cytosolic tail. In most cases, they are probably activated by receptor crosslinking mechanisms.

L-, P-, and E-selectin are cell-surface, C-type lectins responsible for leukocyte homing [24]. Unlike other members of the family, they do not possess a monosaccharide binding specificity. They require at least a tetrasaccharide, sialyl Lewis x (Figure 43.1) for binding, and specific sulfation further enhances binding to L- and P-selectin [25]. The crystal structures of P- and E-selectin, in complex with oligosaccharide/ glycopeptide ligands, have shown the importance of electrostatics in these interactions, a factor thought to be important in the rapid binding kinetics required for leukocyte rolling [26]. Moreover, the structures have provided a rationalization for the specificity differences that ensure that lymphocytes target to lymph nodes and neutrophils reach sites of inflammation. Although the selectins are not known to form oligomers, E-selectin-mediated clustering at contact points between interacting cells has been shown to activate the ERK1/2 signaling pathway [27].

DC-SIGN and DC-SIGNR are also C-type lectins, but in this case they are involved in dendritic cell/T cell interactions [28], as well as the promotion of HIV-1 infection [29]. These lectins possess a mannose-binding specificity, but in addition show a marked increase in affinity for high mannose oligosaccharides [30]. The crystal structures of their CRDs in complex with a mannopentasaccharide show that the increased affinity arises from a further set of interactions in addition to those made with the mannose in the primary binding site [31]. Because these lectins also possess α-helical tetramerization domains, it seems likely that they would be capable of making high-affinity interactions with ICAM-3 and HIV gp120, two of their natural ligands. In fact, it has been suggested that the crosslinking of DC-SIGN tetramers, by the highly multivalent high mannose oligosaccharide containing HIV virus, provides the signal required to promote transport of HIV from the periphery to the T-cell-containing lymph nodes [29].

The hepatic asialoglycoprotein receptor, a member of the C-type lectin family, provides a well-characterized example of the interplay between structure, specificity, and receptor crosslinking. Although an isolated CRD of this receptor binds galactose with a K_d in the millimolar concentration range, the cell-surface form of the receptor can bind the appropriate triantennary N-linked oligosaccharide with nanomolar affinity. Crosslinking studies have shown that the HL-1 subunit forms trimers on the cell surface and that recruitment of an additional HL-2 subunit(s) generates the high-affinity receptor. The galactose terminii of the triantennary oligosaccharides (separated by 15–25 Å) are found to interact with both the HL-1 and HL-2 subunits [32]. Linking receptor specificity to receptor crosslinking in this way may be important for both receptor uptake and signal transduction [33].

The targeting of lysosomal enzymes is also dependent on receptor-mediated endocytosis. In this case, the cation-dependent mannose 6-phosphate receptor (CD-MPR) and the insulin-like growth factor II/cation-independent mannose 6-phosphate receptor (IGF-II/CI-MPR) specifically recognize the mannose-6-phosphate moiety on acid hydrolases destined for lysosomes [34]. Again, multivalency is important; CD-MPR binds mannose 6-phosphate with a dissociation constant in the micromolar concentration range, while the dimeric receptor binds tetrameric β-glucuronidase with nanomolar affinity. Both dimeric and tetrameric forms of the receptor are found in the Golgi membrane, and, based on the crystal structure of the dimeric CD-MPR, a model for its high-affinity interaction with β-glucuronidase has been proposed [35]. The IGF-II/CI-MPR receptor contains two canonical CRDs presumably capable of promoting high-affinity interactions with multivalent lysosomal enzymes, and together with CD-MPR these receptors are responsible for targeting over 50 structurally distinct lysosomal enzymes. Dimerization of the IGF-II/CI-MPR receptor by β-glucuronidase binding increases receptor internalization at the cell surface [36].

The siglecs are a family of sialic acid binding lectins whose canonical CRD is a member of the immunoglobulin (Ig) superfamily. They are particularly important in the immune system, where they function in processes ranging from leukocyte adhesion to hemopoiesis [37]. Members of the family show specificity differences for α2,3- versus α2,6-linked sialic acids, as well as for sialic acids modified with respect to O-acetylation. The crystal structure of the CRD of sialoadhesin in complex with 3′ sialylactose shows that interactions with the bound oligosaccharide are mediated primarily with the terminal sialic acid moiety [38]. Of particular interest are the roles played by cis interactions. CD22 (Siglec-2), for example, is a B-cell-specific receptor which, through interaction with α2,6-linked sialic acid containing glycoproteins on its own cell surface, inhibits B cell receptor signaling. This stable inhibition can be broken by the addition of external competing saccharide and *in vivo* may be controlled by the regulation of sialytransferases and/or sialidase expression levels [39]. The cloning of several CD33-related receptors expressed on myeloid cell progenitors suggests new insight into the significance of their sialic acid binding properties. In all cases, these receptors possess cytoplasmic immunoreceptor tyrosine-based inhibitory motifs (ITIMs), elements now known to be hallmarks of inhibitory receptors central to the initiation, amplification, and termination of immune responses [40]. Through interactions with sialic acid containing self determinants, these receptors may play roles in the control of innate immunity [41].

Serum mannose binding protein (MBP), a component of the vertebrate innate immune system, is also a C-type lectin. Although not membrane bound, it signals activation of the complement cascade though a conformational change initiated by binding the cell surface of a foreign pathogen [42]. Like the asialoglycoprotein receptor, the CRD of MBP also recognizes only a terminal monosaccharide moiety, in this case mannose. The CRDs are also found to form trimers; however, in MBP they are mediated by long, triple-helical, coiled-coil domains that in addition promote the formation of trimer clusters containing 18 CRDs in total [43]. The crystal structures of truncated forms of the trimer show that the mannose binding sites are separated by 45 and 53 Å, respectively, in human [44] and rat [45] MBP. Thus, unlike the asialoglycoprotein receptor, which is designed to recognized the closely spaced galactose determinants of a single N-linked oligosaccharide, MBP is designed to bind the widely spaced mannose determinants typical of the cell surfaces of pathogenic microorganisms [46].

Glycoproteins as Receptors

It has long been known that certain multivalent, soluble plant lectins (e.g., PHA and Con A) can induce mitosis in lymphocytes and oxidative burst in neutrophils. The mechanism for initiation of these signals has generally been assumed to result from the crosslinking of cell-surface glycoproteins. More recently, soluble animal lectins of the galectin type have also been found to induce a variety of signals, including, among others, apoptosis, oxidative burst, cytokine release, and chemotaxis in immune cells [47]. In structural terms, the galectins are either dimeric or contain more than one CRD on a single polypeptide chain and as such they are capable of crosslinking receptors [48]. Recent studies aimed at understanding T cell homeostasis have suggested that CD45, CD43, CD7 [49], and the TCR–CD3 complex [50] are physiologically relevant cell-surface receptors for galectins-1 and -3, respectively.

Glycolipids as Receptors

The role of glycolipids as receptors for microbial lectins has been well studied. Bacterial AB_5 toxins possess a pentameric arrangement of B-subunit lectins which, through multivalent interactions, promote high-affinity binding with host cell-surface gangliosides [51]. In the case of cholera toxin, binding to G_{M1} on the cell surface is followed by retrograde transport and translocation across the ER membrane [52]. Once in the cytosol, the A1 fragment of the A subunit catalyzes the ADP ribosylation of the heterotrimeric Gαs protein, leading to the characteristic chloride and water efflux. In what is a fundamentally different type of interaction, the lectin subunits of the *Escherichia coli* P-fimbriae bind glycolipids in uroepithelial cells leading to ceramide release, activation of ceramide signaling pathways, and ultimately cytokine release through a process that also appears to involve activation of the TLR-4 receptor pathway [53–55]. Although not yet fully characterized, the interactions of glycosphingolipids with various adhesion and signaling receptors found in cell-surface microdomains are being found to mediate signaling events important in cell–cell interactions [56].

Proteoglycans and Glycosaminoglycans

Proteoglycans contain long linear oligosaccharide chains (glycosaminoglycans) made up of disaccharide repeats containing acidic monosaccharides and variable degrees of sulfation. They are found at the cell surface and in the extracellular matrix, where they interact with a wide variety of molecules, including, among others, signaling receptors, growth factors, chemokines, and various enzymes [57–59]. In the well-characterized fibroblast growth factor (FGF)–fibroblast growth factor receptor (FGFR) interaction, heparin/heparan sulfate serves as coreceptor. Two recent crystal structures of ternary complexes have begun to shed light on how the intrinsically multivalent oligosaccharide serves to promote receptor crosslinking in this system [60, 61]. Recent evidence from studies on hepatocyte growth factor/scatter factor suggests that heparan and dermatan sulfate binding serves to promote a conformational change in the growth factor that promotes receptor binding [62]. In some cases, specific sulfation patterns appear to be important determinants of specificity [58, 63]. The syndecans are cell-surface proteoglycans whose core proteins contain cytoplasmic signaling motifs. They have been implicated in the formation of focal adhesions, where interactions with heparin-binding domains and other receptors are proposed to lead to adhesion, crosslinking, and signal transduction [64].

Small Soluble Saccharides

Small nutrient saccharides are often sensed by the receiving cells after entry through a transporter. In mammals, for example, glucose is sensed by an alteration in the adenosine triphosphate (ATP)/adenosine diphosphate (ADP) ratio resulting from glucokinase-initiated glucose metabolism. In microbes, small saccharides are often sensed by specific, non-enzyme cytosolic binding proteins that in turn regulate gene expression (e.g., the Lac-repressor of E. coli). In plants, nutrient sugars are also known to be important mediators of signal transduction [65], and the recognition of small soluble oligosaccharides by membrane and cytoplasmic receptors is important in plant host defense [66]. Although these examples are beyond the scope of this review, it is worth noting that these carbohydrate-mediated signaling mechanisms may be operative in systems yet to be characterized.

Carbohydrates and Lectins in the Nucleocytosolic Compartment

The O-linked glycosylation of serine and threonine residues of nuclear and cytoplasmic proteins by N-acetylglucosamine (O-GlcNAc) is involved in signal transduction in multicellular organisms [67]. This dynamic modification occurs at sites of protein phosphorylation and may serve to transiently block sites of phosphorylation. Although its roles are not yet fully characterized, O-GlcNAc has been found to modulate a wide range of cellular functions, including transcription, translation, nuclear transport, and cytoskeletal assembly [68].

Galectins are also cytosolic and nuclear proteins, but they are not known to bind carbohydrates in these compartments; however, galectins 1 and 3 have been implicated in pre-mRNA splicing, a process inhibited by oligosaccharide binding [69]. The galectins are also secreted from the cytoplasm (by non-classical pathways), and it is at the cell surface that they perform the carbohydrate-mediated processes discussed previously.

For the sake of completeness it is worth noting that well-known second messengers such as cyclic AMP, GDP, GTP, etc. are ribose-containing glycoconjugates and that even more complex saccharide second messengers may be operative in insulin signaling [70].

CONCLUSIONS

The interactions between lectins and carbohydrates are relatively weak in nature and, as such, carbohydrate-mediated interactions may play important roles where weak interactions are required – the leukocyte rolling phenomenon perhaps providing a good example. In many cases, however, type II lectins have employed multivalency as a means of conferring increased affinity and specificity on their binding interactions. The structures of the asialglycoprotein receptor and MBP provide important examples of this principle. Because receptor crosslinking or clustering is a natural outcome of such multivalent interactions, it is clear that lectin-oligosaccharide interactions are inherently well suited to mediating signal transduction by the so-called horizontal mechanisms. In contrast, the higher affinity type-I lectins, typified by the bacterial transport/chemosensory receptors, appear to employ a mechanism more akin to vertical signaling where ligand-induced conformational changes in the receptor lead to signal transduction [71]. Interestingly, the affinity of these receptors for their carbohydrate ligands is close to the minimum affinity ($K_d \sim 10^{-8}$M) thought to be required for vertical signaling though 7TM receptors.

Although similar in some ways, it is clear that protein-carbohydrate interactions differ from protein–protein interactions in ways that might confer on them unique signaling roles or properties. Because glycosylation is a posttranslation modification capable of modifying any molecule with the appropriate acceptor, the subsequent recognition of carbohydrate determinants differs fundamentally from that involving specific protein–protein interactions. The galectins and siglecs, for example, bind β-galactosides and sialic-acid-containing ligands, respectively, and either of them might be expected to interact with more than one receptor type. As such, carbohydrate-mediated interactions may enable the activation of multiple signaling pathways or networks, as described by Bhalla and Iyengar [72]. Alternately, if carbohydrate-mediated interactions lead to heterogeneous crosslinked receptor arrays, this might result in spatial/geometric associations, where the triggering of one receptor type leads to the activation of another [73, 74]. Brewer and colleagues [75] have also provided evidence for the ability of multivalent lectins to form homogeneous crosslinked arrays or lattices, even in the presence of competing ligands. In fact, in recent *in vivo* studies they have shown that galectin-1-induced apoptosis is accompanied by the redistribution and segregation of CD45 and CD43 on T cell surfaces. Galectin-3-mediated crosslinked arrays have also been recently invoked in a model for T cell receptor activation [50]. In a similar vein, it seems likely that the highly multivalent proteoglycans provide scaffolds upon which interacting molecules can be assembled and organized.

In addition to the potential for triggering signaling events, the formation of carbohydrate-mediated crosslinked arrays may also be important in receptor turnover, one way in which signaling events are modulated [76]. In fact, evidence already exists for the ability of galectin-3 to both accelerate [77] and retard the turnover of cell surface receptors (J. Dennis, personal communication). In what might be a variation on this theme, the priming of neutrophil leukocytes with lipopolysaccharide (LPS) leads to galectin responsiveness by inducing the transfer of receptor containing vesicles to the cell surface [78, 79].

Our knowledge of lectin and glycoprotein structures shows that multivalent interactions are a recurring theme. Many lectins are oligomeric and/or membrane bound, and many glycoproteins (and certainly proteoglycans) possess multiple glycosylation sites. Their inherent ability to mediate crosslinks make it certain that new examples of signaling roles will follow from the study of these complex and diverse molecules.

REFERENCES

1. Lis H, Sharon N. Lectins: carbohydrate-specific proteins that mediate cellular recognition. *Chem Rev* 1998;**98**:637–74.
2. Dodd RB, Drickamer K. Lectin-like proteins in model organisms: implications for evolution of carbohydrate-binding activity. *Glycobiology* 2001;**11**:71R–79R.
3. Dennis JW, Granovsky M, Warren CE. Protein glycosylation in development and disease. *Bioessays* 1999;**21**:412–21.
4. Lowe JB. Glycosylation, immunity, and autoimmunity. *Cell* 2001;**104**:809–12.
5. Marth JD, Lowe JB. A genetic approach to mammalian glycan function. *Annu Rev Biochem* 2003;**72**:643–91.
6. Jaeken J, Matthijs G. Congenital disorders of glycosylation. *Annu Rev Genomics Hum Genet* 2001;**2**:129–51.
7. Yoshida A, Kobayashi K, Manya H, Taniguchi K, Kano H, Mizuno M, Inazu T, Mitsuhashi H, Takahashi S, Takeuchi M, Herrmann R, Straub V, Talim B, Voit T, Topaloglu H, Toda T, Endo T. Muscular dystrophy and neuronal migration disorder caused by mutations in a glycosyltransferase, POMGnT1. *Dev Cell* 2001;**1**:717–24.
8. Grewal PK, Holzfeind PJ, Bittner RE, Hewitt JE. Mutant glycosyltransferase and altered glycosylation of alpha-dystroglycan in the myodystrophy mouse. *Nature Genet* 2001;**28**:151–4.
9. Michele DE, Barresi R, Kanagawa M, Saito F, Cohn RD, Satz JS, Dollar J, Nishino I, Kelley RI, Somer H, Straub V, Mathews KD, Moore SA, Campbell KP. Post-translational disruption of dystroglycan-ligand interactions in congenital muscular dystrophies. *Nature* 2002;**418**:417–22.
10. Moloney DJ, Panin VM, Johnston SH, Chen J, Shao L, Wilson R, Wang Y, Stanley P, Irvine KD, Haltiwanger RS, Vogt TF. Fringe is a glycosyltransferase that modifies Notch. *Nature* 2000;**406**:369–75.
11. Bruckner K, Perez L, Clausen H, Cohen S. Glycosyltransferase activity of Fringe modulates Notch–Delta interactions. *Nature* 2000;**406**:411–15.
12. Muller R, Altmann F, Zhou D, Hennet T. The *Drosophila melanogaster* brainiac protein is a glycolipid-specific beta 1,3 N-acetylglucosaminyltransferase. *J Biol Chem* 2002;**277**:32,417–20.
13. Schwientek T, Keck B, Levery SB, Jensen MA, Pedersen JW, Wandall HH, Stroud M, Cohen SM, Amado M, Clausen H. The *Drosophila*

gene brainiac encodes a glycosyltransferase putatively involved in glycosphingolipid synthesis. *J Biol Chem* 2002;**277**:32,421–9.
14. Selleck SB. Genetic dissection of proteoglycan function in *Drosophila* and *C. elegans*. *Semin Cell Dev Biol* 2001;**12**:127–34.
15. Drickamer K, Taylor ME. Evolving views of protein glycosylation. *Trends Biochem Sci* 1998;**23**:321–4.
16. Gagneux P, Varki A. Evolutionary considerations in relating oligosaccharide diversity to biological function. *Glycobiology* 1999;**9**:747–55.
17. Rini JM. Lectin structure. *Annu Rev Biophys Biomol Struct* 1995;**24**:551–77.
18. Weis WI, Drickamer K. Structural basis of lectin-carbohydrate recognition. *Annu Rev Biochem* 1996;**65**:441–73.
19. Elgavish S, Shaanan B. Lectin–carbohydrate interactions: different folds, common recognition principles. *Trends Biochem Sci* 1997;**22**:462–7.
20. Vyas NK. Atomic features of protein–carbohydrate interactions. *Curr Opin Struct Biol* 1991;**1**:732–40.
21. Garcia-Hernandez E, Hernandez-Arana A. Structural basis of lectin–carbohydrate affinities: comparison with protein-folding energetics. *Protein Sci* 1999;**8**:1075–86.
22. Garcia-Hernandez E, Zubillaga RA, Rodriguez-Romero A, Hernandez-Arana A. Stereochemical metrics of lectin– carbohydrate interactions: comparison with protein–protein interfaces. *Glycobiology* 2000;**10**:993–1000.
23. Dam TK, Brewer CF. Thermodynamic studies of lectin–carbohydrate interactions by isothermal titration calorimetry. *Chem Rev* 2002;**102**:387–429.
24. Vestweber D, Blanks JE. Mechanisms that regulate the function of the selectins and their ligands. *Physiol Rev* 1999;**79**:181–213.
25. Kanamori A, Kojima N, Uchimura K, Muramatsu T, Tamatani T, Berndt MC, Kansas GS, Kannagi R. Distinct sulfation requirements of selectins disclosed using cells that support rolling mediated by all three selectins under shear flow. L-selectin prefers carbohydrate 6-sulfation to tyrosine sulfation, whereas P-selectin does not. *J Biol Chem* 2002;**277**:32,578–86.
26. Somers WS, Tang J, Shaw GD, Camphausen RT. Insights into the molecular basis of leukocyte tethering and rolling revealed by structures of P- and E-selectin bound to SLe(X) and PSGL-1. *Cell* 2000;**103**:467–79.
27. Hu Jr. Y, Szente B, Kiely JM, Gimbrone MA Molecular events in transmembrane signaling via E-selectin. SHP2 association, adaptor protein complex formation and ERK1/2 activation. *J Biol Chem* 2001;**276**:48,549–53.
28. Geijtenbeek TB, Torensma R, van Vliet SJ, van Duijnhoven GC, Adema GJ, van Kooyk Y, Figdor CG. Identification of DC-SIGN, a novel dendritic cell-specific ICAM-3 receptor that supports primary immune responses. *Cell* 2000;**100**:575–85.
29. Geijtenbeek TB, Kwon DS, Torensma R, van Vliet SJ, van Duijnhoven GC, Middel J, Cornelissen IL, Nottet HS, KewalRamani VN, Littman DR, Figdor CG, van Kooyk Y. DC-SIGN, a dendritic cell-specific HIV-1-binding protein that enhances trans-infection of T cells. *Cell* 2000;**100**:587–97.
30. Mitchell DA, Fadden AJ, Drickamer K. A novel mechanism of carbohydrate recognition by the C-type lectins DC-SIGN and DC-SIGNR. Subunit organization and binding to multivalent ligands. *J Biol Chem* 2001;**276**:28,939–45.
31. Feinberg H, Mitchell DA, Drickamer K, Weis WI. Structural basis for selective recognition of oligosaccharides by DC-SIGN and DC-SIGNR. *Science* 2001;**294**:2163–66.
32. Lodish HF. Recognition of complex oligosaccharides by the multi-subunit asialoglycoprotein receptor. *Trends Biochem Sci* 1991;**16**:374–7.
33. Parker A, Fallon RJ. c-src Tyrosine kinase is associated with the asialoglycoprotein receptor in human hepatoma cells. *Mol Cell Biol Res Commun* 2001;**4**:331–6.
34. Hille-Rehfeld A. Mannose 6-phosphate receptors in sorting and transport of lysosomal enzymes. *Biochim Biophys Acta* 1995;**1241**:177–94.
35. Roberts DL, Weix DJ, Dahms NM, Kim JJ. Molecular basis of lysosomal enzyme recognition: three-dimensional structure of the cation-dependent mannose 6-phosphate receptor. *Cell* 1998;**93**:639–48.
36. York SJ, Arneson LS, Gregory WT, Dahms NM, Kornfeld S. The rate of internalization of the mannose 6-phosphate/insulin-like growth factor II receptor is enhanced by multivalent ligand binding. *J Biol Chem* 1999;**274**:1164–71.
37. Crocker PR, Varki A. Siglecs in the immune system. *Immunology* 2001;**103**:137–45.
38. May AP, Robinson RC, Vinson M, Crocker PR, Jones EY. Crystal structure of the N-terminal domain of sialoadhesin in complex with 3′ sialyllactose at 1.85-Å resolution. *Mol Cell* 1998;**1**:719–28.
39. Kelm S, Gerlach J, Brossmer R, Danzer CP, Nitschke L. The ligand-binding domain of CD22 is needed for inhibition of the B cell receptor signal, as demonstrated by a novel human CD22-specific inhibitor compound. *J Exp Med* 2002;**195**:1207–13.
40. Ravetch JV, Lanier LL. Immune inhibitory receptors. *Science* 2000;**290**:84–9.
41. Crocker PR, Varki A. Siglecs, sialic acids and innate immunity. *Trends Immunol* 2001;**22**:337–42.
42. Hakansson K, Reid KB. Collectin structure: a review. *Protein Sci* 2000;**9**:1607–17.
43. Hoppe HJ, Reid KB. Collectins – soluble proteins containing collagenous regions and lectin domains-and their roles in innate immunity. *Protein Sci* 1994;**3**:1143–58.
44. Sheriff S, Chang CY, Ezekowitz RA. Human mannose-binding protein carbohydrate recognition domain trimerizes through a triple alpha-helical coiled-coil. *Nature Struct Biol* 1994;**1**:789–94.
45. Weis WI, Drickamer K. Trimeric structure of a C-type mannose-binding protein. *Structure* 1994;**2**:1227–40.
46. Weis WI, Taylor ME, Drickamer K. The C-type lectin superfamily in the immune system. *Immunol Rev* 1998;**163**:19–34.
47. Rabinovich GA, Baum LG, Tinari N, Paganelli R, Natoli C, Liu FT, Iacobelli S. Galectins and their ligands: amplifiers, silencers or tuners of the inflammatory response?. *Trends Immunol* 2002;**23**:313–20.
48. Barondes SH, Cooper DN, Gitt MA, Leffler H. Galectins. Structure and function of a large family of animal lectins. *J Biol Chem* 1994;**269**:20,807–10.
49. Pace KE, Lee C, Stewart PL, Baum LG. Restricted receptor segregation into membrane microdomains occurs on human T cells during apoptosis induced by galectin-1. *J Immunol* 1999;**163**:3801–11.
50. Demetriou M, Granovsky M, Quaggin S, Dennis JW. Negative regulation of T-cell activation and autoimmunity by Mgat5 N-glycosylation. *Nature* 2001;**409**:733–9.
51. Merritt EA, Hol WG. AB5 toxins. *Curr Opin Struct Biol* 1995;**5**:165–171.
52. Tsai B, Rodighiero C, Lencer WI, Rapoport TA. Protein disulfide isomerase acts as a redox-dependent chaperone to unfold cholera toxin. *Cell* 2001;**104**:937–48.
53. Hedlund M, Duan RD, Nilsson A, Svanborg C. Sphingomyelin, glycosphingolipids and ceramide signaling in cells exposed to P-fimbriated *Escherichia coli*. *Mol Microbiol* 1998;**29**:1297–1306.
54. Frendeus B, Wachtler C, Hedlund M, Fischer H, Samuelsson P, Svensson M, Svanborg C. *Escherichia coli* P fimbriae utilize the Toll-like receptor 4 pathway for cell activation. *Mol Microbiol* 2001;**40**:37–51.

55. Hedlund M, Duan RD, Nilsson A, Svensson M, Karpman D, Svanborg C. Fimbriae, transmembrane signaling, and cell activation. *J Infect Dis* 2001;**183**(Suppl. 1):S47–S50.
56. Hakomori S. Inaugural article: the glycosynapse. *Proc Natl Acad Sci USA* 2002;**99**:225–32.
57. Esko JD, Lindahl U. Molecular diversity of heparan sulfate. *J Clin Invest* 2001;**108**:169–73.
58. Esko JD, Selleck SB. Order out of chaos: assembly of ligand binding sites in heparan sulfate. *Annu Rev Biochem* 2002;**71**:435–71.
59. Trowbridge JM, Gallo RL. Dermatan sulfate: new functions from an old glycosaminoglycan. *Glycobiology* 2002;**12**:117–125R.
60. Schlessinger J, Plotnikov AN, Ibrahimi OA, Eliseenkova AV, Yeh BK, Yayon A, Linhardt RJ, Mohammadi M. Crystal structure of a ternary FGF–FGFR–heparin complex reveals a dual role for heparin in FGFR binding and dimerization. *Mol Cell* 2000;**6**:743–50.
61. Pellegrini L, Burke DF, von Delft F, Mulloy B, Blundell TL. Crystal structure of fibroblast growth factor receptor ectodomain bound to ligand and heparin. *Nature* 2000;**407**:1029–34.
62. Lyon M, Deakin JA, Gallagher JT. The mode of action of heparan and dermatan sulfates in the regulation of hepatocyte growth factor/scatter factor. *J Biol Chem* 2002;**277**:1040–6.
63. Kreuger J, Salmivirta M, Sturiale L, Gimenez-Gallego G, Lindahl U. Sequence analysis of heparan sulfate epitopes with graded affinities for fibroblast growth factors 1 and 2. *J Biol Chem* 2001;**276**:30,744–52.
64. Rapraeger AC. Molecular interactions of syndecans during development. *Semin Cell Dev Biol* 2001;**12**:107–16.
65. Sheen J, Zhou L, Jang JC. Sugars as signaling molecules. *Curr Opin Plant Biol* 1999;**2**:410–18.
66. Ebel J. Oligoglucoside elicitor-mediated activation of plant defense. *Bioessays* 1998;**20**:569–76.
67. Wells L, Vosseller K, Hart GW. Glycosylation of nucleocytoplasmic proteins: signal transduction and O-GlcNAc. *Science* 2001;**291**:2376–78.
68. Comer FI, Hart GW. O-glycosylation of nuclear and cytosolic proteins. Dynamic interplay between O-GlcNAc and O-phosphate. *J Biol Chem* 2000;**275**:29,179–82.
69. Vyakarnam A, Dagher SF, Wang JL, Patterson RJ. Evidence for a role for galectin-1 in pre-mRNA splicing. *Mol Cell Biol* 1997;**17**:4730–7.
70. Stralfors P. Insulin second messengers. *Bioessays* 1997;**19**:327–35.
71. Chen J, Sharma S, Quiocho FA, Davidson AL. Trapping the transition state of an ATP-binding cassette transporter: evidence for a concerted mechanism of maltose transport. *Proc Natl Acad Sci USA* 2001;**98**:1525–30.
72. Bhalla US, Iyengar R. Emergent properties of networks of biological signaling pathways. *Science* 1999;**283**:381–7.
73. Gestwicki JE, Kiessling LL. Inter-receptor communication through arrays of bacterial chemoreceptors. *Nature* 2002;**415**:81–4.
74. Thomason PA, Wolanin PM, Stock JB. Signal transduction: receptor clusters as information processing arrays. *Curr Biol* 2002;**12**:R399–R401.
75. Sacchettini JC, Baum LG, Brewer CF. Multivalent protein–carbohydrate interactions. A new paradigm for supermolecular assembly and signal transduction. *Biochemistry* 2001;**40**:3009–15.
76. Vieira AV, Lamaze C, Schmid SL. Control of EGF receptor signaling by clathrin-mediated endocytosis. *Science* 1996;**274**:2086–9.
77. Furtak V, Hatcher F, Ochieng J. Galectin-3 mediates the endocytosis of beta-1 integrins by breast carcinoma cells. *Biochem Biophys Res Commun* 2001;**289**:845–50.
78. Almkvist J, Faldt J, Dahlgren C, Leffler H, Karlsson A. Lipopolysaccharide-induced gelatinase granule mobilization primes neutrophils for activation by galectin-3 and formylmethionyl-Leu-Phe. *Infect Immun* 2001;**69**:832–7.
79. Almkvist J, Dahlgren C, Leffler H, Karlsson A. Activation of the neutrophil nicotinamide adenine dinucleotide phosphate oxidase by galectin-1. *J Immunol* 2002;**168**:4034–41.

Part 2

Ion Channels

Chapter 44

An Overview of Ion Channel Structure

Daniel L. Minor, Jr.
Cardiovascular Research Institute, Departments of Biochemistry & Biophysics and Cellular & Molecular Pharmacology, California Institute for Quantitative Biosciences, University of California, San Francisco

INTRODUCTION

The physical and mental processes you are using to pick up this book, feel its heft or scroll through an electronic copy, read the text, and understand the ideas presented within, rely on electrical signals in your muscles, eyes, and brain. The molecular bases of these bioelectric signals are ion channel proteins, transmembrane proteins that form rapidly activating and inactivating pores that permit ions to flow down their electrochemical gradients and across cell membranes.

Ion channel proteins make macromolecular pores in cell membranes that allow ions to move in response to a range of signals mediated by chemical and protein ligands, membrane potential changes, temperature, and mechanical force [1]. These changes in ion distribution can cause changes in the membrane potential and, in the case of calcium ions [2], directly activate a variety of intracellular signaling cascades.

The signals generated by ion channels are among the fastest found in biological systems, and occur on the timescale of tens of microseconds to hundreds of milliseconds. The flux through a channel pore can be as high as 10^{8-9} ions per second [1]. Because of their central role in nervous system function, channels have been the subjects of intensive biophysical studies for more than 50 years. The past few decades have witnessed the transformation of ion channels from a biophysical idea, into molecules identified by molecular biology and gene cloning approaches, to, finally, three-dimensional structures. The very nature of channels, membrane proteins that have segments within the lipid bilayer as well as large domains on the extracellular and cytoplasmic sides, has been the largest impediment to understanding their architectures at high resolution. Transmembrane proteins remain among the most difficult classes of proteins to obtain in the large quantities (milligram) and high levels of purity that are required for structural studies. The last edition of this chapter was written at the dawn of ion channel structural biology, just a few years after the first high-resolution structures of ion channels, ion channel regulatory domains, and ion channel associated proteins were being uncovered. Since then, our understanding of ion channel structure has progressed rapidly. We now have high-resolution frameworks for the pore domains, regulatory domains, and, in a few cases, complete channels, for examples of many of the well-studied ion channel types. A recent survey of channel structures can be found in Minor, 2007 [3]. Because of the many advances, it is impossible to cover the breadth of this swiftly growing field in a brief chapter. Thus, the focus here will be to present an overview of some key concepts, highlights of important advances, and an accounting of a few of the many outstanding questions.

Ion channels do two basic things. They open and close to regulate the passage of specific ions across the cell membrane (see Chapter 32 of Handbook of Cell Signaling, Second Edition), and they sense and respond to signals that shift the equilibrium between closed and open states (see Chapter 19; also Chapters 31, 33, and 34 of Handbook of Cell Signaling, Second Edition). Their main task is to provide passageway for ions, which are charged, to traverse the formidable hydrophobic barrier presented by the hydrophobic core of the cell membrane. There is a wide variety of ion channel types that respond to a range of signals and that have varied types of ion selectivity. In the human genome, ion channel pore-forming subunit genes are abundant. For example, genes for the pore forming subunits of members of the voltage-gated ion channel family number around 150 [4], placing this class just behind G-protein-coupled receptors, and kinases in terms of the most abundant human genes. Occurrences of genes for other ion channel types (ligand-gated ion channels, ionotropic glutamate receptors, and chloride channels) each number in the 10–30 range. Thus, it is clear that a significant number of gene products in our genome are dedicated to generating electrical signals.

Handbook of Cell Signaling, Three-Volume Set 2 ed.

As many ion channels are composed of multiple pore-forming subunits, and heteromeric channels often have different functional properties than do homomeric ones, the functional diversity available from this genetic complement is even greater than one might imagine from simply counting up the ion channel genes.

The general principle of ion channel architecture is that nearly all are multimeric assemblies of three to six pore-forming subunits that possess cyclic symmetry (Figure 44.1a, b). In this barrel-stave architecture, a fixed number of subunits are arranged around the central axis that forms the pathway for ion conduction. The number of subunits is roughly related to the size and selectivity properties of the pore. For example, potassium channels, which have an exquisite selectivity for potassium over sodium, are tetramers in which four identical or very homologous subunits are arranged around the pore [5]. Pentameric channels, typified by the nicotinic acetylcholine receptor (nAChR) family (see Chapter 33 of Handbook of Cell Signaling, Second Edition), have larger pores and generally are only able to discriminate between positive and negative ions, but not among the ions in these general classes. Hexameric channels, such as gap junctions, allow ions and even small solutes to pass between cells [6]. Despite the prevalence of the barrel-stave architecture among diverse types of ion channels that have different numbers of subunits and different numbers of transmembrane domains, this plan is not the only one used by nature. Voltage-gated chloride channels have two pores that are formed from a dimer of subunits in which each subunit makes its own ion conduction pathway [7] (Figure 44.1c).

OBTAINING THREE-DIMENSIONAL STRUCTURES OF CHANNELS: METHODS AND CHALLENGES

The main goals for understanding ion channel architecture can be simply stated. One would like to see the structures of the same channel in each of its functional states (ex. closed, opened, and inactivated) to understand the molecular rearrangements that underlie function. Nevertheless, these goals remain unattained for any ion channel. Although the full battery of structural approaches, including electron microscopy, nuclear magnetic resonance spectroscopy, and X-ray crystallography has been employed to reveal features of ion channel structure. X-ray crystallography remains the most powerful tool in structural biology studies, as it can yield atomic-level resolution of macromolecules of any size [3]. The challenge for all of these structural methods with respect to ion channels is that the objects of study are membrane proteins that have transmembrane portions that reside in a very hydrophobic environment as well as domains that reside in the aqueous intra- and extracellular spaces. To keep channels soluble upon extraction from the bilayer, a necessary step for purification, reagents such as detergents, lipids, or both, must be used in sample purification and preparation. There is a multitude of detergents and lipids, and it is impossible to predict which combination will work for any particular membrane protein. This complicates purification attempts and the search for conditions that yield diffraction-quality crystals. Further, trapping the channels in some state, be it open or closed, can prove difficult, as pharmacological reagents that are capable of binding a single state are not available for every channel type.

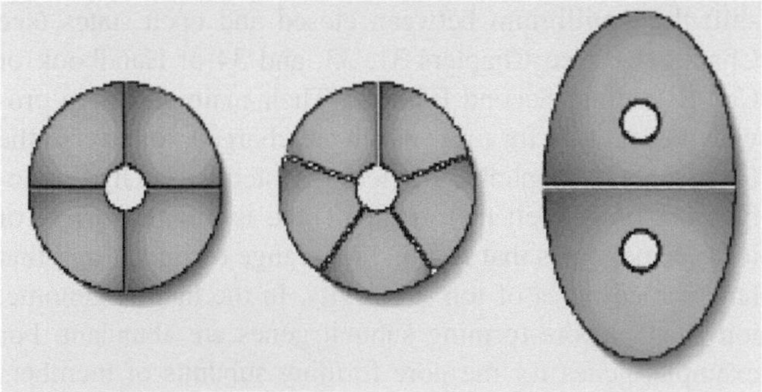

(a) (b) (c)

FIGURE 44.1 Cartoons showing the general architectural plans of ion channels.
(a, b) The "barrel-stave" plans of the voltage-gated cation channel family and nicotinic acetylcholine receptor family, respectively. In each, the channel subunits are arranged around a central pore through which the ions pass. (c) The general achitecture of voltage-gated chloride channels. These channels are dimers in which each subunit contains a pore. Figure adapted from [31] (Jentsch, 2002).

The size of most ion channels has kept them outside the realm of what can be studied by NMR. There have been a number of notable advances in both solution and solid state NMR methods that are beginning to allow study of some ion channels [3], but such efforts remain at the edge of the possible for the moment. Electron microscopy has proven very useful for studying ion channels that can be isolated from high-abundance natural sources. The greatest insights have come from landmark studies of the nAChR from the Torpedo electric organ [8, 9]. While difficult, electron microscopy studies require much less protein than other structural methods, and information can be obtained from two-dimensional crystals, tubular membrane crystals, and single particle analysis.

All of the structural methods require highly purified samples. The principal challenge to all ion channel structural studies is the difficulty in obtaining milligram quantities of pure material. There are two sources of such material; specific organs in which a particular channel is highly expressed (such as the nAChR in the Torpedo electric organ), and heterologously expressed sources. Very few ion channels are found in natively enriched sources. The main advances over the past five years have come from the identification and development of expression systems for archaeal and bacterial ion channels, and from the identification of expression systems for eukaryotic ion channel domains and a few full-length channels. Routine overexpression of prokaryotic or eukaryotic membrane proteins remains a major challenge, and the focus of much current research [3].

PROKARYOTIC ION CHANNELS: GATEWAYS TO FULL LENGTH CHANNEL STRUCTURE

One of the most remarkable discoveries of the past 10 years has been the abundance of ion channels in single-celled organisms [10, 11]. Once thought to be only in the purview of higher organisms, ion channels have been found throughout the tree of life. Although the functions of channels in microorganisms remain poorly understood, the existence of such molecules has been a boon to structural biologists. Prokaryotic channels can often be expressed more readily than their eukaryotic counterparts, and have been the first ion channels to be crystallized and have their structures determined by high-resolution X-ray crystallography. Even though many of the bacterial channels are less complex than in their eukaryotic cousins, the ones that have been characterized structurally contain the core elements that are required for function, and have proven exceptionally instructive for understanding fundamental mechanisms of selectivity and gating. Potassium channels [11], cyclic nucleotide-gated channels [12], voltage-gated chloride channels [7], and ligand-gated ion channels from the nAChR family [13] all have counterparts from the microbial world that have been successfully used for crystallographic structure determination.

One of the most impressive properties of potassium channels is their ability to discriminate potassium ions from sodium ions. Ion channels, including those selective for potassium, pass $\sim 10^{8-9}$ ions per second, leaving only tens of nanoseconds for the channel to interact with each ion. Despite the brief interaction, potassium channels allow potassium ions to pass with a fidelity of ~10,000:1 versus sodium (see Chapter 32 of Handbook of Cell Signaling, Second Edition) [1]. The reason this level of discrimination is so impressive is that sodium is the smaller ion and thus, based on size considerations alone, any pore through which potassium fits should also pass sodium. The way the potassium channels accomplish this amazing feat is through chemistry. All ions have shells of closely associated water molecules in solution [1]. For an ion to enter the filter, it must shed its waters of hydration. The selectivity filter of potassium channels is organized in such a way as to display rings of carbonyl oxygen atoms from the protein backbone that are arranged to coordinate potassium ions. Thus, the waters of hydration surrounding a potassium ion are replaced by oxygen atoms from the protein, creating a perfect chemical and energetic match as the ion enters the selectivity filter (Figure 44.2) [14, 15]. Although the smaller sodium ion can pass through the filter, it is much more energetically costly, since fewer of its lost water ligands can be replaced by the channel. Other selectivity filters may work in similar ways, where the protein makes intimate contact with the permeant ion [16].

OPEN CHANNELS

The key thing that ion channels do is open and close. Structural studies are beginning to reveal the general rearrangements that occur when channels are prompted to move between closed and open states. Electron microscopy studies of the nAChR show that ligand binding to the extracellular domain causes a twisting of the subunits that is propagated some 60 Å away to the narrowest part of the channel pore embedded deep in the membrane. This conformational change widens the narrow constriction or "gate" that prevents ion flow in the closed state [9] (Figure 44.3).

Potassium channels also have a portion of the protein that acts as a gate to prevent ion flow in the closed state. X-ray crystallographic comparisons of the homologous pore regions of open [17] and closed [18] bacterial potassium channels suggest that the lower part of the inner helix moves during gating, to widen the narrow constriction formed by the bundle crossing of the inner helices [19] (Figure 44.3).

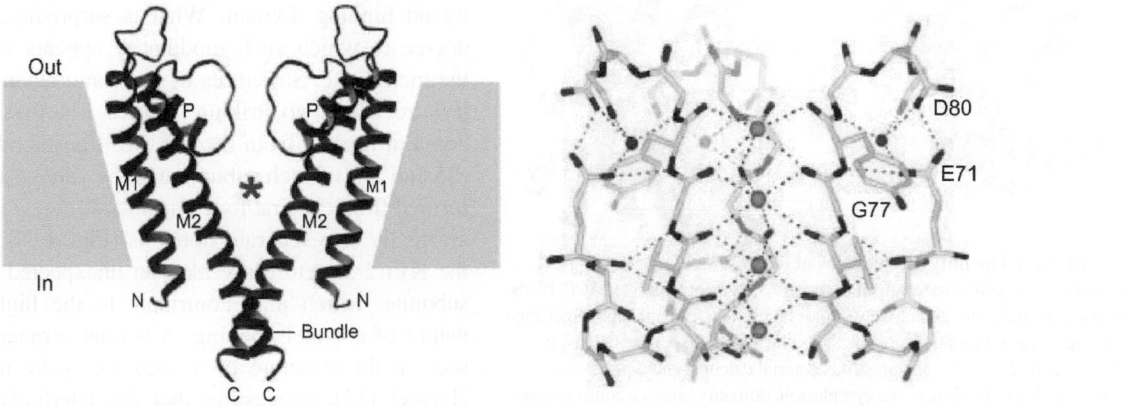

FIGURE 44.2 Structural elements of a potassium channel pore.
Left: two subunits from the bacterial KcsA potassium channel are shown. The transmembrane segments are labeled M1 and M2, and the pore helix is labeled P. M2 subunits cross at the region marked "bundle" and are thought to restrict access to the channel pore. The red star (color shown in online version) marks the inner cavity. Right: close-up view of the network of contacts made between the KcsA selectivity filter oxygens (red) and potassium ions (green spheres). Figure adapted from (19) and (15).

These conformational changes occur below the selectivity filter, which remains largely unchanged. A similar idea has been proposed for how voltage-gated channels work. Instead of movements of the extramembranous domains driving pore rearrangements, the linker segment that connects the transmembane voltage-sensing domain to the pore domain is thought to play a key role [20] (see Chapter 111 of Handbook of Cell Signaling, Second Edition). The details of voltage-sensing [21] as well as ion channel gating mechanisms are far from resolved. The field has yet to see high-resolution structures of the same channel trapped in different states. Such information will be essential for understanding exactly how accurate the current inferences made about gating are, and for elucidating how the movement of domains that sense the signals to which the channel responds (ligand binding domains or voltage sensors) couple to the pore domain.

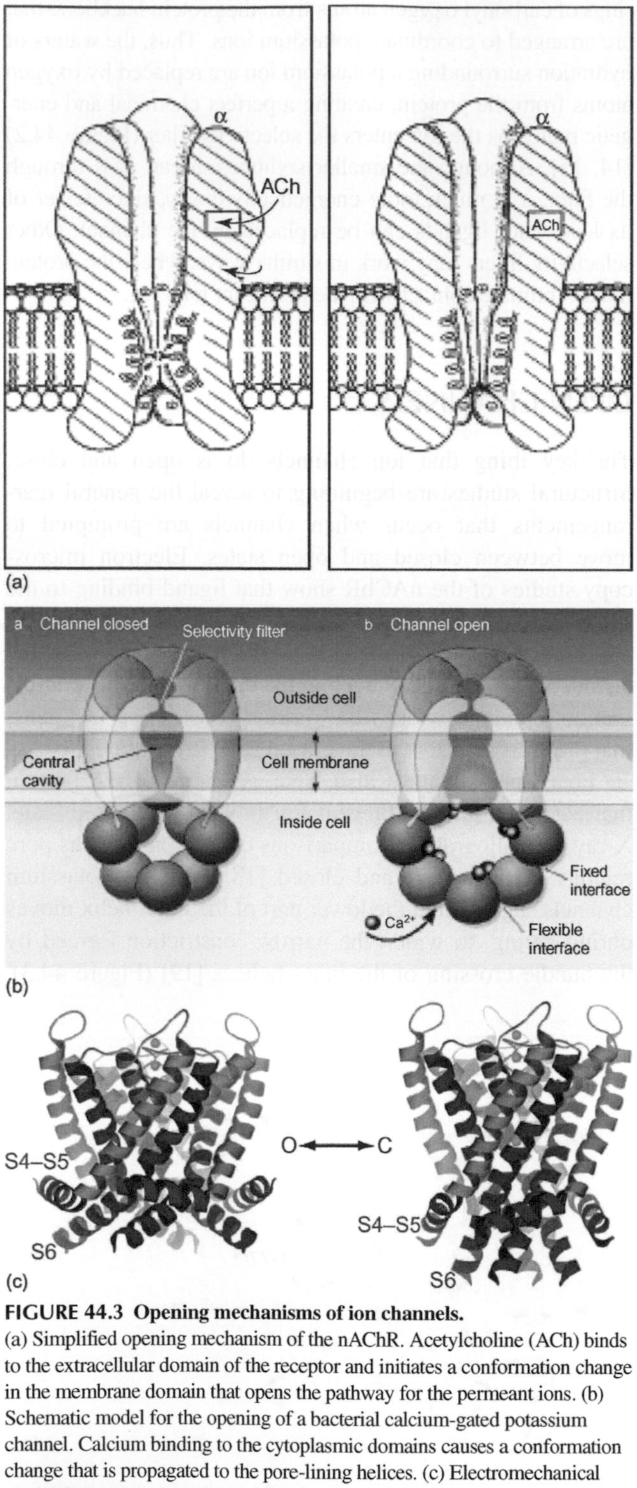

FIGURE 44.3 Opening mechanisms of ion channels.
(a) Simplified opening mechanism of the nAChR. Acetylcholine (ACh) binds to the extracellular domain of the receptor and initiates a conformation change in the membrane domain that opens the pathway for the permeant ions. (b) Schematic model for the opening of a bacterial calcium-gated potassium channel. Calcium binding to the cytoplasmic domains causes a conformation change that is propagated to the pore-lining helices. (c) Electromechanical model for how movements of voltage-sensing domains drive conformational change in the pore region of voltage-gated ion channels through the movement of the S4–S5 segment. Figure adapted from [32] (Unwin, 1998), [33] (Schuhmacher and Adelman, 2002), and [20] (Long *et al.*, 2005).

EUKARYOTIC ION CHANNELS AT HIGH RESOLUTION: WHOLE CHANNELS AND EXPLOITATION OF MODULAR STRUCTURE TO DIVIDE AND CONQUER

While working with bacterial and archaeal membrane proteins has been difficult, the challenges are even greater for eukaryotic membrane proteins. One of the major challenges is the development of robust methods for the overexpression and isolation of eukaryotic membrane proteins [3]. There are only three eukaryotic ion channels that have had their architectures resolved at high resolution (Figures 44.4, 44.5). The first is the Torpedo nicotinic acetylcholine receptor [8]. Unlike in most ion channels, the Torpedo nAChR can be isolated in large quantities from a highly enriched native source. Two channels have been crystallized and had their structures solved from recombinant material; the Kv1.2/Kvβ complex [22, 23] and the Acid Sensing Ion Channel, ASCI-1 [24]. In all cases, the structures show a modular architecture in which defined extramembranous domains are linked to the transmembrane domains that constitute the transmembrane pore.

It is not surprising that channels are built from specific modules such as a transmembrane pore and an extracellular ligand binding domain. What is surprising, though, is the degree to which such modularity appears to be present in the membrane. Structures of the mammalian Kv1.2 channel have revealed two striking features. The first is that the pore domain, formed from the last two transmembrane segments (S5 and S6) of each subunit, and the voltage sensing domain, formed from the first four transmembrane segments (S1–S4), are really two separate domains (Figure 44.5b, c). Further, the Kv1.2 structure revealed an unexpected interlocking of subunits, which may contribute to the highly cooperative nature of channel opening. A similar arrangement has been seen in the structure of a bacterial cyclic nucleotide-gated channel [12], suggesting that the interlocked arrangement is a general feature of the voltage-gated channel family. The consequences for such an architecture arrangement in terms of function and intersubunit cooperativity remain to be explored. Seen schematically, members of the voltage-gated

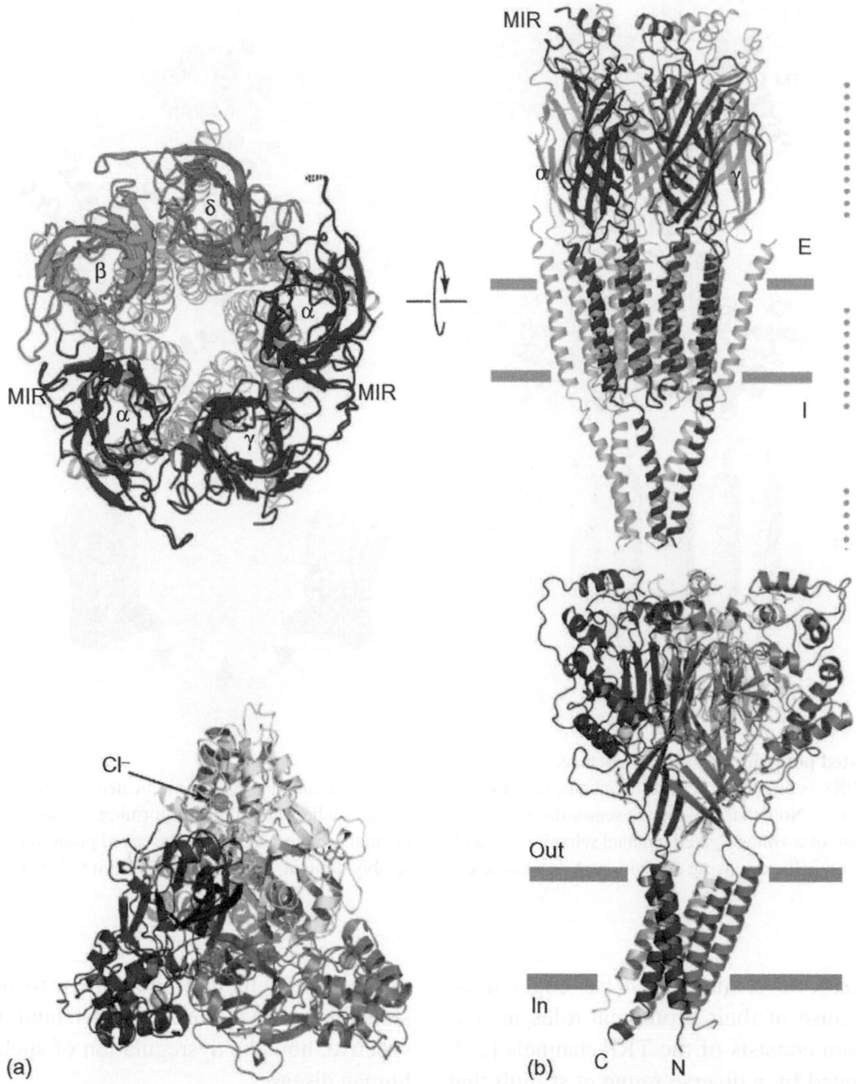

FIGURE 44.4 Structures of nAChR (top) and ASIC (bottom) channels. In each panel, the extracellular view is shown on the left. Figures are from [34] (Unwin, 2005) and [24] (Jasti *et al.*, 2007).

ion channel superfamily are put together from three domains that each have specific roles (Figure 44.5c): a pore-forming domain; a transmembrane module that can be used for voltage sensing; and intracellular domains that may be used to direct assembly, bind modulatory subunits, and serve as a sensor for detecting intracellular signals that modify function such as changes in calcium concentration or phorphorylation. Understanding how channels integrate into cellular signaling networks remains an important future goal, with implications for both basic science and the understanding of disease.

DIVIDE AND CONQUER: EXPLOITATION OF THE MODULAR NATURE OF ION CHANNEL STRUCTURE

In the absence of suitable full-length material for structural work, a second very fruitful approach has been to exploit the modular nature of many ion channels. Thus, one can focus on obtaining structures of isolated extramembranous regions that are known to have important roles in assembly or modulation, and use this information to inform functional studies.

The list of channels that have been successfully dissected is now very long [3]. The most notable success is for the ionotropic glutamate receptors, in which there are now more than 60 structures of the agonist binding domains from various isoforms alone and in complexes with a range of pharmacological agents [25]. The combination of these structures with biochemical and electrophysiological studies has given a deep insight into the likely mechanism of glutamate receptor action, even though high-resolution structure of the entire channel has not been attained. Voltage-gated calcium channels are also yielding structural information as a result of this dissection approach [26].

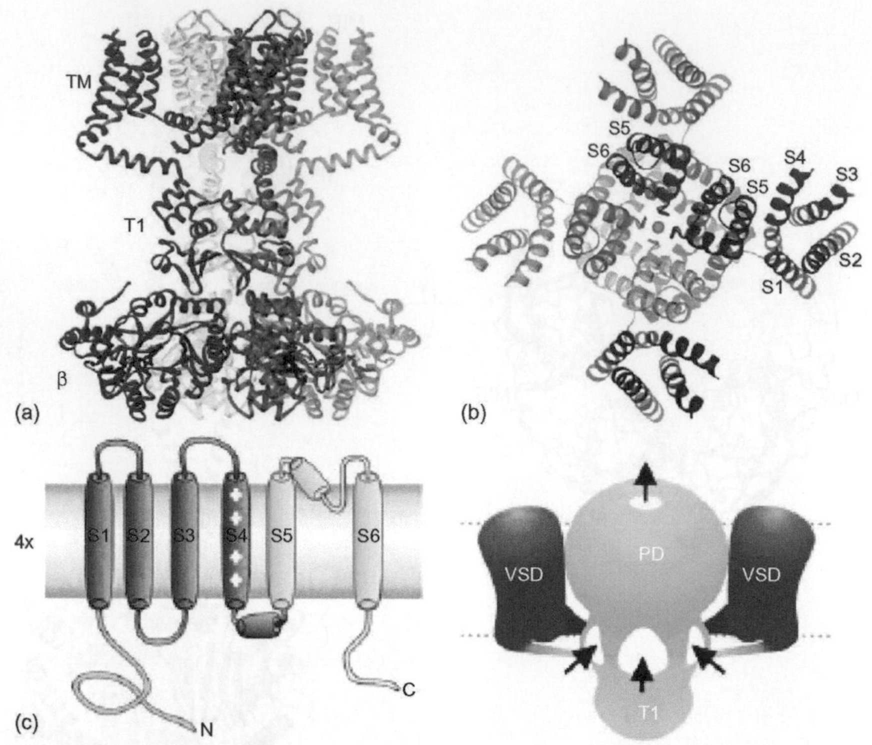

FIGURE 44.5 Voltage-gated potassium channel structure.
(a) Side view of the Kv1.2/Kvβ complex. The transmembrane domains, T1 assembly domain, and Kvβ are indicated by TM, T1, and β, respectively. (b) Extracellular view of Kv1.2 Note that the voltage sensor domain (labeled S1–S4) is adjacent to the pore forming domain (S5–S6) from a different subunit. (c) Cartoon diagram of a voltage-gated channel subunit (left) and the modular architecture of voltage-gated potassium channels (right). VSD, PD, and T1 correspond to the voltage-sensing domain, pore domain, and T1 assembly domain. Figures adapted from [22] (Long *et al.*, 2005) and [35] (Tombola *et al.*, 2005).

One class of eukaryotic channels that has engendered much excitement because of their prominent roles in sensory processes and pain consists of the TRP channels [27]. These channels are gated by a diverse range of stimuli that include pungent compounds such as capsaicin and menthol, and temperature. As close relatives of the voltage-gated ion channels, TRP channels share the six-transmembrane, four-subunit property [28, 29]. Thus far, structural understanding of TRP channels has largely come from the "divide and conquer" approach [30]. In terms of topology, TRP channels look very much like voltage-gated ion channels. Just how far the similarly goes on the structural and mechanistic level is unknown, and provides a fertile area of current research and speculation.

ION CHANNEL COMPLEXES

It is clear that efforts to understand ion channel function at high resolution now have a great deal of momentum. We can anticipate that the coming years will see the determination of more high-resolution structures of a variety of ion channels and ion channel regulatory proteins. As the field matures, it will be ever more important to place the structural efforts in the context of the protein complexes that form around ion channels in cells that allow them to interface with cellular signaling pathways and to understand, from a structural perspective, how the dysregulation of such interactions leads to human diseases.

REFERENCES

1. Hille B. *Ion channels of excitable membranes*. Sunderland, MA: Sinauer Associates, Inc; 2001.
2. Clapham DE. Calcium signaling. *Cell* 2007;**131**:1047–58.
3. Minor Jr DL. The neurobiologist's guide to structural biology: a primer on why macromolecular structure matters and how to evaluate structural data. *Neuron* 2007;**54**:511–33.
4. Yu FH, Catterall WA. The VGL-chanome: a protein superfamily specialized for electrical signaling and ionic homeostasis. *Sci STKE* 2004;**2004**:re15.
5. Gouaux E, Mackinnon R. Principles of selective ion transport in channels and pumps. *Science* 2005;**310**:1461–5.
6. Kovacs JA, Baker KA, Altenberg GA, Abagyan R, Yeager M. Molecular modeling and mutagenesis of gap junction channels. *Prog Biophys Mol Biol* 2007;**94**:15–28.
7. Dutzler R. A structural perspective on ClC channel and transporter function. *FEBS Letts* 2007;**581**:2839–44.
8. Miyazawa A, Fujiyoshi Y, Unwin N. Structure and gating mechanism of the acetylcholine receptor pore. *Nature* 2003;**423**:949–55.

9. Unwin N. Acetylcholine receptor channel imaged in the open state. *Nature* 1995;**373**:37–43.
10. Kung C, Blount P. Channels in microbes: so many holes to fill. *Mol Microbiol* 2004;**53**:373–80.
11. Kuo MM, Haynes WJ, Loukin SH, Kung C, Saimi Y. Prokaryotic K(+) channels: from crystal structures to diversity. *FEMS Microbiol Rev* 2005;**29**:961–85.
12. Clayton GM, Altieri S, Heginbotham L, Unger VM, Morais-Cabral JH. Structure of the transmembrane regions of a bacterial cyclic nucleotide-regulated channel. *Proc Natl Acad Sci USA* 2008;**105**:1511–15.
13. Hilf RJ, Dutzler R. X-ray structure of a prokaryotic pentameric ligand-gated ion channel. *Nature* 2008;**452**:375–9.
14. Morais-Cabral JH, Zhou Y, MacKinnon R. Energetic optimization of ion conduction rate by the K^+ selectivity filter. *Nature* 2001;**414**:37–42.
15. Zhou Y, Morais-Cabral JH, Kaufman A, MacKinnon R. Chemistry of ion coordination and hydration revealed by a K^+ channel-Fab complex at 2.0-Å resolution. *Nature* 2001;**414**:43–8.
16. Shi N, Ye S, Alam A, Chen L, Jiang Y. Atomic structure of a Na^+- and K^+-conducting channel. *Nature* 2006;**440**:570–4.
17. Jiang Y, Lee A, Chen J, Cadene M, Chait BT, MacKinnon R. Crystal structure and mechanism of a calcium-gated potassium channel. *Nature* 2002;**417**:515–22.
18. Doyle DA, Morais Cabral J, Pfuetzner RA, et al. The structure of the potassium channel: molecular basis of K^+ conduction and selectivity. *Science* 1998;**280**:69–77.
19. Jiang Y, Lee A, Chen J, Cadene M, Chait BT, MacKinnon R. The open pore conformation of potassium channels. *Nature* 2002;**417**:523–6.
20. Long SB, Campbell EB, Mackinnon R. Voltage sensor of Kv1.2: structural basis of electromechanical coupling. *Science* 2005;**309**:903–8.
21. Tombola F, Pathak MM, Isacoff EY. How does voltage open an ion channel? *Annu Rev Cell Dev Biol* 2006;**22**:23–52.
22. Long SB, Campbell EB, Mackinnon R. Crystal structure of a mammalian voltage-dependent Shaker family K^+ channel. *Science* 2005; **309**:897–903.
23. Long SB, Tao X, Campbell EB, MacKinnon R. Atomic structure of a voltage-dependent K^+ channel in a lipid membrane–like environment. *Nature* 2007;**450**:376–82.
24. Jasti J, Furukawa H, Gonzales EB, Gouaux E. Structure of acid-sensing ion channel 1 at 1.9-Å resolution and low pH. *Nature* 2007;**449**:316–23.
25. Mayer ML. Glutamate receptor ion channels. *Curr Opin Neurobiol* 2005;**15**:282–8.
26. Van Petegem F, Minor DL. The structural biology of voltage-gated calcium channel function and regulation. *Biochem Soc Trans* 2006;**34**:887–93.
27. Venkatachalam K, Montell C. TRP channels. *Annu Rev Biochem* 2007;**76**:387–417.
28. Moiseenkova-Bell VY, Stanciu LA, Serysheva, II, Tobe BJ, Wensel TG. Structure of TRPV1 channel revealed by electron cryomicroscopy. *Proc Natl Acad Sci USA* 2008;**105**:7451–5.
29. Tsuruda PR, Julius D, Minor Jr DL. Coiled coils direct assembly of a cold-activated TRP channel. *Neuron* 2006;**51**:201–12.
30. Gaudet R. TRP channels entering the structural era. *J Physiol* 2008;**586**:3565–75.
31. Jentsch TJ. Chloride channels are different. *Nature* 2002;**415**:276–7.
32. Unwin N. The nicotinic acetylcholine receptor of the Torpedo electric ray. *J Struct Biol* 1998;**121**:181–90.
33. Schumacher M, Adelman JP. Ion channels: an open and shut case. *Nature* 2002;**417**:501–2.
34. Unwin N. Refined structure of the nicotinic acetylcholine receptor at 4 Å resolution. *J Mol Biol* 2005;**346**:967–89.
35. Tombola F, Pathak MM, Isacoff EY. How far will you go to sense voltage? *Neuron* 2005;**48**:719–25.

Chapter 45

Voltage-Gated Calcium Channels

William A. Catterall

Department of Pharmacology, University of Washington, Seattle, Washington

PHYSIOLOGICAL ROLES OF VOLTAGE-GATED Ca^{2+} CHANNELS

Ca^{2+} channels in many different cell types activate upon membrane depolarization, and mediate Ca^{2+} influx in response to action potentials and sub-threshold depolarizing signals. Ca^{2+} entering the cell through voltage-gated Ca^{2+} channels serves as the second messenger of electrical signaling, initiating many different cellular events (Figure 45.1). In cardiac and smooth muscle cells, activation of Ca^{2+} channels initiates contraction directly by increasing cytosolic Ca^{2+} concentration, and indirectly by activating ryanodine-sensitive Ca^{2+} release channels in the sarcoplasmic reticulum [1–3]. In skeletal muscle cells, voltage-gated Ca^{2+} channels in the transverse tubule membranes interact directly with ryanodine-sensitive Ca^{2+} release channels in the sarcoplasmic reticulum and activate them to initiate rapid contraction [4, 5]. The same Ca^{2+} channels in the transverse tubules also mediate a slow Ca^{2+} conductance that increases cytosolic concentration and thereby regulates the force of contraction in response to high-frequency trains of nerve impulses [4]. In endocrine cells, voltage-gated Ca^{2+} channels mediate Ca^{2+} entry that initiates secretion of hormones [6]. In neurons, voltage-gated Ca^{2+} channels initiate synaptic transmission [7–9]. In many different cell types, Ca^{2+} entering the cytosol via voltage-gated Ca^{2+} channels regulates enzyme activity, gene expression, and other biochemical processes [10]. Thus, voltage-gated Ca^{2+} channels are the key signal transducers of electrical excitability, converting the electrical signal of the action potential in the cell surface membrane to an intracellular Ca^{2+} transient. Signal transduction in different cell types involves different molecular subtypes of voltage-gated Ca^{2+} channels, which mediate voltage-gated Ca^{2+} currents with different physiological, pharmacological, and regulatory properties.

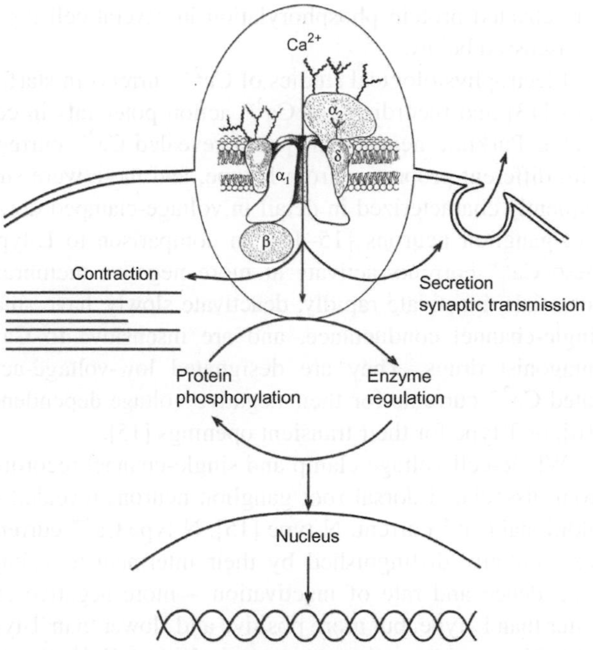

FIGURE 45.1 Signal transduction by voltage-gated Ca^{2+} channels. Ca^{2+} entering cells initiates numerous intracellular events, including contraction, secretion, synaptic transmission, enzyme regulation, protein phosphorylation/dephosphorylation, and gene transcription. *Inset*: Subunit structure of voltage-gated Ca^{2+} channels. The five-subunit complex that forms high voltage activated Ca^{2+} channels is illustrated with a central pore-forming $\alpha 1$ subunit, a disulfide-linker glycoprotein dimer of $\alpha 2$ and δ subunits, an intracellular β subunit, and a transmembrane glycoprotein γ subunit in some Ca^{2+} channel subtypes.

Ca^{2+} CURRENT TYPES DEFINED BY PHYSIOLOGICAL AND PHARMACOLOGICAL PROPERTIES

Since the first recordings of Ca^{2+} currents in cardiac myocytes [1], it has become apparent that there are multiple types of Ca^{2+} currents as defined by physiological and pharmacological criteria [7, 11, 12]. In cardiac, smooth, and skeletal muscle, the major Ca^{2+} currents are distinguished by high voltage of activation, large single channel conductance, slow voltage-dependent inactivation, marked regulation by cAMP-dependent protein phosphorylation pathways, and specific inhibition by Ca^{2+} antagonist drugs including dihydropyridines, phenylalkylamines, and benzothiazepines [1, 7]. These Ca^{2+} currents have been designated L-type, as they have slow voltage-dependent inactivation and therefore are long-lasting when Ba^{2+} is the current carrier and there is no Ca^{2+}-dependent inactivation [7]. L-type Ca^{2+} currents are also recorded in endocrine cells, where they initiate release of hormones [6], and in neurons, where they are important in regulation of gene expression, integration of synaptic input, and initiation of neurotransmitter release at specialized ribbon synapses in sensory cells [7, 10, 12]. L-type Ca^{2+} currents are subject to regulation by second messenger-activated protein phosphorylation in several cell types, as discussed below.

Electrophysiological studies of Ca^{2+} currents in starfish eggs [13] and recordings of Ca^{2+} action potentials in cerebellar Purkinje neurons [14] first revealed Ca^{2+} currents with different properties from L-type, and these were subsequently characterized in detail in voltage-clamped dorsal root ganglion neurons [15–17]. In comparison to L-type, these Ca^{2+} currents activate at more negative membrane potentials, inactivate rapidly, deactivate slowly, have small single-channel conductance, and are insensitive to Ca^{2+} antagonist drugs. They are designated low-voltage-activated Ca^{2+} currents for their negative voltage dependence [16], or T-type for their transient openings [15].

Whole-cell voltage clamp and single-channel recording from dissociated dorsal root ganglion neurons revealed an additional Ca^{2+} current, N-type [15]. N-type Ca^{2+} currents were initially distinguished by their intermediate voltage dependence and rate of inactivation – more negative and faster than L-type, but more positive and slower than T-type [15]. They are insensitive to organic L-type Ca^{2+} channel blockers, but blocked by the cone snail peptide ω-conotoxin GVIA [7, 18]. This pharmacological profile has been the primary method to distinguish N-type Ca^{2+} currents, because the voltage dependence and kinetics of N-type Ca^{2+} currents in different neurons vary considerably.

Analysis of the effects of other peptide toxins revealed three additional Ca^{2+} current types. P-type Ca^{2+} currents, first recorded in Purkinje neurons [19], are distinguished by high sensitivity to the spider toxin ω-agatoxin IVA [20]. Q-type Ca^{2+} currents, first recorded in cerebellar granule neurons [21], are blocked by ω-agatoxin IVA with lower affinity. R-type Ca^{2+} currents in cerebellar granule neurons are resistant to the subtype-specific organic and peptide Ca^{2+} channel blockers [21], and may include multiple channel subtypes [22]. While L-type and T-type Ca^{2+} currents are recorded in a wide range of cell types, N-, P-, Q-, and R-type Ca^{2+} currents are most prominent in neurons.

MOLECULAR PROPERTIES OF Ca^{2+} CHANNELS

Subunit Structure

Ca^{2+} channels purified from skeletal muscle transverse tubules are complexes of α1, α2, β, γ,and δ subunits [23–29]. Analysis of the biochemical properties, glycosylation, and hydrophobicity of these five subunits led to a model comprising a principal transmembrane α_1 subunit of 190kDa in association with a disulfide-linked $\alpha_2\delta$ dimer of 170kDa, an intracellular phosphorylated β subunit of 55kDa, and a transmembrane γ subunit of 33kDa (Figure 45.1) [29].

The α1 subunit is a protein of about 2000 amino acid residues with an amino acid sequence and predicted transmembrane structure like the previously characterized, pore-forming α subunit of sodium channels (Figure 45.2) [30]. The amino acid sequence is organized in four repeated domains (I–IV), which each contain six transmembrane segments (S1–S6) and a membrane-associated loop between transmembrane segments S5 and S6. As expected from biochemical analysis [29], the intracellular β subunit has predicted α-helices but no transmembrane segments (Figure 45.2) [31], while the γ subunit is a glycoprotein with four transmembrane segments (Figure 45.2) [32]. The cloned α2 subunit has many glycosylation sites and several hydrophobic sequences [33], but biosynthesis studies indicate that it is an extracellular, extrinsic membrane protein, attached to the membrane through disulfide linkage to the δ subunit ([34], Figure 45.2). The δ subunit is encoded by the 3′ end of the coding sequence of the same gene as the α2 subunit, and the mature forms of these two subunits are produced by posttranslational proteolytic processing and disulfide linkage (Figure 45.2) [35].

Purification of cardiac Ca^{2+} channels labeled by dihydropyridine Ca^{2+} antagonists identified subunits of the sizes of the α1, α2δ, β, and γ subunits of skeletal muscle Ca^{2+} channels [36–38], whereas immunoprecipitation of Ca^{2+} channels from neurons labeled by dihydropyridine Ca^{2+} antagonists revealed α1, α2δ, and β subunits but no γ subunit [39]. Purification and immunoprecipitation of N-type and P/Q-type Ca^{2+} channels labeled by ω-conotoxin GVIA and ω-agatoxin IVA, respectively, from brain membrane preparations also revealed α1, α2δ, and β subunits [40–43]. More recent experiments have unexpectedly revealed a novel γ subunit (stargazin), which is the target of the *stargazer* mutation in mice [44], and a related series of γ subunits

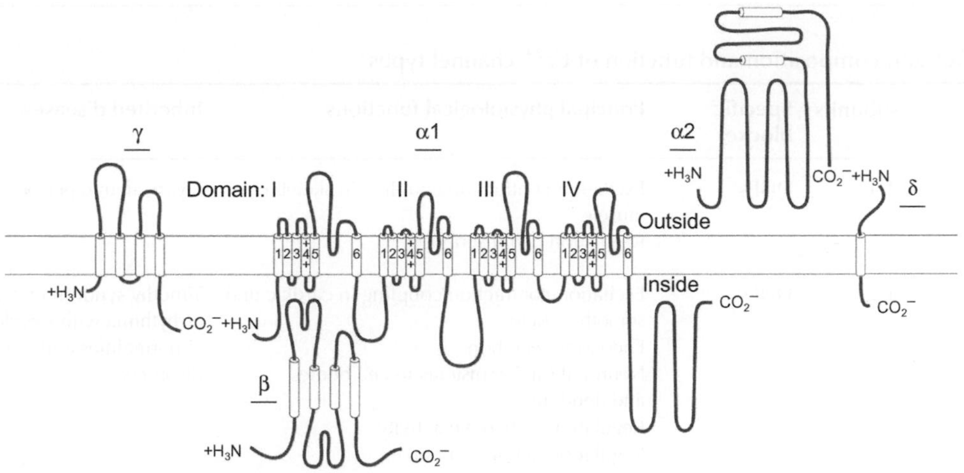

FIGURE 45.2 Subunit structure of Ca^{2+} channels.
(a) The subunit composition and structure of Ca^{2+} channels purified from skeletal muscle are illustrated. The model is updated from our original description of the subunit structure of skeletal muscle Ca^{2+} channels. As described in the text, this model also fits most biochemical and molecular biological results for neuronal Ca^{2+} channels. P, sites of phosphorylation by cAMP-dependent protein kinase. ψ, sites of N-linked glycosylation. (b). Transmembrane folding models for the Ca^{2+} channel subunits. Predicted α-helices are depicted as cylinders. The lengths of lines correlate approximately to the lengths of the polypeptide segments represented.

expressed in brain and other tissues [45]. These γ-subunit-like proteins can modulate the voltage dependence of expressed $Ca_v2.1$ channels, so they may be associated with these Ca^{2+} channels *in vivo*. However, the stargazin-like γ subunits are primary modulators of glutamate receptors in the postsynaptic membranes of brain neurons [46], and it remains to be determined whether they are also associated with voltage-gated Ca^{2+} channels in brain neurons *in vivo*.

Functions of Ca^{2+} Channel Subunits

The initial analyses of functional expression of Ca^{2+} channel subunits were carried out with skeletal muscle Ca^{2+} channels. Expression of the α1 subunit is sufficient to produce functional skeletal muscle Ca^{2+} channels, but with low expression level and abnormal kinetics and voltage dependence of the Ca^{2+} current [47]. Co-expression of the α2δ subunit and especially the β subunit enhanced the level of expression and conferred more normal gating properties [48, 49]. As for skeletal muscle Ca^{2+} channels, co-expression of β subunits has a large effect on the level of expression and the voltage dependence and kinetics of gating of cardiac and neuronal Ca^{2+} channels (reviewed in [50, 51]). In general, the level of expression is increased and the voltage dependence of activation and inactivation is shifted to more negative membrane potentials, and the rate of inactivation is increased. However, these effects are different for the individual β-subunit isoforms. For example, the β2a subunit slows channel inactivation in most subunit combinations. Co-expression of α2δ subunits also increases expression and enhances function of Ca^{2+} channels, but to a lesser extent and in a more channel-specific way than β subunits, whereas γ subunits have much smaller functional effects [52, 53].

Ca^{2+} Channel Diversity

The different types of Ca^{2+} currents are primarily defined by different α1 subunits. The primary structures of 10 distinct Ca^{2+} channel α_1 subunits have been defined by homology screening, and their function has been characterized by expression in mammalian cells or *Xenopus* oocytes. These subunits can be divided into three structurally and functionally related sub families (Ca_v1, Ca_v2, and Ca_v3) [54, 55]. L-type Ca^{2+} currents are mediated by the Ca_v1 type of α1 subunits, which have about 75 percent amino acid sequence identity among them. The Ca_v2 type Ca^{2+} channels form a distinct subfamily with less than 40 percent amino acid sequence identity with Ca_v1 α1 subunits but greater than 70 percent amino acid sequence identity among themselves (Table 45.1). Cloned $Ca_v2.1$ subunits [56, 57] conduct P- or Q-type Ca^{2+} currents, which are inhibited by ω-agatoxin IVA. $Ca_v2.2$ subunits conduct N-type Ca^{2+} currents blocked with high affinity by ω-conotoxin GVIA [58, 59]. Cloned $Ca_v2.3$ subunits form R-type Ca^{2+} channels, which are resistant to both organic Ca^{2+} antagonists specific for L-type Ca^{2+} currents and the peptide toxins specific for N-type or P/Q-type Ca^{2+} currents [60]. T-type Ca^{2+} currents are mediated by the Ca_v3 Ca^{2+} channels [61]. These α1 subunits are only distantly related to the other known homologs, with less than 25 percent amino acid sequence identity. These results reveal a surprising structural dichotomy between the T-type, low-voltage-activated Ca^{2+} channels and the high-voltage-activated Ca^{2+} channels. Evidently, these two lineages of Ca^{2+} channels diverged very early in the evolution of multicellular organisms. Single representatives of the Ca_v1, Ca_v2, and Ca_v3 subfamilies are present in invertebrate genomes, including the flatworm *C. elgans* and the fruit fly *Drosophila*.

TABLE 45.1 Subunit composition and function of Ca^{2+} channel types

Ca^{2+} current type	α1 subunits	Specific blocker	Principal physiological functions	Inherited diseases
L	Cav1.1	DHPs	Excitation-contraction coupling in skeletal muscle Regulation of transcription	Hypokalemic periodic paralysis
	Cav1.2	DHPs	Excitation-contraction coupling in cardiac and smooth muscle Endocrine secretion Neuronal Ca^{2+} transients in cell bodies and dendrites Regulation of enzyme activity Regulation of transcription	Timothy syndrome: cardiac arrhythmia with developmental abnormalities and autism spectrum disorders
	Cav1.3	DHPs	Endocrine secretion Cardiac pacemaking Neuronal Ca^{2+} transients in cell bodies and dendrites Auditory transduction	
	Cav1.4	DHPs	Visual transduction	Stationary night blindness
P/Q	Cav2.1	ω-CTx-GVIA	Neurotransmitter release Dendritic Ca2+ transients	
N	Cav2.2	ω-Agatoxin	Neurotransmitter release Dendritic Ca2+ transients	Inherited migraine Cerebellar ataxia
R	Cav2.3	SNX-482	Neurotransmitter release Dendritic Ca2+ transients	
T	Cav3.1	None	Pacemaking and repetitive firing	
	Cav3.2			
	Cav3.3			

Abbreviations: DHP, dihydropyridine; ω-CTx-GVIA, ω-conotoxin GVIA from the cone snail Conus geographus*; SNX-482, a synthetic version of a peptide toxin from the tarantula* Hysterocrates gigas.

The diversity of Ca^{2+} channel structure and function is substantially enhanced by multiple β subunits. Four β-subunit genes have been identified, and each is subject to alternative splicing to yield additional isoforms (reviewed in [50, 51]). In Ca^{2+} channel preparations isolated from brain, individual Ca^{2+} channel α1 subunit types are associated with multiple types of β subunits, although there is a different rank order in each case [62, 63]. The different β subunit isoforms cause different shifts in the kinetics and voltage dependence of gating, so association with different β subunits can substantially alter the physiological function of an α1 subunit. Genes encoding four α2δ subunits have been described [64], and the α2δ isoforms produced by these different genes have selective effects on the level of functional expression and the voltage dependence of different α1 subunits [65].

Molecular Basis for Ca^{2+} Channel Function

Intensive studies of the structure and function of the related pore-forming subunits of Na^+, Ca^{2+}, and K^+ channels have led to identification of their principal functional domains (reviewed in [66–70]). Each domain of the principal subunits consists of six transmembrane α-helices (S1–S6) and a membrane-associated loop between S5 and S6. The S4 segments of each homologous domain serve as the voltage sensors for activation, moving outward and rotating under the influence of the electric field and initiating a conformational change that opens the pore. The S5 and S6 segments and the membrane-associated pore loop between them form the pore lining of the voltage-gated ion channels. The narrow external pore is lined by the pore loop, which contains a key glutamate residue in each domain that is required for Ca^{2+} selectivity – a structural feature that is unique to Ca^{2+} channels [71]. Remarkably, substitutions that substitute these glutamate residues in the pore loops between the S5 and S6 segments in domains II, III, and IV of sodium channels are sufficient to confer Ca^{2+} selectivity [71, 72]. The inner pore is lined by the S6 segments, which form the receptor sites for the pore-blocking Ca^{2+} antagonist drugs specific for L-type Ca^{2+} channels [73, 74]. All Ca^{2+} channels share these general structural features, but the amino acid residues that confer high affinity for the organic Ca^{2+} antagonists used in therapy of cardiovascular diseases are present only in the Ca_v1 family of Ca^{2+} channels, which conduct L-type Ca^{2+} currents.

Ca^{2+} CHANNEL SIGNALING COMPLEXES

An emerging theme in cell signaling is that Ca^{2+} transients are local, requiring close association between the proteins that generate Ca^{2+} signals and the proteins that serve as effectors and regulators of Ca^{2+} signaling. Two clear examples of this paradigm for voltage-gated Ca^{2+} channels are excitation–contraction coupling, where Ca^{2+} entering muscle cells forms a microdomain that triggers further release of Ca^{2+} from the sarcoplasmic reticulum and initiates contraction of actomyosin; and synaptic transmission, where Ca^{2+} entering nerve terminals forms a microdomain that initiates exocytosis of nearby docked synaptic vesicles. These two examples of localized Ca^{2+} signaling are mediated by Ca^{2+} channel signaling complexes as discussed in the following sections.

A Ca_v1 Channel Signaling Complex in Excitation–Contraction Coupling

As part of the flight-or-flight response, the rate and force of contraction of both skeletal and cardiac muscle is increased through the activity of the sympathetic nervous system. Release of catecholamines stimulates β-adrenergic receptors (βARs), which increases force of skeletal and cardiac muscle contraction and the heart rate. In cardiac muscle, Ca^{2+} influx through $Ca_v1.2$ channels is responsible for initiating excitation–contraction coupling, and increased Ca^{2+} channel activity via the PKA pathway is primarily responsible for the increase in contractility. $Ca_v1.2$ channels are modulated by the β-adrenergic receptor/cAMP signaling pathway [75, 76]. Activation of β-adrenergic receptors increases L-type Ca^{2+} currents through PKA-mediated phosphorylation of the $Ca_v1.2$ channel protein and/or associated proteins [77–80].

PKA Regulation

Both the pore-forming α_1 subunit and the auxiliary β subunits of skeletal muscle $Ca_v1.1$ channels [81–83] and cardiac $Ca_v1.2$ channels [84–87] are phosphorylated by PKA; the α1 subunits are both truncated by proteolytic processing of the C-terminal domain [85, 88–90]; and the primary sites of PKA phosphorylation are located in the distal C-terminus beyond the point of proteolytic cleavage (Figure 45.3) [91–94]. Voltage-dependent potentiation of $Ca_v1.1$ channels on the 50–ms timescale requires PKA phosphorylation [95] as well as PKA anchoring via an A Kinase Anchoring Protein (AKAP) [96, 97], suggesting close association of PKA and Ca^{2+} channels. A novel, plasma-membrane-targeted AKAP (AKAP15) is associated with both $Ca_v1.1$ channels [98, 99] and $Ca_v1.2$ channels [100, 101], and may mediate their regulation by PKA. This AKAP (also known as AKAP18 [102]) binds to the C-terminal domain of $Ca_v1.1$ channels [103] and $Ca_v1.2$ channels [101] via a novel modified leucine-zipper interaction near the primary sites of PKA phosphorylation Block of this interaction by competing peptides prevents PKA regulation of Ca^{2+} currents in intact skeletal and cardiac myocytes [101, 103, 104]. These physiological results suggest that a Ca^{2+} channel signaling complex containing AKAP15 and PKA is formed in both skeletal and cardiac muscle, and this conclusion is supported by specific co-localization of these proteins in both skeletal and cardiac myocytes, and specific co-immunoprecipitation of this complex from both tissues [101, 103, 105]. Remarkably, block of kinase anchoring is as effective as block of kinase activity in preventing $Ca_v1.1$ and $Ca_v1.2$ channel regulation, consistent with the conclusion that PKA targeting via leucine-zipper interactions is absolutely required for regulation of Ca_v1 channels in intact skeletal and cardiac myocytes.

Proteolytic Processing and Regulation Via the C-Terminal Domain

The distal C-terminal domains of skeletal muscle and cardiac Ca^{2+} channels are proteolytically processed *in vivo* (Figure 45.3a) [94, 106]. Nevertheless, the most prominent PKA phosphorylation sites of both proteins are located beyond the site of proteolytic truncation [91, 92], and interaction of AKAP15 and PKA with the distal C-terminal domain through a leucine-zipper motif is required for regulation of cardiac Ca^{2+} channels in intact myocytes [101]. These results imply that the distal C-terminal domain remains associated with the proteolytically processed cardiac $Ca_v1.2$ channel, and this is supported by evidence that the distal C-terminus can bind to the truncated $Ca_v1.1$ and $Ca_v1.2$ channels *in vitro* [107–109] and in transfected cells [103, 104]. Moreover, formation of this complex dramatically inhibits cardiac Ca^{2+} channel function [104]. Deletion of the distal C-terminal near the site of proteolytic processing increases Ca^{2+} channel activity [104, 110]. However, non-covalent association of the cleaved distal C-terminal reduces channel activity more than 10–fold, to a level much below that of channels with an

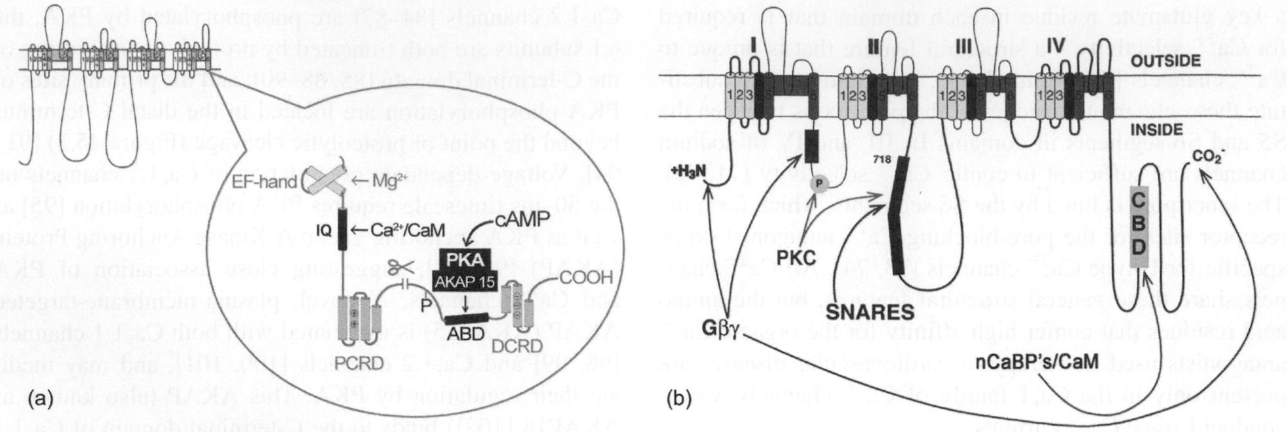

FIGURE 45.3 Ca^{2+} channel signaling complexes.
(a) The cardiac Ca^{2+} channel signaling complex. The C-terminal domain of the cardiac Ca^{2+} channels is shown in expanded presentation to illustrate the regulatory interactions clearly. ABD, AKAP15 binding domain; DCRD, distal C-terminal regulatory domain; PCRD, proximal C-terminal regulatory domain; scissors, site of proteolytic processing. The DCRD binds to the PCRD through a modified leucine zipper interation. (b) The presynaptic Ca^{2+} channel signaling complex. A presynaptic Ca^{2+} channel $\alpha 1$ subunit is illustrated as a transmembrane folding diagram as in Figure 45.2. Sites of interaction of SNARE proteins (the synprint site), G$\beta\gamma$ subunits, protein kinase C (PKC), and CaM and CaS proteins (CBD) are illustrated.

intact C-terminus [104]. Thus, proteolytic processing produces an autoinhibited Ca^{2+} channel complex containing non-covalently bound distal C-terminus, with AKAP15 and PKA associated through a modified leucine-zipper interaction. This autoinhibited complex may be a primary substrate for regulation of cardiac Ca^{2+} channels by the β-adrenergic receptor/PKA pathway *in vivo* [104].

Ca^{2+} Binding Proteins

In addition to their regulation by the PKA/AKAP15 signaling complex, cardiac Ca^{2+} channels have calmodulin bound to their C-terminal domain through an IQ motif (Figure 45.3a), and Ca^{2+} binding to calmodulin causes Ca^{2+}-dependent inactivation [111–113]. Activation of $Ca_v1.2$ channels in the presence of Ba^{2+} as the permeant ion results in inward Ba^{2+} currents that activate rapidly and inactivate slowly via a voltage-dependent inactivation process. In contrast, in the presence of Ca^{2+} as the permeant ion, Ca^{2+} currents are rapidly inactivated via Ca^{2+}/calmodulin-dependent inactivation. The Ca^{2+}-dependent inactivation process is crucial for limiting Ca^{2+} entry during long cardiac action potentials. In light of these results, it is evident that both the cAMP and Ca^{2+} second messenger pathways regulate $Ca_v1.2$ channels locally, dependent on associated regulatory proteins in Ca^{2+} channel signaling complexes.

A Presynaptic Ca^{2+} Channel Signaling Complex in Synaptic Transmission

Presynaptic Ca^{2+} channels conduct P/Q-, N-, and R-type Ca^{2+} currents, which initiate synaptic transmission. The efficiency of neurotransmitter release depends on the third or fourth power of the entering Ca^{2+}. This steep dependence of neurotransmission on Ca^{2+} entry makes the presynaptic Ca^{2+} channel an exceptionally sensitive and important target of regulation. In the nervous system, $Ca_v2.1$ channels conducting P/Q-type Ca^{2+} currents and $Ca_v2.2$ channels conducting N-type Ca^{2+} currents are the predominant pathways for Ca^{2+} entry initiating fast release of classical neurotransmitters like glutamate, acetylcholine, and GABA. Extensive studies indicate that they are controlled by many different protein interactions with their intracellular domains, which serve as a platform for Ca^{2+}-dependent signal transduction (Figure 45.3b).

SNARE Proteins

Ca^{2+} entry through voltage-gated Ca^{2+} channels initiates exocytosis by triggering the fusion of secretory vesicle membranes with the plasma membrane through actions on the SNARE protein complex of syntaxin, SNAP-25, and VAMP/synaptobrevin (reviewed in [114–116]). The function of the SNARE protein complex is regulated by interactions with numerous proteins, including the synaptic vesicle Ca^{2+}-binding protein synaptotagmin. Presynaptic $Ca_v2.1$ and $Ca_v2.2$ channels interact directly with the SNARE proteins through a specific synaptic protein interaction (synprint) site in the large intracellular loop connecting domains II and III (Figure 45.3b) [117, 118]. This interaction is regulated by Ca^{2+} and protein phosphorylation [119–121]. Synaptotagmin also binds to the synprint site of Ca_v2 channels [122–124]. Injection of peptide inhibitors of SNARE protein interactions with Ca_v2 channels into presynaptic neurons inhibits synaptic transmission, consistent with the conclusion that interaction with SNARE proteins is required to position docked synaptic vesicles near Ca^{2+} channels for effective fast exocytosis [125, 126].

These results define a second functional activity of the presynaptic Ca^{2+} channel–targeting docked synaptic vesicles to a source of Ca^{2+} for effective transmitter release.

In addition to this functional role of interaction between Ca^{2+} channels and SNARE proteins in the anterograde process of synaptic transmission, these interactions also have retrograde regulatory effects on Ca^{2+} channel function. Co-expression of the plasma membrane SNARE proteins syntaxin or SNAP-25 with $Ca_v2.1$ or $Ca_v2.2$ channels reduces the level of channel expression and inhibits Ca^{2+} channel activity by shifting the voltage dependence of steady-state inactivation during long depolarizing prepulses toward more negative membrane potentials [127–129]. The inhibitory effects of syntaxin are relieved by co-expression of SNAP-25 and synaptotagmin to form a complete SNARE complex [124, 129, 130], which has the effect of enhancing activation of Ca_v2 channels with nearby docked synaptic vesicles that have formed complete SNARE complexes and are ready for release.

G-protein Modulation

N-type and P/Q-type Ca^{2+} currents are regulated through multiple G-protein-coupled pathways [131–133]. Although there are several G-protein signaling pathways that regulate these channels, one common pathway that has been best studied at both cellular and molecular levels is voltage-dependent and membrane-delimited – that is, a pathway without soluble intracellular messengers whose effects can be reversed by strong depolarization. Inhibition of Ca^{2+} channel activity is typically caused by a positive shift in the voltage dependence and a slowing of channel activation [134]. These effects are relieved by strong depolarization resulting in facilitation of Ca^{2+} currents [134, 135]. Synaptic transmission is inhibited by neurotransmitters through this mechanism. G-protein α subunits are thought to confer specificity in receptor coupling, but $G\beta\gamma$ subunits are responsible for modulation of Ca^{2+} channels. Co-transfection of cells with the Ca^{2+} channel $\alpha1$ and β subunits plus $G\beta\gamma$ causes a shift in the voltage dependence of Ca^{2+} channel activation to more positive membrane potentials and reduces the steepness of voltage dependent activation – effects that closely mimic the actions of neurotransmitters and guanyl nucleotides on N-type and P/Q-type Ca^{2+} currents in neurons and neuroendocrine cells [136]. In contrast, transfection with a range of $G\alpha$ subunits does not have this effect. This voltage shift can be reversed by strong positive prepulses resulting in voltage-dependent facilitation of the Ca^{2+} current in the presence of $G\beta\gamma$, again closely mimicking the effects of neurotransmitters and guanyl nucleotides on Ca^{2+} channels. Similarly, injection or expression of $G\beta\gamma$ subunits in sympathetic ganglion neurons induces facilitation and occludes modulation of N-type channels by norepinephrine, but $G\alpha$ subunits do not [136, 137]. These results point to the $G\beta\gamma$ subunits as the primary regulators of presynaptic Ca^{2+} channels through direct protein–protein interactions (Figure 45.3b).

Possible sites of G-protein $\beta\gamma$ subunit interaction with Ca^{2+} channels have been extensively investigated by construction and analysis of channel chimeras, by G-protein-binding experiments, and by site-directed mutagenesis and expression (Figure 45.3b). Evidence from G-protein-binding and site-directed mutagenesis experiments points to the intracellular loop between domains I and II (L_{I-II}) as a crucial site of G-protein regulation, and peptides from this region of $Ca_v2.2$ prevent inhibition of channel activity by $G\beta\gamma$, presumably by binding to $G\beta\gamma$ and competitively inhibiting its access to Ca^{2+} channels [138–140]. This region of the channel binds $G\beta\gamma$ *in vitro* as well as *in vivo* in the yeast two-hybrid assay [138, 139, 141]. Increasing evidence also points to segments in the N-terminal and C-terminal domains of Ca^{2+} channels that are required for G-protein regulation [142–147]. As the N-terminal and C-terminal domains are likely to interact with each other in the folded channel protein, a second site of interaction for G proteins may be formed at their intersection.

Ca^{2+}-Binding Proteins

Ca^{2+}-dependent facilitation and inactivation of presynaptic Ca^{2+} channels was observed in patch clamp recordings of presynaptic nerve terminals in the rat neurohypophysis [148] and the calyx of Held synapse in the rat brainstem [149]. During tetanic stimulation at this synapse, $Ca_v2.1$ channel currents show both Ca^{2+}-dependent facilitation and inactivation [150–152], which results in facilitation and depression of excitatory postsynaptic responses [149, 151, 152]. Ca^{2+}-dependent facilitation and inactivation are also observed for cloned and expressed $Ca_v2.1$ channels expressed in mammalian cells [153, 154]. A novel CaM-binding site was identified by yeast two-hybrid screening in the C-terminal domain of the pore-forming $\alpha_12.1$ subunit of $Ca_v2.1$ channels [155]. This CaM-binding domain (CBD) (Figure 45.3b) is located on the C-terminal side of the sequence in $\alpha_12.1$ that corresponds to the IQ-domain that is required for CaM modulation of cardiac $Ca_v1.2$ channels [111, 112, 156]. The modified IQ domain of $\alpha_12.1$ contains IM instead of IQ, and has other changes that would be predicted to substantially reduce its affinity for CaM. CaM binding to the CBD is Ca^{2+} dependent. Both Ca^{2+}-dependent facilitation and inactivation are blocked by co-expression of a CaM inhibitor peptide [155], suggesting that Ca^{2+}-dependent modulation of $Ca_v2.1$ channels in neurons is caused by two sequential interactions with CaM or a related Ca^{2+}-binding protein.

The mechanism for Ca^{2+}-dependent facilitation and inactivation of $Ca_v2.1$ channels involves CaM binding to two adjacent subsites – the CBD and the upstream IQ-like motif [157]. The IQ-like motif is required for facilitation, while the CBD is required for inactivation. In addition, the two lobes of CaM are also differentially involved in these two processes. Mutation of the two EF-hands in the C-terminal lobe prevents facilitation, whereas mutation of the EF-hands in the N-terminal lobe prevents inactivation [157–159]. FRET studies indicate that apo-calmodulin can bind to $Ca_v2.1$ channels in intact cells, and binding is enhanced by Ca^{2+} binding to calmodulin [160]. Altogether, these results support a model in which the two lobes of CaM interact differentially with the modified IQ domain and the CBD in order to effect bi-directional regulation, with the C-terminal lobe primarily controlling facilitation through interactions with the IQ-like domain, and the N-terminal lobe primarily controlling inactivation through interactions with the CBD.

CaM is the most well-characterized member of a superfamily of Ca^{2+} sensor (CaS) proteins, many of which differ from CaM in having neuron-specific localization, N-terminal myristoylation, and amino acid substitutions that prevent Ca^{2+} binding to one or two of the EF-hands [161]. The CaS protein CaBP1 binds to the CBD, but not the IQ-like domain, of $\alpha_12.1$, and its binding is Ca^{2+} independent [162]. CaBP1 causes a strong enhancement of the rate of inactivation, a positive shift in the voltage-dependence of activation, and a loss of Ca^{2+}-dependent facilitation of $Ca_v2.1$ channels, which would combine to reduce the activity of these channels. Since it co-immunoprecipitates and colocalizes with $Ca_v2.1$ channels in the brain [162], CaBP1 may be an important determinant of $Ca_v2.1$ channel function in neurons and may contribute to the diversity of function of these channels in the nervous system. Visinin-like protein 2 (VILIP-2) is a neuronal Ca^{2+}-binding protein that is distantly related to CaBP-1 [161]. Consistent with these structural differences, VILIP-2 has opposite effects on $Ca_v2.1$ channels than CaBP-1 [163]. Co-expression of VILIP-2 causes slowed inactivation and enhanced facilitation, but its binding and effects are Ca^{2+} independent, like CaBP-1. VILIP-2 may serve as a positive modulator of synaptic transmission, prolonging Ca^{2+} channel opening and enhancing facilitation. Differential expression of CaBP1 and VILIP-2 at synapses would lead to opposite modulation of synaptic transmission in response to trains of action potentials.

THE EFFECTOR CHECKPOINT MODEL OF Ca^{2+} CHANNEL REGULATION

In closing this chapter on Ca^{2+} signaling via voltage-gated Ca^{2+} channels, it is interesting to introduce an emerging theme that unites several aspects of the localized regulation of these proteins. Ca^{2+} channel signaling complexes are formed when the effectors and regulators of the Ca^{2+} signal bind to the intracellular domains of Ca^{2+} channels in order to effectively receive and respond to the local Ca^{2+} signal. In four cases, binding of the effectors of the Ca^{2+} signal has been shown to enhance the activity of the Ca_v1 and Ca_v2 channels. First, in skeletal muscle, interactions of the plasma membrane $Ca_v1.1$ channel with the ryanodine-sensitive Ca^{2+} release channel in the sarcoplasmic reticulum, which serves as the effector of excitation–contraction coupling, greatly increase the functional activity of the $Ca_v1.1$ channels [164]. Second, as described above, interaction with individual plasma membrane SNARE proteins inhibits the activity of Ca_v2 channels, but formation of a complete SNARE complex containing synaptotagmin, the effector of exocytosis, relieves this inhibition and enhances Ca^{2+} channel activity [124, 128, 129, 165]. Third, binding of Ca^{2+}/CaM-dependent protein kinase II, an effector of Ca^{2+}-dependent regulatory events, to a site in the C-terminal domain of $Ca_v2.1$ channels substantially increases their activity [166]. Finally, binding of RIM, a regulator of SNARE protein function, to the $Ca_v\beta$ subunits substantially increases Ca_v2 channel activity [167]. The common thread in all of these diverse examples of Ca^{2+} channel regulation by interacting proteins is that binding of an effector ready to respond to the Ca^{2+} signal enhances the activity of the Ca^{2+} channel. Thus, this mechanism provides a functional checkpoint of the fitness of a Ca^{2+} channel to carry out its physiological role, and enhances its activity if it passes this checkpoint criterion. This "*effector checkpoint*" mechanism would serve to focus Ca^{2+} entry on the Ca_v channels that are ready to use the resulting Ca^{2+} signal to initiate a physiological intracellular signaling process. It seems likely that further studies will reveal more examples of this form of regulation, and that it may be a unifying theme in Ca^{2+} signaling by Ca_v channels.

REFERENCES

1. Reuter H. Properties of two inward membrane currents in the heart. *Annu Rev Physiol* 1979;**41**:413–24.
2. Tsien RW. Calcium channels in excitable cell membranes. *Annu Rev Physiol* 1983;**45**:341–58.
3. Bers DM. Cardiac excitation–contraction coupling. *Nature* 2002;**415**:198–205.
4. Catterall WA. Excitation–contraction coupling in vertebrate skeletal muscle: a tale of two calcium channels. *Cell* 1991;**64**:871–4.
5. Tanabe T, Mikami A, Niidome T, Numa S, Adams BA, Beam KG. Structure and function of voltage-dependent calcium channels from muscle. *Ann NY Acad Sci* 1993;**707**:81–6.
6. Yang SN, Berggren PO. The role of voltage-gated calcium channels in pancreatic beta-cell physiology and pathophysiology. *Endocr Rev* 2006;**27**:621–76.
7. Tsien RW, Lipscombe D, Madison DV, Bley KR, Fox AP. Multiple types of neuronal calcium channels and their selective modulation. *Trends Neurosci* 1988;**11**:431–8.
8. Dunlap K, Luebke JI, Turner TJ. Exocytotic calcium channels in mammalian central neurons. *Trends Neurosci* 1995;**18**:89–98.
9. Catterall WA, Few AP. Calcium channel regulation and presynaptic plasticity. *Neuron* 2008;**59**:882–901.

10. Flavell SW, Greenberg ME. Signaling mechanisms linking neuronal activity to gene expression and plasticity of the nervous system. *Annu Rev Neurosci* 2008;**31**:563–90.
11. Llinas R, Sugimori M, Hillman DE, Cherksey B. Distribution and functional significance of the P-type, voltage-dependent calcium channels in the mammalian central nervous system. *Trends Neurosci* 1992;**15**:351–5.
12. Bean BP. Classes of calcium channels in vertebrate cells. *Annu Rev Physiol* 1989;**51**:367–84.
13. Hagiwara S, Ozawa S, Sand O. Voltage clamp analysis of two inward current mechanisms in the egg cell membrane of a starfish. *J Gen Physiol* 1975;**65**:617–44.
14. Llinas R, Yarom Y. Electrophysiology of mammalian inferior olivary neurones in vitro. Different types of voltage-dependent ionic conductances. *J Physiol (Lond)* 1981;**315**:569–84.
15. Nowycky MC, Fox AP, Tsien RW. Three types of neuronal calcium channel with different calcium agonist sensitivity. *Nature* 1985;**316**:440–3.
16. Carbone E, Lux HD. A low voltage-activated, fully inactivating Ca channel in vertebrate sensory neurones. *Nature* 1984;**310**:501–2.
17. Fedulova SA, Kostyuk PG, Veselovsky NS. Two types of calcium channels in the somatic membrane of new-born rat dorsal root ganglion neurones. *J Physiol* 1985;**359**:431–46.
18. Olivera BM, Miljanich GP, Ramachandran J, Adams ME. Calcium channel diversity and neurotransmitter release: The omega-conotoxins and omega-agatoxins. *Annu Rev Biochem* 1994;**63**:823–67.
19. Llinás RR, Sugimori M, Cherksey B. Voltage-dependent calcium conductances in mammalian neurons. The P channel. *Ann NY Acad Sci* 1989;**560**:103–11.
20. Mintz IM, Adams ME, Bean BP. P-type calcium channels in rat central and peripheral neurons. *Neuron* 1992;**9**:85–95.
21. Randall A, Tsien RW. Pharmacological dissection of multiple types of calcium channel currents in rat cerebellar granule neurons. *J Neurosci* 1995;**15**:2995–3012.
22. Tottene A, Moretti A, Pietrobon D. Functional diversity of P-type and R-type calcium channels in rat cerebellar neurons. *J Neurosci* 1996;**16**:6353–63.
23. Curtis BM, Catterall WA. Purification of the calcium antagonist receptor of the voltage-sensitive calcium channel from skeletal muscle transverse tubules. *Biochem* 1984;**23**:2113–18.
24. Curtis BM, Catterall WA. Reconstitution of the voltage-sensitive calcium channel purified from skeletal muscle transverse tubules. *Biochem* 1986;**25**:3077–83.
25. Flockerzi V, Oeken HJ, Hofmann F, Pelzer D, Cavalie A, Trautwein W. Purified dihydropyridine-binding site from skeletal muscle t-tubules is a functional calcium channel. *Nature* 1986;**323**:66–8.
26. Leung AT, Imagawa T, Campbell KP. Structural characterization of the 1,4-dihydropyridine receptor of the voltage-dependent calcium channel from rabbit skeletal muscle. Evidence for two distinct high molecular weight subunits. *J Biol Chem* 1987;**262**:7943–6.
27. Striessnig J, Knaus HG, Grabner M, Moosburger K, Seitz W, Lietz H, Glossmann H. Photoaffinity labelling of the phenylalkylamine receptor of the skeletal muscle transverse-tubule calcium channel. *FEBS Letts* 1987;**212**:247–53.
28. Hosey MM, Barhanin J, Schmid A, Vandaele S, Ptasienski J, O'Callahan C, Cooper C, Lazdunski M. Photoaffinity labelling and phosphorylation of a 165 kilodalton peptide associated with dihydropyridine and phenylalkylamine-sensitive calcium channels. *Biochem Biophys Res Commun* 1987;**147**:1137–45.
29. Takahashi M, Seagar MJ, Jones JF, Reber BF, Catterall WA. Subunit structure of dihydropyridine-sensitive calcium channels from skeletal muscle. *Proc Natl Acad Sci USA* 1987;**84**:5478–82.
30. Tanabe T, Takeshima H, Mikami A, Flockerzi V, Takahashi H, Kangawa K, Kojima M, Matsuo H, Hirose T, Numa S. Primary structure of the receptor for calcium channel blockers from skeletal muscle. *Nature* 1987;**328**:313–18.
31. Ruth P, Röhrkasten A, Biel M, Bosse E, Regulla S, Meyer HE, Flockerzi V, Hofmann F. Primary structure of the beta subunit of the DHP-sensitive calcium channel from skeletal muscle. *Science* 1989;**245**:1115–18.
32. Jay SD, Ellis SB, McCue AF, Williams ME, Vedvick TS, Harpold MM, Campbell KP. Primary structure of the gamma subunit of the DHP-sensitive calcium channel from skeletal muscle. *Science* 1990;**248**:490–2.
33. Ellis SB, Williams ME, Ways NR, Brenner R, Sharp AH, Leung AT, Campbell KP, McKenna E, Koch WJ, Hui A, Schwartz A, Harpold MM. Sequence and expression of mRNAs encoding the alpha 1 and alpha 2 subunits of a DHP-sensitive calcium channel. *Science* 1988;**241**:1661–4.
34. Gurnett CA, De Waard M, Campbell KP. Dual function of the voltage-dependent Ca^{2+} channel $\alpha 2\delta$ subunit in current stimulation and subunit interaction. *Neuron* 1996;**16**:431–40.
35. De Jongh KS, Warner C, Catterall WA. Subunits of purified calcium channels. $\alpha 2$ and δ are encoded by the same gene. *J Biol Chem* 1990;**265**:14738–41.
36. Chang FC, Hosey MM. Dihydropyridine and phenylalkylamine receptors associated with cardiac and skeletal muscle calcium channels are structurally different. *J Biol Chem* 1988;**263**:18,929–37.
37. Schneider T, Hofmann F. The bovine cardiac receptor for calcium channel blockers is a 195-kDa protein. *Eur J Biochem* 1988;**174**:369–75.
38. Kuniyasu A, Oka K, Ide-Yamada T, Hatanaka Y, Abe T, Nakayama H, Kanaoka Y. Structural characterization of the dihydropyridine receptor-linked calcium channel from porcine heart. *J Biochem (Tokyo)* 1992;**112**:235–42.
39. Ahlijanian MK, Westenbroek RE, Catterall WA. Subunit structure and localization of dihydropyridine-sensitive calcium channels in mammalian brain, spinal cord, and retina. *Neuron* 1990;**4**:819–32.
40. Witcher DR, De Waard M, Kahl SD, Campbell KP. Purification and reconstitution of N-type calcium channel complex from rabbit brain. *Methods Enzymol* 1995;**238**:335–48.
41. McEnery MW, Snowman AM, Sharp AH, Adams ME, Snyder SH. Purified omega-conotoxin GVIA receptor of rat brain resembles a dihydropyridine-sensitive L-type calcium channel. *Proc Natl Acad Sci USA* 1991;**88**:11,095–9.
42. Martin-Moutot N, Leveque C, Sato K, Kato R, Takahashi M, Seagar M. Properties of omega conotoxin MVIIC receptors associated with alpna-1 A calcium channel subunits in rat brain. *FEBS Letts* 1995;**366**:21–5.
43. Liu H, De Waard M, Scott VES, Gurnett CA, Lennon VA, Campbell KP. Identification of three subunits of the high affinity omega-conotoxin MVIIC-sensitive calcium channel. *J Biol Chem* 1996;**271**:13,804–10.
44. Letts VA, Felix R, Biddlecome GH, Arikkath J, Mahaffey CL, Valenzuela A, Bartlett IFS, Mori Y, Campbell KP, Frankel WN. The mouse stargazer gene encodes a neuronal calcium-channel gamma subunit. *Nat Genet* 1998;**19**:340–7.
45. Klugbauer N, Dai S, Specht V, Lacinova L, Marais E, Bohn G, Hofmann F. A family of gamma-like calcium channel subunits. *FEBS Letts* 2000;**470**:189–97.
46. Nicoll RA, Tomita S, Bredt DS. Auxiliary subunits assist AMPA-type glutamate receptors. *Science* 2006;**311**:1253–6.

47. Perez-Reyes E, Kim HS, Lacerda AE, Horne W, Wei XY, Rampe D, Campbell KP, Brown AM, Birnbaumer L. Induction of calcium currents by the expression of the alpha 1-subunit of the dihydropyridine receptor from skeletal muscle. *Nature* 1989;**340**:233–6.
48. Singer D, Biel M, Lotan I, Flockerzi V, Hofmann F, Dascal N. The roles of the subunits in the function of the calcium channel. *Science* 1991;**253**:1553–7.
49. Lacerda AE, Kim HS, Ruth P, Perez-Reyes E, Flockerzi V, Hofmann F, Birnbaumer L, Brown AM. Normalization of current kinetics by interaction between the alpha-1 and beta subunits of the skeletal muscle dihydropyridine-sensitive calcium channel. *Nature* 1991;**352**:527–30.
50. Hofmann F, Biel M, Flockerzi V. Molecular basis for calcium channel diversity. *Annu Rev Neurosci* 1994;**17**:399–418.
51. Dolphin AC. Beta subunits of voltage-gated calcium channels. *J Bioenerg Biomembr* 2003;**35**:599–620.
52. Davies A, Hendrich J, Van Minh AT, Wratten J, Douglas L, Dolphin AC. Functional biology of the alpha-2-delta subunits of voltage-gated calcium channels. *Trends Pharmacol Sci* 2007;**28**:220–8.
53. Arikkath J, Campbell KP. Auxiliary subunits: essential components of the voltage-gated calcium channel complex. *Curr Opin Neurobiol* 2003;**13**:298–307.
54. Snutch TP, Reiner PB. Calcium channels: diversity of form and function. *Curr Opin Neurobiol* 1992;**2**:247–53.
55. Ertel EA, Campbell KP, Harpold MM, Hofmann F, Mori Y, Perez-Reyes E, Schwartz A, Snutch TP, Tanabe T, Birnbaumer L, Tsien RW, Catterall WA. Nomenclature of voltage-gated calcium channels. *Neuron* 2000;**25**:533–5.
56. Starr TVB, Prystay W, Snutch TP. Primary structure of a calcium channel that is highly expressed in the rat cerebellum. *Proc Natl Acad Sci USA* 1991;**88**:5621–5.
57. Mori Y, Friedrich T, Kim MS, Mikami A, Nakai J, Ruth P, Bosse E, Hofmann F, Flockerzi V, Furuichi T, Mikoshiba K, Imoto K, Tanabe T, Numa S. Primary structure and functional expression from complementary DNA of a brain calcium channel. *Nature* 1991;**350**:398–402.
58. Dubel SJ, Starr TVB, Hell J, Ahlijanian MK, Enyeart JJ, Catterall WA, Snutch TP. Molecular cloning of the alpha-1 subunit of an omega-conotoxin-sensitive calcium channel. *Proc Natl Acad Sci USA* 1992;**89**:5058–62.
59. Williams ME, Brust PF, Feldman DH, Patthi S, Simerson S, Maroufi A, McCue AF, Velicelebi G, Ellis SB, Harpold MM. Structure and functional expression of an omega-conotoxin-sensitive human N-type calcium channel. *Science* 1992;**257**:389–95.
60. Soong TW, Stea A, Hodson CD, Dubel SJ, Vincent SR, Snutch TP. Structure and functional expression of a member of the low voltage-activated calcium channel family. *Science* 1994;**260**:1133–6.
61. Perez-Reyes E, Cribbs LL, Daud A, Lacerda AE, Barclay J, Williamson MP, Fox M, Rees M, Lee JH. Molecular characterization of a neuronal low-voltage-activated T-type calcium channel. *Nature* 1998;**391**:896–900.
62. Pichler M, Cassidy TN, Reimer D, Haase H, Krause R, Ostler D, Striessnig J. Beta subunit heterogeneity in neuronal L-type calcium channels. *J Biol Chem* 1997;**272**:13,877–82.
63. Witcher DR, De Waard M, Liu H, Pragnell M, Campbell KP. Association of native calcium channel beta subunits with the alpha-1 subunit interaction domain. *J Biol Chem* 1995;**270**:18,088–93.
64. Klugbauer N, Lacinová L, Marais E, Hobom M, Hofmann F. Molecular diversity of the calcium channel alpha-2-delta subunit. *J Neurosci* 1999;**19**:684–91.
65. Davies A, Hendrich J, Van Minh AT, Wratten J, Douglas L, Dolphin AC. Functional biology of the $\alpha 2\delta$ subunits of voltage-gated calcium channels. *Trends Pharmacol Sci* 2007;**28**:220–8.
66. Yu FH, Yarov-Yarovoy V, Gutman GA, Catterall WA. Overview of molecular relationships in the voltage-gated ion channel superfamily. *Pharmacol Rev* 2005;**57**:387–95.
67. Bichet D, Haass FA, Jan LY. Merging functional studies with structures of inward-rectifier potassium channels. *Nat Rev Neurosci* 2003;**4**:957–67.
68. Catterall WA. From ionic currents to molecular mechanisms: the structure and function of voltage-gated sodium channels. *Neuron* 2000; **26**:13–25.
69. Catterall WA. Structure and regulation of voltage-gated calcium channels. *Annu Rev Cell Dev Bio* 2000;**16**:521–55.
70. Yi BA, Jan LY. Taking apart the gating of voltage-gated potassium channels. *Neuron* 2000;**27**:423–5.
71. Heinemann SH, Terlau H, Stühmer W, Imoto K, Numa S. Calcium channel characteristics conferred on the sodium channel by single mutations. *Nature* 1992;**356**:441–3.
72. Sather WA, McCleskey EW. Permeation and selectivity in calcium channels. *Annu Rev Physiol* 2003;**65**:133–59.
73. Hockerman GH, Peterson BZ, Sharp E, Tanada TN, Scheuer T, Catterall WA. Construction of a high-affinity receptor site for dihydropyridine agonists and antagonists by single amino acid substitutions in a non-L-type calcium channel. *Proc Natl Acad Sci USA* 1997;**94**:14,906–11.
74. Hockerman GH, Peterson BZ, Johnson BD, Catterall WA. Molecular determinants of drug binding and action on L-type calcium channels. *Annu Rev Pharmacol Toxicol* 1997;**37**:361–96.
75. Reuter H. Calcium channel modulation by neurotransmitters, enzymes and drugs. *Nature* 1983;**301**:569–74.
76. Tsien RW, Bean BP, Hess P, Lansman JB, Nilius B, Nowycky MC. Mechanisms of calcium channel modulation by beta-adrenergic agents and dihydropyridine calcium agonists. *J Mol Cell Cardiol* 1986;**18**:691–710.
77. Reuter H, Scholz H. The regulation of calcium conductance of cardiac muscle by adrenaline. *J Physiol* 1977;**264**:49–62.
78. Tsien RW. Adrenaline-like effects of intracellular iontophoresis of cyclic AMP in cardiac Purkinje fibres. *Nat New Biol* 1973;**245**:120–2.
79. Osterrieder W, Brum G, Hescheler J, Trautwein W, Flockerzi V, Hofmann F. Injection of subunits of cyclic AMP-dependent protein kinase into cardiac myocytes modulates Ca^{2+} current. *Nature* 1982;**298**:576–8.
80. McDonald TF, Pelzer S, Trautwein W, Pelzer DJ. Regulation and modulation of calcium channels in cardiac, skeletal, and smooth muscle cells. *Physiol Rev* 1994;**74**:365–507.
81. Curtis BM, Catterall WA. Phosphorylation of the calcium antagonist receptor of the voltage-sensitive calcium channel by cAMP-dependent protein kinase. *Proc Natl Acad Sci USA* 1985;**82**:2528–32.
82. Flockerzi V, Jaimovich P, Ruth P, Hofmann F. Phosphorylation of purified bovine cardiac sarcolemma and potassium-stimulated calcium uptake. *Eur J Biochem* 1983;**135**:132–42.
83. Takahashi M, Seagar MJ, Jones JF, Reber BF, Catterall WA. Subunit structure of dihydropyridine-sensitive calcium channels from skeletal muscle. *Proc Natl Acad Sci USA* 1987;**84**:5478–82.
84. Hell JW, Yokoyama CT, Wong ST, Warner C, Snutch TP, Catterall WA. Differential phosphorylation of two size forms of the neuronal class C L-type calcium channel $\alpha 1$ subunit. *J Biol Chem* 1993;**268**:19,451–7.

85. De Jongh KS, Murphy BJ, Colvin AA, Hell JW, Takahashi M, Catterall WA. Specific phosphorylation of a site in the full-length form of the α1 subunit of the cardiac L-type calcium channel by cAMP-dependent protein kinase. *Biochemistry* 1996;**35**:10,392–402.
86. Puri TS, Gerhardstein BL, Zhao XL, Ladner MB, Hosey MM. Differential effects of subunit interactions on protein kinase A- and C-mediated phosphorylation of L-type calcium channels. *Biochemistry* 1997;**36**:9605–15.
87. Haase H, Bartel S, Karczewski P, Morano I, Krause EG. In-vivo phosphorylation of the cardiac L-type calcium channel beta-subunit in response to catecholamines. *Mol Cell Biochem* 1996;**163–164**:99–106.
88. De Jongh KS, Merrick DK, Catterall WA. Subunits of purified calcium channels: a 212-kDa form of α1 and partial amino acid sequence of a phosphorylation site of an independent β subunit. *Proc Natl Acad Sci USA* 1989;**86**:8585–9.
89. De Jongh KS, Warner C, Colvin AA, Catterall WA. Characterization of the two size forms of the α1 subunit of skeletal muscle L-type calcium channels. *Proc Natl Acad Sci U S A* 1991;**88**:10,778–82.
90. Hulme JT, Konoki K, Lin TW, Gritsenko MA, Camp DG, Bigelow DJ, Catterall WA. Sites of proteolytic processing and non-covalent association of the distal C-terminal domain of $Ca_v1.1$ channels in skeletal muscle. *Proc Natl Acad Sci USA* 2005; **102**:5274–9.
91. Rotman EI, De Jongh KS, Florio V, Lai Y, Catterall WA. Specific phosphorylation of a COOH-terminal site on the full-length form of the α1 subunit of the skeletal muscle calcium channel by cAMP-dependent protein kinase. *J Biol Chem* 1992;**267**:16,100–5.
92. Rotman EI, Murphy BJ, Catterall WA. Sites of selective cAMP-dependent phosphorylation of the L-type calcium channel α1 subunit from intact rabbit skeletal muscle myotubes. *J Biol Chem* 1995;**270**:16,371–7.
93. Mitterdorfer J, Froschmayr M, Grabner M, Moebius FF, Glossmann H, Striessnig J. Identification of PKA phosphorylation sites in the carboxyl terminus of L-type calcium channel alpha-1 subunits. *Biochem* 1996;**35**:9400–6.
94. De Jongh KS, Murphy BJ, Colvin AA, Hell JW, Takahashi M, Catterall WA. Specific phosphorylation of a site in the full-length form of the alpha-1 subunit of the cardiac L-type calcium channel by cAMP-dependent protein kinase. *Biochem* 1996;**35**:10,392–402.
95. Sculptoreanu A, Scheuer T, Catterall WA. Voltage-dependent potentiation of L-type Ca^{2+} channels due to phosphorylation by cAMP-dependent protein kinase. *Nature* 1993;**364**:240–3.
96. Johnson BD, Brousal JP, Peterson BZ, Gallombardo PA, Hockerman GH, Lai Y, Scheuer T, Catterall WA. Modulation of the cloned skeletal muscle L-type Ca^{2+} channel by anchored cAMP-dependent protein kinase. *J Neurosci* 1997;**17**:1243–55.
97. Johnson BD, Scheuer T, Catterall WA. Voltage-dependent potentiation of L-type Ca^{2+} channels in skeletal muscle cells requires anchored cAMP-dependent protein kinase. *Proc Natl Acad Sci USA* 1994;**91**:11,492–6.
98. Gray PC, Tibbs VC, Catterall WA, Murphy BJ. Identification of a 15-kDa cAMP-dependent protein kinase-anchoring protein associated with skeletal muscle L-type calcium channels. *J Biol Chem* 1997;**272**:6297–302.
99. Gray PC, Johnson BD, Westenbroek RE, Hays LG, Yates JR, Scheuer T, Catterall WA, Murphy BJ. Primary structure and function of an A kinase anchoring protein associated with calcium channels. *Neuron* 1998;**20**:1017–26.
100. Gao T, Yatani A, Dell'Acqua ML, Sako H, Green SA, Dascal N, Scott JD, Hosey MM. cAMP-dependent regulation of cardiac L-type Ca^{2+} channels requires membrane targeting of PKA and phosphorylation of channel subunits. *Neuron* 1997;**19**:185–96.
101. Hulme JT, Lin TW, Westenbroek RE, Scheuer T, Catterall WA. Beta-adrenergic regulation requires direct anchoring of PKA to cardiac $Ca_v1.2$ channels via a leucine zipper interaction with A kinase-anchoring protein 15. *Proc Natl Acad Sci USA* 2003;**100**:13,093–8.
102. Fraser IDC, Tavalin SJ, Lester LB, Langeberg LK, Westphal AM, Dean RA, Marrion NV, Scott JD. A novel lipid-anchored A-kinase anchoring protein facilitates cAMP-responsive membrane events. *EMBO J* 1998;**17**:2261–72.
103. Hulme JT, Ahn M, Hauschka SD, Scheuer T, Catterall WA. A novel leucine zipper targets AKAP15 and cyclic AMP-dependent protein kinase to the C terminus of the skeletal muscle calcium channel and modulates its function. *J Biol Chem* 2002;**277**:4079–87.
104. Hulme JT, Yarov-Yarovoy V, Lin TW-C, Scheuer T, Catterall WA. Autoinhibitory control of the $Ca_v1.2$ channel by its proteolytically cleaved distal C-terminal domain. *J Physiol (Lond)* 2006;**576**:87–102.
105. Hulme JT, Westenbroek RE, Scheuer T, Catterall WA. Phosphorylation of serine 1928 in the distal C-terminal of cardiac $Ca_v1.2$ channels during beta-adrenergic regulation. *Proc Natl Acad Sci USA* 2006;**103**:16,574–9.
106. De Jongh KS, Warner C, Colvin AA, Catterall WA. Characterization of the two size forms of the alpha-1 subunit of skeletal muscle L-type calcium channels. *Proc Natl Acad Sci USA* 1991;**88**:10,778–82.
107. Gao T, Cuadra AE, Ma H, Bunemann M, Gerhardstein BL, Cheng T, Eick RT, Hosey MM. C-terminal fragments of the α1C (Cav1.2) subunit associate with and regulate L-type calcium channels containing C-terminal-truncated α1C subunits. *J Biol Chem* 2001;**276**:21,089–97.
108. Gerhardstein BL, Gao T, Bunemann M, Puri TS, Adair A, Ma H, Hosey MM. Proteolytic processing of the C terminus of the $alpha_{(1C)}$ subunit of L-type calcium channels and the role of a proline-rich domain in membrane tethering of proteolytic fragments. *J Biol Chem* 2000;**275**:8556–63.
109. Hulme JT, Konoki K, Lin TW, Gritsenko MA, Camp DG, Bigelow DJ, Catterall WA. Sites of proteolytic processing and noncovalent association of the distal C-terminal domain of $Ca_v1.1$ channels in skeletal muscle. *Proc Natl Acad Sci USA* 2005;**102**:5274–9.
110. Wei XNA, Lacerda AE, Olcese R, Stefani E, Perez-Reyes E, Birnbaumer L. Modification of Ca^{2+} channel activity by deletions at the carboxyl terminus of the cardiac α1 subunit. *J Biol Chem* 1994;**269**:1635–40.
111. Peterson BZ, DeMaria CD, Yue DT. Calmodulin is the Ca^{2+} sensor for Ca^{2}-dependent inactivation of L-type calcium channels. *Neuron* 1999;**22**:549–58.
112. Zühlke RD, Pitt GS, Deisseroth K, Tsien RW, Reuter H. Calmodulin supports both inactivation and facilitation of L-type calcium channels. *Nature* 1999;**399**:159–62.
113. Qin N, Olcese R, Bransby M, Lin T, Birnbaumer L. Ca^{2+} induced inhibition of the cardiac Ca^{2+} channel depends on calmodulin. *Proc Natl Acad Sci USA* 1999;**96**:2435–8.
114. Bajjalieh SM, Scheller RH. The biochemistry of neurotransmitter secretion. *J Biol Chem* 1995;**270**:1971–4.
115. Sudhof TC. The synaptic vesicle cycle: a cascade of protein–protein interactions. *Nature* 1995;**375**:645–53.
116. Sudhof TC. The synaptic vesicle cycle. *Annu Rev Neurosci* 2004;**27**:509–47.

117. Sheng Z-H, Rettig J, Takahashi M, Catterall WA. Identification of a syntaxin-binding site on N-type calcium channels. *Neuron* 1994;**13**:1303–13.
118. Rettig J, Sheng Z-H, Kim DK, Hodson CD, Snutch TP, Catterall WA. Isoform-specific interaction of the α1A subunits of brain Ca2+ channels with the presynaptic proteins syntaxin and SNAP-25. *Proc Natl Acad Sci USA* 1996;**93**:7363–8.
119. Sheng Z-H, Rettig J, Cook T, Catterall WA. Calcium-dependent interaction of N-type calcium channels with the synaptic core-complex. *Nature* 1996;**379**:451–4.
120. Yokoyama CT, Sheng Z-H, Catterall WA. Phosphorylation of the synaptic protein interaction site on N-type calcium channels inhibits interactions with SNARE proteins. *J Neurosci* 1997;**17**:6929–38.
121. Yokoyama CT, Myers SJ, Fu J, Mockus SM, Scheuer T, Catterall WA. Mechanism of SNARE protein binding and regulation of Ca_v2 channels by phosphorylation of the synaptic protein interaction site. *Mol Cell Neurosci* 2005;**28**:1–17.
122. Charvin N, Lévêque C, Walker D, Berton F, Raymond C, Kataoka M, Shoji-Kasai Y, Takahashi M, De Waard M, Seagar MJ. Direct interaction of the calcium sensor protein synaptotagmin I with a cytoplasmic domain of the α1A subunit of the P/Q-type calcium channel. *EMBO J* 1997;**16**:4591–6.
123. Sheng Z-H, Yokoyama C, Catterall WA. Interaction of the synprint site of N-type Ca^{2+} channels with the C2B domain of synaptotagmin I. *Proc Natl Acad Sci USA* 1997;**94**:5405–10.
124. Wiser O, Tobi D, Trus M, Atlas D. Synaptotagmin restores kinetic properties of a syntaxin-associated N-type voltage sensitive calcium channel. *FEBS Letts* 1997;**404**:203–7.
125. Mochida S, Sheng ZH, Baker C, Kobayashi H, Catterall WA. Inhibition of neurotransmission by peptides containing the synaptic protein interaction site of N-type Ca^{2+} channels. *Neuron* 1996;**17**:781–8.
126. Rettig J, Heinemann C, Ashery U, Sheng ZH, Yokoyama CT, Catterall WA, Neher E. Alteration of Ca^{2+} dependence of neurotransmitter release by disruption of Ca^{2+} channel/syntaxin interaction. *J Neurosci* 1997;**17**:6647–56.
127. Bezprozvanny I, Scheller RH, Tsien RW. Functional impact of syntaxin on gating of N-type and Q-type calcium channels. *Nature* 1995;**378**:623–6.
128. Wiser O, Bennett MK, Atlas D. Functional interaction of syntaxin and SNAP-25 with voltage-sensitive L- and N-type Ca^{2+} channels. *EMBO J* 1996;**15**:4100–10.
129. Zhong H, Yokoyama C, Scheuer T, Catterall WA. Reciprocal regulation of P/Q-type calcium channels by SNAP-25, syntaxin and synaptotagmin. *Nat Neurosci* 1999;**2**:939–41.
130. Tobi D, Wiser O, Trus M, Atlas D. N-type voltage-sensitive calcium channel interacts with syntaxin, synaptotagmin and SNAP-25 in a multiprotein complex. *Recept Channels* 1999;**6**:89–98.
131. Hille B. Modulation of ion-channel function by G-protein-coupled receptors. *Trends Neurosci* 1994;**17**:531–6.
132. Ikeda SR, Dunlap K. Voltage-dependent modulation of N-type calcium channels: role of G protein subunits. *Adv Second Messenger Phosphoprot Res* 1999;**33**:131–51.
133. Jones LP, Patil PG, Snutch TP, Yue DT. G-protein modulation of N-type calcium channel gating current in human embryonic kidney cells (HEK 293). *J Physiol (Lond)* 1997;**498**:601–10.
134. Bean BP. Neurotransmitter inhibition of neuronal calcium currents by changes in channel voltage dependence. *Nature* 1989;**340**:153–6.
135. Marchetti C, Carbone E, Lux HD. Effects of dopamine and noradrenaline on Ca channels of cultured sensory and sympathetic neurons of chick. *Pflügers Arch* 1986;**406**:104–11.
136. Herlitze S, Garcia DE, Mackie K, Hille B, Scheuer T, Catterall WA. Modulation of calcium channels by G protein beta/gamma subunits. *Nature* 1996;**380**:258–62.
137. Ikeda SR. Voltage-dependent modulation of N-type calcium channels by G-protein beta/gamma subunits. *Nature* 1996;**380**:255–8.
138. De Waard M, Liu HY, Walker D, Scott VES, Gurnett CA, Campbell KP. Direct binding of G-protein βγ complex to voltage-dependent calcium channels. *Nature* 1997;**385**:446–50.
139. Zamponi GW, Bourinet E, Nelson D, Nargeot J, Snutch TP. Crosstalk between G proteins and protein kinase C mediated by the calcium channel α_1 subunit. *Nature* 1997;**385**:442–6.
140. Herlitze S, Hockerman GH, Scheuer T, Catterall WA. Molecular determinants of inactivation and G protein modulation in the intracelular loop connecting domains I and II of the calcium channel α_{1A} subunit. *Proc Natl Acad Sci USA* 1997;**94**:1512–16.
141. Garcia DE, Li B, Garcia-Ferreiro RE, Hernández-Ochoa EO, Yan K, Gautam N, Catterall WA, Mackie K, Hille B. G-protein β-subunit specificity in the fast membrane-delimited inhibition of Ca^{2+} channels. *J Neurosci* 1998;**18**:9163–70.
142. Zhang JF, Ellinor PT, Aldrich RW, Tsien RW. Multiple structural elements in voltage-dependent Ca^{2+} channels support their inhibition by G proteins. *Neuron* 1996;**17**:991–1003.
143. Qin N, Platano D, Olcese R, Stefani E, Birnbaumer L. Direct interaction of Gβγ with a C-terminal Gβγ-binding domain of the Ca^{2+} channel α1 subunit is responsible for channel inhibition by G protein-coupled receptors. *Proc Natl Acad Sci USA* 1997;**94**:8866–71.
144. Page KM, Stephens GJ, Berrow NS, Dolphin AC. The intracellular loop between domains I and II of the B-type calcium channel confers aspects of G-protein sensitivity to the E-type calcium channel. *J Neurosci* 1997;**17**:1330–8.
145. Page KM, Cantí C, Stephens GJ, Berrow NS, Dolphin AC. Identification of the amino terminus of neuronal Ca^{2+} channel α1 subunits α1B and α1E as an essential determinant of G-protein modulation. *J Neurosci* 1998;**18**:4815–24.
146. Canti C, Page KM, Stephens GJ, Dolphin AC. Identification of residues in the N terminus of alpha 1B critical for inhibition of the voltage-dependent calcium channel by Gβγ. *J Neurosci* 1999;**19**:6855–64.
147. Li B, Zhong H, Scheuer T, Catterall WA. Functional role of a C-terminal Gβγ-binding domain of $Ca_{(v)}2$. 2 channels. *Mol Pharmacol* 2004;**66**:761–9.
148. Branchaw JL, Banks MI, Jackson MB. Ca^{2+}- and voltage-dependent inactivation of Ca^{2+} channels in nerve terminals of the neurohypophysis. *J Neurosci* 1997;**17**:5772–81.
149. Forsythe ID, Tsujimoto T, Barnes-Davies M, Cuttle MF, Takahashi T. Inactivation of presynaptic calcium current contributes to synaptic depression at a fast central synapse. *Neuron* 1998;**20**:797–807.
150. Forsythe ID, Tsujimoto T, Barnes-Davies M, Cuttle MF, Takahashi T. Inactivation of presynaptic calcium current contributes to synaptic depression at a fast central synapse. *Neuron* 1998;**20**:797–807.
151. Borst JG, Sakmann B. Facilitation of presynaptic calcium currents in the rat brainstem. *J Physiol* 1998;**513**:149–55.
152. Cuttle MF, Tsujimoto T, Forsythe ID, Takahashi T. Facilitation of the presynaptic calcium current at an auditory synapse in rat brainstem. *J Physiol* 1998;**512**:723–9.
153. Lee A, Wong ST, Gallagher D, Li B, Storm DR, Scheuer T, Catterall WA. Calcium/calmodulin binds to and modulates P/Q-type calcium channels. *Nature* 1999;**399**:155–9.
154. Lee A, Scheuer T, Catterall WA. Calcium/calmodulin dependent inactivation and facilitation of P/Q-type calcium channels. *J Neurosci* 2000;**20**:6830–8.

155. Lee A, Wong ST, Gallagher D, Li B, Storm DR, Scheuer T, Catterall WA. Ca^{2+}/calmodulin binds to and modulates P/Q-type calcium channels. *Nature* 1999;**399**:155–9.
156. Qin N, Olcese R, Bransby M, Lin T, Birnbaumer L. Ca^{2+} induced inhibition of the cardiac Ca^{2+} channel depends on calmodulin. *Proc Natl Acad Sci USA* 1999;**96**:2435–8.
157. Lee A, Zhou H, Scheuer T, Catterall WA. Molecular determinants of $Ca^{(2+)}$/calmodulin-dependent regulation of $Ca_{(v)}2$. 1 channels. *Proc Natl Acad Sci USA* 2003;**100**:16,059–64.
158. DeMaria CD, Soong TW, Alseikhan BA, Alvania RS, Yue DT. Calmodulin bifurcates the local Ca^{2+} signal that modulates P/Q-type Ca^{2+} channels. *Nature* 2001;**411**:484–9.
159. Erickson MG, Alseikhan BA, Peterson BZ, Yue DT. Preassociation of calmodulin with voltage-gated $Ca(^{2+})$ channels revealed by FRET in single living cells. *Neuron* 2001;**31**:973–85.
160. Erickson MG, Alseikhan BA, Peterson BZ, Yue DT. Preassociation of calmodulin with voltage-gated calcium channels revealed by FRET in single living cells. *Neuron* 2001;**31**:973–85.
161. Haeseleer F, Palczewski K. Calmodulin and calcium-binding proteins: variations on a theme. *Adv Exp Med Biol* 2002;**514**:303–17.
162. Lee A, Westenbroek RE, Haeseleer F, Palczewski K, Scheuer T, Catterall WA. Differential modulation of $Ca_v2.1$ channels by calmodulin and Ca^{2+}-binding protein 1. *Nat Neurosci* 2002;**5**:210–17.
163. Lautermilch NJ, Few AP, Scheuer T, Catterall WA. Modulation of $Ca_v2.1$ channels by the neuronal calcium-binding protein visinin-like protein-2. *J Neurosci* 2005;**25**:7062–70.
164. Nakai J, Dirksen RT, Nguyen HT, Pessah IN, Beam KG, Allen PD. Enhanced dihydropyridine receptor channel activity in the presence of ryanodine receptor. *Nature* 1996;**380**:72–5.
165. Bezprozvanny I, Scheller RH, Tsien RW. Functional impact of syntaxin on gating of N-type and Q-type calcium channels. *Nature* 1995;**378**:623–6.
166. Jiang X, Lautermilch NJ, Watari H, Westenbroek RE, Scheuer T, Catterall WA. Modulation of $Ca_v2.1$ channels by calcium/calmodulin-dependent protein kinase II bound to the C-terminal domain. *Proc Natl Acad Sci USA* 2008;**105**:341–6.
167. Kiyonaka S, Wakamori M, Miki T, Uriu Y, Nonaka M, Bito H, Beedle AM, Mori E, Hara Y, De Waard M, Kanagawa M, Itakura M, Takahashi M, Campbell KP, Mori Y. RIM1 confers sustained activity and neurotransmitter vesicle anchoring to presynaptic calcium channels. *Nat Neurosci* 2007;**10**:691–701.

Chapter 46

Store-Operated Calcium Channels

James W. Putney, Jr.
Calcium Regulation Section, Laboratory of Signal Transduction, National Institute of Environmental Health Sciences – NIH, Research Triangle Park, North Carolina

STORE-OPERATED OR CAPACITATIVE CALCIUM ENTRY

Many signaling pathways involve the generation of cytoplasmic Ca^{2+} signals. As reviewed in Chapter 115 of Handbook of Cell Signaling, Second Edition, in many instances these Ca^{2+} signals arise as a result of the Ca^{2+}-mobilizing actions of the intracellular messenger, inositol 1,4,5-trisphosphate (IP_3) [1]. IP_3 binds to specific receptor/channels on the endoplasmic reticulum; the binding of IP_3 results in channel opening and release of stored Ca^{2+} to the cytoplasm. In most cell types, this release of Ca^{2+} from intracellular stores is accompanied by an accelerated entry of Ca^{2+} across the plasma membrane. A variety of mechanisms may be responsible for this entry of Ca^{2+} (reviewed in [2, 3]). One mechanism that appears to be ubiquitous in non-excitable cells and is found too in a number of excitable cell types is *capacitative calcium entry*, also known at *store-operated calcium entry* [4–6]. The signal for capacitative calcium entry appears to be the fall in the concentration of Ca^{2+} in the endoplasmic reticulum, or in a specialized sub-compartment of it.

While the physiological mechanism for depleting stores and activating capacitative calcium entry generally involves IP_3-mediated discharge of Ca^{2+} stores, a number of experimental manipulations can bypass receptor activation to empty Ca^{2+} stores. Inhibitors of sarcoplasmic endoplasmic reticulum Ca^{2+} ATPases, such as thapsigargin, cause passive depletion of Ca^{2+} stores and are thus efficient activators of capacitative calcium entry. In electrophysiological studies, utilizing the patch clamp technique to examine whole-cell store-operated membrane currents, IP_3 can be included in the patch pipette, or Ca^{2+} stores can be depleted simply by high concentrations of a Ca^{2+} chelator. The first store-operated current to be described was the Ca^{2+} release-activated Ca^{2+} current (I_{crac}) [7], characteristically found in hematopoetic cells. Noise analysis indicates that the unitary conductance of single CRAC channels is likely too small to measure directly [8]. However, in other cell types the electrophysiological profile of store-operated currents may differ significantly from I_{crac}, and in some instances single channels have been observed (reviewed in [5]). In these instances, the whole cell current resulting from store depletion is always less Ca^{2+}-selective than I_{crac}. This indicates that the molecular composition of store-operated channels differs among cell types, and it is also possible therefore that multiple mechanisms exist for gating these channels.

STORE-OPERATED CHANNELS – TRPs?

For a number of years, the leading contenders for molecular components of stored-operated were members of the *trp* gene superfamily [9]. In *Drosophila*, the *trp* gene encodes a subunit of a cation channel regulated by a light-sensitive phospholipase C [10]. In mammalian cells, within the *trp* superfamily a subgroup of seven mammalian genes with 30–40 percent sequence similarity to *Drosophila trp* have been designated *trpc* for *canonical trp* [11, 12]. The proteins they encode have been designated TRPC1 … TRPC7. In a large number of studies, at one time or another, all of the TRPCs have been reported to give rise to store-operated channels following ectopic expression in various cell systems (reviewed in [5]). In addition, a large number of studies have reported abrogation of store-operated Ca^{2+} entry following knockdown of specific TRPC channels (reviewed in [5]). A general issue is that the channels formed in expression studies are not highly calcium selective, and thus if they form store operated channels, these would likely be less selective channels than those found in non-hematopoetic cells. However, it has been suggested that TRPCs might form calcium-selective channels through association with other proteins, such as Orai, described below [13, 14].

Handbook of Cell Signaling, Three-Volume Set 2 ed.

MAJOR PLAYERS IDENTIFIED: STIM AND ORAI/CRACM

Two independent targeted RNAi screens identified a protein, Stim1 (Stim in *Drosophila*), that is essential for activation of store-operated channels [15, 16]. Stim1 is found in the endoplasmic reticulum, where it appears to function as the initial sensor for Ca^{2+} levels [16, 17]. Knockdown of Stim1 reduces or eliminates store-operated entry in a number of cell types [15, 16, 18]. Stim1 has a calcium-binding EF hand motif near its N-terminus, which would be located in the lumen of the endoplasmic reticulum. Mutation of this EF hand to reduce its Ca^{2+} affinity results in constitutive activation of store-operated entry [16–18]. Stim1 is organized spatially within the endoplasmic reticulum by the microtubule network, which apparently plays a facillatory role in localizing it optimally for signaling purposes [19, 20]. Stim1 is also found in the plasma membrane; its function there is less clear. A homolog of Stim1, Stim2 has been less extensively studied. Knockdown of Stim2 in HeLa cells resulted in a modest decrease in Ca^{2+} entry [16]. In transfection studies, Stim2 has been reported to either inhibit [21] or support [22] store-operated Ca^{2+} entry.

The discovery of Orai (also called CRACM), the second major component of the capacitative calcium entry machinery, was reported almost simultaneously by three laboratories [23–25]. Orai1 is located in the plasma membrane, and has four transmembrane domains. However, Orai1 has no other recognizable signaling or channel-like domains. When Orai1 and Stim1 (or, in the case of *Drosophila*, Orai and Stim) are co-expressed, they synergize to form unusually large I_{crac}-like currents [18, 25–27]. This result indicates that these two proteins can fully recapitulate the fundamental properties of store-operated Ca^{2+} entry and I_{crac}, a finding notably missing in all previous studies of TRP channels. The ability of overexpressed Stim1 and Orai1 to produce huge I_{crac}-like currents does not necessarily mean that no other players are involved; however, to produce the very huge currents, other proteins functioning in a stoichiometric complex with Stim1 and/or Orai1 would have to be constitutively present in considerable excess.

The large I_{crac}-like currents observed with expression of Stim1 and Orai1 immediately led to speculation that Orai1 was likely the molecular component of the channel underlying I_{crac} (CRAC channel). However, there are no obvious channel pore-like sequences in Orai1. Three laboratories focused on a string of acidic residues near the extracellular boundary of the first transmembrane domain [28–30]. The most interesting mutations were substitutions for a Glu in position 106 in mammalian Orai1. Mutation to Ala resulted in a non-functional channel; however, the conservative mutation of this Glu to an Asp (E106D, E180D in *D. melanogaster* Stim) resulted in a functional channel with markedly reduced selectivity for Ca^{2+}. These findings provide strong evidence that this region functions as part of the Ca^{2+}-binding selectivity filter, and indicate that Orai1 is indeed a pore-forming subunit of the CRAC channel. Prakriya and colleagues [28] and Vig *et al.* [30] also investigated a Glu residue at position 190. Mutation of this glutamate to Asp or even Ala had no effect on channel function; however, alteration to a Gln (E190Q) resulted in diminished Ca^{2+} selectivity.

In addition to Orai1, mammalian cells have genes for two additional homologs, Orai2 and Orai3, which are also expressed on the plasma membrane. Orai2 and Orai3 function similarly to Orai1 when expressed in cell lines, although there are some quantitative differences [31–33]. Whether Orai2 and Orai3 form distinct store-operated channels in specific cell types, or function as subunits of heteromeric channels with Orai1, will be a topic of future investigation.

SIGNALING TO STORE-OPERATED CHANNELS

There is very strong evidence that the signaling pathway for CRAC channel activation begins with the Ca^{2+} sensor, Stim1, and culminates in the activation of channels composed partly or wholly of Orai subunits. The obvious question is: how does Stim1 convey information of depleted Ca^{2+} stores to Orai channels? The simplest possibility is that when Stim1 coalesces into punctae and approaches the plasma membrane, it directly interacts with Orai channels there. This would presumably occur through the C-terminal domains of Stim1, since these domains are cytoplasmic and would have access to intracellular domains of Orai. Consistent with this idea, in one study overexpression of a soluble peptide corresponding to the C-terminal cytoplasmic sequence of Stim1 induced constitutive Ca^{2+} entry [34]. In support of a possible direct interaction, Yeromin and colleagues [29] reported that *D. melanogaster* Stim and Orai could be co-immunoprecipitated, and this association was increased by depletion of Ca^{2+} stores. It is clear from the work of Luik and colleagues [35] that communication between Stim1 and Orai1 occurs over very short distances. These investigators observed that Orai1 and Stim1 aggregated at specific sites, and that Ca^{2+} entry was spatially restricted to these plasma membrane sites as well.

Thus, the precise means of communication between Stim1 and Orai channels is presently unknown. A possibility is that although Stim1 and Orai1 communicate over very short distances, this could still involve the generation and release from the endoplasmic reticulum of a diffusible messenger, or calcium influx factor [36, 37]. In addition, one study provided evidence that Orai brings to the complex with Stim1 an additional but mysterious protein that is considerably larger than either Orai or Stim1 [38]. This key issue will no doubt be the focus of intense investigation in the near future.

REFERENCES

1. Berridge MJ. Inositol trisphosphate and calcium signalling. *Nature* 1993;**361**:315–25.
2. Meldolesi J, Clementi E, Fasolato C, Zacchetti D, Pozzan T. Ca^{2+} influx following receptor activation. *Trends Pharmacol Sci* 1991;**12**:289–92.
3. Barritt GJ. Receptor-activated Ca^{2+} inflow in animal cells: a variety of pathways tailored to meet different intracellular Ca^{2+} signalling requirements. *Biochem J* 1999;**337**:153–69.
4. Putney Jr JW. A model for receptor-regulated calcium entry. *Cell Calcium* 1986;**7**:1–12.
5. Parekh AB, Putney Jr JW. Store-operated calcium channels. *Physiol Rev* 2005;**85**:757–810.
6. Berridge MJ. Capacitative calcium entry. *Biochem J* 1995;**312**:1–11.
7. Hoth M, Penner R. Depletion of intracellular calcium stores activates a calcium current in mast cells. *Nature* 1992;**355**:353–5.
8. Zweifach A, Lewis RS. Mitogen-regulated Ca^{2+} current of T lymphocytes is activated by depletion of intracellular Ca^{2+} stores. *Proc Natl Acad Sci USA* 1993;**90**:6295–9.
9. Birnbaumer L, Zhu X, Jiang M, Boulay G, Peyton M, Vannier B, Brown D, Platano D, Sadeghi H, Stefani E, Birnbaumer M. On the molecular basis and regulation of cellular capacitative calcium entry: Roles for Trp proteins. *Proc Natl Acad Sci USA* 1996;**93**:15,195–202.
10. Montell C. TRP channels in Drosophila photoreceptor cells. *J Physiol* 2005;**567**:45–51.
11. Montell C, Birnbaumer L, Flockerzi V, Bindels RJ, Bruford EA, Caterina MJ, Clapham DE, Harteneck C, Heller S, Julius D, Kojima I, Mori Y, Penner R, Prawitt D, Scharenberg AM, Schultz G, Shimizu N, Zhu MX. A unified nomenclature for the superfamily of TRP cation channels. *Mol Cell* 2002;**9**:229–31.
12. Montell C. The TRP superfamily of cation channels. *Sci STKE* 2005:re3.
13. Liao Y, Erxleben C, Yildirim E, Abramowitz J, Armstrong DL, Birnbaumer L. Orai proteins interact with TRPC channels and confer responsiveness to store depletion. *Proc Natl Acad Sci USA* 2007;**104**:4682–7.
14. Ong HL, Cheng KT, Liu X, Bandyopadhyay BC, Paria BC, Soboloff J, Pani B, Gwack Y, Srikanth S, Singh BB, Gill D, Ambudkar IS. Dynamic assembly of TRPC1/STIM1/Orai1 ternary complex is involved in store operated calcium influx: Evidence for similarities in SOC and CRAC channel components. *J Biol Chem* 2007;**282**:9105–16.
15. Roos J, DiGregorio PJ, Yeromin AV, Ohlsen K, Lioudyno M, Zhang S, Safrina O, Kozak JA, Wagner SL, Cahalan MD, Velicelebi G, Stauderman KA. STIM1, an essential and conserved component of store-operated Ca2+ channel function. *J Cell Biol* 2005;**169**:435–45.
16. Liou Jr J, Kim ML, Heo WD, Jones JT, Myers JW, Ferrell JE, Meyer T. STIM is a Ca^{2+} sensor essential for Ca^{2+}-store-depletion-triggered Ca^{2+} influx. *Curr Biol* 2005;**15**:1235–41.
17. Zhang SL, Yu Y, Roos J, Kozak JA, Deerinck TJ, Ellisman MH, Stauderman KA, Cahalan MD. STIM1 is a Ca^{2+} sensor that activates CRAC channels and migrates from the Ca^{2+} store to the plasma membrane. *Nature* 2005;**437**:902–5.
18. Mercer Jr JC, DeHaven WI, Smyth JT, Wedel B, Boyles RR, Bird GS, Putney JW. Large store-operated calcium-selected currents due to co-expression of orai1 or orai2 with the intracellular calcium sensor, stim1. *J Biol Chem* 2006;**281**:24,979–90.
19. Baba Y, Hayashi K, Fujii Y, Mizushima A, Watarai H, Wakamori M, Numaga T, Mori Y, Iino M, Hikida M, Kurosaki T. Coupling of STIM1 to store-operated Ca^{2+} entry through its constitutive and inducible movement in the endoplasmic reticulum. *Proc Natl Acad Sci USA* 2006;**103**:16,704–9.
20. Smyth Jr JT, DeHaven WI, Bird GS, Putney JW. Role of the microtubule cytoskeleton in the function of the store-operated Ca2+ channel activator, Stim1. *J Cell Sci* 2007;**120**:3762–71.
21. Soboloff J, Spassova MA, Hewavitharana T, He LP, Xu W, Johnstone LS, Dziadek MA, Gill DL. STIM2 is an inhibitor of STIM1-mediated store-operated Ca^{2+} entry. *Curr Biol* 2006;**16**:1465–70.
22. Parvez S, Beck A, Peinelt C, Soboloff J, Lis A, Monteilh-Zoller M, Gill DL, Fleig A, Penner R. STIM2 protein mediates distinct store-dependent and store-independent modes of CRAC channel activation. *FASEB J* 2007;**22**:752–61.
23. Feske S, Gwack Y, Prakriya M, Srikanth S, Puppel SH, Tanasa B, Hogan PG, Lewis RS, Daly M, Rao A. A mutation in Orai1 causes immune deficiency by abrogating CRAC channel function. *Nature* 2006;**441**:179–85.
24. Vig M, Peinelt C, Beck A, Koomoa DL, Rabah D, Koblan-Huberson M, Kraft S, Turner H, Fleig A, Penner R, Kinet JP. CRACM1 is a plasma membrane protein essential for store-operated Ca^{2+} entry. *Science* 2006;**312**:1220–3.
25. Zhang SL, Yeromin AV, Zhang XH, Yu Y, Safrina O, Penna A, Roos J, Stauderman KA, Cahalan MD. Genome-wide RNAi screen of Ca^{2+} influx identifies genes that regulate Ca2+ release-activated Ca2+ channel activity. *Proc Natl Acad Sci USA* 2006;**103**:9357–62.
26. Peinelt C, Vig M, Koomoa DL, Beck A, Nadler MJS, Koblan-Huberson M, Lis A, Fleig A, Penner R, Kinet JP. Amplification of CRAC current by STIM1 and CRACM1 (Orai1). *Nat Cell Biol* 2006;**8**:771–3.
27. Soboloff J, Spassova MA, Tang XD, Hewavitharana T, Xu W, Gill DL. Orai1 and STIM reconstitute store-operated calcium channel function. *J Biol Chem* 2006;**281**:20,661–5.
28. Prakriya M, Feske S, Gwack Y, Srikanth S, Rao A, Hogan PG. Orai1 is an essential pore subunit of the CRAC channel. *Nature* 2006;**443**:230–3.
29. Yeromin AV, Zhang SL, Jiang W, Yu Y, Safrina O, Cahalan MD. Molecular identification of the CRAC channel by altered ion selectivity in a mutant of Orai. *Nature* 2006;**443**:226–9.
30. Vig M, Beck A, Billingsley JM, Lis A, Parvez S, Peinelt C, Koomoa DL, Soboloff J, Gill DL, Fleig A. CRACM1 multimers form the ion-selective pore of the CRAC channel. *Curr Biol* 2006;**16**:2073–9.
31. Gwack Y, Srikanth S, Feske S, Cruz-Guilloty F, Oh-hora M, Neems DS, Hogan PG, Rao A. Biochemical and functional characterization of Orai proteins. *J Biol Chem* 2007;**282**:16,232–43.
32. DeHaven Jr WI, Smyth JT, Boyles RR, Putney JW. Calcium inhibition and calcium potentiation of Orai1, Orai2, and Orai3 calcium release-activated calcium channels. *J Biol Chem* 2007;**282**:17,548–56.
33. Lis A, Peinelt C, Beck A, Parvez S, Monteilh-Zoller M, Fleig A, Penner R. CRACM1, CRACM2, and CRACM3 are store-operated Ca2+ channels with distinct functional properties. *Curr Biol* 2007;**17**:794–800.
34. Huang GN, Zeng W, Kim JY, Yuan JP, Han L, Muallem S, Worley PF. STIM1 carboxyl-terminus activates native SOC, Icrac and TRPC1 channels. *Nat Cell Biol* 2006;**8**:1003–10.
35. Luik RM, Wu MM, Buchanan J, Lewis RS. The elementary unit of store-operated Ca2+ entry: local activation of CRAC

channels by STIM1 at ER-plasma membrane junctions. *J Cell Biol* 2006;**174**:815–25.
36. Bolotina VM, Csutora P. CIF and other mysteries of the store-operated Ca^{2+}-entry pathway. *Trends Biochem Sci* 2005;**30**:378–87.
37. Csutora P, Zarayskiy V, Peter K, Monje F, Smani T, Zakharov SI, Litvinov D, Bolotina VM. Activation mechanism for CRAC current and store-operated Ca^{2+} entry: calcium influx factor and Ca^{2+} independent phospholipase A2beta-mediated pathway. *J Biol Chem* 2006;**281**:34,926–35.
38. Varnai P, Toth B, Toth DJ, Hunyady L, Balla T. Visualization and manipulation of plasma membrane–endoplasmic reticulum contact sites indicates the presence of additional molecular components within the STIM1–Orai1 complex. *J Biol Chem* 2007;**282**:29,678–90.

Intracellular Calcium Signaling

Dagmar Harzheim, H. Llewelyn Roderick,* and Martin D. Bootman

*The Babraham Institute, Babraham, Cambridge, England, UK and *Department of Pharmacology, University of Cambridge, Tennis Court Road, Cambridge, England, UK*

THE "CALCIUM SIGNALING TOOLKIT" AND CALCIUM HOMEOSTASIS

Calcium (Ca^{2+}) is a ubiquitous intracellular messenger, controlling a diverse range of cellular processes, such as gene expression, cellular contraction, synaptic transmission, and cell proliferation. It has been presumed that cells began to export or sequester free Ca^{2+} ions early during evolution. There are many possible reasons for this. For example, cells possess substantial levels of phosphate, and some Ca^{2+}-phosphate salts are relatively insoluble. In addition, sustained high levels of Ca^{2+} are deleterious to cells, and can trigger cell death. The use of energy to export Ca^{2+} provided cells with a steep electrochemical gradient of the ions, which could be harnessed for signaling. The concentration of Ca^{2+} in the cytoplasm of resting cells is typically around 100 nM, and it can be up to 10,000-fold higher in the extracellular space. All cells contain tools to shape Ca^{2+} signals in the dimensions of space, time, and amplitude. To generate and interpret the variety of observed Ca^{2+} signals, different cell types employ components selected from a "Ca^{2+} signaling toolkit", which comprises an array of homeostatic and sensory mechanisms [1]. Since many of the molecular components of this toolkit have multiple isoforms with subtly different properties, each specific cell type can exploit this large repertoire of proteins to construct highly versatile Ca^{2+} signaling networks. Thus by mixing and matching components from the toolkit, cells can obtain Ca^{2+} signals that suit their physiology.

Classically, Ca^{2+} is considered to elicit its major signaling function when it is elevated in the cytosolic compartment. From there, it can also diffuse into organelles such as mitochondria and the nucleus. The Ca^{2+} concentration inside cells is regulated by the simultaneous interplay of multiple counteracting processes, which can be divided into Ca^{2+} "on" and "off" mechanisms depending on whether they serve to increase or decrease cytosolic Ca^{2+} [1] (Figure 47.1). The Ca^{2+} "on" mechanisms include channels located at the plasma membrane which regulate the supply of Ca^{2+} from the extracellular space; and channels on the endoplasmic reticulum/sarcoplasmic reticulum (ER/SR respectively), Golgi, secretory granules, and acidic stores (e.g., lysosomes) which release the finite intracellular Ca^{2+} stores. The "off" mechanisms include Ca^{2+}ATPases on the plasma membrane and ER/SR, and exchangers that utilize the electrochemical Na^+ gradient to drive the transport of Ca^{2+} out of the cell. Occasionally, some of the "off" mechanisms contribute to cytosolic Ca^{2+} increases; examples include "slippage" of Ca^{2+} through Ca^{2+}ATPases, and reverse-mode Na^+/Ca^{2+} exchange.

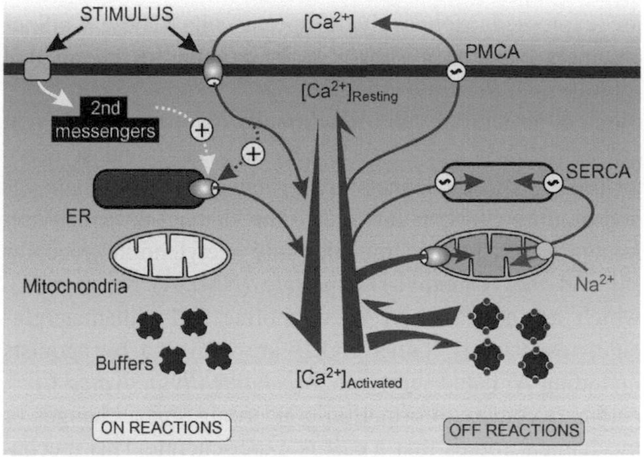

FIGURE 47.1 Calcium "on" and "off" mechanisms in cellular signaling and homeostasis.
Illustration of the various pathways and mechanisms by which cytosolic Ca^{2+} levels can increase or decline. At rest, cells generally have a free cytosolic Ca^{2+} concentration of around 100 nM. This can be increased by activation of channels at the plasma membrane and release from internal stores (denoted "ER"). Cytosolic Ca^{2+} signals are attenuated by passive Ca^{2+} buffering, and actively reversed by mitochondria and Ca^{2+} ATPases on the ER and plasma membrane.

When cells are at rest, the balance lies in favor of the "off" mechanisms, thus yielding an intracellular Ca^{2+} concentration of ~100 nM. However, when cells are stimulated (e.g., by depolarization, mechanical deformation, or hormones), the "on" mechanisms are activated and the cytosolic Ca^{2+} concentration increases to levels of 1 μM or more.

As mentioned above, Ca^{2+} signals can be modulated in their temporal, amplitude, and spatial dimensions. Furthermore, Ca^{2+} signals can arise from different cellular sources, which appear to be regulated by a growing number of messengers [2]. This complex spatiotemporal organization of Ca^{2+} signaling is key to fulfilling its diverse cellular functions. The following sections describe the currently known mechanisms responsible for the generation of Ca^{2+} signals ("on" mechanisms) and examples of how Ca^{2+} signals can be shaped in time and space.

CHANNELS UNDERLYING CA^{2+} INCREASE

Ca^{2+} Influx Channels

Cells utilize several different types of Ca^{2+} influx channels, which can be grouped on the basis of their activation mechanisms. *Voltage-operated Ca^{2+} channels (VOCs)* mediate Ca^{2+} influx in response to depolarization of the plasma membrane [3]. These channels are mainly employed by excitable cell types such as muscle and neuronal cells. According to the characteristics of the corresponding Ca^{2+} current and channel pharmacology, VOCs have been classified as L-, N-, P-, Q-, R-, and T-type channels. Recently, a new class of VOC has been identified, members of which are expressed exclusively in sperm cells. These CatSper channels have been shown to be crucial for sperm cell motility [4]. *Receptor-operated Ca^{2+} channels* comprise a range of structurally and functionally diverse channels that are particularly prevalent in secretory cells and at nerve terminals. These channels are so named because their ligand binding site(s) is integral to the channel protein. Well-known ROCs include the nicotinic acetylcholine receptor [5] and the N-methyl-D-aspartate (NMDA) receptor [6], which are activated by acetylcholine and glutamate/glycine, respectively. Other ROCs are activated by agonists including ATP and serotonin. *Mechanically-activated Ca^{2+} channels* convey information concerning shape changes or mechanical forces that a cell is experiencing. This mechanism is important for the transduction of mechanical senses like hearing and touch. The identity of the underlying channels is still unclear, although members of the transient receptor family (TRP channels) have been implicated, for example, in controlling the myogenic tone and response of the bladder urothelium to stretch [7, 8]. *Second messenger operated channels (SMOCs)* are Ca^{2+} influx channels that open in response to a diverse array of cellular signals. These channels also include members of the TRP family of proteins, which are in many cases stimulated by lipid metabolites, such as diacylglycerol and polyunsaturated fatty acids generated downstream of PLC activation [9].

Store-operated Ca^{2+} channels (SOCs) are activated upon release of Ca^{2+} from intracellular stores. The process of store-operated Ca^{2+} entry has been known and well-characterized for many years, and has been shown to be important in numerous cell functions. In particular, it is necessary in the immune system, where, in the absence of store-operated Ca^{2+} influx, lymphocyte activation is compromised and immune responses cannot be mounted. The molecules that link the depletion of Ca^{2+} stores with Ca^{2+} entry across the plasma membrane have only recently been identified [10]. STIM1 is a single-transmembrane domain protein that senses the Ca^{2+} concentration in the ER via an EF hand in its N-terminus. Following Ca^{2+} store depletion, STIM1 relocates along the endoplasmic reticulum to regions under the plasma membrane, where it forms discrete punctae and activates Orai, the recently-described channel that conducts Ca^{2+} release-activated Ca^{2+} influx (CRAC) [11, 12]. Basal Ca^{2+} entry required to maintain the ER Ca^{2+} store in a replete state is controlled by STIM2, which, due to its lower affinity for Ca^{2+} than STIM1, senses modest decreases in lumenal Ca^{2+} concentration [13].

Ca^{2+} Release Channels

Inositol 1,4,5-trisphosphate receptors ($InsP_3Rs$). $InsP_3$ receptors form a family of abundantly expressed Ca^{2+} release channels located principally in the ER, but also found in the Golgi, nuclear envelope and plasma membrane. $InsP_3Rs$ bind the intracellular messenger inositol 1,4,5-trisphosphate ($InsP_3$), which is generated through phospholipase C (PLC)-mediated hydrolysis of phospholipids. There are 13 different PLC isozymes that can be grouped into six subfamilies (β, γ, δ, ε, η, and ζ). They are activated by several different mechanisms, including binding of hormones and growth factors to specific receptors in the plasma membrane, increases in cytosolic Ca^{2+}, and also by the small GTPase Ras. $InsP_3$ is water-soluble, and diffuses into the cell interior where it can engage $InsP_3Rs$ to release Ca^{2+} into the cytoplasm. Three different isoforms of $InsP_3Rs$ have been described, which appear to differ subtly in their characteristics, such as affinity for $InsP_3$. An important feature of $InsP_3Rs$ is that they are co-regulated by $InsP_3$ and Ca^{2+}. Indeed, it seems that $InsP_3$ may serve to make $InsP_3Rs$ responsive to an activating Ca^{2+} signal. $InsP_3R$ opening is biphasically regulated by Ca^{2+}; 0.1- to 0.5-μM Ca^{2+} increases channel activity, whereas greater Ca^{2+} concentrations inactivate the channels [14, 15]. This dependence of $InsP_3R$ activity on cytosolic Ca^{2+} is crucial in the generation of the complex patterns of Ca^{2+} signals seen in many cells. Although $InsP_3Rs$ are largely thought to control relatively slow Ca^{2+} signaling in non-electrically excitable cells, they are present within cells that have rapid Ca^{2+} changes. Indeed, $InsP_3$-evoked Ca^{2+} signals control a plethora of cellular functions, ranging from secretion in exocrine tissue, to higher brain functions like behavior, learning, and memory. One of the more novel roles for $InsP_3Rs$ is in cardiomyocytes, where they appear to cause the generation of arrhythmogenic Ca^{2+} signals and modulate the activity of certain transcriptional regulators involved in cardiac hypertrophy [16, 17]. The expression of $InsP_3Rs$ in cardiomyocytes is puzzling, since current evidence would suggest their role is more pathological than physiological.

Ryanodine receptors (RyRs) are structurally and functionally analogous to $InsP_3Rs$, although they have approximately twice the conductance and molecular mass of $InsP_3Rs$. RyRs are largely present in excitable cell types, such as muscle and neurons. As with $InsP_3Rs$, RyR subunits are encoded by three genes. RyR1 is the predominant isoform in skeletal muscle, whereas RyR2 is the isoform expressed in cardiac muscle. In neurons, all three isoforms are present. In skeletal and cardiac muscle, RyRs control intracellular Ca^{2+} release and thereby initiate muscle contraction by a process called excitation–contraction coupling (ECC). The first step in ECC is the depolarization of the plasma membrane by an action potential. This leads to opening of L-type voltage-operated Ca^{2+} channels and, thereby, Ca^{2+} influx into the cell. However, there is a fundamental difference in ECC between skeletal and cardiac muscle. In skeletal muscles RyR1 is opened by direct interaction with the VOC through a conformational coupling mechanism, whereas in cardiac muscles Ca^{2+} entering the cell through L-type channels induces opening of RyR2. This mechanism, in which Ca^{2+} acts as the activating ligand for RyRs, is termed *Ca^{2+} induced Ca^{2+} release* (CICR). RyRs are also modulated by the pyrimidine nucleotide cyclic ADP ribose. Recently, it has been shown that inherited mutations in RyR1 cause skeletal muscle diseases, such as central core disease (CCD) and skeletal muscle fatigue. Mutations in RyR2 have been proposed to cause catecholaminergic polymorphic ventricular tachycardia (CPVT) and heart failure. Therefore, RyRs play an important role as potential therapeutic targets [18].

Although RyRs and $InsP_3Rs$ are the best-characterized intracellular Ca^{2+} release channels, a number of other Ca^{2+} release have been described [19]. Notable among these is NAADP-induced Ca^{2+} release from acidic lysosomal organelles, which appears to mobilize Ca^{2+} via "two-pore channels". Ca^{2+} release from the ER is also mediated by polycystin 2, a protein involved in polycystic kidney disease, via a CICR mechanism. Presenilin, a protein mutated in Alzheimer's disease, may promote Ca^{2+} flux by forming a pore, or by interacting with Ca^{2+} channels such as $InsP_3Rs$.

The content of Ca^{2+} within intracellular stores can be an important factor in determining how cells respond to various stimuli. In particular, the release of substantial amounts of Ca^{2+} from internal stores has been consistently linked with the induction of cell death [20]. The concentration of Ca^{2+} within the stores is determined by the competitive activities of the channels that release Ca^{2+} versus the Ca^{2+} pumps that sequester Ca^{2+}. In addition, several other factors have been proposed to modulate the concentration of Ca^{2+} within the lumen of internal stores, and thereby modulate cellular activities. For example, the anti-apoptotic proteins Bcl-2 [21] and Bax inhibitor-1 [22] have been proposed to reduce the Ca^{2+} concentration of intracellular stores. Exactly how this is achieved is not fully understood. One suggestion is that these proteins may induce a new steady-state Ca^{2+} concentration within intracellular stores due to the induction of a leak pathway. However, they may function by modulating the activity of Ca^{2+} pumps or Ca^{2+} release channels [23, 24].

TEMPORAL REGULATION OF CA^{2+} SIGNALS

Ca^{2+} signaling occurs across timescales spanning from microseconds to days. An example where rapid Ca^{2+} signaling controls cellular function is excitation–contraction coupling in striated muscle cells. As described above, this mechanism involves activation of RyR1 in skeletal muscles and RyR2 in cardiac muscles. In cardiomyocytes, opening of VOCs increases the local $[Ca^{2+}]_i$ from 100 nM to 10–20 μM within ~1 ms. In turn, stored Ca^{2+} is released via opening of RyR2 within < 50 ms. Following this CICR event, Ca^{2+} diffuses away from the RyRs to activate the myofilaments and cause contraction. Consequently, Ca^{2+} has to be removed from the cytosol, mainly by the SR Ca-ATPase (SERCA) and the Na^+/Ca^{2+} exchanger (NCX) in the sarcolemma, to enable the initiation of a new heartbeat. In organisms such as the mouse, which has a heart rate of 300 beats per minute, the entire Ca^{2+} cycle has to be completed within 200 ms.

Another important example of temporal Ca^{2+} regulation, but using slower fluxes, is the response of hepatocytes to stimulation with various hormones that regulate glycogen metabolism and mitochondrial respiration [25, 26]. The Ca^{2+} increases that occur in hepatocytes during such stimulation are transient spikes, which arise from the cyclical activation of $InsP_3Rs$. The frequency of the Ca^{2+} spikes is directly proportional to the concentration of hormone applied to the cells, and they can persist for the duration of agonist application. These Ca^{2+} spikes are therefore essentially a frequency-modulated digital read-out of cell stimulation. Ca^{2+}-sensitive processes will be activated in direct proportion to the frequency of the Ca^{2+} transients.

Although the Ca^{2+} signals observed in both hormonally-stimulated hepatocytes and cardiomyocytes take the form of repetitive Ca^{2+} transients, the periodicity and kinetics of these signals are very different. The Ca^{2+} spikes in hepatocytes (and many other non-electrically excitable cells) typically have frequencies in the range of 0.1–0.01 Hz, a time to peak amplitude of several seconds, and a recovery phase lasting tens of seconds [27]. In contrast, cardiac Ca^{2+} signals are triggered at frequencies in the 1- to 10-Hz range (depending on the species of animal), reach peak within a few tens of milliseconds, and persist for only a few hundred milliseconds [28]. The distinct timescales of these Ca^{2+} responses reflect the very different mechanisms by which they are generated, as well as the cell-specific mechanisms for recovery of the Ca^{2+} transients.

Pulsatile Ca^{2+} increases, such as those observed in hepatocytes, are generally considered to have a much higher fidelity of information transfer than simple tonic changes in Ca^{2+} concentrations, since they are much less prone to noisy fluctuations. The major sensors for these Ca^{2+} spikes are Ca^{2+}-binding proteins such as calmodulin [29]. This ubiquitous protein is one of a family of proteins bearing structural Ca^{2+}-binding motifs known as *EF-hands*. The binding of Ca^{2+} to calmodulin has a K_d of around 1 μM, making it an ideal receiver for the rapid transient Ca^{2+} increases seen with each spike. One of the best-known enzymes that uses calmodulin to help it "count" Ca^{2+} spikes is calmodulin-dependent protein kinase II, which can activate other proteins via phosphorylation. This enzyme is composed of 12 subunits that undergo variable degrees of activation, depending on the frequency of Ca^{2+} spikes. Essentially, increasing the frequency or duration of Ca^{2+} spikes maintains this enzyme in an active state by trapping calmodulin and causing autophosphorylation [30,31].

Long-term Ca^{2+} signaling affects cell function by altering gene transcription. One of the most prominent examples of long-term changes in Ca^{2+} homeostasis affecting cellular function is the development of cardiac hypertrophy due to sustained mechanical strain. Increased cardiac afterload and chronic hypertension are two of the main causes of cardiac hypertrophy. On a molecular level, these factors lead to an increase in the cytoplasmic Ca^{2+} concentration, promoting cardiac remodeling that can extend over months and years in humans [32]. As described earlier, intracellular Ca^{2+} release in muscles controls cellular contraction (ECC). However, in cardiomyocytes it has been shown that Ca^{2+} not only initiates contraction but also controls gene transcription promoting cardiac hypertrophy – a process described as excitation–transcription coupling (ETC) [33,34]. The question is, how does the cell distinguish between Ca^{2+} that controls ECC and Ca^{2+} that is involved in ETC? The answer to this question seems to be, by compartmentalization of Ca^{2+} signals. Ca^{2+} signaling controlling transcription has to be directed to the nucleus. Recent reports strongly suggest the involvement of $InsP_3Rs$ in generating local Ca^{2+} signals [16,17]. In this model, $InsP_3$ is generated at the sarcolemma through G_q-coupled receptor activation (e.g., via endothelin-1 or angiotensin II receptors) and diffuses throughout the cytosol to the nucleus. Consequently, it activates $InsP_3Rs$ in the nuclear envelope or peri-nuclear region, thereby specifically increasing Ca^{2+} within the nucleoplasm. One suggested consequence of nuclear $InsP_3R$ activation is phosphorylation of the transcriptional repressor HDAC (histone deacetylase). Phosphorylated HDAC is exported out of the nucleus, which relieves suppression of MEF2-dependent transcription. Moreover, local $InsP_3$-dependent Ca^{2+} release in the cytosol activates calcineurin which dephosphorylates the transcription factor NFAT (nuclear factor of activated T cells). NFAT then translocates into the nucleus, where it stimulates gene transcription together with the transcription factor GATA4. Altogether, this leads to remodeling of gene expression and induction of cardiac hypertrophy [33].

SPATIAL REGULATION OF CA^{2+} SIGNALS

Both Ca^{2+} entry and Ca^{2+} release channels can give rise to brief pulses of Ca^{2+} that form a small plume around the mouth of the channel before diffusing into the cytoplasm [35]. The small plume of Ca^{2+} that forms around the Ca^{2+} channel has been given different names, depending on the type of channel that conducts the Ca^{2+} flux. Ca^{2+} "sparklets" are generated by the opening of single voltage-operated Ca^{2+} channels, and have been observed in cardiomyocytes. Formation of a Ca^{2+} "spark" is due to the opening of clusters of RyRs. The spatial overlap and temporal summation of the Ca^{2+} sparks gives rise to the global responses that ensure synchronized contraction in muscle cells [36, 37]. In neuronal presynaptic endings, Ca^{2+} sparks are called "syntillas". Ca^{2+} "puffs" are local signals that derive from the activation of a cluster of $InsP_3Rs$. They are the analogous events to Ca^{2+} sparks. Typically, Ca^{2+} puffs give a modest elevation of cytosolic Ca^{2+} (~50–600 nM), with a limited spatial spread (~2–6 μm), and are transient (duration of ~1 second). Such events were first described in *Xenopus* oocytes [38], but have subsequently been observed in many other cell types. For example, they have been shown to be important for neocortical glutamatergic presynaptic Ca^{2+} signalling [39]. The temporally and spatially coordinated recruitment of Ca^{2+} puffs is responsible for the generation of repetitive Ca^{2+} waves and oscillations observed during hormonal stimulation – for example, in hepatocytes. Essentially, Ca^{2+} waves reflect the progressive, saltatoric, release of Ca^{2+} by Ca^{2+} puff sites distributed along the ER/SR. Ca^{2+} released by one puff site can diffuse to a neighboring site and activate it (providing $InsP_3$ is bound to the channels). Successive rounds of Ca^{2+} release and diffusion allow the initially local Ca^{2+} puffs to trigger global Ca^{2+} waves and oscillations [37, 40].

It has become increasingly obvious that Ca^{2+} functions in nano/microdomains where both Ca^{2+} channels and target effector molecule are present [40]. An example of this is neurotransmitter release at presynaptic membranes. The fusion of vesicles filled with neurotransmitters at the plasma membrane is caused by proteins such as synaptotagmin, which require a local Ca^{2+} concentration in the order of tens of μM. In the presynaptic membrane, only submembrane regions within the so-called "active zone" provide this substantial Ca^{2+} concentration due to the enrichment of VOCs [41]. Another well-known local Ca^{2+} signal occurs in the apical region of secretory cells such as pancreatic acinar cells [42]. Experimental evidence suggest that the apical Ca^{2+} spikes arise from a stimulus-dependent hierarchical activation of different types of Ca^{2+} release channel [33]. With low levels of cell stimulation, the Ca^{2+} spikes stay restricted to the apical pole of the acinar cells, where they can activate ion channels and trigger limited secretion. Greater stimulation causes the Ca^{2+} spikes to trigger Ca^{2+} waves that propagate towards the basal pole [43]. It appears that the restriction of the Ca^{2+} signal in the apical pole is due in part to a "firewall" of mitochondria that act to buffer Ca^{2+} as it diffuses from the apical pole and prevent the activation of RyRs in the basal pole [44, 45]. Sequestration of Ca^{2+} by mitochondria can lead to acceleration of the citric acid cycle and increased ATP production [46], but exaggerated levels of Ca^{2+} in the mitochondrial matrix can also be a trigger for permeability transition and apoptosis [47].

REFERENCES

1. Berridge MJ, Bootman MD, Roderick HL. Calcium signalling: dynamics, homeostasis and remodelling. *Nat Rev Mol Cell Biol* 2003;**4**(7):517–29.
2. Bootman MD, Berridge MJ, Roderick HL. Calcium signalling: more messengers, more channels, more complexity. *Curr Biol* 2002;**12**(16):R563–5.
3. Catterall WA. Structure and regulation of voltage-gated Ca^{2+} channels. *Annu Rev Cell Dev Biol* 2000;**16**(1):521–55.
4. Carlson AE, Westenbroek RE, Quill T, Ren D, Clapham DE, Hille B, Garbers DL, Babcock DF. CatSper1 required for evoked Ca^{2+} entry and control of flagellar function in sperm. *Proc Natl Acad Sci USA* 2003;**100**(25):14,864–8.
5. Changeux J-P, Taly A. Nicotinic receptors, allosteric proteins and medicine. *Trends Mol Med* 2008;**14**(3):93–102.
6. Cull-Candy S, Brickley S, Farrant M. NMDA receptor subunits: diversity, development and disease. *Curr Opin Neurobiol* 2001;**11**(3):327–35.
7. Welsh DG, Morielli AD, Nelson MT, Brayden JE. Transient receptor potential channels regulate myogenic tone of resistance arteries. *Circ Res* 2002;**90**(3):248–50.
8. Birder LA, Kanai AJ, de Groat WC, Kiss S, Nealen ML, Burke NE, Dineley KE, Watkins S, Reynolds IJ, Caterina MJ. Vanilloid receptor expression suggests a sensory role for urinary bladder epithelial cells. *Proc Natl Acad Sci USA* 2001;**98**(23):13,396–401.
9. Chyb S, Raghu P, Hardie RC. Polyunsaturated fatty acids activate the *Drosophila* light-sensitive channels TRP and TRPL. *Nature* 1999;**397**(6716):255–9.
10. Cahalan MD, Zhang SL, Yeromin AV, Ohlsen K, Roos J, Stauderman KA. Molecular basis of the CRAC channel. *Cell Calcium* 2007;**42**(2):133–44.
11. Feske S, Gwack Y, Prakriya M, Srikanth S, Puppel SH, Tanasa B, Hogan PG, Lewis RS, Daly M, Rao A. A mutation in Orai1 causes immune deficiency by abrogating CRAC channel function. *Nature* 2006;**441**(7090):179–285.
12. Lewis RS. The molecular choreography of a store-operated calcium channel. *Nature* 2007;**446**(7133):247–84.
13. Brandman O, Liou J, Park WS, Meyer T. STIM2 is a feedback regulator that stabilizes basal cytosolic and endoplasmic reticulum Ca^{2+} levels. *Cell* 2007;**131**(7):1327–39.
14. Foskett JK, White C, Cheung K-H, Mak D-OD. Inositol trisphosphate receptor Ca^{2+} release channels. *Physiol Rev* 2007;**87**(2):593–658.
15. Katsuhiko M. IP_3 receptor/Ca^{2+} channel: from discovery to new signaling concepts. *J Neurochem* 2007;**102**(5):1426–46.
16. Higazi DR, Fearnley CJ, Drawnel FM, Talasila A, Corps EM, Ritter O, McDonald F, Mikoshiba K, Bootman MD, Roderick HL. Endothelin-1-stimulated InsP3-induced Ca^{2+} release is a nexus for hypertrophic signaling in cardiac myocytes. *Mol Cell* 2009;**33**:472–82.
17. Wu X, Zhang T, Bossuyt J, Li X, McKinsey TA, Dedman JR, Olson EN, Chen J, Brown JH, Bers DM. Local InsP3-dependent perinuclear Ca^{2+} signaling in cardiac myocyte excitation-transcription coupling. *J Clin Invest* 2006;**116**(3):675–82.
18. Zalk R, Lehnart SE, Marks AR. Modulation of the ryanodine receptor and intracellular calcium. *Annu Rev Biochem* 2007;**76**(1):367–85.
19. Bootman MD, Berridge MJ, Roderick HL. Calcium signalling: more messengers, more channels, more complexity. *Curr Biol* 2002;**12**(16):R563–5.
20. Pinton P, Ferrari D, Rapizzi E, Di Virgilio F, Pozzan T, Rizzuto R. The Ca^{2+} concentration of the endoplasmic reticulum is a key determinant of ceramide-induced apoptosis: significance for the molecular mechanism of Bcl-2 action. *EMBO J* 2001;**20**:2690–701.
21. Ferrari D, Pinton P, Szabadkai G, Chami M, Campanella M, Pozzan T, Rizzuto R. Endoplasmic reticulum, Bcl-2 and Ca^{2+} handling in apoptosis. *Cell Calcium* 2002;**32**:413–20.
22. Xu C, Xu W, Palmer AE, Reed JC. BI-1 regulates endoplasmic reticulum Ca^{2+} homeostasis downstream of Bcl-2 family proteins. *J Biol Chem* 2008;**283**:11,477–84.
23. Hanson CJ, Bootman MD, Distelhorst CW, Wojcikiewicz RJ, Roderick HL. Bcl-2 suppresses Ca^{2+} release through inositol 1,4,5-trisphosphate receptors and inhibits Ca^{2+} uptake by mitochondria without affecting ER calcium store content. *Cell Calcium* 2008;**44**:324–38.
24. Rong YP, Aromolaran AS, Bultynck G, Zhong F, Li X, McColl K, Matsuyama S, Herlitze S, Roderick HL, Bootman MD, Mignery GA, Parys JB, de Smedt H, Distelhorst CW. Targeting Bcl-2-IP3 receptor interaction to reverse Bcl-2's inhibition of apoptotic calcium signals. *Mol Cell* 2008;**31**:255–65.
25. Hajnóczky G, Robb-Gaspers LD, Seitz MB, Thomas AP. Decoding of cytosolic calcium oscillations in the mitochondria. *Cell* 1995;**82**(3):415–24.
26. Hajnoczky G, Thomas AP. Minimal requirements for calcium oscillations driven by the IP3 receptor. *EMBO J* 1997;**16**(12):3533–43.
27. Woods NM, Cuthbertson KSR, Cobbold PH. Repetitive transient rises in cytoplasmic free calcium in hormone-stimulated hepatocytes. *Nature* 1986;**319**(6054):600–2.
28. Mackenzie L, Bootman MD, Berridge MJ, Lipp P. Predetermined recruitment of calcium release sites underlies excitation–contraction coupling in rat atrial myocytes. *J Physiol* 2001;**530**(3):417–29.

29. Chin D, Means AR. Calmodulin: a prototypical calcium sensor. *Trends Cell Biol* 2000;**10**(8):322–8.
30. De Koninck P, Schulman H. Sensitivity of CaM kinase II to the frequency of Ca^{2+} oscillations. *Science* 1998;**279**(5348):227–30.
31. Hudmon A, Schulman H. Neuronal Ca^{2+}/calmodulin-dependent protein kinase II: the role of structure and autoregulation in cellular function. *Annu Rev Biochem* 2002;**71**(1):473–510.
32. Schaub MC, Hefti MA, Zaugg M. Integration of calcium with the signaling network in cardiac myocytes. *J Mol Cell Cardiol* 2006;**41**(2):183–214.
33. Bers DM. Calcium Cycling and signaling in cardiac myocytes. *Annu Rev Physiol* 2008;**70**(1):23–49.
34. Roderick HL, Higazi DR, Smyrnias I, Fearnley C, Harzheim D, Bootman MD. Calcium in the heart: when it's good, it's very very good, but when it's bad, it's horrid. *Biochem Soc Trans* 2007;**035**(5):957–61.
35. Neher E. Vesicle pools and Ca^{2+} microdomains: new tools for understanding their roles in neurotransmitter release. *Neuron* 1998;**20**(3):389–99.
36. Niggli E. Localized intracellular calcium signaling in muscle: calcium sparks and calcium quarks. *Annu Rev Physiol* 1999;**61**(1):311–35.
37. Bootman MD, Lipp P, Berridge MJ. The organisation and functions of local Ca^{2+} signals. *J Cell Sci* 2001;**114**(12):2213–22.
38. Yao Y, Choi J, Parker I. Quantal puffs of intracellular Ca^{2+} evoked by inositol trisphosphate in *Xenopus* oocytes. *J Physiol* 1995;**482**:533–53.
39. Simkus CRL, Stricker C. The contribution of intracellular calcium stores to mEPSCs recorded in layer II neurones of rat barrel cortex. *J Physiol* 2002;**545**(2):521–35.
40. Berridge MJ. Calcium microdomains: organization and function. *Cell Calcium* 2006;**40**(5–6):405–12.
41. Oheim M, Kirchhoff F, Stühmer W. Calcium microdomains in regulated exocytosis. *Cell Calcium* 2006;**40**(5–6):423–39.
42. Ole H, Petersen DBAVT. Polarity in intracellular calcium signaling. *BioEssays* 1999;**21**(10):851–60.
43. Cancela JM, Van Coppenolle F, Galione A, Tepikin AV, Petersen OH. Transformation of local Ca^{2+} spikes to global Ca^{2+} transients: the combinatorial roles of multiple Ca^{2+} releasing messengers. *EMBO J* 2002;**21**(5):909–19.
44. Straub SV, Giovannucci DR, Yule DI. Calcium wave propagation in pancreatic acinar cells: functional interaction of inositol 1,4,5-trisphosphate receptors, ryanodine receptors, and mitochondria. *J Gen Physiol* 2000; **116**(4):547–60.
45. Tinel H, Cancela JM, Mogami H, Gerasimenko JV, Gerasimenko OV, Tepikin AV, Petersen OH. Active mitochondria surrounding the pancreatic acinar granule region prevent spreading of inositol trisphosphate-evoked local cytosolic Ca(2+) signals. *EMBO J* 1999;**18**(18):4999–5008.
46. Jouaville LS, Pinton P, Bastianutto C, Rutter GA, Rizzuto R. Regulation of mitochondrial ATP synthesis by calcium: evidence for a long-term metabolic priming. *Proc Natl Acad Sci USA* 1999;**96**(24):13,807–12.
47. Hanson CJ, Bootman MD, Roderick HL. Cell signalling: IP_3 receptors channel calcium into cell death. *Curr Biol* 2004;**14**(21):R933–5.

Chapter 48

Cyclic Nucleotide-Regulated Cation Channels

Martin Biel

Munich Center for Integrated Protein Science CIPSM and Department of Pharmacy, Center for Drug Research, Ludwig-Maximilians-Universität München, Munich, Germany

INTRODUCTION

Cyclic nucleotides exert their physiological effects by binding to four major classes of cellular receptors: cAMP- and cGMP-dependent protein kinases [1, 2], cyclic GMP-regulated phosphodiesterases [3], cAMP-binding guanine nucleotide exchange factors [4], and cyclic nucleotide-regulated cation channels. Cyclic nucleotide-regulated cation channels are unique among these receptors because their activation is directly coupled to the influx of extracellular cations into the cytoplasm and to the depolarization of the plasma membrane. Two families of channels regulated by cyclic nucleotides have been identified; the cyclic nucleotide-gated (CNG) channels and the hyperpolarization-activated cyclic nucleotide-gated (HCN) channels [5–7]. The two channel classes differ from each other with regard to their mode of activation. CNG channels are opened by direct binding of cAMP or cGMP. In contrast, HCN channels are principally operated by voltage. These channels open at hyperpolarized membrane potentials, and close on depolarization. Apart from their voltage sensitivity, HCN channels are also activated directly by cyclic nucleotides, which act by increasing the channel open probability.

GENERAL FEATURES OF CYCLIC NUCLEOTIDE-REGULATED CATION CHANNELS

Structurally, CNG and HCN channels are members of the superfamily of voltage-gated cation channels [8]. Like other subunits encoded by this large gene family, CNG and HCN channel subunits assemble to tetrameric complexes. The proposed structure and the phylogenetic relationship of mammalian CNG and HCN channel subunits are shown in Figure 48.1. The transmembrane channel core consists of six α-helical segments (S1–S6) and an ion-conducting pore loop between the S5 and S6. The amino- and carboxy-termini are localized in the cytosol. CNG and HCN channels contain a positively-charged S4 helix carrying three to nine regularly spaced arginine or lysine residues at every third position. In HCN channels, as in most other members of the channel superfamily, the S4 helix functions as "voltage-sensor" conferring voltage-dependent gating. However, inward movement of S4 charges through the plane of the cell membrane leads to opening of HCN channels while it triggers the closure of depolarization-activated channels such as the Kv channels [9]. The molecular determinants underlying the different polarity of the gating process in HCN and depolarization-gated channels are not known. In CNG channels which are not gated by voltage, the specific role of S4 remains to be determined.

CNG and HCN channels reveal different ion selectivities. CNG channels pass monovalent cations, such as Na^+ and K^+, but do not discriminate between them. Ca^{2+} is also permeable, but at the same time acts as a voltage-dependent blocker of monovalent cation permeability [10]. By providing an entry pathway for Ca^{2+}, CNG channels control a variety of cellular processes that are triggered by this cation. HCN channels conduct Na^+ and K^+ with permeability ratios of about 1:4, and are blocked by millimolar concentrations of Cs^+ [11–13]. Despite this preference for K^+ conductance, HCN channels carry an inward Na^+ current under physiological conditions. HCN channels can also conduct Ca^{2+}, but not as well as CNG channels. At 2.5-mM external Ca^{2+} the fractional Ca^{2+} current of HCN2 and HCN4 is about 0.5 percent, whereas for native CNG channels it is in the range 10–80 percent [14].

In the carboxy-terminus, CNG and HCN channels contain a cyclic nucleotide-binding domain (CNBD) that has

Handbook of Cell Signaling, Three-Volume Set 2 ed.

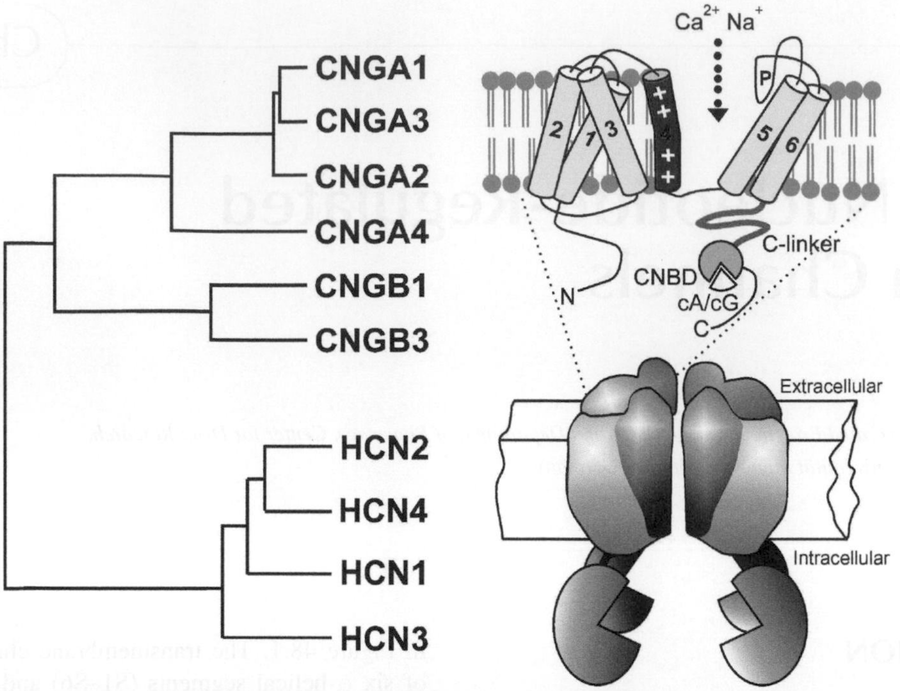

FIGURE 48.1 Phylogenetic tree and structural model of cyclic nucleotide-regulated cation channels. The CNG channel family comprises six members, which are classified into A subunits (CNGA1–4) and B subunits (CNGB1 and CNGB3). The HCN channel family comprises four members (HCN1–4). CNG and HCN channels share a common transmembrane topology, consisting of six α helical segments (1–6) and a pore loop (P). In the cytosolic C-terminus, all subunits carry a cyclic nucleotide-binding domain (CNBD) that is functionally coupled with the transmembrane channel core via the C-linker domain. The S4 segment contains a series of positively-charged residues, and forms the voltage sensor in HCN channels. CNG channels are activated *in vivo* by binding of either cAMP (cA) or cGMP (cG), depending on the channel type. HCN channels activate upon membrane hyperpolarization. Binding of cAMP to the CNBD produces an allosteric conformational change that increases the open probability of the channel pore.

significant sequence similarity to the CNBDs of most other types of cyclic nucleotide receptors. The crystal structure of the CNBD has been determined for HCN2 [15] and a bacterial cyclic nucleotide-regulated potassium channel [16]. In CNG channels, the binding of cyclic nucleotides to the CNBD initiates a sequence of allosteric transitions that leads to the opening of the ion-conducting pore [7]. In HCN channels, the binding of cyclic nucleotides is not required for activation. However, cyclic nucleotides shift the voltage dependence of channel activation to more positive membrane potentials, and thereby facilitate voltage-dependent channel activation [11–13]. Despite the fact that the CNBDs of HCN and CNG channels show significant sequence homology, the two channel classes reveal different selectivities for cyclic nucleotides. HCN channels display an ~10-fold higher affinitiy for cAMP than for cGMP, whereas CNG channels select cGMP over cAMP [5, 7]. Recently, amino acid residues determining this difference have been identified [17, 18]

CNG CHANNELS

CNG channels are expressed in retinal photoreceptors and olfactory neurons, and play a key role in visual and olfactory signal transduction [5, 6]. CNG channels are also found at low density in some other cell types and tissues, such as brain, testis, and kidney [19]. While the function of CNG channels in sensory neurons has been unequivocally demonstrated, the role of these channels in other cell types remains to be established. Based on phylogenetic relationship, the six CNG channels identified in mammals are divided in two subfamilies; the A subunits (CNGA1–4) and the B subunits (CNGB1 and CNGB3) (Figure 48.1). CNG channel A subunits (with the only exception of CNGA4) form functional homomeric channels in various heterologous expression systems. In contrast, B subunits do not give rise to functional channels when expressed alone. However, together with CNGA1–3 they confer novel properties (e.g., single channel flickering, increased sensitivity for cAMP and L-cis diltiazem) that are characteristic of native CNG channels [5]. Recent genetic studies in mice indicate that B subunits, besides being functionally important, also play a key role in principal channel formation and channel targeting in native sensory neurons [20, 21]. The subunit composition is known for three native CNG channels: the rod and cone photoreceptor channels, and the olfactory channel. The CNG channel of rod photoreceptors consists of the CNGA1 subunit and a long isoform of the CNGB1 subunit (CNGB1a) (3 : 1 stoichiometry) [22–24].

The cone photoreceptor channel consists of the CNGA3 and the CNGB3 subunit (2:2 stoichiometry) [25]. CNG channels control the membrane potential and the calcium concentration of photoreceptors. In the dark, both channels are maintained in the open state by a high concentration of cGMP. The resulting influx of Na^+ and Ca^{2+} ("dark current") depolarizes the photoreceptor, and promotes synaptic transmission. Light-induced hydrolysis of cGMP leads to the closure of the CNG channel. As a result, the photoreceptor hyperpolarizes and shuts off synaptic glutamate release. Mutations in the CNGA1 [26] and CNGB1 [27] subunits have been identified in patients suffering from retinitis pigmentosa. The functional loss of either the CNGA3 [28, 29] or the CNGB3 [30] subunit causes total colorblindness (achromatopsia) and degeneration of cone photoreceptors.

The CNG channel expressed in cilia of olfactory neurons consists of three different subunits: CNGA2, CNGA4, and a short isoform of the CNGB1 subunit (CNGB1b) (2:1:1 stoichiometry) [31]. The channel is activated *in vivo* by cAMP, which is synthesized in response to the binding of odorants to their cognate receptors. The olfactory CNG channel mainly conducts Ca^{2+} under physiological ionic conditions [32]. The increase in cellular Ca^{2+} activates a Ca^{2+}-activated Cl^- channel, which further depolarizes the cell membrane. Ca^{2+} is not only a permeating ion of the olfactory CNG channel, but also an important modulator of this channel. By forming a complex with calmodulin, which binds to the CNGB1b and CNGA4 subunits, Ca^{2+} decreases the sensitivity of the CNG channel to cAMP [32]. The resulting inhibition of channel activity is the principal mechanism underlying fast odorant adaptation.

HCN CHANNELS

A cation current that is slowly activated by membrane hyperpolarization (termed I_h, I_f, or I_q) is found in a variety of excitable cells, including neurons, cardiac pacemaker cells, and photoreceptors [33]. The best established function of I_h is to control heart rate and rhythm by acting as "pacemaker current" in the sinoatrial (SA) node [34]. I_h is activated during the membrane hyperpolarization following the termination of an action potential, and provides an inward Na^+ current that slowly depolarizes the plasma membrane. Sympathetic stimulation of SA node cells raises cAMP levels and increases I_h by a positive shift of the current activation curve, thus accelerating diastolic depolarization and heart rate. Stimulation of muscarinic receptors slows down heart rate by the opposite action. In neurons, I_h fulfills diverse functions, including generation of pacemaker potentials, control of membrane potential, generation of rebound depolarizations during light-induced hyperpolarizations of photoreceptors, dendritic integration, and synaptic transmission [35, 36].

HCN channels represent the molecular correlate of the I_h current [11–13]. In mammals, the HCN channel family comprises four members (HCN1–4) that share about 60 percent sequence identity to each other and about 25 percent sequence identity to CNG channels. The highest degree of sequence homology between HCN and CNG channels is found in the CNBD. When expressed in heterologous systems, all four HCN channels generate currents displaying the typical features of native I_h: (1) activation by membrane hyperpolarization; (2) permeation of Na^+ and K^+; (3) positive shift of the voltage-dependence of channel activation by direct binding of cAMP; and (4) channel blockade by extracellular Cs^+. HCN1–4 mainly differ from each other with regard to their speed of activation and the extent by which they are modulated by cAMP. HCN1 is the fastest channel, followed by HCN2, HCN3, and HCN4. Unlike HCN2 and HCN4, whose activation curves are shifted by about +15 mV by cAMP, HCN1 and HCN3 are, if at all, only weakly affected by cAMP. Site-directed mutagenesis experiments have provided initial insight into the complex mechanism underlying dual HCN channel activation by voltage and cAMP. Like in other voltage-gated cation channels, activation of HCN channels is initiated by the movement of the positively charged S4 helix in the electric field [9]. The resulting conformational change in the channel protein is allosterically coupled by other channel domains to the opening of the ion-conducting pore. Major determinants affecting channel activation are the intracellular S4–S5 loop, the S1 segment, and the extracellular S1–S2 loop [37–39]. The CNBD fulfills the role of an autoinhibitory channel domain. In the absence of cAMP, the cytoplasmic carboxy-terminus inhibits HCN channel gating by interacting with the channel core and thereby shifting the activation curve to more hyperpolarizing voltages [40]. Binding of cAMP to the CNBD relieves this inhibition. Differences in the magnitude of the response to cAMP among the four HCN channel isoforms are largely due to differences in the extent to which the CNBD inhibits basal gating. It remains to be determined if the inhibitory effect of the CNBD is conferred by a direct physical interaction with the channel core domain or by some indirect pathway. There is initial evidence that the so-called C-linker, a peptide of about 80 amino acids that connects the last transmembrane helix (S6) to the CNBD, plays an important role in this process. The C-linker was also shown to play a key role in the gating of CNG channels, suggesting that the functional role of this domain has been conserved during channel evolution [7, 41, 42].

HCN channels are found in neurons and heart cells. In mouse and rat brain, all four HCN channel isoforms have been detected [43, 44]. HCN2 is the most abundant channel, and is found almost ubiquitously in the brain. In contrast, HCN1, HCN3, and HCN4 are enriched in specific regions of the brain, such as thalamus (HCN4), hippocampus (HCN1), or olfactory bulb and hypothalamus

(HCN3). HCN channels have also been detected in the retina and some peripheral neurons, such as dorsal root ganglion neurons. In SA node cells, HCN4 represents the predominantly expressed HCN channel isoform. In addition, minor amounts of HCN1 and HCN2 are also present in these cells. Insights into the (patho)physiological relevance of HCN channels have been gained from the anaylsis of mouse lines lacking individual HCN channel isoforms. Disruption of HCN1 impairs motor learning, but enhances spatial learning and memory [45, 46]. Deletion of HCN2 results in absence epilepsy, ataxia, ans sinus node dysfunction [47]. Mice lacking HCN4 die *in utero* because of the failure to generate mature sinoatrial pacemaker cells [48]. The key role of HCN4 in controlling heart rhythmicity is corroborated by genetic data from human patients. Mutations in the human HCN4 gene leading to mutated or truncated channel proteins have been found to be associated with sinus bradycardia [49–52] and complex cardiac arrhythmia [50].

ACKNOWLEDGEMENT

This work was supported by the Deutsche Forschungsgemeinschaft.

REFERENCES

1. Taylor SS, Kim C, Vigil D, Haste NM, Yang J, Wu J, Anand GS. Dynamics of signaling by PKA. *Biochim Biophys Acta* 2005;**1754**:25–37.
2. Hofmann F, Feil R, Kleppisch T, Schlossmann J. Function of cGMP-dependent protein kinases as revealed by gene deletion. *Physiol Rev* 2006;**86**:1–23.
3. Bender AT, Beavo JA. Cyclic nucleotide phosphodiesterases: molecular regulation to clinical use. *Pharmacol Rev* 2006;**58**:488–520.
4. Bos JL. Epac proteins: mulit-purpose cAMP targets. *Trends Biochem Sci* 2006;**31**:680–6.
5. Kaupp UB, Seifert R. Cyclic nucleotide-gated ion channels. *Physiol Rev* 2002;**82**:769–824.
6. Hofmann F, Biel M, Kaupp UB. International Union of Pharmacology. LI. Nomenclature and structure–function relationships of cyclic nucleotide-regulated channels. *Pharmacol Rev* 2005;**57**:455–62.
7. Craven KB, Zagotta WN. CNG and HCN channels: two peas, one pod. *Annu Rev Physiol* 2006;**68**:375–401.
8. Yu FH, Catterall WA. The VGL-chanome: a protein superfamily specialized for electrical signaling and ionic homeostasis. *Sci STKE* 2004;**2004**:re15.
9. Männikkö R, Elinder F, Larsson PH. Voltage-sensing mechanism is conserved among ion channels gated by opposite voltages. *Nature*;**419**:837–41.
10. Frings S, Seifert R, Godde M, Kaupp UB. Profoundly different calcium permeation and blockage determine the specific function of distinct cyclic nucleotide-gated channels. *Neuron* 1995;**15**:169–79.
11. Gauss R, Seifert R, Kaupp UB. Molecular identification of a hyperpolarization-activated channel in sea urchin sperm. *Nature* 1998;**393**:583–7.
12. Ludwig A, Zong X, Jeglitsch M, Hofmann F, Biel M. A family of hyperpolarization-activated mammalian cation channels. *Nature* 1998;**393**:587–91.
13. Santoro B, Liu DT, Yao H, Bartsch D, Kandel ER, Siegelbaum SA, Tibbs GR. Identification of a gene encoding a hyperpolarization-activated pacemaker channel of brain. *Cell* 1998;**93**:717–29.
14. Yu X, Chen XW, Zhou P, Yao L, Liu T, Zhang B, Li Y, Zheng H, Zheng LH, Zhang CX, Bruce I, Ge JB, Wang SQ, Hu ZA, Yu HG, Zhou Z. Calcium influx through If channels in rat ventricular myocytes. *Am J Physiol Cell Physiol* 2007;**292**:C1147–55.
15. Zagotta WN, Olivier NB, Black KD, Young EC, Olson R, Gouaux E. Structural basis for modulation and agonist specificity of HCN pacemaker channels. *Nature* 2003;**425**:200–5.
16. Clayton GM, Silverman WR, Heginbotham L, Morais-Cabral JH. Structural basis of ligand activation in a cyclic nucleotide regulated potassium channel. *Cell* 2004;**119**:615–27.
17. Flynn GE, Black KD, Islas LD, Sankaran B, Zagotta WN. Structure and rearrangements in the carboxy-terminal region of SpIH channels. *Structure* 2007;**15**:671–82.
18. Zhou L, Siegelbaum SA. Gating of HCN channels by cyclic nucleotides: residue contacts that underlie ligand binding, selectivity, and efficacy. *Structure* 2007;**15**:655–70.
19. Richards MJ, Gordon SE. Cooperativity and cooperation in cyclic nucleotide-gated ion channels. *Biochemistry* 2000;**39**:14,003–14,011.
20. Hüttl S, Michalakis S, Seeliger S, Luo D, Acar N, Geiger H, Hudl K, Mader R, Haverkamp S, Moser M, Pfeifer A, Gerstner A, Yau KW, Biel M. Impaired channel targeting and retinal degeneration in mice lacking the cyclic nucleotide-gated channel subunit CNGB1. *J Neurosci* 2005;**25**:130–8.
21. Michalakis S, Reisert S, Geiger H, Wetzel C, Zong X, Bradley J, Spehr M, Hüttl S, Gerstner A, Pfeifer A, Hatt H, Yau KW, Biel M. Loss of CNGB1 protein leads to olfactory dysfunction and subciliary CNG channel trapping. *J Biol Chem* 2006;**281**:35,156–35,166.
22. Zheng J, Trudeau MC, Zagotta WN. Rod cyclic nucleotide-gated channels have a stoichiometry of three CNGA1 subunits and one CNGB1 subunit. *Neuron* 2002;**36**:891–6.
23. Weitz D, Ficek N, Kremmer E, Bauer PJ, Kaupp UB. Subunit stoichiometry of the CNG channel of rod photoreceptors. *Neuron* 2002;**36**:881–9.
24. Zhong H, Molday LL, Molday RS, Yau KW. The heteromeric cyclic nucleotide-gated channel adopts a 3A:1B stoichiometry. *Nature* 2002;**420**:193–8.
25. Peng C, Rich ED, Varnum MD. Subunit configuration of heteromeric cone cyclic nucleotide-gated channels. *Neuron* 2004;**42**:401–10.
26. Dryja TP, Finn JT, Peng YW, McGee TL, Berson EL, Yau KW. Mutations in the gene encoding the alpha subunit of the rod cGMP-gated channel in autosomal recessive retinitis pigmentosa. *Proc Natl Acad Sci USA* 1995;**92**:10,177–10,181.
27. Bareil C, Hamel CP, Delague V, Arnaud B, Demaille J, Claustres M. Segregation of a mutation in CNGB1 encoding the β-subunit of the rod cGMP-gated channel in a family with autosomal recessive retinitis pigmentosa. *Hum Genet* 2001;**108**:328–34.
28. Kohl S, Marx T, Giddings I, Jägle H, Jacobson SG, Apfelstedt-Sylla E, Zrenner E, Sharpe LT, Wissinger B. Total colourblindness is caused by mutations in the gene encoding the alpha-subunit of the cone photoreceptor cGMP-gated cation channel. *Nat Genet* 1998;**19**:257–9.
29. Biel M, Seeliger M, Pfeifer A, Kohler K, Gerstner A, Ludwig A, Jaissle G, Fauser S, Zrenner E, Hofmann F. Selective loss of cone function in mice lacking the cyclic nucleotide-gated channel CNG3. *Proc Natl Acad Sci USA* 1999;**96**:7553–7.

30. Sundin OH, Yang JM, Li Y, Zhu D, Hurd JN, Mitchell TN, Silva ED, Maumenee IH. Genetic basis of total colourblindness among the Pingelapese islanders. *Nat Genet* 2000;**25**:289–93.
31. Zheng J, Zagotta WN. Stoichiometry and assembly of olfactory cyclic nucleotide-gated channels. *Neuron* 2004;**42**:411–21.
32. Bradley J, Reisert J, Frings S. Regulation of cyclic nucleotide-gated channels. *Curr Opin Neurobiol* 2005;**15**:343–9.
33. Pape HC. Queer current and pacemaker: the hyperpolarization-activated cation current in neurons. *Annu Rev Physiol* 1996;**58**:299–327.
34. Baruscotti M, Bucchi A, DiFrancesco D. Physiology and pharmacology of the cardiac pacemaker ("funny") current. *Pharmacol Ther* 2005;**107**:59–79.
35. Robinson RB, Siegelbaum SA. Hyperpolarization-activated cation currents: from molecules to physiological function. *Annu Rev Physiol* 2003;**65**:453–80.
36. Frère SGA, Kuisle M, Lüthi A. Regulation of recombinant and native hyperpolarization-activated cation channels. *Mol Neurobiol* 2004;**30**:279–305.
37. Chen J, Mitcheson JS, Tristani-Firouzi M, Lin M, Sanguinetti MC. The S4-S5 linker couples voltage sensing and activation of pacemaker channels. *Proc Natl Acad Sci USA* 2001;**98**:11,277–11,282.
38. Ishii TM, Takano M, Ohmori H. Determinants of activation kinetics in mammalian hyperpolarization-activated cation channels. *J Physiol* 2001;**537**:93–100.
39. Stieber J, Thomer A, Much B, Schneider A, Biel M, Hofmann F. Molecular basis for the different activation kinetics of the pacemaker channels HCN2 and HCN4. *J Biol Chem* 2003;**278**:33,672–33,780.
40. Wainger BJ, DeGennaro M, Santoro B, Siegelbaum SA, Tibbs GR. Molecular mechanism of cAMP modulation of HCN pacemaker channels. *Nature* 2001;**411**:805–10.
41. Zong X, Zucker H, Hofmann F, Biel M. Three amino acids in the C-linker are major determinants of gating in cyclic nucleotide-gated channels. *EMBO J* 1998;**17**:353–62.
42. Zhou L, Olivier NB, Yao H, Young EC, Siegelbaum SA. A conserved tripeptide in CNG and HCN channels regulates ligand gating by controlling C-terminal oligomerization. *Neuron* 2004;**44**:823–34.
43. Moosmang S, Biel M, Hofmann F, Ludwig A. Differential distribution of four hyperpolarization-activated cation channels in mouse brain. *Biol Chem* 1999;**380**:975–80.
44. Notomi T, Shigemoto R. Immunohistochemical localization of Ih channel subunits, HCN1-4, in the rat brain. *J Comp Neurol* 2004;**471**:241–76.
45. Nolan MF, Malleret G, Lee KH, Gibbs E, Dudman JT, Santoro B, Yin D, Thompson RF, Siegelbaum SA, Kandel ER, Morozov A. The hyperpolarization-activated HCN1 channel is important for motor learning and neuronal integration by cerebellar Purkinje cells. *Cell* 2003;**115**:551–64.
46. Nolan MF, Malleret G, Dudman JT, Buhl DL, Santoro B, Gibbs E, Vronskaya S, Buzsaki G, Siegelbaum SA, Kandel ER, Morozov A. A behavioral role for dendritic integration: HCN1 channels constrain spatial memory and plasticity at inputs to distal dendrites of CA1 pyramidal neurons. *Cell* 2004;**119**:719–32.
47. Ludwig A, Budde T, Stieber J, Moosmang S, Wahl C, Holthoff K, Langebartels A, Wotjak C, Munsch T, Zong X, Feil S, Feil R, Lancel M, Chien KR, Konnerth A, Pape HC, Biel M, Hofmann F. Absence epilepsy and sinus dysrhythmia in mice lacking the pacemaker channel HCN2. *EMBO J* 2003;**15**:216–24.
48. Stieber J, Herrmann S, Feil S, Loster J, Feil R, Biel M, Hofmann F, Ludwig A. The hyperpolarization-activated channel HCN4 is required for the generation of pacemaker action potentials in the embryonic heart. *Proc Natl Acad Sci USA* 2003;**100**:15,235–15,240.
49. Schulze-Bahr E, Neu A, Friederich P, Kaupp UB, Breithardt G, Pongs O, Isbrandt D. Pacemaker channel dysfunction in a patient with sinus node disease. *J Clin Invest* 2003;**111**:1537–45.
50. Ueda K, Nakamura K, Hayashi T, Inagaki N, Takahashi M, Arimura T, Morita H, Higashiuesato Y, Hirano Y, Yasunami M, Takishita S, Yamashina A, Ohe T, Sunamori M, Hiraoka M, Kimura A. Functional characterization of a trafficking-defective HCN4 mutation, D553N, associated with cardiac arrhythmia. *J Biol Chem* 2004;**279**:27,194–27,198.
51. Milanesi R, Baruscotti M, Gnecchi-Ruscone T, DiFrancesco D. Familial sinus bradycardia associated with a mutation in the cardiac pacemaker channel. *N Engl J Med* 2006;**354**:151–7.
52. Nof E, Luria D, Brass D, Marek D, Lahat H, Reznik-Wolf H, Pras E, Dascal N, Eldar M, Glikson M. Point mutation in the HCN4 cardiac ion channel pore affecting synthesis, trafficking, and functional expression is associated with familial asymptomatic sinus bradycardia. *Circulation* 2007;**31**:1,16,463–1,16,470.

Part 3

Immunoglobulin Receptors

Chapter 49

Immunoglobulin–Fc Receptor Interactions

Jenny M. Woof
Division of Medical Sciences, University of Dundee Medical School, Ninewells Hospital, Dundee, Scotland, UK

INTRODUCTION

Immunoglobulins play a central role in immune protection. They serve as flexible adaptor molecules capable of coupling the recognition of antigenic structures on foreign micro-organisms with effector responses that bring about neutralization or elimination of the invading pathogen. There are five classes of immunoglobulin (Ig), namely IgG, IgE, IgA, IgM, and IgD, with differing biological functions and distributions in the body. Through triggering distinct effector functions, each makes a unique contribution to the immune response. IgG, the major antibody class in serum, elicits protective mechanisms against invading bacteria and viruses, whereas IgE has a role in protection against parasitic infections but is responsible also for the symptoms of allergy. IgA, while present at quite high concentrations in serum, carries out its key role at the mucosal surfaces, such as the linings of the respiratory and gastrointestinal tracts, where it is the major antibody class. IgM, found chiefly in serum because its large size precludes ready diffusion into the tissues, provides the primary response to bacterial and fungal invaders. The function of IgD remains less clear. For IgG, IgA, and IgE, which will be the major focus of this chapter, a key element of the Ig effector armory is driven through interaction with specific Fc receptors (FcR) on a variety of immune cells.

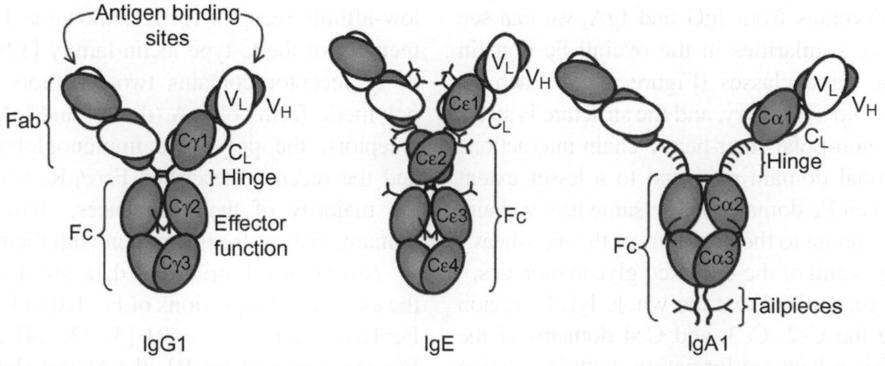

FIGURE 49.1 Schematic representation of human IgG1, IgE and IgA1. Heavy chains are shown in dark gray, and light chains in white.

IMMUNOGLOBULIN STRUCTURE

All Igs share a common structural unit comprising two identical heavy chains and two identical light chains (Figure 49.1). Folded up into globular domains (two per light chain and either four or five per heavy chain) with a characteristic anti-parallel β sheet arrangement, these chains are arranged into two Fab regions, responsible for binding to antigen, and one Fc region in which the effector function elements reside. The N-terminal domains of both types of chain, the so-called variable (V) domains, vary from one antibody to the next and determine the specificity for antigen. In the heavy chain, the remaining domains (constant or C domains) determine the class of the antibody, and are constant within a particular Ig class, or subclass in the case of

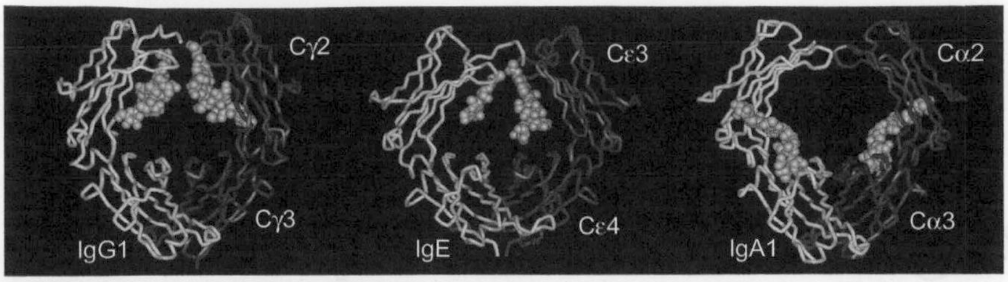

FIGURE 49.2 Structures of the equivalent carboxy-terminal domains of the Fc regions of human IgG1 (PDB code 1IIS), IgE (PDB code 1F6A), and IgA1 (PDB code 1OW0).
Structures adopted by the Fc regions when complexed with cognate Fc receptor are shown. One heavy chain is shown in dark gray, the other in light gray. N-linked oligosaccharides are shown using spheres.

IgG and IgA. In humans there are four subclasses of IgG, known as IgG1, IgG2, IgG3, and IgG4, and two subclasses of IgA, named IgA1 and IgA2.

IgA and IgM have the capacity to polymerize into predominantly dimers and pentamers respectively, which are stabilized by disulfide bonds between subunits and an additional polypeptide called joining (J) chain. IgA exists as both monomers (the predominant serum form) and dimers (the main form in mucosal secretions).

In IgG and IgA, the heavy chain has four domains, and a flexible linker or hinge region links the Fab arms with the Fc region (Figure 49.1). Thus in these Igs the Fc region comprises four domains. In contrast, in IgE and IgM the hinge is replaced by an extra domain pair, and the entire Fc region (in IgE referred to as Fcε2-4) comprises six domains. Recombinant versions of IgE Fc made up of the C-terminal four domains (Cε3-4) have been produced to aid structural analysis.

Crystal structures are now available for the Fc regions of human IgG1, IgE, and IgA1 (Figure 49.2) [1–7]. For IgG Fc and IgE Fc, structures are available for both the uncomplexed state and in complex with the extracellular domains of Fc receptor (FcγRIII and FcεRI respectively). Some differences are apparent – receptor binding to IgG Fc is associated with an asymmetrical opening of the structure by about 7 Å between the tips of the Cγ2 domains, and Fc domain rearrangements also accompany FcεRI binding to IgE Fc [4, 5]. For IgA Fc, only the structures within complexes with either FcαRI or the bacterial IgA binding protein SSL7 have been solved.

When we consider only the Cε3 and Cε4 domains of IgE and the equivalent domains from IgG and IgA, we can see that there are striking similarities in the overall Fc domain arrangements of the three classes (Figure 49.2). Each Fc exhibits pseudo two-fold symmetry, and the structure is stabilized principally by non-polar inter-heavy chain interactions between the C-terminal domain pair, and to a lesser extent by interactions between Fc domains in the same heavy chain. The chief differences relate to the positions of the inter-heavy chain disulfide bridges and of the N-linked glycan moieties.

It has long been recognized that the whole IgE Fc region (Fcε2-4, comprising the Cε2, Cε3, and Cε4 domains of the two heavy chains) has a bent conformation, both in solution and in complex with FcεRI [8,9]. More recently, the X-ray crystal structure for free Fcε2-4 confirmed an acute and asymmetrical bend in the linker region between Cε2 and Cε3 [5]. The Cε3 domains are not symmetrical, and the bend brings one of the Cε2 domains into contact with the Cε3 domain and, to a small extent, the Cε4 domain of the other heavy chain. The other Cε2 domain makes only less extensive contacts with the Cε3 domain of the opposite heavy chain.

Fc RECEPTORS AND THEIR STRUCTURES

Receptors specific for Ig Fc regions are expressed on a range of cell types. Broadly, they fall into four categories. The first group is a family of closely-related receptors comprising the three IgG Fc receptors (FcγRI, the two forms of FcγRII (FcγRIIa and FcγRIIb) and the two forms of FcγRIII), the high-affinity receptor for IgE named FcεRI, and the IgA receptor, FcαRI [10–13]. Each receptor in this class comprises a distinct α−chain either alone, as in the case of FcγRIIa, FcγRIIb, and FcγRIIIb, or associated with β and a γ-chain dimer (FcεRI) or with the γ chains only (FcγRI, FcγRIIIa, FcαRI). The α chains have evolved as part of the Ig gene superfamily, and have two (or three, in the case of FcγRI) Ig-like extracellular domains responsible for binding their respective Fc ligands. Second, we have the so-called neonatal Fc receptor, FcRn, which belongs to the class I major histocompatibility complex (MHC) family and both transports IgG across the placenta and mucosal epithelia, and regulates IgG turnover [14]. The third type of Ig receptor, the low-affinity receptor for IgE known as FcεRII or CD23, is a member of the C-type lectin family [12]. The final category of Ig receptor contains two receptors capable of binding polymeric forms of IgA (dimer) and IgM (pentamer). These receptors, the polymeric immunoglobulin receptor (pIgR) and the recently described Fcα/μR, while very different in the majority of their sequences, share similar N-terminal domains critical for interaction with their ligands [15,16].

Atomic-level structural data are available currently for the extracellular portions of FcγRIIa [17, 18], FcγRIIb [19], FcγRIIIb [20–22], FcεRI [3, 23, 24] and FcαRI [6, 25]. For the Fcγ and FcεRI, the "heart-shaped" structures are remarkably similar. In each case, the two Ig-like domains

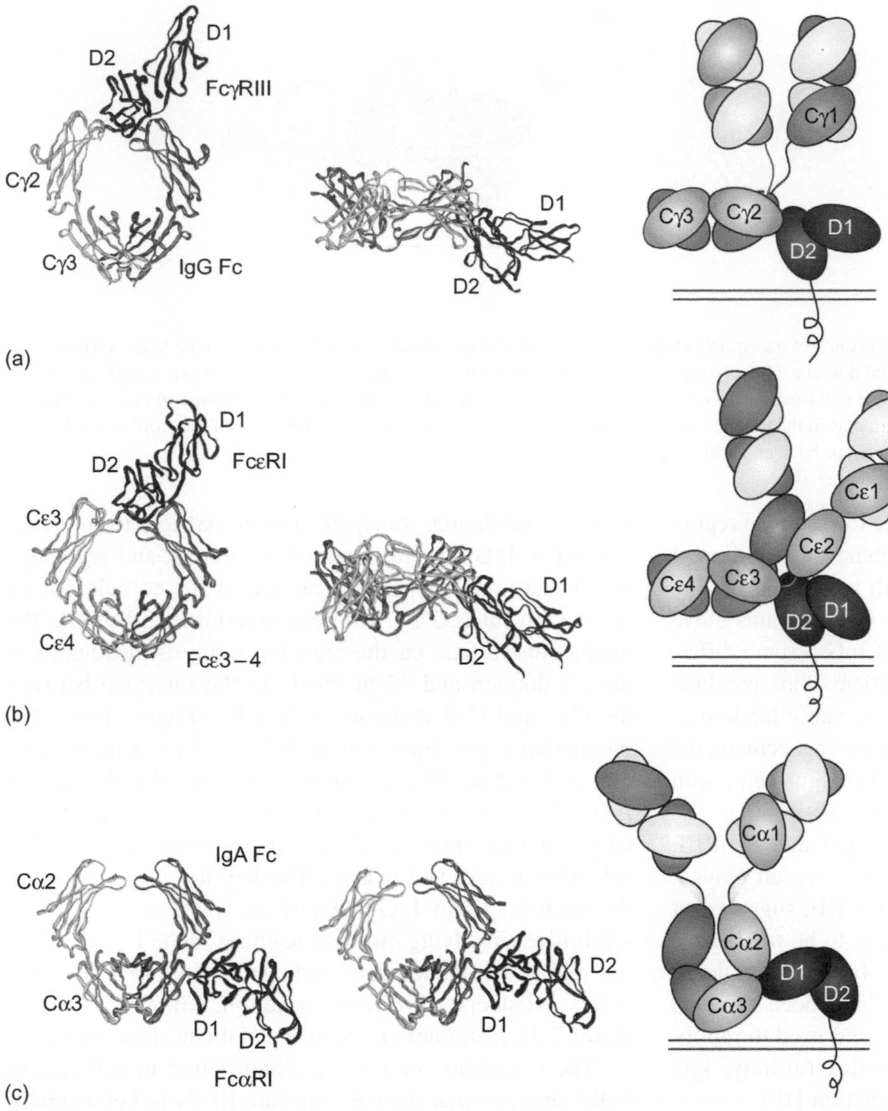

FIGURE 49.3 Complexes of IgG Fc, IgE Fcε3-4, and IgA Fc with their respective Fc receptors.
The left and middle columns show X-ray crystal structures of the Fc–FcR complexes. (a) IgG Fc–FcγRIII complex (PDB code 1E4K); (b) IgE Fcε3-4–FcεRI complex (PDB code 1F6A); and (c) IgA Fc–FcαRI complex (PDB code 1OW0). One Fc heavy chain is colored light gray, the other mid-gray. Extracellular domains of the receptors are shown in dark gray. In the left column, each Fc region is viewed face on. In the middle column, the complexes are orientated so that the C-termini of the D2 domains of each receptor face downwards. The right column shows schematic representations of the receptors in the same orientation as in the middle column, interacting with their intact ligands. Here, light chain domains are shown in the palest gray.

of the α chain share a large interface, and are orientated at an angle of ~70° to each other. In contrast, the two Ig-like domains of FcαRI are arranged very differently relative to each other. Essentially, the domain orientations are inverted compared to those of FcγR and FcεRI.

Structural information is available also for FcεRII [26, 27], for FcRn [28–30], and for domain 1 of pIgR [31]. FcεRII has the topology typical of a C-type lectin domain, with two α-helices and eight β strands forming two anti-parallel β sheets. The structure of FcRn resembles that of a class I MHC molecule, with the three domains of the α chain interacting with β2-microglobulin (β2-m), the obligate subunit of all class I molecules. However, the groove between the two N-terminal α-chain domains is much narrower and unable to hold peptide as in the MHC molecule. The N-terminal domain 1 of pIgR resembles an Ig variable domain, but there are differences in the orientations of the loops analogous to complementarity determining regions (CDR).

We will now examine the nature of receptor–Ig Fc interactions in detail, discussing in turn the receptors specific for IgG, IgE, and IgA.

IgG–RECEPTOR INTERACTIONS

FcγR

Interactions sites on both FcγR receptors [32–35] and IgG [36–38] were originally proposed on the basis of mutagenesis studies. The theme that emerged suggested that the membrane proximal D2 domains of the receptors interacted with the so-called lower hinge regions at the N-terminal ends of the IgG Fc. X-ray crystallographic studies of complexes of IgG Fc and soluble FcγRIII have confirmed this binding mode [20, 22] (Figure 49.3a). Specifically, the main interaction regions comprise a "sandwich" of Pro329 of one IgG heavy chain between Trp87 and Trp110 of the D2 domain

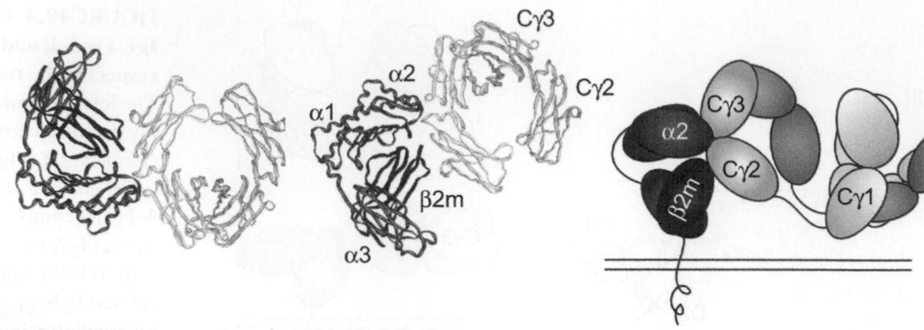

FIGURE 49.4 Complex of IgG Fc and FcRn.
The left and middle images show X-ray crystal structures of the complex between rat IgG Fc and rat FcRn (PDB code 1I1A). One Fc heavy chain is colored light gray, the other mid-gray. Extracellular domains of the receptor (three α-chain domains and β2-microglobulin) are shown in dark gray. In the middle column, the complex is orientated so that the C-termini of the α3 domains of the receptor face downwards. The right column shows a schematic representation of the receptor in the same orientation as in the middle column, interacting with intact IgG. The close approach of the "top" of the Fc region to the membrane suggests that the Fab arms must be bent out of the plane of the Fc.

of the receptor, and the interaction of the lower hinge region (Leu234–Ser239) of both IgG heavy chains with different surfaces of the receptor. Interaction with receptor results in a slight opening of the Fc such that the Cγ2 domains move slightly apart. The two hinge peptides of IgG assume different conformations, and their relative orientations preclude binding of a second receptor molecule by steric hindrance. A 1 : 1 stoichiometry therefore results. Upon engagement, the receptor structure also undergoes a slight adjustment, with the angle between the domains increasing to about 80°.

The similarities between FcγRI, FcγRII and FcγRIII, and the evidence that the IgG lower hinge region plays a key role in interaction with FcγRI and FcγRII, suggest that the mode of interaction with IgG is likely to be fundamentally similar for all three FcγR classes. Indeed, complexes between IgG Fc and FcγRI and FcγRII have been modeled based on the FcγRIII–IgG Fc structure, and used to rationalize the particular binding characteristics (affinity, IgG subclasses specificity, etc.) of each interaction [18].

Modulation of the N-linked oligosaccharides attached to Asn297 in the Cγ2 domain of IgG can have an impact on the affinity of the interaction with FcγR [39], and removal of these glycans has long been known to greatly decrease the capacity to bind FcγR [40, 41]. In the crystal structures of the complex, the oligosaccharides are seen not to contribute directly to receptor binding. Most likely, they play an indirect role, stabilizing the structure of the Fc around the binding region [42].

The location of the FcγR interaction site at the N-terminal end of the Fc means that in order for an IgG molecule to engage both FcγR via its Fc region and antigenic surface via the tips of its Fab arms, some degree of antibody dislocation would appear necessary. Thus the Fab arms would move out of the plane of the Fc region, presumably by virtue of hinge flexibility (Figure 49.3a).

FcRn

It is now appreciated that FcRn has multiple roles. These include transport of ingested IgG from the intestinal lumen into the circulation, transport of maternal IgG to the fetus, transfer of IgG out into mucosal secretions, and regulation of IgG turnover [14]. The structure of the complex of rat FcRn with rat IgG Fc has been determined [29, 30]. The interaction surface on the receptor, comprising regions in the α2 domain and β2-m, binds to the interface between the Cγ2 and Cγ3 domains of IgG Fc (Figure 49.4). The interaction is pH dependent, with low pH favoring association and neutral pH favoring dissociation. Thus binding of IgG to FcRn is driven in acidic environments (e.g., neonatal gut and transport vesicles), and release occurs at neutral pH in the tissues and plasma. The key lies in the nature of the extensive FcRn–IgG Fc interface, which is stabilized by salt bridges involving histidine residues in the Fc and aspartate and glutamine residues on FcRn. Histidine protonation at low pH steers salt bridge formation, while at pH values above 7 the histidines are neutral and the bridges dissolve.

The orientation of IgG Fc when bound to cell surface FcRn suggests that the Fab arms are likely to bend significantly out of the plane of the Fc in order to prevent "steric clashes" with cell membrane components (Figure 49.4).

IgE–RECEPTOR INTERACTIONS

FcεRI

The structure of the complex between the extracellular domains of FcεRI and the IgE fragment comprising the paired Cε3 and Cε4 domains (Fcε3-4) has been solved [3]. The receptor interacts principally with the Cε3 domains and the Cε2–Cε3 linker regions of IgE Fc (Figure 49.3b). The interaction is strikingly reminiscent of the FcγRIII–IgG Fc complex. There are two distinct interaction subsites, one on each Fc heavy chain. In one, a pocket between loops of one of the Cε3 domains and the Cε2–Cε3 linker binds the receptor D2 domain. In the second, the other Cε3 domain sandwiches between two tryptophan residues at the top of the receptor D1–D2 interface. Binding of FcεRI generates

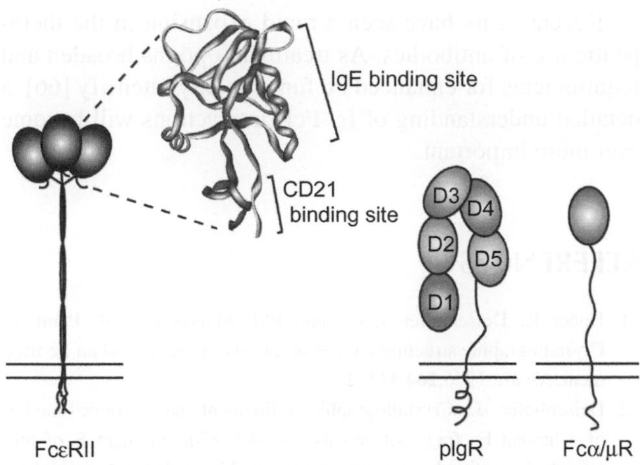

FIGURE 49.5 Schematic representation of FcεRII (left), pIgR (centre) and Fcα/μR (right).
FcεRII comprises C-type lectin domains that mediate IgE binding and α-helical coiled-coil stalk regions that are thought to mediate trimerization of the receptor. The structure of a single lectin domain is shown (PDF code 2H2T) and the regions involved in binding IgE and CD21 are indicated. pIgR has five extracellular Ig-like domains, which are suggested to loop back towards the membrane. Domains D1 and D5 are critical for ligand binding. Fcα/μR has a single Ig-like domain related to D1 of pIgR, which is responsible for binding ligand.

structural asymmetry in the Cε2–Cε3 linker region similar to that seen in the IgG lower hinge region on FcγRIII binding. This asymmetry dictates the 1 : 1 stoichiometry by precluding entry of a second receptor.

While there is little conformational change in FcεRI on ligand binding, it is proposed that the structure of IgE Fc undergoes significant rearrangements upon receptor interaction. Differences between the structure of the Cε3–Cε4 fragment when complexed with FcεRI and that of the whole uncomplexed Fc (Cε2–Cε4 domains) suggest that a conformational change involving both the Cε2 and Cε3 domains occurs upon engagement [5]. The extremely slow dissociation rate of IgE from FcεRI that results, coupled with restricted tissue diffusion that allows rebinding of IgE to receptor, gives rise to an extensive residence time of IgE on tissue mast cells. This long-lasting association of IgE with FcεRI (~14 days) accounts for the immediate induction of the characteristic symptoms of hypersensitivity that is seen when allergen is encountered by allergic patients.

FcεRII

FcεRII, a 45-kDa type II membrane glycoprotein with medium affinity for IgE, is believed to self-associate into a trimer at the cell surface, with the three C-type lectin domains held together above a 15-nm α-helical coiled-coil stalk [26, 27, 43] (Figure 49.5). The lectin domain is known to interact with the Cε3 domain of IgE Fc, at a site distinct from that bound by FcεRI [44–46]. The binding, which has a stoichiometry of 2 : 1 (FcεRII : IgE Fc), does not depend on any lectin-like characteristics of the receptor since IgE glycans are not involved. The NMR analysis located the interaction site for IgE on FcεRII (Figure 49.5), which is quite distinct from that for CD21, another FcεRII ligand [26]. In the trimer arrangement predicted from the NMR data, the three CD21 sites lie close together near to the stalk, while the IgE binding sites occupy the outer edges of the lectin domain trimer. This arrangement means that the formation of extensive B cell surface arrays may be driven by co-crosslinking of surface IgE and CD21 by soluble forms of trimeric FcεRII lectin domains, known to be generated through proteolysis of the FcεRII stalk by endogenous proteases. Thus, through actions as both a cell surface and soluble receptor, FcεRII is believed to play a key role in the regulation of the IgE response [12].

IgA–RECEPTOR INTERACTIONS

FcαRI

The FcαRI interaction site on IgA was originally localized to the interface of the Cα2 and Cα3 domains through a series of mutagenesis studies. One loop in the Cα2 domain (Leu257, Leu258) and another in the Cα3 domain (Pro440–Phe443) were particularly implicated [47, 48]. The interaction site for IgA Fc on the membrane-distal D1 domain of FcαRI was also identified on the basis of mutagenesis studies [49–51]. Subsequently these site localizations were confirmed when the X-ray crystal structure of the complex of IgA1 Fc with the extracellular domains of FcαRI was determined [6] (Figure 49.3c). The IgA Fc–FcαRI interface has a hydrophobic centre, encircled by a region featuring interactions between polar groups. These peripheral interactions may include hydrogen bond interactions, but electrostatic interactions are minor [52, 53]. The N-linked oligosaccharides attached to Asn263 of the IgA Cα2 domains do not play a role in the interaction [54].

Despite structural similarities between both receptors and Fc ligands, the interaction mode of IgA Fc with FcαRI differs significantly from those of IgG Fc with FcγR and IgE Fc with FcεRI (Figure 49.3). Moreover, the interaction stoichiometries differ with the extracellular portions of FcαRI, when free in solution, binding IgA Fc with a 2 : 1 stoichiometry [52]. Under normal physiological conditions at the cell membrane, it is presumed that one IgA molecule does not bind two FcαRI, or such binding is insufficient to initiate a signaling cascade. However, two characteristics of the FcαRI–IgA interaction may relate to the 2 : 1 stoichiometry, namely the ability of cytokine stimulation to enhance affinity without an increase in receptor expression [55, 56], and the inability of secretory IgA to elicit phagocytosis via FcαRI [57]. In the first example, in the unstimulated state, cytoskeletal interactions may restrict movement of FcαRI molecules within the membrane, preventing binding of two receptors to one IgA molecule. Cytokine stimulation may disrupt cytoskeletal interactions and allow reorientation of receptors so that two receptors may bind to the same IgA molecule, resulting in higher avidity. In the second example, the secretory component of secretory IgA (derived from the pIgR that delivers the IgA dimer into the mucosal secretions) is proposed to block access to one or more of the FcαRI sites on the four IgA heavy chains, resulting in the inability of secretory IgA to promote phagocytosis via FcαRI.

pIgR

The 620 amino-acid extracellular portion of pIgR folds into five Ig-like domains, termed D1 to D5 from the N-terminus, followed by a sixth non-Ig-like region (Figure 49.5). The first three pIgR domains are critical for interaction with dimeric IgA, which itself comprises two IgA monomers linked via joining (J) chain. D1 of pIgR plays an essential role in binding, with the CDR-like loops being particularly implicated. In addition, a disulfide bond forms between Cys467 in D5 of pIgR and Cys311 in the IgA Cα2 domain, and direct interaction between J chain and pIgR is required [13]. Mutagenesis studies have highlighted human IgA Fc regions responsible for the non-covalent interaction with human pIgR. The Cα3 domain appears to play the main role, and motifs lying principally across the Cα2-proximal surface of this domain have been particularly implicated [58–61].

Fcα/μR

The membrane distal extracellular portion of this type I receptor shares structural features with Ig-like domains, and has particularly strong identity with D1 of pIgR (Figure 49.5). Indeed, the presence of key conserved residues implies that it adopts a similar structure to that of pIgR D1, and may share a similar mode of interaction with dimeric IgA [31]. It is now known that Fcα/μR interacts only with polymeric forms of IgA and IgM [62, 63]. Recent experiments indicate that the Cα2/Cα3 domain interface of the IgA heavy chain is critical for the interaction [63], suggesting possible overlap with sites that bind pIgR and FcαRI.

CONCLUSIONS

As our knowledge of Ig–FcR interactions grows, certain themes emerge. For example, FcγR and FcεRI bind via analogous sites in their membrane proximal domains to approximately analogous positions in the penultimate Fc domains of IgG and IgE respectively. For other receptors, the Ig Fc interdomain region frequently serves an important role in engagement, presumably due to its highly accessible and hydrophobic nature. Indeed, it has been identified as one of a limited number of regions on the Ig surface particularly suited to protein–protein interaction [64, 65].

Recent years have seen a rapid expansion in the therapeutic use of antibodies. As treatment options broaden and requirements for enhanced Fc functionality intensify [66], a detailed understanding of Ig–FcR interactions will become ever more important.

REFERENCES

1. Huber R, Deisenhofer J, Colman PM, Matsushima M, Palm W. Crystallographic structure studies of an IgG molecule and an Fc fragment. *Nature* 1976;**264**:415–20.
2. Deisenhofer J. Crystallographic refinement and atomic models of a human Fc fragment and its complex with fragment B of protein A from *Staphylococcus aureus* at 2.9- and 2.8-Å resolution. *Biochemistry* 1981;**20**:2361–70.
3. Garman SC, Wurzburg BA, Tarchevskaya SS, Kinet JP, Jardetzky TS. Structure of the Fc fragment of human IgE bound to its high affinity receptor FcεRIα. *Nature* 2000;**406**:259–66.
4. Wurzburg BA, Garman SC, Jardetzky TS. Structure of the human IgE–Fc Cε3–Cε4 reveals conformational flexibility in the antibody effector domains. *Immunity* 2000;**13**:375–85.
5. Wan T, Beavil RL, Fabiane SM, Beavil AJ, Sohi MK, Keown M, Young RJ, Henry AJ, Owens RJ, Gould HJ, Sutton BJ. The crystal structure of IgE reveals an asymmetrically bent conformation. *Nat Immunol.* 2002;**3**:681–6.
6. Herr AB, Ballister ER, Bjorkman PJ. Insights into IgA-mediated immune responses from the crystal structures of human FcαRI and its complex with IgA1-Fc. *Nature* 2003;**423**:614–20.
7. Ramsland PA, Willoughby N, Trist HM, Farrugia W, Hogarth PM, Fraser JD, Wines BD. Structural basis for evasion of IgA immunity by *Staphylococcus aureus* revealed in the complex of SSL7 with Fc of human IgA1. *Proc. Natl Acad. Sci. USA* 2007;**104**:15,051–6.
8. Zheng Y, Shopes B, Holowka D, Baird B. Conformation of IgE bound to its receptor FcεRI and in solution. *Biochemistry* 1991;**30**:9125–32.
9. Zheng Y, Shopes B, Holowka D, Baird B. Dynamic conformation compared for IgE and IgG1 in solution and bound to receptors. *Biochemistry* 1992;**31**:7446–56.
10. Woof JM, Burton DR. Human antibody–Fc receptor interactions illuminated by crystal structures. *Nat Rev. Immunol.* 2004;**4**:89–99.
11. Cohen-Solal JFG, Cassard L, Fridman WH, Sautès-Fridman C. Fcγ receptors. *Immunol. Lett.* 2004;**92**:199–205.
12. Gould HJ, Sutton BJ. IgE in allergy and asthma today. *Nat Rev. Immunol.* 2008;**8**:205–17.
13. Woof JM, Kerr MA. The function of immunoglobulin A in immunity. *J. Pathol.* 2006;**208**:270–82.
14. Roopenian DC, Akilesh S. FcRn: the neonatal Fc receptor comes of age. *Nat Rev. Immunol.* 2007;**7**:715–25.
15. Rojas R, Apodaca G. Immunoglobulin transport across polarized epithelial cells. *Nat Rev. Mol. Cell Biol.* 2002;**3**:944–55.
16. Shibuya A, Honda SI. Molecular and functional characteristics of the Fcα/μR, a novel Fc receptor for IgM and IgA. *Springer Semin. Immun.* 2006;**28**:377–82.
17. Maxwell KF, Powell MS, Hulett MD, Barton PA, McKenzie IF, Garrett TP, Hogarth PM. Crystal structure of the human leukocyte Fc receptor, FcγRIIa. *Nat Struct. Biol.* 1999;**6**:437–42.
18. Sondermann P, Kaiser J, Jacob U. Molecular basis for immune complex recognition: a comparison of Fc-receptor structures. *J. Mol. Biol.* 2001;**309**:737–49.

19. Sondermann P, Huber R, Jacob U. Crystal structure of the soluble form of the human Fcγ-receptor IIb: a new member of the immunoglobulin superfamily at 1.7 Å resolution. *EMBO J.* 1999;**18**:1095–103.
20. Sondermann P, Huber R, Oosthuizen V, Jacob U. The 3.2-Å crystal structure of the human IgG1 Fc fragment–FcγRIII complex. *Nature* 2000;**406**:267–73.
21. Zhang Y, Boesen CC, Radaev S, Brooks AG, Fridman WH, Sautès-Fridman C, Sun PD. Crystal structure of the extracellular domain of a human FcγRIII. *Immunity* 2000;**13**:387–95.
22. Radaev S, Motyka S, Fridman WH, Sautès-Fridman C, Sun PD. The structure of a human type III Fcγ receptor in complex with Fc. *J. Biol. Chem.* 2001;**276**:16,469–77.
23. Garman SC, Kinet JP, Jardetzky TS. Crystal structure of the human high-affinity IgE receptor. *Cell* 1998;**95**:951–61.
24. Garman SC, Sechi S, Kinet JP, Jardetzky TS. The analysis of the human high affinity IgE receptor FcεRIα from multiple crystal forms. *J. Mol. Biol.* 2001;**311**:1049–62.
25. Ding Y, Xu G, Yang M, Yao M, Gao GF, Wang L, Zhang W, Rao Z. Crystal structure of the ectodomain of human FcαRI. *J. Biol. Chem.* 2003;**278**:27,966–70.
26. Hibbert RG, Teriete P, Grundy GJ, Beavil RL, Reljic R, Holers VM, Hannan JP, Sutton BJ, Gould HJ, McDonnell JM. The structure of human CD23 and its interactions with IgE and CD21. *J. Exp. Med.* 2005;**202**:751–60.
27. Wurzburg BA, Tarchevskaya SS, Jardetsky TS. Structural changes in the lectin domain of CD23, the low-affinity IgE receptor, upon calcium binding. *Structure* 2006;**14**:1049–58.
28. Burmeister WP, Gastinel LN, Simister NE, Blum ML, Bjorkman PJ. Crystal structure at 2.2-Å resolution of the MHC-related neonatal Fc receptor. *Nature* 1994;**372**:336–43.
29. Burmeister WP, Huber AH, Bjorkman PJ. Crystal structure of the complex of rat neonatal Fc receptor with Fc. *Nature* 1994;**372**:379–83.
30. Martin WL, West AP, Gan L, Bjorkman PJ. Crystal structure at 2.8 Å of an FcRn/heterodimeric Fc complex: mechanism of pH-dependent binding. *Mol. Cell* 2001;**7**:867–77.
31. Hamburger AE, West AP, Bjorkman PJ. Crystal structure of a polymeric immunoglobulin binding fragment of the human polymeric immunoglobulin receptor. *Structure* 2004;**12**:1925–35.
32. Hulett MD, Witort E, Brinkworth RI, McKenzie IF, Hogarth PM. Identification of the IgG binding site of the human low affinity receptor for IgG FcγRII. Enhancement and ablation of binding by site-directed mutagenesis. *J. Biol. Chem* 1994;**269**:15,287–93.
33. Hulett MD, Witort E, Brinkworth RI, McKenzie IF, Hogarth PM. Multiple regions of human FcγRII (CD32) contribute to the binding of IgG. *J. Biol. Chem* 1995;**270**:21,188–94.
34. Tamm A, Kister A, Nolte KU, Gessner JE, Schmidt RE. The IgG binding site of human FcγRIIIB receptor involves CC′ and FG loops of the membrane-proximal domain. *J. Biol. Chem.* 1996;**271**:3659–66.
35. Hulett MD, Hogarth PM. The second and third extracellular domains of FcγRI (CD64) confer the unique high affinity binding of IgG2a. *Mol. Immunol.* 1998;**35**:989–96.
36. Woof JM, Partridge LJ, Jefferis R, Burton DR. Localisation of the monocyte- binding region on human immunoglobulin G. *Mol. Immunol.* 1986;**23**:319–30.
37. Duncan AR, Woof JM, Partridge LJ, Burton DR, Winter G. Localization of the binding site for the human high-affinity Fc receptor on IgG. *Nature* 1988;**332**:563–4.
38. Lund J, Winter G, Jones PT, Pound JD, Tanaka T, Walker MR, Artymiuk PJ, Arata Y, Burton DR, Jefferis R, Woof JM. Human FcγRI and FcγRII interact with distinct but overlapping sites on human IgG. *J. Immunol.* 1991;**147**:2657–62.
39. Satoh M, Iida S, Shitara K. Non-fucosylated therapeutic antibodies as next-generation therapeutic antibodies. *Expert Opin. Biol Ther.* 2006;**6**:1161–73.
40. Leatherbarrow RJ, Rademacher TW, Dwek RA, Woof JM, Clark A, Burton DR, Richardson N, Feinstein A. Effector functions of a monoclonal aglycosylated mouse IgG2a: binding and activation of complement component C1 and interaction with human monocyte Fc receptor. *Mol. Immunol.* 1985;**22**:407–15.
41. Jefferis R, Lund J, Pound JD. IgG–Fc-mediated effector functions: molecular definition of interaction sites for effector ligands and the role of glycosylation. *Immunol. Rev.* 1998;**163**:59–76.
42. Krapp S, Mimura Y, Jefferis R, Huber R, Sondermann P. Structural analysis of human IgG–Fc glycoforms reveals a correlation between glycosylation and structural integrity. *J. Mol. Biol.* 2001;**325**:979–89.
43. Beavil RL, Graber P, Aubonney N, Bonnefoy JY, Gould HJ. CD23/FcεRII and its soluble fragments can form oligomers on the cell surface and in solution. *Immunology* 1995;**84**:202–6.
44. Vercelli D, Helm BA, Marsh P, Padlan EA, Geha RS, Gould HJ. The B-cell binding site on human immunoglobulin E. *Nature* 1989;**338**:649–51.
45. Shi J, Ghirlando R, Beavil RL, Beavil AJ, Keown MB, Young RJ, Owens RJ, Sutton BJ, Gould HJ. Interaction of the low affinity receptor CD23/FcεRII lectin domain with the Fcε3-4 fragment of human IgE. *Biochemistry* 1997;**36**:2112–22.
46. Sayers I, Housden JEM, Spivey AC, Helm BA. The importance of Lys-352 of human immunoglobulin E in FcεRII/CD23 recognition. *J. Biol. Chem* 2004;**279**:35,320–25.
47. Carayannopoulos L, Hexham JM, Capra JD. Localization of the binding site for the monocyte immunoglobulin (Ig) A-Fc receptor (CD89) to the domain boundary between Cα2 and Cα3 in human IgA1. *J. Exp. Med.* 1996;**183**:1579–86.
48. Pleass RJ, Dunlop JI, Anderson CM, Woof JM. Identification of residues in the CH2/CH3 domain interface of IgA essential for interaction with the human Fcα receptor (FcαR) CD89. *J. Biol. Chem.* 1999;**274**:23,508–14.
49. Wines BD, Hulett MD, Jamieson GP, Trist HM, Spratt JM, Hogarth PM. Identification of residues in the first domain of human Fcα receptor essential for interaction with IgA. *J. Immunol.* 1999;**162**:2146–53.
50. Morton HC, van Zandbergen G, van Kooten C, Howard CJ, van de Winkel JG, Brandtzaeg P. Immunoglobulin binding sites of human FcαRI (CD89) and bovine Fcγ2R are located in their membrane-distal extracellular domains. *J. Exp. Med.* 1999;**189**:1715–22.
51. Wines BD, Sardjono CT, Trist HH, Lay CS, Hogarth PM. The interaction of FcαRI with IgA and its implications for ligand binding by immunoreceptors of the leukocyte receptor cluster. *J. Immunol.* 2001;**166**:1781–9.
52. Herr AB, White CL, Milburn C, Wu C, Bjorkman PJ. Bivalent binding of IgA1 to FcαRI suggests a mechanism for cytokine activation of IgA phagocytosis. *J. Mol. Biol.* 2003;**327**:645–57.
53. Pleass RJ, Dehal PK, Lewis MJ, Woof JM. Limited role of charge matching in the interaction of human immunoglobulin A with the immunoglobulin A Fc receptor (FcαRI) CD89. *Immunology* 2003;**109**:331–5.
54. Mattu TS, Pleass RJ, Willis AC, Kilian M, Wormald MR, Lellouch AC, Rudd PM, Woof JM, Dwek RA. The glycosylation and structure of human serum IgA1, Fab and Fc regions and the role of N-glycosylation on Fcα receptor interactions. *J. Biol. Chem.* 1998;**273**:2260–72.
55. Weisbart RH, Kacena A, Schuh A, Golde DW. GM-CSF induces human neutrophil IgA-mediated phagocytosis by an IgA Fc receptor activation mechanism. *Nature* 1988;**332**:647–8.
56. Bracke M, Dubois GR, Bolt K, Bruijnzeel PL, Vaerman JP, Lammers JW, Koenderman L. Differential effects of the T helper cell type-2 derived

cytokines IL-4 and IL-5 on ligand binding to IgG and IgA receptors expressed by human eosinophils. *J. Immunol.* 1997;**159**:1459–65.
57. van Egmond M, van Garderen E, van Spriel AB, Damen CA, van Amersfoort ES, van Zandbergen G, van Hattum J, Kuiper J, van de Winkel JG. FcαRI-positive liver Kuppfer cells: reappraisal of the function of immunoglobulin A in immunity. *Nature Med.* 2000;**6**: 680–5.
58. Hexham JM, White KD, Carayannopoulos LN, Mandecki W, Brisette R, Yang YS, Capra JD. A human immunoglobulin (Ig)A Cα3 domain motif directs polymeric Ig receptor-mediated secretion. *J. Exp. Med.* 1999;**189**:747–52.
59. White KD, Capra JD. Targeting mucosal sites by polymeric immunoglobulin receptor-directed peptides. *J. Exp. Med.* 2002;**196**: 551–5.
60. Braathen R, Sorensen V, Brandtzaeg P, Sandlie I, Johansen FE. The carboxyl-terminal domains of IgA and IgM direct isotype-specific polymerization and interaction with the polymeric immunoglobulin receptor. *J. Biol. Chem.* 2002;**277**:42,755–62.
61. Lewis MJ, Pleass RJ, Batten MR, Atkin JD, Woof JM. Structural requirements for the interaction of human IgA with the human polymeric Ig receptor. *J. Immunol.* 2005;**175**:6694–701.
62. Kikuno K, Kang DW, Tahara K, Torii I, Kubagawa HM, Ho KJ, Baudine L, Nishizaki N, Shibuya A, Kubagawa H. Unusual biochemical features and follicular dendritic cell expression of human Fcα/μ receptor. *Eur. J. Immunol.* 2007;**37**:3540–50.
63. Ghumra A, Shi J, Mcintosh RS, Rasmussen IB, Braathen R, Johansen FE, Sandlie I, Mongini PK, Areschoug T, Lindahl G, Lewis MJ, Woof JM, Pleass RJ, 2008, Structural requirements for the interaction of human IgM and IgA with the human Fcα/μ receptor. *Eur. J. Immunol.* 2009;**39**:1147–56.
64. Burton DR. Immunoglobulin G: functional sites. *Mol. Immunol.* 1985;**22**:161–206.
65. DeLano WL, Ultsch MH, De Vos AM, Wells JA. Convergent solutions to binding at a protein–protein interface. *Science* 2000;**287**:1279–83.
66. Presta L. Molecular engineering and design of therapeutic antibodies. *Curr. Opin. Immunol.* 2008;**20**:460–70.

Chapter 50

T Cell Receptor/pMHC Complexes

Markus G. Rudolph[1], Robyn L. Stanfield[2], and Ian A. Wilson[2]

[1]*Department of Molecular and Structural Biology, Georg-August University, Goettingen, Germany*

[2]*The Scripps Research Institute, Department of Molecular Biology, and The Skaggs Institute for Chemical Biology, La Jolla, California*

TCR GENERATION AND ARCHITECTURE

T cells bearing clonotypic T cell receptors (TCRs) are generated from a pool of naïve progenitor cells by a two-stage process of positive and negative selection. The TCRs on these cells must recognize self-peptides bound to self, or syngeneic, major histocompatibility complexes (MHCs) before they can differentiate from "double positives" into $CD4^+$- or $CD8^+$- expressing "single positives". However, positively selected T cells that are reactive against self-pMHCs are destroyed by negative selection. Positive selection establishes two sub-classes of TCRs that associate with either of the two co-receptors CD8 (Figure 50.1) and CD4. $CD8^+$ T cells recognize pMHC class Ia molecules, while $CD4^+$ T cells are activated by peptides bound to MHC class II.

TCRs are heterodimeric cell-surface glycoproteins that consist of disulfide-linked α and β or γ and δ chains and have a domain organization similar to antibodies (Figure 50.2).

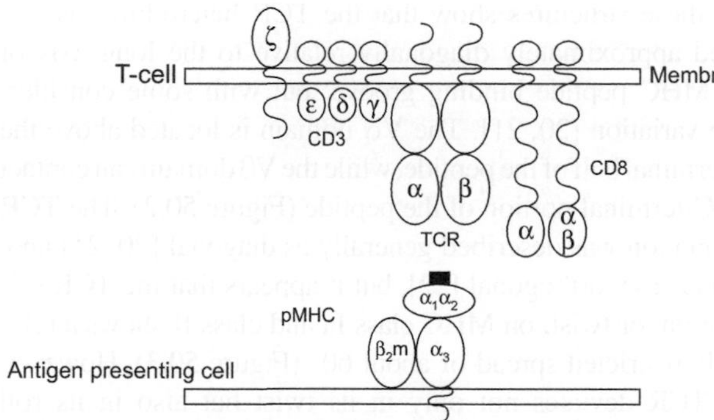

FIGURE 50.1 Schematic representation of the components in a class Ia TCR/pMHC/CD8/CD3 signaling complex.
The heavy chain consists of the α_1–α_3 domains, to which the light chain β_2-microglobulin (β_2m) is non-covalently attached. The peptide–MHC (pMHC) complex is anchored to the plasma membrane of the antigen-presenting cell *via* its α_3 domain while the $\alpha_1\alpha_2$ super-domain binds the peptide (■). The CDR loops of the αβ TCR recognize the pMHC complex, while the co-receptor CD8 binds simultaneously to the α_3 domain either as an αα homodimer or an αβ heterodimer. The signal from the pMHC complex (if any) is then transmitted through the T cell plasma membrane by the CD3 signaling modules. Phosphorylation of the CD3ζ chain by the ZAP70 kinase (not shown) is an early step in this signal transduction cascade.

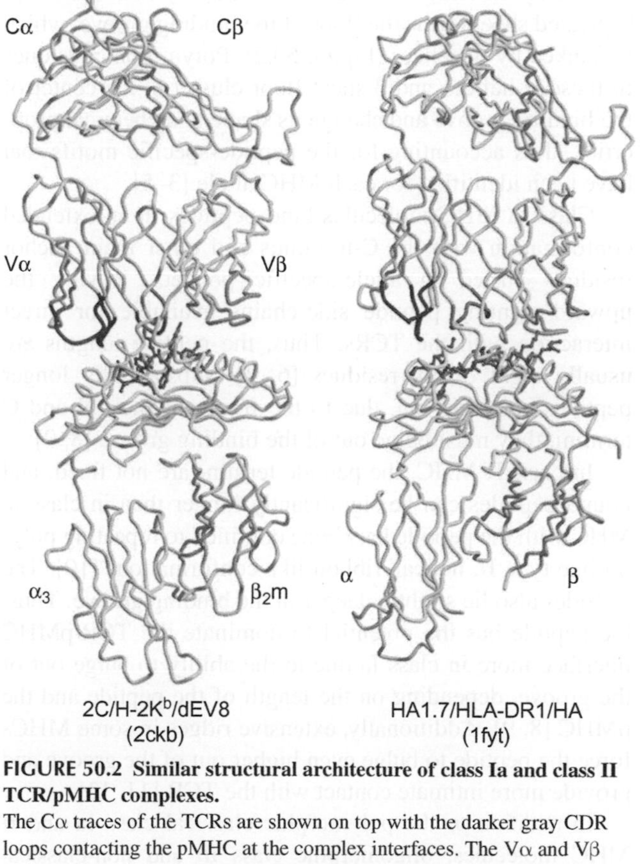

FIGURE 50.2 Similar structural architecture of class Ia and class II TCR/pMHC complexes.
The Cα traces of the TCRs are shown on top with the darker gray CDR loops contacting the pMHC at the complex interfaces. The Vα and Vβ domains are positioned over the N-terminal and C-terminal halves of the peptide, respectively. The peptides are drawn as ball-and-stick representations, and have fixed termini in class Ia MHC, but can project out of the binding groove in class II MHC.

Handbook of Cell Signaling, Three-Volume Set 2 ed.

Each chain is composed of an immunoglobulin (Ig)-like variable (V) and constant (C) domain, a transmembrane region and a short cytoplasmic tail. The C-domains serve to anchor the TCR in the membrane of the T cell and to interact with accessory signaling molecules such as CD3. The variable domains carry the complementarity determining regions (CDRs), with which the TCR binds pMHC antigen with a generally low affinity but moderate specificity. The αβ TCRs recognize peptides bound to MHC molecules, while γδ TCRs, although less well studied than their αβ counterparts, appear to recognize intact proteins or non-peptide ligands bound to either MHC or MHC-like molecules.

PEPTIDE BINDING TO MHC CLASS IA AND II

In the cellular immune response, peptides are displayed to T cells in complex with "classical" class Ia or class II MHC molecules. Both classes of MHC are heterodimers of similar structures; they are composed of three domains, two Ig-like and one α/β domain (MHC fold) that forms the peptide binding site. Whereas in class Ia MHC molecules the peptide binding site is constructed from the heavy chain only, in class II MHC it is formed by both chains. A β-pleated sheet forms the floor of the binding groove, which is flanked by α-helices (Figure 50.2). Polymorphic residues in these α-helices and β-sheet floor cluster at the center of the binding groove and change its shape and chemical properties, thus accounting for the peptide-specific motifs that have been identified for each MHC allele [3–5].

Class Ia MHC molecules bind peptides in an extended conformation with the C-terminus and other main anchor residues buried in allele-specific pockets, leaving the upward-pointing peptide side-chains available for direct interaction with the TCRs. Thus, the peptide lengths are usually eight to ten residues [6, 7]; substantially longer peptides can bind but, due to the fixing of their N and C termini, they must bulge out of the binding groove [8, 9].

In class II MHC, the peptide termini are not fixed, and bound peptides can be significantly longer than in class Ia MHC with the peptide backbone confined to repeating polyproline type II, helical, ribbon-like conformations [10]. The peptides also lie slightly deeper in the binding groove. Thus, the peptide has the potential to dominate the TCR/pMHC interface more in class Ia due to the ability to bulge out of the groove, depending on the length of the peptide and the pMHC [8, 9]. Additionally, extensive ridges in some MHCs force the peptide to bulge even higher out of the groove and provide more intimate contact with the TCR [11, 12].

In addition to the polymorphic classical class Ia and II MHC molecules, oligomorphic class Ib and non-classical MHC molecules are very structurally homologous to their class Ia brethren [13]. The MHC class Ib molecules include HLA-E, HLA-F, and HLA-G which present peptide to TCRs, and the evolutionarily older non-classical MHC molecules, including CD1 molecules that recognize non-peptidic ligands such as hydrophobic bacterial cell-wall components, including glycolipids and lipopeptides [14]. In addition to the CD1 family, other non-classical MHCs function as ligands for natural killer (NK) receptor cells (MICA, MICB, ULBP) and ligands for γδ T cells, such as T10 and T22. Most of this chapter will deal with the much better studied interactions between αβ TCRs and class Ia and II MHC, but we will briefly touch on recent structural results for the less well known γδ TCRs, and Ib and non-classical MHC.

TCR/pMHC INTERACTION

Whereas in humoral immunity antibodies identify antigenic molecules as distinct entities, in the cellular response, TCRs recognize antigenic peptide fragments only when presented by an appropriate MHC molecule. A fundamental difference between antibody/antigen and TCR/pMHC recognition is that the specificity of the former is dependent on high affinity (K_d is nanomolar) for the free antigen, whereas in the latter low affinities predominate (K_d is ~0.1–500 micromolar); thus, specificity must be ensured by a different mechanism. Possible mechanisms are outlined in the following sections.

ORIENTATION OF THE TCR IN TCR/PMHC COMPLEXES

Over 30 independent αβ TCR/pMHC complex structures have been determined to date – for reviews, see [15–19] – and these structures show that the TCR heterodimer is oriented approximately diagonally relative to the long axis of the MHC peptide binding groove, but with some considerable variation [20, 21]. The Vα domain is located above the N-terminal half of the peptide, while the Vβ domain can contact the C-terminal portion of the peptide (Figure 50.2). The TCR orientation was described generally as diagonal [20, 21] and, in one case, orthogonal [22], but it appears that the TCR orientation, or twist, on MHC class Ia and class II shows a relatively restricted spread of about 60° (Figure 50.3). However, the TCR deviates not only in its twist but also in its roll and tilt, which can be gleaned from the angle of the pseudo two-fold axis between the TCR Vα and Vβ domains and the MHC β-sheet floor (Figure 50.3). In addition, the TCRs can differ in their αβ chain pairings, such that the pseudo-Vα/Vβ twofold angle can also contribute to the variation in TCR orientation on the pMHC. As a result of the various TCR orientations, the buried surface for the TCR/pMHC complex can vary extensively between 1240 $Å^2$ and 2020 $Å^2$, with the peptide contributing a relatively restricted range of 19–40 percent to the pMHC side of that interface. Vα can contribute 33–74 percent and Vβ 26–67 percent of the TCR

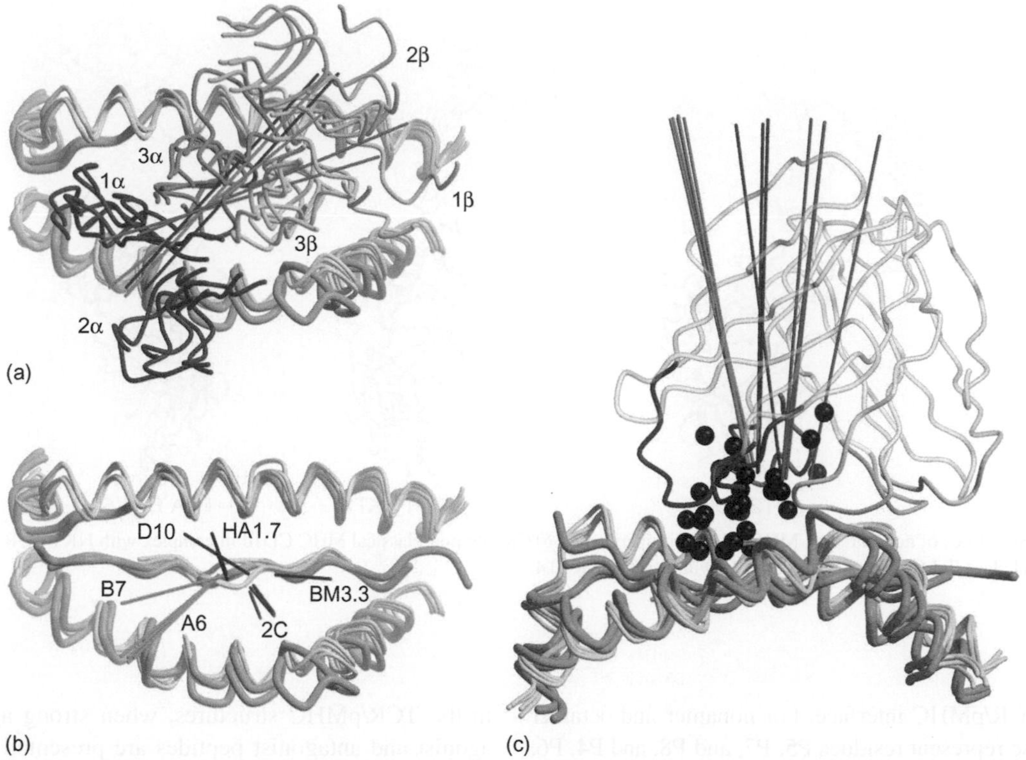

FIGURE 50.3 Relative orientation of the TCR on top of the MHC and comparison of peptide conformations in class Ia and class II TCR/pMHC complexes.
The MHC helices are shown as light and dark gray tubes for class Ia and class II, respectively. The projection of a linear least-squares fit through the centers of gravity of the CDR loops is shown for the six different TCRs. (b, c) Variation in the tilt and roll of TCR/pMHC complexes. The pseudo two-fold axes that relate the Vα and Vβ domains of the TCRs to each other are shown for twelve TCR/pMHC structures. This gives a good estimate of the inclination (roll, tilt) of the TCR on top of the MHC, which is a function of the TCR, not the pMHC ligand. One extreme case is the allogeneic BM3.3 TCR, which is shown as a transparent Cα trace.

buried surface. This bias in chain usage has been noted for antibodies, where V_H usually provides a larger contribution to the antibody–antigen interface [23]. The question of whether any common set(s) of strictly conserved interactions dictate the diagonal orientation is not yet answered, although a recent proposal for germline-derived interaction "codons" that encode pairwise interaction motifs has been put forward for particular TCR Vβ chain segments and MHC haplotypes [1]. Similarly, it has been shown that cross-reactive TCRs maintain conserved contacts to three CDR1 and CDR2 residues while maintaining a diagonal orientation [2].

While αβ TCRs usually bind to MHC–peptide complexes in roughly similar, diagonal binding modes, with the center of the bound peptide recognized by CDRs 3α and 3β, the binding modes seen for the γδ TCR G8 bound to the non-classical MHC T22 (T22 does not present a ligand), and also for the NKT TCR NKT15 bound to the non-classical MHC molecule CD1d-lipid antigen complex, are very different (Figure 50.4). For the G8–T22 complex, the TCR binds to MHC with an acute angle between the major axes of the two molecules, with the TCR β chain making the majority of contacts to T22 [24]. The binding mode of the CD1d in complex with an NKT TCR is also unusual, with the TCR binding to MHC parallel to the long axis of the binding groove, almost perpendicular to the "diagonal" mode seen for TCR binding to classical MHC, with more contacts from the α than the β chain, and a slightly smaller interface than usually found in TCR–pMHC complexes (Figure 50.4). The TCR also hangs off one end of the binding groove [25]. More structures of TCRs in complex with non-classical MHC molecules are needed to see whether these modes of recognition are repeated. The structure for the αβ TCR KK50.4 in complex with the Class Ib MHC HLA-E–peptide complex (Figure 50.4) [26] shows that the mode of binding is similar to the normal "diagonal" orientation, but with more interactions from the CDR2β loop than is normal for TCR to classical MHC.

PEPTIDE RECOGNITION BY THE TCR CDR LOOPS

The suggestion that only a few up-pointing peptide side-chains contribute to the specificity of the TCR/pMHC interaction [27] has been confirmed many times now by TCR/pMHC crystal structures. In class Ia, these interactions are dominated by the peptide residues that extend or bulge most out of the groove and, hence, represent functional hotspots

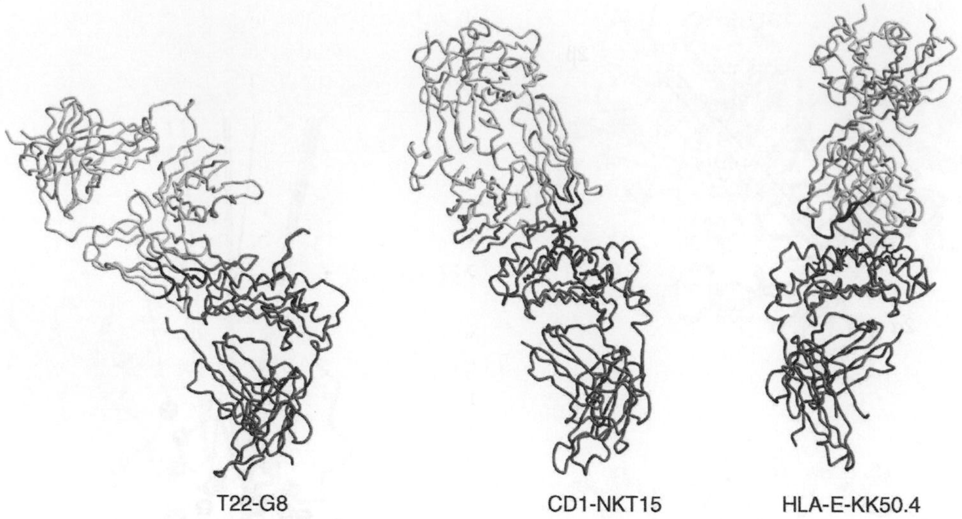

FIGURE 50.4 Structures of non-classical MHC T22 in complex with γδTCR G8; non-classical MHC CD1d in complex with NKT TCR NKT15 and lipid antigen; and class Ib MHC HLA-E in complex with αβTCR KK50.4.

[28] in the TCR/pMHC interface. For nonamer and octamer peptides, these represent residues P5, P7, and P8, and P4, P6, and P7, respectively. For class II peptides, the key side-chain contributions are more uniformly dispersed (P1, P2, P3, P5, P8). On the other hand, the contribution of the peptide backbone to TCR interaction is very modest for both class Ia and class II, where none to only a handful of contacts are made. An exception is for the HLA-A2/Tax complex, where the large P4–P5 bulge includes a glycine at P4 that enables the TCR to access the peptide backbone [21, 29].

Analysis of the number of contacts reveals that CDR1β and CDR2β often make minimal contact with the pMHC compared to CDR3β. In Vα, CDR2α tends to have fewer contacts with the pMHC than CDR3α, although an exception is found in the allogeneic BM3.3 complex where CDR3α has almost no contacts (see above). However, in most cases peptide contacts are made primarily through the central CDR3 loops, which also exhibit the greatest degree of genetic variability. In contrast, the majority of conserved MHC contacts are mediated by the CDR1 and CDR2 loops [30], particularly in Vα.

DISCREPANCY BETWEEN MAGNITUDE OF STRUCTURAL CHANGES AND BIOLOGICAL OUTCOMES

Altered Peptide Ligands: Antagonism and Superagonism

So far, no dramatic structural changes that could account for the magnitude of the different signaling outcomes of various altered peptide ligands (APLs) have been observed in the TCR/pMHC structures, when strong agonist, weak agonist and antagonist peptides are presented by the same MHC to the same TCR [28, 31]. Only slight readjustments occur in the TCR/pMHC interface to accommodate different up-pointing peptide side-chains. In the A6 system, the number of peptide–TCR contacts does not correlate with the degree of agonism and antagonism [21,31]. Similarly, in the 2C system, the buried surface does not change much when weak and strong agonists are compared, but the complementarity [28] and the number of TCR/pMHC contacts increases despite the relatively minor substitution of an arginine (strong agonist) for a lysine (weak agonist) at P4. Again, no gross conformational changes in the TCR or pMHC are observed, but slight rearrangements in the CDR loops accommodate the different peptides [28].

The correlation of complex half-life [32] with the degree of agonism or antagonism is also not clear cut. In both 2C and A6, the strong agonists (SIYR and Tax) have a longer half-life (9.2 and 7.5 s) than weak agonists (3.7 s for H-2K^b-dEV8 and 1.5 s for HLA-A2-V7R). However, by using surface plasmon resonance, agonists have been found in the A6 system that have shorter half-lives than antagonists [33]. An antagonist was converted to an agonist by stepwise filling of a cavity in the TCR/pMHC interface and the biological activity paralleled the TCR/pMHC affinity, not the half-life of the complex [33]. Half-lives of TCR/pMHC complexes on the cell surface could be extended by interaction with the co-receptors CD4 and CD8 [34]. Lateral interactions among the TCR/pMHC signaling complexes or interactions with other co-stimulatory or inhibitory receptors, as in the immunological synapse, may thus form above a certain threshold of TCR/pMHC complex half-life [35].

TCR Confomational Variation and Changes

Sufficient numbers of TCR structures are now available to assess the extent of conformational variation that arises in their antigen combining sites. As expected, the four TCR outer CDRs 1 and 2 adopt canonical conformations [36], as first described for antibodies [37, 38]. A small number of discrete canonical conformations may be able to describe most of the known sequences of the α1,2 and β1,2 loops. At present, three to four canonical structures have been defined for each of these loops [36]. What makes the TCR different from antibodies is the enormous variation seen in both of the central CDR3s (Figure 50.3). In antibodies, CDR L3 adopts a well-defined set of canonical structures, but the equivalent CDR3α loop is, in fact, the most variable in the current set of TCR structures, as well as the CDR3β loop [39]. Thus, the prediction [40] that these CDRs would be most variable and adapt to the pMHC primarily, but not exclusively [41], through contact with the peptide has been borne out. Several examples are available to assess the extent of conformational variation in the CDR loops in the presence of APL. For TCR 2C, only small variations are seen in CDR3β but, for TCR A6, these conformational rearrangements are much larger. Evidence for flexibility in the TCR has also been derived from kinetic and thermodynamic studies [42–44]. Flexible CDRs have been proposed to rearrange upon pMHC binding in a two-step mechanism to facilitate scanning of the diverse peptide complexes [45, 46]. What is consistent so far in the structural and kinetic/thermodynamic experiments is that conformational rearrangements of the CDRs can provide better complementarity of the TCR to both the MHC [47] and the peptide [28, 31] and affect T cell activation [45].

Alloreactivity

Alloreactivity is the phenomenon in which a strong immune response can be generated against foreign pMHC molecules to which one's T cells have not been previously exposed [48, 49]. Thus, an important practical corollary in defining the structural rules of T cell recognition is to explain alloreactivity [50]. So far, several complexes have addressed this issue [41, 51–53]. The complex of the BM3.3 TCR with the allogeneic MHC H-2K^b is perhaps the most structurally distinct so far, but the corresponding syngeneic complex is currently not known. The BM3.3 TCR tilts substantially towards the β-chain side (Figure 50.3), with the α chain making few direct contacts with the MHC. In fact, the long central CDR3α is flared back such that it makes no contacts with the peptide and only two with the MHC. The majority of the interactions are with the β chain, consistent with that proposed for the interaction of H-2L^d with TCR 2C, where an extreme bulge in the C-terminal half of the peptide is likely to increase its interaction with the TCR β chain [12]. Two studies [51, 52] suggest that subtle changes in allogeneic MHCs can alter the peptide conformation and location such that the same peptide is presented differently to the TCR. Thus, these structural studies conclude that TCR interaction with the bound peptide strongly affects the alloresponse.

ROLE OF BOUND WATER IN TCR/pMHC RECOGNITION

Several TCR/pMHC complexes contain bound water molecules in their TCR/pMHC interfaces. The ability of water molecules to provide additional complementarity by filling of cavities in the interface is well documented for antibodies [54]. Typical TCR/pMHC complexes contain 17 (2C/H-2K^{bm3}/dEV8 [52]), 39 (BM3.3/H-2K^b/pBM1 [41]), and 15 (HA1.7/HLA-DR4/HA [51]) waters in their interface with 6, 12, and 6, respectively, mediating contact between the TCR and pMHC. Thus, these high-resolution TCR/pMHC structures indicate a strong involvement of bound water to provide complementarity and specificity to the recognition process. Yet, no specific waters are conserved among these structures, indicating that their presence is dependent on the individual sequences of both the TCR and pMHC. In the allogeneic BM3.3 complex, about 30 interfacial waters are sequestered in a cavity between the Vα and the pMHC, as a result of the TCR Vα domain lifting up from the pMHC surface [41].

Water molecules can also improve complementarity to (and, thus, stability of) pMHC interactions. Small sequence and structure changes in either the peptide (APLs) or the MHC (as in alloreactive complexes) can be amplified on the pMHC surface by redistribution or acquisition of bound waters in the TCR/pMHC interface. A good example is the allogeneic H-2K^{bm8} complex, where water can partially substitute for loss in buried side-chain functional groups [55]. In addition, such buried MHC substitutions, which occur frequently in allogeneic MHC, can transmit their effects by altering the water structure and the electrostatic properties on the surface, even though their mutated residues are not directly "seen" by the TCR [52].

CONCLUSIONS AND FUTURE PERSPECTIVES

The evolution of a common docking mode that enables the $\alpha\beta$ TCR to survey the contents of the MHC binding groove is remarkable. However, the plethora of TCR–pMHC complex structures determined so far have not yet adequately explained the basis for this conserved orientation. No absolutely conserved pairs of interactions are apparent in these different TCR/pMHC complex interfaces that would account for their relatively fixed docking orientations, although some suggestions of key contacts have been made for some alleles and MHC molecules. The variability in the tilt, twist, and roll of the TCR indicates that the docking problem is solved slightly differently in each case to provide sufficient complementarity for binding (K_ds in the micromolar range). With the exception of the alloreactive BM3.3 TCR, where most of the interactions with pMHC are due to the β chain, the TCR Vα interactions with the pMHC seem to predominate, providing some basis for a conserved orientation. Additionally, glycosylation may play a role in facilitating docking, as both the TCR and MHC are highly glycosylated, and, hence, could sterically restrict the range of possible orientations [56, 57].

Another major unresolved issue is how the exceedingly small changes in the TCR/pMHC interface in response to different APLs can lead to such drastically different biological outcomes. Complementarity, buried surface area, or number of contacts in agonist versus antagonist complexes are very similar and are difficult to reconcile with the substantial differences in T cell responses. Therefore, differentiation of strong from weak agonists, or agonists from antagonists, by visual inspection of the crystal structures is not easy. Similarly, while the trend of increased half-life for agonist *versus* antagonist TCR/pMHC complexes is so far maintained, exceptions have been found that belie this as a general rule.

In order to extract all of the general principles that govern TCR/pMHC recognition, further TCR/pMHC complex structures are needed. Although models of the TCR/pMHC/co-receptor (CD4/CD8) complex can be assembled from the component pieces [56], which include the distal globular domains of CD8/pMHC class Ia complexes [58, 59], the low-resolution CD4/pMHC class II complex [60], the CD3εγ and CD3δγ NMR structures [61, 62], and crystal structures of human CD3δγ in complex with antibody [63] and human CD3εγ in complex with antibody [64], perhaps the most important breakthrough would be the determination of a complete αβ TCR signaling complex that includes the αβ TCR, CD3εγ, pMHC, and CD4 or CD8. This more complex assembly would lay open any global changes that may influence TCR signaling events. However, the lack of the membrane-anchoring domains in the constructs used for the current structure determinations will remain a problem until intact membrane proteins can be routinely crystallized. Future studies will also reveal how bulky ligands, such as bulged peptides [8, 9], glycopeptides [65, 66] or glycolipids in the case of CD1 [25, 67] can be accommodated in the TCR/pMHC interface.

REFERENCES

1. Feng D, Bond CJ, Ely LK, Maynard J, Garcia KC. Structural evidence for a germline-encoded T cell receptor–major histocompatibility complex interaction "codon". *Nature Immunol* 2007;**8**:975–83.
2. Dai S, Huseby ES, Rubtsova K, Scott-Browne J, Crawford F, Macdonald WA, Marrack P, Kappler JW. Crossreactive T cells spotlight the germline rules for αβ T cell receptor interactions with MHC molecules. *Immunity* 2008;**28**:324–34.
3. Falk K, Rotzschke O, Stevanovic S, Jung G, Rammensee HG. Allele-specific motifs revealed by sequencing of self-peptides eluted from MHC molecules. *Nature* 1991;**351**:290–6.
4. Rudensky AY, Mazel SM, Yurin VL. Presentation of endogenous immunoglobulin determinant to immunoglobulin-recognizing T cell clones by the thymic cells. *Eur J Immunol* 1990;**20**:2235–9.
5. van Bleek GM, Nathenson SG. The structure of the antigen-binding groove of major histocompatibility complex class I molecules determines specific selection of self-peptides. *Proc Natl Acad Sci USA* 1991;**88**:11032–6.
6. Fremont DH, Matsumura M, Stura EA, Peterson PA, Wilson IA. Crystal structures of two viral peptides in complex with murine MHC class I H-2K^b. *Science* 1992;**257**:919–27.
7. Madden DR, Garboczi DN, Wiley DC. The antigenic identity of peptide–MHC complexes: a comparison of the conformations of five viral peptides presented by HLA-A2. *Cell* 1993;**75**:693–708.
8. Speir JA, Stevens J, Joly E, Butcher GW, Wilson IA. Two different, highly exposed, bulged structures for an unusually long peptide bound to rat MHC class I RT1-A^a. *Immunity* 2001;**14**:81–92.
9. Tynan FE, Borg NA, Miles JJ, Beddoe T, El-Hassen D, Silins SL, van Zuylen WJ, Purcell AW, Kjer-Nielsen L, McCluskey J, Burrows SR, Rossjohn J. High resolution structures of highly bulged viral epitopes bound to major histocompatibility complex class I. Implications for T cell receptor engagement and T cell immunodominance. *J Biol Chem* 2005;**280**:23900–9.
10. Stern LJ, Wiley DC. Antigenic peptide binding by class I and class II histocompatibility proteins. *Structure* 1994;**2**:245–51.
11. Young AC, Zhang W, Sacchettini JC, Nathenson SG. The three-dimensional structure of H-2D^b at 2.4-Å resolution: implications for antigen-determinant selection. *Cell* 1994;**76**:39–50.
12. Speir JA, Garcia KC, Brunmark A, Degano M, Peterson PA, Teyton L, Wilson IA. Structural basis of 2C TCR allorecognition of H-2L^d peptide complexes. *Immunity* 1998;**8**:553–62.
13. Rodgers JR, Cook RG. MHC class Ib molecules bridge innate and acquired immunity. *Nature Rev Immunol* 2005;**5**:459–71.
14. Van Rhijn I, Zajonc DM, Wilson IA, Moody DB. T cell activation by lipopeptide antigens. *Curr Opin Immunol* 2005;**17**:222–9.
15. Rudolph MG, Stanfield RL, Wilson IA. How TCRs bind MHCs, peptides, and co-receptors. *Annu Rev Immunol* 2006;**24**:419–66.
16. Sundberg EJ, Deng L, Mariuzza RA. TCR recognition of peptide/MHC class II complexes and superantigens. *Semin Immunol* 2007;**19**:262–71.
17. Clements CS, Dunstone MA, Macdonald WA, McCluskey J, Rossjohn J. Specificity on a knife-edge: the alphabeta T cell receptor. *Curr Opin Struct Biol* 2006;**16**:787–95.
18. Deng L, Mariuzza RA. Recognition of self-peptide–MHC complexes by autoimmune T cell receptors. *Trends Biochem Sci* 2007;**32**:500–8.
19. Godfrey DI, Rossjohn J, McCluskey J. The fidelity, occasional promiscuity, and versatility of T cell receptor recognition. *Immunity* 2008;**28**:304–14.
20. Garcia KC, Degano M, Stanfield RL, Brunmark A, Jackson MR, Peterson PA, Teyton L, Wilson IA. An αβ T cell receptor structure at 2.5 Å and its orientation in the TCR–MHC complex. *Science* 1996;**274**:209–19.
21. Garboczi DN, Ghosh P, Utz U, Fan QR, Biddison WE, Wiley DC. Structure of the complex between human T cell receptor, viral peptide and HLA-A2. *Nature* 1996;**384**:134–41.
22. Reinherz EL, Tan K, Tang L, Kern P, Liu J, Xiong Y, Hussey RE, Smolyar A, Hare B, Zhang R, Joachimiak A, Chang HC, Wagner G,

Wang J. The crystal structure of a T cell receptor in complex with peptide and MHC class II. *Science* 1999;**286**:1913–21.
23. Wilson IA, Stanfield RL. Antibody–antigen interactions: new structures and new conformational changes. *Curr Opin Struct Biol* 1994;**4**:857–67.
24. Adams EJ, Chien YH, Garcia KC. Structure of a γδ T cell receptor in complex with the nonclassical MHC T22. *Science* 2005;**308**:227–31.
25. Borg NA, Wun KS, Kjer-Nielsen L, Wilce MC, Pellicci DG, Koh R, Besra GS, Bharadwaj M, Godfrey DI, McCluskey J, Rossjohn J. CD1d-lipid-antigen recognition by the semi-invariant NKT T cell receptor. *Nature* 2007;**448**:44–9.
26. Hoare HL, Sullivan LC, Pietra G, Clements CS, Lee EJ, Ely LK, Beddoe T, Falco M, Kjer-Nielsen L, Reid HH, McCluskey J, Moretta L, Rossjohn J, Brooks AG. Structural basis for a major histocompatibility complex class Ib-restricted T cell response. *Nature Immunol* 2006;**7**:256–64.
27. Shibata K, Imarai M, van Bleek GM, Joyce S, Nathenson SG. Vesicular stomatitis virus antigenic octapeptide N52-59 is anchored into the groove of the H-2K^b molecule by the side-chains of three amino acids and the main-chain atoms of the amino terminus. *Proc Natl Acad Sci USA* 1992;**89**:3135–59.
28. Degano M, Garcia KC, Apostolopoulos V, Rudolph MG, Teyton L, Wilson IA. A functional hotspot for antigen recognition in a superagonist TCR/MHC complex. *Immunity* 2000;**12**:251–61.
29. Ding YH, Smith KJ, Garboczi DN, Utz U, Biddison WE, Wiley DC. Two human T cell receptors bind in a similar diagonal mode to the HLA-A2/Tax peptide complex using different TCR amino acids. *Immunity* 1998;**8**:403–11.
30. Garcia KC, Teyton L, Wilson IA. Structural basis of T cell recognition. *Annu Rev Immunol* 1999;**17**:369–97.
31. Ding YH, Baker BM, Garboczi DN, Biddison WE, Wiley DC. Four A6-TCR/peptide/HLA-A2 structures that generate very different T cell signals are nearly identical. *Immunity* 1999;**11**:45–56.
32. Matsui K, Boniface JJ, Reay PA, Schild H, Fazekas de St Groth B, Davis MM. Low affinity interaction of peptide–MHC complexes with T cell receptors. *Science* 1991;**254**:1788–91.
33. Baker BM, Gagnon SJ, Biddison WE, Wiley DC. Conversion of a T cell antagonist into an agonist by repairing a defect in the TCR/peptide/MHC interface: implications for TCR signaling. *Immunity* 2000;**13**:475–84.
34. Garcia KC, Scott CA, Brunmark A, Carbone FR, Peterson PA, Wilson IA, Teyton L. CD8 enhances formation of stable T cell receptor/MHC class I molecule complexes. *Nature* 1996;**384**:577–81.
35. Krummel M, Wulfing C, Sumen C, Davis MM. Thirty-six views of T cell recognition. *Philos Trans R Soc Lond B Biol Sci* 2000;**355**:1071–6.
36. Al-Lazikani B, Lesk AM, Chothia C. Canonical structures for the hypervariable regions of T cell αβ receptors. *J Mol Biol* 2000;**295**:979–95.
37. Chothia C, Lesk AM. Canonical structures for the hypervariable regions of immunoglobulins. *J Mol Biol* 1987;**196**:901–17.
38. Chothia C, Lesk AM, Tramontano A, Levitt M, Smith-Gill SJ, Air G, Sheriff S, Padlan EA, Davies D, Tulip WR. Conformations of immunoglobulin hypervariable regions. *Nature* 1989;**342**:877–83.
39. Reiser JB, Gregoire C, Darnault C, Mosser T, Guimezanes A, Schmitt-Verhulst AM, Fontecilla-Camps JC, Mazza G, Malissen B, Housset D. A T cell receptor CDR3β loop undergoes conformational changes of unprecedented magnitude upon binding to a peptide/MHC class I complex. *Immunity* 2002;**16**:345–54.
40. Bjorkman PJ, Davis MM. Model for the interaction of T cell receptors with peptide/MHC complexes. *Cold Spring Harb Symp Quant Biol* 1989;**54**:365–73.
41. Reiser JB, Darnault C, Guimezanes A, Gregoire C, Mosser T, Schmitt-Verhulst A-M, Fontecilla-Camps JC, Malissen B, Housset D, Mazza G. Crystal structure of a T cell receptor bound to an allogeneic MHC molecule. *Nature Immunol* 2000;**1**:291–7.
42. Davis M, Boniface J, Reich Z, Lyons D, Hampl J, Arden B, Chien Y. Ligand recognition by αβ T cell receptors. *Annu Rev Immunol* 1998;**16**:523–44.
43. Willcox BE, Gao GF, Wyer JR, Ladbury JE, Bell JI, Jakobsen BK, van der Merwe PA. TCR binding to peptide–MHC stabilizes a flexible recognition interface. *Immunity* 1999;**10**:357–65.
44. Boniface JJ, Reich Z, Lyons DS, Davis MM. Thermodynamics of T cell receptor binding to peptide–MHC: evidence for a general mechanism of molecular scanning. *Proc Natl Acad Sci USA* 1999;**96**:11446–51.
45. Krogsgaard M, Davis MM. How T cells "see" antigen. *Nature Immunol* 2005;**6**:239–45.
46. Wu LC, Tuot DS, Lyons DS, Garcia KC, Davis MM. Two-step binding mechanism for T cell receptor recognition of peptide MHC. *Nature* 2002;**418**:552–6.
47. Garcia KC, Degano M, Pease LR, Huang M, Peterson PA, Teyton L, Wilson IA. Structural basis of plasticity in T cell receptor recognition of a self peptide–MHC antigen. *Science* 1998;**279**:1166–72.
48. Lindahl KF, Wilson DB. Histocompatibility antigen-activated cytotoxic T lymphocytes, II. Estimates of the frequency and specificity of precursors. *J Exp Med* 1977;**145**:508–22.
49. Widmer MB, MacDonald HR. Cytolytic T lymphocyte precursors reactive against mutant K^b alloantigens are as frequent as those reactive against a whole foreign haplotype. *J Immunol* 1980;**124**:48–51.
50. Sherman LA, Chattopadhyay S. The molecular basis of allorecognition. *Annu Rev Immunol* 1993;**11**:385–402.
51. Hennecke J, Wiley DC. Structure of a complex of the human alpha/beta T cell receptor (TCR) HA1.7, influenza hemagglutinin peptide, and major histocompatibility complex class II molecule, HLA-DR4 (DRA*0101 and DRB1*0401): insight into TCR cross-restriction and alloreactivity. *J Exp Med* 2002;**195**:571–81.
52. Luz JG, Huang M, Garcia KC, Rudolph MG, Apostolopoulos V, Teyton L, Wilson IA. Structural comparison of allogeneic and syngeneic T cell receptor–peptide–major histocompatibility complex complexes: a buried alloreactive mutation subtly alters peptide presentation substantially increasing V_β interactions. *J Exp Med* 2002;**195**:1175–86.
53. Colf LA, Bankovich AJ, Hanick NA, Bowerman NA, Jones LL, Kranz DM, Garcia KC. How a single T cell receptor recognizes both self and foreign MHC. *Cell* 2007;**129**:135–46.
54. Bhat TN, Bentley GA, Boulot G, Greene MI, Tello D, Dall'Acqua W, Souchon H, Schwarz FP, Mariuzza RA, Poljak RJ. Bound water molecules and conformational stabilization help mediate an antigen–antibody association. *Proc Natl Acad Sci USA* 1994;**91**:1089–93.
55. Rudolph MG, Speir JA, Brunmark A, Mattsson N, Jackson MR, Peterson PA, Teyton L, Wilson IA. The crystal structures of K^{bm1} and K^{bm8} reveal that subtle changes in the peptide environment impact thermostability and alloreactivity. *Immunity* 2001;**14**:231–42.
56. Rudd PM, Wormald MR, Stanfield R, Huang M, Mattsson N, Speir JA, DiGennaro JA, Fetrow JS, Dwek RA, Wilson IA. Roles for glycosylation in the cellular immune system. *J Mol Biol* 1999;**293**:351–66.
57. Rudd PM, Elliott T, Cresswell P, Wilson IA, Dwek RA. Glycosylation and the immune system. *Science* 2001;**291**:2370–6.

58. Gao GF, Tormo J, Gerth UC, Wyer JR, McMichael AJ, Stuart DI, Bell JI, Jones EY, Jakobsen BK. Crystal structure of the complex between human CD8αα and HLA-A2. *Nature* 1997;**387**:630–4.
59. Kern PS, Teng MK, Smolyar A, Liu JH, Liu J, Hussey RE, Spoerl R, Chang HC, Reinherz EL, Wang JH. Structural basis of CD8 co-receptor function revealed by crystallographic analysis of a murine CD8αα ectodomain fragment in complex with H-2K^b. *Immunity* 1998;**9**:519–30.
60. Wang J, Meijers R, Xiong Y, Liu J, Sakihama T, Zhang R, Joachimiak A, Reinherz EL. Crystal structure of the human CD4 N-terminal two domain fragment complexed to a class II MHC molecule. *Proc Natl Acad Sci USA* 2001;**98**:10799–804.
61. Sun ZJ, Kim KS, Wagner G, Reinherz EL. Mechanisms contributing to T cell receptor signaling and assembly revealed by the solution structure of an ectodomain fragment of the CD3εγ heterodimer. *Cell* 2001;**105**:913–23.
62. Sun ZY, Kim ST, Kim IC, Fahmy A, Reinherz EL, Wagner G. Solution structure of the CD3epsilondelta ectodomain and comparison with CD3epsilongamma as a basis for modeling T cell receptor topology and signaling. *Proc Natl Acad Sci USA* 2004;**101**:16867–72.
63. Arnett KL, Harrison SC, Wiley DC. Crystal structure of a human CD3-ε/δ dimer in complex with a UCHT1 single-chain antibody fragment. *Proc Natl Acad Sci USA* 2004;**101**:16268–73.
64. Kjer-Nielsen L, Dunstone MA, Kostenko L, Ely LK, Beddoe T, Mifsud NA, Purcell AW, Brooks AG, McCluskey J, Rossjohn J. Crystal structure of the human T cell receptor CD3εγ heterodimer complexed to the therapeutic mAb OKT3. *Proc Natl Acad Sci USA* 2004;**101**:7675–80.
65. Speir JA, Abdel-Motal UM, Jondal M, Wilson IA. Crystal structure of an MHC class I-presented glycopeptide that generates carbohydrate-specific CTL. *Immunity* 1999;**10**:51–61.
66. Glithero A, Tormo J, Haurum JS, Arsequell G, Valencia G, Edwards J, Springer S, Townsend A, Pao YL, Wormald M, Dwek RA, Jones EY, Elliott T. Crystal structures of two H-2D^b/glycopeptide complexes suggest a molecular basis for CTL cross-reactivity. *Immunity* 1999;**10**:63–74.
67. Moody DB, Besra GS, Wilson IA, Porcelli SA. The molecular basis of CD1-mediated presentation of lipid antigens. *Immunol Rev* 1999;**172**:285–96.

Chapter 51

NK Receptors

Roland K. Strong

Division of Basic Sciences, Fred Hutchinson Cancer Research Center, Seattle, Washington

IMMUNORECEPTORS

Recognition events between the archetypical αβ receptors on T cells (TCRs) and processed peptide fragments of endogenous proteins, presented on target cell surfaces as complexes with major histocompatibility complex (MHC) class I proteins, ultimately mediate activation of T cell cytotoxic responses by the cellular arm of the adaptive immune system [1]. MHC class I proteins are integral-membrane, heterodimeric proteins with ectodomains consisting of a polymorphic heavy chain, comprising three extracellular domains (α1, α2 and α3), associated with a non-polymorphic light chain, β_2-microglobulin (β_2-m) [2]. The α1 and α2 domains together comprise the peptide- and TCR binding "platform" domain; the α3 and β_2-m domains have C-type immunoglobulin (Ig) folds. Crystal structures of TCR/MHC complexes show that the TCR variable domains sit diagonally on the MHC platform domain, making contacts to the peptide and the MHC α1 and α2 domains (Figure 51.1) [3, 4]. Binding studies show that the equilibrium dissociation constants for these interactions range from one to tens of micromolar; the strength of these interactions, including consideration of kinetic and thermodynamic components, are directly correlated with output signal strength [5]. TCR/MHC binding is also highly degenerate, with any TCR capable of recognizing a range of peptides, often in complexes with different MHCs, through "induced-fit" interactions [6, 7].

NATURAL KILLER CELLS

Surveillance against cells undergoing tumorigenesis [8–13] or infection by viruses [14, 15] or internal pathogens [16, 17] is provided by natural killer (NK) cells, components of the innate immune system, thus helping to provide "covering fire" during the period that responses by the adaptive immune system are gearing up [18, 19]. NK cells are defined as $CD56^+/CD16^+$ cells comprising 10–20 percent of PBMC. NK cells also act to regulate innate and acquired immune responses through the release of various immune modulators, chemokines and cytokines, such as tumor necrosis factor α, interferon γ, MIP-1 and RANTES. Unlike T cells, which clonally express unique TCRs, NK cells function through a diverse array of cell-surface inhibitory and activating receptors with varying specificities.

Many NK cell surface receptors (NKRs) are specific for classical (such as HLA-A, -B and -C in humans) and non-classical (such as HLA-E in humans) MHC class I proteins, and occur in paired activating and inhibitory isoforms [20–22]. Different NKRs, with different MHC class I specificities, are expressed on overlapping, but distinct, subsets of NK cells in variegated patterns–where the strength of the inhibitory signals may be stronger than stimulatory signals. Thus, NK cell effector functions are regulated by integrating signals across the array of stimulatory and inhibitory NKRs engaged upon interaction with target cell surface NKR ligands ("KIR-mismatch") [21–23], resulting in the elimination of cells with reduced or altered MHC class I expression ("missing self"), a common consequence of infection or transformation [24, 25]. The developmental mechanisms that govern NKR expression patterns are still not fully understood, but NK cells that express an inhibitory receptor specific for self MHC class I proteins become "licensed", or functionally competent, while those lacking such a receptor are rendered functionally inert [26]. Other NKRs, such as human and murine NKG2D, recognize divergent MHC class I homologs (ULBPs [27], MICA and MICB in humans [28], and RAE-1 and H60 in mice [29, 30]) not involved in conventional peptide antigen presentation. Inhibitory receptors transduce signals through recruitment of tyrosine phosphatases, such as SHP-1 and SHP-2, and

Handbook of Cell Signaling, Three-Volume Set 2 ed.

contain immunoreceptor tyrosine-based inhibitory motifs (ITIMs) in their cytoplasmic domains [31, 32]. Activating receptors associate with immunoreceptor tyrosine-based activation motif (ITAM) -bearing adaptor proteins, either DAP12 [33] or DAP10 [34, 35], through a basic residue in their transmembrane domain. Spontaneous NK effector functions can be activated through triggering receptors, including NKG2D, DNAM-1, and natural cytotoxicity receptors (NCRs: NKp30, NKp44, NKp46); alternatively, NK-mediated antibody-dependent cellular cytotoxicity (ADCC) can be directed against opsonized target cells through antibody Fc/CD16 interactions [18, 36–38]. NKR/ligand affinities span considerable ranges, from hundreds of micromolar to tens of nanomolar (Table 51.1), both within and between NKR families, suggesting signaling mechanisms that respond differentially, comparable to TCRs, likely impacting both activation and developmental pathways. NKRs also display widely varying degrees of specificity, from many KIRs, where binding is determined by the identity of a single residue, to NKG2D, which binds a range of highly polymorphic, structurally divergent ligands.

NKRs can be divided into two broad groups based on structural homologies [39, 40], with some families differentially represented across species. The first group includes

TABLE 51.1 Cell-surface NK receptor/ligand interactions

Type	Receptor	Polymorphism	Signal (+/−)	Ligand	K_D	Ref.	Structures available?
KIRs							
	KIR2DL1-5	High	−	1: HLA-C^{K80}	15−7 μM	[69]	Yes
	(CD158a–f)			2–3: HLA-C^{N80}			
				4: HLA-G			
				5: ?			
	KIR2DS1-5	High	+	1: HLA-C^{K80}	>50−23 μM	[69]	Yes
	(CD158g–j)			2: HLA-C^{N80}			
				3–5: ?			
	KIR3DL1-3	High	−	1: HLA-Bw4	?		No
	(CD158e1, k, z)			2: HLA-A3, -A11			
	KIR3DS1	High	+	?	?		No
	(CD158e2)						
	LILRs	High/low	+/−	MHC class Ia & Ib	100−15 μM	[70]	Yes
	(LIRs, ILTs, CD85)			CMV UL18	2 nM/14 μM	[53]	
KLRs							
	NKG2A/B-CD94	Low	−	Human:	20−0.7 μM	[67, 71]	Yes*
				HLA-E	(varies with peptide)		
				Murine:			
				Qa-1			
	NKG2C/E/H-CD94	Low	+	Human:	120−0.7 μM	[67, 72]	No
				HLA-E	(varies with peptide)		

(Continued)

TABLE 51.1 (Continued)

Type	Receptor	Polymorphism	Signal (+/−)	Ligand	K_D	Ref.	Structures available?
				Murine:			
				Qa-1			
	NKG2D	Low	+	Human:			Yes
				MIC-A/B	0.9−0.3 μM	[63]	
				ULBPs	4.0−1.1 μM	[73]	
				Murine:	1.9−0.35 μM	[74]	
				RAE-1α–δ	0.028 μM	[75]	
				RAE-1ε	0.02 μM	[75]	
				H60		[73, 74]	
	Ly49A-W	Low	+/−	MHC class Ia, ?	100−10 μM	[58, 76]	Yes
				(CMV m157)	0.2 μM	[60]	
NCRs							
	NKp30	Low	+	?			No
	NKp44	Low	+	?			Yes
				(Viral HA?)			
	NKp46	Low	+	?			Yes
				(Viral HA?)			
	NKp80	Low	+	?			No
	CD2	Low	+	LFA-2			Yes
	CD16	Low	+	IgG			Yes
	CD59	Low	+	?			Yes
	2B4	Low	+	CD48			Yes
	(CD244)						
	DNAM-1	Low	+	Nectin-2 *(CD112)*			No
	(CD226)			PVR *(CD155)*			
	LFA-1	Low	+	ICAM			Yes
	TLRs	Low	+	PAMPs			Yes
				(dsRNA, LPS,			
				flagellin, etc.)			

Alternate nomenclature is italicized within parentheses; *: Kaiser et al., personal communication.

the killer cell Ig receptors (KIR, restricted to NK cells) and the leukocyte Ig-like receptors (LILR, found on many cell types), and consists of type I transmembrane glycoproteins with ectodomains containing tandem Ig domains. The second group, including the rodent Ly49 receptor family and the CD94/NKG2x and NKG2D receptor families found in primates and rodents, comprises homo- and heterodimeric type II transmembrane glycoproteins containing C-type lectin-like NK receptor domains (CTLDs) [41]. Ongoing X-ray crystallographic analyses continue to detail NKR/ligand interactions (Figure 51.1).

IG-TYPE NK RECEPTORS: KIR

Two crystal structures of complexes between inhibitory KIR family NKRs and their MHC class I ligands, KIR2DL2/HLA-Cw3 [42] and KIR2DL1/HLA-Cw4 [43], show that the receptor binds in a 1:1 complex with HLA-C, making contacts to both the $\alpha1$ and $\alpha2$ platform domains and the carboxy-terminal end of the bound peptide (Figures 51.1, 51.2). [KIR receptor nomenclature identifies the number of Ig domains (2D(omains) or 3D, specific for HLA-C or HLA-A/B respectively), and whether the receptor is a long (L) form, containing ITIM repeats, or a short (S) form, interacting with ITAM-containing adaptor proteins.] Both complexes have interfaces showing both significant shape and charge complementarity, with the N-terminal KIR domains interacting primarily with the $\alpha1$ domains of HLA-C, the C-terminal KIR domains contacting the $\alpha2$ domains, and with additional contacts provided by the interdomain KIR linker peptides (the "elbow"). The kinetics of binding, rapid on and off rates, are consistent with interactions dominated by charge–charge interactions.

Despite a high degree of conservation of binding surface residues between both KIR2DL2 and KIR2DL1, and HLA-Cw3 and -Cw4, few actual intermolecular interactions are conserved. This recognition flexibility is accomplished through altered side-chain conformations. KIR2D receptors distinguish between HLA-C allotypes on the basis of the residue at position 80: KIR2DL1 recognizes lysine and KIR2DL2 recognizes asparagine, and this

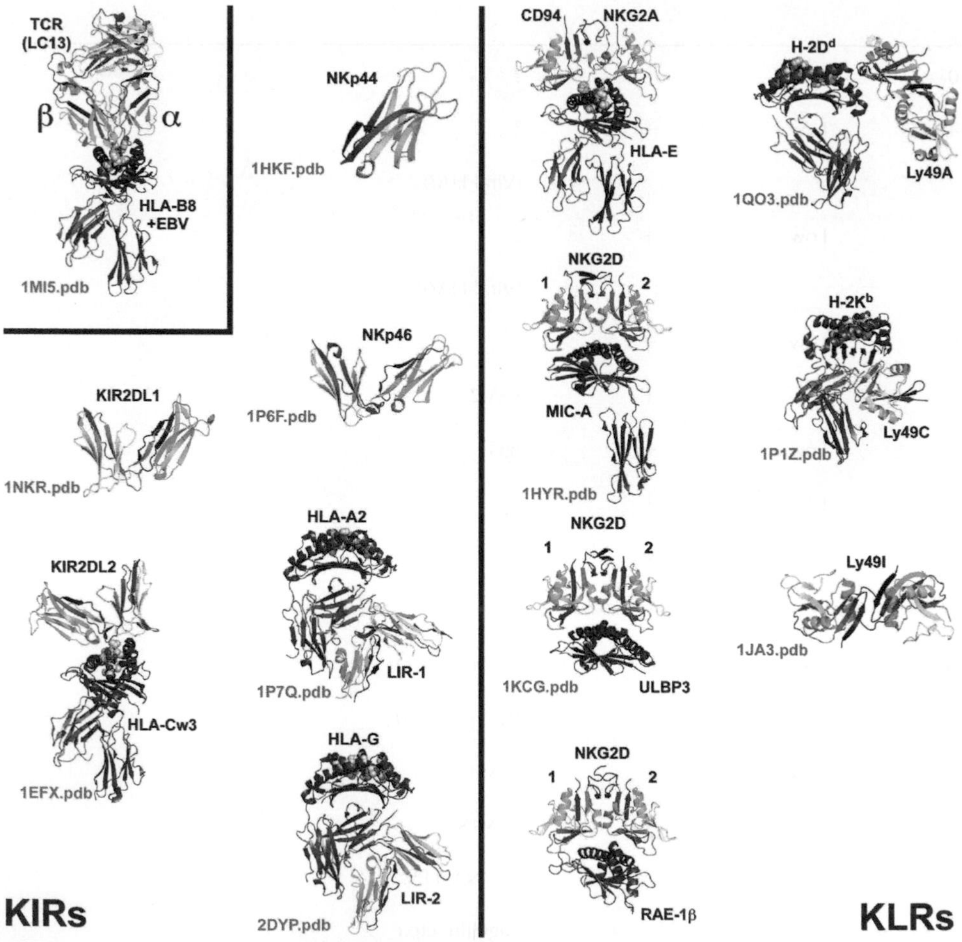

FIGURE 51.1 Ribbon representations of an $\alpha\beta$ TCR/MHC class I complex (upper left) and several NKR and NKR/ligand complex structures are shown. PDB accession codes for the coordinate files used to generate the figure are indicated. Only one half of the 1:2 Ly49C:MHC complex is shown, for simplicity.

specificity is conferred by the identity of the residue at position 44 in the receptor. In KIR2DL1, Lys80 is shape- and charge-matched to a distinct pocket on the surface of the receptor; while Asn80 is sensed through a direct hydrogen bond in the KIR2DL2 complex. Additional structures for isolated KIRs are also available: KIR2DL1 [44], KIR2DL2 [45], KIR2DL3 [46] and KIR2DS2 [47].

OTHER IG-TYPE RECEPTORS ON NK CELLS

Several structures are available for isolated NCRs (NKp44 [48] and NKp46 [49, 50]), but little is currently known about their ligands or the details of NCR/ligand interactions. While LILRs contribute significantly to NK function and are subverted through viral decoys like the MHC class I homolog CMV UL18 [51], they are expressed on many cell types and are, therefore, not a focus of this review. Structures are available for isolated LILRs (LILRA5 [52], LILRB1 [53]) and for the LILRB1/HLA-A2 [54] and LILRB2/HLA-G [55] complexes, which show that the receptor Ig-like ectodomains interact with MHC class Ia and Ib ligands in a similar, peptide-independent manner (Figure 51.1), contacting mostly β_2-m and, to a lesser extent, the MHC class I $\alpha3$ domain. While neighboring to an extent that would result in competition, the Ly49C- and LILR binding sites on MHC class I proteins are distinct.

C-TYPE LECTIN-LIKE NK RECEPTORS: LY49A

Ly49A is a disulfide-linked, symmetric, homodimeric, CTLD-type NKR that is specific for the murine MHC class I protein H-2D^d (the human ortholog is non-functional) [56]. The crystal structure of the D^d/Ly49A complex [57] shows D^d homodimers binding to two distinct sites on the MHC protein (Figure 51.2). The first binding site positions Ly49A on the D^d platform domain, contacting both $\alpha1$ and $\alpha2$ and the N-terminal end of the bound peptide–the opposite end from where KIR2D binds. The second binding site positions Ly49A in the cleft between the underside of the platform domain (the top being the peptide and TCR binding surface), the $\alpha3$ domain and β_2-m. The second site is considerably more extensive than site #1, though less shape-complementary and less dominated by charge–charge interactions, and is likely to be the immunologically relevant interaction on the basis of subsequent mutagenesis studies. Site #2 also overlaps the CD8 binding site on MHC class I proteins (Figure 51.2). As predicted, Ly49A clearly displays a C-type lectin-like fold, though failing to retain any remnant of the divalent cation or carbohydrate binding

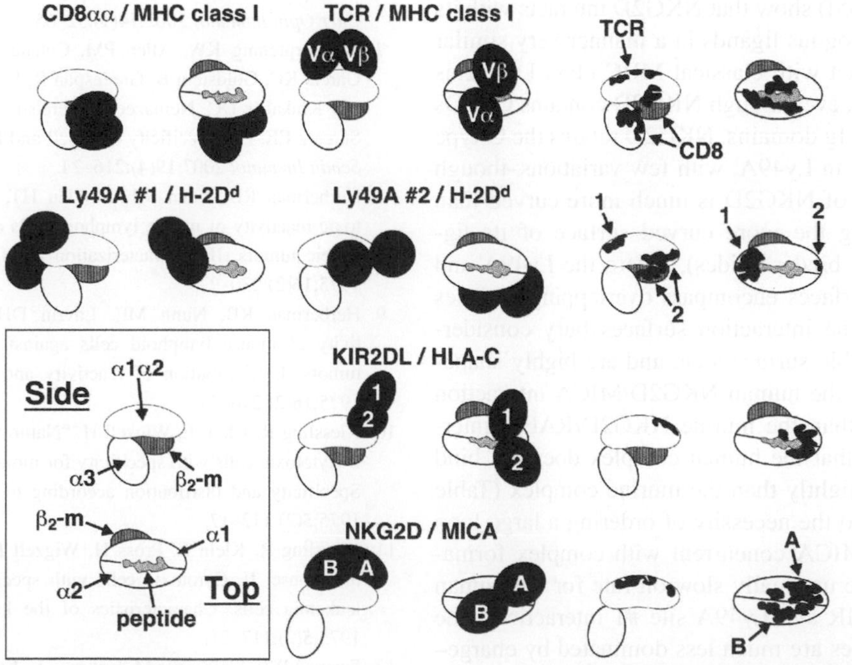

FIGURE 51.2 Structurally-characterized NK receptor/ligand complexes are shown in schematic representations to highlight interaction surfaces. Each row shows two views of a receptor–ligand complex, first showing the organization of domains in the complex (receptor domains in black, labeled where a distinction between domains is significant; MHC class I ligand heavy chains in white and β_2-m in vertical stripes. The arrangement of domains in the ligands is detailed in the inset; the approximate solvent-accessible surface area of the bound peptide, if present, is shown as a cross-hatched area. The right-most columns show approximate footprints of receptors and co-receptors on the ligands as black patches, labeled by receptor component, subsite or domain as appropriate.

sites conserved in true C-type lectins. While the simplest binding mode for a symmetric homodimer is to interact with two monomeric ligands through two identical binding sites, each Ly49A interaction with D^d is with a single monomer because binding of ligand at one site sterically blocks binding at the second, homodimer-related site. Interestingly, the interactions of Ly49C with MHC class I proteins [58] are quite distinct from Ly49A (Figure 51.1). Ly49C makes symmetric interactions with two MHC proteins across the receptor dyad axis of symmetry, not directly contacting the peptide. A crystal structure of isolated Ly49I is also available [59], revealing a distinct dimerization interface from other Ly49 structures. m157 is an MHC class I-like, CMV-encoded decoy ligand that interacts with both Ly49H and I, with much tighter affinities than their true MHC class I ligands (Table 51.1); the structure of m157 shows a compact, minimal MHC molecule which dispenses with peptide and β_2-m association [60].

C-TYPE LECTIN-LIKE NK RECEPTORS: NKG2D

NKG2D is an activating, symmetric, homodimeric, CTLD-type NKR. While highly conserved between primates and rodents, its ligands include very different molecules, both in humans and inrodents. Multiple crystal structures of the receptor alone [61,62] and three complexes (human NKG2D/MICA [63], NKG2D/ULBP3 [64] and murine NKG2D/RAE-1β [65]) show that NKG2D interacts with its MHC class I homologous ligands in a manner very similar to how TCRs interact with classical MHC class I proteins (Figures 51.1, 51.2), even though NKG2D contains CTLDs while TCRs contain Ig domains. NKG2D retains the C-type lectin-like fold seen in Ly49A, with few variations–though the binding surface of NKG2D is much more curved than in Ly49A, matching the more curved surface of its ligands (which do not bind peptides), where the Ly49A and NKG2D binding surfaces encompass overlapping surfaces on the receptors. The interaction surfaces bury considerable solvent-accessible surface area, and are highly shape-complementary, but the human NKG2D/MICA interaction markedly more so than the murine NKG2D/RAE-1 interaction. The reason that the human complex does not bind considerably more tightly than the murine complex (Table 51.1) is likely due to the necessity of ordering a large loop on the surface of MICA concurrent with complex formation, reflected in the unusually slow on-rate for the human complex. Unlike KIR and Ly49A site #1 interactions, the NKG2D binding sites are much less dominated by charge–charge interactions. The stoichiometries of the NKG2D complexes are one homodimer binding to one monomeric ligand. However, unlike Ly49, both homodimer-related binding sites on NKG2D contribute approximately equally to the interactions in both complexes, reflecting a binding site that has evolved to bind multiple target sites without the degree of side-change rearrangements seen in the KIR interactions. The considerable recognition degeneracy of NKG2D, accommodating structurally divergent, polymorphic families of ligands (Table 51.1), is enabled not by a conformationally-plastic binding site (induced-fit recognition), but rather by a "rigid-adaptation" mechanism [7, 66]. The CD94/NKG2A/HLA-E complex structure [68] is quite similar in overall arrangement to NKG2D/ligand and TCR/ligand complexes (Figure 51.1), though peptide sequence differences, which strongly affect receptor affinities, are read out by CD94 and not the NKG2x moiety [67].

REFERENCES

1. Germain RN, Margulies DH. The biochemistry and cell biology of antigen processing and presentation. *Annu Rev Immun* 1993; **11**:403–50.
2. Bjorkman PJ, Parham P. Structure, function and diversity of class I major histocompatibility complex molecules. *Ann Rev Biochem* 1990;**90**:253–88.
3. Garcia KC, Degano M, Speir JA, Wilson IA. Emerging principles for T cell receptor recognition of antigen in cellular immunity. *Rev Immunogen* 1999;1:75–90.
4. Rudolph MG, Stanfield RL, Wilson IA. How TCRs bind MHCs, peptides, and co-receptors. *Annu Rev Immunol* 2006;**24**:419–66.
5. Rudolph MG, Luz JG, Wilson IA. Structural and thermodynamic correlates of T cell signaling. *Annu Rev Biophys Biomol Struct* 2002;**31**:121–49.
6. Rudolph MG, Wilson IA. The specificity of TCR/pMHC interaction. *Curr Opin Immunol* 2002;**14**(1):52–65.
7. Wucherpfennig KW, Allen PM, Celada F, Cohen IR, De Boer R, Garcia KC, Goldstein B, Greenspan R, Hafler D, Hodgkin P, Huseby ES, Krakauer DC, Nemazee D, Perelson AS, Pinilla C, Strong RK, Sercarz EE. Polyspecificity of T cell and B cell receptor recognition. *Semin Immunol* 2007;**19**(4):216–24.
8. Herberman RB, Nunn ME, Holden HT, Lavrin DH. Natural cytotoxic reactivity of mouse lymphoid cells against syngeneic and allogeneic tumors. II. Characterization of effector cells. *Intl J Cancer* 1975;**16**(2):230–9.
9. Herberman RB, Nunn ME, Lavrin DH. Natural cytotoxic reactivity of mouse lymphoid cells against syngeneic acid allogeneic tumors. I. Distribution of reactivity and specificity. *Intl J Cancer* 1975;**16**(2):216–29.
10. Kiessling R, Klein E, Wigzell H. "Natural" killer cells in the mouse. I. Cytotoxic cells with specificity for mouse Moloney leukemia cells. Specificity and distribution according to genotype. *Eur J Immunol* 1975;**5**(2):112–17.
11. Kiessling R, Klein E, Pross H, Wigzell H. "Natural" killer cells in the mouse. II. Cytotoxic cells with specificity for mouse Moloney leukemia cells. Characteristics of the killer cell. *Eur J Immunol* 1975;**5**(2):117–21.
12. Seaman WE, Sleisenger M, Eriksson E, Koo GC. Depletion of natural killer cells in mice by monoclonal antibody to NK-1.1. Reduction in host defense against malignancy without loss of cellular or humoral immunity. *J Immunol* 1987;**138**(12):4539–44.
13. Zamai L, Ponti C, Mirandola P, Gobbi G, Papa S, Galeotti L, Cocco L, Vitale M. NK cells and cancer. *J Immunol* 2007;**178**(7):4011–16.

14. Biron CA, Nguyen KB, Pien GC, Cousens LP, Salazar-Mather TP. Natural killer cells in antiviral defense: function and regulation by innate cytokines. *Annu Rev Immunol* 1999;**17**:189–220.
15. Biron CA, Byron KS, Sullivan JL. Severe herpesvirus infections in an adolescent without natural killer cells. [see comments]. *N Engl J Med* 1989;**320**(26):1731–5.
16. Scharton-Kersten TM, Sher A. Role of natural killer cells in innate resistance to protozoan infections. *Curr Opin Immunol* 1997;**9**(1):44–51.
17. Unanue ER. Studies in listeriosis show the strong symbiosis between the innate cellular system and the T-cell response. *Immunol Rev* 1997;**158**:11–25.
18. Trinchieri G. Biology of natural killer cells. *Adv Immunol* 1989;**47**:187–376.
19. O'Connor GM, Hart OM, Gardiner CM. Putting the natural killer cell in its place. *Immunology* 2005;**117**:1–10.
20. Bakker AB, Wu J, Phillips JH, Lanier LL. NK cell activation: distinct stimulatory pathways counterbalancing inhibitory signals. *Hum Immunol* 2000;**61**(1):18–27.
21. Lanier LL. Face off–the interplay between activating and inhibitory immune receptors. *Curr Opin Immunol* 2001;**13**(3):326–31.
22. Raulet DH, Vance RE, McMahon CW. Regulation of the natural killer cell receptor repertoire. *Annu Rev Immunol* 2001;**19**:291–330.
23. Ruggeri L, Capanni M, Casucci M, Volpi I, Tosti A, Perruccio K, Urbani E, Negrin RS, Martelli MF, Velardi A. Role of natural killer cell alloreactivity in HLA-mismatched hematopoietic stem cell transplantation. *Blood* 1999;**94**(1):333–9.
24. Kärre K, Ljunggren HG, Piontek G, Kiessling R. Selective rejection of H-2-deficient lymphoma variants suggests alternative immune defence strategy. *Nature* 1986;**319**(6055):675–8.
25. Lanier LL. Turning on natural killer cells. *J Exp Med* 2000;**191**(8):1259–62.
26. Kim S, Poursine-Laurent J, Truscott SM, Lybarger L, Song YJ, Yang L, French AR, Sunwoo JB, Lemieux S, Hansen TH, Yokoyama WM. Licensing of natural killer cells by host major histocompatibility complex class I molecules. *Nature* 2005;**436**(7051):709–13.
27. Cosman D, Mullberg J, Sutherland CL, Chin W, Armitage R, Fanslow W, Kubin M, Chalupny NJ. ULBPs, novel MHC class I-related molecules, bind to CMV glycoprotein UL16 and stimulate NK cytotoxicity through the NKG2D receptor. *Immunity* 2001;**14**(2):123–33.
28. Bauer S, Groh V, Wu J, Steinle A, Phillips JH, Lanier LL, Spies T. Activation of NK cells and T cells by NKG2D, a receptor for stress-inducible MICA. *Science* 1999;**285**(5428):727–9.
29. Diefenbach A, Jamieson AM, Liu SD, Shastri N, Raulet DH. Ligands for the murine NKG2D receptor: expression by tumor cells and activation of NK cells and macrophages. *Nat Immunol* 2000;**1**(2):119–26.
30. Cerwenka A, Bakker AB, McClanahan T, Wagner J, Wu J, Phillips JH, Lanier LL. Retinoic acid early inducible genes define a ligand family for the activating NKG2D receptor in mice. *Immunity* 2000;**12**(6):721–7.
31. Ravetch JV, Lanier LL. Immune inhibitory receptors. *Science* 2000;**290**:84–9.
32. Lanier LL. On guard–activating NK cell receptors. *Nature Immunol* 2001;**2**(1):23–7.
33. Lanier LL, Bakker AB. The ITAM-bearing transmembrane adaptor DAP12 in lymphoid and myeloid cell function. *Immunol Today* 2000;**21**(12):611–14.
34. Wu J, Song Y, Bakker AB, Bauer S, Spies T, Lanier LL, Phillips JH. An activating immunoreceptor complex formed by NKG2D and DAP10. *Science* 1999;**285**(5428):730–2.
35. Wu J, Cherwinski H, Spies T, Phillips JH, Lanier LL. DAP10 and DAP12 form distinct, but functionally cooperative, receptor complexes in Natural Killer cells. *J Exp Med* 2000;**192**:1059–67.
36. Bottino C, Moretta L, Pende D, Vitale M, Moretta A. Learning how to discriminate between friends and enemies, a lesson from Natural Killer cells. *Mol Immunol* 2004;**41**(6–7):569–75.
37. Moretta L, Bottino C, Pende D, Vitale M, Mingari MC, Moretta A. Different checkpoints in human NK-cell activation. *Trends Immunol* 2004;**25**(12):670–6.
38. Moretta L, Moretta A. Unravelling natural killer cell function: triggering and inhibitory human NK receptors. *EMBO J* 2004;**23**(2):255–9.
39. Lanier LL. NK cell receptors. *Annu Rev Immunol* 1998;**16**:359–93.
40. Jones EY. Blueprints for life or death. *Nat Immunol* 2001; **2**(2):379–80.
41. Weis WI, Taylor ME, Drickamer K. The C-type lectin superfamily in the immune system. *Immunol Rev* 1998;**163**:19–34.
42. Boyington JC, Motyka SA, Schuck P, Brooks AG, Sun PD. Crystal structure of an NK cell immunoglobulin-like receptor in complex with its class I MHC ligand. *Nature* 2000;**405**(6786):537–43.
43. Fan QR, Long EO, Wiley DC. Crystal structure of the human natural killer cell inhibitory receptor KIR2DL1–HLA-Cw4 complex. *Nat Immunol* 2001;**2**(5):452–60.
44. Fan QR, Mosyak L, Winter CC, Wagtmann N, Long EO, Wiley DC. Structure of the inhibitory receptor for human natural killer cells resembles haematopoietic receptors. *Nature* 1997;**389**(6646):96–100.
45. Snyder GA, Brooks AG, Sun PD. Crystal structure of the HLA-Cw3 allotype-specific killer cell inhibitory receptor KIR2DL2. *Proc Natl Acad Sci USA* 1999;**96**(7):3864–9.
46. Maenaka K, Juji T, Stuart DI, Jones EY. Crystal structure of the human p58 killer cell inhibitory receptor (KIR2DL3) specific for HLA-Cw3-related I. *Structure* 1999;**7**(4):391–8.
47. Saulquin X, Gastinel LN, Vivier E. Crystal structure of the human natural killer cell activating receptor KIR2DS2 (CD158j). *J Exp Med* 2003;**197**(7):933–8.
48. Cantoni C, Ponassi M, Biassoni R, Conte R, Spallarossa A, Moretta A, Moretta L, Bolognesi M, Bordo D. The three-dimensional structure of the human NK cell receptor NKp44, a triggering partner in natural cytotoxicity. *Structure* 2003;**11**(6):725–34.
49. Ponassi M, Cantoni C, Biassoni R, Conte R, Spallarossa A, Pesce A, Moretta A, Moretta L, Bolognesi M, Bordo D. Structure of the human NK cell triggering receptor NKp46 ectodomain. *Biochem Biophys Res Commun* 2003;**309**(2):317–23.
50. Foster CE, Colonna M, Sun PD. Crystal structure of the human natural killer (NK) cell activating receptor NKp46 reveals structural relationship to other leukocyte receptor complex immunoreceptors. *J Biol Chem* 2003;**278**(46):46,081–86,
51. Brown D, Trowsdale J, Allen R. The LILR family: modulators of innate and adaptive immune pathways in health and disease. *Tissue Antigens* 2004;**64**(3):215–25.
52. Shiroishi M, Kajikawa M, Kuroki K, Ose T, Kohda D, Maenaka K. Crystal structure of the human monocyte-activating receptor, "Group 2" leukocyte Ig-like receptor A5 (LILRA5/LIR9/ILT11). *J Biol Chem* 2006;**281**(28):19,536–44.
53. Chapman TL, Heikema AP, West Jr. AP, Bjorkman PJ. Crystal structure and ligand binding properties of the D1D2 region of the inhibitory receptor LIR-1 (ILT2). *Immunity* 2000;**13**(5):727–36.
54. Willcox BE, Thomas LM, Bjorkman PJ. Crystal structure of HLA-A2 bound to LIR-1, a host and viral major histocompatibility complex receptor. *Nat Immunol* 2003;**4**(9):913–19.

55. Shiroishi M, Kuroki K, Rasubala L, Tsumoto K, Kumagai I, Kurimoto E, Kato K, Kohda D, Maenaka K. Structural basis for recognition of the nonclassical MHC molecule HLA-G by the leukocyte Ig-like receptor B2 (LILRB2/LIR2/ILT4/CD85d). *Proc Natl Acad Sci USA* 2006;**103**(44):16,412–17.
56. Natarajan K, Dimasi N, Wang J, Margulies DH, Mariuzza RA. MHC class I recognition by Ly49 natural killer cell receptors. *Mol Immunol* 2002;**38**(14):1023–7.
57. Tormo J, Natarajan K, Margulies DH, Mariuzza RA. Crystal structure of a lectin-like natural killer cell receptor bound to its MHC class I ligand. *Nature* 1999;**402**:623–31.
58. Dam J, Guan R, Natarajan K, Dimasi N, Chlewicki LK, Kranz DM, Schuck P, Margulies DH, Mariuzza RA. Variable MHC class I engagement by Ly49 natural killer cell receptors demonstrated by the crystal structure of Ly49C bound to H-2K(b). *Nat Immunol* 2003;**4**(12):1213–22.
59. Dimasi N, Sawicki MW, Reineck LA, Li Y, Natarajan K, Margulies DH, Mariuzza RA. Crystal structure of the Ly49I natural killer cell receptor reveals variability in dimerization mode within the Ly49 family. *J Mol Biol* 2002;**320**(3):573–85.
60. Adams EJ, Juo ZS, Venook RT, Boulanger MJ, Arase H, Lanier LL, Garcia KC. Structural elucidation of the m157 mouse cytomegalovirus ligand for Ly49 natural killer cell receptors. *Proc Natl Acad Sci USA* 2007;**104**(24):10,128–33.
61. Wolan DW, Teyton L, Rudolph MG, Villmow B, Bauer S, Busch DH, Wilson IA. Crystal structure of the murine NK cell-activating receptor at 1.95 Å. *Nat Immunol* 2001;**2**(3):248–54.
62. McFarland BJ, Kortemme T, Yu SF, Baker D, Strong RK. Symmetry recognizing asymmetry. Analysis of the interactions between the C-type lectin-like immunoreceptor NKG2D and MHC Class I-like ligands. *Structure (Camb.)* 2003;**11**(4):411–22.
63. Li P, Morris DL, Willcox BE, Steinle A, Spies T, Strong RK. Complex structure of the activating immunoreceptor NKG2D and its MHC class I-like ligand MICA. *Nat Immunol* 2001;**2**:443–51.
64. Radaev S, Rostro B, Brooks AG, Colonna M, Sun PD. Conformational plasticity revealed by the cocrystal structure of NKG2D and its class I MHC-like ligand ULBP3. *Immunity* 2001;**15**(6):1039–49.
65. Li P, McDermott G, Strong RK. Crystal structures of RAE-1beta and its complex with the activating immunoreceptor NKG2D. *Immunity* 2002;**16**(1):77–86.
66. McFarland BJ, Strong RK. Thermodynamic analysis of degenerate recognition by the NKG2D immunoreceptor: not induced fit but rigid adaptation. *Immunity* 2003;**19**(6):803–12.
67. Kaiser BK, Barahmand-Pour F, Paulsene W, Medley S, Geraghty DE, Strong RK. Interactions between NKG2x immunoreceptors and HLA-E ligands display overlapping affinities and thermodynamics. *J Immunol* 2005;**174**(5):2878–84.
68. Kaiser BK, Pizarro JC, Kerns J, Strong RK. Structural Basis for NKG2A/CD94 Recognition of HLA-E. *Proc Natl Acad Sci USA* 2008;**105**(18):6696–701.
69. Stewart CA, Laugier-Anfossi F, Vely F, Saulquin X, Riedmuller J, Tisserant A, Gauthier L, Romagne F, Ferracci G, Arosa FA, Moretta A, Sun PD, Ugolini S, Vivier E. Recognition of peptide–MHC class I complexes by activating killer immunoglobulin-like receptors. *Proc Natl Acad Sci USA* 2005;**102**(37):13,224–9.
70. Chapman TL, Heikema AP, Bjorkman PJ. The inhibitory receptor LIR-1 uses a common binding interaction to recognize class I MHC molecules and the viral homolog UL-18. *Immunity* 1999;**11**:603–13.
71. Vales-Gomez M, Reyburn HT, Erskine RA, Lopez-Botet M, Strominger JL. Kinetics and peptide dependency of the binding of the inhibitory NK receptor CD94/NKG2-A and the activating receptor CD94/NKG2-C to HLA-E. *EMBO J* 1999;**18**(15):4250–60.
72. Vance RE, Kraft JR, Altman JD, Jensen PE, Raulet DH. Mouse CD94/NKG2A is a natural killer cell receptor for the nonclassical major histocompatibility complex (MHC) class I molecule Qa-1(b). *J Exp Med* 1998;**188**(10):1841–8.
73. Radaev S, Kattah M, Zou Z, Colonna M, Sun PD. Making sense of the diverse ligand recognition by NKG2D. *J Immunol* 2002;**169**(11):6279–85.
74. O'Callaghan CA, Cerwenka A, Willcox BE, Lanier LL, Bjorkman PJ. Molecular competition for NKG2D: H60 and RAE1 compete unequally for NKG2D with dominance of H60. *Immunity* 2001;**15**:201–11.
75. Carayannopoulos LN, Naidenko OV, Kinder J, Ho EL, Fremont DH, Yokoyama WM. Ligands for murine NKG2D display heterogeneous binding behavior. *Eur J Immunol* 2002;**32**(3):597–605.
76. Dam J, Baber J, Grishaev A, Malchiodi EL, Schuck P, Bax A, Mariuzza RA. Variable dimerization of the Ly49A natural killer cell receptor results in differential engagement of its MHC class I ligand. *J Mol Biol* 2006;**362**(1):102–13.

Chapter 52

Toll-Like Receptors–Structure and Signaling

Istvan Botos and David R. Davies
Laboratory of Molecular Biology, National Institute of Diabetes and Digestive and Kidney Diseases, National Institutes of Health, Bethesda, Maryland

The innate immune system provides a rapid first line of defense against pathogen attack. One of its principal components is the array of pathogen-associated molecular pattern detectors known as the Toll-Like Receptors (TLRs)–germline-encoded type one transmembrane receptors of ancient lineage that are expressed on numerous cell types [1–3]. TLR ectodomains contain about 18–25 leucine-rich repeats (LRRs) [4], and recognize molecules that are as diverse as double-stranded RNA, lipopolysaccharide, peptidoglycan, and unmethylated CpG DNA. TLRs rapidly trigger the production of a variety of cytokines, producing inflammation, and also activate the adaptive immune response by up-regulating co-stimulatory molecules and antigen-presenting cells (Figure 52.1; [5, 6]).

STRUCTURE OF TLR3

The first structure of a TLR ectodomain, TLR3 was determined simultaneously in two laboratories [7, 8]. Human TLR3 (hTLR3) responds to dsRNA, a molecular signature of many types of viruses. It is located mainly intracellularly in endosomes. The two TLR3 ECD structures are very similar, and were determined from crystals grown under very similar conditions, the major difference being the pH of crystallization. TLR3 has 23 LRRs together with N- and C-terminal capping domains. The overall structure is that of a large curved solenoid with an inner diameter of~42 Å, an outer diameter of~90 Å and a thickness of 35 Å (Figure 52.2). The concave inner surface consists of a large parallel β-sheet, with each β-strand roughly perpendicular to the plane of the solenoid and linked to the next strand by an irregular loop, the whole solenoid being held together by the hydrophobic side-chains.

The overall shape of TLR3 differs from that of most LRR structures in that there is practically no twist along the solenoid, resulting in an almost flat molecule. The molecule is heavily glycosylated, with 15 potential glycosylation sites, and Bell and colleagues [8] observed glycans at 11 of these sites. The LRR solenoid is capped at its N- and C-termini by characteristic disulfide-linked capping domains. There are two large loops at positions following the β-strands that extend from the solenoid at LRR12 and LRR20.

THE dsRNA BINDING SITE IN hTLR3

Mutational analysis to identify the dsRNA binding site revealed a small patch of residues in the vicinity of LRR20 where mutations diminished or abrogated binding [9]. These include His 539, Asn 541, and the loop at LRR20. A model linking two TLR3 ectodomains through the RNA brings the two C-terminal domains into sufficient proximity to permit dimerization of the cytoplasmic TIR domains (Figure 52.3).

TLR4

TLR4 couples with MD-2 to react to the presence of lipopolysaccharide from gram-negative bacteria, leading to an inflammatory response. MD-2 is an accessory protein that associates 1 : 1 with the extracellular domain of TLR4 and directly binds LPS. In addition to LPS, TLR4 has been reported to recognize several other ligands, such as: taxol, viral envelope and fusion proteins; and endogenous ligands such as HSP60, HSP70, oligosaccharides of hyaluronic acid and fibrinogen, usually in association with MD-2. The structure of mouse TLR4 (mTLR4) complexed with mouse MD-2 in the absence of lipid has been determined [10].

Handbook of Cell Signaling, Three-Volume Set 2 ed.

The horseshoe-shaped mTLR4 molecule has three regions–N-terminal, central, and C-terminal–each with a different curvature, unlike most other LRR proteins, which have a uniform curvature. mMD-2 interacts with mTLR4 at a highly conserved region of the concave surface near the boundary of the N-terminal and central regions (Figure 52.4). This interaction is mediated by an extensive network of charge-enhanced hydrogen bonds. To increase the probability of crystallization, the authors created a variety of chimaeras of hTLR4, LRR truncations capped at the C-terminus LRRs with a hagfish Variable lymphocyte receptor (VLR) capping motif. From these structures the entire hTLR4 ectodomain was reconstructed. In the same study a fragment of hTLR4 was crystallized with hMD-2 bound to Eritoran, a synthetic lipid antagonist of TLR4/MD-2 [10]. Eritoran has four acyl chains compared with six in lipid A (Figure 52.5a), and occupies

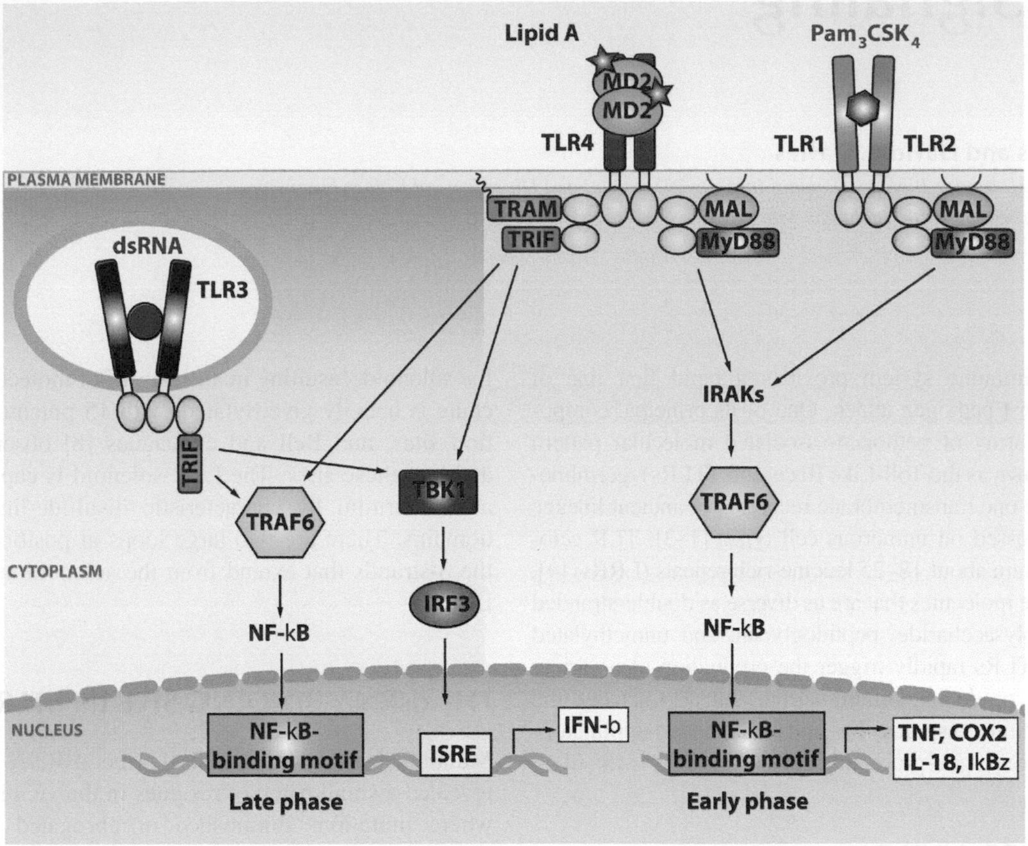

FIGURE 52.1 Two common TLR signaling pathways: MyD88-dependent and TRIF-dependent (adapted from [5–6]). The MyD88-dependent pathway is used by all TLRs except TLR3, which uses the TRIF-dependent pathway. TLRs and adaptor molecules interact through their Toll/Interleukin-1 Receptor (TIR) domains.

Abbreviations: MyD88–Myeloid Differentiation primary response gene 88; MAL–MyD88 Adaptor-Like; TRIF–TIR-domain containing Adaptor Protein; TRAM–TRIF-Related Adaptor Molecule; TRAF–TNF Receptor Associated Factor; IRF–Interferon Regulatory Factor; TBK–TANK binding kinase.

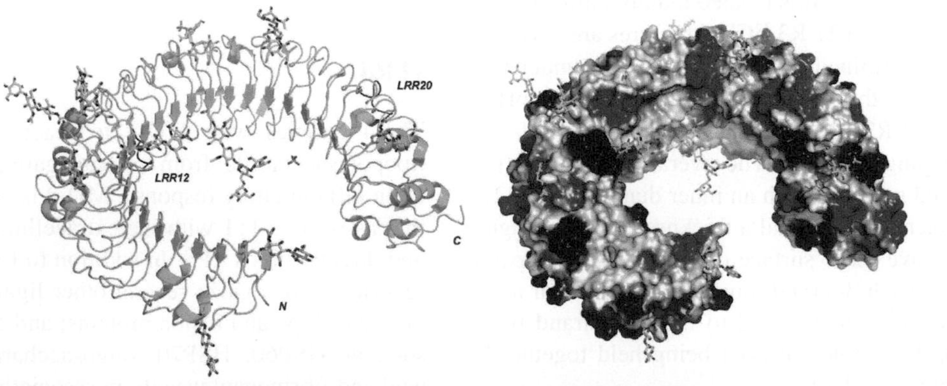

FIGURE 52.2 Ribbon representation and molecular surface of the human TLR3 ectodomain structure.

Positive charges are shown in dark gray, negative charges in lighter gray. Glycosyl moieties are shown in stick representation. All figures were prepared with PyMol (DeLano Scientific, San Carlos, CA).

about 90 percent of the binding pocket of hMD-2. Two of the acyl chains of Eritoran are bent and occupy the empty space in the hydrophobic pocket, whereas the two phosphate groups form ionic bonds with residues at the mouth of the pocket.

Mouse and human MD-2 share 64 percent sequence identity, with almost identical structures, despite their liganded and unliganded forms. The structure of their central regions differs most, due to the binding of mMD-2 that bends this concave region by almost 20 degrees.

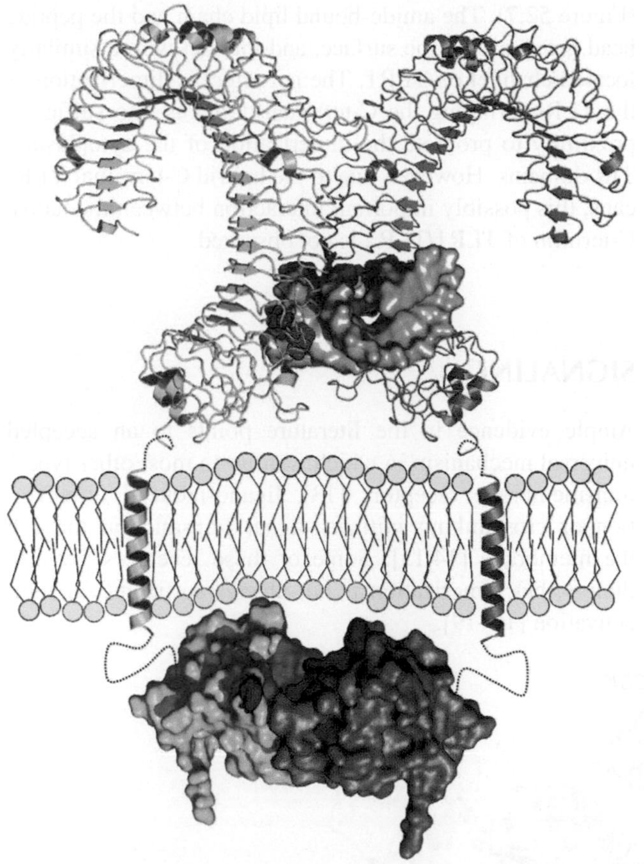

FIGURE 52.3 Model of ligand-induced dimerization of TLR3 triggered by binding of dsRNA (adapted from [8]).

Binding of agonistic LPS induces aggregation of TLR4 and initiates the intracellular signaling cascade, whereas the antagonistic Eritoran does not cause receptor dimerization. LPS binds to the N-terminal TLR4/MD-2 fragment without aggregation. The central and C-terminal regions of TLR4 are required for aggregation. Lee and colleagues [10] proposed a model in which LPS binding induces a structural change in bound MD-2 that in turn promotes interaction between the edge of the MD-2 molecule and a second TLR4 molecule, forming a hetero-tetramer.

MD-2

The structure of human MD2 (hMD2) and its complex with antiendotoxic lipid IVa was recently determined [11]. Like Eritoran, lipid IVa has four acyl chains and acts as an antagonist in human cells (Figure 52.6). The lipid IVa complex with MD-2 is virtually identical to the native MD-2 structure. However, in the native MD-2 there is electron density within the cavity that has been attributed to three myristic acid molecules attached to the MD-2, leaving the possibility that the absence of acyl groups within the pocket could be associated with a conformational change in MD-2.

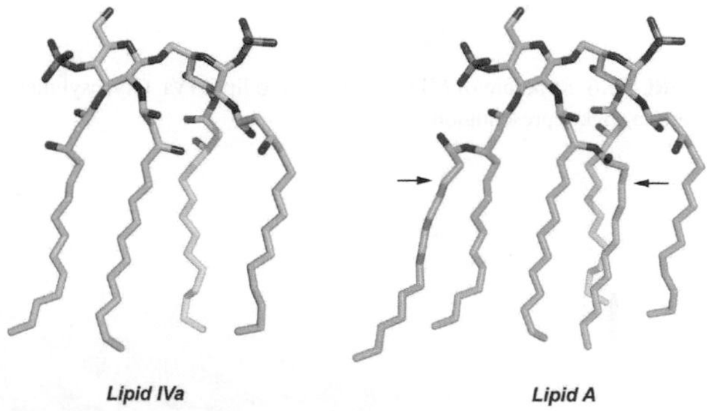

FIGURE 52.5 The antagonist lipid IVa has four acyl chains, whereas the agonist lipid A has six. The two additional chains are marked by arrows.

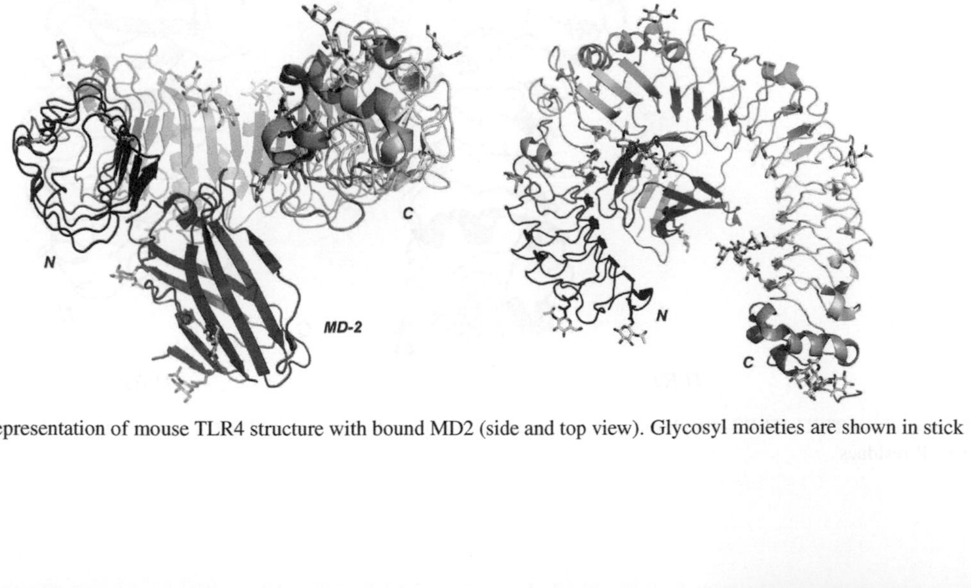

FIGURE 52.4 Ribbon representation of mouse TLR4 structure with bound MD2 (side and top view). Glycosyl moieties are shown in stick representation.

It should be noted that in the case of the complex with TLR4, the MD-2 structure again shows little change upon binding to Eritoran [10]. The authors do not think it likely that lipid A (Figure 52.5b) could be accommodated within the MD-2 cavity without major conformational change.

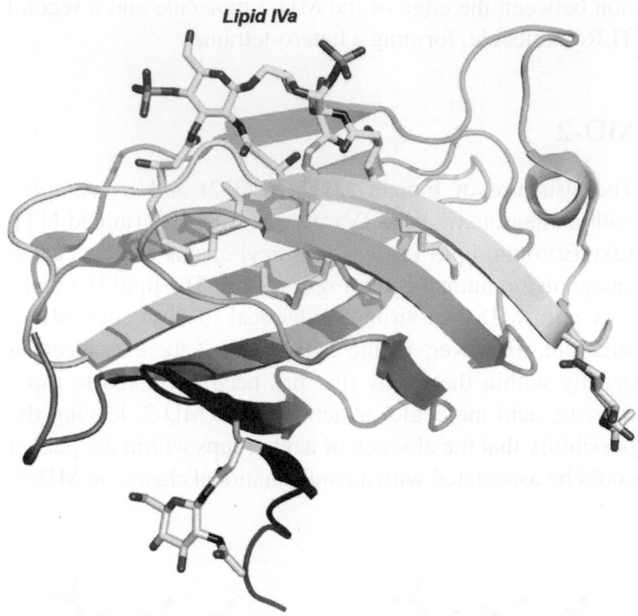

FIGURE 52.6 Structure of MD2 complexed to lipid IVa. Glycosyl moieties are in stick representation.

TLR1–TLR2 DIMERIZATION BY A TRI-ACYLATED LIPOPEPTIDE

TLR2 recognizes triacylated lipopeptide (Pam_3SCK_4) as a complex with TLR1. The structure of a ligand-induced complex between chimaeric TLR1 and TLR2 constructs has been determined [12]. Heterodimerization is caused by this peptide bound to TLR2, with the two ester-bound lipid chains inserted through a crevice at the border between the central and C-terminal regions into a large internal pocket (Figure 52.7). The amide-bound lipid chain and the peptide head group are on the surface, and interact with a similarly located channel on TLR1. The net effect is dimerization of the TLRs, bringing the C-terminal domains into sufficient proximity to promote the dimerization of the cytoplasmic TIR domains. However, due to the hybrid C-terminal VLR-caps, this possibly important interaction between the native C-termini of TLR1/TLR2 is not observed.

SIGNALING

Ample evidence in the literature points to an accepted universal mechanism in which, similar to most other type 1 transmembrane receptors [13], ligand-induced dimerization or crosslinking initiates signal transduction through the membrane [14,15]. Some of these receptors preform dimers, but dimerization is not always sufficient for their activation [16–19].

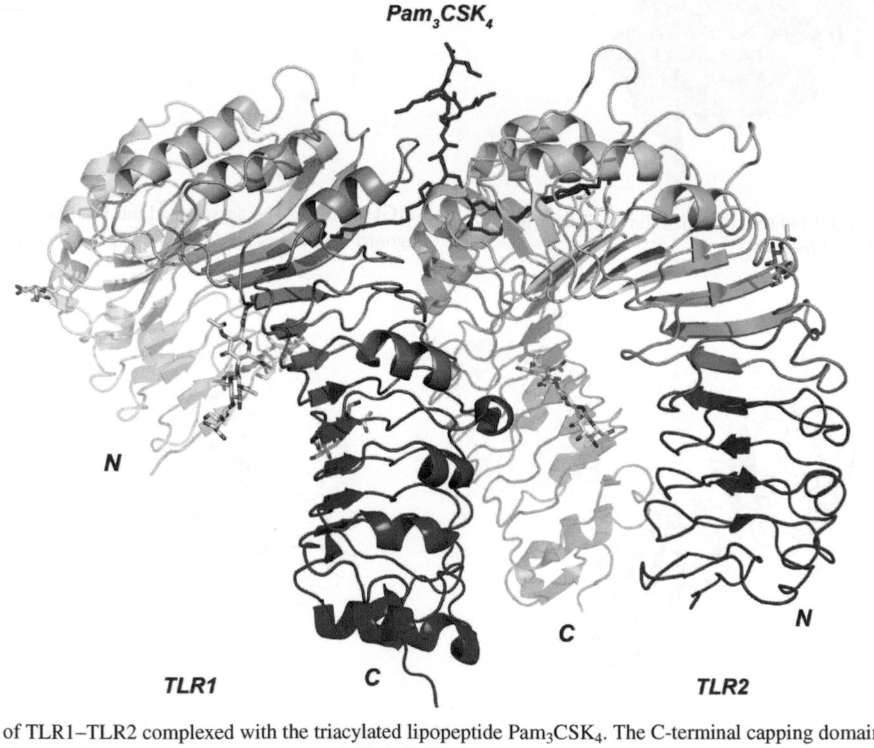

FIGURE 52.7 Structure of TLR1–TLR2 complexed with the triacylated lipopeptide Pam_3CSK_4. The C-terminal capping domains were replaced by the corresponding VLR residues.

It appears that dimerization of the TLR cytoplasmic TIR domains is a prerequisite for the binding of other TIR-domain containing adaptor molecules. However, for some TLR ECDs dimerization may be necessary but not sufficient, and it has been suggested that some TLRs pre-exist as inactive dimers, and ligand binding induces conformational changes necessary for signaling [20]. To date, the only example of a signaling complex structure is the TLR1/TLR2 ECD complex with a tri-acylated lipopeptide [10]. The structure reveals only receptor dimerization upon ligand binding, with no conformational changes in the ECDs, although in this case the C-terminal cap has been replaced by a VLR C-terminus. More structures of signaling complexes are needed to resolve this mechanism.

ACKNOWLEDGEMENTS

We would like to thank Dr David Segal for helpful discussions.

REFERENCES

1. Janeway Jr CA. The immune system evolved to discriminate infectious nonself from noninfectious self. *Immunol Today* 1992;**13**:11–16.
2. Janeway Jr CA, Medzhitov R. Innate immune recognition. *Annu Rev Immunol* 2002;**20**:197–216.
3. Carpenter S, O'Neill LA. How important are Toll-like receptors for antimicrobial responses?. *Cell Microbiol* 2007;**8**:1891–901.
4. Bell JK, Mullen GED, Leifer CA, Mazzoni A, Davies DR, Segal DM. Leucine-rich repeats and pathogen recognition in Toll-like receptors. *Trends Immunol* 2003;**24**:528–33.
5. Liew FY, Xu D, Brint EK, O'Neill LAJ. Negative regulation of Toll-like receptor immune responses. *Nat Rev Immunol* 2005;**5**:446–58.
6. Watters TM, Kenny EF, O'Neill LAJ. Structure, function and regulation of the Toll/IL-1 receptor adaptor proteins. *Immunol Cell Biol* 2007;**85**:411–19.
7. Choe J, Kelker MS, Wilson IA. Crystal structure of human Toll-like Receptor 3 (TLR3) ectodomain. *Science* 2005;**309**:581–5.
8. Bell JK, Botos I, Hall PR, Askins J, Shiloach J, Segal DM, Davies DR. The molecular structure of the Toll-like receptor 3 ligand-binding domain. *Proc Natl Acad Sci USA* 2005;**102**:10,976–10,980.
9. Bell JK, Askins J, Hall PR, Davies DR, Segal DM. The dsRNA binding site of human Toll-like receptor 3. *Proc Natl Acad Sci USA* 2006;**103**:8792–7.
10. Kim HM, Park BS, Kim J-I, Kim SE, Lee J, Oh SC, Enkhbayar P, Matsushima N, Lee H, Yoo OJ, Lee J-O. Crystal structure of the TLR4–MD-2 complex with bound endotoxin antagonist eritoran. *Cell* 2007;**130**:1–12.
11. Ohto U, Fukase K, Miyake K, Satow Y. Crystal structures of human MD-2 and its complex with antiendotoxic lipid IVa. *Science* 2007;**316**:1632–4.
12. Jin MS, Kim SE, Heo JY, Lee ME, Kim HM, Paik S-G, Lee H, Lee J-O. Crystal structure of the TLR1–TLR2 heterodimer induced by binding of a triacylated lipopeptide. *Cell* 2007;**130**:1071–82.
13. Stroud RM, Wells JA. Mechanistic diversity of cytokine receptor signaling across cell membranes. *Sci STKE* 2004;re7.
14. Lemmon MA, Schlessinger J. Regulation of signal transduction and signal diversity by receptor oligomerization. *Trends Biochem Sci* 1994;**19**:459–63.
15. Weber AN, Moncrieffe MC, Gangloff M, Imler JL, Gay NJ. Ligand–receptor and receptor–receptor interactions act in concert to activate signaling in the Drosophila toll pathway. *J Biol Chem* 2005;**280**:22,793–22,799.
16. Remy I, Wilson IA, Michnick SW. Erythropoietin receptor activation by a ligand-induced conformational change. *Science* 1999;**283**:990–3.
17. Guo C, Dower SK, Holowka D, Baird B. Fluorescence resonance energy transfer reveals interleukin (IL)-1-dependent aggregation of IL-1 type I receptors that correlates with receptor activation. *J Biol Chem* 1995;**270**:27,562–27,568.
18. Damjanovich S, Bene L, Matko J, Alileche A, Goldman CK, Sharrow S, Waldmann TA. Preassembly of interleukin 2 (IL-2) receptor subunits on resting Kit 225 K6 T cells and their modulation by IL-2, IL-7, and IL-15: a fluorescence resonance energy transfer study. *Proc Natl Acad Sci USA* 1997;**94**:13,134–13,139.
19. Gadella Jr T, Jovin TM. Oligomerization of epidermal growth factor receptors on A431 cells studied by time-resolved fluorescence imaging microscopy. A stereochemical model for tyrosine kinase receptor activation. *J Cell Biol* 1995;**129**:1543–58.
20. Gay NJ, Gangloff M, Weber AN. Toll-like receptors as molecular switches. *Nat Rev Immunol* 2006;**6**:693–8.

Chapter 53

Toll Family Receptors

Yann Hyvert and Jean-Luc Imler

Centre National de la Recherche Scientifique and Faculté des Sciences de la Vie, Université Louis Pasteur, Institut de Biologie Moléculaire et Cellulaire, Strasbourg, France

INTRODUCTION

Innate immunifty is the first-line host defense mechanism that operates to protect multicellular organisms from infectious microbes. In mammals, it contributes to the activation of adaptive immunity, which involves highly specific antigen receptors expressed by B and T lymphocytes. By contrast, the innate immune response is activated by receptors collectively known as "pattern recognition receptors (PRR)" that broadly sense microbial ligands and trigger cellular activation. The best-characterized family of PRRs is the Toll receptor family, which is described here.

STRUCTURE–FUNCTION OF TOLL RECEPTORS

Toll and Toll-like Receptors in Flies and Mammals

The *Toll* gene was first identified in the early 1980s by Nusslein-Volhard and Anderson during a mutagenesis screen to identify the genes controlling the establishment of the dorso-ventral (DV) axis of the *Drosophila* embryo, and was found to encode a new kind of type I transmembrane receptor [1]. The 11 remaining genes identified in this screen encode factors acting upstream and downstream of Toll in the signaling pathway. Activation of Toll on the ventral side of the embryo results in a ventral to dorsal gradient of nuclear translocation of the transcription factor Dorsal, thus establishing embryonic polarity. It was also through studies in *Drosophila* that Toll was assigned an immune function in the control of the inducible expression of antimicrobial peptides. The transcriptional activation of the genes encoding these peptides requires transcriptional activators of the Rel family, to which NFκB belongs. At the time, Dorsal was the only identified member of this family in *Drosophila*. This prompted analysis of the known mutant strains of the Toll pathway for immunodeficiency phenotypes, which led to the demonstration by Hoffmann and collaborators that the Toll pathway controls the response to fungal and Gram-positive bacterial infections [2,3]. These results were rapidly followed by the description of a mammalian Toll homolog (now known as TLR4) capable of activating NFκB and the synthesis of cytokines and co-stimulatory molecules [4]. Shortly after, Beutler and colleagues showed that mice from the lipopolysaccharide-hyporesponsive strains C57BL/10ScCr and C3H/HeJ carry mutations in their *tlr4* gene, indicating that Toll-like receptors (TLRs) play an important role in the control of infection in mammals [5]. Since these initial discoveries, a family of 10–13 TLRs has been described in mammals (9 members in drosophila), the properties of which are described below.

Structure of Toll-Family Receptors

The 150 amino acid intracytoplasmic domain of Toll bears striking similarities with that of the Interleukin-1 type I receptor (IL-1R), and is referred to as the TIR (Toll/IL-1R) homology domain (Figure 53.1a) [6]. TIR domains are also present in intracytoplasmic signaling molecules such as MyD88 or TIRAP/MAL (see below). Plants also express TIR domain-containing factors [7]. The structures of the TIR domains of human TLR1 and TLR2 have been solved and shown to be composed of a central five-stranded parallel β-sheet surrounded by five α-helices [8]. This domain is devoid of catalytic activity, but rather serves as a homophilic protein–protein interaction domain, mediating the recruitment of cytosolic death-domain adaptors (see below).

Handbook of Cell Signaling, Three-Volume Set 2 ed.

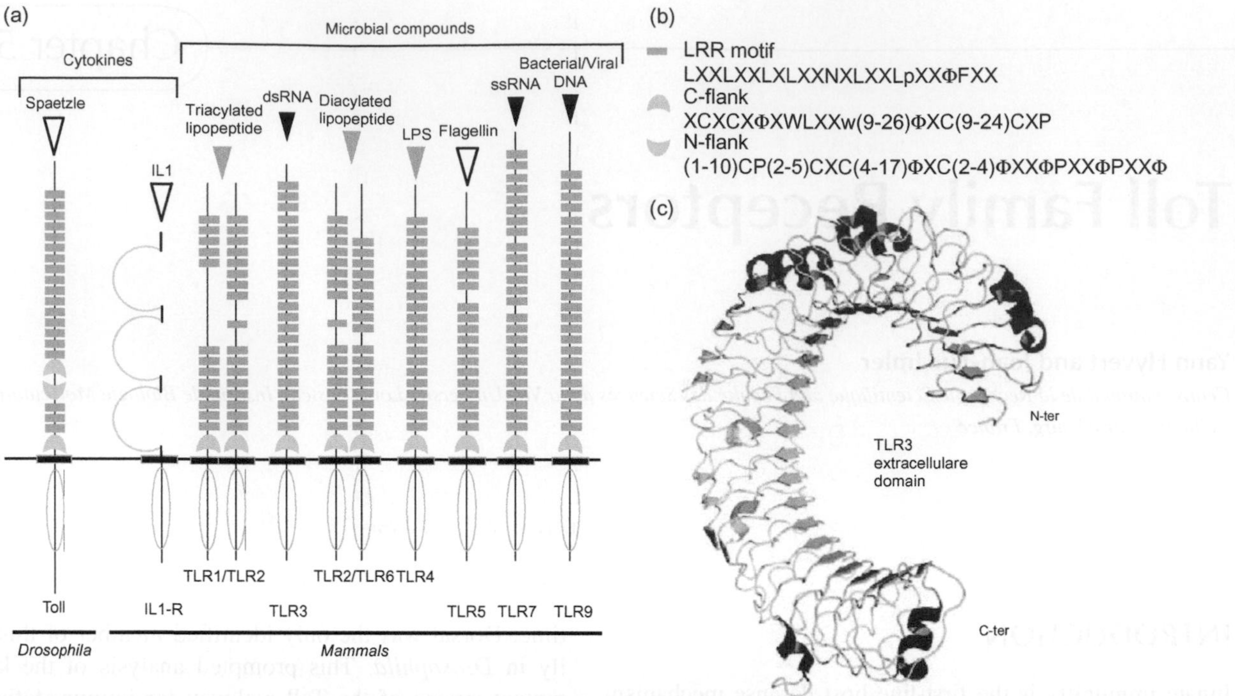

FIGURE 53.1 Structure and activation of TIR domain receptors in *Drosophila* and mammals.
(a, b) Schematic representation of the drosophila Toll receptor and the mammalian TIR domain receptors (IL1-R and TLR1 to 9). The TIR domains are represented as white-gray ovals, and the Ig domains as half-circles. The sequence of LRR, C-flank, and N-flank motifs is shown in panel (b). Peptidic ligands are represented with a white triangle, lipopeptide/lipoprotein ligands with a gray triangle and nucleic acids ligand with a black triangle. (c) Human Toll-like receptor 3 extracellular domain structures : helices are colored in black, β-sheets in gray and turns in white (structure from PDB, ID : 1ziw).

The extracellular domain of Toll family receptors does not contain Ig domains, like the IL-1R, but comprises several leucine-rich repeats (LRRs). A number of other membrane receptors, such as the pattern recognition receptor CD14, the adhesion molecule GpIbα, or the members of the Trk family of neurotrophins receptors, also feature LRRs in their ectodomains. The ectodomains of TLR3 and TLR4 have been crystallized, and their tridimensional structure solved [9–11]. The LRRs are flanked at each end by small domains that cap the exposed hydrophobic residues at the ends of the LRRs. These small domains are stabilized by disulfide bonds involving pairs of conserved cysteine residues (Figure 53.1b). Regarding the LRR motifs, they fold into a short parallel β-sheet, followed by a turn and a more variable region. In TLR3, the 23 tandem copies of LRRs form a curved, solenoidal structure, with the short β-sheets forming the concave (inner) surface of the structure (Figure 53.1c). This horseshoe-shaped structure is characteristic of LRR-containing proteins, and in several cases the concave surface of the solenoid provides the ligand binding site, although the situation is probably more complex for Toll receptors [6, 9, 12].

Comparison of sequences from Toll receptors in invertebrates and mammals reveals that they can be subdivided in two different groups. All mammalian TLRs cluster to one of these subfamilies, together with a single *Drosophila* receptor, Toll-9. The eight other *Drosophila* Toll family members cluster to the second subfamily, together with the *Caenorhabditis elegans Tol-1* gene product [7]. Altogether, these data indicate that most *Drosophila* Tolls have evolved independently from mammalian TLRs, possibly to fulfill different functions.

Activation of Toll-Family Receptors

In agreement with the differences mentioned above between *Drosophila* Toll and mammalian TLRs, these receptors function differently. In flies, Toll is a cytokine receptor, and is activated by the cysteine-knot growth factor Spaetzle, which is structurally related to neurotrophins [13]. Spaetzle is synthesized as an inactive precursor unable to bind to Toll. In response to infection in adult flies, or on the ventral side of the embryo, a cascade of serine-proteases is triggered, leading to activation of Spaetzle processing enzyme (SPE) or Easter, respectively. These proteases cleave the procytokine, releasing a C-terminal fragment of 106 amino acids (C106). C106 binds to Toll with nanomolar affinity, and triggers signaling. Interestingly, the *Drosophila* genome encodes a family of six Spaetzle genes, which may function as ligands for the other members of the Toll family in flies [7, 13].

By contrast, TLRs in mammals appear to be directly activated by microbial ligands. Genetic evidence indicates that these receptors sense a biochemically diverse set of ligands. Some TLRs are activated by lipopeptides and lipoproteins: TLR4 mediates cellular responses to lipopolysaccharides found in the outer cell wall of Gram-negative bacteria; TLR2 forms an heterodimer with TLR6 or TLR1 to sense respectively di- or tri-acylated bacterial lipopeptides. Another subgroup of TLRs detect nucleic acids: TLR3 is involved in the detection of double-stranded RNA; TLR7 recognizes single-stranded RNA enriched in U residues; and TLR9 mediates responses to DNA containing unmethylated CpG motifs. Interestingly, these TLRs functions in endosomes, rather than on the plasma membrane, and the reduced pH found in this environment is required for signaling. Finally, two TLRs, TLR5 and TLR11, sense microbial proteins, namely flagellin and profilin (Figure 53.1a) (reviewed in [14, 15]).

In summary, the common framework provided by the LRRs of the ectodomain offers remarkable plasticity in terms of molecular recognition, and can accommodate chemically very diverse ligands. Of note, some TLRs may interact with more than one type of ligand, as exemplified by TLR4, which senses the G glycoprotein from Vesicular Stomatitis Virus (VSV) in addition to lipopolysaccharides. This plasticity of interaction can be explained by the diversity of the side-chains in the variable amino acids found in the LRR motifs. Another important point to take into account is the existence of co-receptors associating with TLRs. For example, sensing of lipopolysaccharides involves CD14, a GPI-linked LRR protein, and MD2, a secreted 20-kDa protein that binds to TLR4 and lipopolysaccharides [9]. Another example is CD36, which is required for the sensing of lipopeptides by TLR2 [14].

SIGNALING BY TOLL FAMILY RECEPTORS

Upon activation, TLRs induce transcription factors of the NFκB, AP-1, and interferon regulatory factor families, which mediate transcription of many genes encoding antimicrobial molecules, cytokines, and adhesion molecules. The TIR domain of the receptor plays a critical role in the signaling events leading to transcription activation. There are, however, differences in the signaling by different members of the family, depending on the type of TIR domain adaptor that is recruited by the receptor (Figure 53.2).

MyD88-Dependent Signaling

MyD88 (myeloid differentiation clone number 88) was the first TIR domain cytoplasmic molecule discovered. It is composed of an amino-terminal death domain and a C-terminal TIR domain. MyD88 interacts with members of the IL-1R family and with most TLRs (with the notable exception of TLR3) through its TIR domain [14, 15]. The death domain also functions as a homophilic protein–protein interaction domain, and mediates recruitment of serine-threonine kinases of the IRAK family [16]. Toll receptors probably exist as preformed dimers, and ligand binding triggers conformational changes that result in allosteric activation [17, 18]. One consequence of activation is the phosphorylation of IRAK1 and IRAK4, resulting in their release from the receptor complex. IRAK4 then binds to and activates TRAF6. This RING domain containing protein functions as an ubiquitin ligase, and promotes its auto-polyubiquitination. This polymerization of ubiquitin involves the lysine residue in position 63 (K63), and results in signal transduction, unlike the polymers using the lysine 48 residue (K48), which label proteins for degradation by the 26S proteasome. Activated TRAF6 interacts with the MAP3K kinase TAK1, through its associated proteins TAB2 and TAB3. TAK1 subsequently activates the IKK (IκB kinase) complex. The IKKβ subunit phosphorylates IκB, a cytosolic inhibitor of NFκB. Phosphorylation of IκB is followed by K48-polyubiquitination, and degradation, thus allowing the released NFκB to translocate to the nucleus and activate gene expression [19]. TAK1 also phosphorylates the MAP2Ks MKK3 and MKK6, which mediate activation of the MAP kinases JNK and p38, and activation of AP-1 (Figure 53.2).

TRIF-Dependent Signaling

The human genome encodes 4 additional TIR domain-containing cytosolic molecules [20]. The most important of these is TRIF, which mediates MyD88-independent signaling downstream of TLR3 and TLR4. Unlike all other TLRs, TLR3 does not interact with MyD88, and only signals through TRIF. On the other hand, TLR4 signals through both MyD88- and TRIF-dependent pathways. TRIF can activate NFκB and AP-1, like MyD88, and can also induce expression of the antiviral cytokine interferon (IFN)β. The transcription of the gene encoding this cytokine is controlled by the cooperative action of NFκB, AP-1, and two members of the interferon regulatory factor family, IRF3 and IRF7. TRIF does not contain a death domain like MyD88, but a C-terminal RHIM (RIP homotypic interaction motif). This motif mediates interaction with RIP (receptor interacting protein)-1, a kinase initially characterized for its role in activation of NFκB in the TNF (tumor necrosis factor) pathway [21]. TRIF also interacts with TRAF6, and both RIP1 and TRAF6 probably contribute to activation of AP-1 and NFκB. This pathway participates in the full induction of inflammatory cytokines in response to lipopolysaccharides [14, 15].

For induction of IFNβ, induction of IRF3 or IRF7 is required [22]. These molecules are present in an inactive form in the cytosol. TRIF signaling activates two non-canonical IKKs, known as TBK (TANK binding kinase)1 and IKKε. A member of the TRAF family, TRAF3, interacts

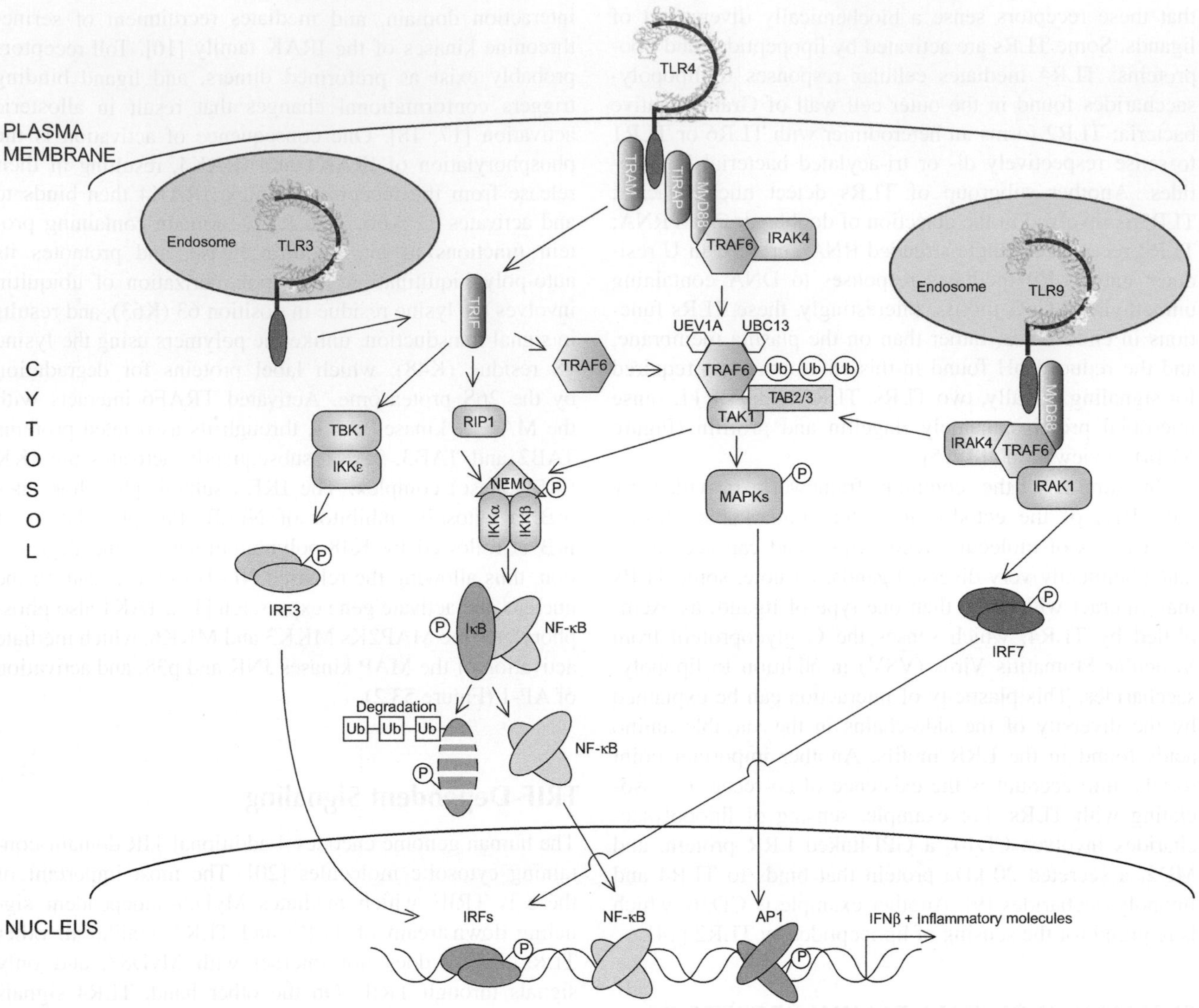

FIGURE 53.2 Signaling by Toll-like receptors.
Most TLRs (and the IL-1R) signal through the adaptor MyD88, and interact directly with this molecule (e.g., TLR9), or through the intermediary of TIRAP (e.g., TLR4). MyD88 recruits kinases of the IRAK family through its death domain. This pathway activates TRAF6, and the MAP3K kinase TAK1, leading to activation of AP-1 and NFκB. TLR3 and TLR4 can also signal through the TIR adaptor TRIF, leading to activation of the transcription factor IRF3 or 7. Induction of the promoters of the genes encoding inflammatory cytokines or type I IFN depends on a combination of transcription factors of the AP-1, NFκB, and IRF families. K48- and K63-mediated polyubiquitinations are indicated by a square and a circle, respectively.

with both MyD88 and TRIF, and is involved in activation of TBK-1/IKKε kinases. These kinases phosphorylate IRF3 and 7, leading to their dimerization, and nuclear translocation (Figure 53.2). Only IRF3 is expressed constitutively in most cell-types. Expression of IRF7 is induced by lipopolysaccharides, dsRNA, and IFNβ, and this factor participates in the amplification of the response.

Two other TIR domain containing cytosolic molecules participate in TLR signaling. The first one is TIRAP (also known as MAL), which is required together with MyD88 for signal transduction by TLR4 and the TLR1/2 and TLR2/6 heterodimers. The second, TRAM, functions as an adaptor between TLR4 and TRIF (Figure 53.2). It is not clear yet whether the fifth and last member of the family, SARM, which is preferentially expressed in neurons, plays a role in Toll signaling.

MyD88-Dependent Activation of IRFs by TLRs

In addition to TLR3 and TLR4, two other TLRs that are activated by ligands of viral origins, TLR7 and TLR9, regulate induction of IFNβ. However, these receptors do not require TRIF, and activate IRF7 in plasmacytoid dendritic cells (a subset of antigen presenting cells that express constitutively IRF7 and play a major role in IFN production). Instead they rely on MyD88, and kinases of the IRAK family. One proposed model is that IRF7 interacts with MyD88 at the receptor level, and is phosphorylated by IRAK1. A role for IKKα in the phosphorylation of IRF7 has also been proposed. Interestingly, another member of the IRF family, IRF5, interacts with MyD88 at the receptor level, and participates together with AP-1 and NFκB, to the induction of several genes encoding inflammatory cytokines [15, 22].

In summary, Toll receptors form an evolutionary ancient family of receptors. They activate cellular responses through a complex set of interactions involving homophilic association between TIR or death domains at the receptor level. These interactions involve members of a small number of families (TIR adaptors, IRAK kinases). At the post-receptor level, a distinct set of molecules triggers posttranslational modifications such as polyubiquitination (TRAFs) and phosphorylation (IKKs, MAPKs) leading to nuclear translocation and/or activation of transcription factors of the NFκB, AP-1, and IRF families. The elucidation of these signaling cascades holds great promise for the development of new drugs acting on inflammation.

REFERENCES

1. Belvin MP, Anderson KV. A conserved signaling pathway: the *Drosophila* toll-dorsal pathway.. *Ann Rev Cell Dev Biol* 1996;**12**:393–416.
2. Lemaitre B, Nicolas E, Michaut L, Reichhart J, Hoffmann J. The dorsoventral regulatory gene cassette spätzle/Toll/cactus controls the potent antifungal response in *Drosophila* adults. *Cell* 1996;**86**:973–83.
3. Hoffmann J. The immune response of *Drosophila*. *Nature* 2003;**426**:33–8.
4. Medzhitov R, Preston-Hurlburt P, Janeway C. A human homologue of the *Drosophila* Toll protein signals activation of adaptive immunity. *Nature* 1997;**388**:394–7.
5. Poltorak A, He X, Smirnova I, Liu M, Huffel C, Du X, Birdwell D, Alejos E, Silva M, Galanos C, Freudenberg M, Ricciardi-Castagnoli P, Layton B, Beutler B. Defective LPS signaling in C3H/HeJ and C57BL/10ScCr mice: mutations in Tlr4 gene. *Science* 1998;**282**:2085–8.
6. Gay NJ, Gangloff M. Structure and function of Toll receptors and their ligands. *Annu Rev Biochem* 2007;**76**:141–65.
7. Imler JL, Zheng L. Biology of Toll receptors: lessons from insects and mammals. *J Leukoc Biol.* 2003;**75**:18–26.
8. Xu Y, Tao X, Shen B, Horng T, Medzhitov R, Manley JL, Tong L. Structural basis for signal transduction by the Toll/interleukin-1 receptor domains. *Nature* 2000;**408**:111–15.
9. Kim HM, Park BS, Kim JI, Kim SE, Lee J, Oh SC, Enkhbayar P, Matsushima N, Lee H, Yoo OJ, Lee JO. Crystal structure of the TLR4–MD-2 complex with bound endotoxin antagonist eritoran. *Cell* 2007;**130**:906–17.
10. Choe J, Kelker MS, Wilson IA. Crystal structure of human toll-like receptor 3 (TLR3) ectodomain. *Science* 2005;**309**:581–5.
11. Bell JK, Botos I, Hall PR, Askins J, Shiloach J, Segal DM, Davies DR. The molecular structure of the Toll-like receptor 3 ligand-binding domain. *Proc Natl Acad Sci USA* 2005;**102**:10,976–10,980.
12. Andersen-Nissen E, Smith KD, Bonneau R, Strong RK, Aderem A. A conserved surface on Toll-like receptor 5 recognizes bacterial flagellin. *J Exp Med* 2007;**204**:393–403.
13. Weber AN, Tauszig-Delamasure S, Hoffmann JA, Lelievre E, Gascan H, Ray KP, Morse MA, Imler JL, Gay NJ. Binding of the *Drosophila* cytokine Spatzle to Toll is direct and establishes signaling. *Nat Immunol* 2003;**4**:794–800.
14. Beutler B, Jiang Z, Georgel P, Crozat K, Croker B, Rutschmann S, Du X, Hoebe K. Genetic analysis of host resistance: Toll-like receptor signaling and immunity at large. *Annu Rev Immunol* 2006;**24**:353–89.
15. Akira S, Uematsu S, Takeuchi O. Pathogen recognition and innate immunity. *Cell* 2006;**124**:783–801.
16. Janssens S, Beyaert R. Functional diversity and regulation of different interleukin-1 receptor-associated kinase (IRAK) family members. *Mol Cell* 2003;**11**:293–302.
17. Latz E, Verma A, Visintin A, Gong M, Sirois CM, Klein DC, Monks BG, McKnight CJ, Lamphier MS, Duprex WP, Espevik T, Golenbock DT. Ligand-induced conformational changes allosterically activate Toll-like receptor 9. *Nat Immunol* 2007;**8**:772–9.
18. Weber AN, Moncrieffe MC, Gangloff M, Imler JL, Gay NJ. Ligand–receptor and receptor–receptor interactions act in concert to activate signaling in the Drosophila toll pathway. *J Biol Chem* 2005;**280**:22,793–22,799.
19. Wullaert A, Heyninck K, Janssens S, Beyaert R. Ubiquitin: tool and target for intracellular NF-kappaB inhibitors. *Trends Immunol* 2006;**27**:533–40.
20. O'Neill LA, Bowie AG. The family of five: TIR-domain-containing adaptors in Toll-like receptor signalling.. *Nat Rev Immunol* 2007;**7**:353–64.
21. Meylan E, Burns K, Hofmann K, Blancheteau V, Martinon F, Kelliher M, Tschopp J. RIP1 is an essential mediator of Toll-like receptor 3-induced NF-kappa B activation. *Nat Immunol* 2004;**5**:503–7.
22. Honda K, Taniguchi T. IRFs: master regulators of signalling by Toll-like receptors and cytosolic pattern-recognition receptors. *Nat Rev Immunol* 2006;**6**:644–58.

Index

A

Abl, 119
AC. *See* Adenylyl cyclase
Acid Sensing Ion Channel (ASIC), 354, 355f
Acidic box, of FGFR, 126
ActA:Follistatin complex, 270
Activation
  of AP-1, 305
  of EGFR, 59, 61, 103–108, 103t, 104f, 104t
    activation-dependent higher-order ErbB oligomers in cancer cells, 107–108, 107t
    biophysical studies of activation at cell surface, 106
    conformations of ECD fragments, 105, 105f
    EGFR in pre-formed dimers, 106
    kinase domain fragment structures, 105–106, 105f
    ligand-induced EGFR tetramers, 106–107, 106f
    new paradigm in ErbB activation and signaling, 108
    signaling network pathways, 104–105, 104f
    structural biology of receptor fragments, 105
  of Eph receptors, 59
  of ErbB receptors, 95–97, 96f
  of Erk, 168–169
  of GH-PRL hormone receptors, 177–181
    cytokine hormones as transcriptional enhancers, 178–179, 178f
    growth hormone and receptor family, 177, 178f
    hormone specificity and cross-reactivity, 181
    receptor activation triggering, 177–178
    receptor homodimerization, 179–181, 179f–180f
  of GPCRs, 213, 237–242
    agonist-induced activation, 240–242
    conformational states during, 238–239, 238f
    conformational states on pathway to activation, 238–239, 238f
    crystal structure of human β_2-AR, 227–228, 239, 240f
    GPCR recognition of diffusible ligands, 237–238
    rhodopsin structure compared with human β_2-AR structure, 239–240, 240f, 241f
  of guanylyl cyclases, 315, 317
  of IR, 59
  of IRFs, 424–425
  of JAK kinases, 163–164, 163f, 164t
  of Jnk, 168–169
  of MAP kinase, 114
  of NFκB
    chemokine activation, 233
    TNFα activation, 299
    TRAF activation, 305
  of p38, 168–169
  of PDGF receptors, 118
  of PDGFβ receptor, 121
  of phospholipase C, 233
  of PI-3K, 97, 104
  of PKC, 233
  of PSKs, 36–38, 37f
  of PTK receptors, 58–61, 60f
    binding of molecules to intracellular parts of PTK receptors, 60
    docking of SH2 and PTB domain signaling proteins, 60, 60f
    homo- and heterodimerization, 59
    inhibition of phosphatases, 60
    ligand-induced receptor dimerization, 58–59
    nuclear function of PTK receptors, 61
    receptor kinase activation, 59
  of PTKs, 36–38, 37f
  of SAP kinase, 168–169
  of Smads, 278
  of STAT transcription factors, 162t, 165–168, 166t
  of TGFβ receptors, 275–276
  of urokinase, 121
  of VEGF, 59
  of VEGFR, 143–147
Active site, of GPCRs, 215–217
Active site dynamics, of PTPs and DUSPs, 43–45, 44f
Activins
  classification of, 273, 274f
  flexibility of, 266f, 269
  receptor I binding of, 266f, 267, 268f, 269, 275
  receptor II binding of, 269, 275
  structure of, 265–266, 266f
Activin-responsive element (ARE), 279
Activin-responsive factor (ARF), 279
Adalimumab (Humira), 300
ADAM10, 151
ADCC. *See* Antibody-dependent cellular cytotoxicity
Adenylyl cyclase (AC), 221–222, 221f
Adherens junctions, 330–331, 331f
Adherent cells, ECM and, 329–330, 330f
Adhesion receptors, 5, 5f
Adrenaline, GPCR interaction with, 228
β-Adrenergic receptor kinase (BARK), 246
Affinity. *See also* Binding affinity
  of drugs, 226
Agonist, types of, 213, 215, 226
Agonist-induced activation, of human β_2-AR, 240–242
Agonist-induced endocytosis, of GPCRs, 247–248, 247f
AGS3, 205
AKAP15, PKA complex with, 363–364, 364f
Akt
  insulin action on, 72
  in IRS→PI-3K→Akt cascade, 85–87, 86f
ALK. *See* Anaplastic lymphoma kinase
Alloreactivity, of TCR/pMHC complexes, 403
Altered peptide ligands (APLs), outcome of TCR/pMHC complexes, 402
γ-Aminobutyric acid B ($GABA_B$), 257
Analgesia
  DOR-KOR in, 259
  MOR-DOR in, 259
Anaplastic lymphoma kinase (ALK), 59
ANP. *See* Atrial natriuretic peptide
Antibody-dependent cellular cytotoxicity (ADCC), 408
Anti-TNF agents, 299–300
AP-1, TRAF activation of, 305
APLs. *See* Altered peptide ligands

β_2-AR. *See* Human β_2-adrenergic receptor
β_2AR–AT1R, in cardiomyocytes, 258
β_2AR–β_1AR, in cardiomyocytes, 258
ARE. *See* Activin-responsive element
ARF. *See* Activin-responsive factor
β_2AR-prostaglandin receptor, in asthma, 258–259
Arrestins, 246–248, 247f, 249, 256–257
AS160, in glucose transport, 72
ASIC. *See* Acid Sensing Ion Channel
Asthma, β_2AR-prostaglandin receptor in, 258–259
AT1R–B2R, in pregnancy-induced pre-eclampsia, 258
Atomic resolution, of heterotrimeric G proteins, 219–222
future structures, 222
Gα and AC, 221–222, 221f
Gβγ, 220–221, 220f
GTP hydrolytic mechanism gained from, 220, 220f
phosducin and Gβγ, 221, 221f
RGS domains and GAPs, 222, 222f
signaling in visual fidelity, 222, 222f
switching mechanism of Gα subunits, 219–220, 220f
Atrial natriuretic peptide (ANP), 317
Autoimmune disease, TNFα role in, 300
Autoinhibition
of PSKs, 36–38, 37f
of PTKs, 36–38, 37f
Autophosphorylation, of PTKs, 59
Avastin. *See* Bevacizumab
Axon guidance, Eph receptor family and, 333–334, 334f

B

B7-1, CTLA-4 complex with, 332–333, 332f
Bacterial AB_5 toxins, 344
BAM22, 256
BARK. *See* β-Adrenergic receptor kinase
Barrel-stave architecture, of ion channels, 352
Bax inhibitor-1, 379
Bcl-2, 379
BDNF. *See* Brain-derived neurotrophic factor
Beta-arrestin, 246, 247f, 256–257
Beta-blockers, GPCR interaction with, 214–215
Bevacizumab (Avastin), 62t
Binding
complex formation, 12, 15–20
interaction between membrane-anchored proteins, 19–20
interaction kinetics, 16–17, 16f–17f
modular structure of protein–protein binding sites, 18–19, 18f
protein complex dissociation, 18–19
thermodynamics, 15–16, 16f
to G protein-coupled receptors, 213–215, 214f, 216f
of GH-PRL hormones, 179–181, 180f
energetics of, 180–181
hormone-receptor binding sites, 179–180
receptor–receptor interactions, 180, 180f
structural and functional coupling of binding sites, 181
to GPCRs, 228, 255–256
to IR, 70–71, 71f
to PTK receptors, 60
structural and energetic basis of, 11–13
binding principles, 11–12
future prospects in, 12–13
non-specific association with membrane surfaces, 12
protein–protein interactions, 12
structural basis of protein-protein recognition, 25–26, 25–26t, 27f
Binding affinity
between membrane-anchored proteins, 19–20
between proteins, 15
Binding "hotspots," 180
of EPOR, 186–188, 188f
Binding sites, modular structure of, 18–19, 18f
BMP2:BMPRIA$_{ec}$, 269
BMP7:ActRII$_{ec}$, 269
BMP7:Noggin complex, 270
BMPS. *See* Bone morphogenetic proteins
BNP. *See* B-type natriuretic peptide
Bone homeostasis, TNFα role in, 299
Bone morphogenetic proteins (BMPS)
classification of, 273, 274f
co-receptors for, 276
receptor I binding of, 266f, 267, 268f, 269, 275
receptor II binding of, 269, 275
structure of, 265–266, 266f
Box-1 domain, of cytokine receptors, 161–163, 185
Box-2 domain, of cytokine receptors, 161–163
BPS region, IR binding to, 78–79, 79f
Brain-derived neurotrophic factor (BDNF), 133–134, 137
therapeutics based on, 138
B-type natriuretic peptide (BNP), 317

C

C106. *See* C-terminal fragment of 106 amino acids
CaBP-1, 366
Cachexia, 293
Cadherins, 330–331, 331f
Calcium channels, 359–366
calcium increase with, 378–379
current types, 360
effector checkpoint regulation of, 366
influx, 378
molecular function of, 362–363
molecular properties of, 360–363
physiological roles of, 359, 359f
signaling complexes, 363–366
in excitation-contraction coupling, 363–364
in synaptic transmission, 364–366, 364f
SNARE proteins and, 364–365, 364f
store-operated, 373–374
subunits of
function, 361
structure, 360–361, 361f
types of, 360–362, 362t
Calcium currents, types of, 360
Calcium homeostasis, 377–378, 377f
Calcium induced calcium release (CICR), 379
Calcium puffs, 380
Calcium release activated calcium (CRAC) channels, 373–374, 378
Calcium release activated calcium current (I_{crac}), 373–374
Calcium sensor proteins (CaS), 366
Calcium signaling
calcium homeostasis, 377–378, 377f
channels for, 378–379
in hepatocytes, 379–380
$InsP_3Rs$ and, 380
intracellular, 377–381
long-term, 380
spatial regulation of, 380–381
temporal regulation of, 379–380
toolkit, 377–378, 377f
Calcium sparks, 380
Calcium-binding proteins, 364–366, 364f
Calmodulin, bound to C-terminal domains, 364
Calmodulin-dependent protein kinase II, 380
CaM, 366
CaM-binding domain (CBD), 364f, 365–366
cAMP, in CNG channels, 383
Cancer
Eph/ephrin interactions in, 153
ErbB receptors in, 99–100, 107–108, 107t, 111
JAK mutation in, 171
loss of RTK ubiquitination and, 67
TNF role in, 300
TNFα as therapy for, 293
CAP. *See* Cbl-associated protein
Capacitative calcium entry. *See* Store-operated calcium channels

Carazolol, GPCR interaction with, 214–215, 214f, 216f, 225, 227–228, 239–240, 240f
Carbohydrates
recognition of
biological roles of, 341
lectins and, 342
signaling and, 341–345
structure and diversity of, 341–342, 342f
Carbohydrate recognition domain (CRD), 342
Carbohydrate-binding proteins. *See* Lectins
Carbohydrate-mediated signaling, 342–344
glycolipids as receptors, 344
glycoproteins as receptors, 343–344
glycosaminoglycans, 344
lectins as receptors, 342–343
and lectins in nucleocytosolic compartment, 344
proteoglycans, 344
small soluble saccharides, 344
Cardiomyocytes
β_2AR–AT1R and β_2AR–β_1AR in, 258
InsP_3Rs in, 379
VOCs opening in, 379
CaS. *See* Calcium sensor proteins
Catalytic mechanism, of PTPs and DUSPs, 43–45, 44f
Catechol, GPCR interaction with, 238, 241
β-Catenin, 330–331
Cation-dependent mannose 6-phosphate receptor (CD-MPR), 343
CatSper channels, 378
Caveolae, EGFR in, 99, 112–115
Caveolin-1, 99, 112–115, 276
CB_1R receptor, 257
CBD. *See* CaM-binding domain
Cbl
insulin action on, 72–73
in PDGF receptor ubiquitination and degradation, 120
RTK endocytosis and, 66–67, 66f
Cbl-associated protein (CAP), 72–73
CC chemokines. *See* β-Chemokines
CCD. *See* Central core disease
CCR4, 232
CCR5, 232, 234
CD27, 50–51
CD40, 50
CD45, 45
CDC25 cell cycle regulators, 41
CDGs. *See* Congenital disorders of glycosylation
CDKs, 38
CD-MPR. *See* Cation-dependent mannose 6-phosphate receptor
Cell positioning, oncogenesis and dysregulation of, 153
Cell signaling
FRET analysis of, 29–32
detection techniques in, 30–32
fluorescent probes for, 29–30
future prospects in, 32
of transmembrane receptors, 3–7
origins of cell signaling research, 3–4
receptor–protein signaling complexes, 25–26t
signaling paradigms, 23–26, 24f
Cell surface
EGFR activation at, 106
TNFRs at, 293–300, 294f, 295t
biological functions of, 298–299
signaling pathways and regulation of, 298
therapeutic expectations for, 299–300
TNF structural features and, 293–298, 296f, 297t
Cell survival signaling, cytokine receptors in, 168
Cell-cell adhesions, 330–331, 331f
Cell–cell communication, Eph/ephrin facilitation of, 153
Cell–cell contacts, Eph/ephrin in disruption of, 181
Cell-surface adhesion receptors, 329–334
axon guidance and neural development, 333–334, 334f
cell-cell adhesions/adherens junctions, 330–331, 331f
mechanosensory mechanisms, 329–330, 330f
T cell co-stimulation, 331–333, 332f
Central core disease (CCD), RyRs in, 379
Cetuximab (Erbitux), 62t, 111
cGMP. *See* Cyclic GMP
CHD. *See* Cytokine homology domain
Chemokine inhibitors, 234
Chemokine receptors, 231–234
ligands of, 232t
Chemokines, 231–234
structure and function of, 231–233, 232t
α-Chemokines, 231, 233
β-Chemokines, 231, 233
Cholesterol, EGFR activity and, 113–114
CICR. *See* Calcium induced calcium release
c-Kit, 118
autoinhibition of, 36
Classical PTPs, 41–43, 42f
CNBD. *See* Cyclic nucleotide-binding domain
CNG channels. *See* Cyclic nucleotide-gated channels
CNP. *See* C-type natriuretic peptide
Common Smad (co-Smad), 276, 285
categorization of, 285
Complex dissociation, 18–19
Complex formation, 12
free energy landscapes in, 15–20
interaction between membrane-anchored proteins, 19–20
interaction kinetics, 16–17, 16f–17f
modular structure of protein–protein binding sites, 18–19, 18f
protein complex dissociation, 18–19
thermodynamics, 15–16, 16f
structural basis of protein-protein recognition, 25–26, 25–26t, 27f
Conformational states, of GPCR activation, 238–239, 238f
Congenital disorders of glycosylation (CDGs), 341
Conventional TNFs, 293–294, 296f
co-Smad. *See* Common Smad
Coupled monoubiquitination, 67
CRAC channels. *See* Calcium release activated calcium channels
CRACM. *See* Orai
CRD. *See* Carbohydrate recognition domain
CRDs. *See* Cysteine-rich domains
Cross-reactivity, of GH and PRL, 181
Cross-regulation, of Smads, 287
Cross-talk
of Ephs and ephrins, 152–153
between PDGF receptors and TGFβ receptor system, 121–122
between PTK receptors and other signaling pathways, 61–62
C-terminal domains
calmodulin bound to, 364
C-terminal TRAF domain, 307, 308f
proteolytic processing and regulation with, 363–364, 364f
C-terminal fragment of 106 amino acids (C106), toll family receptors and, 422
CTLA-4, B7-1 complex with, 332–333, 332f
CTLDs. *See* C-type lectin-like NK receptor domains
C-type lectin-like NK receptor domains (CTLDs), 410–412, 411f
LY49A, 408–409t, 410f, 411–412, 411f
NKG2D, 408–409t, 410f, 411f, 412
C-type natriuretic peptide (CNP), 318
Cubic ternary complex, of GPCRs, 213
CXC chemokines. *See* α-Chemokines
Cyanopindolol, GPCR interaction with, 214–215
Cyclic AMP-dependent protein kinase (PKA)
AKAP15 complex with, 363–364, 364f
GPCR desensitization by, 246–247

Cyclic AMP-dependent protein kinase (PKA) *(Continued)*
regulation with calcium channel signaling complexes, 363, 364f
structural features of, 35–36, 36f
Cyclic GMP (cGMP)
in CNG channels, 383
history of, 315
Cyclic nucleotide-binding domain (CNBD), 383–384
Cyclic nucleotide-gated (CNG) channels, 384–385, 384f
function of, 384
subunit composition of, 384–385
types of, 384
Cyclic nucleotide-regulated cation channels, 383–386
CNG channels, 384–385, 384f
general features of, 383–384, 384f
HCN channels, 385–386
CypA, 179
CypB, 178–179
Cysteine-based PTPs, 41
Cysteine-rich domains (CRDs), of TNFRs, 294–296, 296f
Cytokines, in development of insulin resistance, 87
Cytokine homology domain (CHD), 185
Cytokine hormones, as transcriptional enhancers, 178–179, 178f
CypA as activation switch, 179
CypB as chaperone and activator, 178–179
Cytokine receptors, 5–6, 5f, 159–171
developmental regulation of, 169–170, 170f
dimerization of, 185–186, 188–190, 189f
EPOR as paradigm for signaling by, 185–190
biochemical studies supporting preformed dimers, 188–190, 189f
structural studies on EPOR, 185–188, 187f, 188f
Erk, Jnk, and p38 activation downstream of, 168–169
extracellular domain of, 161–163
generation of high-affinity cytokine-receptor complexes, 161, 162t
in human disease, 170–171, 172t
JAK kinase activation by, 163–165, 163f, 164t
JAK1, 164, 164t
JAK2, 164, 164t
JAK3, 164–165, 164t
Tyk2, 164t, 165
negative regulation of, 169
oligomerization of, 47t, 48–50
overview of, 159–160
in PI-3K pathway, 168
signaling by, 163
STAT recruitment and activation by, 162t, 165–168, 166t
STAT1, 165, 166t
STAT2, 165–167, 166t
STAT3, 166t, 167
STAT4, 166t, 167
STAT5, 166t, 167–168
STAT6, 166t, 168

D

DC-SIGN, 342–343
DC-SIGNR, 342–343
DD. *See* Death domain
DDR receptors, 58
Death domain (DD), of TNFRs, 298
Death-effector domain (DED), of TNFRs, 298
DED. *See* Death-effector domain
Degradation
of cytokine receptors, 169
of GPCRs, 257
of PDGF receptors, 120
of PTK receptors, 61
Dephosphorylation, of PDGF receptor, 120
Desensitization, of GPCRs, 245, 246–248, 247f
agonist-induced endocytosis, 247–248, 247f
endocytosis role in, 247f, 248
functional uncoupling of GPCRs from heterotrimeric G proteins, 246–247, 247f
heterologous desensitization, 246–247
De-ubiquitinating enzymes (DUBs), 65
Developmental regulation, of cytokine receptors, 169–170, 170f
Diabetes
insulin resistance in, 69, 83, 89, 89f, 90
IR as therapeutic target for, 80
Diffusible hormones, GPCR recognition of, 237–238
Diffusive transition state, 17, 17f
Dimeric complexes, of GPCRs, 253–259
heterodimerization and altered receptor function, 255–257, 255f
heterodimerization in physiology and pathology, 257–259, 258t
historical perspective, 253–254
Dimerization
of cytokine receptors, 185–186, 188–190, 189f
of EPOR, 185–186, 188–190, 189f
of ErbB receptors and EGFR, 106
of GH and PRL receptors, 179–181, 179f–180f
of PTKs, 58–59
PTP inactivation via, 44f, 45
Disease
cytokine receptor involvement in, 170–171, 172t
ErbB receptors and EGFR in, 99–100
GPCR heterodimerization in, 257–259, 258t
5HT2AR–mGluR2 in schizophrenia, 259
AT1R–B2R in pregnancy-induced pre-eclampsia, 258
β_2AR–AT1R and β_2AR–β_1AR in cardiomyocytes, 258
β_2AR-prostaglandin receptor in asthma, 258–259
PTK receptors in, 62–63, 62t
Dissociation, of protein complexes, 18–19
Divergent TNFs, 294, 296f
Docking problem, of protein–protein interactions, 12
Dopamine, GPCR interaction with, 238, 241
DOR. *See* δ Opioid receptors
Double-strand RNA (dsRNA) receptors, 305
Downregulation
of GPCRs, 246, 247f, 248–249
of Smads, 287
DR4, 300
DR5, 300
dsRNA receptors. *See* Double-strand RNA receptors
Dual-specificity phosphatases (DUSPs)
catalytic mechanism of, 43–45, 44f
structure of, 41–43, 42f
DUBs. *See* De-ubiquitinating enzymes
DUSPs. *See* Dual-specificity phosphatases
Dynamin, 248

E

E3 ligases, 65–66, 66f
ECC. *See* Excitation-contraction coupling
ECD fragments, conformations of, 105, 105f
ECM. *See* Extracellular matrix
Ectodomains, 5–6
EF-disulfide TNFs, 294, 296f
Effector checkpoint regulation, of calcium channels, 366
Effector proteins
G protein interactions with, 202, 202f
in RTK signaling, ubiquitination of, 67
Efficacy, of drugs, 226
EF-hands, 380
EGF. *See* Epidermal growth factor
EGF-CFC. *See* Epidermal growth factor-cripto-FRL1-cryptic
EGFR. *See* Epidermal growth factor receptor
Electrostatic forces, of binding, 11–12
EMP1, EPOR bound to, 186, 187f
EMP33, EPOR bound to, 186, 187f
EMT. *See* Epithelial-to-mesenchymal transition

Enbrel. *See* Etanercept
Endocrine cells, calcium channels in, 359
Endocrinology, origins of, 3–4
Endocytosis
  of Eph/ephrin, 181
  of ErbB and EGFR molecules, 98–99, 98f
  of GPCRs, 246–248, 247f
    agonist-induced endocytosis, 247–248, 247f
    functional uncoupling of GPCRs from heterotrimeric G proteins, 246–247, 247f
    heterologous desensitization, 246–247
    role in controlling specificity of signal transduction, 249
    role in mediating proteolytic downregulation, 247f, 248–249
    role in rapid desensitization, 247f, 248
    role in resensitization, 247f, 248
  of RTKs, 61
    molecular link between RTK signaling and, 65–66, 66f
    ubiquitination in, 66–67, 66f
Endodomains, 5–6
Endothelial nitric oxide synthase (eNOS), 317
Energetics
  for GHR and PRL binding, 180–181
  of molecular recognition, 11–13
    binding principles, 11–12
    future prospects in, 12–13
    non-specific association with membrane surfaces, 12
    protein–protein interactions, 12
  of protein–protein interactions, 15–20
    interaction between membrane-anchored proteins, 19–20
    interaction kinetics, 16–17, 16f–17f
    modular structure of protein–protein binding sites, 18–19, 18f
    protein complex dissociation, 18–19
    thermodynamics, 15–16, 16f
eNOS. *See* Endothelial nitric oxide synthase
Enthalpy, of protein–protein interactions, 15–16
Entropy, of protein–protein interactions, 15–16
Eph receptors, 58
  activation of, 59
  in axon guidance and neural development, 333–334, 334f
  Eph/ephrin signaling involving, 149–153
    cell–cell communication during vertebrate development, 153
    cross-talk with other signal pathways, 152–153
    disruption of cell–cell contacts and internalization of signaling complexes, 181
    Eph forward signaling, 151–152
    ephrin reverse signaling, 152
    oncogenesis, 153
    protein structures and signaling concepts, 149–151, 150f
    regulation of signaling, 151
Epidermal growth factor (EGF), 111, 159
  preformed dimers of, 190
Epidermal growth factor receptor (EGFR), 57, 95–100, 96t
  activation of, 59, 61, 103–108, 103t, 104f, 104t
    activation-dependent higher-order ErbB oligomers in cancer cells, 107–108, 107t
    biophysical studies of activation at cell surface, 106
    conformations of ECD fragments, 105, 105f
    EGFR in pre-formed dimers, 106
    kinase domain fragment structures, 105–106, 105f
    ligand-induced EGFR tetramers, 106–107, 106f
    new paradigm in ErbB activation and signaling, 108
    signaling network pathways, 104–105, 104f
    structural biology of receptor fragments, 105
  attenuation of signaling by, 98–99, 98f
    receptor endocytosis, 98–99, 98f
    transcription-induced negative regulators, 99
  discovery of, 4
  in disease, 99–100
  domain structure of, 104f
  ligand binding to, 104t
  lipid domain role in EGFR signaling, 111–115
    EGFR and caveolin-1, 99, 112–115
    lipid raft studies, 112
    localization of EGFR in lipid rafts, 112–113
    rafts and EGFR-mediated signaling, 113–114
  phenotypes of mice deficient in, 96t
  regulation of, 36–37, 37f
  signaling pathways induced by, 97
  specificity of signaling by, 97–98
  structure and activation of, 95–97, 96f
  ubiquitination of, 65–67
Epidermal growth factor-cripto-FRL1-cryptic (EGF-CFC), 276
Epinephrine, GPCR interaction with, 228, 237–238
Epithelial-to-mesenchymal transition (EMT), 280–281
EPO. *See* Erythropoietin
EPOR. *See* Erythropoietin receptor
Equilibrium association constant, of protein–protein interactions, 15–16
ErbB receptors, 95–100, 96t, 103t. *See also* Epidermal growth factor receptor
  attenuation of signaling by, 98–99, 98f
    receptor endocytosis, 98–99, 98f
    transcription-induced negative regulators, 99
  biophysical studies of activation at cell surface, 106
  in cancer, 99–100, 107–108, 107t, 111
  conformations of ECD fragments of, 105, 105f
  in disease, 99–100
  domain structure of, 104f
  ligand binding to, 104t
  new paradigm in activation and signaling of, 108
  phenotypes of mice deficient in, 96t
  pre-formed dimers of, 106
  signaling pathways induced by, 97
  specificity of signaling by, 97–98
  structure and activation of, 95–97, 96f
ErbB1. *See* Epidermal growth factor receptor
ErbB2, 67, 95–97, 96f, 96t, 103t
  in cancer, 107–108, 111
ErbB3, 60, 95, 96t, 103t
ErbB4, 61, 95, 96t, 103t
Erbitux. *See* Cetuximab
Erk, cytokine receptors in activation of, 168–169
Erlotinib (Tarceva), 62t, 111
Erythropoietin (EPO), 161, 185
  structure of complex with EPOR, 186, 187f
Erythropoietin receptor (EPOR), 159–160
  dimerization of, 185–186, 188–190, 189f
  extracellular domain of, 161–163
  negative regulation of, 169
  oligomerization of, 49
  as paradigm for cytokine signaling, 185–190
    biochemical studies supporting preformed dimers, 188–190, 189f
    structural studies on EPOR, 185–188, 187f, 188f
  in PI-3K pathway, 168
  signaling by, 163–164, 163f
  structural studies of, 185–188, 187f, 188f
    bound to agonist, 186, 187f
    bound to antagonist, 186, 187f
    bound to EPO, 186, 187f
    EPOR binding hotspots, 186–188, 188f
    unliganded, 186, 187f
Etanercept (Enbrel), 300

Eukaryotic ion channels, 354–355, 355f, 356f
Excitation-contraction coupling (ECC)
calcium channel signaling in, 379
calcium channels signaling complexes in, 363–364
calcium binding proteins, 364, 364f
PKA regulation, 363, 364f
proteolytic processing and regulation, 363–364, 364f
steps in, 379
Expanding-network hypothesis, 51
Extended ternary complex, of GPCRs, 213
Extracellular domain
of cytokine receptors, 161–163
of GH and PRL receptors, 179, 180f
of GPCRs, 213, 214f
Extracellular matrix (ECM)
adherent cells and, 329–330, 330f
integrin signaling and, 337–338
Eya proteins. *See* Eyes absent proteins
Eyes absent (Eya) proteins, 41

F

FADD. *See* Fas-associated DD
Fas, 298–300
Fas-associated DD (FADD), 298
FAST1. *See* Forkhead activin signal transducer 1
Fc receptors (FcR), 391
Fcα/μR, IgA interaction with, 395f, 396
FcαRI, IgA interaction with, 393f, 395–396
FcεRI, IgE interaction with, 393f, 394–395
FcγR, IgG interaction with, 393–394, 393f
FcRn, IgG interaction with, 394, 394f
IgE-FcεRII, IgE interaction with, 395, 395f
pIgR, IgA interaction with, 395f, 396
structures of, 382f, 392–393
Fcα/μR, IgA interaction with, 395f, 396
FcαRI, IgA interaction with, 393f, 395–396
FcεRI, IgE interaction with, 393f, 394–395
FcεRII, IgE interaction with, 395, 395f
FcγR, IgG interaction with, 393–394, 393f
FcR. *See* Fc receptors
FcRn, IgG interaction with, 394, 394f
FERM domain, of JAK kinases, 163
FGF. *See* Fibroblast growth factor
FGFR. *See* Fibroblast growth factor receptor
Fibroblast growth factor (FGF), 57
in oligometric FGF–FGFR–HS complex, 127–128
polypeptides of, 125
Fibroblast growth factor receptor (FGFR), 125–126, 126f
in oligometric FGF–FGFR–HS complex, 127–128
Fibroblast growth factor signaling complex, 125–129
FGF polypeptides in, 125
FGFR in, 125–126, 126f
HS in, 125–127
intracellular signal transduction by, 128–129
klothos in, 126–127, 129
oligometric FGF–FGFR–HS complex, 127–128
Flies, toll family receptors in, 421–423, 422f
FLIM. *See* Fluorescence lifetime imaging microscopy
Fluorescence lifetime imaging microscopy (FLIM), 32
Fluorescence resonance energy transfer (FRET) analysis, 29–32
detection techniques in, 30–32
FLIM, 32
photobleaching methods, 31–32
ratio imaging, 30
sensitized emission measurements, 30–31
fluorescent probes for, 29–30
future prospects in, 32
Fluorescent probes, for FRET analysis, 29–30
Focal adhesions, 329
Forkhead activin signal transducer 1 (FAST1), 279–280
Formoterol, GPCR interaction with, 228
FoxH1. *See* Forkhead activin signal transducer 1
FOXO, in IRS→FOXO cascade, 87–88, 88f
FOXO3a, 168
Fractalkine, 231
Free energy landscapes, in protein–protein interactions, 15–20
interaction between membrane-anchored proteins, 19–20
interaction kinetics, 16–17, 16f–17f
modular structure of protein–protein binding sites, 18–19, 18f
protein complex dissociation, 18–19
thermodynamics, 15–16, 16f
FRET analysis. *See* Fluorescence resonance energy transfer analysis
Full agonist, 213, 215, 226
Functional activity, of GPCRs, 245
Functional complementation analysis, of GPCR dimeric complexes, 254
Functional cooperativity, 181

G

G proteins
atomic resolution of, 219–222
future structures, 222
Gα and AC, 221–222, 221f
Gβγ, 220–221, 220f
GTP hydrolytic mechanism gained from, 220, 220f
phosducin and Gβγ, 221, 221f
RGS domains and GAPs, 222, 222f
signaling in visual fidelity, 222, 222f
switching mechanism of Gα subunits, 219–220, 220f
functional uncoupling of GPCRs from, 246–247, 247f
in GPCR signaling, 211–213
modulation of, 364f, 365
structures of, 199–206, 200t
Gα subunits, 199–202, 201f
Gα–GPCR interactions, 205–206
GAP regulation, 202–204, 202f, 203f
Gβγ dimers, 204–205, 204f, 205f
G-effector interactions, 202, 202f
GTP hydrolysis by G proteins, 202–204, 202f, 203f
in Protein Data Bank, 200t
receptor-independent regulators of G proteins, 205
G protein regulatory (GPR) motifs, 205
G protein-coupled receptor kinases (GRKs), 246, 247f, 248
G protein-coupled receptors (GPCRs), 5–6, 5f, 199
β_2-AR as model of structure and activation of, 237–242
agonist-induced activation, 240–242
conformational states on pathway to activation, 238–239, 238f
crystal structure of human β_2-AR, 227–228, 239, 240f
GPCR recognition of diffusible ligands, 237–238
rhodopsin structure compared with human β_2-AR structure, 239–240, 240f, 241f
desensitization and endocytosis of
functional consequences of, 247f, 248–249
mechanisms of, 246–248, 247f
as drug targets, 214–215, 226, 234, 237
functional roles of dimeric complexes of, 253–259
heterodimerization and altered receptor function, 255–257, 255f
heterodimerization in physiology and pathology, 257–259, 258t
historical perspective, 253–254
Gα interactions with, 205–206
oligomerization of, 47t, 51, 253
in PDGF receptor regulation, 121
regulation of, 245–246

desensitization and resensitization for rapid regulation of functional activity, 245
downregulation and upregulation, 246
sequestration and rapid regulation of subcellular localization of receptors, 245–246
structures of, 209–217, 210–211t, 225–229
active site structure, 215–217
active state structures, 213–215, 214f, 216f, 228–229, 229f
classification based on, 209–211, 212f
extracellular domains, 213, 214f
GPCR signaling and, 211–213
human β_2-AR crystal structure, 227–228, 239, 240f
ligand binding specificity in GPCRs, 228
in Protein Data Bank, 209, 210–211t, 226t
recent advances in structural studies, 225–226, 226t
receptor activation models based on, 213
Gα subunits
AC and, 221–222, 221f
effector interactions with, 202, 202f
GPCR interactions with, 205–206
GTP binding by, 199, 201, 212
structure of, 199–202, 201f
switching mechanism of, 219–220, 220f
$GABA_B$. *See* γ-Aminobutyric acid B
Galectins, 343–344
GAP domains. *See* GTPase activating proteins domains
Gβγ dimers, 204–205, 204f, 205f, 220–221, 221f
GC-A, 316f, 316t, 317–318
oligomerization of, 47t, 50
GC-B, 316f, 316t, 318
GC-C, 316f, 316t, 318–319
GC-D, 316f, 316t, 319
GC-E, 316f, 316t, 319
GC-F, 316f, 316t, 319
GC-G, 316f, 316t, 319–320
GCSF receptor. *See* Granulocyte colony stimulating factor receptor
GDFs. *See* Growth differentiation factors
G-effector interactions, 202, 202f
Gefitinib (Iressa), 62t, 111
GEFs. *See* Guanine nucleotide exchange factors
Gene transcription, long-term calcium signaling and, 380
GFPs. *See* Green fluorescent proteins
GH. *See* Growth hormone
GHR. *See* Growth hormone receptor
Gleevec. *See* Imatinib
GLP1. *See* Glucagon-like peptide-1
Glucagon-like peptide-1 (GLP1), 89
Glucose, 89
Glucose transport, insulin signaling to, 71–73, 72f
Glucose transporter type 4 (GLUT4), 72–73, 72f
GLUT4. *See* Glucose transporter type 4
Glutamate residues, in calcium channels, 362–363
Glycolipids, as receptors in carbohydrate-mediated signaling, 344
Glycoproteins, as receptors in carbohydrate-mediated signaling, 343–344
Glycosaminoglycans, 344
Glycosphingolipids, 344
Glycosyltransferases, 342
GM-CSF. *See* Granulocyte-macrophage colony stimulating factor
GoLoco motifs. *See* G protein regulatory motifs
GPCRs. *See* G protein-coupled receptors
GPR motifs. *See* G protein regulatory motifs
Granulocyte colony stimulating factor (GCSF) receptor, 190
Granulocyte-macrophage colony stimulating factor (GM-CSF), 161
Grb2, 78
Grb14, IR recruitment of, 78–79, 79f
Green fluorescent proteins (GFPs), 29–30
GRKs. *See* G protein-coupled receptor kinases
Growth differentiation factors (GDFs)
classification of, 273, 274f
structure of, 265–266, 266f
Growth factors, discovery of, 4
Growth hormone (GH), 161, 177
role of FOXO in signaling pathways of, 88, 88f
Growth hormone receptor (GHR)
binding of
energetics of, 180–181
hormone-receptor binding sites, 179–180
receptor–receptor interactions, 180, 180f
structural and functional coupling of binding sites, 181
dimerization of, 190
oligomerization of, 49
structural basis for activation and regulation of, 177–181
cytokine hormones as transcriptional enhancers, 178–179, 178f
growth hormone and receptor family, 177, 178f
hormone specificity and cross-reactivity, 181
receptor activation triggering, 177–178
receptor homodimerization, 179–181, 179f–180f
GTP. *See* Guanosine triphosphate
GTPase activating proteins (GAP) domains, 202–204, 202f, 203f, 222, 222f
Guanine nucleotide exchange factors (GEFs), 199
Guanosine triphosphate (GTP)
G protein binding of, 199, 201, 212
G protein hydrolysis of, 202–204, 202f, 203f, 220, 220f
Guanylate cyclase. *See* Guanylyl cyclases
Guanylyl cyclases, 315–320
GC-A/NPR-A/NPR1, 316f, 316t, 317–318
GC-B/NPR-B/NPR2, 316f, 316t, 318
GC-C/StaR, 316f, 316t, 318–319
GC-D, 316f, 316t, 319
GC-E/RET-GC1, 316f, 316t, 319
GC-F/RET-GC2, 316f, 316t, 319
GC-G, 316f, 316t, 319–320
history of, 315–320
mammalian, 315, 316f, 316t
NO activation of, 315, 317
NPR-C/NPR3, 316f, 316t, 318
Guanylyl cyclase receptors, 5–6, 5f
oligomerization of, 47t, 50

H

HCN channels. *See* Hyperpolarization-activated cyclic nucleotide-gated channels
hD2. *See* Human MD-2
HDACs. *See* Histone deacetylases
Heparan sulfate (HS)
in FGF signaling complex, 125–127
in oligometric FGF–FGFR–HS complex, 127–128
Heparin
FGF binding of, 125–127
FGFR binding of, 126, 126f
PDGFα receptor modulation by, 121
Hepatic asialoglycoprotein receptor, 341, 343
Hepatocyte, temporal calcium regulation in, 379–380
Hepatocyte growth factor (HGF) receptors, 58
Heptahelical receptors. *See* G protein-coupled receptors
HER1. *See* Epidermal growth factor receptor
HER2. *See* ErbB2
Herceptin. *See* Trastuzumab
Heterodimerization
of GPCRs, 255–257, 255f
modulation of receptor coupling and signaling by, 256–257
modulation of receptor maturation, trafficking, and localization by, 257
modulation of receptor pharmacology by, 255–256

Heterodimerization *(Continued)*
physiology and pathology response to, 257–259, 258t
of PTKs, 59
Heterologous desensitization, of GPCRs, 246–247
Heterotrimeric G proteins
atomic resolution of, 219–222
future structures, 222
$G\alpha$ and AC, 221–222, 221f
$G\beta\gamma$, 220–221, 220f
GTP hydrolytic mechanism gained from, 220, 220f
phosducin and $G\beta\gamma$, 221, 221f
RGS domains and GAPs, 222, 222f
signaling in visual fidelity, 222, 222f
switching mechanism of $G\alpha$ subunits, 219–220, 220f
functional uncoupling of GPCRs from, 246–247, 247f
in GPCR signaling, 211–213
structures of, 199–206, 200t
$G\alpha$ subunits, 199–202, 201f
$G\alpha$–GPCR interactions, 205–206
GAP regulation, 202–204, 202f, 203f
$G\beta\gamma$ dimers, 204–205, 204f, 205f
G-effector interactions, 202, 202f
GTP hydrolysis by G proteins, 202–204, 202f, 203f
in Protein Data Bank, 200t
receptor-independent regulators of G proteins, 205
HGF receptors. *See* Hepatocyte growth factor receptors
hGH receptors. *See* Human growth hormone receptors
Histone deacetylases (HDACs), 286
HIV, CCR5 interaction with, 232, 234
Homodimerization, of PTKs, 59
Horizontal receptor signaling, 159–160
Hormones. *See also* Diffusible hormones
discovery of, 3–4
"Hot-spot" residues, of protein complexes, 18–19
HS. *See* Heparan sulfate
5HT2AR–mGluR2, in schizophrenia, 259
hTLR3. *See* Human TLR3
hTLR4. *See* Human TLR4
Human adenosine A2A receptor, 209, 213–215, 214f, 216f
Human β_2-adrenergic receptor (β_2-AR)
crystal structures of, 227–228, 239, 240f
desensitization and resensitization of, 245
dimeric complexes of, 254, 256
endocytosis of, 248–249
as model of GPCR structure and activation, 237–242
agonist-induced activation, 240–242
conformational states on pathway to activation, 238–239, 238f
crystal structure of human β_2-AR, 227–228, 239, 240f
GPCR recognition of diffusible ligands, 237–238
rhodopsin structure compared with human β_2-AR structure, 239–240, 240f, 241f
PKA-mediated phosphorylation of, 246–247
rhodopsin *v.*, 239–240, 240f, 241f
structure of, 209, 214–216, 214f, 216f, 225
Human growth hormone (hGH) receptors, 159
Human MD-2 (hD2), 417–418, 417f, 418f
Human TLR3 (hTLR3), 415, 417f
Human TLR4 (hTLR4), 415–417, 417f
Humira. *See* Adalimumab
Hydrophobic forces, of binding, 11–12
Hyperpolarization-activated cyclic nucleotide-gated (HCN) channels, 385–386
function of, 385
locations of, 385–386
types of, 385

I

IA-2, 45
IA-2β, 45
I_{crac}. *See* Calcium release activated calcium current
IgA. *See* Immunoglobulin A
IgD. *See* Immunoglobulin D
IgE. *See* Immunoglobulin E
IGF. *See* Insulin-like growth factor
IGFR. *See* Insulin-like growth factor receptor
IgG. *See* Immunoglobulin G
IgM. *See* Immunoglobulin M
Igs. *See* Immunoglobulins
IL-1. *See* Interleukin-1
IL-2. *See* Interleukin-2
IL-2 receptor, oligomerization of, 49
IL-3. *See* Interleukin-3
IL-4 receptor, oligomerization of, 49
IL-5. *See* Interleukin-5
IL-6 receptor, oligomerization of, 49
IL-8. *See* Interleukin-8
IL-10 receptor, oligomerization of, 49
IL-11 receptor, oligomerization of, 49
Imatinib (Gleevec), 38, 62t
Immune receptor signaling, TRAFs in, 305–309, 306f
biological functions of, 305–306
discovery of, 305, 306t
domain organizations and structures of, 306–307, 307f–308f
C-terminal TRAF domain, 307, 308f
N-terminal RING domain and zinc fingers, 306–307
regulation of TRAF signaling, 309
TRAF signaling and Lys63 linked polyubiquitination, 308–309
TRAF6, 307–308
Immune response, TNFRs in, 299
Immunoglobulins (Igs)
function of, 391
structure of, 391–392, 391f, 392f
Immunoglobulin A (IgA)
FcR interactions
FcαRI, 393f, 395–396
IgA-Fcα/μR, 395f, 396
pIgR, 395f, 396
function of, 391
structure of, 391–392, 391f, 392f
Immunoglobulin D (IgD)
function of, 391
structure of, 392
Immunoglobulin E (IgE)
FcR interactions
FcεRI, 393f, 394–395
IgE-FcεRII, 395, 395f
function of, 391
structure of, 391–392, 391f, 392f
Immunoglobulin G (IgG)
FcR interactions, 393–394
FcγR, 393–394, 393f
FcRn, 394, 394f
function of, 391
structure of, 391–392, 391f, 392f
Immunoglobulin M (IgM)
function of, 391
structure of, 392
Immunoglobulin receptors, 5f, 6
Immunoglobulin-Fc receptor interactions, 391–396
IgA-Fcα/μR, 395f, 396
IgA-FcαRI, 393f, 395–396
IgA-pIgR, 395f, 396
IgE-FcεRI, 393f, 394–395
IgE-FcεRII, 395, 395f
IgG-FcγR, 393–394, 393f
IgG-FcRn, 394, 394f
Immunoreceptor tyrosine-based activation motifs (ITAMs), 408
Immunoreceptor tyrosine-based inhibitory motifs (ITIMs), 407–408
Inducible nitric oxide synthase (iNOS), 317
INFγ. *See* Interferon-γ
INFγR. *See* Interferon-γ receptor

Inflammation
chemokine role in, 234
TNFRs in, 299–300
Infliximab (Remicade), 300
Inhibins, structure of, 265–266, 266f
Inhibitory Smads (I-Smads), 276, 285
downregulation and cross-regulation of, 287
TGFβ signal transduction regulation by, 278–279
iNOS. *See* Inducible nitric oxide synthase
Inositol 1,4,5-trisphosphate ($InsP_3$), 373, 378
Inositol 1,4,5-trisphosphate receptors ($InsP_3Rs$), 378–379
calcium signals and, 380
$InsP_3$. *See* Inositol 1,4,5-trisphosphate
$InsP_3Rs$. *See* Inositol 1,4,5-trisphosphate receptors
Insulin, 69, 75
glucose transport due to, 71–73, 72f
in insulin signaling cascade, 83–84, 84f
Insulin and insulin-like signaling cascade, 83–90
dysregulation of IRS-protein signaling, 87–90, 88f–89f
IRS2 signaling and central control of nutrient homeostasis and lifespan, 90
role of IRS2 signaling in pancreatic β-cells, 88–89, 89f
role of IRS→FOXO cascade, 87–88, 88f
insulin, IGFs, and their receptors, 83–84, 84f
IRS1 and IRS2, 84–85, 85f
IRS→PI-3K→Akt cascade, 85–87, 86f
saccharides in, 344
Insulin receptor (IR), 57, 69–73, 75–80, 76f
activation of, 59
binding determinants of, 70–71, 71f
domain structure of, 69–70, 70f
glucose transport due to insulin signaling, 71–73, 72f
in insulin signaling cascade, 83–84, 84f
oligomerization of, 48
protein recruitment to, 76–79, 77f–79f
Grb14 recruitment, 78–79, 79f
IRS1 and IRS2, 77–78, 78f
PTP1B recruitment, 79, 79f
SH2B2 recruitment, 76–77, 77f
regulation of, 37f
Insulin receptor substrate 1 (IRS1), 83
dysregulation of signaling by, 87–90, 88f–89f
role of IRS→FOXO cascade, 87–88, 88f
in insulin signaling cascade, 84–85, 85f
IR recruitment of, 77–78, 78f
in IRS→PI-3K→Akt cascade, 85–87, 86f
Insulin receptor substrate 2 (IRS2), 83
dysregulation of signaling by, 87–90, 88f–89f
IRS2 signaling and central control of nutrient homeostasis and lifespan, 90
role of IRS2 signaling in pancreatic β-cells, 88–89, 89f
role of IRS→FOXO cascade, 87–88, 88f
in insulin signaling cascade, 84–85, 85f
IR recruitment of, 77–78, 78f
in IRS→PI-3K→Akt cascade, 85–87, 86f
Insulin resistance, 69, 83, 87–90, 88f–89f
IRS2 signaling and central control of nutrient homeostasis and lifespan, 90
role of IRS2 signaling in pancreatic β-cells, 88–89, 89f
role of IRS→FOXO cascade in, 87–88, 88f
Insulin-like growth factor (IGF), in insulin signaling cascade, 83–84, 84f
Insulin-like growth factor receptor (IGFR), in insulin signaling cascade, 83–84, 84f
Insulin-responsive aminopeptidase (IRAP), 72
Integrin
quaternary changes in, 338
structure of, 337–338, 337f
tail interactions, 339
tertiary changes in, 338–339
Integrin signaling, 337–339
Integrin-associated signaling, 329–330, 330f
Interaction kinetics, of protein–protein interactions, 16–17, 16f–17f
Interferon-γ (INFγ), 193–195, 194f
Interferon-γ receptor (INFγR)
oligomerization of, 49
structure of, 193–195, 194f
Interleukin-1 (IL-1), 307–308
Interleukin-2 (IL-2), 161
Interleukin-3 (IL-3), 161
Interleukin-5 (IL-5), 161
Interleukin-8 (IL-8), 231
Internalization
of Eph/ephrin, 181
of ErbB and EGFR molecules, 98–99, 98f
of GPCRs, 246–248, 247f
agonist-induced endocytosis, 247–248, 247f
functional uncoupling of GPCRs from heterotrimeric G proteins, 246–247, 247f
heterologous desensitization, 246–247
of RTKs, 61
molecular link between RTK signaling and, 65–66, 66f
ubiquitination in, 66–67, 66f
Intracellular calcium signaling, 377–381
calcium homeostasis, 377–378, 377f
channels for, 378–379
spatial regulation of, 380–381
temporal regulation of calcium signals, 379–380
Inverse agonist, 213, 215, 226
Ion channels, 5–6, 5f, 23. *See also specific channels*
architecture of, 351–352, 352f
complexes, 356
eukaryotic, 354–355, 355f, 356f
function of, 351–352
modular nature of, 355–356
open channels, 353–354, 354f
prokaryotic, 353, 353f
signals generated by, 351
structure of, 351–356
obtaining three-dimensional, 352
"Ionic lock," of GPCRs, 215, 216f, 229
disruption of, 238, 240, 241f
Ionotropic glutamate receptors, 355
IQ-like motif, 364f, 365–366
IR. *See* Insulin receptor
IRAP. *See* Insulin-responsive aminopeptidase
IRE1, regulation of, 36–38
Iressa. *See* Gefitinib
IRFs, MyD88 activation of, 424–425
IRS1. *See* Insulin receptor substrate 1
IRS2. *See* Insulin receptor substrate 2
I-Smads. *See* Inhibitory Smads
Isoproterenol, GPCR interaction with, 228
ITAMs. *See* Immunoreceptor tyrosine-based activation motifs
ITIMs. *See* Immunoreceptor tyrosine-based inhibitory motifs

J

JAK kinases
cytokine receptor activation of, 163–165, 163f, 164t
JAK1, 164, 164t
JAK2, 164, 164t
JAK3, 164–165, 164t
Tyk2, 164t, 165
cytokine receptor binding with, 161, 162t
in developmental regulation, 170
GHR and PRL in activation of, 177, 178f
in human disease, 170–171, 172t
Janus Homology 1 (JH1) domain, 164
Janus Homology 2 (JH2) domain, 164
JH1 domain. *See* Janus Homology 1 domain
JH2 domain. *See* Janus Homology 2 domain
Jnk. *See* Jun kinase
Jun kinase (Jnk), cytokine receptors in activation of, 168–169

K

KEYY motifs, of JAK kinases, 164
Killer cell Ig receptors (KIR), 408–409t, 410–411, 410f, 411f

Kinase domain fragment structures, of EGFR, 105–106, 105f
Kinase regulatory-loop binding (KRLB) domain, IR binding to, 78, 78f, 85, 85f
Kinetics, of protein–protein interactions, 16–17, 16f–17f
KIR. *See* Killer cell Ig receptors
KIT. *See* Stem cell factor receptor
Klothos, in FGF signaling complex, 126–127, 129
Knuckle epitope
  BMP and activin receptor I binding on, 266f, 267, 268f, 269
  BMP and activin receptor II binding on, 269
KOR. *See* κ Opioid receptors
KRLB domain. *See* Kinase regulatory-loop binding domain
Kv1.2 channel, 354, 356f

L

Lapatinib (Tykerb), 62t, 111
Latent TGFβ binding proteins (LTBPs), 273–274
Lectins
  carbohydrate and, in nucleocytosolic compartment, 344
  carbohydrate recognition, 342
  as receptors in carbohydrate-mediated signaling, 342–343
Leucine-rich repeats (LRRs), in Toll family receptors, 422–423
Leukemia, JAK mutation in, 171
Leukocyte Ig-like receptors (LILR), 408, 410–411, 410f
Lifespan, IRS2 signaling effect on, 90
Ligand, receptor interaction with, 23–24, 24f
Ligand binding. *See* Binding
Ligand traps, TGFβ signaling and, 274
Ligand-induced EGFR tetramers, 106–107, 106f
Ligand-induced receptor dimerization, of PTKs, 58–59
LILR. *See* Leukocyte Ig-like receptors
Lipid A, 417–418, 417f
Lipid domains, EGFR signaling and, 111–115
  EGFR and caveolin-1, 99, 112–115
  lipid raft studies, 112
  localization of EGFR in lipid rafts, 112–113
  rafts and EGFR-mediated signaling, 113–114
Lipid IVa, 417, 417f
LMPTP. *See* Low molecular weight PTP
Localization, of GPCRs, 245–246, 257
Low molecular weight PTP (LMPTP), 41
LRRs. *See* Leucine-rich repeats
LTBPs. *See* Latent TGFβ binding proteins
L-type calcium currents, 360
  mediation of, 361, 362t
LY49A, 408–409t, 410f, 411–412, 411f
Lymphotactin, 231
Lys63 linked polyubiquitination, of TRAF, 308–309
Lysergic acid diethylamide, 259

M

Macrophage-derived chemokine (MDC), 232
MAD. *See* Mothers Against Decapentaplegic
MAD homology domain (MH1), 276–278, 277f
MAD homology domain (MH2), 276–278, 277f
MAD related (MADR), 276
MADR. *See* MAD related
Major histocompatibility complex (MHC), peptide binding to, 399–400, 399f
Malignancy. *See* Cancer
Mammals, toll family receptors in, 421–423, 422f
Man-6-phosphate-mediated intracellular protein targeting, 341
Mannose binding protein (MBP), 343
MAP kinase. *See* Mitogen-activated protein kinase
Maturation, of GPCRs, 257
Maturity onset diabetes of the young (MODY), Pdx1 in, 89, 89f
MBP. *See* Mannose binding protein
MD-2, 417–418, 417f, 418f
MDC. *See* Macrophage-derived chemokine
Mechanical force, focal adhesions and, 329
Mechanically-activated calcium channels, 378
Mechanosensory mechanisms, of cell-surface adhesion receptors, 329–330, 330f
Membrane surfaces, non-specific association with, 12
Membrane-anchored proteins, interaction between, 19–20
Metabotropic glutamate receptor (mGluR), 254
Metal ion-dependent adhesion site (MIDAS), 337f, 338
Metastasis, Eph/ephrin interactions in, 153
N-Methyl-D-aspartate (NMDA), 378
mGluR. *See* Metabotropic glutamate receptor
MH1. *See* MAD homology domain
MH2. *See* MAD homology domain
MHC. *See* Major histocompatibility complex
Microscopy, FLIM, 32
MIDAS. *See* Metal ion-dependent adhesion site
MIS. *See* Mullerian inhibitory substance
Mitogen-activated protein (MAP) kinase
  EGF-induced activation of, 114
  Eph and ephrin down-modulation of, 152–153
  GPCR signaling to, 249
Modular structures, of protein–protein binding sites, 18–19, 18f
MODY. *See* Maturity onset diabetes of the young
Molecular recognition, structural and energetic basis of, 11–13
  binding principles, 11–12
  future prospects in, 12–13
  non-specific association with membrane surfaces, 12
  protein–protein interactions, 12
Molecular sociology, 23–27
  structural basis of protein–protein recognition, 25–26t, 26–27, 27f
  transmembrane signaling paradigms, 23–26, 24f
MOR. *See* μ Opioid receptors
MOR-neurokinin-1 receptor (NK1R), 257
Morphine, 256–257, 259
Mothers Against Decapentaplegic (MAD), 276, 285
Mouse TLR4 (mTLR4), 415–417, 417f
mTLR4. *See* Mouse TLR4
Mullerian inhibitory substance (MIS), 273, 274f
Muscle cells, calcium channels in, 359
  regulation of, 366
MyD88. *See* Myeloid differentiation clone number 88
Myeloid differentiation clone number 88 (MyD88)
  IRFs activation by, 424–425
  signaling, 423, 424f
Myeloproliferative disease, JAK mutation in, 171

N

NAADP-induced calcium release, 379
nAChR. *See* Nicotinic acetylcholine receptor
Na^+/H^+ exchange regulatory factor (NHERF), in stabilization of PDGF receptor dimer formation, 121
Natural cytotoxicity receptors (NCRs), 408–409t, 411
Natural killer cell receptors (NKRs), 407–412
  KIR, 408–409t, 410–411, 410f, 411f
  LILR, 408, 410–411, 410f
  specificity of, 407–408, 408–409t
  types of, 408, 410
Natural killer cells (NK), 407–410, 408–409t, 410f
NCRs. *See* Natural cytotoxicity receptors
Negative regulation, of cytokine receptors, 169
Nerve growth factor (NGF), 66, 133–138

discovery of, 4
neurotrophin signaling excursions and, 137
neurotrophins and, 133–134, 134f, 135t
$p75^{NTR}$ complexes with, 136–137, 137f
$p75^{NTR}$ in signaling of, 133, 136
in ternary receptor complexes, 138
therapeutics based on, 138
Trk receptors in signaling of, 133–135
TrkA complexes with, 135–136, 136f
Neural development, Eph receptor family and, 333–334, 334f
Neuron
calcium channels in, 359
calcium currents in, 360
Neuronal nitric oxide synthase (nNOS), 317
Neuropilin, 146
Neurotransmitter release, calcium signalling and, 380–381
Neurotrophin, 133–134, 134f, 135t
in p75•Trk•neurotrophin ternary complex, 138
signaling excursions of, 137
therapeutics based on, 138
Neurotrophin receptors, 58
Neurotrophin-3 (NT-3), 133–134, 137
Neurotrophin-4/5 (NT-4/5), 133–134
Neurotrophin-6 (NT-6), 133
Neutral antagonist, 213, 215, 226
Nexavar. *See* Sorafenib
NFκB. *See* Nuclear factor κB
NGF. *See* Nerve growth factor
NHERF. *See* Na^+/H^+ exchange regulatory factor
Nicotinic acetylcholine receptor (nAChR), 378
opening of, 353, 353f
structure of, 352, 352f, 354, 355f
in Torpedo electric organ, 352
Nitric oxide (NO), guanylyl cyclase activation by, 315, 317
Nitric oxide synthase (NOS), 317
Nitric oxide synthase 1 (NOS1). *See* Neuronal nitric oxide synthase
Nitric oxide synthase 2 (NOS2). *See* Inducible nitric oxide synthase
Nitric oxide synthase 3 (NOS3). *See* Endothelial nitric oxide synthase
NK. *See* Natural killer cells
NK1R. *See* MOR-neurokinin-1 receptor
NKG2D, 408–409t, 410f, 411f, 412
NKRs. *See* Natural killer cell receptors
NLRs. *See* NOD-like receptors
NMDA. *See* N-Methyl-D-aspartate
NMR. *See* Nuclear magnetic resonance
NMuMG. *See* Normal Murine Mammary Gland cells
nNOS. *See* Neuronal nitric oxide synthase
NO. *See* Nitric oxide
Nodals
classification of, 273, 274f
cripto for, 276
NOD-like receptors (NLRs), 305
Noradrenaline, GPCR interaction with, 228
Norepinephrine, 237, 256
Normal Murine Mammary Gland cells (NMuMG), 281
NOS. *See* Nitric oxide synthase
NOS1. *See* Neuronal nitric oxide synthase
NOS2. *See* Inducible nitric oxide synthase
NOS3. *See* Endothelial nitric oxide synthase
NPR1, 316f, 316t, 317–318
NPR2, 316f, 316t, 318
NPR3, 316f, 316t, 318
NPR-A, 316f, 316t, 317–318
NPR-B, 316f, 316t, 318
NPR-C, 316f, 316t, 318
NT-3. *See* Neurotrophin-3
NT-4/5. *See* Neurotrophin-4/5
NT-6. *See* Neurotrophin-6
N-type calcium currents, 360
mediation of, 361, 362t
regulation of, 364f, 365
in synaptic transmission, 364–366
Nuclear factor κB (NFκB)
chemokine activation of, 233
TNFα activation of, 299
TRAF activation of, 305
Nuclear function, of PTK receptors, 61
Nuclear magnetic resonance (NMR), for obtaining three-dimensional structures of ion channels, 352
Nutrient homeostasis, IRS2 signaling in, 90

O

Obesity, insulin resistance and, 87
Oligomerization
of EGFR, 106–107, 106f
of ErbB receptors in cancer cells, 107–108, 107t
of transmembrane receptors, 47–51, 47t
cytokine receptors, 47t, 48–50
GPCRs, 47t, 51, 253
guanylyl cyclase receptors, 47t, 50
PSKs, 47t, 50
RTKs, 47t, 48
TNF receptors, 47t, 50–51
Oligosaccharides. *See* Carbohydrates
Oncogenesis, Eph/ephrin interactions in, 153
Opiate analgesia, MOR-DOR in, 259
δ Opioid receptors (DOR), 255–257, 259
κ Opioid receptors (KOR), 255–256, 259
μ Opioid receptors (MOR), 256–257, 259
Opsin, 225, 229, 229f, 240, 241f
Orai, store-operated calcium channels and, 374, 378
Orexin receptor, 257
Oxytocin receptors, 256

P

p38, cytokine receptors in activation of, 168–169
p63RhoGEF, 202
$p75^{NTR}$
NGF complexes with, 136–137, 137f
in NGF signaling, 133, 136
in p75•Trk•neurotrophin ternary complex, 138
proneurotrophin binding and, 137
p115RhoGEF, 202, 204
PAMPs. *See* Pathogen-associated molecular patterns
Pancreatic β-cells, role of IRS2 signaling in, 88–89, 89f
Panitumumab (Vectibix), 62t
Partial agonist, 213, 215, 226
Pathogen-associated molecular patterns (PAMPs), 305
Pathology. *See* Disease
Pattern recognition receptors (PRR), 421
PDEγ, G protein binding to, 202f, 203, 222, 222f
PDGF. *See* Platelet-derived growth factor
PDGFα receptor, heparin modulation of, 121
PDK1. *See* Phosphoinositide-dependent protein kinase
Pdx1, in MODY, 89, 89f
PDZ-RhoGEF, 202, 204
Peptide-major histocompatibility complex (pMHC), 399. *See also* TCR/pMHC complexes
Pertuzumab, 111
PH domains. *See* Pleckstrin homology domains
Phosducin, 221, 221f
Phosphatidylinositol 3-kinase (PI-3K), 61–62
cytokine receptors in cell survival signaling involving, 168
ErbB activation of, 97, 104
insulin action on, 72–73, 78
in IRS→PI-3K→Akt cascade, 85–87, 86f
Phosphoinositide-dependent protein kinase (PDK1), insulin action on, 72, 86
Phospholipase C
chemokine activation of, 233
ErbB recruitment of, 97
Phosphorylation
of Eph/ephrin, 151
of GPCRs, 246–247, 247f

Phosphorylation *(Continued)*
of proteins, 35–36
of PTKs, 59
of tyrosine residues, 41
Phosphotyrosine-binding (PTB) domain
IR binding to, 76–78, 84–85, 85f
of PTKs, 60, 60f
Photobleaching, FRET analysis using, 31–32
PI-3K. *See* Phosphatidylinositol 3-kinase
pIgR, IgA interaction with, 395f, 396
PIR region. *See* BPS region
PKA. *See* Cyclic AMP-dependent protein kinase
PKC. *See* Protein kinase C
PL. *See* Placental lactogen
Placental growth factor (PlGF), 143–145, 144f
Placental lactogen (PL), 177
PLAD. *See* Pre-ligand-blinding assembly domain
Plasma membrane, integrin signaling across, 337–338
Platelet-derived growth factor (PDGF), 57, 117–122
activation of, 59
docking of SH2 and PTB domain signaling proteins, 60, 60f
isoforms of, 117, 117f
PDGFR and analogies to VEGFRs, 145–146, 146f
physiological function of, 117
receptor activation and regulation of kinase activity, 118
receptor interaction with downstream signal transduction molecules, 118–120, 119f–120f
regulation and modulation of receptor signaling, 120–122
Cbl mediated ubiquitination and degradation, 120
cross-talk with TGFβ receptor system, 121–122
dephosphorylation of PDGF receptor, 120
GPCR transactivation, 121
heparin modulation of PDGFα receptor, 121
NHERF stabilization of PDGF receptor dimer formation, 121
ROS role in, 122
serine/threonine phosphorylation, 121
urokinase activation of PDGFβ receptor, 121
Pleckstrin homology (PH) domains, IR binding to, 78, 84–85, 85f
PlGF. *See* Placental growth factor
pMHC. *See* Peptide-major histocompatibility complex
Potassium channels
opening of, 353–354, 353f
properties of, 353
structural elements of, 352, 353f, 354–355, 356f
PPI. *See* Prolyl-peptide isomerase
Pregnancy-induced pre-eclampsia, AT1R–B2R in, 258
Pre-ligand-blinding assembly domain (PLAD), of TNFRs, 298
Presenilin, 379
PRL. *See* Prolactin
PRLR. *See* Prolactin receptor
Prokaryotic ion channels, 353, 353f
Prolactin (PRL), 161, 177
Prolactin receptor (PRLR)
dimerization of, 190
structural basis for activation and regulation of, 177–181
cytokine hormones as transcriptional enhancers, 178–179, 178f
growth hormone and receptor family, 177, 178f
hormone specificity and cross-reactivity, 181
receptor activation triggering, 177–178
receptor homodimerization, 179–181, 179f–180f
Prolyl-peptide isomerase (PPI), in cytokine transcriptional enhancement, 178–179
Proneurotrophins, 137
Proteins
calcium-binding, 364–366, 364f
carbohydrate interactions with, 345
Protein Data Bank
G protein structures in, 200t
GPCR structures in, 209, 210–211t, 226t
receptor–protein signaling complexes in, 25–26t
Protein kinase B. *See* Akt
Protein kinase C (PKC), 61, 105
chemokine activation of, 233
Protein kinase inhibitors, 38
Protein phosphorylation, 35–36
Protein recruitment, to IR, 76–79, 77f–79f
Grb14 recruitment, 78–79, 79f
IRS1 and IRS2, 77–78, 78f
PTP1B recruitment, 79, 79f
SH2B2 recruitment, 76–77, 77f
Protein retrotransport, 178
Protein tyrosine kinases (PTKs), 35–38. *See also* Receptor tyrosine kinases
future prospects of, 35–36, 36f
regulation of, 36–38, 37f
structural features of, 35–36, 36f
Protein tyrosine phosphatases (PTPs)
catalytic mechanism of, 43–45, 44f
Eph/ephrin regulation by, 151
inhibition of, 60
large-scale structural analysis of, 41–45
family structural coverage, 41
receptor dimerization, 44f, 45
structural features, 41–43, 42f
PDGF receptor dephosphorylation by, 120
Protein–protein interactions, 12
free energy landscapes in, 15–20
interaction between membrane-anchored proteins, 19–20
interaction kinetics, 16–17, 16f–17f
modular structure of protein–protein binding sites, 18–19, 18f
protein complex dissociation, 18–19
thermodynamics, 15–16, 16f
oligomerization, 47–51, 47t
cytokine receptors, 47t, 48–50
GPCRs, 47t, 51, 253
guanylyl cyclase receptors, 47t, 50
PSKs, 47t, 50
RTKs, 47t, 48
TNF receptors, 47t, 50–51
structural basis of protein-protein recognition, 25–26, 25–26t, 27f
Proteoglycans, 344
Proteolytic processing and regulation, with calcium channels signaling complexes, 363–364, 364f
PRR. *See* Pattern recognition receptors
Pseudokinase domain. *See* Janus Homology 2 domain
Psilocybin, 259
PSKs. *See* Serine/threonine kinases
PTB domain. *See* Phosphotyrosine-binding domain
PTKs. *See* Protein tyrosine kinases
PTP1B, IR recruitment of, 79, 79f
PTPs. *See* Protein tyrosine phosphatases
P-type calcium currents, 360
mediation of, 361, 362t
regulation of, 364f, 365
in synaptic transmission, 364–366

Q

Q-type calcium currents, 360
mediation of, 361, 362t
regulation of, 364f, 365
in synaptic transmission, 364–366
Quinpirole, 256

R

RANTES, 232
Ras, 61–62

Ras/Erk pathway, PDGF receptor interaction in, 119
RasGAP, 119
Ras-like domain, of Gα proteins, 199–201, 219
Ratio imaging, FRET analysis using, 30
Reaction coordinate, of protein–protein interactions, 16
Reactive oxygen species (ROS), in PDGF receptor signaling, 122
Receptor binding domain-2. *See* Kinase regulatory-loop binding domain
Receptor tyrosine kinases (RTKs), 5–6, 5f, 35–38, 57–63, 95
activation mechanism of, 58–61, 60f
binding of molecules to intracellular parts of PTK receptors, 60
docking of SH2 and PTB domain signaling proteins, 60, 60f
homo- and heterodimerization, 59
inhibition of phosphatases, 60
ligand-induced receptor dimerization, 58–59
nuclear function of PTK receptors, 61
receptor kinase activation, 59
antagonists, 62–63, 62t
control of, 61
receptor internalization and degradation, 61
signaling control, 61
cross-talk between signaling pathways, 61–62
disease and, 62–63, 62t
future prospects of, 35–36, 36f
oligomerization of, 47t, 48
PTK subfamilies, 57–58, 58f
regulation of, 36–38, 37f
structural features of, 35–36, 36f
ubiquitination and, 65–67
conjugation of ubiquitin to target proteins, 65
molecular link between RTK signaling and endocytosis, 65–66, 66f
ubiquitination in RTK endocytosis, 66–67, 66f
ubiquitination of effector proteins in RTK signaling, 67
Receptor-independent regulators, of G proteins, 205
Receptor-operated calcium channels (ROCs), 378
Receptor–receptor interactions, of GH-PRL hormone receptors, 180, 180f
Receptor-regulated Smads (R-Smads), 273, 274f, 276, 285
activation of, 278
categorization of, 285
downregulation and cross-regulation of, 287
nucleocytoplasmic shuttling and, 285
receptor regulation of, 285
signaling by, 286
ubiquitin and, 287
Receptors. *See* Transmembrane receptors
Receptor-type phosphatases (RPTPs), 41
catalytic mechanism of, 43–45, 44f
large-scale structural analysis of, 41–45
family structural coverage, 41
receptor dimerization, 44f, 45
structural features, 41–43, 42f
Recombinant TNFα, 300
Recruitment. *See* Protein recruitment
Regulation
with calcium channel signaling complexes, 363, 364f
of calcium channels, 366
of calcium signaling, 379–381
with C-terminal domains, 363–364, 364f
of cytokine receptors, 169–170, 170f
of Eph/ephrin signaling activity, 151
of G proteins, 205
GAP, 202–204, 202f, 203f
of GH-PRL hormone receptors, 177–181
cytokine hormones as transcriptional enhancers, 178–179, 178f
growth hormone and receptor family, 177, 178f
hormone specificity and cross-reactivity, 181
receptor activation triggering, 177–178
receptor homodimerization, 179–181, 179f–180f
of GPCRs, 245–246
desensitization and resensitization for rapid regulation of functional activity, 245
downregulation and upregulation, 246
sequestration and rapid regulation of subcellular localization of receptors, 245–246
of N-type calcium currents, 364f, 365
of PDGF receptor signaling, 120–122
Cbl mediated ubiquitination and degradation, 120
cross-talk with TGFβ receptor system, 121–122
dephosphorylation of PDGF receptor, 120
GPCR transactivation, 121
heparin modulation of PDGFα receptor, 121
NHERF stabilization of PDGF receptor dimer formation, 121
ROS role in, 122
serine/threonine phosphorylation, 121
urokinase activation of PDGFβ receptor, 121
PKA, 363, 364f
of PSKs, 36–38, 37f
of PTK receptors, 61
of PTKs, 36–38, 37f
of P-type calcium currents, 364f, 365
of Q-type calcium currents, 364f, 365
of Smads, 285, 287
of TGFβ receptors, 275–276
of TGFβ signal transduction, 278–279
of TNFRs, 298
of TRAF signaling, 309
transcriptional, 285–287
Regulators of G protein signaling (RGS) domains, 202f, 203, 203f, 222, 222f, 247
Remicade. *See* Infliximab
Resensitization, of GPCRs, 245, 247f, 248
Resistance. *See* Insulin resistance
Resistance to inhibitors of cholinesterase-8 (RIC-8), 205
RET-GC1, 316f, 316t, 319
RET-GC2, 316f, 316t, 319
RGS domains. *See* Regulators of G protein signaling domains
RGS-RhoGEFs, 202, 204
RHIM. *See* RIP homotypic interaction motif
Rhodopsin
functional role of dimeric complexes of, 253
human β_2-AR *v.*, 239–240, 240f, 241f
structure of, 209, 213–216, 214f, 216f, 225–229, 229f
Rhodopsin kinase, 246
RIC-8. *See* Resistance to inhibitors of cholinesterase-8
RIM, binding of, 366
RING domain, of TRAFs, 306–307
RIP homotypic interaction motif (RHIM), 423
RNAKL : RANK system, 299
ROCs. *See* Receptor-operated calcium channels
ROS. *See* Reactive oxygen species
RPTPs. *See* Receptor-type phosphatases
R-Smads. *See* Receptor-regulated Smads
RTKs. *See* Receptor tyrosine kinases
R-type calcium currents, 360
mediation of, 361, 362t
in synaptic transmission, 364–366
Ryanodine receptors (RyRs), 379
RyRs. *See* Ryanodine receptors

S

Saccharides
in carbohydrate-mediated signaling, 344
in insulin signaling, 344

Salbutamol, GPCR interaction with, 228
SAP kinase, cytokine receptors in activation of, 168–169
SARA. *See* Smad anchor for receptor activation
SBEs. *See* Smad binding elements
SCF receptor. *See* Stem cell factor receptor
Schizophrenia
  5HT2AR–mGluR2 in, 259
  DOR-KOR in spinal analgesia, 259
  MOR-DOR in opiate analgesia and tolerance, 259
SCID. *See* Severe combined immunodeficiency
Second messenger operated channels (SMOCs), 378
E-Selectin, 342
L-Selectin, 342
P-Selectin, 342
Sensitized emission measurements, FRET analysis using, 30–31
Sensory neuron-specific receptor-4 (SNSR-4), 256
Sequestration, of GPCRs, 245–246
Serine/threonine kinases (PSKs), 35–38
  future prospects of, 35–36, 36f
  oligomerization of, 47t, 50
  regulation of, 36–38, 37f
  structural features of, 35–36, 36f
Serine/threonine phosphorylation, in regulation of PDGF receptor signaling, 121
Severe combined immunodeficiency (SCID), JAK mutation in, 171
SH2 domain. *See* Src homology-2 domain
SH2B2, IR recruitment of, 76–77, 77f
SHP-1, 61
SHP-2, 61, 78
Siglecs, 343
Signal transducers and activators of transcription (STAT)
  cytokine receptor recruitment and activation of, 162t, 165–168, 166t
    STAT1, 165, 166t
    STAT2, 165–167, 166t
    STAT3, 166t, 167
    STAT4, 166t, 167
    STAT5, 166t, 167–168
    STAT6, 166t, 168
  in developmental regulation, 170
  GHR and PRL in activation of, 177, 178f
  in human disease, 170–171
Signaling. *See* Cell signaling
Smads, 62, 274f, 275f, 276–278, 277f
  activation of, 278
  categorization of, 285
  DNA-binding partners and, 279–280
  as DNA-binding proteins, 279
  downregulation and cross-regulation of, 287
  function *in vivo*, 287–288
  nucleocytoplasmic shuttling and, 285
  receptor regulation of, 285
  TGFβ signal transduction regulation by, 278–279
  transcription co-activators and repressors, 280
  transcriptional regulation by, 285–287
Smad anchor for receptor activation (SARA), 276
Smad binding elements (SBEs), 279
SMOCs. *See* Second messenger operated channels
Smurfs, 287
SNAP23, in glucose transport, 73
SNARE proteins
  in effector checkpoint regulation of calcium channels, 366
  in glucose transport, 73
  in synaptic transmission, 364–365, 364f
SNSR-4. *See* Sensory neuron-specific receptor-4
SOCs. *See* Store-operated calcium channels
Sorafenib (Nexavar), 62t
Sortilin, proneurotrophin interaction with, 137
Spaetzle processing enzyme (SPE), 422
SPE. *See* Spaetzle processing enzyme
Specific transition state, 17, 17f
Specificity
  of GH and PRL, 181
  of GPCR binding, 228
  of GPCR signaling, endocytosis role in, 249
Spinal analgesia, DOR-KOR in, 259
Spinal cord injury, TNF role in, 300
SR141716A, 257
Src homology-2 (SH2) domain
  IR binding to, 76–77, 77f
  of PTKs, 60, 60f
Src-family PTKs, 38
StaR, 316f, 316t, 318–319
STAT. *See* Signal transducers and activators of transcription
Stem cell factor (SCF) receptor, 145–146, 146f
Stim, store-operated calcium channels and, 374, 378
Store-operated calcium channels (SOCs), 373–374
  operation of, 378
  Orai and, 374
  physiological mechanism for, 373
  signaling to, 374
  Stim and, 374
  TRPCs and, 374
STX4. *See* Syntaxin 4
Sulfated sialyl Lewis x, 341–342, 342f
Sunitinib (Sutent), 62t
Sutent. *See* Sunitinib
Sympathetic nervous system. *See* Excitation-contraction coupling
Synaptic transmission
  calcium channels signaling complexes in, 364–366, 364f
    calcium-binding proteins, 364f, 365–366
    G-protein modulation, 364f, 365
    SNARE proteins, 364–365, 364f
  inhibition of, 365
Syntaxin 4 (STX4), in glucose transport, 73
Syntillas, 380

T

T cell receptors (TCRs). *See also* TCR/pMHC complexes
  activation of, 331–332, 332f
  architecture of, 399–400, 399f
  conformational variation and change, 403
  generation of, 399–400, 399f
  peptide recognition, 401–402
  TCR/pMHC complexes orientation of, 399f, 400–401, 401f, 402f
T cells, co-stimulation of, 331–333, 332f
TANK binding kinase (TBK), 423–424
Tarceva. *See* Erlotinib
TBK. *See* TANK binding kinase
TCR/pMHC complexes, 399–404
  alloreactivity, 403
  APLs outcome, 402
  architecture of, 399–400, 399f
  bound water in, 403
  interactions, 400
  orientation of TCR in, 399f, 400–401, 401f, 402f
  peptide recognition by TCR CDR loops, 401–402
  structural changes and biological outcome discrepancies, 402–403
  TCR conformational variation and change, 403
TCRs. *See* T cell receptors
TGFβ. *See* Transforming growth factor β
TGFβ serine/threonine kinase receptor, 58
TGFβ-RII, 268f, 269
Thapsigargin, 373
THD. *See* TNF homology domain
Thermodynamics, of protein–protein interactions, 15–16, 16f
TIE. *See* Transcription inhibitory element
Timolol, GPCR interaction with, 214–215
TLRs. *See* Toll-like receptors
TNF. *See* Tumor necrosis factor

TNF homology domain (THD), 293
TNF receptor associated factors (TRAFs), 50, 298, 305–309, 306f
biological functions of, 305–306
discovery of, 305, 306t
domain organizations and structures of, 306–307, 307f–308f
C-terminal TRAF domain, 307, 308f
N-terminal RING domain and zinc fingers, 306–307
regulation of TRAF signaling, 309
TRAF signaling and Lys63 linked polyubiquitination, 308–309
TRAF6, 307–308
TNF receptor-associated DD (TRADD), 298
TNFα, 293, 299
in development of insulin resistance, 87
TNFRs. *See* Tumor necrosis factor receptors
"Toggle" switch, of GPCRs, 215, 216f, 239, 241
Toll family receptors, 421–425
activation of, 422–423, 422f
in flies and mammals, 421–423, 422f
function of, 421–423
MyD88-dependent activation of IRFs, 424–425
MyD88-dependent signaling, 423, 424f
signaling by, 423–425, 424f
structure of, 421–422, 422f
TRIF-dependent signaling, 423–425, 424f
Toll-like receptors (TLRs), 305, 415–419, 416f
dimerization by tri-acylated lipopeptide, 418, 418f
dsRNA binding site, 415, 417f
human TLR3, 415, 417f
MD-2, 417–418, 417f, 418f
signaling in, 418–419
structure of, 415, 416f, 417f, 418f
TLR1-TLR2 dimerization, 418, 418f
TLR3, 415, 416f
TLR4, 415–417, 417f
Torpedo electric organ, nAChR in, 352
TRADD. *See* TNF receptor-associated DD
TRAF domain, 307, 308f
Trafficking, of GPCRs, 257
TRAFs. *See* TNF receptor associated factors
TRAIL receptors, 300
Transactivation, of GPCRs, PDGF receptor signaling and, 121
Transcription, cytokine hormone enhancement of, 178–179, 178f
CypA as activation switch, 179
CypB as chaperone and activator, 178–179
Transcription co-activators, Smads and, 280
Transcription inhibitory element (TIE), 286
Transcription repressors, Smads and, 280
Transcription-induced negative regulators, of ErbB receptors and EGFR, 99
Transforming growth factor β (TGFβ), 265–271
classification of, 273, 274f
ligands of, 273–274, 274f, 275f
structures of, 265–266, 266f
receptors of, 274–275, 274f
activation and regulation of, 275–276
structures of, 266–267, 267f
receptor-ligand complexes, 267–270
BMP and activin receptor I binding, 266f, 267, 268f, 269
BMP and activin receptor II binding, 269
complete assembly, 268f, 269
extracellular mediators, 270, 270f
TGFβ-RII, 268f, 269
receptors of, 5–6, 5f
oligomerization of, 50
PDGF receptor cross-talk with, 121–122
signal transduction, 273–281
non-Smad signaling pathways, 280–281
Smad regulation of, 278–279
Smads, 274f, 275f, 276–278, 277f
activation of, 278
categorization of, 285
DNA-binding partners and, 279–280
as DNA-binding proteins, 279
downregulation and cross-regulation of, 287
function *in vivo*, 287–288
nucleocytoplasmic shuttling and, 285
receptor regulation of, 285
signal transduction regulation, 278–279
transcription co-activators and repressors, 280
transcriptional regulation by, 285–287
Transition state, diffusive *v.* specific, 17, 17f
7-Transmembrane core, of GPCRs, 209–211, 210–211t, 212f, 213, 214f, 226–227, 237
Transmembrane domains, 5
Transmembrane receptors, 4–6, 5f
classes of, 47t
oligomerization of, 47–51, 47t
cytokine receptors, 47t, 48–50
GPCRs, 47t, 51, 253
guanylyl cyclase receptors, 47t, 50
PSKs, 47t, 50
RTKs, 47t, 48
TNF receptors, 47t, 50–51
receptor–protein signaling complexes, 25–26t
signaling paradigms of, 23–26, 24f
signaling properties of, 3–7, 5f
origins of cell signaling research, 3–4
Transverse tubule membranes, calcium channels in, 359
Trastuzumab (Herceptin), 62t, 67, 111
TRIF-dependent signaling, 423–425, 424f
Trk receptors
antagonists of, 138
internalization of, 66
in NGF signaling, 133–135
in p75•Trk•neurotrophin ternary complex, 138
TrkA, NGF complexes with, 135–136, 136f
TRP channels, 356
TRPCs, store-operated calcium channels and, 374
T-type calcium currents, 360
mediation of, 361, 362t
Tumor formation. *See* Cancer
Tumor necrosis factor (TNF), 293–298, 296f, 297t
Tumor necrosis factor receptors (TNFRs), 5–6, 5f, 159, 293–300, 294f, 295t
biological functions of, 298–299
oligomerization of, 47t, 50–51
preformed dimers of, 190
signaling pathways and regulation of, 298
therapeutic expectations for, 299–300
TNF structural features and, 293–298, 296f, 297t
Tyk2, 164t, 165
in human disease, 171, 172t
Tykerb. *See* Lapatinib
Tyrosine kinases. *See* Protein tyrosine kinases; Receptor tyrosine kinases
Tyrosine residues, phosphorylation of, 41

U

Ubiquitin, 65, 66, 169, 287
Ubiquitination
of cytokine receptors, 169
of PDGF receptors, 120
of RTKs, 65–67
conjugation of ubiquitin to target proteins, 65
molecular link between RTK signaling and endocytosis, 65–66, 66f
ubiquitination in RTK endocytosis, 66–67, 66f
ubiquitination of effector proteins in RTK signaling, 67
of TRAF, 308–309
Upregulation, of GPCRs, 246
Urokinase, PDGFβ receptor activation by, 121

V

Vascular endothelial cell growth factor (VEGF), 57
activation of, 59
co-receptors of, 146
structural characterization of, 143–144, 144f
VEGFR activation by, 143–147

Vascular endothelial cell growth factor receptor (VEGFR)
neuropilins and, 146
PDGFR and analogies to VEGFRs, 145–146, 146f
structural characterization of, 144–145, 145f
VEGF activation of, 143–147
Vasopressin receptors, 256–257
Vectibix. *See* Panitumumab
VEGF. *See* Vascular endothelial cell growth factor
VEGF-A, 143–145, 144f, 146f
VEGF-B, 143–144, 144f
VEGF-C, 143–144
VEGF-D, 143–144
VEGF-E, 143–144, 144f
VEGF-F, 143–144, 144f
VEGFR. *See* Vascular endothelial cell growth factor receptor
Venus Flytrap domain, of GPCRs, 213
Vertebrate development, Eph/ephrin facilitated cell–cell communication during, 153
VILIP-2. *See* Visinin-like protein 2
Visinin-like protein 2 (VILIP-2), 366
Visual arrestin, 246
Visual fidelity, GPCR signaling in, 222, 222f
VOCs. *See* Voltage-operated calcium channels
Voltage-gated ion channels
opening of, 354, 354f
structure of, 352, 352f, 354–355, 356f
Voltage-operated calcium channels (VOCs), 378
cardiomyocyte opening of, 379

W

WD motif, of Gβ subunits, 204, 220
Wnt, Eph cross-talk with, 153
Wrist epitope, BMP and activin receptor I binding on, 266f, 267, 268f, 269

X

X-ray crystallography, for obtaining three-dimensional structures of ion channels, 352

Z

Zinc fingers, of TRAFs, 306–307
ZM241385, GPCR interaction with, 214–215, 214f, 225

Printed and bound by CPI Group (UK) Ltd, Croydon, CR0 4YY
03/10/2024
01040311-0018